Theoretical Plasma Physics

Karl-Heinz Spatschek

Theoretical Plasma Physics

An Introduction to the Fundamentals of Classical Plasmas

Springer

Karl-Heinz Spatschek
Institut für Theoretische Physik I
Heinrich Heine Universität Düsseldorf
Düsseldorf, Nordrhein-Westfalen, Germany

ISBN 978-3-662-72827-7 ISBN 978-3-662-72828-4 (eBook)
https://doi.org/10.1007/978-3-662-72828-4

This book is a translation of the original German edition "Theoretische Plasmaphysik," 2nd edition, by Karl-Heinz Spatschek, published by Springer-Verlag GmbH, DE in 2025. The translation was done with the help of an artificial intelligence machine translation tool. A subsequent human revision was done primarily in terms of content, so that the book will read stylistically differently from a conventional translation. Springer Nature works continuously to further the development of tools for the production of books and on the related technologies to support the authors.

This Springer imprint is published by the registered company Springer-Verlag GmbH, DE, part of Springer Nature.
The registered company address is: Heidelberger Platz 3, 14197 Berlin, Germany

Preface to the 2nd Edition

Many years have passed since the publication of the first edition. During these years, plasma physics has made great progress. Nevertheless, the topics presented in the first edition still form a relevant foundation for subsequent expansions in various current directions. The present edition takes this development into account without having to forgo the basic elements of the first edition. The price, however, is an almost doubling of the page count in order to do even approximate justice to the various newer aspects. The aim, however, is not to present all advances in detail. In keeping with the "philosophy" of the first edition, the idea behind the monograph is to present the established theoretical foundations for the development of plasma physics in such a way that they provide students with an introduction to the "basics."

Plasma physics holds great potential for the future. Currently, progress is evident in magnetic fusion, laser-plasma interactions, plasma technology, and plasma astrophysics. What the more distant future may bring remains, of course, uncertain. However, it is worthwhile to have a reliable "toolbox" for all possible applications, which can then be further developed as needed.

In today's world, the question arises as to whether a "textbook" is still up to date at all. Are there not already AI-based "chatbots" that respond to all questions in detail and with apparent competence? Chatbots are certainly helpful tools, but they should be used with a healthy dose of skepticism. Personal responsibility, critical thinking, and an awareness of the potential weaknesses of the technology are essential. In order to avoid the dangers of falling for false information as much as possible, it is necessary to train one's own critical thinking. But even the ability to ask the right questions must be developed. Universities will have to continue to contribute to this in teaching, and especially in research, as they have in the past. This textbook is intended to help in this regard.

The choice of an appropriate system of units for electrodynamics is widely discussed. In plasma physics, the Gaussian system is often used. Nevertheless, we follow the international recommendation here to use the Maxwell equations in the SI unit system. This recommendation comes from the BIPM (Bureau International des Poids et Mesures). In the publications of the BIPM, particularly the “SI Brochure,” the advantages of the SI system are emphasized.

Finally, a personal word. Perhaps more than the desire to communicate something to others, writing a book serves one’s own self-reflection. In humility and self-imposed discipline, a magnificent field of activity opens up, which—alongside writing original publications—allows me to partake in the fascination of science. I am deeply grateful to all my colleagues in the Department of Physics at Heinrich Heine University Düsseldorf for the opportunities to work and for the trustful conversations. In recent years, I have been able to explore new areas of research with Götz Lehmann—for this, too, a heartfelt thank you. Springer-Verlag encouraged me to undertake the second edition; Caroline Strunz, Claus-Dieter Bachem, and Suresh Syasam, to name just a few, have recently helped me greatly, not only to embark on the project but also to bring it to a—hopefully satisfactory—conclusion. For all shortcomings that surely remain in the second edition, I alone bear full responsibility, not they. So, to sum up briefly: To all who made it possible for me to write the second edition of “Theoretical Plasma Physics,” and especially to my wife Gertrud (Tuta), I extend my very, very heartfelt thanks.

Düsseldorf
in March 2025

K.-H. Spatschek

Preface to the 1st Edition

Physics is "the science of natural phenomena, which can be determined, traced, systematically recorded through observation and measurement, and thus made accessible to mathematical representation." The connection to mathematics is ensured in particular by theoretical physics. The renowned physicist Richard Feynman even claimed, "those who do not understand mathematics will hardly be able to penetrate to the deepest beauties of nature. Physicists have no other language at their disposal, and if you want to learn more about nature, you must learn to understand the language it speaks." Engaging with nature becomes a dialogue when we learn to ask the right questions and have the patience to understand the answers. In this respect, we are sometimes happier than the poet Heinrich Heine, who laments in his "Questions":

The waves murmur their eternal murmur,

The wind blows, the clouds flee,

The stars twinkle, indifferent and cold,

And a fool waits for an answer.

In physics education, students are expected to learn how to "ask questions of nature," to understand established answers, and to remain open to new perspectives. Theoretical physics makes a decisive contribution in this regard. At Heinrich Heine University Düsseldorf, theoretical physics is offered as a core course covering mechanics, electrodynamics, quantum theory I, statistical mechanics and thermodynamics, as well as quantum theory II. In addition, as at other universities, lectures on general relativity, quantum electrodynamics, quantum chromodynamics, statistical physics, and so on, are part of a series of supplementary and specialized lectures that follow the core course. Some of these, such as lectures on theoretical solid-state physics and theoretical plasma physics, occupy a central position, as they significantly enhance the range of elective physics courses—corresponding to the research focus areas. While theoretical solid-state physics is a "standard lecture" in most physics departments, this is not the case for theoretical plasma physics at many universities. The reasons for this are complex; however, they do not stem from a lack of importance of theoretical plasma physics.

There are hardly any books on Theoretical Plasma Physics in the German-speaking world that adequately meet the specific requirements as study and companion texts for advanced studies. Exceptions, such as the excellent presentation by Artsimovich and Sagdeev, only prove the rule—however, they place less emphasis on a systematic theoretical foundation based on the principles of Theoretical Physics ("first principles"). On the other hand, there are numerous monographs in the English-language literature (see the general literature references at the end of the book) that provide a very good overview of plasma physics. I would like to explicitly mention some of these, in particular the works by Ecker, Balescu, and Nicholson, as they have had a significant personal influence on me; I am sure this will also be evident in the following chapters.

The intention of this book is to present plasma physics as a field of theoretical physics that, following the core lectures on mechanics, electrodynamics, quantum theory, as well as statistics and thermodynamics, should have its place in the advanced studies of physics students as a supplementary and specialized course. The book builds upon the fundamental knowledge that is generally acquired after attending the required lecture on statistics and thermodynamics. It addresses (primarily in a classical framework) the plasma as a many-particle system using the methods of theoretical physics. Equal emphasis is placed on equilibrium theory, the kinetic description out of equilibrium, and macroscopic models for transport processes, as well as on the collective effects of nonlinearities. While writing the individual chapters, it became clear to me once again that "theoretical plasma physics" is the result of intensive research by numerous authors. Therefore, the use of the personal pronoun *we* in presenting the facts and their interrelations is to be understood as a "plural of modesty."

Many people deserve thanks. First and foremost, G. Ecker and R. Balescu, from whom I was able to learn essential aspects of plasma physics. Secondly, the colleagues with whom I have been able to work on plasma physics problems in recent years, some of which have also found their way into this book. I would especially like to mention R. Blaha, Th. Eickermann, E.W. Laedke, Chr. Marquardt, H. Pietsch, and H. Wenk. My special thanks also go to Mrs. R. Wohlgemuth, who prepared the manuscript of this book using a text processing system, to Mr. E. Zügge for producing the figures, as well as to the secretaries G. Laufen and E. Gröters, who assisted in preparing the final print version.

Düsseldorf
in June 1990

K.-H. Spatschek

Contents

Part I
General Principles

Overview of Basic Principles

1

Abstract

This first chapter attempts to introduce important plasma phenomena using simple examples. The discussion is not exhaustive and will be presented in more detail and supplemented in later chapters. The aim is to develop an initial “feel” for the special properties of plasmas, which can (hopefully) serve as a guide during the more detailed and systematic derivations that follow.

1.1 Plasma as a Many-Particle System

In this section, we begin by introducing plasma as a many-particle system of electrically charged particles. The constituents and parameters can vary greatly. Accordingly, adjectives are often prefixed to the word *plasma*. Some typical characteristics are mentioned, but the initial overview then focuses on classical, (quasi-) neutral plasmas, which we will use as a working hypothesis for our first approach.

A system of charged particles or quasiparticles (ions, electrons, molecules, quarks, gluons, holes, etc.) is called *plasma* under quite different conditions. In formulating these conditions, the literature shows differences depending on whether one is concerned with ionized gases, solids, fully or partially ionized systems, or macroscopically neutral or non-neutral arrangements. As is often the case, the characteristic properties and their effects only become clear when general knowledge is available that allows insight into fundamentally new phenomena. Therefore, in this introductory chapter, we will start from a

K.-H. Spatschek, *Theoretical Plasma Physics*,
https://doi.org/10.1007/978-3-662-72828-4_1

simple and not overly strict definition of plasma as a many-particle system of charged particles, and only later delve deeper.

A large portion of the matter in the universe exists in the plasma state. It was previously said that about 99% of the matter in the universe is in the plasma state. However, this estimate refers only to baryonic matter, which likely makes up only 4.9% of the total matter content. Baryonic matter includes ordinary matter, from which stars, planets, and all known living beings are made. The "rest" of the universe consists mainly of dark matter and dark energy (dark matter about 26.8% and dark energy about 68.3%). At present, we know little to nothing about dark matter and dark energy.

The term *plasma* was introduced by Langmuir, Tonks, and their collaborators in the 1920s, when they were investigating processes in electronic lamps filled with ionized gases, i.e., low-pressure discharges. The word *plasma* appears to be a misnomer [1]. The Greek $\pi\lambda\acute{\alpha}\sigma\mu\alpha$ means something shaped or molded. However, a plasma generally does not tend to adapt to external influences. On the contrary, due to its collective behavior, it often acts as if it has a will of its own [1].

Modern plasma physics emerged in the 1950s, when the idea of a thermonuclear reactor arose. Fortunately, modern plasma physics has become completely decoupled from weapons development (for example, the hydrogen bomb). The progress of modern plasma physics can be traced in many monographs, e.g., [2–20].

The plasma state is sometimes referred to as the fourth state of matter. After the solid, liquid, and gaseous states, we reach it as the temperature increases, when the thermal energy is sufficient to break bonds and cause ionization.

Simply put, plasmas can be characterized by two parameters, namely the density of charged particles n and the temperature T. The density varies over about 28 orders of magnitude, e.g., from 10^6 to 10^{34} m^{-3}. The kinetic energy $k_B T$, where k_B is the Boltzmann constant, can vary over about seven orders of magnitude, e.g., from 0.1 to 10^6 eV.

Plasmas appear in space and astrophysics [21–23], in laser-matter interaction [24, 25], in technology [26], in nuclear fusion (magnetic or gravitational confinement, laser fusion) [15, 27–30], etc. Technical plasmas, magnetic fusion plasmas, and laser-generated plasmas represent the main applications of plasma physics on Earth. Space plasmas, such as those found in the Earth's magnetosphere, are very important for our life on Earth. For example, a continuous stream of charged particles, mainly electrons and protons, known as the solar wind, encounters the Earth's magnetosphere, which protects us from this radiation. Typical parameters of the solar wind are $n = 5 \times 10^6$ m^{-3}, $k_B T_i = 10$ eV and $k_B T_e = 50$ eV. However, temperatures on the order of $k_B T = 1$ keV can also occur. The drift velocity is about 300 km/s.

We could cite numerous other examples of important plasmas: stellar cores and atmospheres, which are hot enough to be in the plasma state. Free electrons and holes in semiconductors. And so on.

In our further approach, we are guided by two considerations: On the one hand, we must exclude the extremely important questions concerning the structure of individual "particles"—which constitute a subject in their own right—and on the other hand, we aim to distill the characteristic phenomena of a many-particle system with long-range interactions in the simplest possible form. We therefore begin with the working hypothesis that a plasma is a macroscopically neutral gas composed of many electrically charged (and, if applicable, additionally neutral) particles, whose behavior is essentially determined by *collective* degrees of freedom. This definition requires some explanatory remarks; generalizations will follow later.

Since ionized matter is generally present in a plasma, we should know something about the degree of ionization. More on this in the next section. For now, in advance: The degree of ionization can often be roughly estimated using the Saha equation

$$\boxed{\frac{n_i n_e}{n_n} = \frac{2g_1}{g_0}\left(\frac{m_e}{2\pi}\right)^{3/2}\hbar^{-3}(k_B T_e)^{3/2} e^{-E_i/(k_B T_e)}} \tag{1.1}$$

Ionization equilibrium is assumed. In this formula, the following denote: n_e the particle density of electrons, n_n that of neutrals, n_i that of singly ionized species, g_ν the corresponding statistical weights ($g = 2$ for electrons, $g = (2S+1)(2L+1)$ for atoms or ions), k_B the Boltzmann constant $k_B = 1.3807 \times 10^{-16}$ erg/K, T_e the electron temperature, m_e the electron mass, and E_i the ionization energy. For temperature measurement, we note, due to $1eV \mathrel{\widehat{=}} 1.6022 \times 10^{-12}$ erg, the equivalence $1eV \mathrel{\widehat{=}} 1.1605 \times 10^4$ K. Simple calculations show that a degree of ionization close to one is already reached at temperatures (in eV) that are below the ionization energy.

Depending on the degree of ionization, one distinguishes between *fully ionized* or "hot" and *weakly ionized* or "cold" (low-temperature) plasmas. In very hot plasmas, multi-electron atoms are predominantly multiply ionized. According to our definition, in the ensemble of charged particles, the potential energy of a particle due to its interaction with its nearest neighbors should, on average, be much less than its (mean) kinetic energy. These are then *ideal* plasmas, as opposed to *non-ideal* plasmas.

At this point, it already becomes clear why a plasma is not merely a—albeit complicated—practice example for classical electrodynamics. As it is treated in standard lecture courses, electrodynamics is a theory of electromagnetic fields and the motion of particles in external (prescribed) fields. In contrast, the collective effects that arise in the motion of many particles, taking into account long-range interactions, represent new phenomena that define the specific properties of plasma. The electric charges in plasmas generate electromagnetic fields, which in turn exert forces on the charges and influence their dynamics. Therefore, even in the simplest case, the description of a plasma must be carried out in a self-consistent manner using both the mechanical and electrodynamical fundamental equations. It should be noted that, in general, Coulomb forces are not necessarily the only or even the main form of interaction in all plasmas. In general, collective processes in plasmas should always be understood as processes in which a large number of particles participate in an organized manner.

Similar to solid-state physics, where the constituents and the interactions between them are also known, one should not underestimate the range of possible phenomena that make even a many-particle system with Coulomb interaction so fascinating. The appeal, but also the difficulty, in tackling concrete problems in plasma physics lies in the necessity to draw on knowledge from all areas of physics, whether quantum theory or hydrodynamics, thermodynamics and statistics, or nonlinear dynamics, atomic physics, or electrodynamics.

A more detailed treatment of plasmas obviously requires, due to their many-particle nature, methods from statistical physics. Depending on the state, this may involve equilibrium or non-equilibrium statistics. Only a few phenomena can be calculated within the framework of very simple models, such as the single-particle model for the motion of individual charged particles in prescribed electromagnetic fields. In the context of magnetohydrodynamics, the plasma is regarded as a conductive continuous medium that can be described by the equations of hydrodynamics and electrodynamics. The two-fluid model allows for the separate treatment of ions and electrons. In general, however, a kinetic description is appropriate, as it can capture the various new phenomena, such as wave-particle interactions.

The characteristic plasma parameters span many orders of magnitude. The environment of the Earth represents a nearby *natural* plasma. However, plasmas also frequently occur in the *laboratory* or are generated for potential energy production (nuclear fusion).

Collective effects in plasmas compete with elementary processes such as elastic and inelastic collisions or radiation processes. The elementary processes lead to energy transfer between the different particle species and among themselves. If the energy transfer occurs mainly through collisions between the material particles (primarily through electron collisions), the plasma is called *collision-dominated.* A *radiation-dominated* plasma exists when the energy is transferred predominantly by emission and absorption of photons. If, on the other hand, collective effects predominate, the plasma is referred to as a *collisionless* plasma.

A key property of almost all plasmas is quasineutrality. This refers to electrical neutrality down to subvolumes that are small compared to the total plasma volume. Quasineutrality (down to volume elements of size λ_D^3) is based on the fact that any excess charge is quickly neutralized due to the strong electric fields it generates. The Debye length λ_D, to which we will return shortly, plays a decisive role in this context. *Neutral* plasmas are those that are macroscopically neutral. Recently, however, *non-neutral* plasmas have also gained considerable importance. It has been shown that an ensemble of electrons or ions in an electromagnetic trap can quite well represent a form of matter that can be described as a one-component plasma. The latest experiments in microplasmas, consisting of a few charged particles confined in a Paul trap, allow for the systematic study of non-ideal behavior in (strongly coupled) systems.

1.2 Degree of Ionization

In this section, we return to the Saha equation (1.1). We provide guidance on how this equation is derived, discuss its range of applicability, and then generalize it.

A plasma can be partially or fully ionized. The degree of ionization depends on several parameters. First, we determine the degree of ionization based on thermal ionization in a system of hydrogen atoms (H), electrons (e), and protons (p) [ions].

Saha Equation

The Saha equation is named after the astrophysicist Meghnad Saha [31], who first derived it in 1920.

The Saha equation establishes a relationship between the free particles (for example, electrons e and protons p) and the particles bound in atoms (H). To derive the Saha equation, we assume thermodynamic equilibrium and collisional ionization. Let the energy levels be E_n. We set $E = 0$ when the electron is free (not bound) and its velocity is zero, and $E = E_n < 0$ when the electron is in a bound state of the hydrogen atom ($Z = 1$). Using the simple Bohr energy formula for $n = 1$, we ignore the higher n levels. It holds that

$$E_n \approx \frac{Z}{n^2} \times (-13.6)\ \text{eV} \ . \tag{1.2}$$

The first excited state is already close to the free-boundary threshold compared to the ground state. If enough energy is available to excite an electron from the ground state $n = 1$ to the excited state $n = 2$, only a little more (about a third) is needed to ionize it directly. Hence the restriction to $n = 1$.

Example 1.1 (Statistical Justification of the Saha Equation)
For independent particles, we first calculate the *single-particle partition functions* for electrons, protons, and hydrogen atoms, each of which takes the following form[1] [32]

[1] We retain the Boltzmann constant k_B here, as is customary in most books on statistical physics.

$$Z = \sum_n e^{-E(n)/k_B T} \, . \tag{1.3}$$

The summation extends over all states (free or bound) with energies $E(n)$. For free particles, the sums in the partition functions are actually integrals, since the particles have a continuous momentum distribution. The degeneracy of the states (or statistical weights) g_ν with $g_e = g_p = 2$ and $g_H = 4$ (for hydrogen) must be taken into account. Thus, for free electrons and protons (ions) we obtain

$$Z_j = \frac{1}{h^3} \int g_j e^{[-p^2/(2m_j)]/(k_B T_j)} d^3 r d^3 p \quad \text{for} \quad j = e, p \, . \tag{1.4}$$

The integrals can be easily evaluated due to isotropy, and thus $d^3 p = 4\pi p^2 dp$,

$$Z_e = \frac{2V}{h^3} (2\pi m_e k_B T_e)^{3/2} \, , \tag{1.5}$$

$$Z_p = \frac{2V}{h^3} (2\pi m_p k_B T_p)^{3/2} \, . \tag{1.6}$$

Here, m_e and m_p are the electron and proton masses, respectively. In equilibrium, the temperatures satisfy $T = T_e = T_p = T_H$. A similar calculation for freely moving hydrogen atoms, which consist of a bound electron-proton pair (in the ground state), yields the following result

$$Z_H = \frac{4V}{h^3} (2\pi m_H k_B T_H)^{3/2} \, e^{E_i/k_B T_H} \tag{1.7}$$

with $E_0 = -13.6 \text{ eV} \equiv -E_i$, where E_i is the ionization energy.

Let Z be the total partition function (please do not confuse with the atomic number Z) for $N_e \equiv N_p$ *free* electrons (or protons). The total number N is the sum of all electrons and protons (whether free or bound). Thus, $N = N_H + N_p$ holds. For indistinguishable particles in a group, it follows that

$$Z(V, T, N_e, N_p, N_H) = \frac{Z_e^{N_e} \, Z_p^{N_p} \, Z_H^{N_H}}{N_e! \; N_p! \; N_H!} \, . \tag{1.8}$$

From this, the free energy is obtained as

$$F = -k_B T \ln Z \, . \tag{1.9}$$

The particle densities actually realized in nature are those that yield a minimum of the free energy.

To find the most probable state, we therefore differentiate the free energy (1.9). For large N we use Stirling's formula

$$\ln N! \approx N \ln N - N \,, \tag{1.10}$$

which leads to

$$\begin{aligned} -\tfrac{F}{k_BT} \approx N_e \ln Z_e + N_p \ln Z_p + N_H \ln Z_H - N_e \ln N_e \\ +N_e - N_p \ln N_p + N_p - N_H \ln N_H + N_H \end{aligned} \tag{1.11}$$

With $N_H = N - N_p$ and $N_p = N_e$ we find by setting to zero

$$\frac{\mathrm{d}F}{\mathrm{d}N_e} \sim \ln Z_e + \ln Z_p - \ln Z_H - \ln N_e - \ln N_e + \ln(N - N_e) = 0 \,. \tag{1.12}$$

In other words

$$\frac{Z_e Z_p}{Z_H} = \frac{N_e^2}{N - N_e} \,. \tag{1.13}$$

If we use the individual partition functions and $m_H \approx m_p$, it follows

$$\frac{V}{h^3}(2\pi m_e k_B T)^{3/2} e^{-E_i/(k_B T)} = \frac{N_e^2}{N - N_e} \,. \tag{1.14}$$

Next, we introduce particle densities $N_e/V = n_e$, $N_p/V = n_p$, and $N_H/V = n_H \equiv n_n$. The Saha (equilibrium) formula then reads

$$\boxed{\begin{aligned} \frac{n_i n_e}{n_n} &= \frac{2g_1}{g_0}\left(\frac{m_e}{2\pi}\right)^{3/2} \hbar^{-3}(k_B T_e)^{3/2 - E_i/(k_B t_e)} \equiv K(T_e) \\ &\approx 2{,}4 \times 10^{15}\,(T_e[K])^{3/2}\, e^{-E_i/(T_e[eV])} \quad [\mathrm{cm}^{-3}] \\ &\approx 3 \times 10^{21}\,(T_e[eV])^{3/2}\, e^{-E_i/(T_e[eV])} \quad [\mathrm{cm}^{-3}] \end{aligned}} \,. \tag{1.15}$$

Again, n_e is the electron density, n_n the density of hydrogen atoms, n_i is the density of ions (protons), g_ν denotes the statistical weights ($g_1 = 2$ for electrons and protons, $g_0 = 4$ for hydrogen), $k_B = 1.3807 \times 10^{-16}$ erg/K is the Boltzmann constant, T_e is the electron temperature, m_e the electron mass, and $E_i = 13.6$ eV is the ionization energy. Note that 1 eV $\hat{=} 1.6022 \times 10^{-12}$ erg, 1 eV $\hat{=} 1.1605 \times 10^4$ K. ■

For hydrogen, we can also write the Saha equation in the form

$$\frac{\alpha_i \alpha_e}{\alpha_n} = \frac{2g_1}{n g_0} \frac{1}{\lambda_e^3} e^{-E_i/k_B T_e} \tag{1.16}$$

where the degrees of ionization

$$\alpha_e = \frac{N_e}{N} \,, \quad \alpha_i = \frac{N_i}{N} \,, \quad \alpha_n = \frac{N_H}{N} \tag{1.17}$$

have been introduced. We have used the thermal de Broglie wavelength

$$\lambda_e = \sqrt{\frac{h^2}{2\pi m_e k_B T}} \tag{1.18}$$

Of course, the following holds

$$N = N_H + N_i \equiv N_H + N_e \quad , \quad N_e \equiv N_i \quad , \quad n = \frac{N}{V} \; . \tag{1.19}$$

Key parameters are the temperature $T \equiv T_e$ and the density n. To introduce the pressure p, we use the classical equation of state

$$pV = \sum_\mu N_\mu k_B T_\mu = (1 + \alpha_e) N k_B T \; . \tag{1.20}$$

Solving for n gives us

$$n = \frac{p}{(1 + \alpha_e) k_B T} \; , \tag{1.21}$$

so that we obtain

$$\frac{\alpha_i \alpha_e}{\alpha_n (1 + \alpha_e)} = \frac{2g_1}{g_0} \frac{k_B T}{p \lambda_e^3} \, e^{-E_i/k_B T_e} \tag{1.22}$$

For hydrogen, we have

$$\alpha_n \equiv \alpha_H = 1 - \alpha_e \quad , \quad \alpha_i \equiv \alpha_e \; , \tag{1.23}$$

and thus

$$\frac{\alpha_e^2}{(1 - \alpha_e)(1 + \alpha_e)} = \frac{k_B T}{p \lambda_e^3} \, e^{-E_i/k_B T} \equiv \frac{1}{p \mathcal{K}_p(T)} \; . \tag{1.24}$$

The solution of this quadratic equation for α_e is

$$\alpha_e = \frac{1}{\sqrt{1 + p \mathcal{K}_p(T)}} \quad , \quad \mathcal{K}_p(T) = \frac{\lambda_e^3}{k_B T} \, e^{E_i/k_B T} \; . \tag{1.25}$$

The evaluation as a function of p and T is carried out for

$$\begin{aligned} m_e &\approx 9.109 \times 10^{-31}\ \text{kg} \quad , \quad h \approx 6.626 \times 10^{-34}\ \text{Js} \\ 1\, eV &\approx 1.602 \times 10^{-19}\ \text{J} \quad , \quad 1\ \text{bar} = 10^5\ \text{N/m}^2 \\ E_i &\approx 13.6057\ \text{eV} \quad , \quad k_B \approx 1.3807 \times 10^{-16}\ \text{erg/K} \end{aligned}$$

The result is shown in Fig. 1.1.

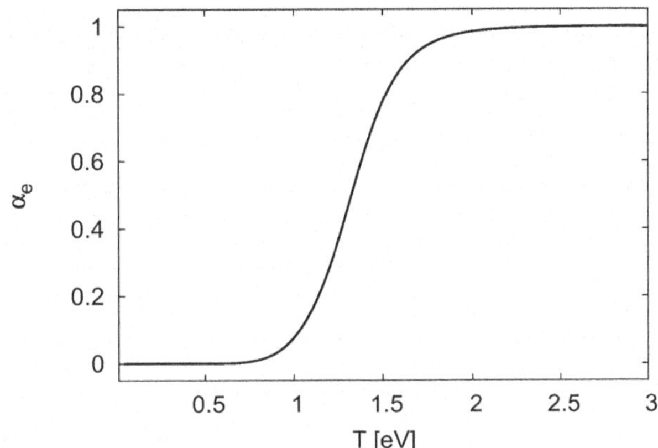

Fig. 1.1 Evaluation of the Saha equation (1.25) for hydrogen ($E_i = 13.6$ eV) at a pressure of $p = 1$ bar

Analogously, we could also evaluate the degree of ionization at a given total density $n = N/V$. The calculation then leads to

$$\alpha_e = \frac{1}{2n\mathcal{K}_n(T)}\left[\sqrt{1 + 4n\mathcal{K}_n(T)} - 1\right] \ , \quad \mathcal{K}_n(T) = \frac{\mathcal{K}_p(T)}{k_B T} \ . \tag{1.26}$$

Ionization and Recombination Coefficient

The Saha equation describes the statistical equilibrium between ionization and recombination under equilibrium conditions. We can represent the processes by collisional ionization

$$H + e^- \rightarrow H^+ + e^- + e^- \tag{1.27}$$

and three-body recombination

$$H^+ + e^- + e^- \rightarrow H + e^- \tag{1.28}$$

Since no further components are involved, these are also referred to as collisional ionization and three-body collisional recombination.

To generalize to non-equilibrium situations, ionization and recombination coefficients are also defined separately.

The cross section for the ionization process by collisions can be easily estimated. The following estimate treats the collision process on the atomic scale (Bohr radius) in a (strictly speaking invalid) classical model. Nevertheless, the results are close to the exact quantum mechanical calculation.

Example 1.2 (Dynamics of an Electron in the Coulomb Field)
When an electron approaches an H atom on atomic dimensions, we consider the scattering at the nucleus (proton). Let us assume high velocities so that the scattering angle θ is small. The electron is accelerated in the Coulomb field of a proton. If ρ is the *constant* impact parameter, the magnitude F of the total force is approximately proportional to $e^2/(\rho^2 + v_e^2 t^2)$, where $v \equiv v_e$ is the characteristic velocity of the electron. The closest approach during the encounter is at $t = 0$. The perpendicular component $F_\perp \equiv F_{perp}$ is (approximately) smaller by a factor of $\rho/(\rho^2 + v_e^2 t^2)^{1/2}$; see Fig. 1.2. Therefore, for the change in the perpendicular velocity component we obtain approximately

$$\frac{dv_\perp}{dt} \approx \frac{1}{4\pi\varepsilon_0}\frac{e^2}{m_e}\frac{\rho}{(\rho^2 + v_e^2 t^2)^{3/2}} \,. \tag{1.29}$$

Integration yields

$$v_\perp \approx \frac{1}{4\pi\varepsilon_0}\frac{e^2}{m_e}\int_{-\infty}^{\infty}\frac{\rho}{(\rho^2 + v_e^2 t^2)^{3/2}}\,dt = \frac{1}{4\pi\varepsilon_0}\frac{2e^2}{m_e v_e \rho} \,. \tag{1.30}$$

The energy gain

$$\Delta E_\perp \equiv \frac{m_e v_\perp^2}{2} \approx \frac{1}{(4\pi\varepsilon_0)^2}\frac{e^4}{E\rho^2} \equiv \varepsilon \,, \quad E = \frac{m_e v_e^2}{2} \tag{1.31}$$

is available for ionization. The last relation can be written as

$$\rho^2 = \frac{1}{(4\pi\varepsilon_0)^2}\frac{e^4}{E\,\varepsilon} \tag{1.32}$$

This is a relation between the impact parameter ρ, initial energy E, and maximum energy transfer ε. ■

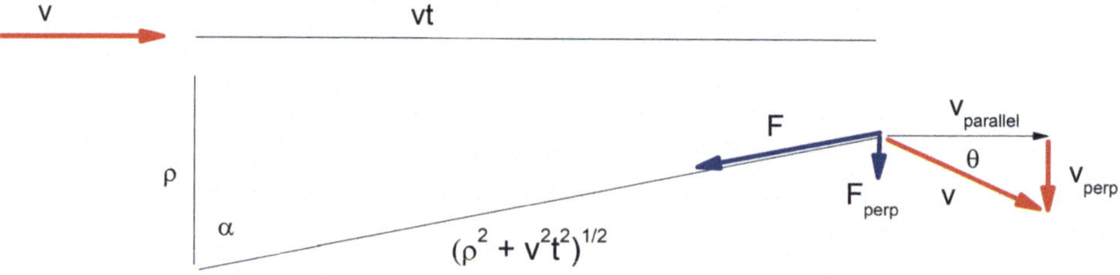

Fig. 1.2 Collision geometry for an electron approaching a proton from infinity with velocity v and being scattered at an angle θ. Left: The situation at times long before the collision. Right: The situation at a much later time

With the results of the previously discussed example of the dynamics of an electron in the Coulomb field of a proton, the differential cross section $d\sigma$ for a given initial energy E can be formulated,

$$d\sigma = |2\pi \rho d\rho| = \frac{1}{(4\pi\varepsilon_0)^2} \frac{\pi e^4}{E\varepsilon^2} d\varepsilon \ . \tag{1.33}$$

For $\varepsilon > E_i$, where E_i is the ionization energy, ionization occurs. By integrating the differential cross section from E_i to E, we find the (total) Thomson cross section.

$$\sigma_T = \frac{1}{(4\pi\varepsilon_0)^2} \frac{\pi e^4}{E^2 E_i} (E - E_i) \ . \tag{1.34}$$

The maximum is reached for $E = 2E_i$. Its value is

$$\sigma_{T,max} = \frac{1}{(4\pi\varepsilon_0)^2} \frac{\pi e^4}{4E_i^2} \approx \pi a_B^2 \approx 10^{-16} \text{ cm}^2 \ . \tag{1.35}$$

Here, a_B is the Bohr radius

$$a_B = \frac{4\pi\varepsilon_0 \hbar^2}{m_e e^2} \approx 0.529 \times 10^{-10} \text{ m} \ . \tag{1.36}$$

From quantum mechanical calculations, we obtain the ionization energy

$$E_i = \frac{m_e e^4}{8\varepsilon_0^2 h^2} \ . \tag{1.37}$$

For large energies, the previously estimated cross section σ_T decreases as $1/E$. More precise calculations yield a proportionality $\ln(E)/E$ for $E \gg E_i$.

If the cross section is multiplied by the velocity and the density of the scattering particles, one obtains the probability of collisional ionization per unit time. Overall, the rate is of course also proportional to the density of the incident particles. For *collisional ionization* and (three-body) recombination, the particle balance is then given by

$$\boxed{\frac{dn_e}{dt} = \alpha n_e n_H - \beta n_e^2 n_p} \ , \tag{1.38}$$

where

$$< \sigma_T v_e > \equiv \alpha \tag{1.39}$$

is called the ionization coefficient due to electrons as a result of collisions (collisional ionization coefficient). The balance (1.38) is not tied to thermodynamic equilibrium.

With β we have denoted the recombination coefficient. It is interesting that we can easily calculate this coefficient for three-body recombination under the assumption of equilibrium (or more generally, a stationary state). In this case, we have

$$\beta = \frac{n_H < \sigma_T v_e >}{n_e n_p} \equiv \frac{\alpha}{K(T)} . \tag{1.40}$$

In the last transformation, $K(T)$ was used as the right-hand side of the Saha equation (1.15).

Coronal Formula

If a plasma is not in thermodynamic equilibrium due to collisions, then in order to calculate its degree of ionization, knowledge of all elementary processes for ionization and recombination is required.

The essential elementary processes are photoionization and ionization by electron impact, as well as recombination accompanied by the emission of photons and by three-body collisions. For each of these four processes, formulas can be given. In contrast to the situation with the Saha equation, both collisional ionizations and photorecombinations can be balanced. By equating their numbers, a new ionization formula results, for example for the solar corona, from which, using the known intensity ratio of the green and red coronal lines, the temperature of the solar corona can be determined.

Let us begin first with photoionization

$$H + \hbar\omega \rightarrow H^+ + e^- \tag{1.41}$$

and photoionization

$$H^+ + e^- \rightarrow H + \hbar\omega . \tag{1.42}$$

These can, together with the processes already discussed, appear in the rate equation

$$\boxed{\frac{dn_e}{dt} = \alpha n_e n_H - \beta n_e^2 n_p + \mu n_H - \gamma n_e n_p} \tag{1.43}$$

Newly appearing are μ for photoionization and γ for radiative recombination.

If we are in thermodynamic equilibrium *and* have a detailed balance between radiative ionization and recombination, then

$$\mu = \gamma \frac{n_e n_p}{n_H} \equiv \gamma K(T) . \tag{1.44}$$

applies. In many (optically thin) systems, radiation can easily escape, and the photon density in the plasma is lower than in equilibrium. In that case, photoionization plays no role. If, in addition, we have a very tenuous plasma, three-body recombination is negligible. In the regime

$$\frac{\mu}{\alpha} \ll n_e \ll \frac{\gamma}{\beta} \tag{1.45}$$

a distribution in equilibrium then follows from the balance of collisional ionization and photorecombination, which is also known as the coronal or Elwert formula

$$\boxed{\frac{n_p}{n_H} = \frac{\alpha}{\gamma}} \tag{1.46}$$

It is suitable for many astrophysical situations. The condition for its application is often written in the form [33]

$$10^{12} t_I^{-1} < n_e\,[\,\mathrm{cm}^{-3}] < 10^{16} (T_e\,[\,\mathrm{eV}])^{7/2} \tag{1.47}$$

where t_I is a normalized ionization time. The photorecombination coefficient can also be approximated by [34]

$$\gamma \approx 2.7 \times 10^{-13}\, T_e^{-1/2} \left[\frac{\mathrm{cm}^3}{\mathrm{sec}}\right] \quad \text{in the region} \quad 1 < T_e\,[\,\mathrm{eV}] < 15 \tag{1.48}$$

while for three-body recombination, it is often

$$\beta \approx 8.75 \times 10^{-27} (T_e\,[\,\mathrm{eV}])^{-4.5} \left[\frac{\mathrm{cm}^6}{\mathrm{s}}\right] \tag{1.49}$$

that is used [33]. More can be found in the book by H. Griem [35].

1.3 Model Zones

In this section, we address the question of which methods of theoretical physics can be used to explore the physics of a plasma. Can a plasma always be described using methods of classical statistics and thermodynamics? When is it necessary to use quantum mechanical calculations? Do we encounter plasmas in the special relativistic or even general relativistic regime? Here, we provide some initial answers based on theoretical physics.

In plasma physics, temperature and density are two characteristic parameters. The terms "hot" and "cold" are, of course, relative when it comes to the temperature in question. The same applies to density when referring to high and low densities. In this section, we aim to develop a better understanding of the relevant orders of magnitude.

In the field of magnetic confinement, for example, "hot" means temperatures high enough to satisfy the Lawson criterion. In this case, we are in the temperature range

$$10 \text{ keV } \leq k_B T \leq 20 \text{ keV} \ . \tag{1.50}$$

The Lawson criterion [36] is[2]

$$\boxed{n k_B T \tau_E \geq 3 \times 10^{21} \text{ m}^{-3} \text{ keV s}} \ . \tag{1.51}$$

Here, τ_E is the energy confinement time.

Relativistic Description

First, we want to address whether a *special relativistic* treatment is required. As a rough estimate for the necessity of a special relativistic description, we postulate

$$\boxed{v_{the}^2/c^2 \geq 0.01} \ , \tag{1.52}$$

with the thermal electron velocity

$$v_{the} = (k_B T_e/m_e)^{1/2} \ . \tag{1.53}$$

The choice of 1% is, of course, somewhat arbitrary. In terms of order of magnitude, we then find the necessity for a relativistic modeling for temperatures

$$k_B T_e \geq 0.01\, m_e c^2 \approx 0.005 \text{ MeV } = 5 \text{ keV } \hat{=} 50\,000\,000 \text{ K} \ . \tag{1.54}$$

This result was obtained using the approximate values

$$m_e c^2 \approx 0.5 \text{ MeV} \ , \quad 1 \text{ eV} \mathrel{\hat{\approx}} 10\,000 \text{ K} \ . \tag{1.55}$$

More precise values can be found, for example, in Ref. [33].

In plasma astrophysics, large systems (mass M, radius R) are often discussed. In such cases, a general relativistic description may become necessary [18]. From General Relativity (GR), it can be estimated that its peculiarities must be taken into account when the

---

[2] John D. Lawson was a British engineer who died on January 15, 2008, at the age of 84. He is particularly known for his fusion criterion from 1955, which he published in 1957: "Some Criteria for a Power Producing Thermonuclear Reactor," Proc. Phys. Soc. 70, 6–10 (1957).

gravitational energy becomes comparable to the rest mass energy:

$$\boxed{\frac{\frac{GM^2}{R}}{Mc^2} \approx 0.7\left(\frac{M}{10^{33}\ \text{g}}\right)\left(\frac{R}{1\ \text{km}}\right)^{-1} \gtrsim \mathcal{O}(1) \quad \rightsquigarrow \quad \text{GR}} \, . \tag{1.56}$$

G is the gravitational constant. Thus, an object with the mass of our Sun would have to have a radius of one kilometer for us to exceed the limits of Newtonian theory.

Quantum Mechanical Description

Quantum mechanical effects become significant when the degeneracy parameter $n_e \lambda_{dB}^3$ becomes greater than 1. In the *non-relativistic* limit, we use the (thermal) de Broglie wavelength for estimation

$$\lambda_{dB} \equiv \lambda_e = h/\sqrt{2\pi m_e k_B T_e} \, , \tag{1.57}$$

so that

$$\boxed{n_e \gg \left(\frac{m_e k_B T_e}{\hbar^2}\right)^{3/2}} \tag{1.58}$$

becomes the criterion for a quantum mechanical description. The mean particle distance (here for electrons) is often estimated by

$$\lambda_n \mathrel{\hat{=}} \lambda_{n_e} \approx n_e^{-1/3} \tag{1.59}$$

Thus, quantum mechanical effects become significant when the de Broglie wavelength approaches or exceeds the order of magnitude of the mean particle distance.

In the *relativistic* case, we use the energy formula

$$E^2 = c^2 p^2 + m_e^2 c^4 \tag{1.60}$$

together with the relations $E \sim k_B T_e$ and $\lambda_{dB} \sim h/p$. From this, we estimate $p = p(T_e)$. In the *ultrarelativistic* case, $p \sim k_B T_e / c$ applies. This then leads to

$$\boxed{n_e \gg \left(\frac{k_B T_e}{\hbar c}\right)^3} \tag{1.61}$$

as the requirement for a quantum mechanical formulation.

Fig. 1.3 shows the classification of plasmas in parameter space, which is (simplified) defined by temperature and density. In addition to the non-relativistic and classical region, the ideal region, which we will now discuss, is also of interest.

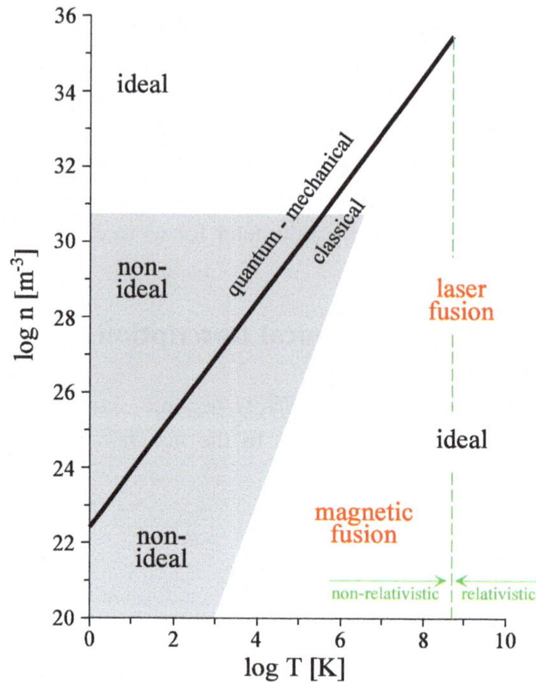

Fig. 1.3 Sketch of the different model regions as a function of temperature T and particle density n

Ideal or Non-Ideal Plasmas

The *ideal* approximation requires $|E_{pot}|/|E_{kin}| \ll 1$, i.e., the interaction energy E_{pot} between the particles must be small compared to their kinetic energy. For Coulomb interaction, we have

$$E_{pot} \sim \frac{1}{4\pi\varepsilon_0} e^2/\lambda_n \sim n_e^{1/3} , \tag{1.62}$$

while classically we calculate the kinetic energy from

$$E_{kin} = \frac{3}{2} k_B T_e \quad \text{[classical]} \tag{1.63}$$

Using the Debye length for electrons

$$\lambda_{De} = (\varepsilon_0 k_B T_e / n_e e^2)^{1/2} \tag{1.64}$$

we obtain for the ratio, and thus for the *classical ideality condition*,

$$\boxed{n_e \lambda_{De}^3 \gg 1} . \tag{1.65}$$

The result is different in the case where a quantum mechanical description is required. In that case, we approximate the kinetic energy by

$$E_{kin} \approx \frac{p_F^2}{2m_e} \sim \frac{h^2 n_e^{2/3}}{m_e}, \tag{1.66}$$

with the Fermi momentum

$$p_F = \sqrt[3]{\frac{3n_e}{8\pi}} h . \tag{1.67}$$

From this follows the *quantum mechanical ideality condition*

$$\boxed{\lambda_B \equiv \frac{e^2 m_e}{4\pi \varepsilon_0 n_e^{1/3} \hbar^2} \ll 1} . \tag{1.68}$$

The expression λ_B on the left-hand side is called the Brueckner parameter.

The difference between (1.65) and (1.68) is of great physical significance. In the classical case, with respect to the density dependence, we have

$$\frac{E_{Coulomb}}{k_B T_e} \sim n_e^{1/3} , \tag{1.69}$$

while in the quantum case

$$\frac{E_{Coulomb}}{p_F^2/2m_e} \sim n_e^{-1/3} \tag{1.70}$$

applies. In contrast to the classical case, quantum mechanically, very dense plasmas become ideal! This is illustrated in Fig. 1.3.

Another important aspect concerns the classification of temperatures. In quantum mechanically degenerate systems, one often speaks of "low" temperatures when

$$k_B T_e \ll \varepsilon_F \equiv \sqrt{m_e^2 c^4 + c^2 p_F^2} \tag{1.71}$$

applies; ε_F is the Fermi energy. Nevertheless, the temperatures can be so high that a relativistic description becomes necessary. This is the case for

$$p_F > m_e c \tag{1.72}$$

In other words, for

$$\boxed{n_e \gg \left(\frac{m_e c}{h}\right)^3} \tag{1.73}$$

we must use a relativistic calculation, even though the temperatures may be referred to as "low".

In Fig. 1.4 we have listed some plasma configurations, from which, together with Fig. 1.3, one can easily determine the appropriate description method. In the following, we show three examples.

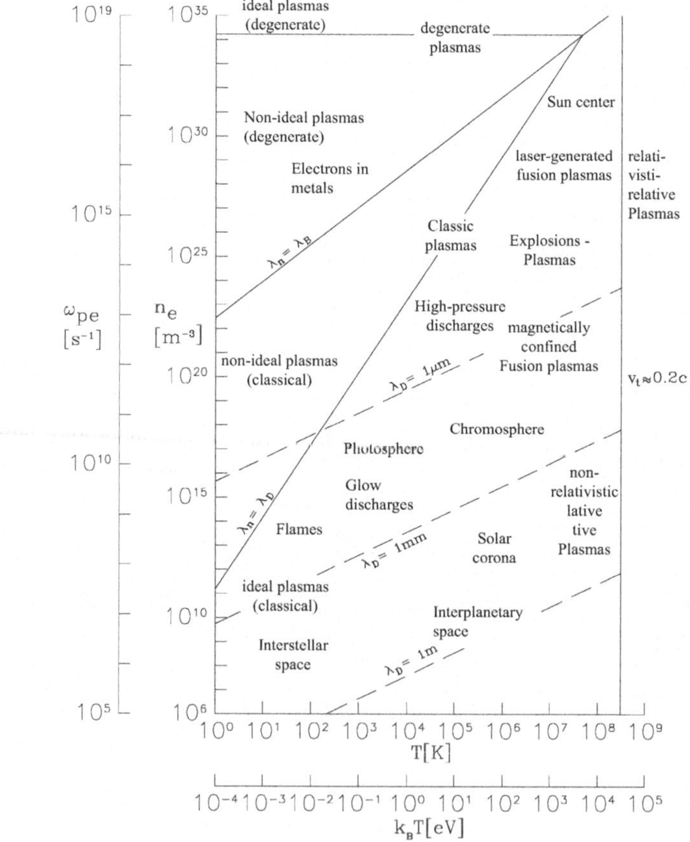

Fig. 1.4 Nomogram for typical plasma scales as a function of particle density and temperature. The electron plasma frequency $\omega_{pe} = (n_e e^2/\varepsilon_0 m_e)^{1/2}$ indicates a characteristic inverse time scale

Example 1.3 (Typical Values in Magnetic Fusion)
When developing a theory, we must define the parameter range to which it should be applicable. Below, we show typical parameters[3] for a magnetic fusion plasma.

$$T \approx 10 \text{ keV} \quad , \quad B \approx 4 \text{ T} \triangleq 4 \times 10^4 \text{ Gauss} ,$$
$$n_e \approx n_i := n \approx 10^{14} \text{ cm}^{-3} . \tag{1.74}$$

For this, the electron Debye length is on the order of

$$\lambda_D \approx 7.43 \times 10^2 \, T^{1/2} \, n^{-1/2} \, [\text{ cm}] \approx 7 \times 10^{-3} \text{ cm} . \tag{1.75}$$

The thermal de Broglie wavelength, the mean particle spacing, and the thermal velocity are, respectively,

$$\lambda_{dB} \approx 2.76 \times 10^{-8} \, T^{-1/2} \, [\text{ cm}] \approx 3 \times 10^{-10} \text{ cm} , \tag{1.76}$$

$$\lambda_n \approx n^{-1/3} \, [\text{ cm}] \approx 10^{-5} \text{ cm} , \tag{1.77}$$

$$v_{the} \approx 4.19 \times 10^7 \, T^{1/2} \, [\text{ cm/s}] \approx 4 \times 10^9 \text{ cm/s} . \tag{1.78}$$

■

Example 1.4 (Temperatures in the Solar Atmosphere)

- The photosphere, from which we receive visible light, is the lowest layer of the solar atmosphere. The mean temperature associated with light emission is 5778 K.
- Above the photosphere, which has a thickness of 400–500 km, lies the chromosphere with a thickness of almost 2000 km. The chromosphere consists predominantly of hydrogen and helium. Its gas density decreases significantly outward.
- Above this is the corona, a fully ionized plasma of hydrogen and helium. Temperatures in the corona are very high; they can exceed 1 million K. Elucidating the heating mechanisms that lead to these high temperatures is one of the main topics of current solar research. ■

Example 1.5 (Values for a White Dwarf)
We now present the example of a white dwarf with typical parameters for a dense system. A characteristic feature of a white dwarf is that quantum effects play an important role.

---

[3] Only in the next formulas should B be used in Gaussian units, as is often done in fusion applications [33]. Unless otherwise specified, in this section T is in eV and B is in Gauss.

- The quantum criterion ($n\lambda_{dB}^3 \geq 1$) is most likely to be fulfilled by the electrons, since $\lambda_{dB} \sim m^{-1/2}$.
- In the degenerate case ($T \to 0$) the pressure P becomes independent of temperature.
- In the quantum case, low temperatures mean $kT \ll \varepsilon_F = \sqrt{m^2c^4 + c^2p_F^2}$ with $p_F = \left(\frac{3n}{8\pi}\right)^{1/3} h$.
- Although the temperatures can be low, an ultrarelativistic treatment may be required. The conditions for this are $c^2p_F^2 \gg m^2c^4$ and $kT \ll cp_F$.

This situation occurs in white dwarfs, which, for example, consist of helium and have a total mass M as well as a mass density ρ. The central temperature is T. White dwarfs are stars with approximately the mass of the Sun, but with a much smaller radius (about by a factor of 10^{-2}). Nuclear fusion has ceased. The pressure that prevents collapse is provided by the degenerate electron gas (i.e., the Fermi pressure). The effective surface temperature is $T_{eff} \approx 8.000$ K, which gives them their white appearance. One of the best-known white dwarfs is Sirius B.

Typical values for white dwarfs are

$$M \approx 10^{33}\ \text{g}\,, \quad \rho \approx 10^7\ \text{g cm}^{-3}, \quad T \approx 10^7\ \text{K}\ . \tag{1.79}$$

At this temperature, helium is almost completely ionized. Each helium atom consists of four nucleons and two electrons. We have $m_{He}c^2 \approx 4\ \text{GeV} \gg k_BT \approx 1\ \text{keV}$, i.e., they are "cold." Similarly, the electrons are also "cold," because $m_ec^2 \approx 511\ \text{keV} \gg 1\ \text{keV}$. With a total of N electrons, the total mass is $M \approx N(m_e + 2m_n) \approx 2m_nN$, where m_n is the mass of a nucleon. From this, the electron density can be determined as a function of the given mass density ρ to the order of magnitude:

$$n_e = \frac{N}{V} \approx \frac{M}{2m_n}\frac{\rho}{M} = \frac{\rho}{2m_n} \approx 3 \times 10^{30}\ \text{cm}^{-3}\ . \tag{1.80}$$

We are therefore in the upper left corner of Fig. 1.3.

Next, we calculate the Fermi momentum

$$p_F = \left(\frac{3n_e}{8\pi}\right)^{1/3} h \approx 5 \times 10^{-17}\ \frac{\text{g cm}}{\text{s}} \approx 0.9\ \frac{\text{MeV}}{c}\ . \tag{1.81}$$

Note that cp_F reaches the same order of magnitude as m_ec^2. This means that relativistic effects become important, even though the electron gas is referred to as "cold." The de Broglie wavelength for helium nuclei is quite small,

$$\lambda_{dBHe} = \left(\frac{h^2}{2\pi mk_BT}\right)^{1/2} \approx 247\ \text{fm} \equiv 247 \times 10^{-13}\ \text{cm}\ . \tag{1.82}$$

By introducing the density of helium nuclei $n = n_e/2$ we find $n\lambda^3_{dBHe} \approx 2.27 \times 10^{-2} \ll 1$. As an important consequence, classical statistics can be applied to the system of helium nuclei, which leads to the helium pressure

$$P_{He} \approx n_{He} k_B T \approx 1.5 \times 10^{-12} \frac{\mathrm{MeV}}{\mathrm{fm}^3} \tag{1.83}$$

On the other hand, the electrons require a quantum mechanical description. Quantum statistics leads for $y_F = p_F/m_e c \approx 2$ to

$$P_e \approx \frac{\pi m_e^4 c^5}{3h^3} A(2) \approx \frac{\pi m_e^4 c^5}{3h^3} 26.7 \approx 10^{-9} \frac{\mathrm{MeV}}{\mathrm{fm}^3} . \tag{1.84}$$

Here, $A(y_F)$ is a function well known in quantum statistics for fermions. In conclusion, we note that the electron pressure dominates, since $P_e \gg P_{He}$. Mainly, it is the electron pressure that counteracts gravitational collapse.

For a detailed discussion, the form of the function $A(y_F)$ must be known. The Chandrasekhar mass results from this relationship and can be easily looked up [22]. ■

1.4 Quasineutrality and Debye Shielding

In this section, we introduce a very important property of a plasma, namely Debye shielding. In a situation close to equilibrium, a charged particle, e.g., an ion, is preferentially surrounded by electrons, which effectively shield it. The plasma is then referred to as quasineutral if (for example, in the case of singly charged ions and $n_e = n_i$) over distances greater than the corresponding Debye length, the Coulomb potentials can no longer be effectively perceived.

Negative charge density fluctuations $\delta\rho = -e\delta n$ (e is the magnitude of the elementary charge) generate electrostatic potential fluctuations $\delta\phi$,

$$\nabla^2 \delta\phi = \frac{1}{\varepsilon_0} e\delta n . \tag{1.85}$$

Roughly estimated, we obtain

$$\nabla^2 \delta\phi \sim \frac{\delta\phi}{l^2} , \tag{1.86}$$

where l is a characteristic fluctuation scale. Thus, it follows that

$$\delta\phi \approx \frac{1}{\varepsilon_0} e\delta n l^2 \,. \tag{1.87}$$

On the other hand, the characteristic potential energy $-e\delta\phi$ cannot be greater than the mean kinetic energy of the particles, which we can approximate by $k_B T$ (we measure the temperature in Kelvin, k_B is the Boltzmann constant, and we ignore numerical factors). Thus,

$$\frac{\delta n}{n} \lesssim \frac{\varepsilon_0 k_B T}{n e^2 l^2} \,. \tag{1.88}$$

We see that a typical length appears, namely the Debye length (further details will be given in the following chapters)

$$\boxed{\lambda_D = \sqrt{\frac{\varepsilon_0 k_B T}{n_{e0} e^2}}} \,, \tag{1.89}$$

so that

$$\frac{\delta n}{n} \lesssim \frac{\lambda_D^2}{l^2} \,. \tag{1.90}$$

A plasma is quasineutral at distances much greater than the Debye radius. If the plasma length is comparable to λ_D, it is not a "real" plasma, but rather a collection of charged particles.

The Debye length is the screening length in a plasma (for now, we are not discussing which species, electrons or ions, dominate the screening process). Let us again start with the Poisson equation

$$\nabla^2\phi = \frac{1}{\varepsilon_0} e(n_e - n_i) \,. \tag{1.91}$$

Assuming that electrons and ions are Boltzmann distributed (let us assume at the same temperature T, which we measure in eV, i.e., $k_B T \to T$), we have

$$n_e = n_{e0} e^{e\phi/T} \approx n_{e0}(1 + e\phi/T) \,, \quad n_i = n_{e0} e^{-e\phi/T} \approx n_{e0}(1 - e\phi/T) \,. \tag{1.92}$$

For spherical symmetry, we obtain

$$\nabla^2\phi = \frac{d^2\phi}{dr^2} + \frac{2}{r}\frac{d\phi}{dr} = \frac{2e^2 n_{e0}}{\varepsilon_0 T}\phi \,. \tag{1.93}$$

This is a homogeneous linear differential equation. The amplitude parameter is free. It is easy to verify that for $r \neq 0$ a solution is given by

$$\Phi = \frac{e}{r} e^{-\sqrt{2}r/\lambda_D} \,. \tag{1.94}$$

Thus, the potential in a plasma is exponentially screened with the Debye length as the screening distance.

Example 1.6 (Debye Potential)
However, the previous calculation is not yet complete. So far, the central charge q that generates the screening cloud is missing. This is reflected in the fact that the amplitude is still undetermined. It is obvious that instead of (1.91) and (1.93), we should solve the following inhomogeneous linearized Poisson-Boltzmann equation:

$$\begin{aligned}\nabla^2\phi &= -\frac{1}{\varepsilon_0}q\,\delta(\mathbf{r}) - \frac{1}{\varepsilon_0}\sum_{s=e,i} q_s\,n_s \\ &\approx \kappa^2\phi - \frac{1}{\varepsilon_0}q\delta(\mathbf{r})\,,\end{aligned} \tag{1.95}$$

where the index s specifies the particle species (electrons or ions). The test charge q is located at $\mathbf{r} = 0$. We define

$$\kappa^2 = \sum_s \frac{n_s q_s^2}{\varepsilon_0 T} \mathrel{\hat{=}} \frac{2}{\lambda_D^2}\,. \tag{1.96}$$

As a solution, we obtain a screened potential q, i.e.,

$$\phi = \frac{1}{4\pi\varepsilon_0}\frac{q}{r}\,e^{-\kappa r}\,. \tag{1.97}$$

The simplest way to obtain the solution is by Fourier transformation

$$\phi_{\mathbf{k}} = \int d^3 r e^{-i\mathbf{k}\cdot\mathbf{r}}\phi(\mathbf{r})\,, \tag{1.98}$$

which leads directly to

$$\phi_{\mathbf{k}} = \frac{1}{\varepsilon_0}\frac{q}{k^2+\kappa^2} \tag{1.99}$$

The inverse transformation yields

$$\begin{aligned}\phi(\mathbf{r}) &= \frac{1}{(2\pi)^3}\int d^3 k e^{i\mathbf{k}\cdot\mathbf{r}}\phi_{\mathbf{k}} = \frac{1}{4\pi\varepsilon_0}\frac{1}{\pi}\int_{-1}^{1} dx \int_0^\infty dk k^2\,\frac{q}{k^2+\kappa^2}\,e^{ikrx} \\ &= \frac{1}{4\pi\varepsilon_0}\frac{2}{\pi}\int_0^\infty dk\,k\,\frac{q}{k^2+\kappa^2}\,\frac{\sin(kr)}{r} = \frac{1}{4\pi\varepsilon_0}\frac{q}{r}\,e^{-\kappa r}\,,\end{aligned} \tag{1.100}$$

since

$$\int_0^\infty \frac{x^{2m+1}\sin(ax)}{(x^2+z)^{n+1}}\,dx = \frac{(-1)^{n+m}}{n!}\,\frac{\pi}{2}\,\frac{d^n}{dz^n}\left(z^m e^{-a\sqrt{z}}\right) . \tag{1.101}$$

We have set the values $n = m = 0,\ a = r,\ x = k$, and $z = \kappa^2$. ∎

Example 1.7 (Induced Space Charge)
Let us now calculate the induced space charge. For this, we decompose:

$$\phi(r) = \frac{1}{4\pi\varepsilon_0}\frac{q}{r}\,e^{-\kappa r} = \phi^{Cb} + \phi^{ind} . \tag{1.102}$$

The "normal" Coulomb potential has been abbreviated as ϕ^{Cb}. Thus, the induced potential is

$$\phi^{ind} = \frac{1}{4\pi\varepsilon_0}\frac{q}{r}\left(e^{-\kappa r} - 1\right) . \tag{1.103}$$

The induced space charge ρ^{ind} is responsible for the screening. It can be calculated from

$$\nabla^2\phi^{ind} \equiv \frac{1}{r^2}\frac{d}{dr}\left(r^2\frac{d\phi^{ind}}{dr}\right) = -\frac{1}{\varepsilon_0}\rho^{ind} \tag{1.104}$$

Differentiation leads directly to

$$\rho^{ind} = -\frac{q\kappa^2}{4\pi r}\,e^{-\kappa r} . \tag{1.105}$$

By integrating over all space, we obtain the expected result

$$\int d^3r\,\rho^{ind} = -q . \tag{1.106}$$

The sign of the screening charge distribution is opposite to that of the test charge q. Due to screening, there is a redistribution of charges compared to the ideal plasma situation (in the latter, the interaction potential is neglected). ∎

The concept of Debye shielding is only valid if there are enough particles in the charge cloud. We can calculate the number N_D of particles in a Debye sphere,

$$N_D = n\,\frac{4}{3}\pi\lambda_D^3 ; \tag{1.107}$$

effective shielding over a Debye length requires

$$N_D \gg 1 . \tag{1.108}$$

A plasma with the characteristic dimension L can be considered quasi-neutral, provided that the dimension L is much greater than the Debye length λ_D. Mathematically expressed:

$$\lambda_D \ll L . \tag{1.109}$$

This means that on scales much larger than the Debye length, the electric charges in the plasma are distributed such that the plasma appears neutral as a whole, even though small local charge inhomogeneities may occur.

Example 1.8 (Maximum Shielding Length)

It is easy to show that the largest spherical volume of a plasma that could spontaneously become free of electrons has a radius of a few Debye lengths. Let us consider a sphere with uniformly distributed ions and density $n_i(\mathbf{r}) = \text{const}$. From the center of the sphere, we introduce the radial coordinate r. The charge enclosed up to radius r is $Q = \frac{4\pi e\, n_i r^3}{3}$, and the electric field has only a radial component

$$E_r = \frac{1}{4\pi\varepsilon_0}\frac{Q}{r^2} = \frac{n_i e r}{3\varepsilon_0} . \tag{1.110}$$

From this, we find the electrostatic field energy W in the sphere of radius r as:

$$W = \frac{\varepsilon_0}{2}\int_0^r E_r^2 4\pi r^2 dr = \frac{2\pi r^5 n_i^2 e^2}{45\varepsilon_0} . \tag{1.111}$$

Let us now consider in more detail the scenario in which the ion-filled sphere was created by removing electrons. Before the electrons left the sphere, they had the kinetic energy

$$E_{kin} = \frac{3}{2} n_e T_e \times \frac{4}{3}\pi r^3 . \tag{1.112}$$

The electrostatic energy W did not exist when the (neutralizing) electrons were originally present in the sphere to balance the ion charge. In other words, W must correspond to the work done by the electrons as they left the sphere. The kinetic energy E_{kin} was available to the electrons. By equating

$$W = E_{kin} \tag{1.113}$$

we find the maximum radius or the largest spherical volume that could spontaneously be depleted of electrons. A short calculation yields

$$r_{\max}^2 = 45\varepsilon_0 \frac{T_e}{n_e e^2} \tag{1.114}$$

or

$$r_{\max} \approx 7\lambda_{De} \,. \tag{1.115}$$

■

1.5 Individual and Collective Effects

In this section, we introduce individual and collective processes using typical examples. Individual effects result, for example, from collision processes. We compare these with collective plasma oscillations.

To understand our initial definition of a (collisionless) plasma, it is already necessary to develop an intuition for collective effects. The latter, in contrast to individual processes of single particles—which we probably have a more elementary understanding of, for example, when we follow the collision process of one particle with another—should be considered separately. Is a plasma merely a collection of rather irregularly wandering particles that collide individually with each other in an indeterminate sequence? By no means, for the plasma is capable of supporting wave motions that require coherent movement over large distances. More on this later.

At this point, let us recall an important effect that we have just discussed and which is significant for the classification of plasmas. If, for example, one solves the Poisson equation for a test particle in an electron-ion system, one finds in a linear calculation under the assumption that electrons and ions are distributed according to Boltzmann statistics, that in the plasma the potential ϕ decreases much more rapidly than in vacuum. The intuitive reason is that, for example, negative charges preferentially accumulate near a positive charge, contributing to screening. Instead of the Coulomb r^{-1} dependence, at large distances one finds the asymptotic form $(q/r)\exp(-r/\lambda_D)$, where q is the charge of the test particle and λ_D is the total Debye length. In formulas, this reads

$$\lambda_D^{-2} = \lambda_{De}^{-2} + \lambda_{Di}^{-2} , \tag{1.116}$$

with the Debye lengths λ_{De} and λ_{Di} for electrons (e) and singly charged ions (i),

$$\lambda_{De,i}^{-2} = \frac{n_{e,i} e^2}{\varepsilon_0 k_B T_{e,i}} \,. \tag{1.117}$$

Here, n denotes the particle density and T the temperature of the electrons or ions. To get a rough idea of the magnitude of the Debye length, we write

$$\lambda_{De,i}[cm] \approx 7.43 \times 10^2 \big(T_{e,i}[\,\text{eV}]/n_{e,i}[\,\text{cm}^{-3}]\big)^{1/2} . \tag{1.118}$$

If, for a plasma, the macroscopic dimension, i.e., its characteristic length L, is much greater than the Debye length,

$$L \gg \lambda_D, \tag{1.119}$$

we speak of a quasineutral or largely neutral system. The screening of an ion by electrons over the characteristic length λ_{De} represents a typical collective process.

In the following, we use classical arguments, i.e., we calculate non-relativistically ($v/c \ll 1$) and neglect quantum effects ($\hbar \to 0$, i.e., the thermal de Broglie wavelength is much smaller than the mean particle distance or the classical interaction radius $r_w = e^2/4\pi\varepsilon_0 k_B T$).

If we compare the mean interparticle distance $\sim n^{-1/3}$ with the Debye length, then collective shielding effects can, of course, only be present if the Debye length is much greater than the mean interparticle distance. In terms of estimation, this means

$$4\pi\varepsilon_0 k_B T_{e,i}/n^{1/3}e^2 \gg 1. \tag{1.120}$$

The left-hand side has a plausible physical meaning. Introducing the plasma parameter

$$\Lambda = \frac{4\pi}{3} n\lambda_D^3, \tag{1.121}$$

with λ_D equal to λ_{De} or λ_{Di}, then the estimation just performed means that the number of particles in the Debye sphere, Λ, must be much greater than one.

A different consideration leads to the same result. If we compare the mean potential energy $\sim n^{1/3}e^2$ of a particle with the mean kinetic energy $\sim k_B T$, we find that for

$$\Lambda \gg 1 \tag{1.122}$$

the kinetic energy component clearly dominates. This closes the loop, as we see that for significant collective effects, the mean kinetic energy must dominate the mean potential energy.

Collective Plasma Oscillations

To further complete the picture, let us discuss characteristic timescales for collective and individual processes. If we take the thermal velocity as the mean particle velocity

$$v_t = (k_B T/m)^{1/2} \tag{1.123}$$

as a basis, where for electrons we use $m = m_e$ and $T = T_e$ (and analogously for ions, using the index i), we obtain from

$$\omega = v_t/\lambda_D \tag{1.124}$$

the two characteristic frequencies

$$\boxed{\omega_{pe} = \left(\frac{n_0 e^2}{\varepsilon_0\, m_e}\right)^{1/2}} \tag{1.125}$$

and

$$\boxed{\omega_{pi} = \left(\frac{n_0 e^2}{\varepsilon_0\, m_i}\right)^{1/2}}\,, \tag{1.126}$$

with $\omega_{pe} \gg \omega_{pi}$. Usually, in these definitions, n_0 (mean particle density) is replaced by n_e (electron number density) or n_i, respectively. Analogous to the total Debye length, a total plasma frequency can be defined by

$$\omega_p^2 = \omega_{pe}^2 + \omega_{pi}^2 \tag{1.127}$$

.

Example 1.9

A "thought experiment" begins with a sphere of radius r, uniformly filled with electrons and protons so that the entire system is globally neutral. However, if only one species, for example electrons, were present, the sphere would have a charge q_e, which would lead to a radial field at the surface $\mathbf{E}$,

$$q_e = -\frac{4\pi}{3} r^3\, e\, n_e\,, \quad \leadsto \quad |E| = \frac{1}{4\pi\varepsilon_0} \frac{q_e}{r^2}\,. \tag{1.128}$$

The field strength could be extremely large, depending on the size of the sphere and the electron density. As already mentioned, we begin our "thought experiment" with homogeneously distributed electrons and protons, so that no electric field exists outside the sphere.

Now we want to expand the electron sphere from radius r to the radius $r+x$ with $x \ll r$, thereby creating an electron shell of thickness x. Since the total number N of electrons and protons remains constant, the electron density is reduced to

$$n_e = n_{e0} + \delta n_e \approx \frac{N}{\frac{4\pi}{3} r^3 \left(1 + 3\frac{x}{r}\right)} \approx n_{e0} - 3 n_{e0} \frac{x}{r} \tag{1.129}$$

Inside the sphere of radius r we now have a positive excess charge

$$\Delta q \approx 4\pi n_{e0} e r^2 x\,, \tag{1.130}$$

which generates an electric field component

$$E \approx \frac{1}{4\pi\varepsilon_0} 4\pi e n_{e0} x \tag{1.131}$$

An electron in the spherical shell experiences the radial force

$$K \equiv m_e \frac{d^2x}{dt^2} = -eE = -\frac{1}{\varepsilon_0} e^2 n_{e0} x \,, \tag{1.132}$$

which pulls it back into the original sphere. An overshoot will occur, leading to oscillations at the electron plasma frequency

$$\omega_{pe} = \sqrt{\frac{e^2 n_e}{\varepsilon_0 m_e}} \tag{1.133}$$

. In this scenario, we have fixed the ions, since their mass is large compared to the electron mass.

The Debye length is related to the plasma frequency by

$$\lambda_D = \frac{v_{th}}{\omega_p} \,. \tag{1.134}$$

■

To get a sense of the order of magnitude of plasma frequencies, we give

$$\omega_{pe}[rad/s] \approx 5.64 \times 10^4 (n_e[cm^{-3}])^{1/2} \tag{1.135}$$

as an example.

Individual Collisions

If we now again ask the question of when collective effects dominate over individual contributions, the characteristic frequency for individual processes is missing for comparison. The latter is usually referred to as the collision frequency, which we understand as the mean inverse time for a (significant) deflection. But how can we define such a frequency precisely in a plasma with long-range and permanent interactions? We will return to this later; here just a few vague remarks, based on the fact that the maximum effective interaction length can be estimated by the Debye length. Because of the low potential energy, we speak of (on average) weak collisions; however, within the Debye zone, many, i.e. Λ, such weak collisions occur. The result, the deflection, is thus a product of a small and a large number. To calculate it, we use the classical results of Rutherford scattering; see Fig. 1.2. If an electron with impact parameter ρ and initial velocity v_0 strikes an ion, the angular momentum $L = m_e r^2 \dot{\theta} = m_e \rho v_0$ is conserved. For the change in the deflection angle θ with time t the following holds

$$dt = \frac{m_e r^2}{L} d\theta \approx \frac{\rho}{v_0} \frac{d\theta}{\sin^2\theta}, \tag{1.136}$$

where, for simplicity, we do not distinguish here between the laboratory and center-of-mass systems (for $m_i \gg m_e$). Furthermore, for weak collisions, in the last step $\rho \approx r \sin\theta$ was used, i.e., it was assumed that (to zeroth order) the trajectory is straight. Here, r measures the distance of the ion from the electron, and we have set $\theta = 0$ at $t = -\infty$. To calculate the velocity components perpendicular to the initial velocity, we use

$$v_\perp = \frac{1}{m_e} \mid \int_{-\infty}^{+\infty} dt\, F_\perp(t) \mid, \tag{1.137}$$

where, due to the Coulomb interaction, ($c \to \infty$) the force component perpendicular to the initial velocity is given by

$$F_\perp = -\frac{1}{4\pi\varepsilon_0} \frac{e^2}{r^2} \sin\theta \tag{1.138}$$

For undisturbed motion, i.e., neglecting the interaction between electron and ion, $\rho \approx r \sin\theta$ holds. Evaluating this, we obtain in the "straight-line approximation"

$$v_\perp \approx \frac{e^2}{4\pi\varepsilon_0 m_e \rho v_0} \int_0^\pi d\theta \sin\theta = \frac{1}{4\pi\varepsilon_0} \frac{2e^2}{m_e v_0 \rho}. \tag{1.139}$$

From this formula, we see that significant changes in velocity with $v_\perp > v_0$ occur at $\rho < \rho_0 \equiv 2e^2/4\pi\varepsilon_0 m_e v_0^2$; however, the derivation is only valid for $v_\perp \ll v_0$. We are therefore stretching the calculation by estimating a portion of the (individual) collision frequency as follows: A *substantial* collision effect occurs at $\rho < \rho_0$. The collision cross section is therefore taken as $\pi\rho_0^2$. How many collisions does an electron experience per unit time? To answer this, we must consider the number of ions in the "cylinder" with "volume" $\pi\rho_0^2 v_0$ as possible collision partners, from which

$$\nu_s \approx \pi\rho_0^2 v_0 n_0 = \frac{1}{(4\pi\varepsilon_0)^2} \frac{4\pi n_0 e^4}{m_e^2 v_0^3} \tag{1.140}$$

follows.

If we estimate v_0 by v_{te}, then, to order of magnitude, we obtain

$$\nu_s \approx \omega_{pe}/\Lambda. \tag{1.141}$$

It is interesting that at least the same order of magnitude results for the total effect of the many small-angle scatterings, if one considers collisions with impact parameters between ρ_0 and the screening length λ_D are taken into account.

Now, some arguments for this. We can use (1.139) for the *small* deviation $\Delta v_\perp$ in a collision ($v_\perp \equiv \Delta v_\perp$) and for dN independent collisions, we set

$$d\langle(\Delta v_\perp)^2\rangle \approx dNv_\perp^2 \approx dN\frac{4e^4}{m_e^2 v_0^2 \rho^2}\frac{1}{(4\pi\varepsilon_0)^2} \tag{1.142}$$

For collisions with impact parameters between ρ and $\rho + d\rho$ we have

$$\nu_c dN = 2\pi\rho d\rho v_0 n_0, \tag{1.143}$$

where ν_c is defined such that in the end $\langle(\Delta v_\perp)^2\rangle \approx v_0^2 \approx v_{te}^2$ should result. Thus, from (1.142) [by integration, where on the right-hand side (1.143) the integration variable p is introduced] we obtain

$$\nu_c \approx \frac{1}{(4\pi\varepsilon_0)^2}\frac{8\pi n_0 e^4}{m_e^2 v_0^3}\int_{\rho_0}^{\lambda_D}\frac{1}{p}dp, \tag{1.144}$$

or, due to

$$\ln(\lambda_D/\rho_0) \approx \ln\Lambda \tag{1.145}$$

$$\boxed{\nu_c \approx \frac{1}{(4\pi\varepsilon_0)^2}\frac{8\pi n_0 e^4}{m_e^2 v_0^3}\ln\Lambda}\,. \tag{1.146}$$

If we compare with (1.140), we find as a difference the factor $2\ln\Lambda > 1$. This result is not all that surprising, since according to our definition of a plasma, only for a small fraction of the particles is the potential energy comparable to the kinetic energy, which is required for individual large deflections. It turns out that many small deflections add up to a greater effect than a few strong deflections. Later, more precise kinetic calculations will support this heuristic argument.

Next, we will discuss the individual collision frequencies in more detail. But first, a few general remarks on collision frequency and cross section using an example.

Example 1.10 (Collision Frequency and Cross Section)
Binary collisions correspond to the classical two-body problem. At this point, it should be mentioned that the two-body problem with Coulomb interaction can be reduced to an effective one-particle problem if we move to the center-of-mass system (COM). The reduced mass then appears. If one mass is much larger than the other (or if we fix the position of the

scatterer, which effectively means we assume an infinitely large mass for the scatterer), the difference between the COM and laboratory system disappears. The following considerations can be interpreted as calculations for a fixed scatterer, once again repeating calculations for the collision frequency and aiming at the cross section.

A charge $q = +e$ scatters an electron beam with charge $-e$. The incoming particle current density is $j = n_e v_e$. Per unit time, $2\pi\rho d\rho j$ particles are scattered as they pass through an annular area $2\pi\rho d\rho$. Each particle is scattered at a specific angle θ. Asymptotically, the change in the *longitudinal* (i.e., in the direction of the original propagation) momentum of a single particle is

$$\Delta p_0 = -m_e v_e (1 - \cos\theta) \; . \tag{1.147}$$

We can define the total force on the *beam* (in the presence of a fixed scatterer). Its magnitude is

$$F_e = -\int_0^\infty m v_e (1 - \cos\theta) j \, 2\pi\rho d\rho \mathrel{\hat=} -m_e v_e j \sigma \; . \tag{1.148}$$

Here, we have assumed fixed initial velocities of magnitude v_e. The expression

$$\sigma = \int_0^\infty (1 - \cos\theta) \, 2\pi\rho d\rho \tag{1.149}$$

is called the transport cross section (for momentum transfer). To evaluate the latter, we need the functional dependence $\theta = \theta(\rho)$. This can be found in standard textbooks on classical mechanics [37]. It holds that

$$\tan\frac{\theta}{2} = \frac{1}{4\pi\varepsilon_0} \frac{e^2}{m_e \rho v_e^2} \; . \tag{1.150}$$

Incidentally, this relation can be easily justified for small scattering angles $\theta \ll 1$. In this case, the straight-line approximation can be used in evaluating the integrals. For example, for the transverse momentum change we obtain

$$\begin{aligned} \Delta p_{0\perp} \approx m_e v_e \theta &= \int_{-\infty}^{\infty} F_\perp dt \approx \frac{1}{4\pi\varepsilon_0} \int_{-\infty}^{\infty} \frac{e^2 \rho}{(\rho^2 + v_e^2 t^2)^{3/2}} dt \\ &= \frac{1}{4\pi\varepsilon_0} \frac{2e^2}{\rho v_e} , \end{aligned} \tag{1.151}$$

where we have approximated $\sin\theta \approx \theta$ and $\cos\alpha \approx \rho/\sqrt{\rho^2 + \mathbf{v}_e^2 t^2}$. The angle θ is the scattering angle and α is the angle between the total force and the perpendicular component.

To evaluate the scattering cross section (1.149), we set $1 - \cos\theta \approx \theta^2/2$ and obtain

$$\sigma = \frac{1}{(4\pi\varepsilon_0)^2} \frac{4\pi e^4}{m_e^2 v_e^4} \int_0^\infty \frac{1}{\rho} d\rho \; . \tag{1.152}$$

Obviously, the integral is logarithmically divergent. We define as the lower limit

$$\rho_{min} = \frac{1}{4\pi\varepsilon_0}\frac{e^2}{m_e v_e^2} , \tag{1.153}$$

which corresponds to 90°-scattering. Note that $\tan(\pi/4) = 1$. Further details can be found in the chapter on transport theory. For very small impact parameters (strong collisions), a classical treatment will be inadequate. The upper limit $\rho_{max} = \lambda_D$ is postulated as the Debye length, due to the screening of the scatterer's potential. Therefore, instead of (1.152), we calculate

$$\sigma = \frac{1}{(4\pi\varepsilon_0)^2}\frac{4\pi e^4}{m_e^2 v_e^4}\int_{\rho_{min}}^{\rho_{max}} \frac{1}{\rho} d\rho . \tag{1.154}$$

Again, the expression

$$\ln\Lambda = \ln\frac{\rho_{max}}{\rho_{min}} = \ln 16\pi^2\varepsilon_0^{5/2}\frac{3(k_B T)^{3/2}}{e^3\sqrt{n}} \tag{1.155}$$

is proportional to the logarithm of the number of particles in the Debye sphere and is referred to as the Coulomb logarithm. In magnetic fusion plasmas, it is on the order of 10 to 20. In (1.155) we have used

$$\frac{1}{2}m_e v_e^2 \approx \frac{3}{2}k_B T \tag{1.156}$$

Of course, within an approximate treatment, slightly different estimates are also used. The result for the cross section is now

$$\sigma = \frac{1}{(4\pi\varepsilon_0)^2}\frac{4\pi e^4}{m_e^2 v_e^4}\ln\Lambda . \tag{1.157}$$

As an order of magnitude, we obtain

$$\sigma \sim \frac{10^{-12}}{(E[eV])^2} cm^2 , \tag{1.158}$$

with the kinetic energy $E = \frac{1}{2}m_e v_e^2$. ■

Collision frequencies for momentum and energy transfer

The transport cross section allows us to determine the *momentum transfer rate* between particles. Let us consider an electron beam in a cold plasma. We further assume that the scatterers are considered fixed. Let n_i be the ion density. If we have n_i scatterers per unit volume, then the average force acting on a single particle of the beam is

$$\mathbf{F}_e = -m_e \mathbf{v}_e j \sigma n_i \frac{1}{n_e} = -m_e v_e \sigma n_i \mathbf{v}_e \ . \quad (1.159)$$

Due to this force, the mean velocity of the electrons is reduced,

$$\frac{d\mathbf{v}_e}{dt} = \frac{\mathbf{F}_e}{m_e} = -n_i \sigma v_e \mathbf{v}_e \equiv -\nu_{ei} \mathbf{v}_e \ . \quad (1.160)$$

Here, $\mathbf{v}_e$ is the relative velocity between electron and ion. For a fixed scatterer, the total energy of an electron remains unchanged. The longitudinal velocity component of the scattered electron changes according to (1.147), or approximately as

$$\Delta v_e = -v_e(1 - \cos\theta) \approx -v_e \frac{\theta^2}{2} \ . \quad (1.161)$$

Thus, the angular broadening increases,

$$\frac{d\theta^2}{dt} = 2\sigma n_i v_e \ . \quad (1.162)$$

The characteristic time

$$\tau_{ei} = \frac{1}{n_i \sigma v_e} \equiv \frac{1}{\nu_{ei}} \quad (1.163)$$

is inverse to ν_{ei}.

The collision frequency follows from the collision cross section after it has been multiplied by $n_i v_e$. The approximate result is therefore

$$\nu_{ei} = \frac{1}{(4\pi\varepsilon_0)^2} \frac{4\pi e^4 n_i}{m_e^2 v_e^3} \ln\Lambda \ . \quad (1.164)$$

The ratio between the individual collision frequency and the collective plasma frequency is proportional to the reciprocal of the number of particles in a Debye sphere,

$$\frac{\nu_{ei}}{\omega_{pe}} \sim \frac{1}{n\lambda_{De}^3} \ , \quad (1.165)$$

for $n_i \approx n_e \approx n$. Note that after the time $T \approx \nu_{ei}^{-1}$ the quantity θ^2 changes significantly. We can calculate characteristic values using

$$\nu_{ei} \sim 6 \times 10^{-5} \frac{n_i[\,\mathrm{cm}^{-3}]}{(E[\,\mathrm{eV}])^{3/2}} \ \mathrm{s}^{-1} \ . \quad (1.166)$$

The mean free path of the electrons is

$$\lambda_{mfpe} = \frac{1}{n_i \sigma} \sim 10^{12} \frac{(E[\,\mathrm{eV}])^2}{n_i[\,\mathrm{cm}^{-3}]}\ \mathrm{cm}\ . \tag{1.167}$$

For a Maxwell distribution, the mean square energy is given by

$$< E^2 >= \frac{15}{4}(k_B T)^2\ . \tag{1.168}$$

This allows us to determine the temperature dependencies of the characteristic quantities.

If we measure lengths in cm, densities in cm^{-3} and temperatures in eV, we obtain the characteristic values

$$\sigma \sim \frac{3 \times 10^{-13}}{T^2}\ \mathrm{cm}^2\ , \tag{1.169}$$

$$\nu_{ei} \sim 3 \times 10^{-5} \frac{n}{T^{3/2}}\ \mathrm{s}^{-1}\ , \tag{1.170}$$

$$\lambda_{mfpe} \sim 3 \times 10^{12} \frac{T^2}{n}\ \mathrm{cm}\ . \tag{1.171}$$

Example 1.11 (Calculation in the Center-of-Mass System)
If the masses of the scatterer (m_2) and the scattered particle (m_1) are similar or $m_1 \gg m_2$, we transform to the center-of-mass system with the center of mass $\mathbf{R}$. Using $\mathbf{r} = \mathbf{r}_2 - \mathbf{r}_1$ for the difference of the position vectors and the reduced mass $m_{12} = m_1 m_2/(m_1 + m_2)$, we have

$$\mathbf{r}_1 = \mathbf{R} - \frac{m_2}{m_1 + m_2}\mathbf{r}\ , \quad \mathbf{r}_2 = \mathbf{R} + \frac{m_1}{m_1 + m_2}\mathbf{r} \tag{1.172}$$

and from this we find

$$m_{12}\ddot{\mathbf{r}} = -Z_1 Z_2 e^2 \frac{\mathbf{r}}{r^3}\ . \tag{1.173}$$

In the following, we assume that the scatterer has charge $Z_2 e$, while the scattered particle has charge $-Z_1 e$. This yields an effective two-body problem, similar to the treatment for fixed scatterers. The mass must be replaced by the reduced mass, and the position of the scatterer is replaced by the relative distance. We can immediately transfer the previous formulas; for example, (1.157) can now be written as

$$\sigma = \frac{1}{(4\pi\varepsilon_0)^2} \frac{4\pi Z_1^2 Z_2^2 e^4}{m_{12}^2 v^4} \ln \Lambda\ , \tag{1.174}$$

and the general formula for the collision frequency (to deflect the incoming particle 1 by an angle of 90°, in the presence of n_2 scatterers per unit volume) is

$$\nu_{12} = v\sigma n_2 = \frac{1}{(4\pi\varepsilon_0)^2}\frac{4\pi Z_1^2 Z_2^2 e^4 n_2 \ln\Lambda}{m_{12}^2 v^3} \,. \tag{1.175}$$

Note that we have calculated the frequency for velocity deflections. Thus, ν_{12} directly gives the frequency for momentum transfer in the center-of-mass system. We will denote the *momentum* scattering frequencies as ν_{ee}, ν_{ii}, ν_{ei} and ν_{ie} for the various possible interactions between the species. The reciprocals are denoted by $\tau \sim \nu^{-1}$. ■

As a typical velocity, we take $v \sim v_{th}$. We reference all collision frequencies to ν_{ee}. Note that for the reduced mass $m_{ee} \sim m_e/2 \sim m_e$ holds, and $v \sim T^{1/2}/m_e^{1/2}$ applies in this case. When we calculate ν_{ei}, we have $m_{ei} \sim m_e$ and $v \sim T^{1/2}/m_e^{1/2}$, i.e., (apart from a factor of 2) the same values, and therefore

$$\boxed{\nu_{ei} \sim \nu_{ee}} \,. \tag{1.176}$$

Next, we calculate ν_{ii}. Now, $m_{ii} \sim m_i/2$ holds and $v \sim T^{1/2}/m_i^{1/2}$. Therefore,

$$\boxed{\nu_{ii} \sim \sqrt{\frac{m_e}{m_i}}\,\nu_{ee}} \,. \tag{1.177}$$

In all these cases, the differences between the laboratory system and the center-of-mass system are tolerable, since in the current estimates we are ignoring factors of order 2.

Special care, however, must be taken when calculating ν_{ie}. The transformation to the laboratory system is necessary, straightforward, but not immediately obvious. A simpler method for estimating ν_{ie} is momentum conservation in the laboratory system, which leads to $m_i\Delta\mathbf{v}_i = -m_e\Delta\mathbf{v}_e$, where Δ denotes the change in magnitude as a result of the collision. For a head-on collision, we have approximately $\Delta\mathbf{v}_e \approx 2\mathbf{v}_i$, and therefore $|\Delta\mathbf{v}_i|/|\mathbf{v}_i| \approx 2m_e/m_i$ holds. Thus, in order to achieve $|\Delta\mathbf{v}_i|/|\mathbf{v}_i|$ of order unity, m_i/m_e collisions are required. Therefore,

$$\boxed{\nu_{ie} \sim \frac{m_e}{m_i}\nu_{ee}} \,. \tag{1.178}$$

The energy scattering is characterized by the time required for an incoming particle to transfer its kinetic energy to the target particle. The frequencies for energy transfer in collisions are denoted as ν_{ee}^E, ν_{ii}^E, ν_{ei}^E and ν_{ie}^E.

Now let us consider *energy* changes. When a moving electron undergoes a head-on collision with a stationary electron, the incoming electron comes to a stop, while the originally stationary electron flies off with the same momentum and energy that the incoming electron had. From this we conclude that

$$\boxed{\nu_{ee}^E \sim \nu_{ee}} \, . \tag{1.179}$$

It is similar when an ion collides with an ion,

$$\boxed{\nu_{ii}^E \sim \nu_{ii} \sim \sqrt{\frac{m_e}{m_i}} \, \nu_{ee}} \, . \tag{1.180}$$

Finally, let us compare the energy changes during electron-ion and ion-electron collisions. The change in momentum of the electron is $-2m_e\mathbf{v}_e$ during a collision of an electron with an ion. From conservation of momentum it follows that $m_i\mathbf{v}_i = 2m_e\mathbf{v}_e$. The energy transferred to the ion is $\frac{1}{2}m_i v_i^2 = 4(m_e/m_i)m_e v_e^2/2$. An electron must undergo m_i/m_e collisions in order to transfer all its energy to the ions. Therefore,

$$\boxed{\nu_{ei}^E \sim \frac{m_e}{m_i} \, \nu_{ee}} \, . \tag{1.181}$$

Similarly, in an ion-electron collision, an initially stationary electron will fly off with twice the velocity of the incoming ion. The electron gains energy $\frac{1}{2}m_e v_e^2 \sim 2(m_e/m_i)m_i v_i^2$. Again, about m_i/m_e collisions are necessary for the ion to transfer all its energy to the electrons, that is,

$$\boxed{\nu_{ie}^E \sim \frac{m_e}{m_i} \, \nu_{ee}} \, . \tag{1.182}$$

These rough estimates have an important consequence, namely that even in nonequilibrium situations we can describe an electron-ion plasma using a two-fluid model. Relaxation to approximate Maxwell distributions occurs rapidly within each component (most rapidly for the electrons), while exchange between the components takes place on a much slower timescale.

Macroscopic Friction Force

We now consider a thermal electron-ion plasma. The velocity distribution functions are assumed to be

$$f_i(\mathbf{v}') = \left(\frac{m_i}{2\pi T_i}\right)^{3/2} \exp\left(-\frac{m_i(\mathbf{v}' - \mathbf{u}_i)^2}{2T_i}\right) \tag{1.183}$$

and

$$f_e(\mathbf{v}') = \left(\frac{m_e}{2\pi T_e}\right)^{3/2} \exp\left(-\frac{m_e(\mathbf{v}' - \mathbf{u}_e)^2}{2T_e}\right) . \tag{1.184}$$

In this configuration, the current density is

$$\mathbf{j} = n_i e \mathbf{u}_i - n_e e \mathbf{u}_e . \tag{1.185}$$

It is useful to transform into the frame of the mean velocity $\mathbf{u}_e$ of the electrons, thereby defining a relative velocity:

$$\mathbf{u}_{rel} = \mathbf{u}_i - \mathbf{u}_e \equiv \mathbf{u} . \tag{1.186}$$

In the new frame, the distribution functions are

$$f_i(\mathbf{v}) = \left(\frac{m_i}{2\pi T_i}\right)^{3/2} \exp\left(-\frac{m_i(\mathbf{v} - \mathbf{u}_{rel})^2}{2T_i}\right) \tag{1.187}$$

and

$$f_e(\mathbf{v}) = \left(\frac{m_e}{2\pi T_e}\right)^{3/2} \exp\left(-\frac{m_e v^2}{2T_e}\right) . \tag{1.188}$$

Since the thermal velocity of the ions is much smaller than that of the electrons, the velocity distribution of the ions is much narrower than that of the electrons, and the ions can be approximately regarded as a monoenergetic beam. When they encounter n_e electrons per unit volume, we calculate the net force on the ions by using the previous expressions and applying the *actio* = *reactio* principle. Using (1.159), we switch for a moment to the frame in which the ions are at rest, so that the electrons have the velocity$\mathbf{v}_e - \mathbf{u}$. Then, $\mathbf{F}_e$ can be used in the form (1.159). If we set $\mathbf{F}_i = -\mathbf{F}_e$, we find

$$\mathbf{F}_i = -m_e|\mathbf{u} - \mathbf{v}_e|\sigma n_e(\mathbf{u} - \mathbf{v}_e) , \tag{1.189}$$

with $(Z_1 = 1, Z_2 = Z)$

$$\sigma \approx \frac{1}{(4\pi\varepsilon_0)^2} \frac{4\pi Z^2 e^4}{m_e^2|\mathbf{u} - \mathbf{v}_e|^4} \ln \Lambda . \tag{1.190}$$

Incidentally, similar formulas would apply for a fast electron component in a thermal plasma. Suppose we have a fraction of scatterers in a certain velocity range. Then we replace

$$n_e \to dn_e = n_e f_e(\mathbf{v}')d^3v' \,. \tag{1.191}$$

We rewrite the friction force for dn_e scatterers per unit volume. Taking into account the relative velocity $\mathbf{u} - \mathbf{v}'$ of the scattered particles, we obtain

$$d\mathbf{F}_i = -\frac{1}{(4\pi\varepsilon_0)^2}\frac{4\pi Z^2 \ln\Lambda e^4 n_e}{m_e}\frac{\mathbf{u}-\mathbf{v}'}{|\mathbf{u}-\mathbf{v}'|^3} f_e(\mathbf{v}')d^3v' \,. \tag{1.192}$$

When averaging over the possible velocities of the scatterers, one must compute the integral

$$\mathbf{I} = \int \frac{\mathbf{u}-\mathbf{v}'}{|\mathbf{u}-\mathbf{v}'|^3} f_e(\mathbf{v}')d^3v' \tag{1.193}$$

Its form is reminiscent of the integral over a charge distribution that appears when solving the Poisson equation [38]. If the distribution function f is isotropic, the integral corresponds to an electric field produced by a spherically symmetric charge distribution with "radius vector" $\mathbf{u}$. We know that the radial electric field at a distance u is produced only by the charge inside the sphere of "radius" u, i.e.,

$$\mathbf{I} = \frac{\mathbf{u}}{u^3}\int_0^u f_e(v')4\pi v'^2 dv' \,. \tag{1.194}$$

Example 1.12
For readers not familiar with the discussion of the Poisson integral, we present a "calculation for beginners." Let us start with

$$\int \frac{\mathbf{u}-\mathbf{v}'}{|\mathbf{u}-\mathbf{v}'|^3} f_e(\mathbf{v}')d^3v' = -\nabla_u \int \frac{1}{|\mathbf{u}-\mathbf{v}'|} f_e(\mathbf{v}')d^3v' \,. \tag{1.195}$$

Next, we introduce spherical coordinates for the integration on the right-hand side, perform the integration over the azimuthal angle, and after a simple transformation obtain

$$\begin{aligned}
&\int \frac{1}{|\mathbf{u}-\mathbf{v}'|} f_e(\mathbf{v}')d^3v' \\
&= 2\pi \int_0^\infty dv' v'^2 \int_{-1}^{+1} d(\cos\vartheta')\frac{d}{\cos\vartheta'}\sqrt{u^2+v'^2-2uv'\cos\vartheta'}\left(-\frac{f_e(\mathbf{v}')}{uv'}\right) \\
&= 4\pi\left\{\frac{1}{u}\int_0^u v'^2 f_e(\mathbf{v}')dv' + \int_u^\infty v' f_e(\mathbf{v}')dv'\right\} .
\end{aligned} \tag{1.196}$$

By applying the operator $-\nabla_u$ to the right-hand side, we obtain (1.194). ■

For small relative velocities compared to the thermal electron velocity, $u \ll v_{th}$, we obtain

$$\int_0^u f_e(v')4\pi v'^2 dv' \approx f_e(0)\frac{4\pi}{3}u^3 , \tag{1.197}$$

and thus for a Maxwell distribution

$$\mathbf{I} \approx \frac{\sqrt{2}}{3\sqrt{\pi}}\left(\frac{m_e}{T_e}\right)^{3/2} \mathbf{u} . \tag{1.198}$$

On the other hand, for large velocities u

$$\mathbf{I} \sim \frac{\mathbf{u}}{u^3} . \tag{1.199}$$

In summary, for small velocities compared to the thermal electron velocity

$$\mathbf{F}_i = -\frac{1}{(4\pi\varepsilon_0)^2}\frac{4\sqrt{2\pi}}{3}\frac{Z^2 \ln\Lambda e^4 n_e}{m_e}\left(\frac{m_e}{T_e}\right)^{3/2} \mathbf{u} , \tag{1.200}$$

i.e., the frictional force increases with velocity. However, when the drift velocity becomes greater than the thermal velocity, we obtain a decreasing frictional force, $F \sim u^{-2}$. Here, u is the mean velocity difference. The frictional force reaches a maximum near the thermal electron velocity.

We now apply these results to the problem of current flow through plasma. In equilibrium, the driving force produced by the electric field must be balanced by friction, i.e., for the ions

$$\mathbf{F}_i + Ze\mathbf{E} = 0 . \tag{1.201}$$

However, if the drift velocity becomes much greater than the thermal electron velocity, collisions can no longer stop the acceleration of the particles, and the particles run away. In other words, if the applied electric field becomes too large,

$$E > \frac{F_{max}}{Ze} , \tag{1.202}$$

we obtain "runaways." The critical field is called the Dreicer field:

$$E_{Dr} \approx \frac{1}{(4\pi\varepsilon_0)^2}\frac{\ln\Lambda\, n_e e^3 Z}{T_e} \sim \ln\Lambda \frac{e}{\lambda_D^2} . \tag{1.203}$$

A more detailed kinetic calculation yields [16]

$$E_{Dr} \approx \frac{0.43}{(4\pi\varepsilon_0)^2}\frac{2\pi\ln\Lambda\, n_e e^3 Z}{T_e} \approx 5.6\times 10^{-18} n_e Z \frac{\ln\Lambda}{T_e}\,\frac{V}{m}\,. \tag{1.204}$$

To illustrate the typical behaviors for $u \ll v_{th}$ and $u \gg v_{th}$, we solve the momentum equation

$$m_i \frac{d\mathbf{u}}{dt} = \mathbf{F}_i + Ze\mathbf{E} \tag{1.205}$$

in two regimes. In both cases, we choose a one-dimensional description.

In the regime of small velocities, the momentum equation takes the form

$$\frac{du}{dt} = -a\,u + b\,, \tag{1.206}$$

(with constants a and b, which are easy to determine). Its solution for $u = u_0 = 0$ at $t = t_0 = 0$ is

$$u(t) = \frac{b}{a}\left(1 - e^{-at}\right) \;\rightarrow\; \frac{b}{a} \quad \text{for} \quad t \rightarrow \infty\,. \tag{1.207}$$

For small velocities (fields), there is a stationary conduction state

$$j \sim enu \sim ne\frac{eE}{m\nu_{ei}} \sim \frac{\varepsilon_0\omega_p^2}{\nu_{ei}}E \sim \sigma E\,. \tag{1.208}$$

The conductivity is inversely proportional to the collision frequency. Since the latter decreases with increasing temperature, we find

$$\sigma \sim T_e^{3/2}\,, \tag{1.209}$$

i.e., the electrical conductivity of hot plasmas becomes very large. The resistance η is inversely proportional to the conductivity σ,

$$\eta = \frac{1}{\sigma}\,, \tag{1.210}$$

and thus decreases with temperature. The value obtained from kinetics [16] is

$$\eta \approx \frac{1}{(4\pi\varepsilon_0)^2}\frac{8\sqrt{\pi}\ln\Lambda\, Ze^2 m_e^{1/2}}{3\sqrt{2}T_e^{3/2}}\,. \tag{1.211}$$

For practical applications, we can use

$$\eta \approx 1.03 \times 10^{-4} \frac{Z \ln \Lambda}{T_e^{3/2}} \text{ Ohm m} , \tag{1.212}$$

where T_e is measured in eV.

In the regime of high velocities, the momentum equation takes the form

$$\frac{du}{dt} = -\frac{g}{u^2} + b , \tag{1.213}$$

(with constants g and b, which in turn are easy to determine). Its solution for $u = u_0 > \sqrt{\frac{g}{b}}$ for $t = t_0 = 0$ is

$$u - u_0 + \frac{1}{2}\sqrt{\frac{g}{b}} \ln \left[1 - 2\frac{\sqrt{\frac{g}{b}}}{u + \sqrt{\frac{g}{b}}} \right]_{u_0}^{u} = bt . \tag{1.214}$$

It is easy to see that t is a monotonically increasing function of u for $u > \sqrt{\frac{g}{b}}$. In other words,

$$u \to bt \quad \text{for} \quad t \to \infty . \tag{1.215}$$

This asymptotic behavior reflects the so-called "runaway phenomenon."

Summary and Outlook

For the ratio of individual to collective effects, we find

$$\boxed{\frac{\nu_c}{\omega_{pe}} \approx \frac{\ln \Lambda}{\Lambda} \approx \frac{1}{\Lambda} \ll 1} , \tag{1.216}$$

and once again, as a consistency condition, $\Lambda \gg 1$ appears.

In the following chapters, we will examine systems of charged particles with large numbers (Λ) of particles in the Debye zone. We already know that, under this condition, collective processes play an essential role. We will conduct our investigations using the methods—and standards—of theoretical physics, that is, we will start more or less from "first principles" and strive as far as possible for a closed mathematical formulation or solution. As already indicated, however, we cannot always claim completeness; the problems of plasma and its associated applications are too complex for that. The selection of topics is based on various considerations: One essential aspect is that the material presented here should form the basis of a lecture on "Theoretical Plasma Physics" in the theoretical physics curriculum, also for future "non-plasma physicists." Another, and certainly not the least important, is guided by personal preference. In this sense, we will first apply our knowledge of classical mechanics and electrodynamics to a plasma, in order to have, as a necessary prerequisite for later considerations, an understanding of the trajectories of charged particles in external electromagnetic fields. We will then discuss, under the assumption of thermodynamic equilibrium, the statistical behavior of a plasma. The main part of the book will deal with nonequilibrium phenomena. Within the framework of a theoretical physics course, this part can be regarded as an extension or continuation of the lecture "Thermodynamics and Statistics." Issues concerning the kinetic description of a many-particle system will be addressed, as well as the possibility of a macroscopic description and its consequences for transport processes. Greater depth—with the aim of examining specific plasma processes in more detail—will be sought in the chapters on Vlasov systems. There, we will essentially address the important questions of waves and instabilities. Ultimately, nonlinear effects must not be neglected in such a presentation. They have recently become increasingly important, both with regard to basic research and in the interpretation of experimental observations.

Motion of Individual Particles 2

Abstract

Particles move in a plasma under the influence of fields. The latter can be externally imposed, i.e., in the form of so-called external electric and magnetic fields. In addition, there is also an interaction between the particles themselves due to so-called internal fields, e.g., Coulomb fields in the electrostatic approximation. We ignore the internal fields in this chapter. In this chapter, we consider the motion of particles in prescribed electromagnetic fields. The latter can be spatially inhomogeneous. Any temporal dependence, if present at all, is assumed to be weak when considering drift motion. The specific aspects of the (relativistic) motion of a particle in an electromagnetic wave are treated in a separate section. Special aspects of Fermi acceleration are treated separately. The chapter concludes with the motion of a particle under the influence of stochastic forces, i.e., within the framework of the Langevin approach.

2.1 Overview of Drifts

In this section, we discuss—physically, that is, not always with the utmost mathematical rigor—how individual charged plasma particles move in given external fields. The fields are usually not simple, as they can be spatially inhomogeneous or curved. Temporal variations are also allowed.

We begin with a very simple example. When a charged particle with charge q and mass m moves in a constant magnetic field $\mathbf{B} = B\hat{z}$, the non-relativistic equation of motion is:

K.-H. Spatschek, *Theoretical Plasma Physics*,
https://doi.org/10.1007/978-3-662-72828-4_2

$$m\frac{d\mathbf{v}}{dt} = q\mathbf{v} \times \mathbf{B} \,. \tag{2.1}$$

The solution for the z-component is trivial, namely $v_z = \text{const}$. If we introduce the gyrofrequency

$$\boxed{\Omega = \frac{qB}{m}} \tag{2.2}$$

the other components follow from

$$\frac{d^2 v_x}{dt^2} = -\Omega^2 v_x, \quad \frac{d^2 v_y}{dt^2} = -\Omega^2 v_y \tag{2.3}$$

with the solutions

$$v_x = a_x \sin(\Omega t + \delta_x) \,, \quad v_y = a_y \cos(\Omega t + \delta_y) \,. \tag{2.4}$$

The solutions contain integration constants a and δ, which are determined by the initial conditions. If we postulate for $t = 0$ that $v_x = 0$ (and thus $\delta_x = 0$) and

$$v_y(t=0) = a_y \cos\delta_y \equiv \pm v_\perp \,, \quad \text{with } v_\perp = const = (v_x^2 + v_y^2)^{1/2} \tag{2.5}$$

hold, the solutions become simple. Because

$$\frac{dv_x}{dt} = \Omega v_y \tag{2.6}$$

we can determine the constants in

$$v_x = a_x \sin\Omega t \,, \quad v_y = (\pm v_\perp)\cos\Omega t - (\pm v_\perp)\tan\delta_y \sin\Omega t \tag{2.7}$$

as $a_x = \pm v_\perp$ and $\delta_y = 0$. This leads to

$$v_x = \pm v_\perp \sin\Omega t \,, \quad v_y = \pm v_\perp \cos\Omega t. \tag{2.8}$$

The result is easy to interpret. Particles with positive charge have $\Omega > 0$, and we obtain—looking in the direction of the magnetic field—a left-handed helical motion, regardless of the sign of $\pm v_\perp$. Integration yields

$$x = x_0 \mp \frac{v_\perp}{\Omega}\cos\Omega t \,, \quad y = y_0 \pm \frac{v_\perp}{\Omega}\sin\Omega t. \tag{2.9}$$

The Larmor radius

$$r_L = v_\perp/|\Omega| \tag{2.10}$$

determines the width of the helix.

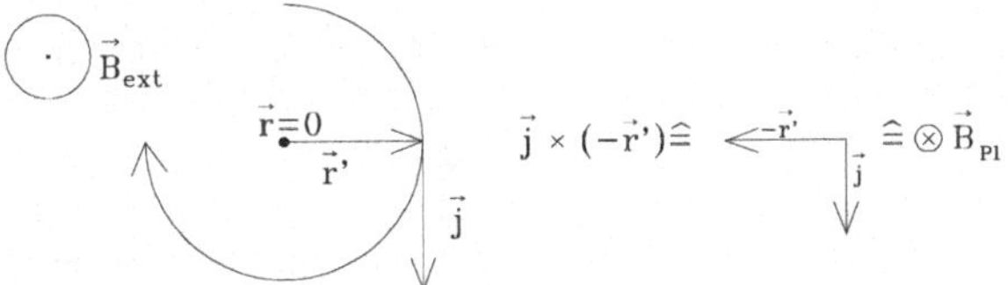

Fig. 2.1 Sketch of the direction of the current density of charged particles moving in an external magnetic field pointing out of the plane of the image. Using the integrand of the Biot–Savart law, one finds the direction of the generated magnetic field, which points into the plane of the image

The circular currents in the x, y plane are responsible for the diamagnetic behavior of the plasma. The Biot–Savart law

$$\mathbf{B}_{Pl}(\mathbf{r}) = \frac{\mu_0}{4\pi} \int \frac{\mathbf{j}(\mathbf{r}') \times (\mathbf{r} - \mathbf{r}')}{|\mathbf{r} - \mathbf{r}'|^3} dV' \tag{2.11}$$

allows one ($\mathbf{j}$ is the current density) to determine the direction of the generated magnetic field $\mathbf{B}_{Pl}$. This is oriented opposite to the original external magnetic field $\mathbf{B} = \mathbf{B}_{ext}$; see Fig. 2.1.

Let us now include an external electric field $\mathbf{E}$. Its component $E_{\|} \neq 0$ along $\mathbf{B}$ causes an acceleration. In the plane perpendicular to $\mathbf{B}$, the motion is governed by

$$\frac{d\mathbf{v}_\perp}{dt} = \frac{q}{m}\mathbf{v}_\perp \times \mathbf{B} + \frac{q}{m}\mathbf{E}_\perp \, . \tag{2.12}$$

We decompose the velocity $\mathbf{v}_\perp$ into $\mathbf{v}_\perp := \mathbf{w}_D + \mathbf{u}$, where

$$\boxed{\mathbf{w}_D = \frac{\mathbf{E}_\perp \times \mathbf{B}}{B^2}} \, . \tag{2.13}$$

After a straightforward calculation, we obtain

$$\frac{d\mathbf{u}}{dt} = \frac{q}{m}\mathbf{u} \times \mathbf{B} \, . \tag{2.14}$$

The velocity $\mathbf{w}_D$ is referred to as the $E \times B$ velocity. Its direction is independent of the sign of the charge. This is illustrated in Fig. 2.2.

Since (2.13) is proportional to $1/B$, one might assume that it diverges for $B \to 0$. However, when $\mathbf{w}_D \to c$, a relativistic calculation becomes necessary, which prevents the divergence.

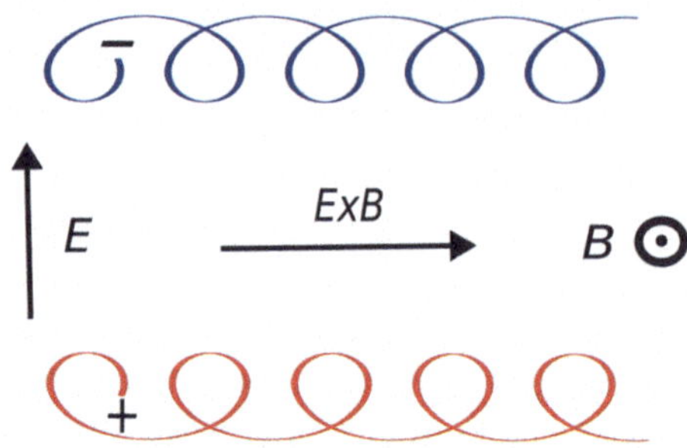

Fig. 2.2 Sketch of the $E \times B$ drift of charged particles. The particles start at + and −, respectively

The splitting into two velocities is the starting point for the "guiding center approximation" [39]. The center of the (rapid) circular motion is referred to as the "guiding center" (gyration center). It moves in the plane perpendicular to the magnetic field with the $E \times B$ drift velocity (in addition to the motion along the magnetic field). Relative to the gyration center, a rotation with velocity $\mathbf{u}$ occurs, which is generally faster than the drift. In the case of homogeneous and constant straight magnetic and electric fields, the decomposition is exact.

Now let us consider the motion of a particle in an inhomogeneous magnetic field. First, we assume that the magnetic field lines remain straight (parallel to the z-axis). No electric field is present. The spatial inhomogeneity introduces a new length scale, characterized by the inhomogeneity length L,

$$\frac{1}{L} \sim |\nabla \ln B| \,. \tag{2.15}$$

The following calculations make the assumption

$$r_L/L \ll 1 \,. \tag{2.16}$$

This is the central assumption in the "guiding center approximation." For $\mathbf{B} = B(x)\hat{z}$ we can assume the vector potential in the form $\mathbf{A} = (0, A_y(x), 0)$. Then $Q_x \equiv x$ is the only non-cyclic coordinate. Starting from the Hamiltonian (Hamilton function), we introduce the canonical momentum $\mathbf{P} = m\mathbf{v} + q\mathbf{A}$, so that the Hamiltonian for the particle motion can be written in the form

$$H = \frac{P_x^2}{2m} + V(Q_x) \tag{2.17}$$

where

$$V(Q_x) = \frac{[P_y - qA_y(Q_x)]^2}{2m} + \frac{P_z^2}{2m} \,. \tag{2.18}$$

We have introduced the constant components P_y and P_z, which are conjugate to the cyclic coordinates y and z, respectively. The motion thus becomes effectively one-dimensional. The solutions depend on $A_y(x)$. Let us assume that the latter has a minimum at $x = 0$. Depending on the value of $H = \text{const}$, the motion remains confined to a finite range in x. Of course, there is also motion along the z- and y-axes. Note that $P_y = \text{const}$ does not mean that $\dot{y} = \text{const}$ is in the y-direction. As we will see, the motion in the y-direction is a drift, the so-called ∇B drift.

Let us now calculate the averaged forces in the x and y directions. The averaging is performed over the (assumed fast) circular motion. Starting with the y component, we obtain from the Lorentz force $\mathbf{F} = q\mathbf{v} \times \mathbf{B}$

$$F_y = -qv_x B_z(x) \; . \tag{2.19}$$

By substituting the exact position x and the velocity with those of the circular motion, we obtain after a Taylor expansion

$$F_y \approx -qv_\perp \sin \Omega t \left[B_0 - \frac{v_\perp}{\Omega} \cos \Omega t \frac{\partial B}{\partial x} \right] , \tag{2.20}$$

where $B_0 = B_z(0)$. Time averaging yields $\langle F_y \rangle \approx 0$. The calculation of F_x leads to

$$F_x \approx qv_\perp \cos \Omega t \left[B_0 - \frac{v_\perp}{\Omega} \cos \Omega t \frac{\partial B}{\partial x} \right] , \tag{2.21}$$

and after averaging

$$\langle F_x \rangle \approx -\frac{qv_\perp^2}{2\Omega} \frac{\partial B}{\partial x} \; . \tag{2.22}$$

A mean force acts on the particle in the plane perpendicular to the magnetic field. Similar to the calculation for a constant electric field, which leads to the $E \times B$ drift, we now find by substituting $q\mathbf{E} \to \langle F_x \rangle \hat{x}$

$$\mathbf{w}_D \approx \frac{1}{q} \frac{\langle \mathbf{F} \rangle \times \mathbf{B}}{B^2} \approx -\frac{v_\perp^2}{2\Omega B^2} \nabla B \times \mathbf{B} \; , \tag{2.23}$$

which can also be written as

$$\boxed{\mathbf{v}_{\nabla B} = \text{sign}\,(q) \frac{1}{2} |v_\perp| |r_L| \frac{\mathbf{B} \times \nabla B}{B^2}} \tag{2.24}$$

This velocity is called the ∇B drift; see Fig. 2.3.

Now let us consider curved magnetic fields. If we assume that the particles move approximately along the magnetic field lines, they experience a centrifugal force

$$\mathbf{F} = m\omega \times (\mathbf{r} \times \omega) \; , \tag{2.25}$$

Fig. 2.3 Sketch of the ∇B drift of a charged particle

which leads to

$$\mathbf{F} \approx \frac{mv_\parallel^2}{R_c^2}\mathbf{R}_c \tag{2.26}$$

Here, R_c is the local radius of curvature, and $v_\parallel$ is the local velocity along a magnetic field line. Similar to the calculation of the $E \times B$ drift (where $q\mathbf{E}_\perp$ is replaced by $\mathbf{F}$), we obtain

$$\boxed{\mathbf{v}_R = \frac{mv_\parallel^2}{qB^2}\frac{\mathbf{R}_c \times \mathbf{B}}{R_c^2}} \,. \tag{2.27}$$

This velocity is called curvature drift.

In general, the curvature drift occurs together with the ∇B drift. Let us illustrate this using the example of a magnetic field that, in cylindrical coordinates, has only a θ component. The z component of $\nabla \times \mathbf{B} = 0$ leads to

$$(\nabla \times \mathbf{B})_z \equiv \frac{1}{r}\frac{\partial}{\partial r}(rB_\theta) \stackrel{!}{=} 0\,, \quad \rightarrow \quad B_\theta \sim \frac{1}{r}\,. \tag{2.28}$$

Since the strength of the magnetic field varies with the radius, this leads to

$$\frac{\nabla B}{B} = -\frac{\mathbf{R}_c}{R_c^2}\,. \tag{2.29}$$

A ∇B drift then follows in the form

$$\mathbf{v}_{\nabla B} = \frac{1}{2}\frac{m}{q}v_\perp^2\frac{\mathbf{R}_c \times \mathbf{B}}{R_c^2 B^2}\,. \tag{2.30}$$

Together with the curvature drift, this leads to

$$\mathbf{v}_R + \mathbf{v}_{\nabla B} = \frac{m}{q}\frac{\mathbf{R}_c \times \mathbf{B}}{R_c^2 B^2}\left[v_\parallel^2 + \frac{1}{2}v_\perp^2\right]. \tag{2.31}$$

Example 2.1 (Force in mirror configurations)
An example originating from mirror machines considers a cylindrically symmetric magnetic field with $B_\theta = 0$. For a specific form of motion, we can assume that $\partial_\theta = 0$. Under this assumption, $\nabla \cdot \mathbf{B} = 0$ leads to

$$\frac{1}{r}\frac{\partial}{\partial r}(rB_r) + \frac{\partial B_z}{\partial z} = 0 . \tag{2.32}$$

For a magnetic bottle with

$$|B_z| \gg |B_r| \tag{2.33}$$

and only a weak dependence of B_z on r, we find approximately

$$B_r \approx -\frac{1}{2}r\frac{\partial B_z}{\partial z}\bigg|_{r=0} \neq 0 . \tag{2.34}$$

From the r component of $\mathbf{B}$ we obtain a z component of the Lorentz force

$$F_z \approx -qv_\theta B_r \approx \frac{1}{2}qv_\theta r\frac{\partial B_z}{\partial z}\bigg|_{z=0} . \tag{2.35}$$

To estimate the order of magnitude, we approximate $v_\theta \approx \pm v_\perp$ and $r \approx r_L$ and find

$$F_\| \approx -\frac{\mu}{B}[\mathbf{B} \cdot \nabla \mathbf{B}]_\| , \quad \mu = \frac{1}{2}\frac{mv_\perp^2}{B} . \tag{2.36}$$

The magnetic moment μ appears as a factor. ■

Finally, we consider inhomogeneous electric fields. We begin with spatial inhomogeneity. If we assume

$$\mathbf{E} = E_x\hat{x} \equiv E_0 \cos kx\,\hat{x} \tag{2.37}$$

we can consider the field address as a Fourier component of a more general form. For $\mathbf{B} = B\hat{z}$, the equations of motion follow in the form

$$\frac{dv_x}{dt} = \Omega v_y + \frac{q}{m}E_x(x) , \quad \frac{dv_y}{dt} = -\Omega v_x . \tag{2.38}$$

We can summarize,

$$\frac{d^2v_y}{dt^2} = -\Omega^2 v_y - \Omega^2\frac{1}{B}E_x(x) ; \tag{2.39}$$

however, in this formulation we require the trajectory $x = x(t)$. For strong magnetic fields, we use to lowest order

$$x \approx x_0 - \frac{v_\perp}{\Omega} \cos \Omega t \ . \tag{2.40}$$

An exact solution is nevertheless difficult. Qualitatively, we expect a $E \times B$ motion plus circular motion. When averaging over the rapid circular motion, the averaged velocity component v_x should vanish, and thus

$$\left\langle \frac{d^2 v_x}{dt^2} \right\rangle \stackrel{!}{=} 0 = -\Omega^2 \langle v_x \rangle + \frac{\Omega}{B} \left\langle \frac{dE_x}{dt} \right\rangle \approx -\Omega^2 \langle v_x \rangle + \frac{\Omega}{B} \frac{\partial E_x}{\partial x} \langle v_x \rangle \ , \tag{2.41}$$

$$\left\langle \frac{d^2 v_y}{dt^2} \right\rangle \approx -\Omega^2 \langle v_y \rangle - \Omega^2 \frac{E_0}{B} \left\langle \cos\left[k x_0 - \frac{k v_\perp}{\Omega} \cos \Omega t \right] \right\rangle \stackrel{!}{=} 0 \ . \tag{2.42}$$

Using the addition theorems and the assumption of small gyroradii, we find

$$\langle v_y \rangle \approx -\frac{E_0}{B} \cos(k x_0) \left[1 - \frac{1}{4} k^2 r_L^2 \right] , \tag{2.43}$$

which is known as the "finite Larmor radius correction" to the $E \times B$ drift, sometimes written as

$$\boxed{\mathbf{v}_D = \left(1 + \frac{1}{4} r_L^2 \nabla^2 \right) \frac{\mathbf{E} \times \mathbf{B}}{B^2}} \ . \tag{2.44}$$

For time-dependent electric fields, we choose

$$\mathbf{E} = E_0 e^{i \omega t} \hat{x} + c.c. \tag{2.45}$$

In the x, y plane, the equations of motion are

$$\frac{d^2 v_x}{dt^2} = -\Omega^2 (v_x - v_p) \ , \quad \frac{d^2 v_y}{dt^2} = -\Omega^2 (v_y - v_E) \ . \tag{2.46}$$

Here we have defined

$$v_p = \frac{i \omega}{\Omega} \frac{E_x}{B} \ , \quad v_E = -\frac{E_x}{B} \ . \tag{2.47}$$

For $\omega^2 \ll \Omega^2$ the solutions are

$$v_x \approx v_\perp \cos \Omega t + v_p \ , \quad v_y \approx -v_\perp \sin \Omega t + v_E \ . \tag{2.48}$$

In the y direction, a $E \times B$ drift occurs, while in the x direction we find the polarization drift.

$$\boxed{\mathbf{v}_p = \frac{1}{\Omega B} \frac{d\mathbf{E}}{dt}} \ . \tag{2.49}$$

In summary, in the presence of strong magnetic fields, the motion can be decomposed into a rapid circular motion $\mathbf{v}_\perp$ and a drift motion $\mathbf{w}$, i.e., $\mathbf{v} = \mathbf{w} + \mathbf{v}_\perp$.

For weak inhomogeneities, we have a dominant $E \times B$ drift $\mathbf{w}_D \equiv \mathbf{v}_{E\times B}$ as well as gradient and polarization drifts

$$\boxed{\mathbf{w}_\perp = \mathbf{v}_{E\times B} + \frac{\mu}{qB^3}\mathbf{B} \times \nabla\frac{B^2}{2} + \frac{m}{qB^2}\mathbf{B} \times \dot{\mathbf{v}}_{E\times B}} \,. \tag{2.50}$$

The dot above $\mathbf{v}_{E\times B}$ denotes the time derivative. The motion parallel to the magnetic field obeys

$$\boxed{\frac{dv_\parallel}{dt} = \frac{q}{m}E_\parallel - \frac{\mu}{mB}[(\mathbf{B}\cdot\nabla)\mathbf{B}]_\parallel} \,. \tag{2.51}$$

Example 2.2 (Invariance of the Magnetic Moment)
We determine a “conserved quantity” using the drift approximation.

When we derive conserved quantities within the guiding center approximation, these may not be exact, but only valid for the approximated system. Such conserved quantities are called adiabatic invariants.

We now discuss the magnetic moment as an adiabatic invariant. We begin with energy conservation

$$\frac{d}{dt}\left[\frac{1}{2}mv^2\right] = q\mathbf{v}\cdot\mathbf{E} \tag{2.52}$$

and average over the rapid circular motion. We note that $\langle v^2\rangle \approx \langle v_\parallel^2\rangle + \langle v_\perp^2\rangle$. The term

$$q\langle \mathbf{v}_\perp \cdot \mathbf{E}\rangle = \frac{q}{T}\int \mathbf{E}\cdot\mathbf{v}_\perp dt = \frac{q}{T}\oint \mathbf{E}\cdot d\mathbf{s} \tag{2.53}$$

can be simplified to

$$\frac{q}{T}\int \frac{\partial \mathbf{B}}{\partial t}\cdot d\mathbf{F} \approx \mu\frac{\partial B}{\partial t} \tag{2.54}$$

Here, the area was approximated as $F \approx \pi r_L^2$ and the period of the circular motion $T = 2\pi/\Omega$ was used. The averaged energy conservation equation (2.52) now reads

$$\left\langle \frac{d}{dt}\left(\frac{1}{2}mv_\perp^2\right)\right\rangle \approx \mu\frac{\partial B}{\partial t} - mv_\parallel\left\langle\frac{dv_\parallel}{dt} - \frac{q}{m}E_\parallel\right\rangle$$
$$\approx \mu\frac{\partial B}{\partial t} + \mu\mathbf{v}\cdot\nabla B \equiv \mu\frac{dB}{dt}\,. \tag{2.55}$$

From

$$\frac{d}{dt}(B\mu) = \mu\frac{dB}{dt} \tag{2.56}$$

we find

$$\boxed{\frac{d\mu}{dt} = 0}\,, \tag{2.57}$$

i.e., the invariance of the magnetic moment. ■

2.2 Physically Motivated Time Averaging

As we have seen, a magnetic field significantly alters particle trajectories. We now introduce more systematic methods to determine the drifts. Here, we will first focus on the *physically* motivated systematic averaging over the rapid gyro-motion. Another, more *mathematically* based method, which relies on Lie transformations, will be presented in the next section.

As previously discussed, the motion of a particle can be decomposed into an averaged drift motion and a rapid gyro-motion. It is useful to introduce the vector $\boldsymbol{\rho}$ from the Larmor center to the particle. Its magnitude corresponds to the Larmor radius (also referred to as r_L), see Fig. 2.4. With the gyrofrequency (the sign is determined by the charge), we write

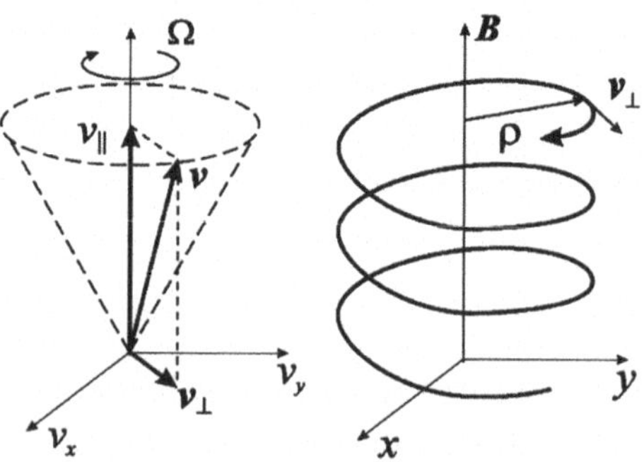

Fig. 2.4 Nomenclature for gyration and drift. Left: in velocity space; right: in configuration space

$$\boldsymbol{\rho} = -\frac{1}{\Omega}\mathbf{v} \times \mathbf{b} \,, \tag{2.58}$$

where $\mathbf{b} = \mathbf{B}/B$ is. We also introduce the radius vector $\mathbf{R}$ of the guiding center of the particles (center of the Larmor helix),

$$\mathbf{R} = \mathbf{r} - \boldsymbol{\rho} = \mathbf{r} + \frac{1}{\Omega}\mathbf{v} \times \mathbf{b} \,. \tag{2.59}$$

In this equation, we must take Ω and $\mathbf{b}$ at the position of the particle. We calculate $d\mathbf{R}/dt$:

$$\dot{\mathbf{R}} = \dot{\mathbf{r}} + \frac{1}{\Omega}\dot{\mathbf{v}} \times \mathbf{b} + \frac{1}{\Omega}\mathbf{v} \times \dot{\mathbf{b}} - \frac{\dot{\Omega}}{\Omega^2}\mathbf{v} \times \mathbf{b} \,. \tag{2.60}$$

Using Newton's equation of motion with the Lorentz force, it follows that

$$\dot{\mathbf{R}} = \mathbf{v} + \frac{1}{\Omega}\left(\frac{q}{m}\,\mathbf{E} \times \mathbf{b} + \Omega(\mathbf{v} \times \mathbf{b}) \times \mathbf{b}\right) + \frac{1}{\Omega}\mathbf{v} \times \dot{\mathbf{b}} - \frac{\dot{\Omega}}{\Omega^2}\mathbf{v} \times \mathbf{b} \,. \tag{2.61}$$

Since $(\mathbf{v} \times \mathbf{b}) \times \mathbf{b} = -\mathbf{v} + (\mathbf{v} \cdot \mathbf{b})\mathbf{b} = -\mathbf{v}_\perp$, it follows that

$$\dot{\mathbf{R}} = v_\parallel \mathbf{b} + \frac{q}{\Omega m}\mathbf{E} \times \mathbf{b} + \frac{1}{\Omega}\mathbf{v} \times \dot{\mathbf{b}} - \frac{\dot{\Omega}}{\Omega^2}\mathbf{v} \times \mathbf{b} \equiv I + II + III + IV \,. \tag{2.62}$$

This equation is exact. It contains terms that vary slowly and terms that oscillate rapidly. To extract drifts, we must average this equation over the rapid Larmor motion. We begin with the first term $I = v_\parallel \mathbf{b}$. It does not contain a small parameter ε if we define the smallness parameter as the ratio of the Larmor radius r_L to the characteristic length L for the field variations. We expand all fields around the guiding center in a Taylor series.

$$\mathbf{b}(\mathbf{r}) = \mathbf{b}(\mathbf{R} + \boldsymbol{\rho}) \approx \mathbf{b}(\mathbf{R}) + (\boldsymbol{\rho} \cdot \nabla)\mathbf{b}(\mathbf{R}) \,. \tag{2.63}$$

The last term here is of the order of ε. However, if we average over the Larmor radius, it vanishes, since $\langle \boldsymbol{\rho} \rangle = 0$ is. It follows that

$$\langle I \rangle = v_\parallel \mathbf{b}(\mathbf{R}) \,. \tag{2.64}$$

The second term $II = \frac{q}{m\Omega}\mathbf{E} \times \mathbf{b}$ is already small, since the electric field in a well-conducting plasma cannot be too large. Therefore, we can consider the fields at the gyrocenter, and we obtain

$$\langle II \rangle = \frac{1}{B^2(\mathbf{R})}\mathbf{E}(\mathbf{R}) \times \mathbf{B}(\mathbf{R}) \,. \tag{2.65}$$

Now we move on to the third term $III = \Omega^{-1}\mathbf{v} \times d\mathbf{b}/dt$. Here, the following holds

$$\dot{\mathbf{b}} = \frac{d\mathbf{b}}{dt} = \frac{\partial \mathbf{b}}{\partial t} + (\mathbf{v} \cdot \nabla)\mathbf{b} \, . \tag{2.66}$$

For time-independent magnetic fields, we write

$$\mathbf{III} = \frac{1}{\Omega}\mathbf{v} \times (\mathbf{v} \cdot \nabla)\mathbf{b} \, . \tag{2.67}$$

For averaging, we use index notation with summation over repeated indices and introduce the fully antisymmetric third-order tensor $\varepsilon_{\alpha\beta\gamma}$ in order to find:

$$III_\alpha = \frac{1}{\Omega}\varepsilon_{\alpha\beta\gamma} v_\beta v_\nu \frac{\partial b_\gamma}{\partial x_\nu} \, . \tag{2.68}$$

This expression already contains a small spatial derivative at the position of the guiding center, so we only need to average the velocities.

$$\langle III_\alpha \rangle = \frac{1}{\Omega}\varepsilon_{\alpha\beta\gamma} \frac{\partial b_\gamma}{\partial x_\nu} \langle v_\beta v_\nu \rangle \, . \tag{2.69}$$

The averaging here must be invariant under rotations about the direction **b**. Then, the result should be a linear combination of invariant tensors and the components of **b**, namely

$$\langle v_\beta v_\nu \rangle = A\delta_{\beta\nu} + B b_\beta b_\nu + C\varepsilon_{\beta\nu\mu} b_\mu \, . \tag{2.70}$$

The constant C must be zero, since the mean value cannot depend on the sign of the magnetic field. The factors A and B are found by contracting with $\delta_{\beta\nu}$ and $b_\beta b_\nu$, which leads to $v^2 = 3A + B$, $v_\parallel^2 = A + B$. Then, we have

$$\langle v_\beta v_\nu \rangle = \frac{1}{2} v_\perp^2 \left\{ \delta_{\beta\nu} - b_\beta b_\nu \right\} + v_\parallel^2 b_\beta b_\nu \, , \tag{2.71}$$

and finally

$$\langle III_\alpha \rangle = \frac{v_\perp^2}{2\Omega}\varepsilon_{\alpha\beta\gamma} \left\{ \frac{\partial b_\gamma}{\partial x_\beta} - b_\beta b_\nu \frac{\partial b_\gamma}{\partial x_\nu} \right\} + \frac{v_\parallel^2}{\Omega}\varepsilon_{\alpha\beta\gamma} b_\beta b_\nu \frac{\partial b_\gamma}{\partial x_\nu} \, . \tag{2.72}$$

Returning to vector notation, we write

$$\langle III \rangle = \frac{v_\perp^2}{2\Omega}\{\nabla \times \mathbf{b} - [\mathbf{b}, (\mathbf{b} \cdot \nabla)\mathbf{b}]\} + \frac{v_\parallel^2}{\Omega}[\mathbf{b}, (\mathbf{b} \cdot \nabla)\mathbf{b}] \, . \tag{2.73}$$

To make the representation more transparent, we have introduced the notation $\mathbf{a} \times \mathbf{b} \equiv [\mathbf{a}, \mathbf{b}]$. The expression can be somewhat simplified. Because

$$[\mathbf{b}, \nabla \times \mathbf{b}] = \underbrace{\nabla \frac{b^2}{2}}_{=0} - (\mathbf{b} \cdot \nabla)\mathbf{b} = -(\mathbf{b} \cdot \nabla)\mathbf{b} \, , \tag{2.74}$$

it follows that

$$\langle III \rangle = \frac{v_\perp^2}{2\Omega}\mathbf{b}(\mathbf{b}\cdot\nabla\times\mathbf{b}) + \frac{v_\parallel^2}{\Omega}[\mathbf{b}, (\mathbf{b}\cdot\nabla)\mathbf{b}] \,. \tag{2.75}$$

Let us now turn to the last, fourth term$IV = -\frac{\dot{\Omega}}{\Omega^2}\mathbf{v}\times\mathbf{b}$. The following holds

$$\dot{\Omega} = \frac{d\Omega}{dt} = \frac{\partial\Omega}{\partial t} + (\mathbf{v}\cdot\nabla)\Omega = (\mathbf{v}\cdot\nabla)\Omega \,. \tag{2.76}$$

Similar to before, we calculate

$$\begin{aligned}\langle IV \rangle &= -\left\langle \frac{1}{\Omega^2}\, v_\nu \frac{\partial\Omega}{\partial x_\nu}\, \varepsilon_{\alpha\beta\gamma} v_\beta b_\gamma \right\rangle = -\frac{1}{\Omega^2}\,\frac{\partial\Omega}{\partial x_\nu}\,\varepsilon_{\alpha\beta\gamma} b_\gamma \langle v_\nu v_\beta\rangle \\ &= -\frac{1}{\Omega^2}\,\frac{\partial\Omega}{\partial x_\nu}\,\varepsilon_{\alpha\beta\gamma} b_\gamma \left\{ \frac{1}{2}\, v_\perp^2 \delta_{\beta\nu} + \left(v_\parallel^2 - \frac{1}{2}\, v_\perp^2 \right) b_\beta b_\nu \right\} . \end{aligned} \tag{2.77}$$

Because of $\varepsilon_{\alpha\beta\gamma} b_\beta b_\gamma = [\mathbf{b}, \mathbf{b}]_\alpha = 0$ we finally obtain

$$\langle IV \rangle = -\frac{v_\perp^2}{2\Omega^2}\,[\nabla\Omega, \mathbf{b}] \,. \tag{2.78}$$

In summary, we obtain

$$\dot{\mathbf{R}} = \left(v_\parallel + \frac{v_\perp^2}{2\Omega}\,\mathbf{b}\cdot\nabla\times\mathbf{b} \right)\mathbf{b} + \frac{1}{B}\,[\mathbf{E}, \mathbf{b}] + \frac{v_\parallel^2}{\Omega}[\mathbf{b}, (\mathbf{b}\cdot\nabla)\mathbf{b}] + \frac{v_\perp^2}{2\Omega B}\,[\mathbf{b}, \nabla B] \,. \tag{2.79}$$

The correction to the parallel velocity $v_\parallel$ in the first term is usually small and can be neglected. The last three terms describe drifts: the electric, centrifugal, and gradient drifts, respectively.

For the centrifugal drift, we note that

$$(\mathbf{b}\cdot\nabla)\mathbf{b} = \frac{\partial\mathbf{b}}{\partial s} = \kappa \tag{2.80}$$

is the curvature vector of a magnetic field line. We have $|\kappa| = 1/R$ with the radius of curvature R; see Fig. 2.5. With the curvature vector κ the centrifugal drift is written as

$$\mathbf{v}_{cd} = \frac{v_\parallel^2}{\Omega}\,\mathbf{b}\times\kappa \,. \tag{2.81}$$

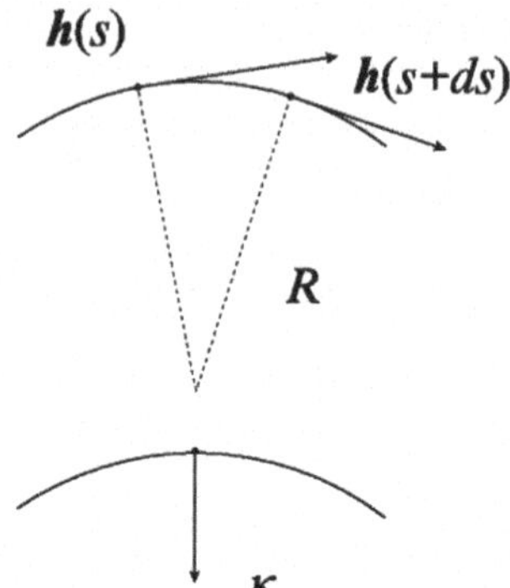

Fig. 2.5 On the definition of the curvature vector of a magnetic field

If $\mathbf{n}$ is the normal vector to the particle trajectory, the centrifugal drift remains directed along the binormal $\tilde{\mathbf{b}} = \mathbf{b} \times \mathbf{n}$. This drift depends on the sign of the charge.

In vacuum, where $\nabla \times \mathbf{B} = 0$ holds, the centrifugal and gradient drifts are similar. To prove this, we use the equation $[\mathbf{b}, \nabla \times \mathbf{b}] = -\kappa$. and

$$\nabla \times \mathbf{b} = \nabla \times \frac{\mathbf{B}}{B} = \frac{1}{B} \underbrace{\nabla \times \mathbf{B}}_{=0} + \left[\nabla \frac{1}{B}, \mathbf{B} \right] . \tag{2.82}$$

Furthermore,

$$-\kappa = [\mathbf{b}, \nabla \times \mathbf{b}] = -\left[\mathbf{b}, \left[\frac{\nabla B}{B}, \mathbf{b} \right] \right] = -\frac{\nabla B}{B} + \mathbf{b} \left(\mathbf{b} \cdot \frac{\nabla B}{B} \right) , \tag{2.83}$$

and after applying $\mathbf{b}\times$

$$[\mathbf{b}, \nabla B] = B[\mathbf{b}, \kappa] , \tag{2.84}$$

so that gradient and centrifugal drifts can be combined,

$$\mathbf{v}_{\text{grad}} + \mathbf{v}_{\text{cd}} = \frac{v_\parallel^2 + \frac{1}{2} v_\perp^2}{\Omega} [\mathbf{b}, \kappa] = \frac{v_\parallel^2 + \frac{1}{2} v_\perp^2}{\Omega R} [\mathbf{b}, \mathbf{n}] . \tag{2.85}$$

Overall, it can be said that the motion of the gyration center to first order is given by

$$\dot{\mathbf{R}} = v_\parallel \mathbf{b} + \frac{1}{B}[\mathbf{E}, \mathbf{b}] + \frac{v_\parallel^2}{\Omega} [\mathbf{b}, \kappa] + \frac{v_\perp^2}{2\Omega B} [\mathbf{b}, \nabla B] \tag{2.86}$$

2.3 Mathematical Background

In this section, we proceed somewhat more systematically in the *mathematical* sense by replacing time averaging with a Lie transformation. The calculations are based on the work of Littlejohn and Balescu [40–46].

From a theoretical standpoint, the question arises as to whether the decomposition of a particle's motion into a gyration and a drift of the gyration center, as presented in the previous section, can also be justified mathematically in an exact way. Within the framework of classical mechanics, the relevant formalism for such a question was developed in the form of canonical transformations. Can we perform a corresponding transformation here, in which the coordinates of the gyration center appear as new coordinates? In this precise form, the question of the validity of the "guiding center approximation" was posed, mainly by Littlejohn and Balescu. Earlier studies were content with averaging over the gyration phases, with the consequence that the necessary prerequisites for a constructive statistical theory were lacking.

In the following, we show that the introduction of the coordinates of a guiding center can indeed be incorporated into the Hamiltonian formalism—albeit with extended transformations, the so-called pseudocanonical transformations.

In Hamiltonian systems, a dynamical function f evolves due to the motion of the system according to

$$\dot{f} = [f, H]. \tag{2.87}$$

In classical mechanics, the Lie bracket $[\ldots, \ldots]$ is identical to the Poisson bracket. Since with Eq. (2.87) we are aiming for a rather important generalization, a few clarifying words are in order. The properties of a Lie algebra are determined by algebraic structures and the operation in the form of the so-called Lie bracket. An algebraic structure is imparted to the set of all $a, b, \ldots \in D$ by the fact that $\alpha a, \alpha a + \beta b, a \cdot b$ and a^{-1} are defined and are also to belong to the set D. Furthermore, $[a, b] \in D$ must hold, where the Lie bracket satisfies the following relations: $[a, b] = -[b, a]$, $[\alpha a, b] = \alpha[a, b]$, $[\alpha a + \beta b, c] = \alpha[a, c] + \beta[b, c]$, $[a \cdot b, c] = a \cdot [b, c] + b \cdot [a, c]$, $[[a, b], c] + [[b, c], a] + [[c, a], b] = 0$. Any Lie bracket can be computed if one knows the so-called fundamental Lie brackets $[q_i, q_j]$, $[q_i, p_j]$, $[p_i, p_j]$ for all i and j of the coordinates q_i and momenta p_j. This statement is manifested in the formula

$$\begin{aligned}
[a(q,p), b(q,p)] = \sum_{i,j} \left\{ \frac{\partial a}{\partial q_i} \frac{\partial b}{\partial q_j} [q_i, q_j] + \frac{\partial a}{\partial q_i} \frac{\partial b}{\partial p_j} [q_i, p_j] \right. \\
\left. + \frac{\partial a}{\partial p_i} \frac{\partial b}{\partial q_j} [p_i, q_j] + \frac{\partial a}{\partial p_i} \frac{\partial b}{\partial p_j} [p_i, p_j] \right\}.
\end{aligned} \tag{2.88}$$

If we define the fundamental Lie brackets by means of

$$[q_i, q_j] = 0, \qquad \text{for all i, j,} \tag{2.89}$$

$$[p_i, p_j] = 0, \qquad \text{for all i, j,} \tag{2.90}$$

$$[q_i, p_j] = \delta_{ij}, \qquad \text{for all i, j,} \tag{2.91}$$

then (2.88) coincides with the Poisson bracket, i.e.,

$$[a, b] = \sum_i \left\{ \frac{\partial a}{\partial q_i} \frac{\partial b}{\partial p_i} - \frac{\partial a}{\partial p_i} \frac{\partial b}{\partial q_i} \right\}. \tag{2.92}$$

The variables q_i, p_i are then called canonically conjugate.

A transformation $Q_i = Q_i(q, p)$, $P_i = P_i(q, p)$ is called canonical if, for the new variables Q_i, P_i the fundamental Lie brackets also take the form

$$[Q_i, Q_j] = [P_i, P_j] = 0, \qquad [Q_i, P_j] = \delta_{ij} \tag{2.93}$$

From the lecture courses it is then known that the canonical equations also hold in the new variables, albeit with a new Hamiltonian function K, which, in the absence of explicit time dependence, is simply obtained from the old one: $K(Q, P) = H(q(Q, P), p(Q, P))$.

In a pseudocanonical transformation, the new variables do not have to satisfy (2.93), however, the transformation equations should still be invertible. Instead of (2.93), the following apply

$$[Q_i, Q_j] = F_{ij}(Q, P), \qquad \text{for all i, j,} \tag{2.94}$$

$$[P_i, P_j] = G_{ij}(Q, P), \qquad \text{for all i, j,} \tag{2.95}$$

$$[Q_i, P_j] = H_{ij}(Q, P), \qquad \text{for all i, j,} \tag{2.96}$$

with known functions F_{ij}, G_{ij}, and H_{ij}.

Based on these fundamental Lie brackets, any Lie bracket can be calculated, including

$$\dot{Q}_i = [Q_i, K(Q, P)], \tag{2.97}$$

$$\dot{P}_i = [P_i, K(Q, P)], \tag{2.98}$$

Equations (2.97) and (2.98) take the place of the canonical equations. For example, (2.97) reads explicitly

$$\dot{Q}_i = \sum_j \left\{ F_{ij}(Q, P) \frac{\partial K(Q, P)}{\partial Q_j} + H_{ij}(Q, P) \frac{\partial K(Q, P)}{\partial P_j} \right\}. \tag{2.99}$$

In addition to the difference already mentioned, there is another distinction from canonical transformations. The functional determinant is not equal to one. Instead, the following holds

$$J^2 \equiv \left| \frac{\partial(q, p)}{\partial(Q, P)} \right|^2 = \frac{1}{\mid \det \Sigma \mid}, \tag{2.100}$$

where the elements Σ_{kl} are defined by

$$\Sigma_{kl} = [z_k, z_l] \tag{2.101}$$

with $\mathbf{z} = (Q_1 \ldots, P_1 \ldots)$.

Example 2.3 (Pseudocanonical Transformation)
To mention a simple example: One can show that the transformation from $\mathbf{r}$ and $\mathbf{p}$ (coordinates and momenta of a particle of mass m and charge e in an external magnetic field) to $\mathbf{r}$ and $\mathbf{v}$ (coordinates and velocities, with $m\mathbf{v} = \mathbf{p} - (e/c)\mathbf{A}(\mathbf{x})$, where $\mathbf{A}$ is the vector potential) is not canonical, but pseudocanonical. ■

After this excursion into the fundamentals of classical mechanics, let us now return to the problem of the motion of a particle in external fields. It is the great merit of Littlejohn and Balescu to have pointed out that the separation into gyration and drift can be cleanly carried out within the framework of a Hamiltonian theory using pseudocanonical transformations.

Some geometry is necessary to carry out the calculation in detail. First, we start with a space-fixed coordinate system, which, however, is not particularly suitable for more complicated magnetic field geometries. Instead, it is more appropriate to use a local coordinate system adapted to the magnetic field lines $\mathbf{x}(s)$. In the general case, at a given point, it has the directions of the tangent $\hat{b}$,, normal $\hat{N}$, and binormal $\hat{\beta}$ (see Fig. 2.6). The sense of rotation is defined by

$$\mathbf{e}_1 \equiv \hat{N} = \hat{\beta} \times \hat{b}, \tag{2.102}$$

$$\mathbf{e}_2 \equiv \hat{\beta} = \hat{b} \times \hat{N}, \tag{2.103}$$

$$\mathbf{e}_3 \equiv \hat{b} = \hat{N} \times \hat{\beta} \tag{2.104}$$

It must be noted that these vectors are position-dependent. To now represent the velocities, we go to the position $\mathbf{q}$ of the particle, construct—by means of the magnetic field line passing through $\mathbf{q}$—a local coordinate system, and project the particle trajectory onto the plane perpendicular to $\mathbf{e}_3$. Let us call the unit tangent vector to the particle velocity at the position $\mathbf{q}$ $\mathbf{n}_1(\mathbf{q})$, and $\mathbf{n}_2(\mathbf{q})$ the vector orthogonal to it in the positive sense of rotation, then in the plane perpendicular to $\mathbf{e}_3$ we obtain a simple relationship between $\mathbf{n}_\nu$ and $\mathbf{e}_\nu$ (see Fig. 2.7). Trivially, the following holds

$$\mathbf{n}_1(\mathbf{q}, \varphi) = -\sin\varphi\, \mathbf{e}_1(\mathbf{q}) - \cos\varphi\, \mathbf{e}_2(\mathbf{q}), \tag{2.105}$$

$$\mathbf{n}_2(\mathbf{q}, \varphi) = \cos\varphi\, \mathbf{e}_1(\mathbf{q}) - \sin\varphi\, \mathbf{e}_2(\mathbf{q}). \tag{2.106}$$

The velocity of the particle is thus given by

$$\mathbf{v} = v_\parallel \mathbf{e}_3(\mathbf{q}) + v_\perp \mathbf{n}_1(\mathbf{q}, \varphi) \tag{2.107}$$

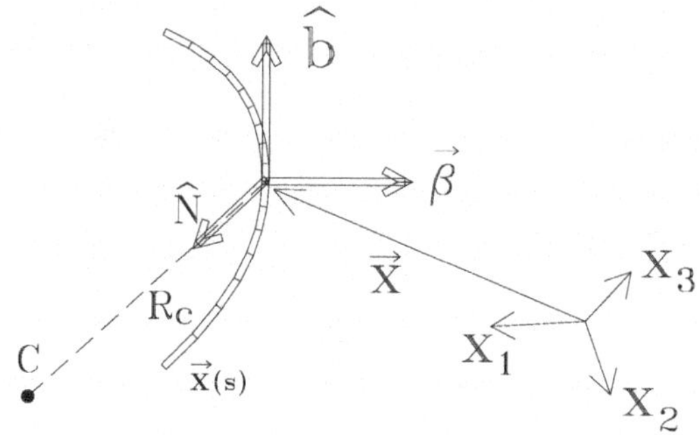

Fig. 2.6 Local coordinate system at a point on the curved magnetic field line $\mathbf{x}(s)$

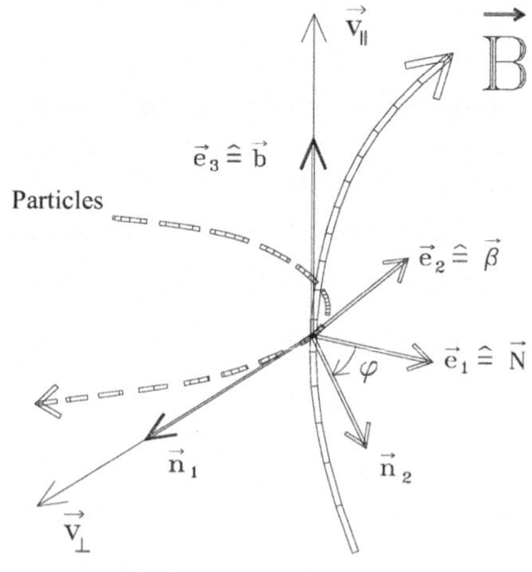

Fig. 2.7 Local coordinate system at the position $\mathbf{q}$ of the particle

The aim of the following considerations is, starting from the variables $\mathbf{q}, v_\parallel, v_\perp$ and φ for the position and velocity of a particle, to specify pseudocanonical transformations that lead to the "gyration coordinates." To begin with, we are of course immediately able to write down the Hamiltonian function

$$H = \frac{m}{2}(v_\parallel^2 + v_\perp^2) + eV(\mathbf{q}) \tag{2.108}$$

and the dynamical equations for $\mathbf{q}, v_\parallel, v_\perp$ and φ. The latter follow from the fundamental Lie brackets (which, up to this point, are still the Poisson brackets with the variables $\mathbf{q}$ and $\mathbf{p}$), e.g., $[\mathbf{q}, v_\parallel] = \mathbf{b}/m$ etc. The result is [see, for example, the excellent book by R. Balescu, "Transport Processes in Plasmas" (North-Holland, Amsterdam 1988)]

$$\dot{\mathbf{q}} = v_\parallel \mathbf{b} + v_\perp \mathbf{n}_1, \tag{2.109}$$

$$\dot{v}_\parallel = v_\perp \mathbf{n}_2 \cdot \mathbf{D} + \frac{e}{m}\mathbf{b} \cdot \mathbf{E}, \tag{2.110}$$

$$\dot{v}_\perp = -v_\parallel \mathbf{n}_2 \cdot \mathbf{D} + \frac{e}{m}\mathbf{n}_1 \cdot \mathbf{E}, \tag{2.111}$$

$$\dot{\varphi} = \frac{eB}{m} + \mathbf{b} \cdot \mathbf{D} - \frac{v_\parallel}{v_\perp}\mathbf{n}_1 \cdot \mathbf{D} - \frac{e}{mv_\perp}\mathbf{n}_2 \cdot \mathbf{E}\,, \tag{2.112}$$

with

$$\mathbf{D} := v_\parallel \nabla \times \mathbf{b} + v_\perp \nabla \times \mathbf{n}_1. \tag{2.113}$$

The transformation from $\mathbf{q}, \mathbf{p}$ to $\mathbf{q}, v_\parallel, v_\perp, \varphi$ is not canonical; the functional determinant is $|J| = v_\perp$.

The derivation of the equations of motion is, in principle, straightforward; let us verify them for the case of constant and homogeneous $\mathbf{E}$ and $\mathbf{B}$ fields. In this case, we can neglect all contributions from $\mathbf{D}$ on the right-hand sides. The reason for this is that $\nabla \times \mathbf{b} = 0$ holds, and in this case $\mathbf{n}_1$ is only a function of φ and thus does not depend on $\mathbf{q}$. From the general equations, we immediately obtain

$$\dot{\mathbf{q}} = v_\parallel \mathbf{b} + v_\perp \mathbf{n}_1, \tag{2.114}$$

$$\dot{v}_\parallel = \frac{e}{m}\mathbf{b} \cdot \mathbf{E}, \tag{2.115}$$

$$\dot{v}_\perp = \frac{e}{m}\mathbf{n}_1 \cdot \mathbf{E}, \tag{2.116}$$

$$\dot{\varphi} = \frac{eB}{m} - \frac{e}{mv_\perp}\mathbf{n}_2 \cdot \mathbf{E}. \tag{2.117}$$

From these equations, we can easily obtain the familiar form of the equations of motion,

$$\begin{aligned}\ddot{\mathbf{q}} &= \frac{e}{m}(\mathbf{b}\cdot\mathbf{E})\mathbf{b} + \frac{e}{m}\mathbf{n}_1(\mathbf{n}_1 \cdot \mathbf{E}) - \mathbf{n}_2 v_\perp \left[\frac{eB}{m} - \frac{e}{mv_\perp}\mathbf{n}_2 \cdot \mathbf{E}\right] \\ &= \frac{e}{m}\mathbf{E} - \mathbf{n}_2 v_\perp \frac{eB}{m} = \frac{e}{m}[\mathbf{E} + \mathbf{v} \times \mathbf{B}].\end{aligned} \tag{2.118}$$

Let us return to the general equations. A closer examination reveals that the gyration term

$$\dot{\varphi} \approx \frac{eB}{m} \tag{2.119}$$

plays a special role. Compared to the other contributions, we assign this term a higher order. We state: within the validity range of the drift approximation, this term should dominate. If we compare, for example, with the contributions arising from **D**, we find that the gyration term is larger by a factor of

$$L/r_L \gg 1 \tag{2.120}$$

In addition, the electric field causes an acceleration and a change in the angular velocity. We will require that these effects of the electric field should not be too large. (Physically, this means, for example, that the $E \times B$ drift must not reach the same order of magnitude as the instantaneous particle velocity, i.e., $v_{E\times B} \ll v_t$.) The condition (2.120) or, equivalently, the requirement

$$\omega/\Omega \ll 1, \tag{2.121}$$

where ω is to be a characteristic frequency, together with the restriction to not excessively large electric fields, constitutes the essential prerequisite for the drift approximation to be carried out below. Formally, we ensure these conditions by replacing Ω with Ω/ε, where ε is to be a smallness parameter (see, for example, the already cited book by Balescu).

Taking into account what has just been said, the equations of motion have the following structure:

$$\frac{dX}{dt} = S(X, \varphi), \tag{2.122}$$

$$\frac{d\varphi}{dt} = \frac{1}{\varepsilon}\Omega + R(X, \varphi), \tag{2.123}$$

where X summarizes the variables $\mathbf{q}$, $v_\parallel$ and $v_\perp$. The functions R and S, whose exact form does not concern us at the moment, depend on all variables; in any case, they are of lower order of magnitude than the gyration term. At this point, it becomes clear once again that, from a physical perspective, averaging over the rapidly varying gyration phase is a natural approach. Such an averaging procedure, which can indeed be successfully applied to calculate the various drifts, is theoretically unsatisfactory, as it does not fit into the canonical formalism of mechanics and, at the latest, leads to difficulties in the statistical description of a plasma. This is the reason why, in the following, we will present the systematic approach, consistent with Hamiltonian transformation theory, for eliminating the rapid phase dependence.

Variable transformations from X, φ to Y, ϕ are to be carried out in such a way that the equations of motion successively take the form, in the various orders in ε,

$$\frac{dY}{dt} = S_0(Y) + \varepsilon S_1(Y) + \ldots, \tag{2.124}$$

$$\frac{d\phi}{dt} = \frac{1}{\varepsilon}\Omega(Y) + R_0(Y) + \varepsilon R_1(Y) + \ldots, \tag{2.125}$$

with the essential condition that the functions $S_0, S_1, \ldots, R_0, R_1, \ldots$ do not depend on ϕ. If we succeed in this, we can interpret part of Y as "drift coordinates" (guiding center coordinates).

Example 2.4 (Elimination of the Phase)

It is best to look at the basic procedure using a simple example. We choose the clear case of constant **E** and **B** fields. We have already seen that the variables $\mathbf{q}$, $v_\parallel$, $v_\perp$ and φ can be related via a pseudocanonical transformation with $\mathbf{q}$ and $\mathbf{p}$. The equations of motion have the desired structure; in addition, we imagine $\Omega = eB/m$ replaced by Ω/ε. Now we perform a transformation

$$\mathbf{Q} = \mathbf{q} + \varepsilon\kappa(\mathbf{q}, v_\parallel, v_\perp, \varphi) + \mathcal{O}(\varepsilon^2), \tag{2.126}$$

$$U = v_\parallel + \varepsilon \nu_\parallel(\mathbf{q}, v_\parallel, v_\perp, \varphi) + \mathcal{O}(\varepsilon^2), \tag{2.127}$$

$$W = v_\perp + \varepsilon \nu_\perp(\mathbf{q}, v_\parallel, v_\perp, \varphi) + \mathcal{O}(\varepsilon^2), \tag{2.128}$$

$$\phi = \varphi + \varepsilon\psi(\mathbf{q}, v_\parallel, v_\perp, \varphi) + \mathcal{O}(\varepsilon^2), \tag{2.129}$$

which, at least up to order ε, is supposed to yield the structure (2.124) and (2.125) (note: Y combines the new variables $\mathbf{Q}$, U and W). In the new variables, we aim for a form of the Hamiltonian function

$$H = H(\mathbf{Q}, U, W) = \frac{m}{2}(U^2 + W^2) - e\mathbf{E}\cdot\mathbf{Q} + \mathcal{O}(\varepsilon^2). \tag{2.130}$$

Substituting the above transformation equations yields, for this form of H, the condition that $\dot{v}_{\parallel}$ and $\dot{v}_{\perp}$ have a very similar functional form, namely

$$\dot{v}_{\perp} = -\frac{v_{\parallel}}{v_{\perp}}\dot{v}_{\parallel} + \frac{e}{mv_{\perp}}\dot{\kappa} \cdot \mathbf{E}. \tag{2.131}$$

We achieve this by choosing

$$\dot{v}_{\parallel} = v_{\perp}\beta(\mathbf{q}, v_{\parallel}, v_{\perp}, \varphi), \tag{2.132}$$

$$\dot{v}_{\perp} = -v_{\parallel}\beta(\mathbf{q}, v_{\parallel}, v_{\perp}, \varphi) + \frac{e}{mv_{\perp}}\dot{\kappa} \cdot \mathbf{E}, \tag{2.133}$$

with a single function β.

Next, we proceed to calculate the various Lie brackets. The main objective here is to suppress the φ-dependence in the equations of motion for $\mathbf{Q}$, U and W up to and including order ε^1. The calculation of the new equations of motion presents an interesting technical detail. In a systematic approach, we start (for constant $\mathbf{E}$ and $\mathbf{B}$ fields) from the Lie brackets

$$[q_i, q_j] = 0, \quad [\mathbf{q}, v_{\parallel}] = \frac{1}{m}\mathbf{b}, \quad [\mathbf{q}, v_{\perp}] = \frac{1}{m}\mathbf{n}_1, \tag{2.134}$$

$$[\mathbf{q}, \varphi] = -\frac{1}{mv_{\perp}}\mathbf{n}_2, \quad [v_{\parallel}, v_{\perp}] = 0, \quad [v_{\parallel}, \varphi] = 0, \tag{2.135}$$

$$[v_{\perp}, \varphi] = -\frac{1}{\varepsilon}\frac{\Omega}{mv_{\perp}}, \tag{2.136}$$

which, in their simplest form, follow from the Poisson brackets for $\mathbf{q}$ and $\mathbf{p}$. For the calculation of the new Lie brackets $[Q_i, Q_j]$, $[\mathbf{Q}, U]$, $[\mathbf{Q}, W]$, $[Q, \phi]$, $[U, W]$, $[U, \phi]$ and $[W, \phi]$, we use the rule (2.88). ■

Example 2.5 (Calculation of Lie Brackets)
Let us consider an example:

$$\begin{aligned}
[\mathbf{Q}, W] &= [\mathbf{q} + \varepsilon\kappa, v_{\perp} - \varepsilon v_{\parallel}\beta + \varepsilon\frac{e}{mv_{\perp}}\kappa \cdot \mathbf{E}] \\
&= [\mathbf{q}, v_{\perp}] + \varepsilon\left\{[\kappa, v_{\perp}] - [\mathbf{q}, v_{\parallel}\beta] + \frac{e}{m}\left[\mathbf{q}, \frac{1}{v_{\perp}}\kappa \cdot \mathbf{E}\right]\right\} + \ldots \\
&= [\mathbf{q}, v_{\perp}] + \varepsilon\frac{\partial\kappa}{\partial\varphi}[\varphi, v_{\perp}] + \mathcal{O}(\varepsilon),
\end{aligned} \tag{2.137}$$

when we consider only terms up to order ε^0. Here, we must take into account that $[\varphi, v_{\perp}] \sim \mathcal{O}(\frac{1}{\varepsilon})$. Since $[\mathbf{q}, v_{\perp}] = \frac{1}{m}\mathbf{n}_1(\varphi)$ depends explicitly on φ, but on the other hand, no φ-dependence is desired at lowest order, we suppress the explicit φ-dependence at lowest order by choosing

$$\kappa = -\frac{v_\perp}{\Omega}\mathbf{n}_2 \tag{2.138}$$

It then follows that

$$[\mathbf{Q}, W] = 0 + \mathcal{O}(\varepsilon). \tag{2.139}$$

■

In a similar manner, the functions β and ψ are defined by determining the Lie brackets in the new variables $\mathbf{Q}$, U, W and ϕ such that, to lowest order, they do not possess any oscillatory component. One possible choice is

$$\beta = 0, \tag{2.140}$$

$$\psi = -\frac{e}{m\Omega v_\perp}\mathbf{n}_1 \cdot \mathbf{E}. \tag{2.141}$$

It is quite interesting and important to note that such an approach suppresses the angular dependence even at the next order. To make this clear: The transformations

$$\mathbf{Q} = \mathbf{q} - \varepsilon\frac{v_\perp}{\Omega}\mathbf{n}_2(\varphi) + \mathcal{O}(\varepsilon^2), \tag{2.142}$$

$$U = v_\parallel + \mathcal{O}(\varepsilon^2), \tag{2.143}$$

$$W = v_\perp - \varepsilon\frac{e}{m\Omega}\mathbf{E} \cdot \mathbf{n}_2(\varphi) + \mathcal{O}(\varepsilon^2), \tag{2.144}$$

$$\phi = \varphi - \varepsilon\frac{e}{m\Omega v_\perp}\mathbf{E} \cdot \mathbf{n}_1(\varphi) + \mathcal{O}(\varepsilon^2) \tag{2.145}$$

lead to the new fundamental Lie brackets

$$[Q_i, Q_j] = -\varepsilon\frac{c}{eB}\varepsilon_{ijk}b_k + \mathcal{O}(\varepsilon^2), \tag{2.146}$$

$$[\mathbf{Q}, U] = \frac{1}{m}\mathbf{b} + \mathcal{O}(\varepsilon^2), \tag{2.147}$$

$$[\mathbf{Q}, W] = 0 + \mathcal{O}(\varepsilon^2), \tag{2.148}$$

$$[\mathbf{Q}, \phi] = 0 + \mathcal{O}(\varepsilon^2), \tag{2.149}$$

$$[U, W] = 0 + \mathcal{O}(\varepsilon^2), \tag{2.150}$$

$$[U,\phi]=0+\mathcal{O}(\varepsilon^2), \tag{2.151}$$

$$[W,\phi]=-\frac{1}{\varepsilon}\frac{\Omega}{mW}+\mathcal{O}(\varepsilon). \tag{2.152}$$

From this, the equations of motion then follow from the new Hamiltonian

$$H=\frac{1}{2}m[U^2+W^2]-e\mathbf{E}\cdot\mathbf{Q}+\mathcal{O}(\varepsilon^2) \tag{2.153}$$

as

$$\dot{\mathbf{Q}}=U\mathbf{b}+\varepsilon\frac{\mathbf{E}\times\mathbf{B}}{B^2}+\mathcal{O}(\varepsilon^2), \tag{2.154}$$

$$\dot{U}=\frac{e}{m}\mathbf{E}\cdot\mathbf{b}+\mathcal{O}(\varepsilon^2), \tag{2.155}$$

$$\dot{W}=0+\mathcal{O}(\varepsilon^2)\ , \tag{2.156}$$

$$\dot{\phi}=\frac{1}{\varepsilon}\Omega+\mathcal{O}(\varepsilon). \tag{2.157}$$

This result is remarkable for several reasons, but also quite understandable:

1. The $E\times B$ drift follows, within the framework of a systematic expansion, to first order in ε.
2. $\mathbf{Q}$ can be regarded as the “intuitively understandable” position of the gyration center.
3. Within the framework of transformation theory, it turns out that we may not identify W with $v_\perp$ or ϕ with φ. Here too, a simple geometric consideration provides intuitively clear reasons.

We should remember that the calculation just presented, which is by no means trivial, is only valid for the case of constant electric and magnetic fields. While it does illustrate the basic procedure in a straightforward way, it must be significantly generalized for the general case of inhomogeneous fields. The main credit for this goes to Littlejohn; Balescu’s clear presentation also deserves mention. In the following, we will limit ourselves to summarizing the results.

In the general case of stationary but spatially inhomogeneous fields, the transformations

$$\begin{aligned}\mathbf{Q}=\mathbf{q}-\frac{\varepsilon}{\Omega}v_\perp\mathbf{n}_2+\left(\frac{\varepsilon}{\Omega}\right)^2v_\perp^2\Bigg\{&\frac{3}{8}\mathbf{b}[\mathbf{n}_2\cdot(\nabla\times\mathbf{n}_1)+\mathbf{n}_1\cdot(\nabla\times\mathbf{n}_2)]\\&+\frac{v_\parallel}{v_\perp}(\mathbf{n}_2\mathbf{b}+2\mathbf{b}\mathbf{n}_2)\cdot(\nabla\times\mathbf{b})-\frac{1}{4B}(\mathbf{n}_2\mathbf{n}_2-\mathbf{n}_1\mathbf{n}_1)\cdot\nabla B\Bigg\},\end{aligned} \tag{2.158}$$

$$U = v_{\parallel} + \frac{\varepsilon}{\Omega} v_{\perp}^2 \left\{ \frac{1}{4} [\mathbf{n}_1 \cdot (\nabla \times \mathbf{n}_1) - \mathbf{n}_2 \cdot (\nabla \times \mathbf{n}_2) + 2\mathbf{b} \cdot (\nabla \times \mathbf{b})] + \frac{v_{\parallel}}{v_{\perp}} \mathbf{n}_1 \cdot (\nabla \times \mathbf{b}) \right\}, \tag{2.159}$$

$$W = v_{\perp} + \frac{\varepsilon}{\Omega} v_{\parallel} v_{\perp} \left\{ -\frac{1}{4} [\mathbf{n}_1 \cdot (\nabla \times \mathbf{n}_1) - \mathbf{n}_2 \cdot (\nabla \times \mathbf{n}_2) + 2\mathbf{b} \cdot (\nabla \times \mathbf{b})] - \frac{v_{\parallel}}{v_{\perp}} \mathbf{n}_1 \cdot (\nabla \times \mathbf{b}) - \frac{v_{\perp}}{v_{\parallel}} \frac{e}{m} \mathbf{n}_2 \cdot \mathbf{E} \right\}, \tag{2.160}$$

$$\phi = \varphi + \frac{e}{\Omega} \left\{ \frac{v_{\parallel}^2}{v_{\perp}} \mathbf{n}_2 \cdot (\nabla \times \mathbf{b}) - v_{\perp} \mathbf{b} \cdot (\nabla \times \mathbf{n}_2) + \frac{1}{4} v_{\parallel} [\mathbf{n}_2 \cdot (\nabla \times \mathbf{n}_1) + \mathbf{n}_1 \cdot (\nabla \times \mathbf{n}_2)] + v_{\perp} \frac{1}{B} \mathbf{n}_1 \cdot \nabla B - \frac{e}{m v_{\perp}} \mathbf{n}_1 \cdot \mathbf{E} \right\} \tag{2.161}$$

lead to the goal. Here, $\nabla = \partial/\partial \mathbf{q}$, and all fields and vectors are to be formed at the location $\mathbf{q}$ with the phase φ. The new Hamiltonian is

$$H = \frac{m}{2}(U^2 + W^2) + eV(Q) + \mathcal{O}(\varepsilon^2). \tag{2.162}$$

For the equations of motion, one needs the fundamental Lie brackets (where now only the relevant orders are given and, for consistency, the fields are formed at the location $\mathbf{Q}$ with $\nabla = \partial/\partial \mathbf{Q}$)

$$[Q_i, Q_j] = -\varepsilon (m\Omega)^{-1} \varepsilon_{ijk} b_k, \tag{2.163}$$

$$[\mathbf{Q}, U] = m^{-1} \mathbf{b}^*, \tag{2.164}$$

$$[\mathbf{Q}, W] = \varepsilon (W/2mB\Omega) \mathbf{b} \times \nabla B, \tag{2.165}$$

$$[\mathbf{Q}, \phi] = \varepsilon (m\Omega)^{-1} \mathbf{b} \times (\nabla \mathbf{n}_1 \cdot \mathbf{n}_2), \tag{2.166}$$

$$[U, W] = -(W/2mB) \mathbf{b}^* \cdot \nabla B, \tag{2.167}$$

$$[U, \phi] = -m^{-1} \mathbf{b}^* \cdot (\nabla \mathbf{n}_1 \cdot \mathbf{n}_1) + \frac{1}{2m} \mathbf{b} \cdot (\nabla \times \mathbf{b}), \tag{2.168}$$

$$[W, \phi] = -\varepsilon^{-1} \frac{\Omega}{mW} + \varepsilon \frac{W}{2mB\Omega} (\nabla B) \cdot (\nabla \times \mathbf{b}) \ , \tag{2.169}$$

with

$$\mathbf{b}^* = \mathbf{b} + \frac{\varepsilon}{\Omega} U \mathbf{b} \times (\mathbf{b} \cdot \nabla)\mathbf{b}. \tag{2.170}$$

With these results, the equations of motion can be easily calculated to lowest order.

$$\dot{\mathbf{Q}} = U\mathbf{b} + \frac{\varepsilon}{\Omega}\left\{\frac{W^2}{2B}\mathbf{b} \times \nabla B + U^2 \mathbf{b} \times (\mathbf{b} \cdot \nabla)\mathbf{b} + \frac{e}{m}\mathbf{E} \times \mathbf{b}\right\}, \tag{2.171}$$

$$\begin{aligned} \dot{U} &= -\frac{W^2}{2B}\mathbf{b} \cdot \nabla B + \frac{e}{m}\mathbf{E} \cdot \mathbf{b} \\ &\quad - \varepsilon \frac{U}{\Omega}[\mathbf{b} \times (\mathbf{b} \cdot \nabla)\mathbf{b}] \cdot \left[\frac{W^2}{2B}\nabla B - \frac{e}{m}\mathbf{E}\right], \end{aligned} \tag{2.172}$$

$$\dot{W} = \frac{UW}{2B}\mathbf{b} \cdot \nabla B + \varepsilon \frac{W}{2B\Omega}\left\{U^2[\mathbf{b} \times (\mathbf{b} \cdot \nabla)\mathbf{b}] - \frac{e}{m}\mathbf{b} \times \mathbf{E}\right\} \cdot \nabla B, \tag{2.173}$$

$$\dot{\phi} = \frac{1}{\varepsilon}\Omega + U\mathbf{b} \cdot \mathbf{R} - \frac{1}{2}U\mathbf{b} \cdot (\nabla \times \mathbf{b}), \tag{2.174}$$

where $\mathbf{R} = \nabla \mathbf{e}_1 \cdot \mathbf{e}_2 = \nabla \mathbf{n}_1 \cdot \mathbf{n}_2$ is. A simple check shows that

1. for homogeneous fields, the special case already explicitly discussed follows;
2. $\dot{\mathbf{Q}}$ in order ε systematically yields the known drift velocities;
3. U is the parallel velocity of the guiding center, where
4. the "parallel force component" $m\dot{U}$ is supplied not only by the electric field, but also by ∇B and centrifugal contributions.

At this point, we will not go into further detail, but rather return to the original question. If we summarize the results, we can say that a systematic and theoretically satisfactory introduction of the coordinates and drift velocities of guiding centers is achieved. The essential requirement that the motion of the guiding centers becomes ϕ-independent can, however, be realized in various ways; the one just presented is one of several possibilities. The set of variables $\mathbf{Q}$, U, W is referred to as natural for a Cartesian choice of coordinates. Also very common is a system in which, instead of W

$$M = \frac{m}{2}\frac{W^2}{B(\mathbf{Q})} \tag{2.175}$$

is introduced. In another representation, for example, W can be replaced by M and U by

$$E = \frac{m}{2}(U^2 + W^2) + eV(\mathbf{Q}) \tag{2.176}$$

These equivalent formulations are advantageous because constants of motion or adiabatic invariants are introduced. To clarify this statement, we calculate

$$\dot{M} = [M, H]. \tag{2.177}$$

A simple calculation yields

$$\begin{aligned}
\dot{M} &= \frac{m}{2}\left[\frac{W^2}{B(\mathbf{Q})}, \frac{m}{2}(U^2 + W^2) + eV\right] \\
&= \frac{m^2}{4}\frac{1}{B}[W^2, U^2] + \frac{m^2}{4}W^2[B^{-1}(\mathbf{Q}), U^2] + \frac{m^2}{4}W^2[B^{-1}(\mathbf{Q}), W^2] \\
&\quad + \frac{m}{2}\frac{e}{B}[W^2, V(\mathbf{Q})] + \frac{em}{2}W^2[B^{-1}(\mathbf{Q}), V(\mathbf{Q})] \\
&= \frac{m^2}{B}UW\frac{W}{2mB}\mathbf{b}^* \cdot \nabla B - \frac{m^2}{2}W^2U\frac{1}{B}\nabla B \cdot \frac{1}{m}\mathbf{b}^* \\
&\quad - \varepsilon\frac{m^2}{2}W^3\frac{W}{2mB\Omega}\nabla B \cdot (\mathbf{b} \times \nabla B) - \frac{em}{B}W\nabla V \cdot \varepsilon\frac{W}{2mB\Omega}(\mathbf{b} \times \nabla B) \\
&\quad + \frac{em}{2}W^2\frac{1}{B^2}\left(-\frac{\partial B}{\partial Q_i}\right)\frac{\partial V}{\partial Q_j}\left(-\varepsilon\frac{\varepsilon_{ijk}b_k}{m\Omega}\right) + \mathcal{O}(\varepsilon^2) \\
&= 0 + \mathcal{O}(\varepsilon^2).
\end{aligned} \tag{2.178}$$

We also see that in the limit $\varepsilon \to 0$ the quantity M coincides with the magnetic moment μ.

For later applications in statistics, it is useful to conclude by comparing the three possibilities with the variable sets $\mathbf{Q}, U, W, \phi$ or $\mathbf{Q}, U, M, \phi$ or $\mathbf{Q}, E, M, \phi$, including the results for the functional determinants.

For $\mathbf{Q}, U, W, \phi$ the corresponding Hamiltonian is

$$H = \frac{m}{2}(U^2 + W^2) + eV(\mathbf{Q}), \tag{2.179}$$

as well as the functional determinant

$$|J| = \left[1 + \frac{\varepsilon}{\Omega}U\mathbf{b} \cdot (\nabla \times \mathbf{b})\right]W. \tag{2.180}$$

For $\mathbf{Q}, U, M, \phi$ we have

$$H = \frac{m}{2}U^2 + MB(\mathbf{Q}) + eV(\mathbf{Q}) \tag{2.181}$$

as well as

$$|J| = \frac{B}{m}\left[1 + \frac{\varepsilon}{\Omega}U\mathbf{b} \cdot (\nabla \times \mathbf{b})\right]. \tag{2.182}$$

Finally, for $\mathbf{Q}, E, M, \phi$ we obtain

$$H = E \tag{2.183}$$

as well as

$$|J| = \frac{B}{m^2}\left[1 + \frac{\varepsilon}{\Omega} U \mathbf{b} \cdot (\nabla \times \mathbf{b})\right] / |U|, \tag{2.184}$$

with

$$|U|^2 = \frac{2}{m}[E - eV(\mathbf{Q})] - \frac{2M}{m} B(\mathbf{Q}). \tag{2.185}$$

For all further details, we refer to the specialized literature and the excellent presentation by Balescu.

2.4 Adiabatic Invariants

In this section, we return to the topic of adiabatic invariants and summarize the three best-known cases: magnetic moment, action integral of trapped particles, and magnetic flux in a "drift shell".

The adiabatic invariants are approximate constants of motion. They are more nearly constant the better the drift approximation holds. The adiabatic invariants can be systematically calculated.

Magnetic Moment as the "First" Adiabatic Invariant

Let us begin with the magnetic moment. Energy conservation is used (we return to the charge q, instead of e, which was used in the previous section):

$$\boxed{\frac{1}{2} \frac{d}{dt} v^2 = \mathbf{v} \cdot \dot{\mathbf{v}} = \mathbf{v} \cdot \left(\frac{q}{m} \mathbf{E} + \frac{q}{m} \mathbf{v} \times \mathbf{B}\right) = \frac{q}{m} \mathbf{v} \cdot \mathbf{E}}\,. \tag{2.186}$$

After averaging over the fast gyration, we obtain to lowest order

$$\frac{1}{2} \frac{d}{dt} \langle v^2 \rangle \approx \frac{q}{m} v_{\parallel}\, \mathbf{b} \cdot \mathbf{E}\,. \tag{2.187}$$

The electric field must be evaluated at the location of the guiding center. The change in the parallel velocity component $v_{\parallel}$ is given by

$$\dot{v}_{\parallel} = \frac{d}{dt}(\mathbf{b} \cdot \mathbf{v}) = \mathbf{v} \cdot \left[\frac{\partial \mathbf{b}}{\partial t} + \mathbf{v} \cdot \nabla \mathbf{b} \right] + \frac{q}{m}\, \mathbf{E} \cdot \mathbf{b}\,. \tag{2.188}$$

The first term on the right-hand side can be neglected if the magnetic field is not explicitly time-dependent. The second term can be simplified as follows. We average over the fast rotation

$$\langle \mathbf{v}[\mathbf{v} \cdot \nabla \mathbf{b}] \rangle = \langle v_\alpha\, v_\beta \rangle \frac{\partial b_\alpha}{\partial x_\beta} \tag{2.189}$$

and apply Einstein's summation convention. We now introduce the representation

$$\langle v_\alpha\, v_\beta \rangle = \frac{v_\perp^2}{2}\, \delta_{\alpha\beta} + \frac{2v_\parallel^2 - v_\perp^2}{2}\, b_\alpha\, b_\beta \tag{2.190}$$

and can then further simplify.

This form has already been derived previously. Another way to understand the representation is to first show the equivalence by assuming a local coordinate system $\mathbf{b} \equiv \hat{z}$. Then, the representation proves to be correct by simply calculating both sides for $\alpha, \beta = x, y, z$. Any other rectangular coordinate system follows by rotation. The tensor representation shown above is rotation invariant.

If we substitute the representation (2.190) into (2.189), we obtain two terms on the right-hand side. The first is

$$\frac{v_\perp^2}{2} \frac{\partial b_\alpha}{\partial x_\alpha} = \frac{v_\perp^2}{2}\, \nabla \cdot \mathbf{b}\,. \tag{2.191}$$

The second vanishes because

$$b_\alpha\, b_\beta\, \frac{\partial b_\alpha}{\partial x_\beta} = b_\beta \frac{\partial}{\partial x_\beta} \frac{b^2}{2} = 0\,. \tag{2.192}$$

Finally, we calculate

$$\nabla \cdot \mathbf{b} = \nabla \cdot \frac{\mathbf{B}}{B} = -\frac{\mathbf{B} \cdot \nabla B}{B^2} = -\frac{1}{B}\, \mathbf{b} \cdot \nabla B\,. \tag{2.193}$$

By summarizing, we arrive at

$$\dot{v}_\parallel = \frac{q}{m}\, \mathbf{E} \cdot \mathbf{b} - \frac{1}{2} \frac{v_\perp^2}{B}\, \mathbf{b} \cdot \nabla B\,. \tag{2.194}$$

If we use energy conservation (2.186), we can express the change in the square of the perpendicular velocity component as follows:

$$\frac{d}{dt} \frac{v_\perp^2}{2} \equiv \frac{d}{dt} \left(\frac{v^2}{2} - \frac{v_\parallel^2}{2} \right) = \frac{1}{2}\, v_\perp^2 v_\parallel\, \frac{1}{B}\, \mathbf{b} \cdot \nabla B\,. \tag{2.195}$$

The right-hand side requires further consideration. If we calculate the change in the magnetic field strength B during the motion of the gyration center, we obtain in the lowest (averaged) order in stationary magnetic fields

$$\frac{dB}{dt} = \frac{\partial B}{\partial t} + \mathbf{v} \cdot \nabla B \approx v_{\parallel} \mathbf{b} \cdot \nabla B \, . \tag{2.196}$$

This means

$$\frac{d}{dt} v_{\perp}^2 \approx \frac{v_{\perp}^2}{B} \frac{dB}{dt} \tag{2.197}$$

or

$$\frac{d}{dt} \left(\frac{v_{\perp}^2}{B} \right) \approx 0 \, . \tag{2.198}$$

The magnetic moment of a particle remains (approximately) invariant along its trajectory,

$$\mu = \frac{m v_{\perp}^2}{2B} \approx const \, . \tag{2.199}$$

Example 2.6 (Adiabatic Heating)
If the magnetic field is (explicitly) time-dependent, then according to Maxwell's equation

$$\nabla \times \mathbf{E} = -\frac{\partial \mathbf{B}}{\partial t} \tag{2.200}$$

an induced electric field arises. Let us consider the simple case where the magnetic field is spatially homogeneous. Then, the work done by the electric field during one gyration cycle is

$$q \oint \mathbf{E} \cdot d\mathbf{l} = q \int \nabla \times \mathbf{E} \cdot d\mathbf{F} \approx q \pi \rho^2 \left(-\frac{\partial \mathbf{B}}{\partial t} \right) \cdot \hat{n} \, . \tag{2.201}$$

Note for gyrating particles of charge q that sign $\{q \, \hat{n} \cdot \mathbf{B}\} = -1$ holds. Therefore, the energy change per unit time is

$$-\frac{|\Omega|}{2\pi} q \pi \rho^2 \frac{\partial \mathbf{B}}{\partial t} \cdot \hat{n} = \frac{|q|}{2} \frac{|q| \mathbf{B}}{m} \cdot \frac{\partial \mathbf{B}}{\partial t} \left(\frac{m v_{\perp}}{qB} \right)^2 = \mu \frac{\partial B}{\partial t} \, . \tag{2.202}$$

This work can be used to heat a plasma (adiabatic heating). ■

The "Second" Adiabatic Invariant

Adiabatic invariants are discussed because perfect symmetry is almost never achieved in reality. This leads to the practical question of how conserved quantities behave when symmetries are "good" but not perfect.

Example 2.7 (Adiabatic Invariant of the Simple Pendulum)
A good exercise to investigate this question is the simple pendulum when time symmetry is not exact. We realize this problem by introducing a slowly time-varying resonance frequency $\omega(t)$, i.e., the equation of motion reads

$$\frac{d^2x}{dt^2} + \omega^2(t)x = 0 \,. \tag{2.203}$$

Slow variation in this context means

$$\frac{1}{\omega}\frac{d\omega}{dt} \ll \omega \,, \tag{2.204}$$

that is, the frequency changes slowly over a period. To solve the equation of motion, we make the ansatz

$$x(t) = Re\Big[A(t)e^{i\int^t \omega(t')dt'}\Big] \,, \tag{2.205}$$

which is based on the solution for constant ω. Substituting this ansatz into the pendulum equation leads to

$$i\frac{d\omega}{dt}A + 2i\omega\frac{dA}{dt} + \frac{d^2A}{dt^2} = 0 \,. \tag{2.206}$$

The last term on the left-hand side is neglected for slow variations. Then we have

$$-\frac{2}{A}\frac{dA}{dt} \approx \frac{1}{\omega}\frac{d\omega}{dt} \,; \tag{2.207}$$

i.e., the amplitude A also varies only weakly

$$A \sim \frac{1}{\sqrt{\omega(t)}} \,, \tag{2.208}$$

and the energy is not an exact constant of motion. The action integral

$$S = \oint v dx \tag{2.209}$$

however, when the integration is performed over one oscillation period. Later we will show that, more generally, we can formulate $\oint PdQ = S =$ const . Let τ be the interval between two times at which $dx/dt = 0$ and d^2x/dt^2 have the same sign. We calculate

$$S = \int_{t_0}^{t_0+\tau} v \frac{dx}{dt} dt = x \frac{dx}{dt}\bigg|_{t_0}^{t_0+\tau} - \int_{t_0}^{t_0+\tau} x \frac{d^2x}{dt^2} dt = \int_{t_0}^{t_0+\tau} \omega^2 x^2 dt \ . \tag{2.210}$$

If we insert

$$x(t) = x(t_0)\sqrt{\frac{\omega(t_0)}{\omega(t)}} \cos\left(\int_{t_0}^{t} \omega(t')dt'\right) \tag{2.211}$$

here, then after a straightforward calculation we obtain

$$S = x^2(t_0)\omega(t_0) \int_0^{2\pi} \cos^2 \xi \ d\xi = \pi x^2(t_0)\omega(t_0) = \text{const} \ , \tag{2.212}$$

where $\xi = \int_{t_0}^{t'} \omega(t'')dt''$ was used. ■

Now we turn to the generalization for a Hamiltonian H with a slowly varying parameter $\lambda(t)$,

$$H(P, Q; \lambda(t)) = E(t) \ . \tag{2.213}$$

We have

$$\begin{aligned} \frac{dS}{dt} &= \frac{d}{dt} \oint PdQ = \frac{d}{dt} \int_{Q(t)}^{Q(t+\tau)} P(E(t), Q, \lambda(t))dQ \\ &= \underbrace{P \frac{dQ}{dt}\bigg|_{Q(t)}^{Q(t+\tau)}}_{=0} + \int_{Q(t)}^{Q(t+\tau)} \frac{\partial P}{\partial t}\bigg|_Q dQ \ . \end{aligned} \tag{2.214}$$

Note

$$\frac{\partial P}{\partial t}\bigg|_Q = \frac{\partial P}{\partial E}\bigg|_{Q,\lambda} \frac{dE}{dt} + \frac{\partial P}{\partial \lambda}\bigg|_{Q,E} \frac{d\lambda}{dt} \tag{2.215}$$

and Therefore, it follows

$$\frac{dS}{dt} = \oint \left(\frac{\partial H}{\partial P}\right)^{-1} \left[\frac{dE}{dt} - \frac{\partial H}{\partial \lambda} \frac{d\lambda}{dt}\right] dQ \ . \tag{2.217}$$

From the Hamiltonian equations, it follows

$$\frac{dE}{dt} = \frac{\partial H}{\partial P}\frac{dP}{dt} + \frac{\partial H}{\partial Q}\frac{dQ}{dt} + \frac{\partial H}{\partial \lambda}\frac{d\lambda}{dt} = \frac{\partial H}{\partial \lambda}\frac{d\lambda}{dt}\,, \tag{2.218}$$

and thus

$$\boxed{\frac{dS}{dt} = 0}\,. \tag{2.219}$$

Next, we apply this to the case of a stationary magnetic field with spatial inhomogeneity, where the field lines are compressed together. The density of the field lines is proportional to the strength of the magnetic fields. Since magnetic field lines have no divergence, they are endless and must bend when compressed. For $\partial B_z/\partial z \neq 0$ a radial component B_r is required, so that $\mathbf{B} = B_z\hat{z} + B_r\hat{r}$. In a reference frame moving with the gyration velocity $\mathbf{v}_{\perp g}$, there is only a single perpendicular velocity component due to the gyration, and $\mathbf{v}_{\perp g} \perp \mathbf{B}$. The parallel velocity is not affected by this change of reference frame. If we introduce s as the distance along the magnetic field, we can write

$$\dot{s} = v_\parallel\,, \quad \dot{\mu} = 0\,, \quad \frac{d}{dt}\frac{mv^2}{2} = m\left(v_\parallel \dot{v}_\parallel + v_\perp \dot{v}_\perp\right). \tag{2.220}$$

Because $\dot{\mu} = 0$ holds

$$v_\perp \dot{v}_\perp = \frac{1}{2} v_\perp^2 \frac{\dot{B}}{B}\,, \tag{2.221}$$

where

$$\dot{B} = v_\parallel \frac{\partial B}{\partial s}\,. \tag{2.222}$$

From energy conservation, it follows

$$mv_\parallel \dot{v}_\parallel = qE_\parallel v_\parallel - v_\parallel\, \mu \frac{\partial B}{\partial s}\,. \tag{2.223}$$

If we introduce

$$\psi := -\int^{s} E_\parallel\, ds \tag{2.224}$$

then in the non-relativistic case

$$\dot{p}_\parallel = -q\frac{\partial \psi}{\partial s} - \mu\frac{\partial B}{\partial s}\,. \tag{2.225}$$

Here, $p_\parallel = mv_\parallel$ in the absence of currents along the magnetic field, since then $A_\parallel = 0$. Therefore, the Hamiltonian in the variables s and $p_\parallel$

$$\mathcal{H} = \mathcal{H}(s, p_\parallel) = \frac{p_\parallel^2}{2m} + q\psi + \mu B \tag{2.226}$$

leads to the canonical equations

$$\dot{s} = \frac{\partial \mathcal{H}}{\partial p_\parallel} = v_\parallel \,, \tag{2.227}$$

$$\dot{p}_\parallel = -\frac{\partial \mathcal{H}}{\partial s} = -qE_\parallel - \mu \frac{\partial B}{\partial s} \,. \tag{2.228}$$

Referring to the result for S in a general Hamiltonian formulation, the quantity

$$\mathcal{J} = \oint p_\parallel ds \tag{2.229}$$

should be invariant, provided that any time dependence of a magnetic "potential well" is slow compared to the oscillation frequency of the trapped particles, and any spatial inhomogeneity of the magnetic field is so gradual that the oscillation trajectory of the particle changes only slightly from one reflection to the next. Note that here we are calculating

$$p_\parallel = \sqrt{2m(\mathcal{H} - q\psi - \mu B)} \,. \tag{2.230}$$

We will use this invariance when we discuss the Fermi acceleration process.

The "Third" Adiabatic Invariant

The first adiabatic invariant μ refers to the rapid gyration of a charged particle, the second to the oscillation in a magnetic well. Now we consider the motion of particles in a cylindrically symmetric configuration with an axial field. The cylinder is assumed to be symmetric in the poloidal angle θ. The angular momentum is a conserved quantity. We assume that A_θ is the only nonzero component of $\mathbf{A}$. The θ-component of the exact equation of motion in cylindrical coordinates is

$$m(r\ddot{\theta} + 2\dot{r}\dot{\theta}) = -\frac{q}{r}\left(r\dot{A}_\theta + \dot{r}A_\theta + r\dot{r}\,\frac{\partial A_\theta}{\partial r} + r\dot{z}\frac{\partial A_\theta}{\partial z} \right) . \tag{2.231}$$

The right-hand side can be rewritten as a total derivative, so that

$$m(r\ddot{\theta} + 2\dot{r}\dot{\theta}) = -\frac{q}{r}\,\frac{d}{dt}\,(rA_\theta) \,. \tag{2.232}$$

Multiplying by r and rearranging slightly leads to

$$\frac{d}{dt}\left(mr^2\dot{\theta} + q\, rA_\theta\right) \equiv \frac{d}{dt}\left[r(mv_\theta + q\, A_\theta)\right] \equiv \frac{dP_\theta}{dt} = 0\,. \tag{2.233}$$

The axial field is

$$B_z = (\nabla \times \mathbf{A}) \cdot \hat{z} = \frac{1}{r}\frac{\partial}{\partial r}\,(rA_\theta)\,. \tag{2.234}$$

By integration we obtain

$$A_\theta(r) = \frac{1}{r}\int_0^r r' B_z(r') dr'\,. \tag{2.235}$$

Comparing the terms that contribute to P_θ, we find

$$\frac{mv_\theta}{|qA_\theta|} \sim \frac{|mv_\perp|}{|qrB_z|} \sim \frac{\rho}{r} \ll 1\,. \tag{2.236}$$

Therefore, for distances greater than the Larmor radius $\rho \equiv r_L$

$$2\pi\, rA_\theta = 2\pi \int r' B_z(r') dr' \equiv \int_F \mathbf{B} \cdot d\mathbf{F} \approx \text{ const}\,. \tag{2.237}$$

The magnetic flux within a drift shell is adiabatically constant.
The surfaces $\psi \equiv rA_\theta = \text{const}$ satisfy

$$\mathbf{B} \cdot \nabla\psi \equiv \mathbf{B} \cdot \nabla(rA_\theta) = 0 \tag{2.238}$$

and are known as magnetic flux surfaces.

That the magnetic field lines lie within the surfaces $\psi = \text{const}$ follows from the following calculations:

$$\begin{aligned} \mathbf{B} \cdot \nabla\psi &= B_r \frac{\partial(rA_\theta)}{\partial r} + B_z \frac{\partial(rA_\theta)}{\partial z} \\ &= -\frac{\partial A_\theta}{\partial z}\frac{\partial(rA_\theta)}{\partial r} + \frac{1}{r}\frac{\partial}{\partial r}(rA_\theta)\frac{\partial(rA_\theta)}{\partial r} = 0\,. \end{aligned} \tag{2.239}$$

Thus, due to the conservation of angular momentum, the particles are forced to move on the magnetic flux surfaces, except for excursions on the order of the Larmor radius.

Selected Applications

Electron cyclotron emission

An accelerated charged particle emits radiation. Obviously, circular motion is an accelerated motion. Therefore, a charged particle in a strong magnetic field emits radiation. The radiation frequency is on the order of the gyrofrequency (and its harmonics). Assuming that the emitted intensity follows a Planck distribution, we estimate that an electron in a magnetic field of 3 T at a temperature of $k_B T_e \approx 1\ keV$ radiates at a frequency below the maximum of the Planck distribution. Therefore, we can use the Rayleigh-Jeans law $I \sim \omega^2 T_e$. Since the intensity is proportional to the temperature, if the magnetic field strength is known, a measurement of I can be used to determine the temperature T_e.

Drift in a Simple Torus

If we bend a cylinder into a torus, one might think that it would be easy to construct a magnetic confinement device for plasma in this way. We introduce magnetic field lines that run parallel to the original axis of the cylinder. The simplest way to achieve this "confinement field" is to use line currents along the Z-axis; see Fig. 2.8. Then, the magnetic field strength decreases with distance from the torus axis Z, i.e., $\sim 1/R$. As shown in the figure, due to the ∇B drift, electrons and ions move in opposite directions, creating a space charge field parallel to the torus axis. Ultimately, together with the magnetic field, this results in an $E \times B$ motion of the entire plasma outward. Therefore, the simple device will not work. One solution to this dilemma is a helical field structure of the magnetic field lines, as realized in tokamaks.

Magnetic Mirror

As an example of trapped drift orbits, let us consider particles that are trapped in a magnetic bottle consisting of two magnetic mirrors, as sketched in Fig. 2.9.

A particle that starts at the magnetic field minimum $B_{\min}$ and initially has $v_{\|0}$ and $v_{\perp 0}$, moves in the direction of the magnetic field maximum. Since $\mu = $ const , its perpendicular velocity increases while the parallel velocity decreases. The turning point occurs at the position where $v_\| = 0$ is satisfied. The condition for the magnetic field strength B at the turning point is

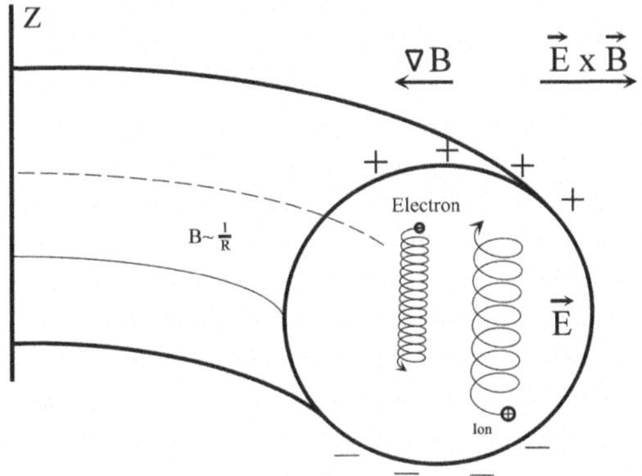

Fig. 2.8 Charge separation in a "simple" torus

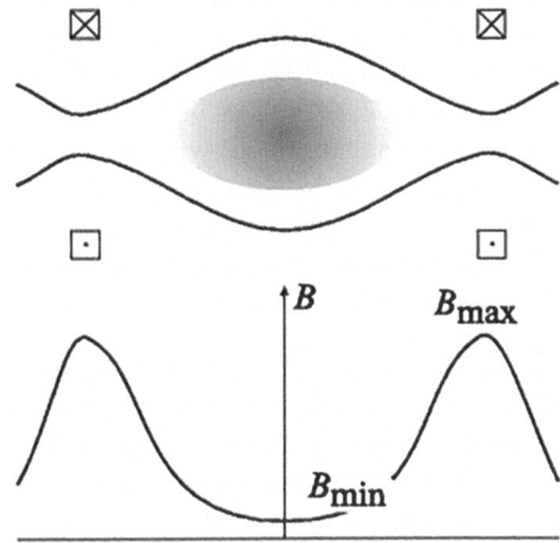

Fig. 2.9 The magnetic mirror uses the invariance of the magnetic moment μ

$$\mu \sim \frac{v_{\perp 0}^2}{B_{\min}} = \frac{v_\perp^2}{B} \stackrel{!}{=} \frac{v_0^2}{B} . \tag{2.240}$$

Thus,

$$B = \frac{v_0^2}{v_{\perp 0}^2} B_{\min} \equiv \frac{B_{\min}}{\sin^2 \theta} . \tag{2.241}$$

The angle θ is called the pitch angle and is defined by

$$\sin \theta = \frac{v_{\perp 0}}{v_0} . \tag{2.242}$$

We can define a cone angle θ_K by

$$\sin^2 \theta_K = \frac{B_{\min}}{B_{\max}} \tag{2.243}$$

As shown in Fig. 2.9 and 2.10, all particles within the cone

$$\theta < \theta_K \tag{2.244}$$

cannot be trapped by the magnetic bottle.

Van Allen Belt

Our Earth possesses a dipole-like magnetic field. The field lines are strongly deformed by the solar wind. As one approaches the poles, the magnetic field strength increases. Particles originating from the solar wind can be scattered into the magnetic structure by collisions. They are then trapped in the Earth's magnetic field and oscillate back and forth between the poles. Further collisions can act on the particles and scatter them into the loss cone, which ultimately causes the auroras. The curvature of the magnetic field lines leads to a curvature drift. Electrons and ions move in opposite directions due to this curvature drift, for example along the equator. In a closed system with rotational symmetry, there is on average no space charge field. However, a current flows from east to west. This current generates a magnetic field, which in turn modifies the original Earth's magnetic field.

Fermi Acceleration Mechanism

The second adiabatic invariant, in combination with mirror trapping/release, forms the basis of the Fermi mechanism for accelerating cosmic ray particles to ultrarelativistic speeds. Let us consider a particle that is initially trapped in a magnetic mirror configuration. Let $\theta > \theta_K$ initially be satisfied for these particles. Now let us assume that the distance between the magnetic mirrors is slowly reduced. As a result, the oscillation path L of the trapped particles gradually shortens. Due to the (adiabatic) invariance of the second adiabatic invariant, the parallel velocity of the particle increases with each oscillation. The continuous increase of $v_\parallel$ means that the velocity angle decreases. Eventually, $\theta < \theta_K$ holds, and the particle can escape at one end of the mirror with high parallel velocity. This mechanism leads to a slow pumping up to very high energies, followed by a sudden and automatic ejection of high-energy particles.

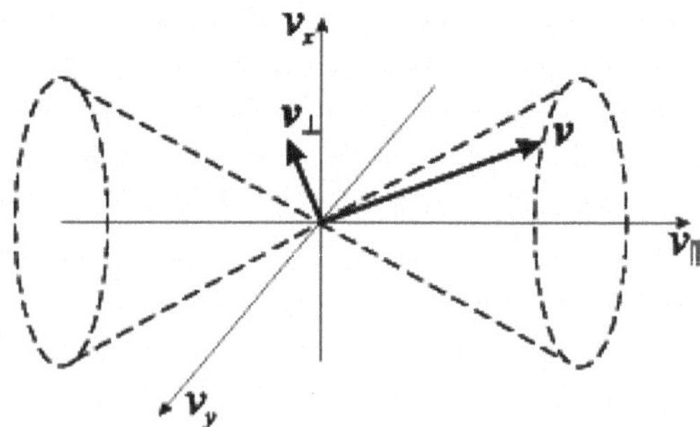

Fig. 2.10 Scattering of particles out of a magnetic mirror configuration (magnetic bottle)

Because of its particular importance in plasma astrophysics, we devote a separate (subsequent) section to Fermi acceleration.

2.5 Fermi Acceleration

Acceleration Models
As early as 1949, Fermi [47] proposed a mechanism for the acceleration of cosmic particles. The basis is energy gain through interaction with moving magnetized plasma regions. Today, we distinguish whether the effect arises from randomly distributed plasma clouds (Mechanism I) or whether reflection at strong shock waves (Mechanism II) is responsible for the energy gain.

Type I

Figure 2.11 shows particles in the laboratory frame encountering magnetized plasma clouds. The velocities $\mathbf{u}_i$ of the plasma clouds are assumed to be isotropically distributed. Let us select one with velocity $\mathbf{u}$. The particle encounters this cloud with velocity $\mathbf{v}_1$, as shown in Fig. 2.11. Using polar coordinates, we define the angle θ_1:

$$\mathbf{u} \cdot \mathbf{v}_1 = uv_1 \cos\theta_1 \ . \tag{2.245}$$

Obviously, $\theta_1 = \pi$ applies for antiparallel incidence and $\theta_1 = 0$ for parallel incidence; $\cos\theta_1$ then varies from -1 to $+1$.

For (ultra-)relativistic particles, the following holds

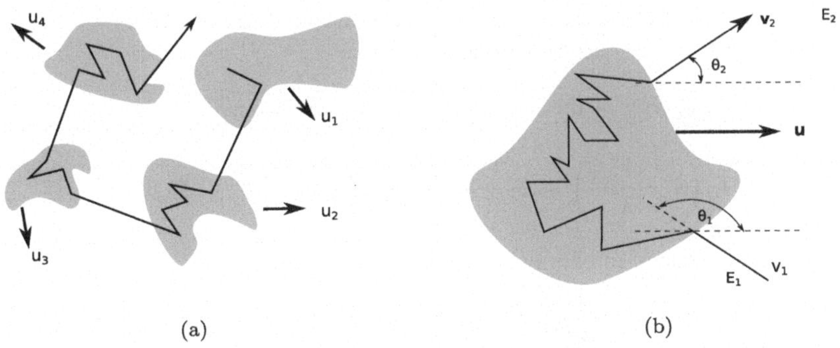

Fig. 2.11 Isotropic scattering of particles by stochastically distributed plasma clouds. **a** Successive interactions with different plasma clouds. **b** Geometry of scattering at a plasma cloud

$$E_1 = \sqrt{c^2 p_1^2 + m^2 c^4} \approx p_1 c \quad \text{with} \quad \gamma \gg 1 \,. \tag{2.246}$$

After multiple isotropic scatterings within the magnetized plasma cloud (note: isotropic scattering in the rest frame of the cloud!), the particle exits at angle θ_2 with energy E_2. Unprimed quantities refer to the laboratory frame.

Using the Lorentz transformation, we transform into the rest frame of the cloud. Energy and momentum form a four-vector $(E, \mathbf{p}c)$, which transforms as a Lorentz vector. Since $v_1 \approx c$, we obtain for the energy in the (primed) cloud frame

$$E_1' \approx \gamma E_1 (1 - \beta \cos\theta_1) \,, \quad \beta = \frac{u}{c} \,, \quad \gamma = \frac{1}{\sqrt{1-\beta^2}} \,. \tag{2.247}$$

Scattering by the magnetic fields in the cloud system is assumed to be elastic, so that $E_1' \approx E_2'$ holds. Furthermore, the scattering angle θ_2' in the cloud system is assumed to be isotropically distributed. Under this assumption, transforming back to the laboratory frame yields

$$E_2 = \gamma E_2' (1 + \beta \cos\theta_2') = \gamma^2 E_1 (1 + \beta \cos\theta_2')(1 - \beta \cos\theta_1) \,. \tag{2.248}$$

Now we average over different realizations of the scattering in the cloud system. Isotropy leads to

$$\langle \cos\theta_2' \rangle = 0 \,. \tag{2.249}$$

The same does not apply to the θ_1-averaging, because the collision probability depends on the relative velocity

$$v_{rel} = v - u \cos\theta_1 \tag{2.250}$$

Averaging with weighting by the relative velocity leads to

$$\langle \cos\theta_1 \rangle = \frac{\int_{-1}^{+1} \cos\theta_1 (v - u\cos\theta_1) d\cos\theta_1}{\int_{-1}^{+1} (v - u\cos\theta_1) d\cos\theta_1} = -\frac{u}{3v} \approx -\frac{\beta}{3} \,. \tag{2.251}$$

Thus, for the mean energy of the scattered particle, we obtain

$$\langle E_2 \rangle = \gamma^2 E_1 \left(1 + \frac{1}{3}\beta^2 \right) = E_1 \frac{1 + \frac{1}{3}\beta^2}{1 - \beta^2} \approx E_1 \left(1 + \frac{4}{3}\beta^2 + \mathcal{O}(\beta^4) \right) \,. \tag{2.252}$$

The mean relative change in energy is calculated according to

$$\boxed{\left\langle \frac{\Delta E}{E} \right\rangle = \frac{\langle E_2 \rangle - E_1}{E_1} \approx \frac{4}{3}\beta^2 \,.} \tag{2.253}$$

Thus, particles gain energy. The energy gain here is of second order in the velocity of the cloud.

Type II

A more effective energy transfer than in the interaction with plasma clouds occurs through scattering at a shock wave. A shock wave is a pressure wave that propagates at supersonic speed (or super-Alfvén speed in MHD) u. Here, the reference speed is determined by the restoring force. If the magnetic pressure is much greater than the kinetic gas pressure, disturbances propagate not at the speed of sound

$$v_S = \sqrt{\frac{\gamma P}{\rho}} , \tag{2.254}$$

but at the Alfvén speed

$$v_A = \sqrt{\frac{B}{\mu_0 \rho}} \tag{2.255}$$

The gas in front of and behind the shock front is characterized by pressure P, mass density ρ, and temperature T.

Example 2.8
Three shock conditions follow from mass, energy, and momentum continuity. For the flow velocities, see Table 2.1. We assume a sharp (discontinuous) transition between the upstream and downstream regions.

From the continuity equation for mass flux [48]

$$\frac{\partial \rho}{\partial t} + \frac{\partial (\rho v)}{\partial x} = 0 \tag{2.256}$$

it follows (in the frame moving with the shock front)

$$\rho_1 v_1 = \rho_2 v_2 \qquad \text{[condition 1]} . \tag{2.257}$$

The momentum flux equation

$$\rho \frac{dv}{dt} = -\frac{\partial P}{\partial x} \tag{2.258}$$

in the stationary shock front frame, i.e.

$$\rho v \frac{dv}{dx} + \frac{dP}{dx} = 0 , \tag{2.259}$$

Table 2.1 Shock parameters, on the one hand in the system in which the incoming gas is at rest (laboratory system), and on the other hand in the system of the moving shock front

Variable	Laboratory system	Shock front system
Front velocity	$\mathbf{u}$	0
Density ahead of the front	ρ_1	ρ_1
Density behind the front	ρ_2	ρ_2
Pressure ahead of the front	P_1	P_1
Pressure behind the front	P_2	P_2
Temperature ahead of the front	T_1	T_1
Temperature behind the front	T_2	T_2
Gas velocity ahead of the front	$\mathbf{u}_1 = 0$	$\mathbf{v}_1 = -\mathbf{u}$
Gas velocity behind the front	$\mathbf{u}_2$	$\mathbf{v}_2 = \mathbf{u}_2 - \mathbf{u}$

yields, with $d(\rho v)/dx = 0$

$$P_1 + \rho_1 v_1^2 = P_2 + \rho_2 v_2^2 \quad \text{[condition 2]} \ . \tag{2.260}$$

The energy density of the gas is

$$\mathcal{E} = \frac{1}{2}\rho v^2 + \varepsilon + P \tag{2.261}$$

with the internal energy density

$$\varepsilon = \frac{1}{\gamma - 1}P \rightarrow \frac{3}{2}P \ , \quad P = nkT = \frac{\rho}{m}kT \tag{2.262}$$

and the enthalpy density

$$\varepsilon + P = \frac{\gamma}{\gamma - 1}P \ . \tag{2.263}$$

Altogether, from the conservation of energy flux, we obtain

$$\frac{1}{2}v_1^2 + \frac{\varepsilon_1 P_1}{\rho_1} = \frac{1}{2}v_2^2 + \frac{\varepsilon_2 P_2}{\rho_1} \quad \text{[condition 3]} \ . \tag{2.264}$$

■

The three shock conditions, together with the relationships between the physical quantities shown in the example, yield solutions for the ratios

$$\frac{\rho_2}{\rho_1} = \frac{v_1}{v_2} = \frac{(\gamma + 1)M_1^2}{(\gamma - 1)M_1^2 + 2} \ , \tag{2.265}$$

Table 2.2 Solutions for the relative shock values in the limit of large Mach numbers and (additionally) $\gamma = \frac{5}{3}$

Ratios	$M_1^2 \gg 1$	$M_1^2 \gg 1$ und $\gamma = \frac{5}{3}$
$\frac{P_2}{P_1}$	$\frac{2\gamma}{\gamma+1} M_1^2$	$\frac{5}{4} M_1^2$
$\frac{\rho_2}{\rho_1} = \frac{v_1}{v_2}$	$\frac{\gamma+1}{\gamma-1}$	4
$\frac{T_2}{T_1}$	$\frac{2\gamma(\gamma-1)}{(\gamma+1)^2} M_1^2$	$\frac{5}{16} M_1^2$

$$\frac{P_2}{P_1} = \frac{2\gamma M_1^2}{\gamma+1} - \frac{\gamma-1}{\gamma+1} , \tag{2.266}$$

$$\frac{T_2}{T_1} = \frac{[2\gamma M_1^2 - \gamma(\gamma-1)][(\gamma-1)M_1^2 + 2]}{(\gamma+1)^2 M_1^2} \tag{2.267}$$

Here, we have introduced the Mach number

$$M_1 = \frac{v_1}{\sqrt{\gamma P_1/\rho_1}} \tag{2.268}$$

For a "strong" shock ($M_1 \gg 1$) and the adiabatic index 5/3, the results in Table 2.2 are simplified. If we take into account

$$v_1 = -u \approx 4v_2 = 4(u_2 - u) , \tag{2.269}$$

then for a strong shock we obtain $u_2 = \frac{3}{4}u$.

When we calculate the scattering of particles at a shock front with velocity $\mathbf{u}$, we must take into account two differences from the previous calculation. First, we have to replace $u \to \frac{3}{4}u$.

Secondly, and this is fundamentally more important, incident angles only occur between π and $\pi/2$, while exit angles only occur between 0 and $\pi/2$. This is best seen in Fig. 2.12. In other words:

$$\frac{\pi}{2} \leq \theta_1 \leq \pi , \quad 0 \leq \theta_2' \leq \frac{\pi}{2} . \tag{2.270}$$

Within these limits, the angles are uniformly distributed. So if we use the formula that also applies here (2.248), i.e.

$$E_2 = \gamma^2 E_1 (1 + \beta \cos\theta_2')(1 - \beta \cos\theta_1) , \tag{2.271}$$

we again perform averaging, namely

$$\langle E_2 \rangle = \gamma^2 E_1 (1 + \beta \langle \cos\theta_2' \rangle)(1 - \beta \langle \cos\theta_1 \rangle) . \tag{2.272}$$

However, now

$$\langle\cos\theta_1\rangle = \frac{\int_{-1}^{0}\cos\theta_1 d\cos\theta_1}{\int_{-1}^{0} d\cos\theta_1} = -\frac{1}{2}\,, \tag{2.273}$$

$$\langle\cos\theta_2'\rangle = \frac{\int_{0}^{+1}\cos\theta_2' d\cos\theta_2'}{\int_{0}^{+1} d\cos\theta_2'} = \frac{1}{2}\,, \tag{2.274}$$

applies, so that

$$\langle E_2\rangle = \gamma^2 E_1\left(1+\frac{1}{2}\beta\right)^2 = E_1\frac{(1+\frac{1}{2}\beta)^2}{1-\beta^2} \approx E_1(1+\beta) \tag{2.275}$$

results. Note

$$u_2 = \frac{3}{4}u \rightarrow \beta = \frac{3}{4}\frac{u}{c}\,. \tag{2.276}$$

For the relative energy gain, this leads to

$$\boxed{\left\langle\frac{\Delta E}{E}\right\rangle = \frac{\langle E_2\rangle - E_1}{E_1} \approx \beta \approx \frac{3}{4}\frac{u}{c}\,.} \tag{2.277}$$

In contrast to the result for scattering by plasma clouds, in shock wave acceleration the energy gain is first order in β. For this reason, this process is considered more effective. Among the various possible sources of cosmic particle acceleration (shock waves from supernovae, pulsars, active galactic nuclei, black holes, gamma-ray bursts, etc.), shock waves from supernovae are favored.

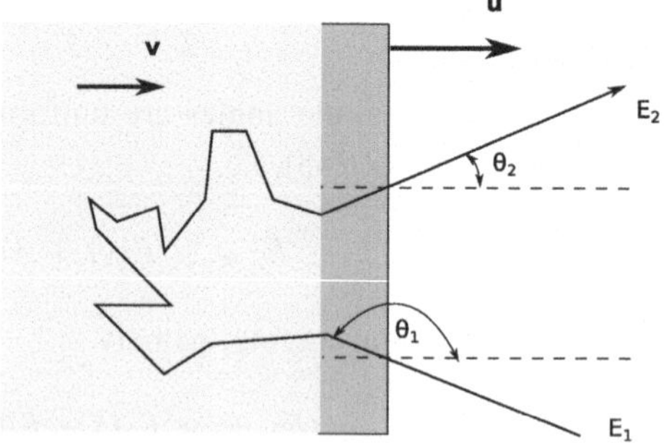

Fig. 2.12 For type II Fermi acceleration at a shock front, the geometry of the scattering process is shown. A particle strikes the shock front head-on, is scattered in the shock, and then leaves the shock with a higher energy E_2

2.6 Motion of an Electron in an Electromagnetic Wave

A particularly interesting case, which is also important for applications, is the motion of electrons in an electromagnetic wave. Here, we are thinking of a laser as the wave. Is the electron preferentially pushed in a certain direction? Does it absorb energy from the wave, and if so, for how long? These are the questions we will discuss here.

A plane light wave propagating in the x-direction can be described by a vector potential:

$$\mathbf{A}(\mathbf{r}, t) = \Re\{\mathbf{A}_0 e^{i\psi}\} , \tag{2.278}$$

where $\mathbf{r}$ and t are spatial and temporal coordinates, $\mathbf{A}_0 = A_0\hat{e}_y$ for linear polarization (LP) and $\mathbf{A}_0 = A_0(\hat{e}_y \pm i\hat{e}_z)$ for circular polarization (CP). The plus and minus signs denote right- and left-circular polarization, respectively. The phase $\psi = \mathbf{k}\cdot\mathbf{r} - \omega t$ contains the wave vector $\mathbf{k} = k\hat{e}_x$. The dispersion relation in vacuum is $\omega = kc$ with $k = |\mathbf{k}| = 2\pi/\lambda$; c is the speed of light in vacuum. The electric and magnetic fields are then obtained as

$$\mathbf{E} = \Re\left\{i\omega\mathbf{A}_0 e^{i\psi}\right\}, \quad \mathbf{B} = \Re\{i\mathbf{k}\times\mathbf{A}_0 e^{i\psi}\} , \tag{2.279}$$

where the Poynting vector $\mathbf{S} = \mathbf{E}\times\mathbf{H}$ is. The latter determines the intensity (energy per unit area and per unit time) of the light

$$I = |\mathbf{S}| = \frac{\omega k}{2\mu_0} A_0^2 \times \begin{cases} [1-\cos(2\psi)], & \text{for LP} \\ 2, & \text{for CP} \end{cases} . \tag{2.280}$$

It should be noted that for linear polarization, the intensity oscillates with double the phase, due to $\sin^2 x = [1-\cos(2x)]/2$. On the other hand, for circular polarization, the intensity is independent of the phase. This leads to a significant difference in the interaction with matter. For the intensity $I_0 = \langle I\rangle$ averaged (over the oscillations), we have

$$I_0\lambda^2 = \zeta\frac{\omega k\lambda^2}{2\mu_0} A_0^2 = \zeta\frac{2\pi^2}{\mu_0} cA_0^2 , \tag{2.281}$$

with $\zeta = 1$ for linear and $\zeta = 2$ for circular polarization.

The relativistic threshold intensity is reached when electrons, which are captured by the light wave, attain (almost) the speed of light. For non-relativistic electrons with $|\mathbf{v}| \ll c$ the equation of motion is

$$m\frac{d\mathbf{v}}{dt} = -e(\mathbf{E} + \mathbf{v}\times\mathbf{B}) \approx -e\mathbf{E} \tag{2.282}$$

with the approximate solutions

$$\mathbf{v} \approx \Re\left\{\frac{e\mathbf{E}}{im\omega}\right\} = \frac{eA_0}{m}\begin{cases}\hat{e}_y \cos\psi & \text{for LP} \\ (\hat{e}_y \cos\psi \mp \hat{e}_z \sin\psi) & \text{for CP}\end{cases}, \tag{2.283}$$

$$\mathbf{r} \approx \Re\left\{\frac{e\mathbf{E}}{m\omega^2}\right\} = -\frac{eA_0}{m\omega}\begin{cases}\hat{e}_y \sin\psi & \text{for LP} \\ (\hat{e}_y \sin\psi \mp \hat{e}_z \cos\psi) & \text{for CP}\end{cases}. \tag{2.284}$$

With the dimensionless light amplitude

$$a_0 = \frac{eA_0}{mc}, \tag{2.285}$$

we can write (2.281) in the form

$$\boxed{I_0\lambda^2 = \zeta\frac{2\pi^2}{\mu_0}P_0a_0^2 = \zeta\left[1.37\times10^{18}\frac{W}{cm^2}\,\mu m^2\right]a_0^2} \tag{2.286}$$

By definition, the relativistic threshold is reached at $a_0 = 1$, i.e., when the oscillation velocity approaches the speed of light c. Naturally, the electron trajectories then differ from the simple transverse oscillation derived above. We will derive these in detail in the following. Eq. (2.286) contains the relativistic power unit

$$P_0 = \frac{4\pi}{\mu_0}\frac{m^2c^3}{e^2} = 8.67\ \text{GW}. \tag{2.287}$$

This can be written as the product of the voltage $\frac{mc^2}{e} = 511$ kV, corresponding to the rest energy of the electron, and the electric current unit $J_0 = \frac{4\pi}{\mu_0}\frac{mc}{e} = 17$ kA, which is related to the Alfvén current $J_A = J_0\beta\gamma$, where $\beta = \frac{v}{c}$ and $\gamma = \frac{1}{\sqrt{1-\beta^2}}$ are. Currents greater than J_A cannot be transported in vacuum due to magnetic self-interaction.

Example 2.9 (Current Units)
Let us briefly compare the electric current unit J_0 in the SI system (ampere) with the corresponding cgs unit. If we construct a unit for electric current from the physical constants e (let this be the unit for a charge Q measured in units M for mass, L for length, and T for time), m (unit M for mass), and the speed c, we find in the cgs system for charge divided by time

$$[J] = \frac{M^{1/2}L^{3/2}}{T^2} \sim m^X c^Y e^Z \sim M^X\left(\frac{L}{T}\right)^Y\left(\frac{M^{1/2}L^{3/2}}{T}\right)^Z, \tag{2.288}$$

which leads to

$$X + \frac{1}{2}Z = \frac{1}{2}, \quad Y + \frac{3}{2}Z = \frac{3}{2}, \quad Y + Z = 2. \tag{2.289}$$

The solution $X = 1$, $Y = 3$, $Z = -1$ leads to $J_0 = \frac{mc^3}{e}$ which agrees with the above formula for $\mu_0 = 4\pi/c^2$. ■

Example 2.10 (Vacuum Impedance)
Another remark concerns the relation (2.281). For circular polarization ($\zeta = 2$) we write it in the form

$$I_0 = \frac{1}{\mu_0}\frac{k}{\omega}E^2 = \frac{1}{\mu_0 c}E^2 = \sqrt{\frac{\varepsilon_0}{\mu_0}}E^2 \equiv \frac{E^2}{Z_0}\,, \tag{2.290}$$

with the impedance of vacuum

$$Z_0 = \sqrt{\frac{\mu_0}{\varepsilon_0}} = 376.73\ \Omega\,. \tag{2.291}$$

■

An exact analytical description of electron trajectories is possible for individual electrons in a plane light wave of arbitrary amplitude. For a clear derivation, it is important to exploit suitable symmetries and invariants of the problem. The relativistic Lagrangian of a particle with charge q, moving in electromagnetic potentials $\mathbf{A}$ and ϕ, is given by

$$L(\mathbf{r}, \mathbf{v}, t) = -mc^2\sqrt{1 - \frac{v^2}{c^2}} + q\,\mathbf{v}\cdot\mathbf{A} - q\phi\,. \tag{2.292}$$

For a plane electromagnetic wave, $\Phi = 0$ holds. From the Euler-Lagrange equations

$$\frac{d}{dt}\frac{\partial L}{\partial \mathbf{v}} - \frac{\partial L}{\partial \mathbf{r}} = 0\,, \tag{2.293}$$

the equations of motion follow

$$\frac{d\mathbf{p}}{dt} = q(\mathbf{E} + \mathbf{v}\times\mathbf{B})\,. \tag{2.294}$$

The canonical momentum is $\mathbf{p}^{\text{can}} \equiv \frac{\partial L}{\partial \mathbf{v}} = m\gamma\mathbf{v} + q\mathbf{A} \equiv \mathbf{p} + q\mathbf{A}$ with $\gamma = \frac{1}{\sqrt{1-\frac{v^2}{c^2}}}$.

For a plane light wave, there are two symmetries that imply two conserved quantities. The planar symmetry yields $\frac{\partial L}{\partial \mathbf{r}_\perp} = 0$ and consequently the conservation of canonical momentum in the transverse direction.

$$\boxed{\partial L/\partial \mathbf{v}_\perp = \mathbf{p}_\perp + q\,\mathbf{A}_\perp = \text{const}}\,. \tag{2.295}$$

The second invariant results from the waveform of $\mathbf{A} \equiv \mathbf{A}(t - x/c)$. We use the relation $\frac{dH}{dt} = -\frac{\partial L}{\partial t}$ for the Hamiltonian function $H(\mathbf{x}, \mathbf{P}, t) = E(t)$, which expresses the time-dependent energy of the particle. We obtain

$$\frac{dE}{dt} = -\frac{\partial L}{\partial t} = c\frac{\partial L}{\partial x} = c\frac{d}{dt}\frac{\partial L}{\partial v_x} = c\frac{dp_x^{can}}{dt} = c\frac{dp_x}{dt} \,. \tag{2.296}$$

Taking into account that $A_x = 0$ holds for a plane light wave, we find

$$\boxed{E - cp_x = C} \,, \tag{2.297}$$

where C is a constant. For electrons that are initially, i.e., when no wave is present, at rest, we have $C = mc^2$. If no potential energy ($\Phi = 0$) is present, the kinetic energy is the total energy minus the rest energy, and one finds

$$E_{kin} \equiv E - mc^2 = p_x c \,. \tag{2.298}$$

Note that due to the famous Einstein formula $E = \gamma mc^2$ (m is the rest mass), the following also holds for the kinetic energy

$$E_{kin} \equiv E - mc^2 = (\gamma - 1)mc^2 \,. \tag{2.299}$$

The two relations for the kinetic energy become more plausible when one considers the canonical equations. Starting from the relativistic Lagrangian, we find the Hamiltonian by first defining the canonical momenta via

$$\frac{\partial L}{\partial \mathbf{v}} = \mathbf{p}^{can} \equiv \mathbf{P} \tag{2.300}$$

The Hamiltonian function $H(\mathbf{P}, \mathbf{q}, t)$ follows via the Legendre transformation

$$H = \sum_k P_k \dot{q}_k - L \,. \tag{2.301}$$

We obtain

$$\mathbf{P}_\perp = \mathbf{p}_\perp + q\mathbf{A}_\perp \,, \quad \mathbf{P}_\parallel = \mathbf{p}_\parallel = p_x\hat{x} \,, \quad \mathbf{A}_\parallel \equiv A_x\hat{x} = 0 \,, \tag{2.302}$$

$$H = \frac{mc^2}{\sqrt{1 - \frac{v_\perp^2}{c^2} - \frac{v_\parallel^2}{c^2}}} \equiv mc^2\sqrt{1 + \frac{p_\perp^2}{m^2c^2} + \frac{p_\parallel^2}{m^2c^2}} \,. \tag{2.303}$$

The Hamiltonian H does not depend on the transverse coordinates; therefore, $\mathbf{P}_\perp = \text{const}$ holds. Since this value is constant, we can determine it from the initial condition, e.g., that the electron is at rest before the wave arrives. Then we set $\mathbf{P}_\perp = 0$. Equivalent formulations of the Hamiltonian are

$$\boxed{H = mc^2\sqrt{1+\frac{q^2A_\perp^2}{m^2c^2}+\frac{P_\parallel^2}{m^2c^2}} \equiv mc^2\gamma} \,. \tag{2.304}$$

The latter formulations show in a very clear way that the relevant variables in the Hamiltonian $x - ct$ and $P_x \equiv P_\parallel$ are, i.e., $H = H(x - ct, P_x)$. The energy $E \equiv H = mc^2\gamma$ is the total energy. If the electron is initially, i.e., before the wave arrives, at rest, the energy corresponds to the rest energy mc^2. This means that $H - mc^2$ is the kinetic energy, as used above. Two canonical equations still need to be considered. The first,

$$\frac{dx}{dt} = \frac{\partial H}{\partial P_x}, \tag{2.305}$$

leads to the well-known relation $p_x = \gamma m v_x$. The other,

$$\frac{dP_x}{dt} = -\frac{\partial H}{\partial x}, \tag{2.306}$$

is equivalent to

$$\frac{dp_x}{dt} = -\frac{\partial H}{\partial x} = -\frac{q^2}{2m\gamma}\frac{\partial A^2}{\partial x} \,. \tag{2.307}$$

Next, we calculate

$$\frac{dH}{dt} = \frac{1}{2\gamma}\frac{q^2}{m}\frac{\partial A^2}{\partial t} + \frac{1}{2\gamma}\frac{q^2}{m}\frac{\partial A^2}{\partial x}\dot{x} + \frac{p_x}{\gamma m}\frac{dp_x}{dt} \,. \tag{2.308}$$

Substituting on the right-hand side of (2.307) yields

$$\frac{dH}{dt} = \frac{1}{2\gamma}\frac{q^2}{m}\frac{\partial A^2}{\partial t} \,. \tag{2.309}$$

Since all fields depend only on $\xi = x - ct$, we replace $\frac{\partial}{\partial t} \hat{=} -c\frac{\partial}{\partial x}$ and obtain

$$\frac{dH}{dt} = c\frac{dp_x}{dt} \,. \tag{2.310}$$

This equation can be integrated, with the result

$$\boxed{H - cp_x = C} \,, \tag{2.311}$$

where the constant $C = mc^2$ applies to particles that are initially at rest.

In summary, for an electron in a plane wave of finite duration, the relativistic equation of motion can be integrated exactly. For an electron that is initially at rest, the conserved quantities yield

$$\tilde{\mathbf{p}}_\perp \equiv \frac{\mathbf{p}_\perp}{mc} = \mathbf{a} \equiv \frac{e\mathbf{A}_\perp}{mc} = (0, a_y, a_z) \;, \quad \gamma - 1 = \tilde{p}_x \;. \tag{2.312}$$

The γ factor can also be written as

$$\gamma = \frac{1}{\sqrt{1-\frac{v^2}{c^2}}} = \sqrt{1+\frac{p^2}{m^2c^2}} \;, \tag{2.313}$$

since $\mathbf{p} = \gamma m \mathbf{v}$. Starting from

$$\gamma - 1 = \frac{p_x}{mc} \tag{2.314}$$

we obtain from the square

$$\gamma = 1 + \frac{p_\perp^2}{2m^2c^2} \;. \tag{2.315}$$

Therefore,

$$\gamma - 1 = \tilde{p}_x = \frac{\tilde{p}_\perp^2}{2} = \frac{a^2}{2} \;. \tag{2.316}$$

One immediate and very important observation at this point is that $E_{kin} \sim \tilde{E}_{kin} \sim \gamma - 1 \sim a^2$ is directly coupled to the light amplitude a and drops to zero as soon as the electron leaves the light field.

The electron cannot gain net energy in a plane light wave. A symmetry breaking (departure from the strict assumption of a plane wave) is required to achieve a net energy gain. This typically occurs in experimental configurations, for example, due to a finite beam diameter or additional interactions.

From the last equations, with $\beta = \mathbf{v}/c$ and $q = -e$, we obtain the "equations of motion"

$$\tilde{p}_x = \gamma\beta_x = \frac{\gamma}{c}\frac{dx}{dt} = a^2/2 \;, \tag{2.317}$$

$$\tilde{p}_y = \gamma\beta_y = \frac{\gamma}{c}\frac{dy}{dt} = a_y \;, \tag{2.318}$$

$$\tilde{p}_z = \gamma\beta_z = \frac{\gamma}{c}\frac{dz}{dt} = a_z \; . \tag{2.319}$$

Since $\gamma = 1 + a^2/2$, we obtain for $a \gg 1$

$$\beta_x = \frac{a^2/2}{1+a^2/2} \to 1, \;\; \beta_y = \frac{a_y}{1+a^2/2} \to 0, \;\; \beta_z = \frac{a_z}{1+a^2/2} \to 0 \tag{2.320}$$

and also $\tan\theta = p_\perp/p_x \to 0$. This means that although the electron oscillates transversely to the direction of propagation at low field strengths $|\mathbf{a}| \ll 1$, at relativistic laser intensities with $|a| \gg 1$ it moves increasingly in the direction of light propagation.

The integration of the equations of motion for a given light pulse $\mathbf{a}(t - x/c)$ is straightforward to carry out in the variable "proper time" $\tau = t - x(t)/c$. We have

$$\gamma\frac{d}{dt} = \gamma\frac{d\tau}{dt}\frac{d}{d\tau} = \gamma\left(1 - \frac{1}{c}\frac{dx}{dt}\right)\frac{d}{d\tau} = \left(1 + \frac{a^2}{2} - \frac{a^2}{2}\right)\frac{d}{d\tau} = \frac{d}{d\tau} \; , \tag{2.321}$$

and therefore $d\tau = dt/\gamma$. Thus, the equations take the simple form

$$\frac{dx}{d\tau} = c\frac{a^2}{2} \; , \quad \frac{dy}{d\tau} = ca_y \; , \quad \frac{dz}{d\tau} = ca_z \; . \tag{2.322}$$

Example 2.11 (Electron motion under circular polarization)
For *circular* polarization with

$$\mathbf{a}(\mathbf{r}, t) = \Re\left\{a_0(\hat{e}_y \pm i\hat{e}_z)e^{-i\omega\tau}\right\} \tag{2.323}$$

the motion of the electron becomes particularly simple. Since $a^2 = a_y^2 + a_z^2 = a_0^2/2$ and $\gamma = 1 + a_0^2/2$, the kinetic energy $E_{kin} = a_0^2/2$ depends on time only through the envelope $a_0(\tau)$, but not through the rapidly oscillating phase $\psi = -\omega\tau$ of the laser pulse. For *constant* a_0 we obtain $\tau = t/\gamma$, and thus the trajectory

$$x(t) = (ca_0^2/2)\tau = \frac{a_0^2/2}{1 + a_0^2/2}\, ct \; , \tag{2.324}$$

$$y(t) = \frac{ca_0}{\omega}\,\sin(\omega t/\gamma) \; , \tag{2.325}$$

$$z(t) = \mp\frac{ca_0}{\omega}\,\cos(\omega t/\gamma) \; . \tag{2.326}$$

This describes an electron moving at constant velocity along a helix. ■

For *linear* polarization, the motion of the electron is more complex. Suppose we have a rectangular pulse (over N periods with constant amplitude a_0), i.e.

$$a_y = a_0 \cos(\omega\tau) \quad \text{for} \quad 0 < \tau < N(2\pi/\omega) \ , \tag{2.327}$$

$$a_z \equiv 0, \quad a^2 = a_y^2 = a_0^2 \cos^2(\omega\tau) \ . \tag{2.328}$$

The electron can initially be at rest and located at $x = y = z = 0$. The trajectories are then obtained in the form

$$x(\tau) = \frac{ca_0^2}{2} \int_0^{\tau} \cos^2(\omega\tilde{\tau})d\tilde{\tau} = \frac{ca_0^2}{4} \left[\tau + \frac{1}{2\omega} \sin(2\omega\tau)\right] , \tag{2.329}$$

$$y(\tau) = ca_0 \int_0^{\tau} \cos(\omega\tilde{\tau})d\tilde{\tau} = \frac{ca_0}{\omega} \sin(\omega\tau) \tag{2.330}$$

The parameter τ must be determined implicitly via $\tau = t - x(t(\tau))/c$ as a function of t. A note on the latter relation: From $\gamma d\tau = dt$ and

$$\gamma = 1 + \frac{a^2}{2} = 1 + \frac{a_0^2}{2} \cos^2(\omega\tau) \ , \tag{2.331}$$

we find by direct integration

$$t = \tau + \frac{a_0^2}{4}\left[\tau + \frac{1}{2\omega} \sin(2\omega\tau)\right] . \tag{2.332}$$

Note that t is a *monotonically increasing* function of τ; see Fig. 2.13. Thus, the motion evidently consists of a general drift in the x-direction when using the non-oscillating part $t \approx \tau(1 + a_0/4)$,

$$x_d(t) \approx \frac{a_0^2}{a_0^2 + 4} ct \ , \tag{2.333}$$

and a superimposed figure-8 motion in the drift reference frame,

$$y(\tau) = \frac{a_0}{k} \sin(\omega\tau) \ , \tag{2.334}$$

$$x(\tau) - x_d \approx \frac{ca_0^2}{4}\tau - \frac{ca_0^2}{a_0^2 + 4}t + \frac{1}{8\omega}ca_0^2 \sin(2\omega\tau) = \frac{1}{2k}\frac{a_0^2}{4 + a_0^2} \sin(2\omega\tau) \ , \tag{2.335}$$

for $k = \omega/c$.

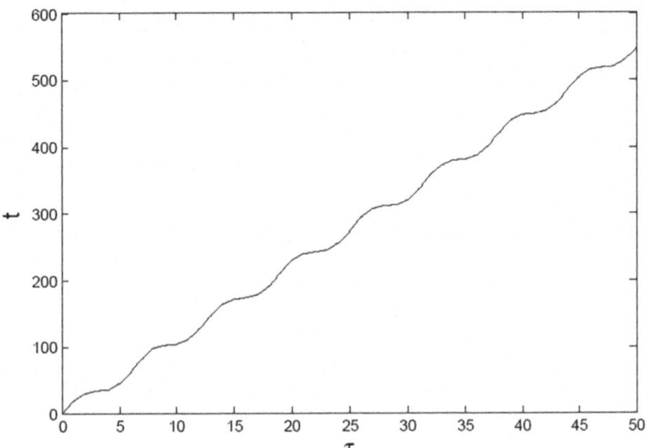

Fig. 2.13 The relation (2.332) between the time t and τ for typical parameters

In summary, for relativistic laser amplitudes, the motion of an electron is predominantly in the direction of the laser. If an electron is initially at rest and is then overtaken by a pulse of electromagnetic radiation of finite length, the electron comes to rest again after the end of the laser pulse, but has experienced a displacement. Importantly, no net energy has been transferred.

2.7 Langevin Approximation

In this section, we discuss some fundamental effects of collisions on the motion of individual particles. Much can be learned from the so-called one-dimensional (1D) Langevin equations, which model the motion of a single particle in a fluid composed of many light particles. The collisions with the constituents of the fluid are described by a stochastic force $f(t)$.

The effect of many weak collisions can be modeled by a stochastic force. We denote the stochastic force by $f(t)$. If we assume that its mean value $\langle f(t)\rangle$ is zero, we explicitly exclude the friction force $-m\zeta v$, where ζ is the friction coefficient (corresponding to the collision frequency). Newton's law for the motion of the (heavy) particle with mass m under the influence of many light collisions then reads:

$$\boxed{m\dot{v} = -m\zeta v + f(t)}\,. \tag{2.336}$$

Here we assume that no external field is present (hence Eq. (2.336) is called the free Langevin equation). The stochastic force is characterized by its mean value $\langle f(t)\rangle$ and its correlation, e.g.,

$$\langle f(t)f(t')\rangle \equiv \phi(t-t') = \lambda\delta(t-t') \tag{2.337}$$

for a delta correlation in the case where the correlation time τ_c is very small (white noise). The solution of (2.336) is

$$v(t) = v_0 e^{-\zeta t} + \frac{1}{m}\, e^{-\zeta t} \int\limits_0^t d\tau\, e^{\zeta\tau} f(\tau). \tag{2.338}$$

It can be easily found using the method of variation of constants. Since we are not interested in the exact motion, but only in the average behavior, we calculate $\left\langle [v(t)]^2 \right\rangle$. For $t \gg \zeta^{-1}$ one obtains

$$\langle v^2\rangle \to \frac{\lambda}{2\zeta m^2}\,. \tag{2.339}$$

Note that for $t \gg \zeta^{-1}$ the influence of the initial velocity v_0 has disappeared. After many collisions, a single particle should be thermalized, $(t \to \infty)$

$$\frac{1}{2} m \langle v^2\rangle \approx \frac{T}{2}\,. \tag{2.340}$$

This leads directly to the Einstein relation

$$\boxed{\lambda = 2\zeta m T}\,, \tag{2.341}$$

i.e., the friction coefficient is proportional to the quadratic amplitude of the stochastic force. The velocity correlation function also follows directly,

$$\begin{aligned} \langle v(t)v(t')\rangle &= \frac{\lambda}{2\zeta m^2}\, e^{-\zeta|t-t'|} + \left(v_0^2 - \frac{\lambda}{2\zeta m^2}\right) e^{-\zeta(t+t')} \\ &\to \frac{\lambda}{2\zeta m^2}\, e^{-\zeta|t-t'|} \quad \text{for} \quad t, t' \gg \zeta^{-1}\,. \end{aligned} \tag{2.342}$$

The correlation function shows an exponential decay for long times with the characteristic time ζ^{-1}.

For transport, the mean square deviation $\left\langle [x(t)]^2 \right\rangle$ is of fundamental importance. Since x follows from temporal integration of v, it can be directly related to the velocity correlation function (2.342). In the long-time limit, we obtain

$$\langle [x(t)]^2 \rangle \approx \int_0^t d\tau \int_0^t d\tau' \, \frac{\lambda}{2\zeta m^2} \, e^{-\zeta|\tau-\tau'|} \, . \tag{2.343}$$

If we split the integral into two parts, we have

$$\begin{aligned}
&\int_0^t d\tau \int_0^\tau d\tau' e^{-\zeta(\tau-\tau')} + \int_0^t d\tau \int_\tau^t d\tau' e^{-\zeta(\tau'-\tau)} \\
&= \int_0^t d\tau \, \frac{1}{\zeta} \, e^{-\zeta\tau}\left[e^{\zeta\tau} - 1\right] + \int_0^t d\tau \, \frac{1}{(-\zeta)} \, e^{\zeta\tau}\left[e^{-\zeta t} - e^{-\zeta\tau}\right] \\
&= \frac{t}{\zeta} + \frac{1}{\zeta^2}\left[e^{-\zeta t} - 1\right] - \frac{e^{-\zeta t}}{\zeta^2}\left[e^{\zeta t} - 1\right] + \frac{t}{\zeta} \\
&\sim \frac{2}{\zeta} t \quad \text{for large t} \, .
\end{aligned} \tag{2.344}$$

Therefore, it follows that

$$\langle [x(t)]^2 \rangle \approx \frac{\lambda}{\zeta^2 m^2} \, t \, . \tag{2.345}$$

The "standard" definition of the (position-independent) diffusion coefficient D is

$$\langle x^2 \rangle = 2Dt \, , \tag{2.346}$$

and thus

$$D = \frac{\lambda}{2\zeta^2 m^2} = \frac{T}{\zeta m} \equiv \mu T \, , \tag{2.347}$$

where $\mu = 1/\zeta m$ is the mobility. Note that for an (unmagnetized) 1D system, the diffusion coefficient is proportional to the square of the thermal velocity and inversely proportional to the collision frequency.

An important remark concerns Eq. (2.346). Why do we call D the diffusion coefficient? To illustrate this, let us first consider a very simple 1D model. The model consists of a linear chain, where particles can jump to the left or right at times $\Delta t, 2\Delta t, \ldots$ with a constant step length Δx and equal probabilities of $\frac{1}{2}$ to the left or right. The probability of having made r steps in one direction after a total of n jumps is

$$P_n(r) = \frac{n!}{r!(n-r)!}\left(\frac{1}{2}\right)^n . \tag{2.348}$$

We now introduce the "physical" variables $t = n\Delta t$ and $x = r\Delta x - (n-r)\Delta x$ in order to calculate the mean square

$$\langle x^2 \rangle = 4(\Delta x)^2 \sum_{r=0}^{n}\left(r - \frac{n}{2}\right)^2 P_n(r) \tag{2.349}$$

In evaluating the terms on the right-hand side, we can use $p = q = 1/2$.

$$\sum_{r=0}^{n} P_n(r) = (p+q)^n = 1 , \tag{2.350}$$

$$\sum_{r=0}^{n} rP_n(r) = p\,\frac{\partial}{\partial p}\,(p+q)^n = \frac{n}{2} , \tag{2.351}$$

$$\sum_{r=0}^{n} r^2 P_n(r) = p\,\frac{\partial}{\partial p}\,p\,\frac{\partial}{\partial p}\,(p+q)^n = \frac{n}{2} + \frac{n(n-1)}{4} , \tag{2.352}$$

where we now more generally set

$$P_n(r) = \frac{n!}{r!(n-r)!}\,p^r\,q^{n-r} . \tag{2.353}$$

The calculation yields

$$\langle x^2 \rangle = \frac{(\Delta x)^2}{\Delta t}\,t \equiv 2Dt . \tag{2.354}$$

Note that $\langle x \rangle = 0$ holds, and therefore the last formula describes the mean squared displacement, which grows linearly with time.

The net particle flux $\Gamma = nv$ at a given point x_0 results from the balance of fluxes in the positive and negative directions. Here, $n(x)$ describes the (initial) distribution of the particles. We count the number of jumps in both directions, and from this it follows that

$$\Gamma = \Gamma_+ - \Gamma_- = \frac{1}{2\Delta t}\left[\int_{x_0-\Delta x}^{x_0} n(x)\,dx - \int_{x_0}^{x_0+\Delta x} n(x)\,dx\right] \tag{2.355}$$

in the continuum approximation. A Taylor expansion of n to first order, i.e., in the form

$$n(x) \approx n(x_0) + \frac{dn}{dx_0}x , \tag{2.356}$$

is used in the integrand. This leads to

$$\Gamma \approx -\frac{(\Delta x)^2}{2\Delta t} \left. \frac{dn}{dx} \right|_{x_0} \equiv -D\,\frac{dn}{dx}\,. \tag{2.357}$$

The 1D continuity equation now appears in the form

$$\frac{\partial n}{\partial t} = -\frac{\partial \Gamma}{\partial x} = \frac{\partial}{\partial x}\left(D\,\frac{\partial n}{\partial x}\right). \tag{2.358}$$

Example 2.12 (3*D* Generalization)
More generally (3*D*), we write

$$\mathbf{j}(x) = -D\,\nabla\, n\,(x) \hat{=} -D\,\frac{\partial n}{\partial x} \tag{2.359}$$

for the 1*D* formulation used so far, which generalizes to

$$\frac{\partial n}{\partial t} + \nabla \cdot \mathbf{j} = 0 \tag{2.360}$$

■

In the spatially one-dimensional case, we obtain the diffusion equation

$$\frac{\partial n}{\partial t} - D\,\frac{\partial^2 n}{\partial x^2} = 0\,, \tag{2.361}$$

where, for simplicity, we (usually) assume a constant diffusion coefficient.

Example 2.13 (Solution of the 1*D* Diffusion Equation)
Starting at time $t = 0$, when all particles are concentrated at the point $x = 0$, i.e., $n(x, t = 0) = N\,\delta(x)$, the solution to Eq. (2.361) is

$$n(x,t) = \frac{N}{\sqrt{4\pi\,Dt}}\, e^{-\frac{x^2}{4Dt}}\,. \tag{2.362}$$

This solution is obtained for $n(x, t = 0) = N\delta(x)$ when we introduce the Fourier transform:

$$\tilde{n}(k,t) = \int_{-\infty}^{\infty} dx\; n(x,t) e^{ikx}\,. \tag{2.363}$$

After spatial Fourier transformation, the diffusion equation reads

$$\frac{\partial \tilde{n}}{\partial t} = -Dk^2 \tilde{n}\,, \tag{2.364}$$

with the obvious solution

$$\tilde{n}(k,t) = A e^{-Dk^2 t} \ . \tag{2.365}$$

The integration constant A can be determined from the initial condition. Thus, we obtain $A = N$. Applying the inverse transformation leads to

$$n(x,t) = \frac{N}{2\pi} \int_{-\infty}^{\infty} dk\ e^{-ikx} e^{-Dk^2 t} = \frac{N}{\sqrt{4\pi D t}}\ e^{-x^2/4Dt} \ . \tag{2.366}$$

■

Let us now generalize the above results for the $1D$ Langevin equation (without external fields) to a Langevin equation that is more suitable for expectations in hot and magnetized systems. For this, we must include the Lorentz force, and the problem essentially becomes $3D$. An appropriate generalization with a magnetic field in the z-direction is

$$\frac{d\mathbf{q}(t)}{dt} = \mathbf{v}(t) \ , \tag{2.367}$$

$$\frac{d\mathbf{v}_\perp(t)}{dt} = \Omega\ (\mathbf{v}_\perp \times \hat{e}_z) - \nu\ \mathbf{v}_\perp + \mathbf{a}(t) \ , \tag{2.368}$$

$$\frac{dv_z(t)}{dt} = -\nu\ v_z + a_z(t) \ . \tag{2.369}$$

We will return to this system in the chapter on stochastic transport. Since the present generalization is still a linear (inhomogeneous) system, it can be solved explicitly. Similar to the simple $1D$ case, for long times t, a characteristic behavior can be observed when solving the velocity correlation function and the mean squared displacement. The collision-induced force has a mean component $-m\nu\mathbf{v}(t)$ and a random, fluctuating component $m\mathbf{a}(t)$. Here, the collisions are modeled as a force that, on average, produces a friction proportional to the velocity. We have already mentioned this. In the more detailed description, the instantaneous intensity and direction of the friction force are random, as described by $m\mathbf{a}(t)$. The constant ν is identified with the collision frequency. The function $\mathbf{a}(t)$ is defined only by its statistical properties. A Gaussian process is completely defined by its first two moments. We assume that $\mathbf{a}(t)$ is a stationary, delta-correlated Gaussian process (white noise), so that the first and second moments are $\langle a_r(t)\rangle = 0$ and $\langle a_r(t) a_s(t+\tau)\rangle = A\delta_{rs}\delta(\tau)$, for $r, s = x, y, z$. The constant will be specified later.

Suppose the velocity of the test particle at $t = 0$ has the value $\mathbf{v}(0) = \mathbf{v}_0$ with probability 1. We first treat the Langevin equation as an ordinary differential equation for a given realization $\mathbf{a}(t)$. In doing so, we introduce a propagator,

$$G(t) = \begin{pmatrix} e^{-\nu t}\cos(\Omega t) & e^{-\nu t}\sin(\Omega t) & 0 \\ e^{-\nu t}\sin(\Omega t) & e^{-\nu t}\cos(\Omega t) & 0 \\ 0 & 0 & e^{-\nu t} \end{pmatrix} . \tag{2.370}$$

For $t \geq 0$ we can write the solution in the form

$$\mathbf{v} = G(t) \cdot \mathbf{v}_0 + \int_0^t d\theta G(\theta) \cdot \mathbf{a}(t-\theta) \tag{2.371}$$

The first moment is obtained by simple averaging,

$$\langle \mathbf{v}(t) \rangle = G(t) \cdot \mathbf{v}_0 . \tag{2.372}$$

The velocity autocorrelation functions, for $t_1 > t_2$, are obtained, for example, as

$$\begin{aligned} R_{xx}(t_1, t_2) &\equiv \langle v_x(t_1) v_x(t_2) \rangle \\ &= e^{-\nu(t_1+t_2)}\left[\cos(\Omega t_1)\cos(\Omega t_2) v_{0x}^2 + \sin(\Omega t_1)\sin(\Omega t_2) v_{0y}^2\right] \\ &+ \frac{A}{2\nu}\, e^{-\nu(t_1-t_2)}\left[1 - e^{-2\nu t_2}\right]\cos[\Omega(t_1-t_2)] . \end{aligned} \tag{2.373}$$

The autocorrelation function R_{yy} is obtained in a similar way. The parallel autocorrelation function R_{zz} follows from R_{xx} by setting $\Omega = 0$. Next, we average over different initial velocities $\mathbf{v}_0$. Let us assume that these are Maxwell-distributed with the velocity distribution function

$$f(\mathbf{v}_0) = \frac{1}{(2\pi)^{3/2} v_{th}^3} \exp\left(-\frac{v_0^2}{2 v_{th}^2}\right) , \tag{2.374}$$

where $v_{th} = \sqrt{\frac{T}{m}}$. We define

$$\bar{R}_{xx}(t_1, t_2) = \int d^3 v_0 f(\mathbf{v}_0) R_{xx}(t_1, t_2) \tag{2.375}$$

and after some algebra we obtain

$$\bar{R}_{xx}(t_1, t_2) = \frac{1}{2}\left\{\frac{A}{\nu}\, e^{-\nu|t_1-t_2|} + \left(2v_{th}^2 - \frac{A}{\nu}\right) e^{-\nu(t_1+t_2)}\right\} \cos[\Omega(t_1-t_2)] . \tag{2.376}$$

In this form, the autocorrelation function is not stationary unless $A = 2v_{th}^2\, \nu$.

We summarize and in doing so define $\tau = t_1 - t_2$:

$$\bar{R}_{xx}(\tau) = v_{th}^2 e^{-\nu|\tau|} \cos(\Omega\tau) , \quad \bar{R}_{zz}(\tau) = v_{th}^2 e^{-\nu|\tau|} . \tag{2.377}$$

The instantaneous position

$$x(t) = x(0) + \int_0^t d\theta v_x(\theta) \equiv x(0) + \delta x(t) \tag{2.378}$$

then yields for the mean square deviation

$$\langle \delta x^2(t) \rangle = \int_0^t dt_1 \int_0^t dt_2 \, \bar{R}_{xx}(t_1 - t_2) . \tag{2.379}$$

For the inner integral on the right-hand side, we use for fixed t_1 the new variable $\tau = t_1 - t_2$, in order to obtain by partial integration

$$\langle \delta x^2(t) \rangle = -\int_0^t dt_1 \int_{t_1}^{t_1 - t} d\tau \, \bar{R}_{xx}(\tau) = \int_0^t dt_1 (t - t_1) \bar{R}_{xx}(t_1) + \int_0^t dt_1 t_1 \bar{R}_{xx}(t_1 - t) . \tag{2.380}$$

The last term on the right-hand side yields the same result as the first. This becomes evident if one introduces the variable $\bar{t}_1 = t - t_1$ and uses the fact that $\bar{R}_{xx}(-\bar{t}_1) = R_{xx}(\bar{t}_1)$ holds. Thus, it follows that

$$\langle \delta x^2(t) \rangle = 2 \int_0^t d\tau (t - \tau) \bar{R}_{xx}(\tau) . \tag{2.381}$$

We are now in a position to determine the diffusion coefficient

$$D_\perp(t) = \frac{1}{2} \frac{d}{dt} \langle \delta x^2(t) \rangle = \int_0^t d\tau \bar{R}_{xx}(\tau) \tag{2.382}$$

One finds

$$D_\perp(t) = v_{th}^2 \frac{\nu + [\Omega \sin(\Omega t) - \nu \cos(\Omega t)] e^{-\nu t}}{\nu^2 + \Omega^2} . \tag{2.383}$$

In the asymptotic limit, it follows

$$\boxed{D_\perp = \lim_{t \to \infty} D_\perp(t) = v_{th}^2 \frac{\nu}{\nu^2 + \Omega^2}} . \tag{2.384}$$

The parallel, still time-dependent diffusion coefficient is

$$D_{\parallel}(t) = v_{th}^2 \frac{1 - e^{-\nu t}}{\nu} , \tag{2.385}$$

with the asymptotic value

$$\boxed{D_{\parallel} = \frac{v_{th}^2}{\nu}} . \tag{2.386}$$

In summary, we have

$$D_{\parallel} \approx \frac{T}{\nu m} \tag{2.387}$$

and

$$D_{\perp} = \frac{T}{m} \frac{\nu}{\Omega^2 + \nu^2} , \tag{2.388}$$

where $\Omega = qB/m$ is the cyclotron frequency. For strong magnetic fields $\Omega \gg \nu$ the following approximation holds

$$D_{\perp} \approx \frac{T}{m} \frac{\nu}{\Omega^2} \sim \frac{T\nu}{B^2} . \tag{2.389}$$

Note that the perpendicular diffusion coefficient is now proportional to the collision frequency and inversely proportional to the square of the magnetic field strength. This prediction suggests good magnetic confinement by strong magnetic fields. Unfortunately, reality is different.

Equilibrium Statistics of a Plasma

3

Abstract

The many-particle system plasma can be described in thermodynamic equilibrium using the well-established methods of equilibrium statistics and thermodynamics. In this respect, the calculations in this chapter are "merely" an application of the principles developed in the corresponding course lectures. However, even in the classical case, the evaluations themselves are by no means trivial; on the contrary: in systems with internal interactions, one quickly encounters very significant mathematical difficulties, the resolution of which remains the subject of intensive research to this day. Quantum statistics are not addressed in this chapter. They are, however, of particular importance in the context of complex plasmas.

3.1 Fundamentals

In this section, we begin with some fundamentals of statistical description, where we are not interested in instantaneous values of random variables, but rather in the "average" (thermodynamic) behavior of the system. We introduce the partition function and recall how the thermodynamic variables (in equilibrium) can be derived from the partition function.

We write the mean value $\langle \dots \rangle$ of a quantity $F(q, p)$, which depends on all positions $\mathbf{q}_i$ and canonically conjugate momenta $\mathbf{p}_i$ of the particles $i = 1, \dots, N$ (abbreviated: q, p) as

$$\langle F \rangle = \int d^{3N} q \, d^{3N} p \, F(q, p) \rho(q, p; t). \tag{3.1}$$

K.-H. Spatschek, *Theoretical Plasma Physics*,
https://doi.org/10.1007/978-3-662-72828-4_3

This expression requires some explanatory remarks. We average with a weighting function ρ, which in turn depends on all coordinates and momenta. The specification of ρ will occupy us intensively in the following. In (3.1), the integration is carried out over the Γ space, which is spanned by the coordinates and momenta of all particles. For N particles, this is thus a $6N$-dimensional space. In this high-dimensional space, the instantaneous (microscopic) state is represented by a point (the ergode). The temporal evolution of the system thus corresponds to the motion of the ergode in Γ space.

We arrive at a statistical description of the entire system by considering a multitude of points in Γ space, i.e., an ensemble of systems. Conceptually, the following distinction must be clear: A system consists of many (N) interacting particles. Two systems of the ensemble do *not* interact with each other. The question of which systems are even allowed in the constructed ensemble is discussed in detail in the lecture "Statistics and Thermodynamics." Here, just two remarks: the internal dynamics of the individual systems are determined by one and the same Hamiltonian function, and all systems must be compatible with the given macroscopic quantities. The ensemble considered is often called a virtual ensemble or Gibbs ensemble. We use its density for averaging. At this point, the "principle of equal a priori probability" comes into play: Every microscopic state that is allowed has the same weight. For the density ρ in Γ space, due to the absence of interaction between ergodes belonging to different systems, a continuity equation holds

$$\frac{\partial \rho}{\partial t} + \sum_{i=1}^{3N} \left[\frac{\partial(\rho \dot{q}_i)}{\partial q_i} + \frac{\partial(\rho \dot{p}_i)}{\partial p_i} \right] = 0 \,. \tag{3.2}$$

The density is incompressible, since

$$\nabla \cdot \mathbf{v}_\Gamma \equiv \sum_{i=1}^{3N} \left(\frac{\partial \dot{q}_i}{\partial q_i} + \frac{\partial \dot{p}_i}{\partial p_i} \right) = 0 \tag{3.3}$$

is fulfilled due to the canonical equations of mechanics.

If we use this in (3.2), it follows that

$$\frac{d\rho}{dt} = 0 \tag{3.4}$$

or

$$\boxed{\frac{\partial \rho}{\partial t} = \{H, \rho\}}\,. \tag{3.5}$$

This equation, known as the Liouville equation, in the given form holds only for canonically conjugate variables with the Poisson bracket $\{\ldots, \ldots\}$. The quantum mechanical generalization with the density matrix $\hat{\rho}$ is the von Neumann equation

$$i\hbar\frac{\partial\hat{\rho}}{\partial t} = [H, \hat{\rho}]; \tag{3.6}$$

The quantum mechanical averaging is given by $\hat{\rho}$ by means of taking the trace

$$\langle\hat{A}\rangle = \mathrm{Sp}\,(\hat{A}\hat{\rho}). \tag{3.7}$$

A discussion of the classical limit, for N indistinguishable particles, establishes the transition

$$\mathrm{Sp}\,(\hat{A}\hat{\rho}) \longrightarrow \int\ldots\int\frac{d^{3N}pd^{3N}q}{h^{3N}N!}A\rho \tag{3.8}$$

Through the ergodic hypothesis, we identify the ensemble average with the measurement average and the time average.

The Liouville equation—despite its simple form in the compact notation (3.5)—is an extremely complicated partial differential equation, whose general solution is practically impossible. A first simplification is possible if one restricts oneself to equilibrium states in which there is no *explicit* time dependence in ρ is to occur,

$$\frac{\partial\rho}{\partial t} = \dot{\rho} = 0, \tag{3.9}$$

so that in the classical case

$$\{H, \rho\} = 0 \tag{3.10}$$

holds. This equation for ρ is satisfied not only by any function of H, but in general by any function of the conserved quantities. Which constants of motion are ultimately relevant for a statistical description cannot be discussed in detail here; however, it is clear that energy plays an outstanding role.

Various ensembles, among which the microcanonical, canonical, and grand canonical ensembles are of particular importance, can be used for ρ. As examples, the canonical ensemble with

$$\rho = \frac{1}{h^{3N}N!Z}e^{-\beta H} \tag{3.11}$$

and the grand canonical ensemble with

$$\rho = \frac{1}{h^{3N} N! Z_g} e^{-\beta H - \alpha N} \tag{3.12}$$

are given here as examples. The "normalizations" or partition functions

$$\boxed{Z = \frac{1}{h^{3N} N!} \int \dots \int d^{3N} p \, d^{3N} q \, e^{-\beta H}}\,, \tag{3.13}$$

$$Z_g = \sum_{N=0}^{\infty} \frac{1}{h^{3N} N!} \int \dots \int d^{3N} p \, d^{3N} q \, e^{-\beta H - \alpha N} \tag{3.14}$$

play a central role in the transition to the thermodynamic description. The canonical partition function Z is related to the free energy F by

$$\boxed{F = -\theta \ln Z} \tag{3.15}$$

where $\theta = k_B T = 1/\beta$ holds. A justification for this relation can perhaps most easily be recalled by the following consideration: The free energy results from the internal energy U and the entropy S,

$$F = U - TS, \tag{3.16}$$

and thus, for the differential, it follows

$$dF = dU - TdS - SdT. \tag{3.17}$$

For F, the natural variables are temperature T, volume V, and particle number N (where, in order to allow for more general forces, one often introduces the notation $\mathbf{a}$ instead of V, with the consequence that the pressure p is replaced by $-\mathbf{A}$ and thus $pdV \widehat{=} -\mathbf{A} \cdot d\mathbf{a}$ holds). To rewrite the right-hand side of (3.17) into a differential form with dT, dV, and dN, we make use of the fundamental laws of thermodynamics. According to these, we have

$$dS = dQ/T \tag{3.18}$$

and

$$dQ = dU + pdV - \mu dN, \tag{3.19}$$

where μ is the chemical potential, which is related to α via $\alpha = -\beta\mu$. Substituting into (3.17) yields

$$dF = \frac{F - U}{T} dT - pdV + \mu dN. \tag{3.20}$$

Comparing this with the differential form of (3.15),

$$dF = -\ln Z\,d\theta - \theta \frac{1}{Z} dZ, \tag{3.21}$$

where

$$dZ = \frac{\partial Z}{\partial N} dN + \frac{1}{h^{3N} N!} \int \ldots \int d^{3N}p\, d^{3N}q \left[\frac{H}{\theta^2} d\theta - \frac{1}{\theta} \frac{\partial H}{\partial V} dV \right] e^{-H/\theta}$$

$$= \frac{\partial Z}{\partial N} dN + \left[\frac{1}{\theta^2} U d\theta + \frac{1}{\theta} p dV \right] \tag{3.22}$$

applies. We will return to the definition of pressure later. A simple rearrangement,

$$d(-\theta \ln Z) = \frac{(-\theta \ln Z - U)}{T} dT - p dV + \frac{\partial}{\partial N}(-\theta \ln Z) dN, \tag{3.23}$$

and comparison with (3.20) lead to the identification (3.15). Similarly, for example, $pV = \theta \ln Z_g$ follows for the grand canonical partition function Z_g and the grand canonical potential pV, with the natural variables T, V, and μ. However, if we remain with the canonical ensemble (as we know: in the thermodynamic limit, the results $N \to \infty$, $N/V = const$ must not depend on the choice of ensemble; we can choose whichever approach is most convenient for us!), then the remaining thermodynamic quantities are obtained from

$$S = -\left.\frac{\partial F}{\partial T}\right|_{N,V}, \quad p = -\left.\frac{\partial F}{\partial V}\right|_{T,N}, \tag{3.24}$$

$$\mu = \left.\frac{\partial F}{\partial N}\right|_{T,V}, \quad c_V = \left.\frac{\partial (F + TS)}{\partial T}\right|_{N,V}. \tag{3.25}$$

This outlines the further procedure within the framework of an equilibrium statistical treatment of the plasma: we must calculate the partition function. If we do this here in a purely classical manner, we neglect the effects arising from bound states. This approach is based on the idea that the partition function factorizes into two parts; one part describes the free particles and the other the bound states. Justifying such a model is difficult, and for this we must refer to the specialized literature. Here, we define particles as free, e.g., an ion and a neighboring electron, if their distance r exceeds a certain threshold. Usually, one requires $r^2 \geq \lambda_B \cdot r_w$, with the de Broglie wavelength λ_B and the classical interaction radius r_w. We will take the lower limit of the separation into account when we write the Hamiltonian for an electron-ion system in the form

$$H = \frac{1}{2}\sum_{j=1}^{N}\left[\frac{p_j^2}{m} + \frac{P_j^2}{M}\right] + \frac{k}{2}\sum_{\substack{i,j \\ i \neq j}}^{N}\left[\frac{e^2}{|\mathbf{R}_i - \mathbf{R}_j|} + \frac{e^2}{|\mathbf{r}_i - \mathbf{r}_j|}\right] - k\sum_{i,j}^{N}\frac{e^2}{|\mathbf{R}_i - \mathbf{r}_j|} \tag{3.26}$$

As a shortcut for the SI unit system, we have temporarily introduced the so-called Coulomb constant $k = \frac{1}{4\pi\varepsilon_0}$. The canonical partition function for N singly charged ions (mass M) at positions $\mathbf{R}_i$ with momenta $\mathbf{P}_i$ and N electrons (mass m) at positions $\mathbf{r}_i$ with momenta $\mathbf{p}_i$ is then

$$Z = \frac{1}{h^{6N}(N!)^2}\int \ldots \int d^{3N}p\, d^{3N}P\, d^{3N}r\, d^{3N}R\, e^{-\beta H}, \tag{3.27}$$

where $H = H(\mathbf{r}, \mathbf{p}, \mathbf{R}, \mathbf{P})$ must be used. We see that the Gibbs factor $\exp(-\beta H)$ splits into a momentum and a position part. To calculate the momentum part, integrals of the type

$$\int_0^\infty x^2 e^{-a^2x^2}\, dx = \frac{\sqrt{\pi}}{4a^3} \tag{3.28}$$

must be evaluated.

Example 3.1 (Ideal Gas Approximation)
As an exercise, let us now discuss the case in which the interaction in the Hamiltonian can be neglected. This approximation is known as the ideal gas approximation. In this case, it follows that

$$Z = \frac{1}{h^{6N}(N!)^2}\left[\frac{2\pi m^{1/2}M^{1/2}}{\beta}\right]^{3N} V^{2N}, \tag{3.29}$$

and thus the free energy can be very simply calculated via

$$F = -k_BT \ln Z \tag{3.30}$$

If, for example, we are interested in the pressure, only the volume dependence is important due to

$$p = -\left.\frac{\partial F}{\partial V}\right|_{T,N} \tag{3.31}$$

The straightforward calculation yields

$$p = \frac{2Nk_BT}{V} \tag{3.32}$$

for $2N$ particles in the volume V with (electron and ion) temperature T. Similarly, it follows that

$$E = -\frac{\partial}{\partial \beta} \ln Z = \frac{3}{2}(2N)k_B T. \tag{3.33}$$

These formulas, as well as others for entropy, specific heat, etc., are known from the statistical mechanics of ideal gases. We can regard them as starting points for calculations including interactions. ■

At this point, it is already possible to make the expectations regarding the results of an exact calculation somewhat more concrete: Since the (average) potential energy of a particle is supposed to be much smaller than its (average) kinetic energy, the contributions of the potential energy in the Gibbs factor are unlikely to play a significant role with respect to the thermodynamic variables. This expectation suggests that deviations from the ideal gas law will probably only be corrections of the order of Λ^{-1} (inverse particle number in the Debye zone).

At this point, we can gain further insight for the subsequent calculation by recalling some results from classical statistics for non-ideal (imperfect) gases. The total partition function for a system of N particles (mass m) with potential energy ϕ is split into two parts,

$$Z = \frac{1}{N!h^{3N}} \int \dots \int d^{3N}p\, d^{3N}q\, e^{-H(p,q)/k_B T}$$

$$= J(T)Q(T, V), \tag{3.34}$$

where

$$H = \frac{1}{2m} \sum_i \mathbf{p}_i^2 + \phi(\mathbf{q}_1, \dots, \mathbf{q}_N) \tag{3.35}$$

and $\phi = \sum_{i<j} \phi_{ij}$ apply. The functions J and Q are defined as

$$J(T) = \frac{1}{h^{3N}} \left[\int_{-\infty}^{\infty} \exp(-p^2/2mk_B T) dp \right]^{3N} = \left[\frac{2\pi m k_B T}{h^2} \right]^{3N/2}, \tag{3.36}$$

$$Q(T, V) = \frac{1}{N!} \int \dots \int d^3q_1 \dots d^3q_N \exp(-\phi/k_B T) \tag{3.37}$$

defined. The factor $\exp(-\phi/k_B T)$ is commonly referred to as the generalized Boltzmann factor.

A key role in the calculation of the deviations is played by the two-particle distribution function

$$n^{(2)}(\mathbf{r}_1, \mathbf{r}_2) := \frac{N(N-1)\int \ldots \int d^3r_3 \ldots d^3r_N \exp(-\phi/k_BT)}{\int \ldots \int d^3r_1 \ldots d^3r_N \exp(-\phi/k_BT)}, \tag{3.38}$$

which gives the probability of finding two particles at the locations $\mathbf{r} = \mathbf{r}_1$ and $\mathbf{r} = \mathbf{r}_2$. It is easy to see that, for example, the total energy can be written exactly as

$$E = -\frac{\partial \ln Z}{\partial \beta} = \frac{3}{2}Nk_BT + (V/2)\int d^3(|\mathbf{r}_1 - \mathbf{r}_2|)\, \phi_{12}\, n^{(2)}(\mathbf{r}_1, \mathbf{r}_2) \tag{3.39}$$

provided that ϕ_{12} depends only on the distance between the two particles. The first term on the right-hand side represents the contribution of the kinetic energy, the second that of the potential energy. Our expectation that, in a plasma, the second term is smaller than the first by a factor of $1/\Lambda$ now becomes quite evident for the energy.

Finally, a few clarifying words about the equation of state that follows from

$$\boxed{p = k_BT \left.\frac{\partial \ln Q}{\partial V}\right|_{T,N}} \tag{3.40}$$

To better understand the differentiation with respect to volume, let us imagine the system enclosed in a cube with edge length L; $V = L^3$. We switch to dimensionless spatial variables by normalizing with L. As a result, the integration limits become constant (0 and 1), and thus play no role in the differentiation with respect to volume. We have

$$Q(T, V) = \frac{L^{3N}}{N!}\int_0^1 \cdots \int_0^1 d^3\Big(\frac{r_1}{L}\Big)\ldots d^3\Big(\frac{r_N}{L}\Big)$$

$$\times \exp\Bigg[-\frac{1}{k_BT}\sum_{i<j}\phi_{ij}\Big(\frac{|\mathbf{r}_i - \mathbf{r}_j|L}{L}\Big)\Bigg]. \tag{3.41}$$

Because

$$\frac{\partial}{\partial V} = \frac{1}{3V}L\frac{\partial}{\partial L} \tag{3.42}$$

we calculate

$$L\frac{\partial \ln Q}{\partial L} = 3N - \frac{1}{k_BT}\sum_{j<k}\frac{\int \ldots \int d^3(\frac{r_1}{L})\ldots d^3(\frac{r_N}{L})|\mathbf{r}_j - \mathbf{r}_k|\phi'_{jk}e^{-\beta\phi}}{N!Q/L^{3N}}$$

$$= 3N - \frac{N(N-1)}{2k_BT}\,\frac{\int \ldots \int d^3r_1 \ldots d^3r_N|\mathbf{r}_1 - \mathbf{r}_2|\phi'_{12}e^{-\beta\phi}}{\int \ldots \int d^3r_1 \ldots d^3r_N e^{-\beta\phi}}. \tag{3.43}$$

Therefore, it follows that

$$pV = Nk_BT - \frac{1}{6}V \int d^3(|\mathbf{r}_1 - \mathbf{r}_2|)\, |\mathbf{r}_1 - \mathbf{r}_2|\, \phi_{12}^{'}\, n^{(2)}(\mathbf{r}_1, \mathbf{r}_2), \tag{3.44}$$

and again, it can be expected that in a plasma (within the approximation chosen here) the first term on the right-hand side dominates. The calculation just performed also allows us to understand in a classical way how the pressure is related to the change in energy eigenvalues with variations in volume. In the next section, we will take a closer look at the evaluations for a classical plasma.

3.2 Cluster Integrals

In this section, we deal with the calculation of the interaction contribution

$$Q' := \int d^{3N}r\, d^{3N}R \exp\left[-\beta \sum_{i,j}^{N} \phi_{ij}\right] \tag{3.45}$$

for a system described by the Hamiltonian function (3.26)

$$H = \frac{1}{2}\sum_{j=1}^{N}\left[\frac{p_j^2}{m} + \frac{P_j^2}{M}\right] + \frac{k}{2}\sum_{\substack{i,j \\ i\neq j}}^{N}\left[\frac{e^2}{|\mathbf{R}_i - \mathbf{R}_j|} + \frac{e^2}{|\mathbf{r}_i - \mathbf{r}_j|}\right] - k\sum_{i,j}^{N}\frac{e^2}{|\mathbf{R}_i - \mathbf{r}_j|} \tag{3.46}$$

where ϕ_{ij} is used as a shorthand for the (Coulomb) interaction energy between two particles i and j. In the summation, the particles may belong to the same or to different species (electrons and ions).

It is easy to see that, due to the pairwise interaction, the exponential function in the interaction energy can be written as a product of very many exponential functions. For demonstration purposes, let us now single out the contribution to Q' that arises from the interaction between identical particles, i.e., with the Coulomb constant k

$$Q'' := \int d^{3N}r \exp\left[-\beta \sum_{i,j}^{N} \phi_{ij}\right], \tag{3.47}$$

$$\phi_{ij} = k\frac{e^2}{|\mathbf{r}_i - \mathbf{r}_j|}\,; \tag{3.48}$$

i and j belong to the same particle species (e.g., electrons), and the summation in the exponent of Q'' extends over all possible interaction contributions within the species. This summation rule can also be specified more precisely by $\frac{1}{2}\sum_{i,j}^{N}{}'$, where the prime ($'$) means that $i = j$ is excluded, and the factor $\frac{1}{2}$ ensures that, when summing independently over i and j, interactions are not counted twice. Equivalently, the restriction $i > j$. applies. However, in some intermediate calculations, we will not write out this restriction explicitly, so as not to unnecessarily complicate the notation.

Within the framework of statistics, one now writes Q'' by means of f-bonds,

$$Q'' = \int d^{3N}r \prod_{i,j}(1 + f_{ij}), \tag{3.49}$$

with the so-called Mayer function

$$f_{ij} = \exp(-\beta\phi_{ij}) - 1. \tag{3.50}$$

For the product over i and j, a similar restriction applies as just discussed, i.e. $i > j$. Initially, f_{ij} in contrast to $\exp(-\beta\phi_{ij})$ has an advantageous property: the Mayer function vanishes for $|\mathbf{r}_i - \mathbf{r}_j| \to \infty$. In the region of repulsion, $f_{ij} < 0$. holds. The introduction of f_{ij} has the advantage that we can recognize the asymptotic behavior of the integrands that arise somewhat more quickly and easily. Otherwise, we have gained nothing so far.

There are various methods for evaluating the partition function, such as the cumulant expansion, the virial expansion, and so on. In the following, we describe a graphical method that leads to a virial expansion. The aim is to organize the various contributions to Q'' by defining the so-called cluster integrals,

$$b_1 = \frac{1}{V}\int d^3r_1 = 1, \tag{3.51}$$

$$b_2 = \frac{1}{2V}\int d^3r_1 d^3r_2 f_{12} \equiv \frac{1}{2}\int_0^\infty 4\pi r^2 dr f(r), \tag{3.52}$$

$$b_3 = \frac{1}{3!V}\int d^3r_1 d^3r_2 d^3r_3 [f_{31}f_{21} + f_{32}f_{31} + f_{32}f_{12} + f_{32}f_{31}f_{12}] \tag{3.53}$$

and so on. The procedure becomes somewhat clearer if we illustrate the formulas step by step. In a cluster b_l we select l particles that interact with each other, for example, in the case of $l = 2$ two particles, which we label 1 and 2; for $l = 3$, three particles labeled 1, 2, and 3, and so on. In general, a cluster integral is defined by

$$\boxed{b_l = \frac{1}{l!V}\int \cdots \int d^3r_1 \ldots d^3r_l \sum \prod_{l \geq i > j \geq 1} f_{ij}} \tag{3.54}$$

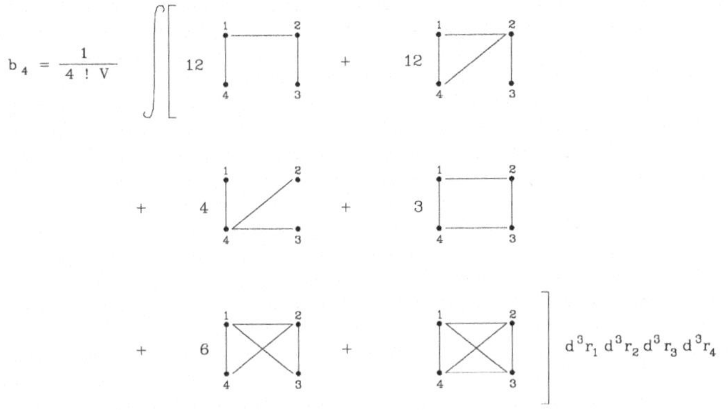

Fig. 3.1 Schematic representation of the cluster integral b_4

In b_l, the sum and the product are to be understood such that, for l selected particles $1, \ldots, l$, all possible connections are represented in diagrams (the product extends over all bonds in a diagram), and then one sums over all diagrams. The reader may follow this using the example of b_4 (see Fig. 3.1). It is important that each particle is "connected" to every other, at least via bonds with other particles.

The following properties of the cluster integrals are important. Starting from a fixed index l, some contributions can be easily summed. For demonstration, if we choose b_4, as we have already indicated in Fig. 3.1, there are a total of twelve diagrams of the first type that yield the same result. The corresponding integrands are

$$\begin{aligned}
&f_{41}f_{12}f_{23}, \quad f_{12}f_{23}f_{34}, \quad f_{23}f_{34}f_{41}, \\
&f_{34}f_{41}f_{12}, \quad f_{31}f_{12}f_{24}, \quad f_{12}f_{24}f_{43}, \\
&f_{24}f_{43}f_{31}, \quad f_{43}f_{31}f_{12}, \quad f_{41}f_{13}f_{32}, \\
&f_{13}f_{32}f_{24}, \quad f_{32}f_{24}f_{41}, \quad f_{24}f_{41}f_{13}.
\end{aligned} \tag{3.55}$$

Similar considerations apply to the other diagram types. (When summing, it is important to note that changing the "sense of rotation" of a closed particle arrangement does not lead to new configurations!) Furthermore, it is important that the cluster integrals b_l are volume-independent. This can be explained by the fact that, when performing the integrations, we introduce one particle position ($\mathbf{r}_1$) as a reference point and, instead of the remaining particle coordinates, we can choose the distances to the reference point. Then, integration over $\mathbf{r}_1$ yields a volume factor V (which cancels with the prefactor $1/V$); assuming the Mayer functions are localized (with an effective range λ_{De})), the remaining integrations over the difference coordinates no longer yield any volume dependence. We will return to this assumption further below.

Finally, we should point out that in the integrand of Q'' the product

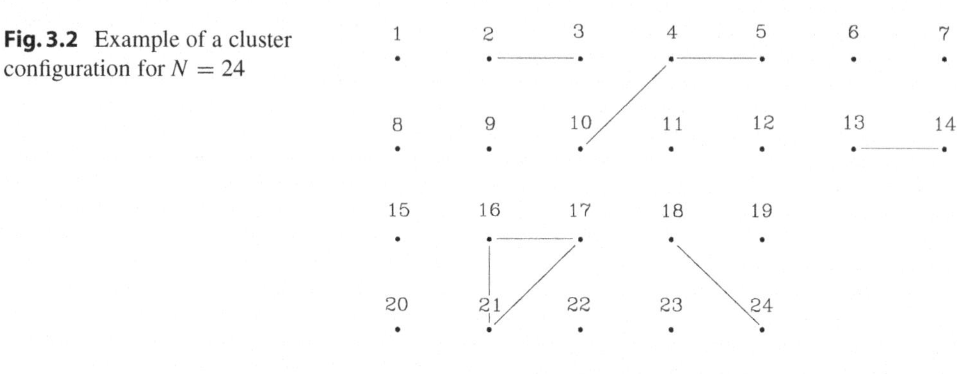

Fig. 3.2 Example of a cluster configuration for $N = 24$

$$\prod_{N\geq i>j\geq 1} (1+f_{ij}) = 1 + \sum_{i>j} f_{ij} + \sum_{i,j}\sum_{k,l} f_{ij}f_{kl} + \ldots \tag{3.56}$$

gives exactly the contributions that we have summarized in the cluster integrals b_l. This raises the question of how $Q'' \equiv N!Q$ (for N electrons) can be calculated using the b_l. The result

$$Q'' = N! \sum_{m_l,\,\Sigma l m_l = N} \prod_l (Vb_l)^{m_l}/m_l! \tag{3.57}$$

will now be justified.

To this end, a few preliminary remarks. If we are dealing with a *specific* cluster, this means that certain particles, e.g., particles 2, 3, 4, 5, 10, 13, 14, 16, 17, 18, 21, and 24 in Fig. 3.2 for $N = 24$, out of the N particles are participating in the "interaction process." Physically, of course, every particle interacts with every other. The statement just made is to be understood in the sense that we represent the product terms appearing in (3.56) of the $f's$ by f-bonds, and call an f-bond f_{23} an "interaction" of particle 2 with particle 3. This type of "interaction" thus has nothing in common with the (physical) Coulomb interaction.

In a specific cluster, individual ($l = 1$) particles appear without any binding, as well as pairs ($l = 2$), that do not "interact" with any other particle, only with each other, and similarly, triplets ($l = 3$) that are only internally connected, and so on. We denote by l the number of particles in a subgroup (subcluster), where the l particles are only connected among themselves, but not with any other particles. The number of subclusters present in a specific cluster, each consisting of l particles, is denoted by m_l. In the example of Fig. 3.2, we thus have $m_1 = 12, m_2 = 3, m_3 = 2, m_4 = 0$ and so on.

With regard to the integrations appearing in the partition function, for indistinguishable particles the particle number itself is not of particular importance; rather, what matters is the characterization of a term by m_l subclusters of particle number l. For the derivation of formula (3.57), the following approach now suggests itself: First, we select all diagrams that group the same specific particles into clusters. Within these specific clusters, the "binding scheme" can differ.

From these special diagrams, we select those that contain a particular specific cluster C of the particles $1, \dots,$ s. For this, we write (by "factoring out" C)

$$\sum \prod f_{ij} = \left[\prod_{C} f_{ij} \right] \cdot R, \tag{3.58}$$

where R denotes the "remainder." Now we sum over all possible specific clusters that contain the specified points $1, \dots,$ s and integrate over the coordinates of $1, \dots,$ s, in order to obtain

$$b_l l! V \cdot R \tag{3.59}$$

We extend this procedure to the remainder R. Thus, for the contribution of all diagrams that connect the same particles into clusters, we obtain

$$\prod_{l} (b_l l! V)^{m_l}, \tag{3.60}$$

where we have further assumed that the distribution in m_1 isolated particles, m_2 pairs, $\dots, m_l$ l subclusters, etc., is given. There are now

$$N! / \prod_{l} m_l! (l!)^{m_l} \tag{3.61}$$

ways to partition the particles accordingly. To justify this, we conduct the following thought experiment. Imagine m_1 "boxes" set up, each of which fits exactly one particle, m_2 "boxes" that fit exactly 2 particles, $\dots, m_l$ "boxes" for exactly l particles, and so on. We imagine the "boxes" lined up in a row and arrange the particles alongside them in a chain. There are exactly $N!$ ways to arrange the N particles in a chain, and thus to distribute them among the "boxes," provided the particles are distinguishable. However, permutations of particles within a "box" are already accounted for in the summation and integration for the b_l, so that $(l!)^{m_l}$ possibilities are omitted. This explains the last factor in (3.61). Furthermore, the individual "boxes" for l particles are indistinguishable, so an additional factor $m_l!$ appears. Altogether, we thus obtain the prefactor (3.61). The expressions (3.60) and (3.61) together yield (3.57).

Although we have succeeded in expressing the contribution Q'' to the partition function in terms of the b_l, we are still far from quantitative results. At this point, two different approaches (yielding the same results) present themselves: On the one hand, the transition to the grand partition function Z_g brings some formal simplifications here; on the other hand, in the limit $N \to \infty$ (3.57) can be evaluated using the standard methods of statistics. We choose the second (practical) approach and approximate $Q''/N!$ by the maximal term in the sum (3.57) over the m_l. This procedure is justified because the number ν of terms in the sum is of the order of magnitude

$$\ln \nu \approx \pi \left(\frac{2}{3} N \right)^{\frac{1}{2}} \tag{3.62}$$

while the maximal term T_m (see below) is of the order $\exp(N)$. From

$$T_m \leq \frac{Q''}{N!} \leq \nu T_m \tag{3.63}$$

it then follows in the limit $N \to \infty$

$$\ln \frac{Q''}{N!} \approx \ln T_m. \tag{3.64}$$

In the next step, we calculate the maximum of

$$\sum_l \ln\left[(Vb_l)^{m_l}/m_l!\right] \tag{3.65}$$

subject to the constraint

$$\sum_l m_l l = N \tag{3.66}$$

To do this, we vary

$$L := \sum_l \left\{ \ln\left[(Vb_l)^{m_l}/m_l!\right] + (\ln F) m_l l \right\} \tag{3.67}$$

with the Lagrange multiplier $\lambda = -\ln F$. As a result, due to Stirling's formula, we obtain

$$\ln m_l! \approx m_l \ \ln \ m_l - m_l \tag{3.68}$$

the Euler equations

$$\ln[Vb_l] - \ln m_l = -\ln F^l. \tag{3.69}$$

We therefore obtain

$$m_l = Vb_l F^l, \tag{3.70}$$

and for the maximal term

$$\ln T_m = N \left(\sum_{l=1}^{N} \frac{V}{N} b_l F^l - \ln F \right). \tag{3.71}$$

F itself follows from the constraint

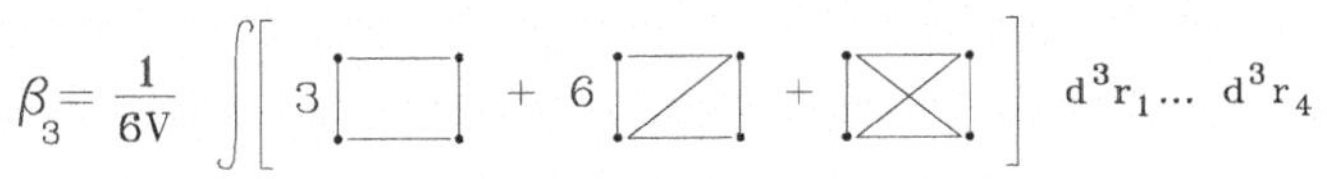

$$\beta_3 = \frac{1}{6V} \int \left[3 \square + 6 \boxslash + \boxtimes \right] d^3r_1 \dots d^3r_4$$

Fig. 3.3 Schematic representation of the irreducible cluster integral β_3

$$\sum_{l=1}^{N} l \frac{V}{N} b_l F^l = 1. \tag{3.72}$$

Thus, we have

$$\boxed{\ln \frac{Q''}{N!} \approx N \left(\sum_{l=1}^{N} \frac{V}{N} b_l F^l - \ln F \right)}. \tag{3.73}$$

We can further simplify this expression considerably. Each of the integrals b_l contains individual diagrams with different connections between the l particles. For example, b_4 contains an integral over $f_{21}f_{31}f_{41}$, which, however, can be simplified. If we choose coordinate 1 as the reference point, the integrations over coordinates 2, 3, and 4 can be performed independently. This portion of b_4 thus reduces to Vb_2^3. Clusters that can be decomposed in this way are called reducible. The integral (in b_4) with the integrand $f_{21}f_{32}f_{31}$ cannot be reduced to b_l with $l < 4$. Such diagrams are called irreducible. We therefore define the irreducible cluster integrals

$$\beta_k = \frac{1}{k!V} \int \sum_{(irr)} \prod_{k+1 \geq j > i \geq 1} f_{ij} d^3r_1 \dots d^3r_{k+1}, \tag{3.74}$$

where the sum is taken only over irreducible representations. This formal definition becomes immediately clear, for example, when one constructs $\beta_2, \beta_3, \dots$ step by step. The irreducible cluster integral β_3 is shown schematically in Fig. 3.3.

Now the interesting point: We have

$$b_l = \frac{1}{l^2} \sum_{n_k} \prod_{k} (l\beta_k)^{n_k} / n_k!\,, \tag{3.75}$$

where the running indices k and n_k are defined by

$$\sum_{k=1}^{l-1} k n_k = l - 1 \tag{3.76}$$

For $l = 4$ we have, for example, the statements

$$\begin{aligned} n_1 &= 0,\ n_2 = 0,\ n_3 = 1; \\ n_1 &= 1,\ n_2 = 1,\ n_3 = 0; \\ n_1 &= 3,\ n_2 = 0,\ n_3 = 0; \end{aligned} \tag{3.77}$$

which, due to the definition (3.75), lead to

$$\begin{aligned} b_4 &= \frac{1}{16}\left\{\frac{(4\beta_1)^3}{3!} + 4\beta_1 4\beta_2 + 4\beta_3\right\} \\ &= \frac{2}{3}\beta_1^3 + \beta_1\beta_2 + \frac{1}{4}\beta_3. \end{aligned} \tag{3.78}$$

This relationship is illustrated in Fig. 3.4. This also clarifies once again how the schematic representations of the cluster integrals are to be understood. The relation (3.75) has an important consequence. We can now explicitly state the solution of (3.72):

$$F = \frac{1}{v}\exp\left(-\sum_k \beta_k v^{-k}\right), \tag{3.79}$$

with

$$v = V/N. \tag{3.80}$$

Moreover, it is possible to significantly simplify (3.73):

$$\ln\frac{Q''}{N!} = N\left[1 + \sum_{k=1}^{N}\frac{1}{k+1}\beta_k v^{-k} + \ln v\right]. \tag{3.81}$$

This result represents a tremendous success. For example, if we calculate pV, we obtain

$$pV = Nk_BT\left[1 - \sum_{k=1}^{N}\frac{k}{k+1}\beta_k v^{-k}\right], \tag{3.82}$$

the virial expansion of the equation of state. For comparison with the literature, we introduce W via

$$Q'' = V^N \exp(NW) \tag{3.83}$$

Then (3.81) yields

$$W = \sum_{k=1}^{N}\frac{\beta_k}{k+1}n^k \tag{3.84}$$

with $n = v^{-1} = N/V$ in the limit $N \to \infty$. The equation of state (3.82) can also be written as

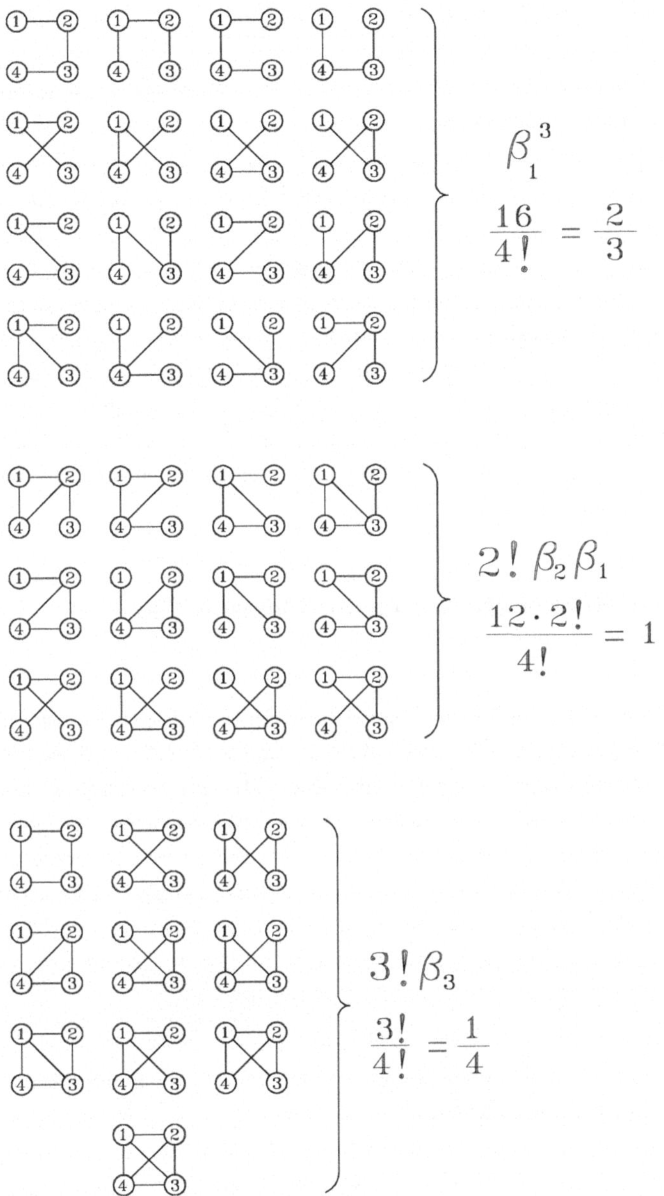

Fig. 3.4 Relationship between reducible and irreducible cluster integrals, using the example of b_4

$$\frac{pV}{Nk_BT} = 1 - n\frac{\partial W}{\partial n} \tag{3.85}$$

We might actually think we have reached our goal, were it not for a surprise at the end. If we try, for example, to calculate

$$\beta_1 = \frac{1}{V}\int d^3r_1 d^3r_2 f_{12} = 4\pi \int dr\, r^2\Big[e^{-\beta\phi_{12}(r)} - 1\Big] \tag{3.86}$$

we find, due to $\phi_{12}(r) \sim r^{-1}$ at $r \to \infty$, divergences, unless we invoke Debye screening. (We will not discuss the specific problems in two-component systems here!) Fortunately, part of the problem is self-inflicted. The divergence proportional to r is remedied in neutral systems by summing over the different components. The divergences caused by integrands proportional to r^0 and r^{-1} can be avoided by rearrangements and convergence-generating factors resulting from partial summations. We cannot reproduce the extensive literature on this topic here, but will limit ourselves to illustrating the basic procedure with a simple example in the next section.

3.3 Debye-Hückel Theory for an Electron Gas

In this section, we demonstrate using the example of an electron gas (i.e., an effectively "one-component" plasma with a "smeared-out" ionic background) that divergences can be overcome by summing over different graphs. Essentially, we use results from Abé and Mayer to justify summations over prototype graphs, though we will not go into all the details here. Thus, the approach in the following is not always fully justified; after this section, however, it should at least be clearer that some of the divergence problems described can be overcome by a renormalization of the bonds. The Debye interaction results from a summation of Coulomb contributions; it guarantees the finiteness of the integrals.

We start with the Fourier representation of the interaction potential ϕ_{ij}, omitting the Coulomb prefactor $k \equiv \frac{1}{4\pi\varepsilon_0}$, and use

$$\frac{e^2}{r} = \frac{1}{V}\sum_{\mathbf{q}\neq 0}\frac{4\pi e^2}{q^2}e^{i\mathbf{q}\cdot\mathbf{r}}, \tag{3.87}$$

which we will briefly justify below.

Example 3.2 (Fourier Representation of the Interaction)

The connection with the integral representation is immediately clear because of

$$\sum_{\mathbf{q}\neq 0} \longrightarrow \frac{V}{(2\pi)^3}\int d^3q \tag{3.88}$$

and $dq_x = (2\pi/L_x)dn_x$, etc. (In the summation over $\mathbf{q}$ we have used $dn_x = dn_y = dn_z = 1$.) Thus, for the right-hand side of (3.87) we obtain

$$\begin{aligned}\frac{1}{(2\pi)^3}\int \frac{4\pi e^2}{q^2} e^{i\mathbf{q}\cdot\mathbf{r}} d^3q &= -\frac{e^2}{\pi}\int e^{iqr\cos\delta} d(\cos\delta)dq \\ &= \frac{2}{\pi}\frac{e^2}{r}\int_0^\infty d(qr)\frac{\sin qr}{qr} \\ &= \frac{e^2}{r},\end{aligned} \tag{3.89}$$

which immediately yields (3.87). ■

For f_{ij} we use the formal expansion

$$f_{ij} = e^{-\beta\phi_{ij}} - 1 = \sum_{\nu=1}^{\infty}\frac{1}{\nu!}(-\beta\phi_{ij})^\nu, \tag{3.90}$$

with which we proceed in

$$W = \sum_{k=1}^{\infty}\frac{\beta_k}{k+1}n^k \tag{3.91}$$

This yields

$$W = \sum_{k=1}^{\infty}\frac{n^k}{(k+1)!}\sum_{(irr)}\int\cdots\int\prod_{k+1\geq i>j\geq 1}\sum_{\nu_{ij}=1}^{\infty}\frac{(-\beta\phi_{ij})^{\nu_{ij}}}{\nu_{ij}!}d^3r_2\ldots d^3r_{k+1}. \tag{3.92}$$

A fundamental change has now taken place: the former "f_{ij} bonds" are now replaced by the more realistic "ϕ_{ij} bonds". The product over specific ϕ_{ij}, ϕ_{kl}, etc., is again illustrated by clusters, which arise from rearrangements and partial summations of the previous ones. We now focus on a particular type of cluster, the so-called ring clusters (see Fig. 3.5). These describe integrals over products of the form $\phi_{12}\,\phi_{23}\,\phi_{34}\ldots\phi_{l1}$.. Let us consider the example

$$I_l := \int\ldots\int\phi_{12}\phi_{23}\ldots\phi_{l1}d^3r_2\ldots d^3r_l$$

$$= \frac{1}{V^l} \int \left\{ \sum_{\mathbf{q}_1} \nu(\mathbf{q}_1) e^{i\mathbf{q}_1 \cdot (\mathbf{r}_1 - \mathbf{r}_2)} \ldots \sum_{\mathbf{q}_l} \nu(\mathbf{q}_l) e^{i\mathbf{q}_l \cdot (\mathbf{r}_l - \mathbf{r}_1)} \right\} d^3 r_2 \ldots d^3 r_l$$

$$= \frac{1}{V} \sum_{\mathbf{q}} \nu^l(\mathbf{q}) \tag{3.93}$$

where $\nu(\mathbf{q}) = 4\pi e^2/q^2$. The integrations over $\mathbf{r}_2, \ldots, \mathbf{r}_l$ yield $\mathbf{q}_1 = \mathbf{q}_2 = \cdots = \mathbf{q}_l$ and volume factors. Returning to W, as written in (3.92), we consider in the sum over ν_{ij} only the first term each time $\nu_{ij} = 1$. Products of ϕ_{ij} arise due to the product formation over i and j with $k+1 \geq i > j \geq 1$. Here, we want to include only the ring diagrams, that is, for example, for $k = 3$ the term $\phi_{12}\,\phi_{23}\,\phi_{34}$, taking into account all permutations of the electrons $1, 2, 3, 4 = k+1$. To simplify the terminology, we define

$$l = k + 1 \tag{3.94}$$

and note that l particles $(l-1)!/2$ can form different rings. The factor $\frac{1}{2}$ arises here because the direction of rotation does not matter when arranging in a ring. We see that under these assumptions

$$\sum_{(irr)} \int \cdots \int \prod_{k+1 \geq i > j \geq 1} \sum_{\nu_{ij}=1}^{\infty} \frac{(-\beta\phi_{ij})^{\nu_{ij}}}{\nu_{ij}!} d^3 r_2 \ldots d^3 r_{k+1}$$

$$\longrightarrow (-\beta)^l \frac{(l-1)!}{2} I_l \tag{3.95}$$

results and

$$W_{ring} = \frac{1}{V} \sum_{l=2}^{\infty} \frac{1}{2l} (-1)^l \beta^l n^{l-1} \sum_{\mathbf{q} \neq 0} \nu^l(\mathbf{q}) \tag{3.96}$$

is obtained. In the following, we approximate W by W_{ring}, i.e., $W \approx W_{ring}$. The rest is algebra: On the right-hand side of (3.96) one easily recognizes the series expansion of ln, i.e.,

$$W \approx \frac{1}{2nV} \sum_{\mathbf{q} \neq 0} \{-\ln[1 + \beta n \nu(\mathbf{q})] + \beta n \nu(\mathbf{q})\}. \tag{3.97}$$

The transition to the continuum approximation yields for N electrons

$$W \approx \frac{1}{2n} \frac{1}{(2\pi)^3} \int d^3 q \{-\ln[1 + \beta n \nu(\mathbf{q})] + \beta n \nu(\mathbf{q})\}$$

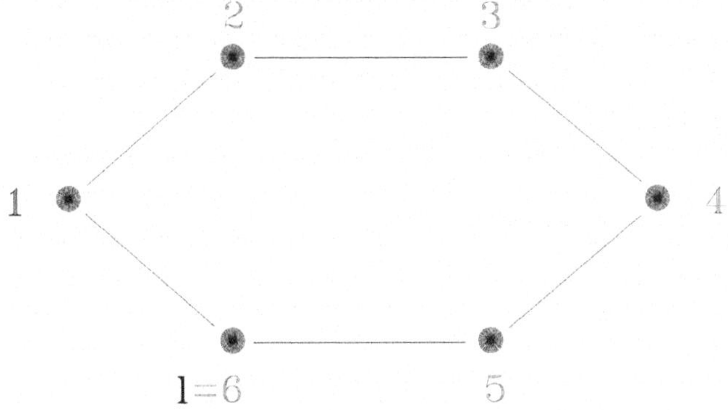

Fig. 3.5 Example of a ring cluster

$$= \frac{1}{2n_e} \frac{4\pi}{(2\pi)^3} \kappa_{De}^3 \int_0^\infty dx\, x^2 \left\{ \frac{1}{x^2} - \ln\left(1 + \frac{1}{x^2}\right) \right\}. \tag{3.98}$$

The evaluation of the integral yields

$$W = \frac{\kappa_{De}^3}{12\pi n_e} \tag{3.99}$$

with $\kappa_{De}^2 = \lambda_{De}^{-2} = n_e e^2 \beta / \varepsilon_0$ and $n_e = n$. For the equation of state, this means

$$\boxed{\frac{pV}{Nk_B T_e} \approx 1 - \frac{1}{24\pi} \frac{1}{n_e \lambda_{De}^3}}, \tag{3.100}$$

i.e., exactly the result we expected in terms of order of magnitude.

Naturally, the question arises as to how we can justify neglecting the remaining contributions that are not captured by the ring approximation. To answer this question, we can only cite the results from the rather extensive specialized literature, e.g., [49–51]. It turns out that all further contributions are corrections to the deviation just given, of at least the order of

$$\frac{r_w}{\lambda_{De}} \sim \frac{1}{n_e \lambda_{De}^3} \sim \frac{1}{\Lambda} \tag{3.101}$$

A few explanatory words on this. At the beginning of this section, we moved from the "f-clusters" to the "ϕ -clusters." It should be noted that irreducible clusters remain irreducible, but now, due to the expansion of the exponential function, multiple bonds can occur between two points. All diagrams are classified into single diagrams (which do not contribute in neutral systems), rings (whose contributions we have just calculated), and a remainder, in which at least two points have more than two connections. Each graph in the remainder can be associated with a prototype graph in which all points are nodes of at least third order (with three connections). In each prototype graph, we can "inflate" bonds into chains to obtain all graphs of the remainder. Conversely, by "shrinking" chains in graphs of the remainder, we arrive at prototype graphs. The extremely interesting point is that all contributions of the remainder, which result from "shrinking" the bonds to a prototype graph, can simply be calculated using the prototype graph with Debye-screened bonds. Here is a plausibility argument: If we let the rings used in the Debye-Hückel approximation "shrink" to a double bond with screening, we obtain an effective contribution to W of the order of

$$W_{eff} = \frac{n_e}{2}\frac{\beta^2}{2}e^4 \int_0^\infty \frac{e^{-2\kappa_{De} r}}{r^2} 4\pi r^2 \, dr$$

$$= \frac{1}{32\pi n_e \lambda_{De}^3} . \tag{3.102}$$

This roughly agrees with (3.99). The prototype graphs between two points (but with at least three bonds) yield for W_{ring} a correction of the order of

$$W^{(2)}_{proto} \approx \frac{r_w}{\lambda_{De}} W_{ring} , \tag{3.103}$$

the contributions of the prototype graphs with three particles (and again at least three bonds) yield

$$W^{(3)}_{proto} \approx \frac{r_w^2}{\lambda_{De}^2} W_{ring} , \tag{3.104}$$

and so on.

This sufficiently suggests that the Debye-Hückel result (3.100) correctly describes the corrections up to and including order $r_w/\lambda_{De} \sim 1/n_e\lambda_{De}^3$.

Heuristic determination of an equation of state

After all the discussions of a strongly mathematically oriented graph theory, we now turn to a more physically motivated derivation of thermodynamic relationships for a classical plasma.

We have already solved the screening problem of a stationary test charge in the introduction. The sign of the screening charge distribution is opposite to the sign of the test charge q_a. The screening causes a redistribution of charges compared to the ideal plasma situation (in the latter, the interaction potential is neglected). The correlation energy is obtained from the general formula for the electrostatic energy of a charge distribution

$$W = \frac{1}{2}\int d^3r\rho(\mathbf{r})\phi(\mathbf{r}) \ . \tag{3.105}$$

We insert into this formula:

$$\rho(\mathbf{r}) = \rho_a^{ind}(r) \quad , \quad \phi(\mathbf{r}) = \phi_a^{Cb}(r) \ . \tag{3.106}$$

The use of ϕ_a^{ind} is obvious, since we want to calculate the energy change due to the redistribution of charges. In general, $\phi = \phi_a^{Cb}+\phi_a^{ind}$ will appear in the integrand. However, the second term leads to the self-energy. For the energy shift, we therefore have to evaluate the integral

$$\Delta E_a = -\frac{1}{2}\frac{1}{4\pi\varepsilon_0}\frac{q_a^2\kappa^2}{4\pi}\int d^3r\,\frac{1}{r^2}\,e^{-\kappa r} = -\frac{1}{2}\frac{1}{4\pi\varepsilon_0}q_a^2\kappa^2\int\limits_0^\infty dre^{-\kappa r} = -\frac{1}{2}\frac{1}{4\pi\varepsilon_0}q_a^2\kappa \tag{3.107}$$

The energy shift of the plasma, which consists of N_e electrons and $N_i = N_e = N/2$ protons, is

$$U^{int} = -\frac{1}{2}\frac{1}{4\pi\varepsilon_0}\sum_s N_s q_s^2\kappa = -\frac{N}{2}\frac{1}{4\pi\varepsilon_0}e^2\kappa \sim N\sqrt{\frac{N}{TV}} \ , \tag{3.108}$$

where we recall $\kappa^2 = \frac{2}{\lambda_D^2}$. The thermodynamic potential, which refers to the natural variables N, T, and V, is the free energy $F = U - TS$. We have

$$U = F + TS = F - T\left(\frac{\partial F}{\partial T}\right)_{N,V} = -T^2\frac{\partial}{\partial T}\left(\frac{F}{T}\right)_{N,V} \ . \tag{3.109}$$

With the splitting of a thermodynamic quantity A into

$$A = A^{ideal} + A^{interaction} \equiv A^{id} + A^{int} \ , \tag{3.110}$$

we find

$$\frac{F^{int}}{T} = -\int^T dT\,\frac{U^{int}}{T^2} \ . \tag{3.111}$$

Now we have

$$U^{int} = \frac{C}{\sqrt{T}} \tag{3.112}$$

as a function of T. Therefore, it follows that

$$\frac{F^{int}}{T} = -C \int^{T} \frac{dT}{T^{5/2}} = \frac{2}{3} \frac{C}{T^{3/2}} = \frac{2}{3} \frac{U^{int}}{T} . \tag{3.113}$$

The result has important consequences. The free energy is modified by the term

$$F^{int} = -\frac{1}{3} \frac{1}{4\pi\varepsilon_0} \sum_s N_s q_s^2 \kappa = -\frac{N}{3} \frac{1}{4\pi\varepsilon_0} e^2 \sqrt{\frac{Ne^2}{\varepsilon_0 k_B T V}} \tag{3.114}$$

where an electron-proton plasma $N = N_e + N_i$ is present. The pressure correction follows from

$$p^{int} = -\left.\frac{\partial F^{int}}{\partial V}\right|_{T,N} = -\frac{1}{6} \frac{1}{4\pi\varepsilon_0} \sum_s n_s q_s^2 \kappa = -\frac{1}{24\pi} \kappa^3 k_B T . \tag{3.115}$$

This is exactly the result obtained using the diagram technique in the ring approximation.

Example 3.3 (Corrections to Other Thermodynamic Quantities)

The present method is not only applicable for calculating the pressure correction. Since we have obtained the correction to the free energy (compared to the ideal gas approximation), we also have access to the corrections of the other thermodynamic variables. The chemical potential must be corrected by

$$\mu_a^{int} = \left.\frac{\partial F^{int}}{\partial N_a}\right|_{T,V} = -\frac{1}{2} \frac{1}{4\pi\varepsilon_0} q_a^2 \kappa . \tag{3.116}$$

For the entropy correction, the following applies accordingly

$$S^{int} = -\left.\frac{\partial F^{int}}{\partial T}\right|_{V,N} = -\frac{1}{6} \frac{1}{4\pi\varepsilon_0} \sum_a N_a \frac{q_a^2 \kappa}{T} . \tag{3.117}$$

If we proceed with the corresponding densities, the following expressions are available:

$$f^{int} \equiv \frac{F^{int}}{V} = -\frac{k_B T}{12\pi} \kappa^3 , \tag{3.118}$$

$$u^{int} \equiv \frac{U^{int}}{V} = -\frac{k_B T}{8\pi} \kappa^3 , \tag{3.119}$$

$$s^{int} \equiv \frac{S^{int}}{V} = -\frac{k_B}{24\pi} \kappa^3 . \tag{3.120}$$

■

We conclude the discussion of the thermodynamic variables here and emphasize once again that this introduction can only provide an impression of the problems encountered in more precise calculations.

As a result of the estimates shown here, we find that in the limiting case where the number of particles in the Debye sphere approaches infinity, the properties of the ideal plasma become increasingly applicable.

3.4 Debye Shielding and Microfields

Since up to now we have not yet rigorously justified the shielding of the Coulomb potential by charge clouds with the result of Debye shielding between two particles, we turn to this set of problems in this section. In addition, we calculate the electric microfield distribution in a plasma, which is of great importance for diagnostic purposes via the Stark effect.

To calculate exactly the effective potential in a plasma, for example of an ion at the position $\mathbf{R}$, it is necessary to solve the Poisson equation

$$\nabla^2 \varphi = -\frac{1}{\varepsilon_0} e\delta(\mathbf{r} - \mathbf{R}) + \frac{1}{\varepsilon_0} e[n_e(\mathbf{r}) - n_i(\mathbf{r})] \tag{3.121}$$

In principle, the further evaluation of the Poisson equation is clear: in this exact notation, we must insert the sum over Dirac delta functions for $n_e(\mathbf{r})$ and $n_i(\mathbf{r})$. This makes the problem formally solvable in full generality, but the solution contains all particle positions and is therefore far too complicated. We should keep in mind that we are not interested in a microscopic and detailed solution that includes all fluctuations. We therefore change the question by asking for the mean potential.

Let us begin with the phase space density ρ. In the electrostatic case, ρ factorizes into a position-dependent and a momentum-dependent part. If we are interested in the probability of finding a particle at a certain position, independent of the current particle velocities, we can integrate over all momenta and obtain from ρ

$$P_N(\mathbf{r}_1, \ldots, \mathbf{r}_N) = \frac{\exp\left(-\sum \phi_{ij}/2k_BT\right)}{\int d^3r_1 \ldots d^3r_N \exp\left(-\sum \phi_{ij}/2k_BT\right)} \,. \tag{3.122}$$

This notation applies to a system of N particles; in a plasma, they belong to different species (electrons and ions).

The integration over momentum space naturally leads to a loss of information. In this context, however, this is not particularly problematic for two reasons. First, the momentum distribution is not of interest for solving the Poisson equation anyway, and second, the velocity distribution in equilibrium is trivially a Maxwell distribution.

The distribution P_N still contains far too much information (and too many coordinates); for a direct calculation, (3.122) is not suitable. Therefore, P_N, is reduced by integrating over the coordinates of entire sets of particles. For example, the specific molecular distribution function P_s (with $s < N$) according to the prescription

$$P_s(\mathbf{r}_1, \ldots, \mathbf{r}_s) = \int P_N(\mathbf{r}_1, \ldots, \mathbf{r}_N) d^3 r_{s+1} \ldots d^3 r_N. \tag{3.123}$$

Physically, this reduction means that we are now only asking about the probabilities of finding the first s particles at certain locations, regardless of where the remaining $N - s$ particles are located. If it were now possible to derive a simple expression for $P_1(\mathbf{r}_1)$ [for electrons or ions], we would have made significant progress in determining the potential φ. We want to illustrate the difficulties that arise using the example of P_s, initially ignoring complications due to different particle species. Because of the interaction terms, the definition equation for P_s cannot be easily evaluated, so we attempt to derive a differential equation for P_s. Let us begin with $s = 1$ and denote the gradient with respect to $\mathbf{r}_1$ by ∇_1. Then, based on the definition (3.123), it follows that

$$\nabla_1 P_1(\mathbf{r}_1) = -\frac{1}{k_B T} \sum_{j=2}^{N} \int (\nabla_1 \phi_{1j}) P_2(\mathbf{r}_1, \mathbf{r}_j) d^3 r_j. \tag{3.124}$$

No further simplification is possible. Eq. (3.124) displays the typical hierarchical character: we do not obtain a closed differential equation for P_1; rather, P_1 is determined by P_2. If we proceed to set up a differential equation for P_2, then P_3 appears, and so on. A straightforward calculation leads for $s = 1, 2, \ldots, N - 1$ to

$$\boxed{k_B T \, \nabla_i \ln P_s = -\sum_{j=1}^{s}{}' (\nabla_i \phi_{ij}) - \sum_{j=s+1}^{N} \int (\nabla_i \phi_{ij}) \frac{P_{s+1}}{P_s} d^3 r_j}\,, \tag{3.125}$$

where $i = 1, \ldots, s$. In this formula, P_s depends on the coordinates $\mathbf{r}_1, \ldots, \mathbf{r}_s$ and P_{s+1} additionally on the coordinate $\mathbf{r}_j$. The bar over the summation symbol excludes $j = i$ and ∇_i denotes differentiation with respect to the ith coordinate.

Eq. (3.125) has a simple physical meaning. If we write the exact formula for the force on the ith particle (with $i \leq s$)

$$\mathbf{F}_i = -\sum_{j=1}^{N}{}' (\nabla_i \phi_{ij}) \tag{3.126}$$

and average over all possible positions of the particles $s+1, \ldots, N$, that is, we multiply by the generalized Boltzmann factor P_N and integrate over the particle positions $\mathbf{r}_{s+1}, \ldots, \mathbf{r}_N$, then, with appropriate normalization, we obtain

$$\langle \mathbf{F}_i \rangle_s = k_B T \, \nabla_i \, \ln \, P_s . \tag{3.127}$$

Thus, $-k_B T \ln P_s \equiv \langle W_i \rangle_s$ has the intuitive meaning of the potential of the mean force. Furthermore, Eq. (3.125) can now be read as stating that the mean force on the ith particle in the s-configuration consists of the exact force caused by the other particles in the s-configuration and the averaged contribution from the remaining particles.

Note that $P_{s+1}(\ldots, \mathbf{r}_j)/P_s(\ldots)$ gives the conditional probability for particle j to be at position $\mathbf{r}_j$, given that the first s particles are at positions $\mathbf{r}_1, \ldots \mathbf{r}_s$.

If we again specialize to $s = 1$, we obtain from (3.127) the Boltzmann distribution

$$P_1(\mathbf{r}_1) = \frac{\exp[-\langle W \rangle / k_B T]}{\int d^3 r_1 \, \exp[-\langle W \rangle / k_B T]} , \tag{3.128}$$

with $\langle W \rangle \equiv \langle W_1 \rangle_1$ as the potential of the mean force on a particle, where the averaging is performed over all other particle positions.

We now answer the question of whether the potential of the mean force coincides with the mean potential energy. For $s = 1$ we obtain

$$\begin{aligned} \langle \phi_1 \rangle_1 \equiv \langle \phi \rangle &= \frac{\displaystyle\int e^{-\Sigma \phi_{ij}/2k_B T} \sum_{j=2}^{N} \phi_{1j} d^3 r_2 \ldots d^3 r_N}{\displaystyle\int e^{-\Sigma \phi_{ij}/2k_B T} \, d^3 r_2 \ldots d^3 r_N} \\ &= \sum_{j=2}^{N} \int \phi_{1j} \frac{P_2(\mathbf{r}_1, \mathbf{r}_j)}{P_1(\mathbf{r}_1)} d^3 r_j , \end{aligned} \tag{3.129}$$

and thus, in general, *not* $\langle \phi \rangle = \langle W \rangle$, since on the right-hand side ∇_1 would not act only on ϕ_{1j} [cf. (3.125) and (3.127)]. However, we can immediately provide an approximation in which $\langle \phi \rangle = \langle W \rangle$ holds, namely

$$\boxed{P_2(\mathbf{r}_1, \mathbf{r}_j) = P_1(\mathbf{r}_1) P_1(\mathbf{r}_j)} \, . \tag{3.130}$$

This so-called single-particle approximation gives us for the probability density $P_1(\mathbf{r}_1)$ a Boltzmann distribution in which the mean potential appears in the exponent. This finally allows a closed and simple formulation of the averaged Poisson equation (3.121), which is known as the Poisson-Boltzmann equation. Before we discuss it explicitly, two remarks. The single-particle approximation physically means that particles 1 and j are completely uncorrelated; particle j occupies position $\mathbf{r}_j$ independently of where particle 1 is located. Another point: So far, we have assumed the particles to be distinguishable and interpret $P_1(\mathbf{r}_1)$ as the probability of finding particle number 1 at position $\mathbf{r}_1$. In fact, in the averaged Poisson equation, we are only interested in the overall distribution of electrons and ions; whether electron number 1 or electron number 27 (if we can even distinguish them) is at position $\mathbf{r}_1$ is actually of no interest to us. This can be formalized mathematically by moving to the general distribution function, e.g.

$$P^{(1)} = \frac{N}{2} P_1. \tag{3.131}$$

For the two-particle distribution function, in principle, we must take into account whether the pair configuration consists of two electrons, two ions, or an electron and an ion. In the limit $N \to \infty$ however, the combinatorial factors become equal, and therefore we define

$$P^{(2)} = \left(\frac{N}{2}\right)^2 P_2 \tag{3.132}$$

for $N/2$ electrons and $N/2$ ions. It is clear that we must use the respective general single-particle distribution functions in the Poisson equation.

For the averaged electric potentials

$$\varphi \equiv \langle \phi \rangle / q_\nu \tag{3.133}$$

— the index ν stands for ions or electrons ($\nu = e, i$) — we obtain the Poisson-Boltzmann equation

$$\boxed{\nabla^2 \varphi = -\frac{1}{\varepsilon_0} \sum_{\nu=e,i} \frac{q_\nu N_\nu \exp(-q_\nu \varphi / k_B T_\nu)}{\int \exp(-q_\nu \varphi / k_B T_\nu) d^3 r}}, \tag{3.134}$$

where, for clarity, we have introduced $N_i = N_e = N/2$ and use $q_i = e$ and $q_e = -e$ to denote the charge of the ions and electrons, respectively. If we choose the zero point of the averaged potential appropriately, we obtain

$$\nabla^2 \varphi = -\frac{1}{\varepsilon_0} \varrho \left[e^{-q_i \varphi / k_B T_i} - e^{-q_e \varphi / k_B T_e} \right], \tag{3.135}$$

with the charge density

$$\varrho = e \frac{N}{2V}. \tag{3.136}$$

The simplest way to see this is by linearization, which shows that the mean potential decays with the Debye length as the characteristic length scale.

Physically, this result means that space charge fields vanish over distances on the order of a Debye length, and in a neutral plasma, significant field effects occur only within a narrow boundary layer.

For a physical interpretation in which the effective interaction potential between two particles is the Debye potential, our calculation so far is not sufficient. For this purpose, we single out two particles (1 and 2), set $s = 2$ and calculate the mean potential energy. Instead of (3.129), we now have, for example, for $i = 1$ ($i = 2$ follows analogously)

$$\langle\phi_i\rangle_2 = \phi_{i2} + \sum_{j=3}^{N} \int \phi_{ij} \frac{P_3(\mathbf{r}_i, \mathbf{r}_2, \mathbf{r}_j)}{P_2(\mathbf{r}_i, \mathbf{r}_2)} d^3 r_j. \tag{3.137}$$

The mean potential energy of the $s = 2$ configurations ($\mathbf{r}_i = \mathbf{r}_1$ and $\mathbf{r}_i = \mathbf{r}_2$),

$$\langle\phi\rangle^{(2)} = \phi_{12} + \sum_{i=1,2} \sum_{j=3}^{N} \int \phi_{ij} \frac{P_3(\mathbf{r}_1, \mathbf{r}_2, \mathbf{r}_j)}{P_2(\mathbf{r}_1, \mathbf{r}_2)} d^3 r_j, \tag{3.138}$$

also depends on $\mathbf{r}_1$ and $\mathbf{r}_2$, more precisely on $r = |\mathbf{r}_1 - \mathbf{r}_2|$. The interaction energy of the particles that do not belong to the ($s = 2$) configuration has not been written out. The following holds

$$\sum_{i=1,2} \nabla_i \langle\phi\rangle^{(2)} \approx \sum_{i=1,2} \nabla_i \langle\phi_i\rangle_2, \tag{3.139}$$

when three-particle correlations are neglected.

Example 3.4 (Justification of (3.139))
To understand this transformation, we write in the pair approximation

$$P_2(\mathbf{r}_1, \mathbf{r}_2) \approx P_1(\mathbf{r}_1) P_1(\mathbf{r}_2) \tag{3.140}$$

and for the conditional probability

$$\frac{P_3(\mathbf{r}_1, \mathbf{r}_2, \mathbf{r}_j)}{P_2(\mathbf{r}_1, \mathbf{r}_2)} \approx P_1(\mathbf{r}_j) \prod_{i=1,2} \left[1 + g_{ij}(\mathbf{r}_i, \mathbf{r}_j)\right], \tag{3.141}$$

neglecting all higher correlations (including products of g functions). After applying the operator $\sum_{i=1,2} \nabla_i$ to (3.137) and (3.138), we obtain in the approximation (3.140) and (3.141) equality of the respective right-hand sides, if

$$\sum_{j=3}^{N} \int \phi_{1j}\, P_1(\mathbf{r}_j)\nabla_2\, g_{2j}\, d^3r_j + \sum_{j=3}^{N} \int \phi_{2j}\, P_1(\mathbf{r}_j)\nabla_1\, g_{1j} d^3r_j \stackrel{!}{=} 0 \tag{3.142}$$

holds. If we introduce in the first integral on the left-hand side of (3.142) $\tilde{\mathbf{r}}_j := \mathbf{r}_j - \mathbf{r}_1$ and in the second integral $\tilde{\mathbf{r}}_j := \mathbf{r}_j - \mathbf{r}_2$ as new integration variables, then under the assumption of homogeneity $[P_1(\mathbf{r}_j) \approx const]$ and for isotropy $[g_{2j}(\mathbf{r}_2 - \mathbf{r}_1 - \tilde{\mathbf{r}}_j) = g_{2j}(|\mathbf{r}_2 - \mathbf{r}_1 - \tilde{\mathbf{r}}_j|);\ g_{1j}(\mathbf{r}_1 - \mathbf{r}_2 - \tilde{\mathbf{r}}_j) = g_{1j}(|\mathbf{r}_1 - \mathbf{r}_2 - \tilde{\mathbf{r}}_j|) = g_{1j}(|\mathbf{r}_2 - \mathbf{r}_1 + \tilde{\mathbf{r}}_j|)]$ the relation (3.142) follows.

Starting from (3.125), we can derive another relation if we use (3.137) in the pair approximation. Note that for $i = 1$ or $i = 2$

$$\int \phi_{ij} \nabla_i \frac{P_3(\mathbf{r}_1, \mathbf{r}_2, \mathbf{r}_j)}{P_2(\mathbf{r}_1, \mathbf{r}_2)} d^3r_j = const \tag{3.143}$$

follows with similar arguments as those just used. The constants can later [see (3.144)] be determined from the asymptotics; we may, without loss of generality, set them to zero. Thus, we find

$$-k_BT \sum_{i=1,2} \nabla_i \ln P_2 \approx \sum_{i=1,2} \nabla_i \langle\phi_i\rangle_2 \approx \sum_{i=1,2} \nabla_i \langle\phi\rangle^{(2)}\ . \tag{3.144}$$

■

The relation (3.144) shows us that

$$P_2 \sim \exp[-\langle\phi\rangle^{(2)}/k_BT] \tag{3.145}$$

is a reasonable approach. Within the framework of the approximation (3.140), it further follows from (3.144)

$$P_1(\mathbf{r}_2) \sim \exp[-\langle\phi_2\rangle_2/k_BT]\ . \tag{3.146}$$

We are now in a position to arrive at a closed equation for $\langle\phi_2\rangle_2$. The subsequent interpretation assumes that we set $i = 2$ and consider particle 1 with charge q_1 as a test particle. Eq. (3.137), after division by q_2, yields the mean electric potential $\langle\phi_2\rangle_2/q_2$ of the test particle 1 at the position $\mathbf{r}_2$. For the calculation, we need the relation

$$\nabla_i^2 \phi_{ij} = -\frac{1}{\varepsilon_0} q_i q_j \delta(\mathbf{r}_j - \mathbf{r}_i)\ , \tag{3.147}$$

after applying the Laplace operator with respect to the coordinate $\mathbf{r}_i = \mathbf{r}_2$ to (3.137). On the right-hand side of (3.137), we then use

$$\nabla_2^2\left[\phi_{2j}\frac{P_3(\mathbf{r}_1,\mathbf{r}_2,\mathbf{r}_j)}{P_2(r_1,\mathbf{r}_2)}\right] \approx [\nabla_2^2\phi_{2j}]P_1(\mathbf{r}_j) \sim \delta(\mathbf{r}_2-\mathbf{r}_j)P_1(\mathbf{r}_2)$$
$$\sim \delta(\mathbf{r}_2-\mathbf{r}_j)\exp[-\langle\phi_2\rangle_2/k_BT] \qquad (3.148)$$

and thus obtain from (3.137) a generalized Poisson-Boltzmann equation for $\varphi(\mathbf{r}) = \langle\phi_2\rangle_2/q_2$ at the position $\mathbf{r}_2-\mathbf{r}_1=\mathbf{r}$. For simplicity, we write this equation for a screening electron gas, whereby for physical reasons we can immediately specify the proportionality constant in (3.148). We have ($q_2=-e$)

$$\nabla^2\varphi = -\frac{1}{\varepsilon_0}q_1\delta(\mathbf{r}) + \frac{1}{\varepsilon_0}e[n_{e0}\exp(e\varphi/k_BT_e) - n_{e0}]. \qquad (3.149)$$

It can now be shown that the linearization in $e\varphi/k_BT_e$ is consistent with the neglect of higher correlations, and as a solution of

$$-\nabla^2\varphi + \frac{n_{e0}e^2}{\varepsilon_0 k_BT_e}\varphi = \frac{1}{\varepsilon_0}q_1\delta(\mathbf{r}) \qquad (3.150)$$

one obtains the Debye potential

$$\boxed{\varphi(\mathbf{r}) = \frac{1}{4\pi\varepsilon_0}\frac{q_1}{r}\exp(-r/\lambda_{De})}\,. \qquad (3.151)$$

Thus, within the framework of equilibrium statistics, we have established Debye screening. Of course, it must be mentioned that, in particular, we have not consistently pursued the consistent neglect of higher correlations. The corresponding calculation, especially when distinguishing several components in the plasma, becomes rather confusing; however, it does not lead to any new or altered insights.

However, a word must still be said about the physical interpretation of neglecting the correlations. It must, of course, not be the case that (3.140) and (3.141)—in the form used for the approximation (3.148)—exclude any correlation between particle 1 and the "screening" particles $j=3,4,\ldots$. In fact, we only require $g_{2j}(\mathbf{r}_2,\mathbf{r}_j)\approx 0$, since we apply the Laplace operator with respect to the coordinate $\mathbf{r}_2$. The physical picture, in which the correlation between test particle 1 and the screening particles $j=3,4,\ldots$ is present, but the correlations among the screening particles themselves are neglected, certainly makes sense.

After having calculated the mean electric field of a particle via the mean potential, we are next interested in the distribution of the fields around the mean value. The problem is quite difficult in full generality. It is a matter of determining the probability distribution of a dependent quantity $\mathbf{E}$, which is composed of many individual contributions, i.e.

$$\mathbf{E}(\mathbf{r}) \equiv \mathbf{E}(\mathbf{r}; \mathbf{r}_1, \ldots) = \sum_{j=1}^{N} \mathbf{E}_j(\mathbf{r}, \mathbf{r}_j), \tag{3.152}$$

to calculate. In general, the following holds

$$\boxed{W(\mathbf{E}) = \int \ldots \int \delta[\mathbf{E} - \mathbf{E}(\mathbf{r})] \rho \, d^{3N} r \, d^{3N} p} \, . \tag{3.153}$$

The difficulties with the Dirac delta function can be circumvented by switching to the Fourier transform[1]

$$W(\xi) = \int e^{-i\xi \cdot \mathbf{E}} W(\mathbf{E}) d^3 E$$

$$= \int \ldots \int e^{-i\xi \cdot \mathbf{E}(\mathbf{r})} \rho \, d^{3N} r \, d^{3N} p \tag{3.154}$$

It is immediately apparent that for electrostatic fields the integration over momentum space becomes trivial and thus

$$\boxed{W(\xi) = \int \ldots \int e^{-i\xi \cdot \Sigma \mathbf{E}_j} P_N \, d^{3N} r} \tag{3.155}$$

follows. In evaluating the right-hand side, similar difficulties arise as in the calculation of the partition function. A cluster-like expansion was successfully applied by Baranger and Mozer [52]. We will not go into this type of evaluation in detail here, but will only discuss the older approaches, which nevertheless already provide considerable insight.

To present the results of Holtsmark [53], we introduce

$$\varepsilon_j = \exp(-i\xi \cdot \mathbf{E}_j) - 1 \tag{3.156}$$

and obtain from (3.155)

$$W(\xi) = \int \ldots \int \prod_{j=1}^{N} (1 + \varepsilon_j) P_N \, d^{3N} r. \tag{3.157}$$

Furthermore, in the Holtsmark theory we use the single-particle approximation

$$P_N = P_1(\mathbf{r}_1) \ldots P_1(\mathbf{r}_N). \tag{3.158}$$

[1] Note that ξ is a vector, even though the Greek letter often does not appear in vector notation in print.

If we take a closer look at the product over all possible factors $(1+\varepsilon_j)$ from $j = 1$ to $j = N$ more closely, we see that it can be represented as a sum over terms of the form $\varepsilon_1\varepsilon_2\ldots\varepsilon_s$. Since we assume indistinguishable particles, i.e., we do not have to pay attention to the indices $1, 2, \ldots$ individually, we only need the information about how many terms contain exactly s factors ε_1 to ε_s contained. The corresponding combinatorial factor is of course

$$\frac{N!}{(N-s)!s!}\,. \tag{3.159}$$

In the single-particle approximation, we can write the integrals over the individual terms as

$$\int\ldots\int \varepsilon_1\ldots\varepsilon_s P_N\, d^{3N}r = \left[\int \varepsilon_1 P_1 d^3r_1\right]^s \tag{3.160}$$

Altogether, we thus obtain

$$W(\xi) = \sum_{s=0}^{N} \frac{N!}{(N-s)!s!}\left[\int \varepsilon_1 P_1 d^3r_1\right]^s. \tag{3.161}$$

If we completely neglect particle correlations and set

$$P_1 = 1/V \tag{3.162}$$

we obtain in the approximation $N \to \infty$,

$$\frac{N!}{(N-s)!} \approx N^s, \tag{3.163}$$

the result

$$\boxed{W(\xi) \approx \exp\left\{n\int\left(e^{-i\xi\cdot\mathbf{E}_1}-1\right)d^3r_1\right\}}\,, \tag{3.164}$$

where we have set $n = \frac{N}{V}$. To calculate the integral in the exponent of (3.164), we use the field of a single ion (position vector $\mathbf{r}_1$)

$$\mathbf{E}_1 = -\frac{1}{4\pi\varepsilon_0}\frac{e\mathbf{r}_1}{r_1^3} \equiv -\frac{e^*\mathbf{r}_1}{r_1^3} \tag{3.165}$$

for the "low-frequency" microfield distribution, where for simplicity we have placed the field point at the origin. This yields

$$\int\left(e^{-i\xi\cdot\mathbf{E}_1}-1\right)d^3r_1 = -\frac{4}{15}(2\pi e^*\xi)^{3/2}. \tag{3.166}$$

The inverse transformation

$$W(\mathbf{E}) = \frac{1}{(2\pi)^3}\int e^{i\xi\cdot\mathbf{E}} W(\xi) d^3\xi$$

$$= \frac{1}{2\pi^2 E}\int_0^\infty \sin(\xi E)\exp\left\{-\frac{4n}{15}(2\pi e^*\xi)^{3/2}\right\}\xi d\xi \tag{3.167}$$

can no longer be carried out fully analytically. Instead of the distribution of the electric field vector **E**, we can ask for the distribution of the magnitude of **E**, and for this, due to the isotropy of the system, we define

$$W(E) = 4\pi E^2 W(\mathbf{E}) \tag{3.168}$$

We then normalize with a mean field strength

$$E_0 = e^*/r_0^2, \tag{3.169}$$

where $r_0 \approx 0{,}465 r_n$ is. If we introduce $\beta = E/E_0$, we obtain

$$\boxed{W(\beta) = \frac{2}{\pi\beta}\int_0^\infty x\sin x\, e^{-(x/\beta)^{3/2}} dx}\,. \tag{3.170}$$

The asymptotic forms

$$W(\beta) \sim \begin{cases} \frac{4}{3\pi}\beta^2 & \text{for } \beta \to 0, \\ \frac{3}{2}\beta^{-5/2} & \text{for } \beta \to \infty \end{cases}$$

are also known. The complete integral (3.170) is shown in Fig. 3.6. Of course, Holtsmark's calculation [53] is only valid under very rough assumptions. It is clear that these assumptions can only be applicable in very dilute plasmas.

Since we have already discussed in the previous section that *one* plasma effect is the screening of the Coulomb fields, it is natural to improve the Holtsmark calculation by using Debye-screened fields instead of the Coulomb fields (3.165). Such an evaluation—with all other assumptions of the previous calculation—was carried out very successfully by Ecker and Müller [54], who found deviations from Holtsmark [53].

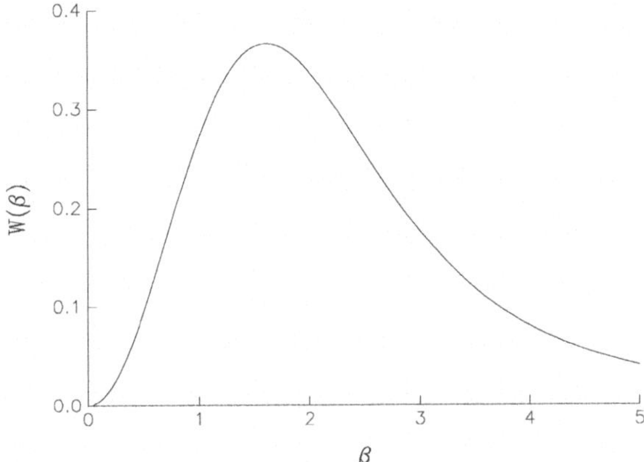

Fig. 3.6 Holtsmark distribution (3.170)

An even more advanced calculation comes from Baranger and Mozer [52], who performed a cluster-like expansion. Such a calculation allows a separation into a high-frequency and a low-frequency component of the microfield distribution. It is clear that effective screening requires sufficient time for the formation of the screening clouds; for rapidly changing processes, our previous calculation must therefore be fundamentally modified. In this sense, in this section we have presented only the low-frequency microfield component.

4 Kinetic Description of a Plasma

Abstract

Kinetics of a many-particle system refers in physics to the dynamic description of a system composed of many particles based on statistical methods. Instead of considering each individual particle, a probability distribution (distribution function) of the particle states is used. In the kinetic theory of gases (neutral particles), the distribution is determined by the Boltzmann equation. Collisions between particles play a central role in the evolution of the distribution function. From the latter, macroscopic properties such as pressure, temperature, and transport phenomena can be derived. The charges of plasma particles lead to particular features, which we address in this chapter. We present the various kinetic equations that are in use.

4.1 BBGKY Hierarchy

In this section, we begin with the fundamental problem of deriving a self-contained determining equation (kinetic equation) for the single-particle distribution function of a many-particle system. The hierarchy problem is fundamental and can only be solved mathematically by physically motivated assumptions.

The starting point for a nonequilibrium description of the plasma is the Liouville equation

$$\boxed{\frac{\partial \rho}{\partial t} = \{H, \rho\}} \tag{4.1}$$

K.-H. Spatschek, *Theoretical Plasma Physics*,
https://doi.org/10.1007/978-3-662-72828-4_4

in canonically conjugate position and momentum variables q and p. In equilibrium, we set the left-hand side of (4.1) to zero; we will no longer do this here. However, the method of reduction, which was successfully applied in the previous chapter, is also suitable here. To distinguish the results from the previous chapter, we choose a new notation for the reduced single-particle distribution function

$$f^{\alpha}(\mathbf{q}, \mathbf{p}; t) \equiv f_1^{\alpha}(\mathbf{q}_1, \mathbf{p}_1; t) := N_{\alpha} \int d^3q_2 \ldots d^3q_N d^3p_2 \ldots d^3p_N \, \rho, \tag{4.2}$$

where the index α indicates the particle species to which the selected particle (1) belongs. Analogously, two-particle distribution functions are defined as

$$f_2^{\alpha\alpha}(\mathbf{q}_1, \mathbf{p}_1, \mathbf{q}_2, \mathbf{p}_2; t) = N_{\alpha}(N_{\alpha} - 1) \int d^3q_3 \ldots d^3q_N d^3p_3 \ldots d^3p_N \rho, \tag{4.3}$$

$$f_2^{\alpha\beta}(\mathbf{q}_1, \mathbf{p}_1, \mathbf{q}_2, \mathbf{p}_2; t) = N_{\alpha}N_{\beta} \int d^3q_3 \ldots d^3q_N d^3p_3 \ldots d^3p_N \, \rho, \tag{4.4}$$

again depending on the particle species of the selected particles. Of course, due to the large numbers $N_{\alpha} - 1 \approx N_{\alpha}$ we will use, the asymmetry between the definitions (4.3) and (4.4) disappears.

As long as we consider a plasma without an external magnetic field, it is not necessary to generalize the formalism to the extent that pseudo-canonical transformations must be applied. The transformation of momenta $\mathbf{p}_j$ to velocities $\mathbf{v}_j$ via

$$\mathbf{p}_j = m\mathbf{v}_j \tag{4.5}$$

is so straightforward in the absence of a magnetic field that we can directly perform the transition

$$f_1^{\alpha}(\mathbf{q}_1, \mathbf{v}_1; t) = m_{\alpha}^3 f_1^{\alpha}(\mathbf{q}_1, \mathbf{p}_1 = m\mathbf{v}_1; t) \tag{4.6}$$

with the correct normalization.

If we now want to derive an equation for the single-particle distribution function from the Liouville equation by reduction, we encounter a hierarchy problem similar to that in equilibrium.

To be able to carry out the calculation in detail, we write the Liouville equation as

$$\boxed{\frac{\partial \rho}{\partial t} = L\rho} \tag{4.7}$$

with the Liouville operator

$$
\begin{aligned}
L &= \sum_{j=1}^{N} \left[\{H, \mathbf{q}_j\} \cdot \frac{\partial}{\partial \mathbf{q}_j} + \{H, \mathbf{p}_j\} \cdot \frac{\partial}{\partial \mathbf{p}_j} \right] \\
&= -\sum_{j=1}^{N} \left[\frac{\partial H}{\partial \mathbf{p}_j} \cdot \frac{\partial}{\partial \mathbf{q}_j} - \frac{\partial H}{\partial \mathbf{q}_j} \cdot \frac{\partial}{\partial \mathbf{p}_j} \right].
\end{aligned}
\tag{4.8}
$$

The Hamiltonian function (with $\mathbf{p}_j = m_j \mathbf{v}_j$) is given by

$$
H = H(\mathbf{q}, \mathbf{v}) = \sum_{j=1}^{N} \left[\frac{m_j}{2} \mathbf{v}_j^2 + e_j \phi(\mathbf{q}_j) \right] + \sum_{i<j} \phi_{ij}(\mathbf{q}_i, \mathbf{q}_j), \tag{4.9}
$$

where $\phi(\mathbf{q}_j)$ denotes the scalar potential of an external electric field $\mathbf{E}_0$ at the position of the j-th particle (with charge e_j) and

$$
\phi_{ij} = \frac{1}{4\pi\varepsilon_0} \frac{e_i e_j}{|\mathbf{q}_i - \mathbf{q}_j|} \tag{4.10}
$$

represents the potential energy for Coulomb interaction. If we insert (4.9) into the Liouville operator, we obtain

$$
\boxed{
\begin{aligned}
L &= -\sum_{j=1}^{N} \left[\mathbf{v}_j \cdot \frac{\partial}{\partial \mathbf{q}_j} - \frac{e_j}{m_j} \frac{\partial \phi}{\partial \mathbf{q}_j} \cdot \frac{\partial}{\partial \mathbf{v}_j} \right] + \sum_{j=1}^{N} \sum_{i=1}^{j-1} \frac{1}{m_j} \frac{\partial \phi_{ij}}{\partial \mathbf{q}_j} \cdot \frac{\partial}{\partial \mathbf{v}_j}, \\
&= L^{(1)} + L^{(2)},
\end{aligned}
}
\tag{4.11}
$$

i.e., this yields a (first) term $L^{(1)}$, which is to be understood merely as a single-particle propagator, and an interaction term $L^{(2)}$. When reducing (4.7) with respect to the functions (4.2), $L^{(1)}$ is relatively easy to handle, especially if we stipulate that the probability densities for $|\mathbf{q}_j|, |\mathbf{p}_j| \to \infty$ vanish. We single out a particle of type $\alpha = e, i$ and integrate over all positions and velocities of the other particles. Without loss of generality, we choose as arguments of the single-particle distribution function $\mathbf{q}_1, \mathbf{v}_1$ and t; we denote the distribution function itself with the index α. From the term

$$
\left[\frac{\partial \rho}{\partial t} \right]^{(1)} = L^{(1)} \rho \tag{4.12}
$$

it follows that

$$
\left[\frac{\partial}{\partial t} f^{\alpha}(\mathbf{q}_1, \mathbf{v}_1; t) \right]^{(1)} = L_1^{\alpha} f^{\alpha}(\mathbf{q}_1, \mathbf{v}_1; t), \tag{4.13}
$$

with

$$L_j^\alpha = -\mathbf{v}_j \cdot \frac{\partial}{\partial \mathbf{q}_j} - \frac{e_\alpha}{m_\alpha}\mathbf{E}_0(\mathbf{q}_j) \cdot \frac{\partial}{\partial \mathbf{v}_j} . \tag{4.14}$$

That in the calculation of

$$N_\alpha \int d^3q_2 \ldots d^3p_N L^{(1)} \rho = L_1^\alpha f^\alpha \tag{4.15}$$

in fact only the part L_1^α of $L^{(1)}$ (i.e., $j = 1$) contributes, is due to the fact that, as a result of possible partial integrations,

$$\int d^3q_2 \ldots d^3p_N \, L_j^\alpha \, \rho = -\int \ldots d^3q_j d^3p_j \ldots \left[\frac{\partial H}{\partial \mathbf{p}_j} \cdot \frac{\partial \rho}{\partial \mathbf{q}_j} - \frac{\partial H}{\partial \mathbf{q}_j} \cdot \frac{\partial \rho}{\partial \mathbf{p}_j}\right]$$

$$= 0, \text{ for} j \neq 1 \tag{4.16}$$

holds; all remaining terms vanish.

Let us allow a further side note here: in the presence of an external magnetic field, in addition to the external electric field $\mathbf{E}_0$, the operator L_1^α must be extended by the Lorentz term, so that

$$\boxed{L_j^\alpha = -\mathbf{v}_j \cdot \frac{\partial}{\partial \mathbf{q}_j} - \frac{e_\alpha}{m_\alpha}\left[\mathbf{v}_j \times \mathbf{B}(\mathbf{q}_j) + \mathbf{E}_0(\mathbf{q}_j)\right] \cdot \frac{\partial}{\partial \mathbf{v}_j}} \tag{4.17}$$

applies. More on this later.

For the complete reduction of the Liouville equation (4.7), we now calculate the contribution from $L^{(2)}$. Now the definitions (4.3) and (4.4) come into play; we only need to pay attention to whether the interaction occurs between particles of the same species or between particles of different species. The factors $1/(N_\alpha N_\beta)$ or $1/N_\alpha^2$ cancel against the number of identical terms, so that

$$\left[\frac{\partial}{\partial t} f^\alpha(\mathbf{q}_1, \mathbf{v}_1; t)\right]^{(2)} = \sum_{\beta=e,i} \int d^3q_2 \, d^3v_2 \, L_{12}^{\alpha\beta} f_2^{\alpha\beta}(\mathbf{q}_1, \mathbf{v}_1, \mathbf{q}_2, \mathbf{v}_2; t) \tag{4.18}$$

follows, with

$$L_{jk}^{\alpha\beta} = \frac{\partial \phi_{jk}}{\partial \mathbf{q}_j} \cdot \left[\frac{1}{m_\alpha}\frac{\partial}{\partial \mathbf{v}_j} - \frac{1}{m_\beta}\frac{\partial}{\partial \mathbf{v}_k}\right]. \tag{4.19}$$

Note that the particle at position $\mathbf{r}_j$ with velocity $\mathbf{v}_j$ is supposed to belong to species α and the other particle (k) to species β, where $\alpha = \beta$ is also allowed. In the operator $L_{12}^{\alpha\beta}$ the indices 1 and 2 are meant to denote different particles. In general, for the operators defined here L_1^α and $L_{12}^{\alpha\beta}$, the lower indices 1 and 1 2 indicate the coordinates with respect to which differentiation is performed.

If we combine the contributions (4.13) and (4.18), we obtain the first equation of the BBGKY (Bogoliubov, Born, Green, Kirkwood, Yvon) hierarchy,

$$\partial_t f^\alpha(\mathbf{q}_1, \mathbf{v}_1; t) = L_1^\alpha f^\alpha(\mathbf{q}_1, \mathbf{v}_1; t) + \sum_{\beta=e,i} \int d^3q_2\, d^3v_2\, L_{12}^{\alpha\beta} f^{\alpha\beta}(\mathbf{q}_1, \mathbf{v}_1, \mathbf{q}_2, \mathbf{v}_2; t). \tag{4.20}$$

For the functions f^α and $f^{\alpha\beta}$ we have omitted the lower indices 1 and 1 2, since the notation is unambiguous even without them. Furthermore, we abbreviate

$$\frac{\partial}{\partial t} = \partial_t, \qquad \frac{\partial}{\partial \mathbf{q}} = \partial_{\mathbf{q}}, \qquad \frac{\partial}{\partial \mathbf{v}} = \partial_{\mathbf{v}} \tag{4.21}$$

to save some writing. Likewise,

$$d^6 1 \equiv d^3q_1\, d^3v_1, \qquad d^6 2 \equiv d^3q_2\, d^3v_2, \qquad \text{etc.} \tag{4.22}$$

applies wherever this unconventional abbreviation does not lead to confusion.

Equation (4.20) confronts us with a hierarchy problem that we have already encountered in equilibrium: for the single-particle distribution function, we do not obtain a closed equation. The two-particle distribution function, which appears on the right-hand side of (4.20), is itself determined by $f_3^{\alpha\beta\gamma}(\mathbf{q}_1, \mathbf{v}_1, \mathbf{q}_2, \mathbf{v}_2, \mathbf{q}_3, \mathbf{v}_3; t)$ and so on.

In order to nevertheless be able to draw useful conclusions from (4.20), we must resort to physical arguments. If we define the pair correlation function $g^{\alpha\beta}$ via

$$\boxed{f^{\alpha\beta}(\mathbf{q}_1, \mathbf{v}_1, \mathbf{q}_2, \mathbf{v}_2; t) = f^\alpha(\mathbf{q}_1, \mathbf{v}_1; t) f^\beta(\mathbf{q}_2, \mathbf{v}_2; t) + g^{\alpha\beta}(\mathbf{q}_1, \mathbf{v}_1, \mathbf{q}_2, \mathbf{v}_2; t)}, \tag{4.23}$$

then we can initially only assume that $g^{\alpha\beta}$ depends solely on the distance and vanishes for $|\mathbf{q}_1 - \mathbf{q}_2| \to \infty$. Furthermore, in the limit $N \to \infty$ the normalizations are chosen such that

$$\int d^3q_1 d^3q_2 d^3v_1 d^3v_2\; g^{\alpha\beta} = 0 \tag{4.24}$$

holds. The three-particle correlation $g^{\alpha\beta\gamma}$ is analogously defined via

$$\boxed{f^{\alpha\beta\gamma} = f^\alpha f^\beta f^\gamma + f^\alpha g^{\beta\gamma} + f^\beta g^{\alpha\gamma} + f^\gamma g^{\alpha\beta} + g^{\alpha\beta\gamma}} \tag{4.25}$$

The arguments $\mathbf{q}_1, \mathbf{v}_1, \mathbf{q}_2, \ldots$ are no longer written out explicitly; the assignment follows from the upper indices α, β and γ, where particle 1 belongs to species α,, particle 2 to species β, and particle 3 to species γ.

The essential assumption that we make in all further calculations is

$$g^{\alpha\beta\gamma}(\mathbf{q}_1, \mathbf{v}_1, \mathbf{q}_2, \mathbf{v}_2, \mathbf{q}_3, \mathbf{v}_3; t) \approx 0, \tag{4.26}$$

i.e., we neglect all three-particle correlations. This approximation, though born out of necessity, proves to be quite good in "dilute" plasmas. Recall the convention that, in a plasma, the mean potential energy of a particle should be much smaller than its mean kinetic energy. We have expressed this fact by $\Lambda \gg 1$. Since $\Lambda \sim n^{-1/2}$ holds, the number of particles in the Debye zone becomes large for sufficiently low densities. When the particles are, on average, far apart, the coupling between three particles can be neglected ("weak coupling approximation").

Eqs. (4.20), (4.23), (4.25), and (4.26) allow for a closed formulation for f^α and $g^{\alpha\beta}$,

$$\begin{aligned}\partial_t f^\alpha(\mathbf{q}_1, \mathbf{v}_1; t) = L_1^\alpha f^\alpha(\mathbf{q}_1, \mathbf{v}_1; t) + \sum_{\beta=e,i} \int d^6 2\, L_{12}^{\alpha\beta} f^\alpha(\mathbf{q}_1, \mathbf{v}_1; t) f^\beta(\mathbf{q}_2, \mathbf{v}_2; t) \\ + \sum_{\beta=e,i} \int d^6 2\, L_{12}^{\alpha\beta} g^{\alpha\beta}(\mathbf{q}_1, \mathbf{v}_1, \mathbf{q}_2, \mathbf{v}_2; t),\end{aligned} \tag{4.27}$$

$$\begin{aligned}&\partial_t g^{\alpha\beta}(\mathbf{q}_1, \mathbf{v}_1, \mathbf{q}_2, \mathbf{v}_2; t) = (L_1^\alpha + L_2^\beta) g^{\alpha\beta}(\mathbf{q}_1, \mathbf{v}_1, \mathbf{q}_2, \mathbf{v}_2; t) + L_{12}^{\alpha\beta} g^{\alpha\beta}(\mathbf{q}_1, \mathbf{v}_1, \mathbf{q}_2, \mathbf{v}_2; t) \\ &+ \sum_{\gamma=e,i} \int d^6 3 \big[L_{13}^{\alpha\gamma} f^\alpha(\mathbf{q}_1, \mathbf{v}_1; t) g^{\beta\gamma}(\mathbf{q}_2, \mathbf{v}_2, \mathbf{q}_3, \mathbf{v}_3; t) \\ &+ L_{23}^{\beta\gamma} f^\beta(\mathbf{q}_2, \mathbf{v}_2; t) g^{\alpha\gamma}(\mathbf{q}_1, \mathbf{v}_1, \mathbf{q}_3, \mathbf{v}_3; t) \\ &+ (L_{13}^{\alpha\gamma} + L_{23}^{\beta\gamma}) f^\gamma(\mathbf{q}_3, \mathbf{v}_3; t) g^{\alpha\beta}(\mathbf{q}_1, \mathbf{v}_1, \mathbf{q}_2, \mathbf{v}_2; t) \big] + L_{12}^{\alpha\beta} f^\alpha(\mathbf{q}_1, \mathbf{v}_1; t) f^\beta(\mathbf{q}_2, \mathbf{v}_2; t).\end{aligned} \tag{4.28}$$

The system just given, for whose derivation the only assumption made was the neglect of the three-particle correlation (4.26), is still far too complicated to be amenable to simple mathematical treatment.

Before we address the problem of reducing (4.27) and (4.28) to tractable kinetic equations, let us return once more to the question of how an external magnetic field $\mathbf{B}$ affects the operators L_1^α and $L_{12}^{\alpha\beta}$. Part of the answer was already given in (4.17); here now is the systematic approach.

If we continue to use $\mathbf{q}$ and $\mathbf{v}$ as coordinates, we start from (4.8) in the form

$$L = \sum_{j=1}^{N}\left[[H, \mathbf{q}_j] \cdot \frac{\partial}{\partial \mathbf{q}_j} + [H, \mathbf{v}_j] \cdot \frac{\partial}{\partial \mathbf{v}_j}\right] \tag{4.29}$$

With regard to the part that can be written as a sum of single-particle contributions, not much changes. The calculation of the Lie brackets proceeds in the usual manner and leads directly to (4.17), since $[H, \mathbf{q}_j] = -\mathbf{v}_j$ remains unchanged and

$$[H, \mathbf{v}_j] = -\frac{e_j}{m_j}\left[\mathbf{v}_j \times \mathbf{B}(\mathbf{q}_j) + \mathbf{E}_0(\mathbf{q}_j)\right] \tag{4.30}$$

applies for the non-interacting part. The operator component $L_{12}^{\alpha\beta}$ is obtained from the purely position-dependent interaction potential in H. Thus, $L_{12}^{\alpha\beta}$ is not changed compared to (4.19) by an external **B** field.

The Lie formalism, of course, only reveals its full strength when we leave the variables $\mathbf{q}$ and $\mathbf{v}$ behind and instead move to the coordinates of the guiding centers and drift velocities. In this case, the calculation, especially regarding the definition of f_1^α and the reduction from the Liouville density, requires some more extensive considerations, to which we will return in the section "Drift-Kinetic Equation."

In the following sections, we will deal with Eqs. (4.27) and (4.28), more precisely, with their simplifications and approximate solutions. Two steps are central to this:

In the first step, Eq. (4.28) is simplified. Essentially, three approximations are known in the literature. The *first* approximation is due to Vlasov; in it, one sets $g^{\alpha\beta} = 0$. In the *second* approximation, according to Landau, for weak coupling in (4.1.28) one omits ($g^{\alpha\beta} \ll f^\alpha f^\beta$) all terms in which particle 3 contributes. This thus leads to

$$\left[\partial_t - L_1^\alpha - L_2^\beta\right] g^{\alpha\beta}(\mathbf{q}_1, \mathbf{v}_1, \mathbf{q}_2, \mathbf{v}_2; t) = L_{12}^{\alpha\beta} f^\alpha(\mathbf{q}_1, \mathbf{v}_1; t) f^\beta(\mathbf{q}_2, \mathbf{v}_2; t). \tag{4.31}$$

This is called a binary collision assumption (in analogy to the Boltzmann approach).

In the *third* approximation, due to Balescu and Lenard, contributions from the third particle are partially taken into account:

$$\left[\partial_t - L_1^\alpha - L_2^\beta\right] g^{\alpha\beta}(\mathbf{q}_1, \mathbf{v}_1, \mathbf{q}_2, \mathbf{v}_2; t) = L_{12}^{\alpha\beta} f^\alpha(\mathbf{q}_1, \mathbf{v}_1; t) f^\beta(\mathbf{q}_2, \mathbf{v}_2; t)$$
$$+ \sum_{\gamma=e,i} \int d^6 3 \left[L_{13}^{\alpha\gamma} f^\alpha(\mathbf{q}_1, \mathbf{v}_1; t) g^{\beta\gamma}(\mathbf{q}_2, \mathbf{v}_2, \mathbf{q}_3, \mathbf{v}_3; t)\right. \tag{4.32}$$
$$\left. + L_{23}^{\beta\gamma} f^\beta(\mathbf{q}_2, \mathbf{v}_2; t) g^{\alpha\gamma}(\mathbf{q}_1, \mathbf{v}_1, \mathbf{q}_3, \mathbf{v}_3; t)\right].$$

If one compares with (4.28), it becomes clear that in all cases the contributions

$$\left[L_{12}^{\alpha\beta} + \sum_{\gamma=e,i} \int d^6 3 [L_{13}^{\alpha\gamma} + L_{23}^{\beta\gamma}] f^\gamma(\mathbf{q}_3, \mathbf{v}_3; t)\right] g^{\alpha\beta}(\mathbf{q}_1, \mathbf{v}_1, \mathbf{q}_2, \mathbf{v}_2; t) \tag{4.33}$$

are neglected as higher-order corrections in the development of $g^{\alpha\beta}(\mathbf{q}_1, \mathbf{v}_1, \mathbf{q}_2, \mathbf{v}_2; t)$. In contrast to (4.31), however, Balescu and Lenard take into account dynamic screening effects.

If, in the second step, the equation for $g^{\alpha\beta}$ were to be solved exactly (let us assume for the moment that we could do this!), then $g^{\alpha\beta}$ would in general ("non-Markovian") depend on the entire temporal history. However, within the framework of a kinetic theory, it is assumed that $g^{\alpha\beta}$ depends on time only functionally via f,

$$\boxed{g^{\alpha\beta}(t) \approx g^{\alpha\beta}[f(t)]}\,. \tag{4.34}$$

We will discuss this assumption, which defines the so-called kinetic regime, in more detail later.

The goal in the next section is to derive and discuss a single kinetic equation of the form

$$\partial_t f^\alpha(\mathbf{q}_1, \mathbf{v}_1; t) = L_1^\alpha f^\alpha(\mathbf{q}_1, \mathbf{v}_1; t) + \sum_{\beta=e,i} \int d^3 q_2 d^3 v_2 L_{12}^{\alpha\beta} f^\alpha(\mathbf{q}_1, \mathbf{v}_1; t) f^\beta(\mathbf{q}_2, \mathbf{v}_2; t)$$
$$+ K^\alpha\{f^\alpha(t)\} \tag{4.35}$$

from (4.27) and (4.28) or their simplifications. Here, the binary collision term K^α summarizes the effects due to close encounters, which influence the kinetics via $g^{\alpha\beta}$. The second term on the right-hand side of (4.35) accounts for the influence of particle 1 by the mean field of all the other particles. To see this more clearly, we rearrange:

$$
\begin{aligned}
&\sum_{\beta=e,i} \int d^3q_2 d^3v_2 L_{12}^{\alpha\beta} f^\alpha f^\beta \\
&= \sum_{\beta=e,i} \int d^3q_2 d^3v_2 \frac{\partial \phi_{12}^{\alpha\beta}}{\partial \mathbf{q}_1} \cdot \left[\frac{1}{m_\alpha} \frac{\partial}{\partial \mathbf{v}_1} - \frac{1}{m_\beta} \frac{\partial}{\partial \mathbf{v}_2} \right] f^\alpha(\mathbf{q}_1, \mathbf{v}_1; t) f^\beta(\mathbf{q}_2, \mathbf{v}_2; t) \\
&= \left[\frac{\partial}{\partial \mathbf{q}_1} \sum_{\beta=e,i} \int d^3q_2 d^3v_2\, \phi_{12}^{\alpha\beta} f^\beta(\mathbf{q}_2, \mathbf{v}_2; t) \right] \cdot \frac{1}{m_\alpha} \frac{\partial f^\alpha(\mathbf{q}_1, \mathbf{v}_1; t)}{\partial \mathbf{v}_1} \\
&\equiv -\frac{e_\alpha}{m_\alpha} \langle \mathbf{E}(\mathbf{q}_1; t) \rangle \cdot \frac{\partial f^\alpha}{\partial \mathbf{v}_1}.
\end{aligned}
\tag{4.36}
$$

Here, we have defined the mean field $\langle \mathbf{E} \rangle$, generated by all the other particles at the location $\mathbf{q}_1$ via

$$
\langle \mathbf{E}(\mathbf{q}_1; t) \rangle = -\frac{\partial}{\partial \mathbf{q}_1} \langle \phi(\mathbf{q}_1; t) \rangle, \tag{4.37}
$$

$$
\langle \phi(\mathbf{q}_1; t) \rangle = \sum_{\beta=e,i} \frac{1}{4\pi\varepsilon_0} e_\beta \int d^3q_2 d^3v_2 \frac{1}{|\mathbf{q}_1 - \mathbf{q}_2|} f^\beta(\mathbf{q}_2, \mathbf{v}_2; t) \tag{4.38}
$$

The interpretation follows directly from this.

In particular, from the form of the kinetic equation for $f^\alpha \equiv f^\alpha(\mathbf{q}_1, \mathbf{v}_1; t)$,

$$
\begin{aligned}
&\frac{\partial f^\alpha}{\partial t} + \mathbf{v}_1 \cdot \frac{\partial f^\alpha}{\partial \mathbf{q}_1} + \frac{e_\alpha}{m_\alpha} \left[\mathbf{v}_\alpha \times \mathbf{B}(\mathbf{q}_1) + \mathbf{E}_0(\mathbf{q}_1) \right] \cdot \frac{\partial f^\alpha}{\partial \mathbf{v}_1} + \frac{e_\alpha}{m_\alpha} \langle \mathbf{E}(\mathbf{q}_1; t) \rangle \cdot \frac{\partial f^\alpha}{\partial \mathbf{v}_1} \\
&\quad = K^\alpha\{f^\alpha(t)\},
\end{aligned}
\tag{4.39}
$$

we see that the total electric field

$$
\mathbf{E}(\mathbf{q}_1) = \mathbf{E}_0(\mathbf{q}_1) + \langle \mathbf{E}(\mathbf{q}_1; t) \rangle \tag{4.40}
$$

acts on a particle 1. In addition, there are the effects of close collisions.

4.2 Vlasov Equation and Landau Damping

In this section, we discuss properties of the Vlasov equation. In particular, after linearization, we find the dynamical behavior near a stationary state. Linearization around a Maxwell distribution leads to Landau damping.

To gain further insight into the collective effects in a plasma, in this section we neglect the influence of binary collisions and set

$$\boxed{K^{\alpha} = 0}\,. \tag{4.41}$$

This approximation is known as the Vlasov approximation. The Vlasov equation for a system of particles in an external electric field $\mathbf{E}_0$

$$\begin{aligned} &\frac{\partial f^{\alpha}(\mathbf{q}_1, \mathbf{v}_1; t)}{\partial t} + \mathbf{v}_1 \cdot \frac{\partial f^{\alpha}(\mathbf{q}_1, \mathbf{v}_1; t)}{\partial \mathbf{q}_1} \\ &+ \frac{e_{\alpha}}{m_{\alpha}} \left[\mathbf{v}_1 \times \mathbf{B}(\mathbf{q}_1) + \mathbf{E}_0 + \mathbf{E}(\mathbf{q}_1; t)\right] \cdot \frac{\partial f^{\alpha}(\mathbf{q}_1, \mathbf{v}_1; t)}{\partial \mathbf{v}_1} = 0, \end{aligned} \tag{4.42}$$

where in the electrostatic approximation ($\nabla \times \mathbf{E} = 0, \mathbf{B}$ externally imposed magnetic field) $\mathbf{E}$ is given by the Poisson equation

$$\nabla \cdot \mathbf{E} = \frac{1}{\varepsilon_0} \sum_{\beta=e,i} e_{\beta} \int d^3 v_1 f^{\beta}(\mathbf{q}_1, \mathbf{v}_1; t) \tag{4.43}$$

This form is identical to (4.39) and (4.40), where for the self-consistent field E we have omitted the averaging brackets. Eq. (4.43) corresponds to the relation (4.38). In the general case, in addition to (4.70), we must use the full set of Maxwell's equations

$$\nabla \cdot \mathbf{B} = 0, \tag{4.44}$$

$$\nabla \times \mathbf{E} = -\frac{\partial \mathbf{B}}{\partial t}, \tag{4.45}$$

$$\nabla \times \mathbf{B} = \mu_0 \mathbf{j} + \mu_0 \varepsilon_0 \frac{\partial \mathbf{E}}{\partial t}, \tag{4.46}$$

with

$$\mathbf{j} = \sum_{\beta=e,i} e_{\beta} \int d^3 v_1 f^{\beta}(\mathbf{q}_1, \mathbf{v}_1; t)\, \mathbf{v}_1 \tag{4.47}$$

Because of the self-consistent fields **E** and **B**, the Vlasov equation is a nonlinear integro-differential equation, whose solution is by no means trivial. As we have already indicated earlier, neglecting K^α in a plasma with many particles in the Debye zone is appropriate as long as specific questions of transport are not under discussion. To better understand this important point, we will consider other justifications for the Vlasov equation.

Example 4.1 (Klimontovich Formulation)
The (mean) probability of finding a particle, described by f^α, can be obtained not only from the Liouville equation by reduction, but also from the so-called Klimontovich equations by averaging. The Klimontovich equations describe the exact dynamics of a system via the *exact* particle density (for point particles) in Γ-space, with

$$F = \sum_{j=1}^{N} F_j \equiv \sum_{j=1}^{N} \delta(\mathbf{q} - \mathbf{q}_j(t))\delta(\mathbf{v} - \mathbf{v}_j(t)). \tag{4.48}$$

If one calculates the time derivative of F_j, the exact dynamics of each individual particle j must be taken into account:

$$\dot{\mathbf{q}}_j(t) = \mathbf{v}_j(t), \tag{4.49}$$

$$m_j \dot{\mathbf{v}}_j(t) = e_j \mathbf{E}[\mathbf{q}_j(t); t] + e_j \mathbf{v}_j \times \mathbf{B}[q_j(t); t], \tag{4.50}$$

where **E** and **B** denote the exact electric and magnetic fields experienced by particle j along its trajectory. Due to the properties of the delta function, the following equation results

$$\boxed{\partial_t F_j(\mathbf{q}, \mathbf{v}; t) + \mathbf{v} \cdot \partial_{\mathbf{q}} F_j(\mathbf{q}, \mathbf{v}; t) + \frac{e_j}{m_j}(\mathbf{E} + \mathbf{v} \times \mathbf{B}) \cdot \partial_{\mathbf{v}} F_j(\mathbf{q}, \mathbf{v}; t) = 0}\,. \tag{4.51}$$

Formally, this equation resembles the Vlasov equation (4.42), but its meaning is entirely different. Eq. (4.51) is exact and still contains the complete dynamics of the entire system. This is also reflected in the fact that the single-particle distribution function F_j, as a microscopic density, contains all fluctuations and, via Maxwell's equations, is still coupled to the exact motions of all other particles. If one wishes to move away from this overly detailed information, one averages F_j, to obtain a "smoothed" distribution function. The specific method (whether by ensemble, time, or measurement averaging) does not concern us at the moment. We write

$$F_j(\mathbf{q}, \mathbf{v}; t) = \langle F_j \rangle + \delta F_j \equiv f^\alpha + \delta f^\alpha \tag{4.52}$$

and obtain from (4.51) by averaging

$$\frac{\partial f^\alpha(\mathbf{q},\mathbf{v};t)}{\partial t}+\mathbf{v}\cdot\frac{\partial f^\alpha(\mathbf{q},\mathbf{v};t)}{\partial \mathbf{q}}$$
$$+\frac{e_\alpha}{m_\alpha}\left[\mathbf{v}\times\langle\mathbf{B}(\mathbf{q};t)\rangle+\mathbf{E}_0+\langle\mathbf{E}(\mathbf{q};t)\rangle\right]\cdot\frac{\partial f^\alpha(\mathbf{q},\mathbf{v};t)}{\partial \mathbf{v}} \tag{4.53}$$
$$=-\frac{e_\alpha}{m_\alpha}\left\langle(\delta\mathbf{E}+\mathbf{v}\times\delta\mathbf{B})\cdot\frac{\partial\delta f^\alpha}{\partial\mathbf{v}}\right\rangle.$$

If we now compare with the Vlasov equation (4.42), we find agreement on the left-hand sides. Obviously, the right-hand side of (4.53) is identical to K^α. The expression to be averaged

$$(\delta\mathbf{E}+\mathbf{v}\times\delta\mathbf{B})\cdot\frac{\partial\delta f^\alpha}{\partial\mathbf{v}} \tag{4.54}$$

represents a product of quantities that fluctuate strongly due to the discrete particle structure. The influence of such collisions on the mean behavior was estimated in the introduction as an effect of order $1/\Lambda$. Here is the complementary argument. ■

Example 4.2 (Discreteness Effects)
In a thought experiment, let us imagine each of the N_α particles being broken up into fragments. The total number of particles then increases according to the rule $N_\alpha\to\infty, m_\alpha\to 0, e_\alpha\to 0$,

$$N_\alpha e_\alpha = const,\quad e_\alpha/m_\alpha = const,\quad v_\alpha = const. \tag{4.55}$$

This procedure leaves the plasma frequencies ω_{pe} and ω_{pi} as well as the Debye lengths λ_{De} and λ_{Di} unaffected and requires $T_\alpha\to 0$ and $\Lambda\to\infty$. In every arbitrarily small but finite volume, there are now infinitely many particles ("fission products"), and according to the general results of statistics, the relative strength of the fluctuations vanishes, e.g., $\delta f^\alpha/f^\alpha$, because

$$\delta f^\alpha \sim N_\alpha^{1/2}\sim\Lambda^{1/2}. \tag{4.56}$$

On the other hand, due to Maxwell's equations

$$\delta\mathbf{E},\delta\mathbf{B}\sim e_\alpha\delta f^\alpha\sim N_\alpha^{-1}N_\alpha^{1/2}\sim N_\alpha^{-1/2}\sim\Lambda^{-1/2}, \tag{4.57}$$

so that the product (4.54), and thus K^α, remains unchanged in the thought experiment. However, if we consider the left-hand side of (4.53), its order of magnitude increases proportionally to $N_\alpha\sim\Lambda$.

This explains the statement that in a plasma, collisional effects are smaller than collective effects by a factor of $1/\Lambda$ and that the Vlasov equation, as the collisionless

Boltzmann equation, precisely accounts for the collective effects. For our real physical system, this means that in the limit $\Lambda \to \infty$ the Vlasov equation becomes exact.

■

Returning to the Vlasov equation (4.42) in the electrostatic approximation. In order to work out some decisive phenomena as simply as possible, we adopt the following approximations:

$$\mathbf{E}_0 = \mathbf{B} = 0 \tag{4.58}$$

and a smeared-out, neutralizing ion background. The last assumption is only valid in a collisionless plasma ($\Lambda \to \infty, \omega \gg \nu_{ei} \sim \omega_{pe}/\Lambda \to 0$) when high-frequency processes are considered. In high-frequency processes ($\omega \simeq \omega_{pe} \gg \omega_{pi}$) the ions, due to their large inertial masses, cannot follow the rapidly changing fields directly. However, after we have evaluated this simple model, we will be able to generalize to the corresponding models with ion dynamics relatively easily.

We linearize the (nonlinear) Vlasov equation (4.42) around a homogeneous stationary state $f_0^e(\mathbf{v})$ and, in what follows, omit the index $\alpha = e$ ($e_e = -e$),

$$\boxed{f(\mathbf{q}, \mathbf{v}; t) = f_0(\mathbf{v}) + f_1(\mathbf{q}, \mathbf{v}; t)}\,. \tag{4.59}$$

For f_0 we can assume a function of the constants of motion of a single particle. In the absence of external fields, the kinetic energy and the momentum components are possible integrals. Thus, f_0 can be an arbitrary function of v_x, v_y and v_z. Using the *linearized* equation, we want to follow the (initial) temporal evolution of a *small* perturbation.

$$\partial_t f_1 + \mathbf{v} \cdot \nabla f_1 - \frac{e}{m_e}\mathbf{E}_1 \cdot \partial_{\mathbf{v}} f_0 = 0, \tag{4.60}$$

$$\nabla \cdot \mathbf{E}_1 = -\frac{1}{\varepsilon_0} e \int d^3v f_1 \tag{4.61}$$

The linearized equations for the electron component with a smeared-out ion background are then

$$n_i = \int f^i(\mathbf{q}, \mathbf{v}; t) d^3v = \int f_0(\mathbf{v}) d^3v. \tag{4.62}$$

We now attempt a naive normal mode ansatz ($q_x = x$)

$$\boxed{f_1 \sim e^{ikx - i\omega t}}\,, \tag{4.63}$$

and thus obtain from (4.60) and (4.61)

$$f_1 = \frac{ieE_{1x}}{m_e} \frac{\partial f_0/\partial v_x}{\omega - kv_x}, \tag{4.64}$$

$$ik\varepsilon_0 E_{1x} = -e \int d^3v f_1. \tag{4.65}$$

We can combine these two equations into the "dispersion relation"

$$\boxed{1 = -\frac{\omega_{pe}^2}{k} \int_{-\infty}^{+\infty} du \frac{dg(u)/du}{\omega - ku}} \tag{4.66}$$

where

$$g(u) = \frac{1}{n_{e0}} \int dv_y dv_z f_0(u = v_x, v_y, v_z) \tag{4.67}$$

has been introduced as an abbreviation. For a given $f_0 g$ thus follows directly. Eq. (4.66) represents a relation between ω and k, albeit with some issues that still need to be resolved.

We will initially set aside the problem of resonant particles with

$$v_x = \omega/k \tag{4.68}$$

by considering only special situations with

$$\left.\frac{dg(u)}{du}\right|_{u=\omega/k} = 0 \tag{4.69}$$

Example 4.3 (Beam Ansatz)
Let us consider the simple example

$$f_0(\mathbf{v}) = \sum_{\nu=1}^{4} n_\nu\, \delta(\mathbf{v} - \mathbf{V}_\nu), \tag{4.70}$$

which leads to

$$\frac{e^2}{\varepsilon_0 m_e} \sum_{\nu=1}^{4} \frac{n_\nu}{(kU_\nu - \omega)^2} = 1 \tag{4.71}$$

with $U_\nu = V_{\nu x}$. The left-hand side of (4.71) is a positive function ($n_\nu > 0$) with four poles at $\omega = kU_\nu$, $\nu = 1, \ldots, 4$. For $\omega \to \pm\infty$ this function approaches zero. From this we can conclude that for each (fixed) k there exist at least two and at most eight different real ω values that satisfy (4.71). Now, the ansatz (4.70) is quite special: we have constructed f_0 from four beams with the "sharp" velocities $\mathbf{V}_\nu$, $\nu = 1, \ldots, 4$. Had we taken significantly more beams instead of just four, then for certain k values, correspondingly many (real) ω

values could exist. If we consider, in particular, a continuous velocity distribution f_0 as the limiting case of infinitely many beams, then for certain values of k infinitely many ω can exist. Eq. (4.66) is therefore (still) not a (well-defined) dispersion relation! In a naive normal mode analysis, we obtain many uninteresting ("false") modes. In a real system, we cannot construct such an artificial initial condition; there is always a phase mixing between the "beams".

A more detailed analysis by Landau shows that this problem, as well as that of the resonant particles, can be solved by a correct treatment of the initial value problem. ■

We therefore return once again to (4.60) and (4.61) and make only the ansatz

$$f_1 \sim f_1(\mathbf{v}; t)e^{ikx}, \tag{4.72}$$

which leads to

$$\partial_t f_1 + ikv_x f_1 - \frac{e}{m_e} E_{1x} \frac{\partial f_0}{\partial v_x} = 0, \tag{4.73}$$

$$ikE_{1x} = -\frac{1}{\varepsilon_0} e \int f_1 \, d^3v, \tag{4.74}$$

where, due to $\nabla \times \mathbf{E} \approx 0$, $E_{1y} = E_{1z} = 0$ holds.

We solve the linear system of integro-differential equations (4.73) and (4.74) by Laplace transformation:

$$F(p, \mathbf{v}) = \int_0^\infty f_1(\mathbf{v}; t)e^{-pt} dt, \tag{4.75}$$

$$f_1(\mathbf{v}; t) = \frac{1}{2\pi i} \int_{\sigma - i\infty}^{\sigma + i\infty} F(p, \mathbf{v})e^{pt} dp. \tag{4.76}$$

In the inverse transformation (4.76) in the complex p-plane, it must be noted that the integration path (parallel to the imaginary p-axis) lies to the right of all singularities of the function $F(p, \mathbf{v})$ runs (see Fig. 4.1). If we multiply both sides of (4.73) by e^{-pt} and integrate over t from 0 to ∞,, we obtain

$$(p + ikv_x)F(p, \mathbf{v}) - \frac{e}{m_e} E_{1x}(p) \frac{\partial f_0}{\partial v_x} = f_1(\mathbf{v}; 0) \tag{4.77}$$

and correspondingly from the second Eq. (4.74)

$$ikE_{1x}(p) = -\frac{1}{\varepsilon_0} e \int F(p, \mathbf{v}) d^3v. \tag{4.78}$$

With the abbreviation (4.67) and accordingly

$$G(u) = \int dv_y dv_z f_1(u = v_x, v_y, v_z; 0) \tag{4.79}$$

it then follows from (4.77) and (4.78)

$$E_{1x}(p) = -\frac{e}{ik\varepsilon_0} \frac{\displaystyle\int_{-\infty}^{+\infty} \frac{G(u)}{p + iku} du}{1 - \dfrac{\omega_{pe}^2}{k} \displaystyle\int \frac{dg}{du} \frac{du}{ku - ip}}. \tag{4.80}$$

From this we obtain $E_{1x}(t)$ by inverse transformation

$$\boxed{E_{1x}(t) = \frac{1}{2\pi i} \int_{\tilde{\sigma} - i\infty}^{\tilde{\sigma} + i\infty} E_{1x}(p)\, e^{pt}\, dp}. \tag{4.81}$$

The contour must be chosen such that $\tilde{\sigma}$ is greater than all values of Re p_ν, where p_ν denotes the singularities of $E_{1x}(p)$. According to (4.80), the singularities of $E_{1x}(p)$ are generally determined by the zeros of the denominator on the right-hand side of (4.80). For the following discussion, we assume stable distribution functions for which Re $p_\nu < 0$ holds, so that $\tilde{\sigma} = 0$ can be chosen.

The inverse transformation (4.81) can be carried out relatively easily using methods from complex analysis. For this, we analytically continue $E_{1x}(p)$ into the entire p-plane.

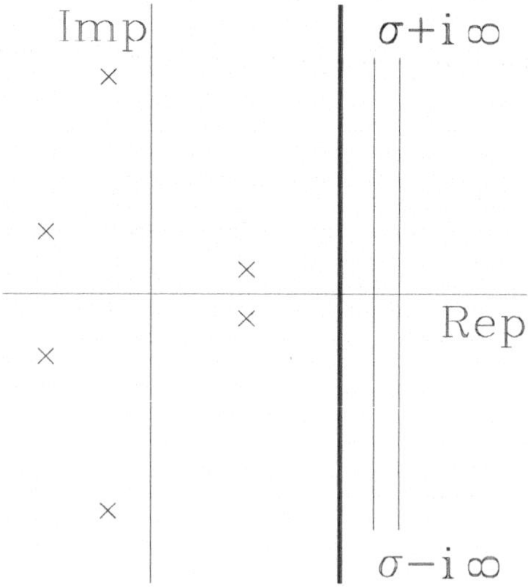

Fig. 4.1 Contour Re $p = \sigma$ for the inverse transformation (4.76) in the complex p-plane. The crosses indicate the singularities of $F(p, \mathbf{v})$

Note that initially $E_{1x}(p)$ is only defined for Re $p \geq 0$. Accordingly, we first assume that the function in the denominator on the right-hand side of (4.80), namely

$$\boxed{H(p) := \frac{\omega_{pe}^2}{k^2}\int_{-\infty}^{+\infty}\frac{dg}{du}\frac{du}{u-\frac{ip}{k}}}\,, \tag{4.82}$$

is defined for Re $p \geq 0$. If we examine this integral more closely, we can imagine $g(u)$ as being analytically continued from the real u-axis into the complex u-plane; the integrand then has a pole at $u = ip/k$ in the upper half of the u-plane. If we now wish to analytically continue $E_{1x}(p)$ into the region Re $p < 0$, we must do the same with $H(p)$ and ensure that no discontinuities arise when passing from Re $p > 0$ to Re $p < 0$. We achieve this by deforming the integration path in the complex u-plane according to Landau (see Fig. 4.2): the pole at $u = ip/k$ must not cross the integration contour. If we use the Plemelj formula

$$\frac{1}{x \pm i0} = \mathcal{P}\frac{1}{x} \mp i\pi\,\delta(x) \tag{4.83}$$

and the integration formula

$$\int_{-\infty}^{+\infty}\frac{f(x)}{x \pm i0}dx = \lim_{\varepsilon\to 0}\left[\int_{|x|\geq\varepsilon}\frac{f(x)}{x}dx + \int_{C_\pm}\frac{f(z)}{z}dz\right], \tag{4.84}$$

where C_+ (or C_-)) is a small semicircular arc of radius ε in the upper (lower) complex z-half-plane, and start from the definition (4.82) for Re $p > 0$, these considerations yield

$$H(p) = \frac{\omega_{pe}^2}{k^2}\int_{-\infty}^{+\infty}\frac{dg/du}{u-\frac{ip}{k}}du \quad \text{for Re } p > 0\,, \tag{4.85}$$

$$H(p) = \frac{\omega_{pe}^2}{k^2}\left[\mathcal{P}\int_{-\infty}^{+\infty}\frac{dg/du}{u-\frac{ip}{k}}du + i\pi\frac{dg}{du}\bigg|_{\frac{ip}{k}}\right] \text{for Re } p = 0\,, \tag{4.86}$$

$$H(p) = \frac{\omega_{pe}^2}{k^2}\left[\int_{-\infty}^{+\infty}\frac{dg/du}{u-\frac{ip}{k}}du + 2\pi i\frac{dg}{du}\bigg|_{\frac{ip}{k}}\right] \text{for Re } p < 0\,. \tag{4.87}$$

With this clarification, we can return to the evaluation of formula (4.60) and also allow $\tilde{\sigma} > 0$, provided unstable distribution functions are present. The definitions (4.85)–(4.87) then generalize in an obvious way.

We choose a path as shown in Fig. 4.3; it is obtained from the contour of $\tilde{\sigma} - i\infty$ to $\tilde{\sigma} + i\infty$ by shifting in the direction of negative real parts of p by $\tilde{\sigma} + s$, but only in the region $p_- \leq$ Im $p \leq p_+$ for $|p_-|, |p_+| \to \infty$. We then obtain from the poles p_j, with

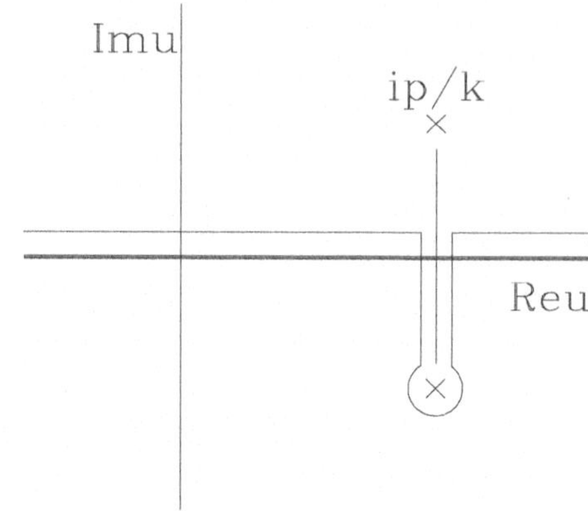

Fig. 4.2 Deformation of the integration path in the complex u-plane according to Landau

$$\boxed{H(p_j) = 1}, \tag{4.88}$$

the contribution

$$E_{1x}(t) = \sum_j \exp(p_j t) \text{ Res } (p_j). \tag{4.89}$$

For $t \to \infty$ the summand originating from the pole with the largest real part dominates, and all other contributions (path segments I, II, and III in Fig. 4.3) can be neglected. The contributions from path segments I are proportional to $\exp(-st)$. For path segments II, it should be noted that, for $|p| \to \infty$, $|E_{1x}(p)| \sim 1/|p|$ becomes very small. The integrands along paths III oscillate very strongly and the integrals vanish for $|p_+|, |p_-| \to \infty$. All this together leads, in the limit $t \to \infty$ to the result

$$E_{1x}(t) \approx \text{ Res } (p) \exp(pt), \tag{4.90}$$

where p is the pole with the largest real part. Thus, for the calculation of a collisionless damping rate, the solution of the dispersion relation (4.66), or

$$\boxed{\varepsilon(k, \omega = ip) := 1 - \frac{\omega_{pe}^2}{k^2} \int \frac{dg}{du} \frac{du}{u - \frac{\omega}{k}} = 0} \tag{4.91}$$

with the Landau prescription, plays the decisive role.

Example 4.4 (Landau damping for small rates)
Before we address this problem in detail, let us present a simple calculation for weak damping and large phase velocities. By weak damping we mean

$$\gamma/\omega_r \ll 1, \tag{4.92}$$

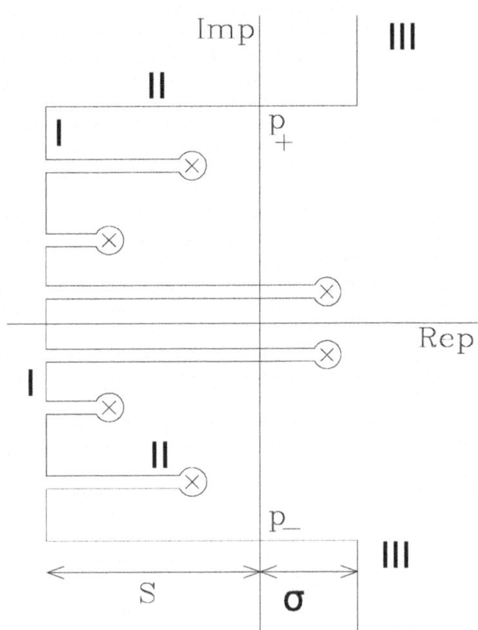

Fig. 4.3 Deformation of the contour $\Re p = \tilde{\sigma}$ for the evaluation of Landau damping

where $\gamma = p_r =$ Im ω has been set and $\omega_r =$ Re ω denotes. We can then perform a Taylor expansion in (4.91) about the point $\omega = \omega_r$. Furthermore, since Im $\varepsilon \sim \gamma$ for small γ, the first-order Taylor expansion in γ reads as

$$\text{Re } \varepsilon(k, \omega_r) + i \text{ Im } \varepsilon(k, \omega_r) + i\gamma \left. \frac{\partial \text{ Re } \varepsilon(k, \omega)}{\partial \omega} \right|_{\omega_r} = 0 \,, \tag{4.93}$$

from which

$$\text{Re } \varepsilon(k, \omega_r) \approx 0 \tag{4.94}$$

and

$$\gamma = -\frac{\text{Im } \varepsilon(k, \omega_r)}{\partial \text{ Re } \varepsilon(k, \omega)/\partial\omega|_{\omega_r}} \tag{4.95}$$

follow. Let us first discuss $\varepsilon(k, \omega_r)$, for which we can refer back to expression (4.86). We obtain

$$\varepsilon(k, \omega_r) = 1 - \frac{\omega_{pe}^2}{k^2} \mathcal{P} \int_{-\infty}^{+\infty} \frac{dg/du}{u - \frac{\omega_r}{k}} du - i\pi \frac{\omega_{pe}^2}{k^2} \left. \frac{dg}{du} \right|_{u=\frac{\omega_r}{k}} , \tag{4.96}$$

from which

$$\text{Re } \varepsilon(k, \omega_r) = 1 - \frac{\omega_{pe}^2}{k^2} \mathcal{P} \int_{-\infty}^{+\infty} \frac{dg/du}{u - \frac{\omega_r}{k}} du \tag{4.97}$$

and

$$\text{Im } \varepsilon(k, \omega_r) = -\pi \frac{\omega_{pe}^2}{k^2} \frac{dg}{du}\bigg|_{u=\frac{\omega_r}{k}} \tag{4.98}$$

follow. The approximate calculation of $\partial \text{ Re } \varepsilon/\partial\omega|_{\omega_r}$ is carried out according to the formula

$$\frac{\partial \text{ Re } \varepsilon(k, \omega)}{\partial \omega}\bigg|_{\omega_r} = \frac{\partial \text{ Re } \varepsilon(k, \omega_r)}{\partial \omega_r} = -\frac{\omega_{pe}^2}{k^2} \frac{\partial}{\partial \omega_r} \mathcal{P} \int_{-\infty}^{+\infty} \frac{dg/du}{u - \frac{\omega_r}{k}} du \tag{4.99}$$

We can only perform concrete evaluations after specifying the distribution functions f_0 or g. Without, for example, assuming a Maxwell distribution

$$f_0 = \frac{n_0}{(2\pi)^{3/2} v_{te}^3} \exp\left[-(v_x^2 + v_y^2 + v_z^2)/2v_{te}^2\right] \tag{4.100}$$

or

$$\boxed{g(u) = \frac{1}{(2\pi)^{1/2} v_{te}} \exp\left[-u^2/2v_{te}^2\right]} \tag{4.101}$$

only a small step can be taken. Namely, if we move into the regime of large phase velocities ω_r/k (this statement means that we only consider those solutions ω_r and k for which the quotient ω_r/k corresponds to a large velocity, for which we can already set $g(u = \omega_r/k)$ to almost zero), then (4.97) yields after partial integration

$$\begin{aligned} \text{Re } \varepsilon(k, \omega_r) &\approx 1 - \frac{\omega_{pe}^2}{k^2} \int_{-\infty}^{+\infty} \frac{g(u)}{\left(u - \frac{\omega_r}{k}\right)^2} du \\ &\approx 1 - \frac{\omega_{pe}^2}{\omega_r^2} \int_{-\infty}^{+\infty} du\, g(u) \left[1 + \frac{2uk}{\omega_r} + \frac{3u^2k^2}{\omega_r^2} + \ldots\right] \\ &\approx 1 - \frac{\omega_{pe}^2}{\omega_r^2} - \frac{3k^2 v_{te}^2 \omega_{pe}^2}{\omega_r^4}. \end{aligned} \tag{4.102}$$

Here we have assumed

$$\int du\, u\, g(u) = 0 \tag{4.103}$$

If we then use

$$\omega_r^2 \approx \omega_{pe}^2 + 3k^2 v_{te}^2 \tag{4.104}$$

as a solution of (4.94) and substitute it into (4.95)

$$\left.\frac{\partial \text{ Re } \varepsilon}{\partial \omega}\right|_{\omega_r} \approx \frac{2}{\omega_{pe}} \tag{4.105}$$

it follows that

$$\gamma \approx \frac{\pi \omega_{pe}^3}{2k^2} \left.\frac{dg(u)}{du}\right|_{u=\frac{\omega_r}{k}} . \tag{4.106}$$

With the Maxwell distribution (4.101), this yields

$$\boxed{\gamma \approx -\omega_{pe}\left(\frac{\pi}{8}\right)^{1/2} (k\lambda_{De})^{-3} \exp\left[-\frac{1}{2}(k\lambda_{De})^{-2} - \frac{3}{2}\right]}. \tag{4.107}$$

■

The dependence of the damping rate $|\gamma|/\omega_{pe}$ on $k\lambda_{De}$ is shown in Fig. 4.4. We observe a maximum at $k\lambda_{De} \approx 0.5$, but should keep in mind when interpreting this that the previous derivation is only valid for $|\gamma/\omega_{pe}| \ll 1$ and $k\lambda_{De} \ll 1$.

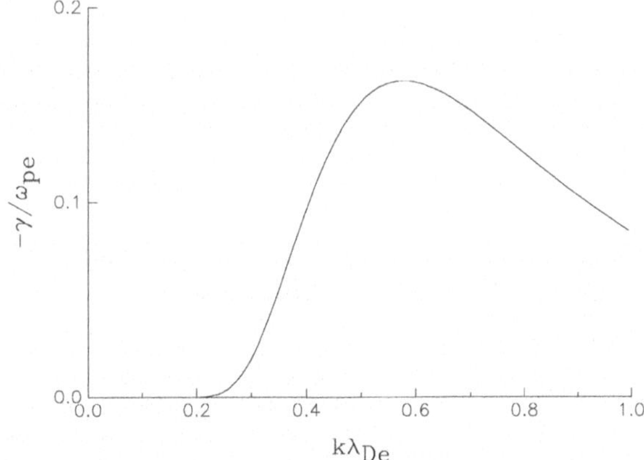

Fig. 4.4 Landau damping rate γ [in ω_{pe}] as a function of $k\lambda_{De}$

The result (4.107), obtained from the Maxwell distribution (4.101), suggests the interpretation that particles with velocities greater than the phase velocity in the wave are decelerated, and conversely, particles with velocities less than the phase velocity in the wave are accelerated. A monotonically decreasing distribution function ($g'(u) < 0$) represents more particles with lower than with higher velocities (relative to a given phase velocity), and thus leads to a net energy loss of the wave (due to a net energy gain of the particles).

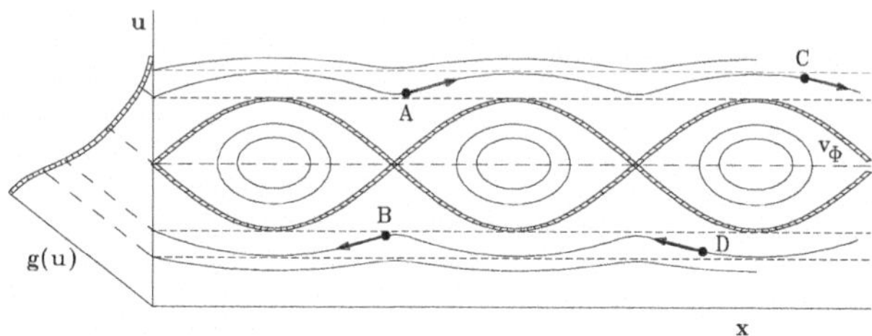

Fig. 4.5 Trajectory in the x, u phase space for the discussion of Landau damping

However, this interpretation is not entirely correct, as can be seen from a phase portrait. In Fig. 4.5 we have depicted, for a one-dimensional situation, "trajectories" in the "x, u phase space." For the potential energy (at the time $t = 0$), a distribution proportional to $\cos(kx)$ is assumed. It then holds (in the reference frame moving with the phase velocity v_ϕ)

$$E = \frac{1}{2} m_e u^2 + a \cos(kx), \tag{4.108}$$

and different curves in Fig. 4.5 correspond to different parameter values of E. The wave (and thus the entire structure) moves to the right with the phase velocity

$$v_\phi = \frac{\omega}{k} \tag{4.109}$$

The symmetry line in Fig. 4.5 is $u = v_\phi$, and one should keep in mind that, due to the monotonically decreasing distribution function $g(u)$, fewer particles have a higher than a lower velocity, i.e.,

$$g(u_1) > g(v_\phi) > g(u_2) \quad \text{for} \quad u_1 < v_\phi < u_2. \tag{4.110}$$

Let us now consider typical constellations (e.g., A, B, C, and D in Fig. 4.5), we see that even particles with $u > v_\phi$ (e.g., at A) gain kinetic energy in the initial phase! Due to $g(u_D) > g(u_B)$ and $g(u_A) > g(u_C)$, there is, overall, an effective gain in kinetic energy in the initial phase.

Example 4.5 (Landau Damping from Energy Balance)
To estimate this effect somewhat more quantitatively, we determine from the linearized force balance

$$m_e\left(\frac{\partial u_1}{\partial t}+u\frac{\partial u_1}{\partial x}\right)=-eE_1\sin(kx-\omega t) \tag{4.111}$$

and the linearized continuity equation

$$\frac{\partial n_1}{\partial t}+u\frac{\partial n_1}{\partial x}=-n_u\frac{\partial u_1}{\partial x} \tag{4.112}$$

the perturbations n_1 and u_1. In this calculation, we are to understand that the entire particle arrangement is conceived as being composed of beams with velocities u, with the corresponding particle densities n_u,. The probability distribution of the beams is determined by $g(u)$. With the appropriate solutions

$$u_1=-\frac{eE_1}{m_e}\,\frac{\cos(kx-\omega t)-\cos(kx-kut)}{\omega-ku}, \tag{4.113}$$

$$n_1=-n_u\frac{eE_1k}{m_e}\,\frac{\cos(kx-\omega t)-\cos(kx-kut)-(\omega-ku)t\sin(kx-kut)}{(\omega-ku)^2} \tag{4.114}$$

the force density on a unit volume of a beam can be calculated as

$$F_u\approx-eE_1\sin(kx-\omega t)[n_u+n_1], \tag{4.115}$$

The change in energy density per unit time

$$\frac{dW}{dt}\approx(u+u_1)F_u \tag{4.116}$$

is, after spatial averaging, given by

$$\begin{aligned}\left\langle\frac{dW}{dt}\right\rangle_u\approx\frac{e^2E_1^2}{2m_e}n_u\Bigg[&\frac{\sin(\omega t-kut)}{\omega-ku}\\&+ku\frac{\sin(\omega t-kut)-(\omega-ku)t\cos(\omega t-kut)}{(\omega-ku)^2}\Bigg].\end{aligned} \tag{4.117}$$

We then sum over all beam contributions,

$$\begin{aligned}\sum_u\left\langle\frac{dW}{dt}\right\rangle_u&\mathrel{\widehat{=}}n_0\int\frac{g(u)}{n_u}\left\langle\frac{dW}{dt}\right\rangle_u du\\&\approx\frac{\varepsilon_0}{2}E_1^2\,\omega_{pe}^2\int_{-\infty}^{+\infty}g(u)\frac{d}{du}\left[u\frac{\sin(\omega t-kut)}{\omega-ku}\right]du.\end{aligned} \tag{4.118}$$

In the case of electrostatic plasma oscillations, the total wave energy density

$$W_w=\frac{\varepsilon_0}{2}E_1^2 \tag{4.119}$$

is composed in equal parts of the electrostatic field energy density and the kinetic oscillation energy density (see the next section). Thus, (4.119) can also be written in the form

$$\frac{dW_w}{dt} = -W_w\,\omega_{pe}^2 \int_{-\infty}^{+\infty} g(u)\frac{d}{du}\left[u\frac{\sin(\omega t - kut)}{\omega - ku}\right]du \tag{4.120}$$

After partial integration and the transition

$$\delta\Big(u - \frac{\omega}{k}\Big) = \frac{k}{\pi}\lim_{t\to\infty}\frac{\sin(\omega - ku)t}{\omega - ku} \tag{4.121}$$

it follows that

$$\frac{dW_w}{dt} = W_w\pi\frac{\omega_{pe}^3}{k^2}g'\Big(\frac{\omega}{k}\Big) \approx 2\gamma W_w. \tag{4.122}$$

The last expression follows because of $W_w \sim E_1^2$ and $E_1 \sim \exp(\gamma t)$. The damping rate thus calculated

$$\boxed{\gamma \approx \frac{\pi}{2}\frac{\omega_{pe}^3}{k^2}g'\Big(u = \frac{\omega}{k}\Big)} \tag{4.123}$$

agrees with the expression (4.107). ■

Despite these remarkable agreements, this does not yet exhaust the physics of Landau damping. In all further interpretations, phase mixing plays a prominent role. This aspect, which we will not discuss in detail here, was studied in depth especially by van Kampen.

4.3 Z-Function and Dispersive Behavior

In the previous section, we focused on demonstrating collective damping effects. We derived the dispersion relation (4.91) exactly and then evaluated it approximately. In this section, we will improve upon the previous approximate evaluation by introducing the plasma dispersion function, which provides a clear and universally applicable calculation method.

We assume a Maxwell distribution for g. From the previous discussion, we already know that solutions of $\varepsilon = 0$, with ε according to (4.91) exist, for which

$$\mathrm{Re}\ p < 0 \tag{4.124}$$

holds. In this section, we restrict ourselves to this branch of solutions. Using the Landau prescription for the integration contour, we write ($\theta_e = k_B T_e$)

$$\begin{aligned}\frac{k^2}{\omega_{pe}^2} = -\frac{(m_e/\theta_e)^{3/2}}{(2\pi)^{1/2}} \Bigg\{ \int_{-\infty}^{+\infty} \frac{u \exp(-m_e u^2/2\theta_e)}{u - ip/k} du \\ + 2\pi i \frac{ip}{k} \exp\left[-\frac{m_e}{2\theta_e} \left(\frac{ip}{k} \right)^2 \right] \Bigg\}.\end{aligned} \tag{4.125}$$

This expression can be written more simply in the dimensionless variables

$$t := u \left(\frac{m_e}{2\theta_e} \right)^{1/2}, \qquad \zeta := \frac{ip}{k} \left(\frac{m_e}{2\theta_e} \right)^{1/2} \tag{4.126}$$

namely as

$$\boxed{-k^2 \lambda_{De}^2 = \frac{1}{\sqrt{\pi}} \int_{-\infty}^{+\infty} \frac{t e^{-t^2}}{t - \zeta} dt + 2i\sqrt{\pi}\, \zeta e^{-\zeta^2}}\,. \tag{4.127}$$

The integrand in the first term on the right-hand side of this equation can be rewritten as follows:

$$\frac{t}{t - \zeta} = \frac{t(t + \zeta)}{t^2 - \zeta^2} \to \frac{t^2}{t^2 - \zeta^2} = \frac{t^2 - \zeta^2 + \zeta^2}{t^2 - \zeta^2}, \tag{4.128}$$

where the odd terms in t vanish upon integration. After some calculation steps, we obtain

$$\begin{aligned}-k^2 \lambda_{De}^2 = 1 + \frac{1}{\sqrt{\pi}} \zeta^2 e^{-\zeta^2} \left\{ \int_{-\infty}^{+\infty} \frac{e^{-(t^2 - \zeta^2)}}{t^2 - \zeta^2} dt + \frac{2\pi i}{\zeta} \right\} \\ \equiv J(\zeta).\end{aligned} \tag{4.129}$$

With the representation

$$\frac{e^{-(t^2 - \zeta^2)}}{t^2 - \zeta^2} = -\int_0^1 e^{-(t^2 - \zeta^2)s} ds + \frac{1}{t^2 - \zeta^2} \tag{4.130}$$

and the formulas

$$\int_{-\infty}^{+\infty} \frac{1}{t^2 - \zeta^2} dt = -\frac{i\pi}{\zeta} \tag{4.131}$$

as well as

$$\int_{-\infty}^{+\infty} e^{-(t^2-\zeta^2)s} dt = \sqrt{\pi}\frac{1}{\sqrt{s}} e^{\zeta^2 s} \tag{4.132}$$

further simplifications can be achieved. With $\nu = i\zeta\sqrt{s}$ one finally obtains

$$\begin{aligned} J(\zeta) &= 1 + 2i\zeta e^{-\zeta^2}\left\{\frac{1}{2}\sqrt{\pi} + \int_0^{i\zeta} e^{-\nu^2} d\nu\right\} \\ &= 1 + 2i\zeta e^{-\zeta^2}\int_{-\infty}^{i\zeta} e^{-\nu^2}\, d\nu. \end{aligned} \tag{4.133}$$

In the literature, the abbreviation

$$Z(\zeta) := 2ie^{-\zeta^2}\int_{-\infty}^{i\zeta} e^{-\nu^2} d\nu \tag{4.134}$$

is common. Using it, the relation (4.129) can be written as

$$k^2\lambda_{De}^2 = -[1 + \zeta Z(\zeta)] \equiv \frac{1}{2}Z'(\zeta). \tag{4.135}$$

For the sake of completeness, we now establish a connection with the G function, which is also commonly used in the literature. We have

$$\boxed{G(\zeta) \equiv Z(-\zeta) = \frac{1}{\sqrt{\pi}}\int_{-\infty}^{+\infty} \frac{e^{-p^2}}{\zeta - p} dp}\,. \tag{4.136}$$

In the last relation, the integration path runs along the real p axis above the pole at $\text{Im}\ \zeta < 0$. For $\text{Im}\ \zeta > 0$ the integration path must be deformed as explained in detail in the previous section.

Example 4.6 (Relationship between Z- and G-function)
To prove (4.136), we use the following relations:

$$G(z) = \frac{1}{\sqrt{\pi}}\int \frac{e^{-\xi^2}}{z - \xi} d\xi, \tag{4.137}$$

$$zG(z) - 1 = \frac{1}{\sqrt{\pi}}\int \frac{\xi e^{-\xi^2}}{z - \xi} d\xi, \tag{4.138}$$

$$G'(z) = -\frac{2}{\sqrt{\pi}} \int \frac{\xi e^{-\xi^2}}{z - \xi} d\xi, \tag{4.139}$$

from which the differential equation

$$\frac{1}{2}G' + zG - 1 = 0 \tag{4.140}$$

follows. The solution to the homogeneous equation is immediately found to be

$$G = ce^{-z^2}. \tag{4.141}$$

Variation of the constant c leads to the particular solution

$$G(z) = 2e^{-z^2} \int_0^z e^{p^2} dp \tag{4.142}$$

of (4.140). If we combine from (4.141) and (4.142) the solution that satisfies $G(z = 0) = i\sqrt{\pi}$, then we have

$$\begin{aligned} G(z) &= \left[2 \int_0^z e^{p^2} dp + i\sqrt{\pi} \right] e^{-z^2} \\ &= 2e^{-z^2} \int_{-i\infty}^z e^{p^2} dp. \end{aligned} \tag{4.143}$$

With $p = i\nu$ and $z = -\zeta$ it finally follows

$$G(-\zeta) = 2ie^{-\zeta^2} \int_{-\infty}^{i\zeta} e^{-\nu^2} d\nu = Z(\zeta), \tag{4.144}$$

i.e., the missing relationship (4.136). ■

Note that the *Z*-function defined in this way or according to (4.136) coincides, for example, with the one given by D.L. Book in the "NRL Plasma Formulary." For many applications, it is not necessary to have complete knowledge of the entire *G*-function or *Z*-function. For small arguments ($|z| \ll 1$) the following holds

$$G(z) \approx i\sqrt{\pi}\, e^{-z^2} + 2z\left(1 - \frac{2}{3}z^2 + \frac{4}{15}z^4 - \dots\right), \tag{4.145}$$

whereas for large arguments ($|z| \gg 1$)

$$G(z) \approx i\sigma\sqrt{\pi}\, e^{-z^2} + \frac{1}{z}\left(1 + \frac{1}{2z^2} + \frac{3}{4z^4} + \dots\right) \tag{4.146}$$

can be used as an approximation. Here, σ for $z = x + iy$ is introduced as an abbreviation for

$$\sigma = \begin{cases} 0 \text{ for } \ y \ > 1/x, \\ 1 \text{ for } |y| < 1/x, \\ 2 \text{ for } -y > 1/x \end{cases} \tag{4.147}$$

There are tables in which the G- or Z-function is listed for all relevant argument ranges. However, modern personal computers allow for the direct calculation of all desired parameters using simple programs. In Figs. 4.6, 4.7, and 4.8 we provide three examples.

Now that we have provided the necessary mathematical tools, we can proceed to address concrete physical problems. Within the framework of linear (electrostatic) Vlasov theory, the "density response" $j = e, i)$ for the species δn_j due to an electric field E [cf. (4.64)] is given by

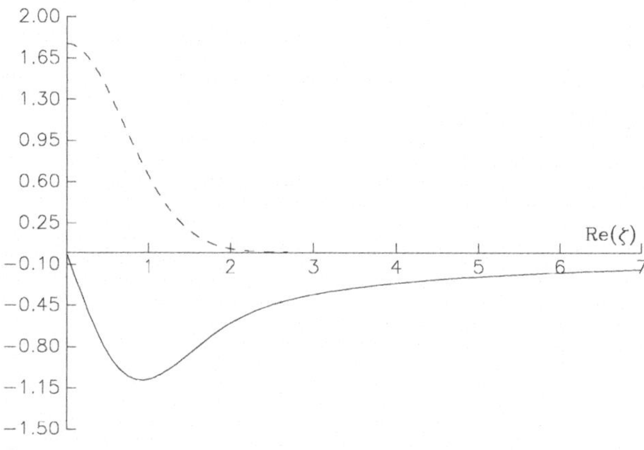

Fig. 4.6 Real part (solid line) and imaginary part (dashed line) of the Z-function for Im $\zeta = 0$

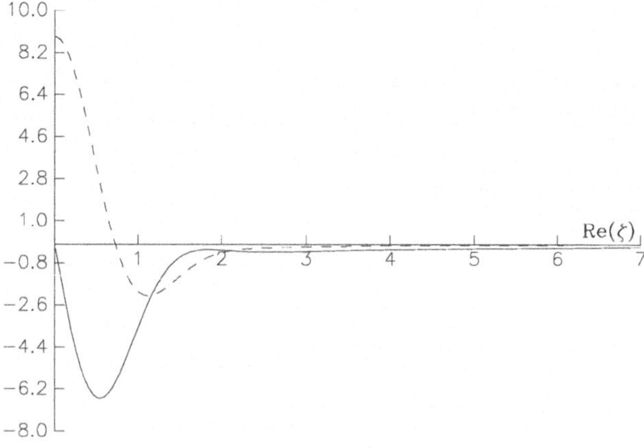

Fig. 4.7 Real part (solid line) and imaginary part (dashed line) of the Z-function for Im $\zeta = -1$

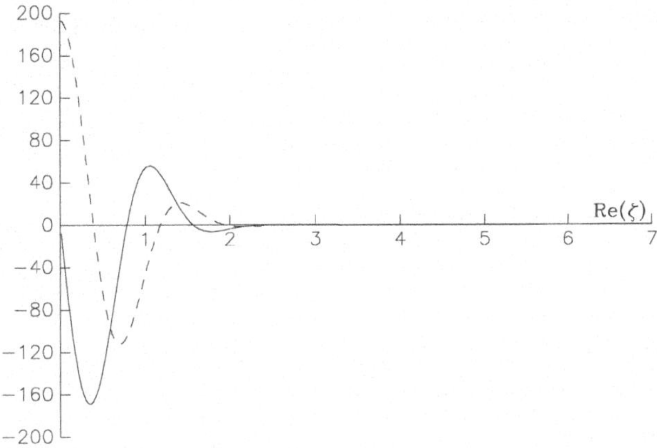

Fig. 4.8 Real part (solid line) and imaginary part (dashed line) of the Z-function for Im $\zeta = -2$

$$\delta n_j = \frac{iq_j E n_{j0}}{2km_j v_{tj}^2} Z'(\zeta_j), \tag{4.148}$$

where the prime (′) Differentiation of the Z-function with respect to the argument

$$\zeta_j = \frac{ip}{k}\left[\frac{m_j}{2\theta_j}\right]^{1/2} \tag{4.149}$$

means, and $v_{tj} = (\theta_j/m_j)^{1/2}$ is. If we insert this "response" into the Poisson equation, it follows from

$$\nabla \cdot \mathbf{E} = \frac{1}{\varepsilon_0}\sum_j q_j \delta n_j \tag{4.150}$$

for an electron-ion plasma

$$\boxed{k^2 = \frac{\omega_{pe}^2}{2v_{te}^2} Z'(\zeta_e) + \frac{\omega_{pi}^2}{2v_{ti}^2} Z'(\zeta_i)} \tag{4.151}$$

as the dispersion relation [cf. (4.91)]. We immediately see that for high-frequency waves ($\omega \geq \omega_{pe}$) the ion contribution can be neglected to a good approximation ($\omega_{pi} \approx 0$) and the dispersion relation becomes

$$k^2 \lambda_{De}^2 = \frac{1}{2} Z'(\zeta_e) \tag{4.152}$$

Let us first consider the simplest case of large phase velocities, i.e., small k values with

$$\frac{\omega}{k} \gg v_{te}, \tag{4.153}$$

then, for the calculation of the real part of the frequencies, damping is negligible and we obtain from (4.152)

$$k^2\lambda_{De}^2 \approx \frac{k^2 v_{te}^2}{\omega^2} + 3\frac{k^4 v_{te}^4}{\omega^4}, \tag{4.154}$$

with the approximate solution

$$\boxed{\omega^2 \approx \omega_{pe}^2\left(1 + 3k^2\lambda_{De}^2\right)}\,. \tag{4.155}$$

For small values of $k\lambda_{De}$ the calculation of the damping decrement proceeds similarly to what was shown in the previous section. We can use the asymptotic expansion (4.145) and start from

$$k^2\lambda_{De}^2 \approx \frac{k^2 v_{te}^2}{\omega^2} - i\sqrt{\pi}\,\frac{\omega}{k}\,\frac{1}{\sqrt{2}}\,\frac{1}{v_{te}}e^{-\omega^2/k^2 2v_{te}^2} \tag{4.156}$$

With (4.155) for the real part of ω, Re $\omega \approx \omega_{pe}$, and $\omega =$ Re $\omega + i\gamma$ the electron Landau damping follows to lowest order

$$\boxed{\frac{\gamma}{\omega_{pe}} \approx -\left(\frac{\pi}{8}\right)^{1/2}(k\lambda_{De})^{-3}\exp\left[-\frac{1}{2}(k\lambda_{De})^{-2} - \frac{3}{2}\right]}\,, \tag{4.157}$$

in complete agreement with the expression (4.107). For larger values of $k\lambda_{De}$ ($k\lambda_{De} \geq 0.2$) however, the approximate expression (4.157) becomes inaccurate. In fact, the solution of (4.152) continues to increase monotonically, with a characteristic value of $\gamma/\omega_{pe} \approx 1$ for $k\lambda_{De} \approx 1$. Thus, the behavior shown in Fig. 4.4 is only valid for $k\lambda_{De} \ll 1$.

Another aspect is also interesting in the solution of (4.152). The equation is not linear in either the real part or the imaginary part of ω. Accordingly, we obtain several solution branches. However, the other branches, besides (4.155), are so strongly damped that we cannot attribute any physical significance to them.

In the low-frequency range, however, there exists another relevant solution branch. To calculate it, we return to (4.151) and investigate solutions for

$$\zeta_e \ll 1, \tag{4.158}$$

so that we can set

$$Z'(\zeta_e) \approx -2 \tag{4.159}$$

Then, the dispersion relation (4.151) is approximately

$$1 + k^2\lambda_{De}^2 \approx \lambda_{De}^2 \frac{\omega_{pi}^2}{2v_{ti}^2} Z'(\zeta_i). \tag{4.160}$$

Analogous to the calculation of electron plasma oscillations, we first assume $\zeta_i \gg 1$ for small k in order to allow for an analytical solution. Using the expansion (4.146), we easily find for the real part

$$1 + k^2\lambda_{De}^2 \approx \lambda_{De}^2 \omega_{pi}^2 \left(\frac{k^2}{\omega^2} + 3\frac{k^4}{\omega^4} v_{ti}^2 \right) \tag{4.161}$$

with the approximate solution

$$\boxed{\omega^2 \approx \frac{k^2 c_s^2}{1 + k^2\lambda_{De}^2}} \tag{4.162}$$

for ion acoustic waves. Here,

$$c_s = [k_B(T_e + 3T_i)/m_i]^{1/2} \tag{4.163}$$

is referred to as the ion sound speed. For $T_e \gg T_i$ we obtain the ion sound speed at the electron temperature

$$\boxed{c_s \approx (k_B T_e/m_i)^{1/2}}\,. \tag{4.164}$$

The calculation of the damping again requires a more precise consideration of the imaginary part. Instead of (4.161), we use

$$\frac{k^2 v_{ti}^2}{\omega^2}\left(1 + 3\frac{T_i}{T_e}\right) - i\left(\frac{\pi}{2}\right)^{1/2} \frac{\omega}{k}\frac{1}{v_{ti}} e^{-\omega^2/2k^2v_{ti}^2} \approx \frac{T_i}{T_e}\left(1 + k^2\lambda_{De}^2\right). \tag{4.165}$$

If we again set $\omega = \text{Re}\ \omega + i\gamma$ and use for the real part of ω (4.162), then a calculation analogous to that for electron Landau damping yields

$$\boxed{\frac{\gamma}{\text{Re}\ \omega} \approx -\left(\frac{\pi}{8}\right)^{1/2} \frac{T_e}{T_i}\left(3 + \frac{T_e}{T_i}\right) e^{-(3+T_e/T_i)/2}}\,. \tag{4.166}$$

This formula applies for $k\lambda_{De} \ll 1$ and $T_e/T_i \gg 1$; the last condition is ensured by $\zeta_i \gg 1$, under which we were able to carry out the analytical calculation. A better numerical solution of (4.160) is therefore necessary for the quite frequently occurring situation $1 \leq T_e/T_i \leq 10$. It shows that the relative damping rate (4.166) also increases monotonically with T_i/T_e in this regime.

We conclude this section with some simple remarks and interpretations of dispersion in plasmas.

Example 4.7 (Susceptibilities)
The dispersion relation

$$\varepsilon = 0 \tag{4.167}$$

with

$$\varepsilon = 1 - \sum_j \frac{\omega_{pj}^2}{k^2} \mathbf{k} \cdot \int \frac{\partial f_j / \partial \mathbf{v}}{\mathbf{k} \cdot \mathbf{v} - \omega} d^3 v \tag{4.168}$$

is deliberately written with ε as the dielectric constant; f_j here is the single-particle distribution function for a particle of species j, normalized to one (when integrating over velocities). From electrodynamics, we know that it is often useful to introduce susceptibilities χ_j, where

$$\boxed{\varepsilon = 1 + \sum_j \chi_j} \tag{4.169}$$

applies. By comparison with (4.158), it follows directly

$$\chi_j = -\frac{\omega_{pj}^2}{k^2} \mathbf{k} \cdot \int \frac{\partial f_j / \partial \mathbf{v}}{\mathbf{k} \cdot \mathbf{v} - \omega} d^3 v = -\frac{\omega_{pj}^2}{2 v_{tj}^2 k^2} Z'(\zeta_j); \tag{4.170}$$

where again $\mathbf{k} = k\hat{x}$ and $v_x = u$. ■

Example 4.8 (Polarization)
After these mathematical transformations, a simple physical interpretation may be appropriate, which we provide for electron oscillations. An electron oscillates in an electric field

$$\mathbf{E} = \mathbf{E}_0 e^{i(\mathbf{k} \cdot \mathbf{r} - \omega t)} ; \tag{4.171}$$

from the equation of motion, we obtain for the displacements $\delta \mathbf{r}$

$$\begin{aligned} \delta \ddot{\mathbf{r}} &= \frac{q_e}{m_e} \mathbf{E}_0 \, e^{-i(\omega - \mathbf{k} \cdot \mathbf{v})t} \, e^{i\mathbf{k} \cdot (\mathbf{r}_0 + \delta \mathbf{r})} \\ &\approx \frac{q_e}{m_e} \mathbf{E}_0 \, e^{i[\mathbf{k} \cdot \mathbf{r}_0 - (\omega - \mathbf{k} \cdot \mathbf{v})t]} . \end{aligned} \tag{4.172}$$

Here, the Doppler shift $\omega - \mathbf{k} \cdot \mathbf{v}$ appears due to $\mathbf{r} = \mathbf{r}_0 + \mathbf{v}t + \delta \mathbf{r}$. From the approximate solution

$$\delta \mathbf{r} \approx -\frac{q_e}{m_e} \frac{\mathbf{E}}{(\omega - \mathbf{k} \cdot \mathbf{v})^2} \tag{4.173}$$

we can calculate the dipole moment

$$\mathbf{p} = q_e \delta \mathbf{r} \approx -\frac{q_e^2 \mathbf{E}}{m_e(\omega - \mathbf{k} \cdot \mathbf{v})^2} \tag{4.174}$$

and, by summation, the polarization

$$\boxed{\mathbf{P} = -\frac{n_{e0} q_e^2}{m_e} \mathbf{E} \int \frac{f_e d^3 v}{(\omega - \mathbf{k} \cdot \mathbf{v})^2}} \tag{4.175}$$

The well-known definition equation $\mathbf{D} = \mathbf{E} + 4\pi \mathbf{P} = \varepsilon \mathbf{E}$, with $\varepsilon = 1 + \chi$, then leads to the susceptibility given in (4.170). ■

It is furthermore interesting that, with the help of the dielectric constant ε, the total wave energy density in a ω mode can be calculated by means of

$$\boxed{U = \frac{\varepsilon_0 E^2}{4} \frac{\partial}{\partial \omega} (\varepsilon \omega)} \tag{4.176}$$

Example 4.9 (Energy Density)
Let us now look at a justification for this relationship. For example, let us focus on a high-frequency mode and start with the (one-dimensional) Vlasov equation for electrons. We multiply by $m_e v$ or $\frac{1}{2} m_e v^2$ and then average over position in the form

$$\langle \ldots \rangle := \frac{k}{2\pi} \int_0^{2\pi/k} dx \ldots, \tag{4.177}$$

i.e., over one wavelength, where we consider oscillations of the form

$$E = E(t) \sin(kx - \omega t) \tag{4.178}$$

with slowly varying amplitude $E(t)$. This procedure leads to

$$\partial_t \langle \int m_e v f dv \rangle \equiv \frac{dP}{dt} = q_e \langle E \int f dv \rangle, \tag{4.179}$$

$$\partial_t \langle \int \frac{1}{2} m_e v^2 f dv \rangle \equiv \frac{dK}{dt} = q_e \langle E \int v f dv \rangle. \tag{4.180}$$

To calculate the averages on the right-hand sides, we need to approximately determine the perturbed distribution $f = f_0 + f_1$. From

$$\frac{df_1}{dt} = -\frac{q_e}{m_e} f_0' \, E(t) \sin(kx - \omega t) \tag{4.181}$$

we find by integration

$$
\begin{aligned}
\frac{m_e}{q_e} f_1 = & f_0' E(t) \frac{\cos(kx - \omega t)}{kv - \omega} - f_0' \frac{dE}{dt} \frac{\sin(kx - \omega t)}{(kv - \omega)^2} \\
& + \int_{-\infty}^{t} \frac{d^2 E}{d\tau^2} f_0' \frac{\sin(kx_0 + kv\tau - \omega\tau)}{(kv - \omega)^2} d\tau.
\end{aligned} \tag{4.182}
$$

If we use this in (4.179) and (4.180), then, neglecting the contribution proportional to d^2E/dt^2, it follows that

$$
P \approx -\frac{\omega_{pe}^2}{n_{e0}} \frac{\varepsilon_0 E^2}{4} \int \frac{f_0' \, dv}{(kv - \omega)^2}, \tag{4.183}
$$

$$
K \approx -\frac{\omega_{pe}^2}{n_{e0}} \frac{\varepsilon_0 E^2}{4} \int \frac{f_0' \, v dv}{(kv - \omega)^2}. \tag{4.184}
$$

The total energy of an oscillation consists of the electrostatic and kinetic energy contributions,

$$
U = \frac{\varepsilon_0 E^2}{4} + K. \tag{4.185}
$$

A short calculation yields

$$
\begin{aligned}
U & \approx \frac{\varepsilon_0 E^2}{4} \left[1 - \frac{\omega_{pe}^2}{k n_{eo}} \int \frac{f_0' (kv - \omega)}{(kv - \omega)^2} dv - \frac{\omega}{k} \frac{\omega_{pe}^2}{n_{e0}} \int \frac{f_0' \, dv}{(kv - \omega)^2} \right] \\
& \approx \frac{\varepsilon_0 E^2}{4} \frac{\partial}{\partial \omega} (\varepsilon \omega),
\end{aligned} \tag{4.186}
$$

which proves (4.176). The approach just outlined can and should only sketch the exact calculation procedure. We will not include additional components (ions) and electromagnetic contributions at this point. ■

Finally, it should be noted that there is a relationship between the energy density U and the momentum density P of the wave given by

$$
\frac{U}{\omega} = \frac{P}{k} \tag{4.187}
$$

4.4 Landau-Fokker-Planck Equation

We now return to the original system of equations (4.27) and (4.28) and apply the approximation (4.31). This leads to a kinetic equation named after Landau, Fokker, and Planck.

For the sake of clarity, we will write out the relevant equations once more:

$$\begin{aligned}\partial_t f^\alpha(1;t) &= L_1^\alpha f^\alpha(1;t) + \sum_\beta \int d^6 2\, L_{12}^{\alpha\beta} f^\alpha(1;t) f^\beta(2;t) \\ &\quad + \sum_\beta \int d^6 2\, L_{12}^{\alpha\beta}\, g^{\alpha\beta}(1,2;t),\end{aligned} \tag{4.188}$$

$$\partial_t g^{\alpha\beta}(1,2;t) - (L_1^\alpha + L_2^\beta) g^{\alpha\beta}(1,2;t) = L_{12}^{\alpha\beta} f^\alpha(1;t) f^\beta(2;t). \tag{4.189}$$

The symbols 1 and 2 in the arguments abbreviate the coordinates $\mathbf{q}_1, \mathbf{v}_1$ and $\mathbf{q}_2, \mathbf{v}_2$, respectively. The next task is to solve (4.189) for $g^{\alpha\beta}$, insert the result into the first Eq. (4.188), and obtain a kinetic equation of the form (4.39). In doing so, we treat the right-hand side of (4.189) as an inhomogeneity and solve it using the propagator $U_{12}^{\alpha\beta}$.. The latter is defined by the operator equation

$$\boxed{\partial_t U_{12}^{\alpha\beta}(t) - (L_1^\alpha + L_2^\beta) U_{12}^{\alpha\beta}(t) = 0} \tag{4.190}$$

with the initial condition

$$U_{12}^{\alpha\beta}(0) = 1 \tag{4.191}$$

It solves

$$g^{\alpha\beta}(1,2;t) = U_{12}^{\alpha\beta}(t) g^{\alpha\beta}(1,2;0) \tag{4.192}$$

(4.189) without the inhomogeneity term. It is just as easy to see that

$$\boxed{\begin{aligned} g^{\alpha\beta}(1,2;t) &= U_{12}^{\alpha\beta}(t) g^{\alpha\beta}(1,2;0) \\ &\quad + \int_0^t d\tau\, U_{12}^{\alpha\beta}(t-\tau) L_{12}^{\alpha\beta}\, f^\alpha(1;\tau) f^\beta(2;\tau)\end{aligned}} \tag{4.193}$$

fully satisfies Eq. (4.189). This expression clearly shows how, in order to calculate the correlation function at time t, contributions from all preceding times must be "collected." If we insert $g^{\alpha\beta}$ from (4.193) into (4.188), we obtain as the contribution of the correlation function to the time evolution of f^α

$$K^\alpha = \sum_\beta \int d^6 2 \int_0^t d\tau \, L_{12}^{\alpha\beta} \, U_{12}^{\alpha\beta}(t-\tau) L_{12}^{\alpha\beta} f^\alpha(1;\tau) f^\beta(2;\tau)$$
$$+ \sum_\beta \int d^6 2 \, L_{12}^{\alpha\beta} \, U_{12}^{\alpha\beta}(t) g^{\alpha\beta}(1,2;0). \tag{4.194}$$

We cannot accept this expression in a kinetic equation (in the kinetic regime yet to be defined) for two reasons:

1. Through the single-particle distribution functions, it contains contributions from *all* previous times $0 \le \tau \le t$.
2. The initial correlation $g(1, 2; 0)$ still appears explicitly.

We now want to discuss how these difficulties can be circumvented in physically relevant situations. To do so, we must take a closer look at the individual contributions in (4.194).

First, from the structure of (4.190), we recognize that we can use a product ansatz of the form

$$\boxed{U_{12}^{\alpha\beta}(t) = U_1^\alpha(t) U_2^\beta(t)} \tag{4.195}$$

where for U_1^α and U_2^β the equations

$$\partial_t U_1^\alpha(t) = L_1^\alpha \, U_1^\alpha(t), \tag{4.196}$$

$$\partial_t U_2^\beta(t) = L_2^\beta \, U_2^\beta(t) \tag{4.197}$$

hold. If we now want to study the effect of U_1^α on the distribution function f^α in the terms on the right-hand side of (4.194), we only need to consider the effects to lowest order. Thus, not in general, but in the treatment of K^α we use

$$U_1^\alpha(t) f^\alpha(1;0) = f^\alpha(1;t) = f^\alpha(\mathbf{q}_1(-t), \mathbf{v}_1(-t);\, 0). \tag{4.198}$$

This conclusion follows from a comparison of (4.196) with the leading terms of (4.188). The relation (4.198) further states that the operator $U_1^\alpha(t)$ produces a time shift by $-t$ in the coordinates and velocities according to the equations of motion in external fields.

Let us now consider the second term on the right-hand side of (4.194):

$$\sum_\beta \int d^6 2 \, L_{12}^{\alpha\beta} \, U_{12}^{\alpha\beta} \, g^{\alpha\beta}(1,2;0)$$

$$= \sum_{\beta} \int d^3r\, d^3v_2\, L_{12}^{\alpha\beta}\, U_1^{\alpha}\, U_2^{\beta}\, g^{\alpha\beta}(\mathbf{q}_1, \mathbf{r}, \mathbf{v}_1, \mathbf{v}_2; 0)$$

$$= \sum_{\beta} \int d^3r\, d^3v_2\, L_{12}^{\alpha\beta}\, g^{\alpha\beta}[\mathbf{q}_1(-t), \mathbf{r}(-t), \mathbf{v}_1(-t), \mathbf{v}_2(-t); 0]. \tag{4.199}$$

We have introduced the difference coordinate $\mathbf{r} = \mathbf{q}_2 - \mathbf{q}_1$ and performed the time shift. In an external magnetic field, for example, the position of a particle of type α is given by

$$\mathbf{q}(-t) = \begin{pmatrix} q_x - \Omega_\alpha^{-1} v_y(\cos\Omega_\alpha t - 1) - \Omega_\alpha^{-1} v_x \sin\Omega_\alpha t \\ q_y + \Omega_\alpha^{-1} v_x(\cos\Omega_\alpha t - 1) - \Omega_\alpha^{-1} v_y \sin\Omega_\alpha t \\ q_z - v_z t \end{pmatrix}. \tag{4.200}$$

Of particular interest is the z-component; for example, we obtain $r_z(-t) = q_{2z} - q_{1z} - (v_{2z} - v_{1z})t \sim t$. On the other hand, we know that the correlation function $g^{\alpha\beta}$ becomes very small for distances $r \geq \lambda_D$. Thus, for times

$$t \geq \frac{\lambda_D}{v_{2z} - v_{1z}} \sim \frac{\lambda_D}{v_t} \sim \omega_p^{-1} \tag{4.201}$$

the influence of the initial distribution $g^{\alpha\beta}(1, 2; 0)$. disappears. More precisely, since the characteristic correlation times (= mean duration of a collision) for electrons and ions are different, we set

$$t \geq \tau_c = \max(\omega_{pe}^{-1}, \omega_{pi}^{-1}). \tag{4.202}$$

The range (4.202) defines the kinetic regime with

$$U_{12}(t) g^{\alpha\beta}(1, 2; 0) \approx 0. \tag{4.203}$$

Let us now consider the first term on the right-hand side of (4.194). We first substitute for $t - \tau$ again τ and obtain

$$\sum_{\beta} \int d^6 2 \int_0^t d\tau\, L_{12}^{\alpha\beta}\, U_{12}^{\alpha\beta}(\tau)\, L_{12}^{\alpha\beta}\, f^{\alpha}(1; t - \tau) f^{\beta}(2; t - \tau). \tag{4.204}$$

We then use (4.195) and (4.198) in the form

$$f^{\alpha}(1; t - \tau) = U_1^{\alpha}(-\tau) f^{\alpha}(1; t) \tag{4.205}$$

(and analogously for particle 2), where of course

$$U_1^\alpha(\tau)U_1^\alpha(-\tau) = 1 \tag{4.206}$$

applies. A relatively straightforward calculation then leads to

$$\begin{aligned} K^\alpha \approx \sum_\beta \int d^3v_2 \int d^3r \int_0^t d\tau \frac{1}{m_\alpha}\frac{\partial}{\partial \mathbf{v}_1} \cdot [\nabla\phi_{\alpha\beta}(\mathbf{r}(\tau))][\nabla\phi_{\alpha\beta}(\mathbf{r}(-\tau))] \\ \cdot \left\{U_1^\alpha(\tau)\frac{1}{m_\alpha}\frac{\partial}{\partial \mathbf{v}_1}U_1^\alpha(-\tau) - U_2^\beta(\tau)\frac{1}{m_\beta}\frac{\partial}{\partial \mathbf{v}_2}U_2^\beta(-\tau)\right\} f^\alpha(1;t)f^\beta(2;t). \end{aligned} \tag{4.207}$$

The gradient ∇ is taken with respect to the difference coordinate $\mathbf{r}$. Due to the two factors $\phi_{\alpha\beta}(\mathbf{r}(\tau))$ and $\phi_{\alpha\beta}(\mathbf{r}(-\tau))$ in the integrand and the fact that these act as weighting factors in the integration, restricting the range to

$$\boxed{0 \le r \le r_c}, \tag{4.208}$$

$$\boxed{0 \le \tau \le \tau_c} \tag{4.209}$$

only a small region contributes to the integral. This is important for the following estimates, in which the correlation length $r_c \approx \max(\lambda_{De}, \lambda_{Di})$ as well as $\tau_c \approx \omega_{pi}^{-1}$ are used as integration limits.

Example 4.10 (Evaluation in the Kinetic Regime)
One finds

$$U_1^\alpha(\tau)\frac{\partial}{\partial \mathbf{v}_1}U_1^\alpha(-\tau)f_1^\alpha(1;t) \approx \frac{\partial}{\partial \mathbf{v}_1}f_1^\alpha(1;t) \tag{4.210}$$

in the following way. The displacement $U_1^\alpha(-\tau)$ acts with respect to the coordinates as shown in (4.200) [except for the sign]. Furthermore, we have

$$\mathbf{v}(-t) = \begin{pmatrix} v_x \cos\Omega_\alpha t - v_y \sin\Omega_\alpha t \\ v_y \cos\Omega_\alpha t + v_x \sin\Omega_\alpha t \\ v_z \end{pmatrix}. \tag{4.211}$$

With this, we can estimate

$$\begin{aligned} &U_1^\alpha(\tau)\frac{\partial}{\partial \mathbf{v}_1}U_1^\alpha(-\tau)f^\alpha(1;t) \\ &= \left\{\hat{x}\left[\frac{1}{\Omega_\alpha}\sin\Omega_\alpha\tau\,\frac{\partial}{\partial x} + \frac{1}{\Omega_\alpha}(\cos\Omega_\alpha\tau - 1)\frac{\partial}{\partial y} + \cos\Omega_\alpha\tau\frac{\partial}{\partial v_x} - \sin\Omega_\alpha\tau\frac{\partial}{\partial v_y}\right]\right. \\ &+ \hat{y}\left[-\frac{1}{\Omega_\alpha}(\cos\Omega_\alpha\tau - 1)\frac{\partial}{\partial x} + \frac{1}{\Omega_\alpha}\sin\Omega_\alpha\tau\frac{\partial}{\partial y} + \sin\Omega_\alpha\tau\,\frac{\partial}{\partial v_x} + \cos\Omega_\alpha\tau\frac{\partial}{\partial v_y}\right] \end{aligned} \tag{4.212}$$

$$+\hat{z}\left[\tau\frac{\partial}{\partial z}+\frac{\partial}{\partial v_z}\right]\Big\}f^\alpha(1;t)\approx\left[\frac{\partial}{\partial \mathbf{v}_1}+\tau\frac{\partial}{\partial \mathbf{q}_1}\right]f^\alpha(1;t)+\mathcal{O}(\Omega_\alpha\tau_c)$$

Without a magnetic field ($\Omega_\alpha = 0$) naturally only the first two terms on the right-hand side remain. Even with a magnetic field, we neglect the contributions of the magnetic field under the assumption

$$|\Omega_\alpha|\tau_c \ll 1. \tag{4.213}$$

Due to the restriction of the τ-integration to $\tau \leq \tau_c$ we have

$$\tau\frac{\partial}{\partial q_1}f^\alpha : \frac{\partial}{\partial v_1}f^\alpha \sim \frac{\tau_c v_{t\alpha}}{L_H} \ll 1. \tag{4.214}$$

Initially, under the assumption that the single-particle distribution function depends on position only through the hydrodynamic moments (density, temperature, etc.), we introduced a hydrodynamic length scale

$$\boxed{L_H^{-1} \sim |\nabla \ln f^\alpha|} \tag{4.215}$$

for which we assume

$$L_H \gg r_c \sim \lambda_{De} \tag{4.216}$$

The characteristic hydrodynamic time scale

$$\boxed{\tau_H^\alpha = L_H/v_{t\alpha}} \tag{4.217}$$

is then, of course, much greater than the correlation time τ_c, from which the estimate (4.214) follows. The calculation (4.212) thus yields (4.210). ■

In a similar manner, one proves

$$\begin{aligned}
f^\alpha(1;t)f^\beta(2;t) &= f^\alpha(\mathbf{q}_1,\mathbf{v}_1;t)f^\beta(\mathbf{q}_1+\mathbf{r},\mathbf{v}_2;t)\\
&\approx f^\alpha(\mathbf{q}_1,\mathbf{v}_1;t)\left[f^\beta(\mathbf{q}_1,\mathbf{v}_2;t)+\mathbf{r}\cdot\frac{\partial}{\partial \mathbf{q}_1}f^\beta(\mathbf{q}_1,\mathbf{v}_2;t)+\ldots\right]\\
&\approx f^\alpha(\mathbf{q}_1,\mathbf{v}_1;t)f^\beta(\mathbf{q}_1,\mathbf{v}_2;t),
\end{aligned} \tag{4.218}$$

where terms of order r_c/L_H have been neglected. [Recall: The estimate is only necessary for the integrand in the integration domain (4.208) and (4.209).] Finally, in the same way, one also finds

$$\phi_{\alpha\beta}(\mathbf{q}_2(-\tau)-\mathbf{q}_1(-\tau)) \approx \phi_{\alpha\beta}(\mathbf{r}-(\mathbf{v}_2-\mathbf{v}_1)\tau), \tag{4.219}$$

where terms of order $\Omega_\alpha\tau_c$ have been neglected.

If we substitute all this into (4.207), we obtain

$$\boxed{\begin{aligned}K^\alpha \approx \sum_\beta \int d^3 v_2 \frac{1}{m_\alpha} \frac{\partial}{\partial v_{1\nu}} \mathcal{T}^{\alpha\beta}_{\nu\mu}(\mathbf{g}) \\ \times \left(\frac{1}{m_\alpha} \frac{\partial}{\partial v_{1\mu}} - \frac{1}{m_\beta} \frac{\partial}{\partial v_{2\mu}} \right) f^\alpha(\mathbf{q}_1, \mathbf{v}_1; t) f^\beta(\mathbf{q}_1, \mathbf{v}_2; t),\end{aligned}} \tag{4.220}$$

with the designation

$$\mathbf{g} = \mathbf{v}_2 - \mathbf{v}_1 \tag{4.221}$$

and the summation convention over the doubly occurring indices ν and μ. In the tensor

$$\mathcal{T}^{\alpha\beta}(\mathbf{g}) := \int d^3 r \int_0^\infty d\tau \left[\nabla \phi_{\alpha\beta}(\mathbf{r}) \right] \left[\nabla \phi_{\alpha\beta}(\mathbf{r} - \mathbf{g}\tau) \right] \tag{4.222}$$

the domain rule (4.208), (4.209) applies. The calculation of $\mathcal{T}^{\alpha\beta}$ uses the Coulomb interaction (see, however, the later discussion regarding screening) with

$$\phi_{\alpha\beta}(\mathbf{r}) = \int d^3 k \, e^{i\mathbf{k}\cdot\mathbf{r}} \tilde{\phi}_{\alpha\beta}(\mathbf{k}) \tag{4.223}$$

and

$$\tilde{\phi}_{\alpha\beta} = \frac{1}{2\pi^2} \frac{1}{4\pi\varepsilon_0} \frac{e_\alpha e_\beta}{k^2}. \tag{4.224}$$

The short calculation

$$\begin{aligned}\mathcal{T}^{\alpha\beta}(\mathbf{g}) &= \int d^3 r \int_0^\infty d\tau \int e^{i(\mathbf{k}+\mathbf{k}')\cdot\mathbf{r} - i\mathbf{k}'\cdot\mathbf{g}\tau} (i\mathbf{k})(i\mathbf{k}') \tilde{\phi}_{\alpha\beta}(\mathbf{k}) \tilde{\phi}_{\alpha\beta}(\mathbf{k}') d^3 k \, d^3 k' \\ &= 8\pi^4 \int d^3 k \, \mathbf{k}\,\mathbf{k} \, \tilde{\phi}^2_{\alpha\beta}(\mathbf{k}) \delta(\mathbf{k}\cdot\mathbf{g}) \\ &= 8\pi^4 \int_0^\infty dk \, k^4 \tilde{\phi}^2_{\alpha\beta}(k) \int_{-1}^1 d\cos\theta \int_0^{2\pi} d\varphi \\ &\quad \times \delta(kg\cos\theta) \begin{pmatrix} \sin\theta \, \cos\varphi \\ \sin\theta \, \sin\varphi \\ \cos\theta \end{pmatrix} \begin{pmatrix} \sin\theta \, \cos\varphi \\ \sin\theta \, \sin\varphi \\ \cos\theta \end{pmatrix} \\ &= [1 - \hat{z}\hat{z}] \, T^{\alpha\beta},\end{aligned} \tag{4.225}$$

with

$$T^{\alpha\beta} = 8\pi^5 \int_0^\infty dk \, k^4 \, \tilde{\phi}^2_{\alpha\beta}(k) \int_{-1}^1 d\mu (1-\mu^2) \delta(kg\mu), \tag{4.226}$$

quickly leads to essential simplifications. First, we note that the expression (4.226) can be further evaluated,

$$T^{\alpha\beta} = \frac{8\pi^5}{g} \int_0^\infty dk\, k^3\, \tilde{\phi}^2_{\alpha\beta}(k) \equiv \frac{A^{\alpha\beta}}{g}. \tag{4.227}$$

Thus, the tensor $\mathcal{T}^{\alpha\beta}$ in a system with $\mathbf{g} = g\hat{z}$ is written as

$$\boxed{\mathcal{T}^{\alpha\beta}_{\nu\mu}(\mathbf{g}) = \left(\delta_{\nu\mu} - \frac{g_\nu g_\mu}{g^2}\right)\frac{A^{\alpha\beta}}{g}}, \tag{4.228}$$

where for the Coulomb interaction

$$\boxed{A^{\alpha\beta}_{Coulomb} = 2\pi\left(\frac{1}{4\pi\,\varepsilon_0}\right)^2 e^2_\alpha e^2_\beta \int_0^\infty dk\, \frac{1}{k}} \tag{4.229}$$

diverges logarithmically at both the upper and lower limits. This was, in essence, to be expected.

Large $k \to \infty$ correspond to small distances, where the assumption of weak coupling breaks down. Small $k \to 0$ correspond to large distances, where the Coulomb interaction does not come into play due to Debye screening. The latter issue was successfully addressed by Balescu and Lenard, in whose equation for the correlation function $g^{\alpha\beta}$ the screening by a third particle was taken into account. We will return to this in the next section.

It should be noted here only that the general dynamic screening in the limit $v_1 \to 0$ reduces to the effective static interaction

$$\phi^{effective}_{\alpha\beta} = \frac{1}{4\pi\,\varepsilon_0} e_\alpha e_\beta \frac{e^{-\lambda_D r}}{r} \tag{4.230}$$

with

$$\tilde{\phi}^{effective}_{\alpha\beta} = \frac{e_\alpha e_\beta}{2\pi^2}\,\frac{1}{4\pi\,\varepsilon_0}\,\frac{1}{k^2 + \kappa_D^2} \tag{4.231}$$

The problem of close collisions is generally difficult to handle. Here, we use a cutoff procedure that restricts k to the range

$$k < k_{max} \approx \frac{12\pi\varepsilon_0 k_B T_\alpha}{e_\alpha^2} \tag{4.232}$$

where $e_\alpha^2/12\pi\varepsilon_0 k_B T_\alpha$ is the impact parameter for 90^0 deflection.

With these restrictions, we find

$$\begin{aligned} A^{\alpha\beta} &\approx 8\pi^5 \int_0^{k_{max}} dk\, k^3 \left[\tilde{\phi}_{\alpha\beta}^{effective}\right]^2 \\ &= \pi \left(\frac{1}{4\pi\varepsilon_0}\right)^2 e_\alpha^2 e_\beta^2 \left[\frac{\kappa_D^2}{k_{max}^2 + \kappa_D^2} - 1 + \ln \frac{k_{max}^2 + \kappa_D^2}{\kappa_D^2}\right] \\ &\approx 2\pi \left(\frac{1}{4\pi\varepsilon_0}\right)^2 e_\alpha^2 e_\beta^2 \ln \Lambda_B \end{aligned} \tag{4.233}$$

because

$$\ln \Lambda \sim \ln \frac{k_{max}}{\kappa_D} = \ln \frac{12\pi\varepsilon_0 \lambda_D k_B (T_e + T_i)}{2e^2} = \ln \Lambda_B\, . \tag{4.234}$$

To summarize: Within the framework of the Landau approximation, the collision term K^α is written as

$$\begin{aligned} K^\alpha = &\sum_{\beta=e,i} 2\pi \left(\frac{1}{4\pi\varepsilon_0}\right)^2 e_\alpha^2 e_\beta^2 \ln \Lambda_B \int d^3 v_2 \frac{1}{m_\alpha} \frac{\partial}{\partial v_{1\nu}} G_{\nu\mu}(\mathbf{g}) \\ &\times \left[\frac{1}{m_\alpha} \frac{\partial}{\partial v_{1\mu}} - \frac{1}{m_\beta} \frac{\partial}{\partial v_{2\mu}}\right] f^\alpha(\mathbf{q}_1, \mathbf{v}_1; t) f^\beta(\mathbf{q}_1, \mathbf{v}_2; t) \end{aligned} \tag{4.235}$$

with the Landau tensor

$$G_{\nu\mu}(\mathbf{g}) = \frac{g^2 \delta_{\nu\mu} - g_\nu g_\mu}{g^3}\, . \tag{4.236}$$

In (4.235),$\ln \Lambda_B$ is an averaged Coulomb logarithm, as defined on the right-hand side of (4.234). We will not go into further detail here regarding the differences that arise from a more precise analysis of screening (by electrons and ions).

A justification for the cutoff at small k values can be found in the later chapter on the Lenard-Balescu equation.

In the following, we will first discuss the Landau equation for a single-component electron plasma. Because

$$\partial_{\mathbf{v}_1} \partial_{\mathbf{v}_1} g = \frac{g^2 \underline{\underline{1}} - \mathbf{g}\mathbf{g}}{g^3} \tag{4.237}$$

we then obtain

$$\frac{df(\mathbf{v}_1)}{dt} = \left(\frac{1}{4\pi\varepsilon_0}\right)^2 \frac{2\pi e^4 \ln\Lambda_B}{m_e^2} \partial_{\mathbf{v}_1} \cdot \left[(\partial_{\mathbf{v}_1} f(\mathbf{v}_1)) \cdot \partial_{\mathbf{v}_1} \partial_{\mathbf{v}_1} \int d^3 v_2\, g f(\mathbf{v}_2) \right.$$
$$\left. - f(\mathbf{v}_1) \int d^3 v_2 \partial_{\mathbf{v}_1} (\partial_{\mathbf{v}_1} \cdot \partial_{\mathbf{v}_1}) g f(\mathbf{v}_2) \right]. \tag{4.238}$$

Here, in the second term on the right-hand side, partial integration was performed and then $\partial_{\mathbf{v}_2} g = -\partial_{\mathbf{v}_1} g$ was used. Furthermore, we have abbreviated the Vlasov term on the left-hand side as d/dt and have written only the relevant variables. Now we move the second operator $\partial_{\mathbf{v}_1}$ in the first term on the right-hand side to the front,

$$\frac{df}{dt} = \left(\frac{1}{4\pi\varepsilon_0}\right)^2 \frac{2\pi e^4 \ln\Lambda_B}{m_e^2} \left\{ \partial_{\mathbf{v}_1} \partial_{\mathbf{v}_1} : \left[f(\mathbf{v}_1) \partial_{\mathbf{v}_1} \partial_{\mathbf{v}_1} \int d^3 v_2\, g f(\mathbf{v}_2) \right] \right.$$
$$\left. - 2\partial_{\mathbf{v}_1} \cdot \left[f(\mathbf{v}_1) \int d^3 v_2 \partial_{\mathbf{v}_1} (\partial_{\mathbf{v}_1} \cdot \partial_{\mathbf{v}_1}) g f(\mathbf{v}_2) \right] \right\}. \tag{4.239}$$

Finally, we also use

$$(\partial_{\mathbf{v}_1} \cdot \partial_{\mathbf{v}_1}) g = \frac{2}{g} \tag{4.240}$$

to obtain the Landau equation in Fokker-Planck form.

The Landau-Fokker-Planck equation is then given by

$$\frac{df(\mathbf{v}_1)}{dt} = -\partial_{\mathbf{v}_1} \cdot [\mathbf{A} f(\mathbf{v}_1)] + \frac{1}{2} \partial_{\mathbf{v}_1} \partial_{\mathbf{v}_1} : [\mathcal{B} f(\mathbf{v}_1)] . \tag{4.241}$$

Two coefficients appear. The coefficient of dynamical friction is

$$\mathbf{A} = \left(\frac{1}{4\pi\varepsilon_0}\right)^2 \frac{8\pi e^4 \ln\Lambda_B}{m_e^2} \partial_{\mathbf{v}_1} \int d^3 v_2 \frac{f(\mathbf{v}_2)}{|\mathbf{v}_1 - \mathbf{v}_2|}, \tag{4.242}$$

while the diffusion coefficient is

$$\mathcal{B} = \left(\frac{1}{4\pi\varepsilon_0}\right)^2 \frac{4\pi e^4 \ln\Lambda_B}{m_e^2} \partial_{\mathbf{v}_1} \partial_{\mathbf{v}_1} \int d^3 v_2 |\mathbf{v}_1 - \mathbf{v}_2| f(\mathbf{v}_2) \tag{4.243}$$

In this standard form of a Fokker-Planck equation, the friction term describes the reduction of particle motion in the original direction due to the many weak collisions, while the diffusion term accounts for the growth of the transverse component.

Note that from (4.242), the inverse relaxation time can be estimated as $\tau_R^{-1} \approx \omega_{pe} \ln \Lambda_B/\Lambda_B$.

Example 4.11 (Plausibility Consideration for the Fokker-Planck Form)
For understanding equation (4.241), a basic plausibility consideration is helpful. If we assume a statistical description of many Coulomb collisions with small deflections, we define the probability $W(\mathbf{v}, \Delta\mathbf{v})$ that after a time Δt the velocity of a particle has changed by $\Delta\mathbf{v}$ from $\mathbf{v}$ to $\mathbf{v} + \Delta\mathbf{v}$. For the single-particle distribution function, this leads to

$$f(\mathbf{q}, \mathbf{v}; t + \Delta t) = \int d^3\Delta v f(\mathbf{q}, \mathbf{v} - \Delta\mathbf{v}; t) W(\mathbf{v} - \Delta\mathbf{v}, \Delta\mathbf{v}) \tag{4.244}$$

If we now determine the change in the distribution function due to Coulomb collisions,

$$\frac{df}{dt}\Delta t \approx f(\mathbf{q}, \mathbf{v}; t + \Delta t) - f(\mathbf{q}, \mathbf{v}; t), \tag{4.245}$$

we use in (4.244) a Taylor expansion

$$f(\mathbf{q}, \mathbf{v} - \Delta\mathbf{v}; t) W(\mathbf{v} - \Delta\mathbf{v}, \Delta v) \approx f(\mathbf{q}, \mathbf{v}; t) W(\mathbf{v}, \Delta\mathbf{v})$$

$$- \frac{\partial (fW)}{\partial \mathbf{v}} \cdot \Delta\mathbf{v} + \sum_{\nu,\mu} \frac{1}{2} \frac{\partial^2 (fW)}{\partial v_\nu \partial v_\mu} \Delta v_\nu \Delta v_\mu - \dots . \tag{4.246}$$

Now we introduce

$$\int d^3\Delta v\, W(\mathbf{v}, \Delta\mathbf{v}) = 1 , \tag{4.247}$$

$$\int d^3\Delta v\, W(\mathbf{v}, \Delta\mathbf{v}) \Delta\mathbf{v} = \langle \Delta\mathbf{v} \rangle_t \Delta t , \tag{4.248}$$

$$\int d^3\Delta v\, W(\mathbf{v}, \Delta\mathbf{v}) \Delta v_\nu \Delta v_\mu = \langle \Delta v_\nu \Delta v_\mu \rangle_t \Delta t \tag{4.249}$$

The first relation follows from the normalization of the probability. The two further definitions exploit the statistical independence of the individual collisions, with the result that the overall effect is proportional to Δt. (If we consider, for example, the product $\Delta v_\nu \Delta v_\mu = \sum_i \sum_j \Delta v_\nu^i \Delta v_\mu^j \approx \sum_i \Delta v_\nu^i \Delta v_\mu^i$ with the contributions from collisions i and j, we see that only the product contributions from the individual collisions contribute. The result is proportional to the number of collisions and thus proportional to Δt.) From this, (4.245) yields

$$\frac{df}{dt} = -\frac{\partial}{\partial \mathbf{v}} \cdot \left[\langle \Delta\mathbf{v} \rangle_t f \right] + \frac{1}{2} \sum_{\nu,\mu} \frac{\partial^2}{\partial v_\nu \partial v_\mu} [\langle \Delta v_\nu \Delta v_\mu \rangle_t f], \tag{4.250}$$

in strict analogy to (4.241). ■

Example 4.12 (Evaluation for a Maxwell Distribution)
It is also of interest to evaluate the coefficients **A** and $\mathcal{B}$ for a Maxwell distribution

$$f = \frac{n_{e0} b^3}{\pi^{3/2}} \exp(-b^2 v_1^2) \tag{4.251}$$

We have

$$\mathbf{A} = -\left(\frac{1}{4\pi\varepsilon_0}\right)^2 \frac{8\pi e^4 n_{e0} \ln \Lambda_B}{m_e^2 v_1^3} \mathbf{v}_1 \phi_1(b v_1), \tag{4.252}$$

$$B_{\nu\mu} = \left(\frac{1}{4\pi\varepsilon_0}\right)^2 \frac{4\pi e^4 n_{e0} \ln \Lambda_B}{m_e^2} \frac{\partial^2}{\partial v_{1\nu} \partial v_{1\mu}} \left[\left(1 + \frac{1}{2b^2 v_1^2}\right)\phi(b v_1) + \frac{1}{\pi^{1/2} b v_1} e^{-b^2 v_1^2}\right], \tag{4.253}$$

with

$$\phi(x) = \frac{2}{\sqrt{\pi}} \int^x e^{-x^2} dx, \tag{4.254}$$

$$\phi_1(x) = \phi(x) - x\frac{d\phi}{dx}. \tag{4.255}$$

If we use this to calculate the right-hand side of (4.241), we obtain

$$\frac{df}{dt} = 0 . \tag{4.256}$$

From this we conclude that the electron-electron collision term tends toward Maxwellization; this statement is confirmed by numerical calculations. ■

We now return at the end of this section to the more general collision term K^α. It is not difficult, for

$$K^\alpha = K^{\alpha e} + K^{\alpha i} \tag{4.257}$$

to prove the relations

$$\int d^3 v_1 K^\alpha = 0 , \tag{4.258}$$

$$\sum_{\alpha} m_\alpha \int d^3v_1 \, \mathbf{v}_1 \, K^\alpha = 0, \tag{4.259}$$

$$\sum_{\alpha} \frac{1}{2} m_\alpha \int d^3v_1 \, v_1^2 \, K^\alpha = 0 \tag{4.260}$$

These physically meaningful results express the conservation of particle number, total momentum, and total energy.

Also of interest are the further results for collisions within a single component

$$\int d^3v_1 \, K^{\alpha\alpha} = 0, \text{for}\, \alpha = e, i, \tag{4.261}$$

$$m_\alpha \int d^3v_1 \, \mathbf{v}_1 \, K^{\alpha\alpha} = 0, \text{for}\, \alpha = e, i, \tag{4.262}$$

$$\frac{1}{2} m_\alpha \int d^3v_1 \, v_1^2 \, K^{\alpha\alpha} = 0, \text{for}\, \alpha = e, i. \tag{4.263}$$

If we combine these detailed balances with the previously obtained global ones, we can conclude that the transfer rates for momentum and energy due to electron-ion and ion-electron collisions (the transfer is always from the first to the second component) are exactly balanced.

With regard to the electron-ion (K^{ei}) and ion-electron (K^{ie}) collision terms, there are approximate forms due to Lorentz, to which we will return later.

We can define a friction force density and a heat exchange density,

$$\mathbf{R}^\alpha = m_\alpha \int d^3 \, v \, \mathbf{v} \, K^\alpha \,, \tag{4.264}$$

$$Q^\alpha = \frac{1}{2} m_\alpha \int d^3 \, v \, |\mathbf{v} - \mathbf{u}^\alpha|^2 K^\alpha \,. \tag{4.265}$$

These satisfy the following relations

$$\mathbf{R}^\mathbf{e} = \mathbf{R}^{ei} = -\mathbf{R}^{ie} = -\mathbf{R}^i \,, \tag{4.266}$$

$$Q^e = Q^{ei} = -Q^{ie} - \left(\mathbf{u}^e - \mathbf{u}^i\right) \cdot \mathbf{R}^{ei} \,, \tag{4.267}$$

$$Q^i = Q^{ie} \approx -3 n_i \frac{Z}{\tau_e} \frac{m_e}{m_i} (T_i - T_e) \,. \tag{4.268}$$

These relations express conservation laws, but additionally show the weak energy exchange between electrons and ions. The latter is the reason why electron and ion temperatures are introduced separately as plasma dynamic variables.

If we express the friction force in terms of collision frequencies $\nu \sim 1/\tau$, we can write

$$\mathbf{R}^{ei} = -\nu_{ei} m_e n_e (\mathbf{u}_e - \mathbf{u}i) \ , \quad \mathbf{R}^{ie} = -\nu_{ie} m_i n_i (\mathbf{u}_i - \mathbf{u}_e) \ . \tag{4.269}$$

The collision frequencies can be easily determined from the linearized collision term. Collisions between like particles conserve the number, momentum, and energy of each species; they distribute momentum and energy very efficiently among themselves and are responsible for local (nearly) equilibrium states (after a time τ_α). Collisions between unlike particles exchange momentum and energy between the species; they transfer energy between ions and electrons extremely slowly [characteristic time $(m_i/m_e)\tau_\alpha$]. Since for the relaxation times ($\tau \sim 1/\nu$) the following holds

$$\boxed{\tau_{ee} \sim \tau_{ei} \sim \tau_e \quad (Z=1) \ , \quad \tau_{ie} \gg \tau_{ii} \sim \tau_i} \ , \tag{4.270}$$

only two times are involved, namely [33]

$$\boxed{\tau_e = \frac{3(4\pi\varepsilon_0)^2}{4\sqrt{2\pi}} \frac{m_e^{1/2} T_e^{3/2}}{n_i Z^2 e^4 \ln\Lambda}} \ , \quad \boxed{\tau_i = \frac{3(4\pi\varepsilon_0)^2}{4\sqrt{\pi}} \frac{m_i^{1/2} T_i^{3/2}}{n_i Z^4 e^4 \ln\Lambda}} \ . \tag{4.271}$$

Here, $\ln\Lambda \equiv \ln(r_{\max}/r_{\min})$ is the Coulomb logarithm, as defined in [33]. Typically, $\ln\Lambda \approx 10-20$ holds. More precise values are given in [33]. Note that we use a slightly different definition of the Coulomb logarithm, namely

$$\ln\Lambda_B = 4\pi\varepsilon_0 \frac{3(T_e+T_i)\lambda_D}{2Ze^2} \ , \quad \lambda_D = \left[\frac{Ze^2(n_e T_e + n_i T_i)}{\varepsilon_0 T_e T_i (1+Z)}\right]^{-12} \ . \tag{4.272}$$

The characteristic times are obtained from the solutions of the kinetic equations. The characteristic times and collision frequencies can be found in formula collections such as [33]. From the orders of magnitude, we conclude that a fusion plasma is not necessarily collision-dominated, especially in the hot core. However, due to the confinement and capture geometries, the particles, although their mean free paths may be much longer than the transverse machine dimensions, can remain in the plasma long enough before reaching the walls. Therefore, it can be expected that even weak collisions contribute to transport.

The derivation of the relaxation times is one of the main tasks of kinetic theory. Considering the parameter dependencies, we find that the relaxation times increase with temperature. From a very simple model, we can then estimate the specific resistivity,

$$\eta \equiv \frac{1}{\sigma} \sim \frac{m_e}{e^2 n \tau_e} . \tag{4.273}$$

Therefore, the specific resistivity of a plasma decreases as the temperature increases. The question of how more accurate results for the plasma (transport) coefficients can be obtained belongs to transport theory. We will explain the basic approach using the following example.

Example 4.13 (Resistivity of a Plasma)

Assumethe system is spatially homogeneous and stationary in the presence of a small electric field **E**. We approximate the Landau-Fokker-Planck equation by

$$-\frac{e}{m_e}\mathbf{E}\cdot\frac{\partial f_{e0}}{\partial \mathbf{v}} \approx \frac{n_i Z^2 e^4 \ln \Lambda_B}{8\pi \varepsilon_0^2 m_e^2 v^3}\left[\frac{1}{\sin\theta}\frac{\partial}{\partial\theta}\left(\sin\theta\frac{\partial f_e}{\partial\theta}\right) + \frac{1}{\sin^2\theta}\frac{\partial^2 f_e}{\partial\phi^2}\right] . \tag{4.274}$$

Here, we have used a polar coordinate system in velocity space, with the z-axis pointing in the direction of the electric field. The assumption of azimuthal symmetry ($\partial/\partial\phi = 0$) is natural. Note also that we treat the electric field as a perturbation, which allows us to assume the zeroth-order distribution function f_{e0} to be a Maxwellian distribution; $f_e = f_{e0} + f_{e1}$. This leads to

$$\frac{eEv}{T_e} f_{e0}\cos\theta \approx \frac{n_i Z^2 e^4 \ln \Lambda_B}{8\pi \varepsilon_0^2 m_e^2 v^3}\frac{1}{\sin\theta}\frac{\partial}{\partial\theta}\left(\sin\theta\frac{\partial f_{e1}}{\partial\theta}\right) . \tag{4.275}$$

The solution f_{e1} quickly follows as

$$f_{e1} \approx -\frac{4\pi\varepsilon_0^2 m_e^2 E v^4 f_{e0}\cos\theta}{n_i Z^2 e^3 T_e \ln \Lambda_B} . \tag{4.276}$$

The calculation of the z-component of the electric current density yields for $n_e = Zn_i$:

$$\begin{aligned} j_z = -e \int f_{e1} v \cos\theta d^3 v &\approx \frac{8\pi\varepsilon_0^2 m_e^2 E}{n_i Z^2 e^2 T_e \ln \Lambda_B} \int_0^\infty v^7 f_{e0} dv \int_0^\pi \cos^2\theta \sin\theta d\theta \\ &= \frac{32\sqrt{\pi}\varepsilon_0^2 E (2T_e)^{32}}{\sqrt{m_e} Z e^2 \ln \Lambda_B} . \end{aligned} \tag{4.277}$$

From this, we can read off the resistivity:

$$\eta = \frac{\sqrt{m_e} Z e^2 \ln \Lambda_B}{32\sqrt{\pi}\varepsilon_0^2 (2T_e)^{32}} . \tag{4.278}$$

This simple model already leads to a fairly accurate result. Despite the simplifications, such as assuming a Maxwellian for the zeroth-order distribution function and treating the electric field as a perturbation, this model provides a realistic estimate for the conductivity and electrical resistance of the plasma. ■

The resulting plasma properties, such as the decrease in specific resistance with increasing temperature, agree well with experimental observations. This shows that even simple models in plasma physics are often sufficient to provide essential physical insights and relatively accurate predictions.

4.5 Lenard-Balescu Equation

In this section, we improve the Landau-Fokker-Planck equation by taking into account the screening by a third particle.

To this end, we return to the original equations (4.27) and (4.28) and apply the approximation (4.32). To avoid unnecessary writing, we restrict ourselves to a spatially homogeneous system without external fields. Furthermore, we present the essential calculation steps only for a single-component plasma. These restrictions are justified, since our focus here is solely on the new effect of screening, and all contributions not considered now have already been discussed in detail in the previous section. We therefore start from the equations

$$\boxed{d_t f(\mathbf{v}_1, t) = \int d^6 2\, L_{12}\, g(1,2),} \tag{4.279}$$

$$\boxed{\begin{aligned} \partial_t g(1,2) + \mathbf{v}_1 \cdot \partial_{\mathbf{q}_1} g(1,2) + \mathbf{v}_2 \cdot \partial_{\mathbf{q}_2} g(1,2) &= L_{12}\, f(\mathbf{v}_1; t) f(\mathbf{v}_2; t) \\ &+ \int d^6 3 \left[L_{13}\, f(\mathbf{v}_1; t) g(2,3) + L_{23}\, f(\mathbf{v}_2; t) g(1,3) \right], \end{aligned}} \tag{4.280}$$

with the abbreviations $d_t f = \partial_t f - L_1 f$ as well as

$$g(1,2) = g(\mathbf{q}_1 - \mathbf{q}_2, \mathbf{v}_1, \mathbf{v}_2; t), \tag{4.281}$$

$$g(2,3) = g(\mathbf{q}_2 - \mathbf{q}_3, \mathbf{v}_2, \mathbf{v}_3; t), \tag{4.282}$$

$$g(1,3) = g(\mathbf{q}_1 - \mathbf{q}_3, \mathbf{v}_1, \mathbf{v}_3; t). \tag{4.283}$$

Due to spatial homogeneity, the correlation functions g depend only on the distances, i.e., not individually on $\mathbf{q}_1$ and $\mathbf{q}_2$, but only on $\mathbf{q}_1 - \mathbf{q}_2$.

For a Coulomb system (consisting of electrons with a smeared-out ionic background), we can furthermore replace the operator

$$L_{jk}^{\alpha\beta} = \frac{\partial \phi_{jk}}{\partial \mathbf{q}_j} \cdot \left[\frac{1}{m_\alpha} \frac{\partial}{\partial \mathbf{v}_j} - \frac{1}{m_\beta} \frac{\partial}{\partial \mathbf{v}_k} \right] \tag{4.284}$$

by

$$L_{jk} = -\left(\mathbf{a}_{jk} \cdot \frac{\partial}{\partial \mathbf{v}_j} + \mathbf{a}_{kj} \cdot \frac{\partial}{\partial \mathbf{v}_k} \right) \tag{4.285}$$

where

$$\mathbf{a}_{jk} = \frac{1}{4\pi \varepsilon_0} \frac{e^2}{m_e \mid \mathbf{q}_j - \mathbf{q}_k \mid^3} (\mathbf{q}_j - \mathbf{q}_k) \tag{4.286}$$

is.

We already know that the correlation functions g also vary on a fast time scale and that no "exact" solution of the system (4.279) and (4.280) is required. Here, too, we will restrict ourselves to the kinetic regime and accordingly, in the equation for f, consider only the term $g(\ldots; t = \infty)$.

Analogous to the procedure used in Landau damping, we first perform a Fourier transform in space. In doing so, the new variables $\mathbf{k}_1$ (for $\mathbf{q}_1$) and $\mathbf{k}_2$ (for $\mathbf{q}_2$) appear. However, due to homogeneity, we only obtain contributions for $\mathbf{k}_2 = -\mathbf{k}_1$. From (4.279) and (4.280) it then follows that

$$d_t f(\mathbf{v}_1; t) = -\frac{i(2\pi)^3}{m_e} \partial_{\mathbf{v}_1} \int d^3 v_2 \, d^3 k_1 \mathbf{k}_1 \varphi(k_1) g(\mathbf{k}_1, \mathbf{v}_1, \mathbf{v}_2; t = \infty), \tag{4.287}$$

$$\begin{aligned} \partial_t g(\mathbf{k}_1, \mathbf{v}_1, \mathbf{v}_2; t) &+ i\mathbf{k}_1 \cdot (\mathbf{v}_1 - \mathbf{v}_2) g(\mathbf{k}_1, \mathbf{v}_1, \mathbf{v}_2; t) \\ &= \frac{(2\pi)^3}{m_e} i\mathbf{k}_1 \cdot \partial_{\mathbf{v}_1} f(\mathbf{v}_1) \varphi(k_1) \int d^3 v_3 \, g(\mathbf{k}_1, \mathbf{v}_2, \mathbf{v}_3; t) \\ &- \frac{(2\pi)^3}{m_e} i\mathbf{k}_1 \cdot \partial_{\mathbf{v}_2} f(\mathbf{v}_2) \varphi(k_1) \int d^3 v_3 \, g(\mathbf{k}_1, \mathbf{v}_1, \mathbf{v}_3; t) \\ &+ \frac{\varphi(k_1)}{m_e} i\mathbf{k}_1 \cdot \left(\partial_{\mathbf{v}_1} - \partial_{\mathbf{v}_2} \right) f(\mathbf{v}_1) f(\mathbf{v}_2). \end{aligned} \tag{4.288}$$

The abbreviation $\varphi(k_1)$ was introduced via

$$\mathbf{a}_{12}(\mathbf{k}_1) = \frac{-i\mathbf{k}_1}{m_e}\varphi(k_1) = -\frac{1}{4\pi\varepsilon_0}\frac{i\mathbf{k}_1 e^2}{2\pi^2 k_1^2 m_e} \tag{4.289}$$

[cf. (4.224)]. In the following, our presentation is based on the original works of Balescu and Lenard as well as a summary by Nicholson [6]. The problem consists in finding, in the kinetic regime, an approximate solution to (4.288) that can be inserted into (4.287). To avoid unwieldy long formulas, we use a formal notation by abbreviating (4.288) in the form

$$\partial_t g + V_1 g + V_2 g = S(\mathbf{k}_1, \mathbf{v}_1, \mathbf{v}_2) \tag{4.290}$$

where S represents the last term on the right-hand side of (4.288), i.e., the source term for g. With respect to the dynamics of g, it can be assumed to be constant in time. The meaning of the operators V_1 and V_2 is obtained by direct comparison with (4.288), e.g., V_1 contains, in addition to the convective term $ik_1 \cdot \mathbf{v}_1$, a contribution from the first term on the right-hand side of (4.288). It should be emphasized that V_1 and V_2 are operators, and when we treat them as numbers in the following, for example by performing "divisions" with them, this is only meant as a shorthand for inversions, etc. We will carry out these inversions explicitly below.

Laplace transformation of (4.290) leads to ($p = -i\omega$)

$$\begin{aligned} &- g(\mathbf{k}_1, \mathbf{v}_1, \mathbf{v}_2; t = 0) - i\omega g(\mathbf{k}_1, \mathbf{v}_1, \mathbf{v}_2; \omega) + V_1\, g(\mathbf{k}_1, \mathbf{v}_2, \mathbf{v}_3; \omega) \\ &+ V_2\, g(\mathbf{k}_1, \mathbf{v}_1, \mathbf{v}_3; \omega) = -\frac{1}{i\omega} S(\mathbf{k}_1, \mathbf{v}_1, \mathbf{v}_2). \end{aligned} \tag{4.291}$$

Formally solved for g, we obtain

$$g(\ldots; \omega) = \frac{g(\ldots; t = 0) - S/i\omega}{-i\omega + V_1 + V_2}. \tag{4.292}$$

By inverse transformation we obtain

$$g(\ldots; t) = \frac{1}{2\pi}\int_C d\omega \frac{g(\ldots; t = 0) - S/i\omega}{-i\omega + V_1 + V_2} e^{-i\omega t}, \tag{4.293}$$

where the contour C in the complex ω-plane runs parallel to the real ω-axis above all singularities.

When evaluating the right-hand side of (4.293), it turns out that all zeros of $-i\omega + V_1 + V_2$ in the complex ω-plane have a negative imaginary part, provided the distribution function f is stable (see below). What remains from (4.293) is $\omega = 0$ as the singularity with the largest imaginary part. By similar considerations as in the evaluation of Landau damping, it follows for $t \to +\infty$

$$\boxed{g(\ldots;t=\infty) = \lim_{\omega\to 0} \frac{S}{-i\omega + V_1 + V_2}}\,. \tag{4.294}$$

We now wish to carry out the inversion in (4.294) explicitly and for this we use the "trick" given by Nicholson:

$$\begin{aligned}
\frac{1}{-i\omega + V_1 + V_2} &= \int_0^\infty dt\, e^{(i\omega - V_1 - V_2)t} \\
&= \int_0^\infty dt\, e^{i\omega t} \int_{C_1} \frac{d\omega_1}{2\pi} \frac{e^{-i\omega_1 t}}{-i\omega_1 + V_1} \int_{C_2} \frac{d\omega_2}{2\pi} \frac{e^{-i\omega_2 t}}{-i\omega_2 + V_2} \\
&= \int_{C_1} \frac{d\omega_1}{2\pi} \int_{C_2} \frac{d\omega_2}{2\pi} \frac{1}{-i\omega_1 + V_1} \frac{1}{-i\omega_2 + V_2} \frac{1}{-i(\omega - \omega_1 - \omega_2)},
\end{aligned} \tag{4.295}$$

which allows the operators V_1 and V_2 to be treated individually (instead of $V_1 + V_2$). In (4.295), the contours C_1 and C_2 are to be chosen parallel to the real ω_1- and ω_2-axes, respectively, such that $\mathrm{Im}\ \omega > \mathrm{Im}\ \omega_1 + \mathrm{Im}\ \omega_2$ is satisfied.

First, we calculate

$$\alpha := \frac{1}{-i\omega_1 + V_1} S, \tag{4.296}$$

i.e., we solve

$$\begin{aligned}
S(\mathbf{k}_1, \mathbf{v}_1, \mathbf{v}_2) &= (-i\omega_1 + i\mathbf{k}_1 \cdot \mathbf{v}_1)\alpha(\mathbf{k}_1, \mathbf{v}_1, \mathbf{v}_2) \\
&\quad - i\frac{(2\pi)^3}{m_e} \mathbf{k}_1 \cdot \partial_{\mathbf{v}_1} f(\mathbf{v}_1)\varphi(k_1) \int d^3v_3\, \alpha(\mathbf{k}_1, \mathbf{v}_3, \mathbf{v}_2).
\end{aligned} \tag{4.297}$$

To begin with, the result for the integrated solution is obtained very easily

$$\int d^3v_3\, \alpha(\mathbf{k}_1, \mathbf{v}_3, \mathbf{v}_2) = \frac{1}{\varepsilon(\mathbf{k}_1, \omega_1)} \int d^3v_3 \frac{S(\mathbf{k}_1, \mathbf{v}_3, \mathbf{v}_2)}{-i\omega_1 + i\mathbf{k}_1 \cdot \mathbf{v}_3}, \tag{4.298}$$

where, for the dielectric constant ε, the integration path must be chosen according to Landau. With (4.298), the solution to (4.297) can be immediately given as

$$\begin{aligned}
\alpha(\mathbf{k}_1, \mathbf{v}_1, \mathbf{v}_2) &= \frac{1}{-i\omega_1 + i\mathbf{k}_1 \cdot \mathbf{v}_1} \\
&\quad \times \left[S(\mathbf{k}_1, \mathbf{v}_1, \mathbf{v}_2) + i\frac{(2\pi^3)\mathbf{k}_1 \cdot \partial_{\mathbf{v}_1} f(\mathbf{v}_1)\varphi(k_1)}{m_e \varepsilon(\mathbf{k}_1, \omega_1)} \int d^3v_3 \frac{S(\mathbf{k}_1, \mathbf{v}_3, \mathbf{v}_2)}{-i\omega_1 + i\mathbf{k}_1 \cdot \mathbf{v}_3} \right].
\end{aligned} \tag{4.299}$$

Example 4.14 (Propagator Inversion)
The next inversion $(-i\omega_2 + V_2)^{-1}\alpha(\mathbf{k}_1, \mathbf{v}_1, \mathbf{v}_2)$ can be carried out in a similar way; ultimately, we obtain

$$\int d^3v_2\, g(\mathbf{k}_1, \mathbf{v}_1, \mathbf{v}_2; t=\infty) = \lim_{\omega\to 0} \int_{C_1} \frac{d\omega_1}{2\pi} \int_{C_2} \frac{d\omega_2}{2\pi}$$
$$\times \frac{1}{\varepsilon(-\mathbf{k}_1, \omega_2)} \frac{1}{-i(\omega - \omega_1 - \omega_2)} \int d^3v_2 \frac{1}{-i\omega_2 - i\mathbf{k}_1\cdot\mathbf{v}_2} \alpha(\mathbf{k}_1, \mathbf{v}_1, \mathbf{v}_2), \tag{4.300}$$

where we still need to insert the expression (4.299) for α. ■

We then perform the ω_2 integration; here, we note that $\omega - \omega_1$ lies above the integration contour C_2 and that the integration contour can be closed in the upper half-plane (the additional contribution is negligible, since the integrand $\sim \omega_2^{-2}$ vanishes). Thus, only the pole at $\omega_2 = \omega - \omega_1$ contributes. From (4.299) and (4.300) it then follows that

$$\int d^3v_2\, g(\mathbf{k}_1, \mathbf{v}_1, \mathbf{v}_2; t=\infty) = \lim_{\omega\to 0} \int d^3v_2 \int_{C_1} \frac{d\omega_1}{2\pi} \frac{1}{\varepsilon(-\mathbf{k}_1, \omega - \omega_1)}$$
$$\times \frac{1}{-i(\omega-\omega_1) - i\mathbf{k}_1\cdot\mathbf{v}_2} \frac{1}{-i\omega_1 + i\mathbf{k}_1\cdot\mathbf{v}_1} \left[\frac{i\mathbf{k}_1}{m_e}\varphi(k_1)\cdot(\partial_{\mathbf{v}_1} - \partial_{\mathbf{v}_2}) f(\mathbf{v}_1) f(\mathbf{v}_2)\right.$$
$$\left. - \frac{i(2\pi)^3\mathbf{k}_1\cdot\partial_{\mathbf{v}_1} f(\mathbf{v}_1)\varphi^2(k_1)}{m_e^2\,\varepsilon(\mathbf{k}_1, \omega_1)} \int d^3v_3 \frac{\mathbf{k}_1\cdot(\partial_{\mathbf{v}_3} - \partial_{\mathbf{v}_2})}{\omega_1 - \mathbf{k}_1\cdot\mathbf{v}_3} f(\mathbf{v}_3) f(\mathbf{v}_2)\right]. \tag{4.301}$$

A somewhat more extensive calculation, in which essentially only the definition of the dielectric constant ε is used, leads to

$$\int d^3v_2\, g(\mathbf{k}_1, \mathbf{v}_1, \mathbf{v}_2; t=\infty) = \lim_{\omega\to 0} \int_{C_1} \frac{d\omega_1}{2\pi} \frac{1}{-i\omega_1 + i\mathbf{k}_1\cdot\mathbf{v}_1}$$
$$\times \left\{\left[1 - \frac{1}{\varepsilon(-\mathbf{k}_1, \omega-\omega_1)}\right]\left[-\frac{f(\mathbf{v}_1)}{(2\pi)^3} + \frac{\mathbf{k}_1\cdot\partial_{\mathbf{v}_1} f(\mathbf{v}_1)\varphi(k_1)}{m_e\varepsilon(\mathbf{k}_1, \omega_1)} \int d^3v_2 \frac{f(\mathbf{v}_2)}{\omega_1 - \mathbf{k}_1\cdot\mathbf{v}_2}\right]\right.$$
$$\left. + \frac{i\mathbf{k}_1\cdot\partial_{\mathbf{v}_1} f(\mathbf{v}_1)\varphi(k_1)}{m_e\varepsilon(\mathbf{k}_1, \omega_1)\varepsilon(-\mathbf{k}_1, \omega-\omega_1)} \int d^3v_2 \frac{f(\mathbf{v}_2)}{-i(\omega-\omega_1) - i\mathbf{k}_1\cdot\mathbf{v}_2}\right\}. \tag{4.302}$$

The individual contributions on the right-hand side of (4.302) can be evaluated using complex analysis. It should be noted that the values $\omega_1 = \omega + \mathbf{k}_1\cdot\mathbf{v}_2$ and the zero of $\varepsilon(-\mathbf{k}_1, \omega-\omega_1) = 0$ lie above the integration contour C_1 in the complex ω plane. Below the contour C_1 — with the largest imaginary part — are the values $\omega_1 = \mathbf{k}_1\cdot\mathbf{v}_1$ and $\omega_1 = \mathbf{k}_1\cdot\mathbf{v}_2$. The contribution

$$-\int_{C_1} \frac{d\omega_1}{2\pi} \frac{1}{-i\omega_1 + i\mathbf{k}_1\cdot\mathbf{v}_1}\left[1 - \frac{1}{\varepsilon(-\mathbf{k}_1, \omega-\omega_1)}\right]\frac{f(\mathbf{v}_1)}{(2\pi)^3}$$

$$\approx -\frac{f(\mathbf{v}_1)}{(2\pi)^3}\left[1 - \frac{1}{\varepsilon(-\mathbf{k}_1, \omega - \mathbf{k}_1 \cdot \mathbf{v}_1)}\right] \tag{4.303}$$

was evaluated by closing the contour C_1 in the lower half-plane. Furthermore,

$$\int_{C_1} \frac{d\omega_1}{2\pi} \frac{1}{-i\omega_1 + \mathbf{k}_1 \cdot \mathbf{v}_1} \int d^3v_2 \frac{1}{\varepsilon(\mathbf{k}_1, \omega_1)} \frac{f(\mathbf{v}_2)}{\omega_1 - \mathbf{k}_1 \cdot \mathbf{v}_2} = 0\,; \tag{4.304}$$

also holds; this is most easily seen if one closes C_1 in the upper half-plane. There remains further

$$\begin{aligned}
&\lim_{\omega\to 0} \int_{C_1} \frac{d\omega_1}{2\pi} \frac{1}{-i\omega_1 + i\mathbf{k}_1 \cdot \mathbf{v}_1} \frac{1}{\varepsilon(-\mathbf{k}_1, \omega - \omega_1)} \frac{1}{\varepsilon(\mathbf{k}_1, \omega_1)} \\
&\times \mathbf{k}_1 \cdot \partial_{\mathbf{v}_1} f(\mathbf{v}_1) \frac{\varphi(k_1)}{m_e} \int d^3v_2 f(\mathbf{v}_2)\left[\frac{-1}{\omega_1 - \mathbf{k}_1 \cdot \mathbf{v}_2} + \frac{1}{\omega_1 - \omega - \mathbf{k}_1 \cdot \mathbf{v}_2}\right].
\end{aligned} \tag{4.305}$$

In the limit $\omega \to 0$ we can bring the contour C_1 close to the real ω axis. Moreover, on the right-hand side of (4.287) we only need the imaginary part of g, i.e., we are only interested in the imaginary parts of (4.303) and (4.305). The imaginary part of (4.305) is obtained via the Plemelj formula and the resulting relation

$$\begin{aligned}
&\mathrm{Re} \lim_{\omega\to 0} \frac{1}{\omega_1 - \mathbf{k}_1 \cdot \mathbf{v}_1}\left[\frac{-1}{\omega_1 - \mathbf{k}_1 \cdot \mathbf{v}_2} + \frac{1}{\omega_1 - \omega - \mathbf{k}_1 \cdot \mathbf{v}_2}\right] \\
&= 2\pi^2 \delta(\omega_1 - \mathbf{k}_1 \cdot \mathbf{v}_1)\delta(\omega_1 - \mathbf{k}_1 \cdot \mathbf{v}_2).
\end{aligned} \tag{4.306}$$

A short calculation leads to the result

$$\pi \mathbf{k}_1 \cdot \partial_{\mathbf{v}_1} f(\mathbf{v}_1) \frac{\varphi(k_1)}{m_e} \int d^3v_2 \frac{\delta[\mathbf{k}_1 \cdot (\mathbf{v}_1 - \mathbf{v}_2)] f(\mathbf{v}_2)}{|\varepsilon(\mathbf{k}_1, \mathbf{k}_1 \cdot \mathbf{v}_1)|^2} \tag{4.307}$$

for the imaginary part of (4.305). A calculation of the imaginary part of (4.303) yields

$$\begin{aligned}
&\frac{f(\mathbf{v}_1)}{(2\pi)^3} \frac{\mathrm{Im}\ \varepsilon(\mathbf{k}_1, \mathbf{k}_1 \cdot \mathbf{v}_1)}{|\varepsilon(\mathbf{k}_1, \mathbf{k}_1 \cdot v_1)|^2} \\
&= -\frac{\pi f(\mathbf{v}_1)\varphi(k_1)}{m_e |\varepsilon(\mathbf{k}_1, \mathbf{k}_1 \cdot \mathbf{v}_1)|^2} \int d^3v_2 [\mathbf{k}_1 \cdot \partial_{\mathbf{v}_2} f(\mathbf{v}_2)] \delta[\mathbf{k}_1 \cdot (\mathbf{v}_1 - \mathbf{v}_2)].
\end{aligned} \tag{4.308}$$

If we use the results (4.302)–(4.308) in (4.287), we obtain the Balescu-Lenard equation (the order of the names varies in the literature)

$$d_t f(\mathbf{v}_1, t) = -\frac{8\pi^4}{m_e^2} \partial_{\mathbf{v}_1} \int d^3k_1\, d^3v_2\, \varphi^2(k_1) \frac{\mathbf{k}_1 \mathbf{k}_1}{|\varepsilon(\mathbf{k}_1, \mathbf{k}_1 \cdot \mathbf{v}_1)|^2}$$

$$\times\, \delta[\mathbf{k}_1 \cdot (\mathbf{v}_1 - \mathbf{v}_2)]\{f(\mathbf{v}_1)\partial_{\mathbf{v}_2} f(\mathbf{v}_2) - f(\mathbf{v}_2)\partial_{\mathbf{v}_1} f(\mathbf{v}_1)\}\,. \tag{4.309}$$

To facilitate comparison with the Landau equation, we write the collision term, i.e., the right-hand side of (4.309), as

$$\overline{K} = \frac{1}{m_e^2}\partial_{\mathbf{v}} \cdot \int d^3v' \mathcal{Q}(\mathbf{v}, \mathbf{v}\,') \cdot (\partial_{\mathbf{v}} - \partial_{\mathbf{v}\,'})f(\mathbf{v})f(\mathbf{v}\,'), \tag{4.310}$$

with $\mathbf{v} = \mathbf{v}_1$ and $\mathbf{v}\,' = \mathbf{v}_2$. Here, for the tensor, we have

$$\boxed{\mathcal{Q} = 8\pi^4 \int d^3k \frac{\mathbf{k}\mathbf{k}\,\varphi^2(k)}{|\varepsilon(\mathbf{k}, \mathbf{k}\cdot\mathbf{v})|^2}\delta[\mathbf{k}\cdot(\mathbf{v} - \mathbf{v}')]}\,. \tag{4.311}$$

We compare this result with (4.220) and (4.225). We find complete agreement except for the screening factor $|\varepsilon|^2$ in the denominator on the right-hand side of (4.311). For small values of $\mathbf{k}\cdot\mathbf{v}$, i.e., in the so-called static limit, we have $\varepsilon \approx 1 + 1/k^2\lambda_{De}^2$, and the Landau approximation with Debye fields becomes understandable. For large velocities, $\varepsilon \approx 1$, and the calculation from the previous paragraph becomes relevant. We refrain from a detailed transformation of (4.311) for all velocity ranges at this point, since an analogous calculation was carried out in the previous section.

The fundamental difference now is that, due to screening, we no longer obtain a divergence for small k values; the divergence for large k values (small distances) still remains. In particular, we see that in the static approximation

$$\varepsilon \approx 1 + \frac{1}{k^2\lambda_{De}^2} \tag{4.312}$$

an effective potential appears, as was already used in (4.231). Thus, the results of Lenard and Balescu lead in an elegant way to a satisfactory completion of the kinetic equations, at least as far as convergence at small wavenumbers is concerned.

To improve our physical understanding, we now turn to the question of why $\varphi^2/|\varepsilon|^2$ correctly describes the effective interaction between two particles with their respective screening clouds. For this, we employ the so-called test particle model, in which the motion of a single particle in the Vlasov plasma can be studied. (We have already treated the problem of the screening of a stationary particle earlier; there, the potential was recognized as a Debye potential.)

Let the test particle move with velocity $\mathbf{v}\,'$ in a plasma with the charge density

$$\rho_p = \sum_\alpha q_\alpha \int d^3v f^\alpha(\mathbf{v}), \tag{4.313}$$

which, together with the charge density of the point charge, enters the Poisson equation

$$\nabla^2 \phi = -\frac{1}{\varepsilon_0} q\delta(\mathbf{r} - \mathbf{r}'_0 - \mathbf{v}'t) - \frac{1}{\varepsilon_0}\rho_p \tag{4.314}$$

We calculate the collective response of the plasma using the linearized Vlasov equation

$$\frac{\partial \delta f^\alpha}{\partial t} + \mathbf{v} \cdot \nabla \delta f^\alpha = \frac{q_\alpha}{m_\alpha}(\nabla \phi) \cdot \partial_{\mathbf{v}} f_0^\alpha. \tag{4.315}$$

Laplace transformation in time and Fourier transformation in space lead to

$$(p + i\mathbf{k} \cdot \mathbf{v})\delta f_k^\alpha = \frac{q_\alpha}{m_\alpha}\phi_k i\mathbf{k} \cdot \partial_{\mathbf{v}} f_0^\alpha, \tag{4.316}$$

$$k^2 \phi_k = \frac{q \exp(-i\mathbf{k} \cdot \mathbf{r}'_0)}{\varepsilon_0 (p + i\mathbf{k} \cdot \mathbf{v}')(2\pi)^3} + \frac{1}{\varepsilon_0}\sum_\alpha q_\alpha \int \delta f^\alpha d^3 v. \tag{4.317}$$

Combining (4.316) and (4.317) yields

$$\phi_k = \frac{q \exp(-i\mathbf{k} \cdot \mathbf{r}'_0)}{(2\pi)^3 \varepsilon_0 k^2 (p + i\mathbf{k} \cdot \mathbf{v}')\varepsilon(\mathbf{k}, ip)}. \tag{4.318}$$

In this calculation, we have assumed the undisturbed plasma to be quasi-neutral with a Maxwellian distribution. The trajectory of the test particle is assumed to be a straight line; deviations from the straight path (in higher-order approximations) are neglected. The response of the plasma is well described up to order $1/\Lambda$ by the Vlasov approximation. From (4.318) we calculate $\phi(\mathbf{r}, t)$ by inverse transformation. Since we are currently not interested in transient processes, we consider the time-asymptotic limit. First, from (4.318) it follows that

$$\begin{aligned} \phi_k(\mathbf{k}, t) = {} & \frac{q \exp(-i\mathbf{k} \cdot \mathbf{r}'_0) \exp(-i\mathbf{k} \cdot \mathbf{v}'t)}{\varepsilon_0 (2\pi)^3 k^2 \varepsilon(\mathbf{k}, \mathbf{k} \cdot \mathbf{v}')} \\ & + \sum_j \frac{q \exp(-i\mathbf{k} \cdot \mathbf{r}'_0) \exp(p_j t)}{(2\pi)^3 \varepsilon_0 k^2 (p_j + i\mathbf{k} \cdot \mathbf{v}') d\varepsilon/dp\,|_{p_j}} \end{aligned} \tag{4.319}$$

when evaluated using the residue theorem. The zeros of $\varepsilon(\mathbf{k}, ip)$ are denoted by p_j. The second term on the right-hand side of (4.319) is negligible in the limit $t \to \infty$ for a stable Vlasov plasma. Ultimately, from (4.319) we obtain

$$\boxed{\phi(\mathbf{r}, t) \approx \frac{q}{(2\pi)^3 \varepsilon_0} \int \frac{\exp[i\mathbf{k} \cdot (\mathbf{r} - \mathbf{r}'_0 - \mathbf{v}'t)]}{k^2 \varepsilon(\mathbf{k}, \mathbf{k} \cdot \mathbf{v}')} d^3 k}\,. \tag{4.320}$$

This expression sufficiently explains the occurrence of the screened potentials in (4.311). We will now explicitly consider some limiting cases.

Example 4.15 (Screening of a Test Particle)
For small velocities of the test particle ($v' \ll v_{ti}$) we can approximate

$$\varepsilon(\mathbf{k}, \mathbf{k} \cdot \mathbf{v}') \approx 1 + \frac{1}{k^2 \lambda_D^2} \tag{4.321}$$

and obtain in a Maxwellian plasma the expected Debye potential

$$\phi(\mathbf{r}, t) = \frac{1}{4\pi\varepsilon_0} \frac{q}{|\mathbf{r} - \mathbf{r}'_0 - \mathbf{v}'t|} \exp\{-|\mathbf{r} - \mathbf{r}'_0 - \mathbf{v}t'|/\lambda_D\} \tag{4.322}$$

with screening by electrons and ions. At high velocities ($v' \gg v_{te}$) we have $\varepsilon \approx 1$ and

$$\phi(\mathbf{r}, t) = \frac{1}{4\pi\varepsilon_0} \frac{q}{|\mathbf{r} - \mathbf{r}'_0 - \mathbf{v}'t|} \tag{4.323}$$

remains unscreened. A screening charge cloud cannot form in such rapidly changing processes. In the intermediate velocity range ($v_{ti} \ll v' \ll v_{te}$) only the electrons contribute to screening, since

$$\varepsilon(\mathbf{k}', \mathbf{k} \cdot \mathbf{v}') \approx 1 + \frac{1}{k^2 \lambda_{De}^2} \tag{4.324}$$

can be approximated. The potential ϕ corresponds to (4.322), but with the electron Debye length as the screening length. ■

In summary: The Balescu-Lenard equation solves, in a self-consistent manner, the problem of the cutoff at large interaction lengths. For this reason, it is of great theoretical significance.

For practical estimates, however, (extreme) simplifications, for example in the form

$$K^e = \nu_{ee}(f_{e0} - f_e) + \nu_{ei}(\overline{f}_{e0} - f_e) , \tag{4.325}$$

$$K^i = \nu_{ie}(\overline{f}_{i0} - f_i) + \nu_{ii}(f_{i0} - f_i) , \tag{4.326}$$

with Maxwellian distributions (e.g., also with $T_i = \overline{T}_e$, $\overline{T}_i = T_e$, $u^i = \overline{u}^e$, $\overline{u}^i = u^e$, where u^α characterizes the mean velocities) are very useful. Here, the frequencies $\nu_{\alpha\beta}$ can be identified approximately with the relaxation rates for momentum ("slowing down rates") $\nu_s^{\alpha\beta} = (1 + m_\alpha/m_\beta)\psi[m_\beta(u^\alpha)^2/2k_B T_\beta]\nu_0^{\alpha\beta}$, where $\nu_0^{\alpha\beta} \approx (n\lambda_{D\alpha}^3)^{-1}\omega_{p\alpha} \ln \Lambda_B$ and $\psi(x) = 2\pi^{-1/2}\int_0^x dt\, t^{1/2} e^{-t}$ apply. We refrain here from a more detailed discussion of these relationships—as well as from a calculation of the relaxation times for temperature, etc.—and refer to the specialized literature or the following chapter.

4.6 Boltzmann Equation and Lorentz Collision Term

In the previous sections, we have become acquainted with kinetic equations of the Fokker-Planck type, which correctly describe the dynamics in fully ionized plasmas for not too large values of Λ. A key point in the derivation was that large deflections of the charged particles do not occur as a result of single collisions, but rather as the result of many small-angle deflections. We now want to address two subsequent questions: First, what is—in the kinetic description—the fundamental difference compared to a weakly ionized plasma? Second, can one specify a somewhat simpler, approximate form of the collision term for estimates?

In the case of a weakly ionized plasma, the individual collisions between charged particles can be neglected, while the short-range interaction of the charged particles with the neutrals dominates. The kinetic equation for such conditions, at least as far as the collision term is concerned, initially has no specific plasma properties. It can be borrowed from the kinetic theory of gases and is known as the Boltzmann equation. We use it here, after a brief plausibility consideration, to characterize the essential differences from the fully ionized case. In the notation (4.39), i.e.,

$$\boxed{\frac{\partial f^\alpha}{\partial t} + \mathbf{v} \cdot \nabla f^\alpha + \frac{1}{m_\alpha} \mathbf{F} \cdot \partial_{\mathbf{v}} f^\alpha = K^\alpha,} \tag{4.327}$$

the collision term K^α ($\alpha = e, i$) now has to be recalculated. The binary collision assumption and molecular chaos are the two most important cornerstones in defining K^α, when one wants to derive the Boltzmann equation from the BBGKY hierarchy (see, for example, the already cited book by G. Ecker). In addition, walls are not taken into account and the influence of external forces on the collision process is neglected. The velocity of a particle and its position are assumed to be uncorrelated. Here, since the question of derivation from the BBGKY hierarchy is not central to our interest, we follow Boltzmann's original derivation.

To calculate the dynamics of the scatterer and the scattered particle, we consider an impact area $b db d\varphi = (d\sigma/d\Omega) d\Omega$, where b is the impact parameter and φ denotes the azimuthal angle. The flux of particles of type α onto this area (with velocities in the volume element of velocity space d^3v around $\mathbf{v}$) is

$$f^\alpha(\mathbf{r}, \mathbf{v}; t)|\mathbf{v} - \mathbf{v}_1| d^3v \; b \, db \, d\varphi. \tag{4.328}$$

If one takes into account the number of scattering particles with a velocity $\mathbf{v}_1$ (in the volume element d^3v_1), i.e., $f^\beta(\mathbf{r}, \mathbf{v}_1; t)d^3v_1$, then, per unit time and per unit volume,

$$f^\alpha(\mathbf{r}, \mathbf{v}; t)|\mathbf{v} - \mathbf{v}_1|f^\beta(\mathbf{r}, \mathbf{v}_1; t)\, b\, db\, d\varphi\, d^3v_1 d^3v \tag{4.329}$$

collisions occur. Here, for the first time, the assumption of molecular chaos enters: The number of particle pairs in a volume element d^3r, whose velocities are in d^3v_1 around $\mathbf{v}_1$ and d^3v around $\mathbf{v}$, is given precisely by $f^\alpha(\mathbf{r}, \mathbf{v}; t)f^\beta(\mathbf{r}, \mathbf{v}_1; t)d^3r\, d^3r\, d^3v\, d^3v_1$. The change in f^α occurs because, through binary collisions, on the one hand, particles are "scattered out" of the considered velocity range, and on the other hand, particles from the velocity range ($\mathbf{v}', \mathbf{v}' + d\mathbf{v}'$ or $\mathbf{v}_1', \mathbf{v}_1' + d\mathbf{v}_1'$) are "scattered into" the considered velocity range. Let us first calculate the loss

$$-d^3v \sum_\beta \int d^3v_1 |\mathbf{v} - \mathbf{v}_1| f^\alpha(\mathbf{r}, \mathbf{v}; t) f^\beta(\mathbf{r}, \mathbf{v}_1; t)\, b\, db\, d\varphi, \tag{4.330}$$

by integrating (4.329) over the velocities of the scatterers and summing over all types of scatterers. The gain from the velocity range ($\mathbf{v}', \mathbf{v}' + d\mathbf{v}'$) is formally written in a similar way, namely as

$$+d^3v' \sum_\beta \int d^3v_1' |\mathbf{v}' - \mathbf{v}_1'| f^\alpha(\mathbf{r}, \mathbf{v}'; t) f^\beta(\mathbf{r}, \mathbf{v}_1'; t) b' db' d\varphi'. \tag{4.331}$$

In order to combine the two expressions (4.330) and (4.331), we must exploit symmetries and conservation laws in the scattering process. We are dealing with binary collisions of the type $\{\mathbf{v}, \mathbf{v}_1\} \rightarrow \{\mathbf{v}', \mathbf{v}_1'\}$ or $\{\mathbf{v}', \mathbf{v}_1'\} \rightarrow \{\mathbf{v}, \mathbf{v}_1\}$. These collisions are inverse to each other, i.e., we can use the symmetry of the differential scattering cross section,

$$\boxed{b\, db\, d\varphi = b' db' d\varphi'}\,. \tag{4.332}$$

From the conservation laws for momentum and energy we have

$$\boxed{m_\alpha \mathbf{v} + m_\beta \mathbf{v}_1 = m_\alpha \mathbf{v}' + m_\beta \mathbf{v}_1',} \tag{4.333}$$

$$\boxed{\frac{1}{2} m_\alpha |\mathbf{v}|^2 + \frac{1}{2} m_\beta |\mathbf{v}_1|^2 = \frac{1}{2} m_\alpha |\mathbf{v}'|^2 + \frac{1}{2} m_\beta |\mathbf{v}_1'|^2.} \tag{4.334}$$

If we introduce the new variables

$$\mathbf{V} = \left(m_\alpha \mathbf{v} + m_\beta \mathbf{v}_1\right) / \left(m_\alpha + m_\beta\right), \tag{4.335}$$

$$\mathbf{u} = \mathbf{v}_1 - \mathbf{v} \tag{4.336}$$

and analogously define the variables $\mathbf{V}'$ and $\mathbf{u}'$, then the conservation laws (4.333) and (4.334) can be written in the form

$$\mathbf{V} = \mathbf{V}' , \tag{4.337}$$

$$|\mathbf{u}| = |\mathbf{u}'| \tag{4.338}$$

A collision (in the center-of-mass system) merely rotates $\mathbf{u}$ into $\mathbf{u}'$, without changing its magnitude. The scattering angles are θ and the previously introduced azimuthal angle φ. We imagine that $\mathbf{V}$ and $\mathbf{u}$ at constant θ and φ in $\mathbf{V} + d\mathbf{V}$ and $\mathbf{u} + d\mathbf{u}$ are modified. Thus, we consider two collisions with slightly altered initial conditions. Then $\mathbf{V}'$ and $\mathbf{u}'$ change accordingly to $\mathbf{V}' + d\mathbf{V}'$ and $\mathbf{u}' + d\mathbf{u}'$. Since $\mathbf{V} = \mathbf{V}'$ holds, we have $d\mathbf{V} = d\mathbf{V}'$. For fixed θ it must also be $d\mathbf{u} = d\mathbf{u}'$ Thus, we have

$$d^3 V d^3 u = d^3 V' d^3 u' , \tag{4.339}$$

since we can write each volume element as a product, e.g., $d^3 V = dV_x dV_y dV_z$ With (4.335) and (4.336), the functional determinant can easily be calculated and thus

$$d^3 V d^3 u = d^3 v \, d^3 v_1 \tag{4.340}$$

proven. The same applies to the primed quantities. Thus, if one varies $\mathbf{v}$ and $\mathbf{v}_1$ while keeping the scattering angles constant, then $\mathbf{v}'$ and $\mathbf{v_1}'$ change in such a way that

$$d^3 v \, d^3 v_1 = d^3 v' d^3 v_1' \tag{4.341}$$

holds. Conservation of energy and momentum thus yield

$$|\mathbf{v}' - \mathbf{v}_1'| d^3 v' d^3 v_1' = |\mathbf{v} - \mathbf{v}_1| d^3 v \, d^3 v_1 . \tag{4.342}$$

With this, we obtain

$$K^\alpha = \sum_\beta \int d^3 v_1 b \, db \, d\varphi |\mathbf{v} - \mathbf{v}_1| [f^\alpha(\mathbf{r}, \mathbf{v}'; t) f^\beta(\mathbf{r}, \mathbf{v}_1'; t) - f^\alpha(\mathbf{r}, \mathbf{v}; t) f^\beta(\mathbf{r}, \mathbf{v}_1; t)] \tag{4.343}$$

for the collision term in the Boltzmann equation. In understanding this term, we must keep in mind that $\mathbf{v}'$ and $\mathbf{v}_1'$ can be related to the collision process for a given interaction with $\mathbf{v}$ and $\mathbf{v}_1$. In the weakly ionized case and for $\alpha = e$ the summation over β, is omitted, since only neutral particles β can be considered as collision partners.

It is also clear that the binary collision assumption is not valid in a fully ionized plasma with many particles in the Debye sphere during the effective interaction process.

For simple calculations, the collision term K^α is often approximated by

$$\boxed{K^\alpha \approx \frac{f_0^\alpha - f^\alpha(\mathbf{r}, \mathbf{v}; t)}{\tau_\alpha}}\,, \tag{4.344}$$

where τ_α is the relaxation time or mean free flight time. In general, τ_α is velocity-dependent; however, in the approximation (4.344), an averaged τ_α is usually used.

The so-called Krook approach (4.344) is intended to express nothing more than the relaxation of the distribution function f^α towards an equilibrium distribution f_0^α, where f_0^α is generally taken as a local Maxwell distribution

$$f_0^\alpha = n_\alpha(\mathbf{r}, t)\left(\frac{m_\alpha}{2\pi k_B T_\alpha}\right)^{3/2} \exp\left(-m_\alpha v^2/2k_B T_\alpha\right) \tag{4.345}$$

This specification, together with

$$n_\alpha(\mathbf{r}, t) = \int f^\alpha(\mathbf{r}, \mathbf{v}; t) d^3 v, \tag{4.346}$$

at least guarantees particle conservation. An improved model, which conserves not only the particle number but also energy and momentum, was proposed by Bhatnagar, Gross, and Krook. Models in which quasi-stationary distributions of the scatterers are assumed for the collision processes are also referred to as Lorentz models.

We do not wish to go into these particularities in weakly ionized plasmas any further here, but rather ask whether similar simplifications are also possible in fully ionized plasmas. In this case, we are dealing with two components of quite different masses, and we cannot expect the simplified forms of the collision terms to be identical for both components. For collisions of electrons with ions, we can, by analogy with collisions of electrons with neutral particles, assume the mass of the ions to be very large and thus treat the ions as quasi-stationary. For the distribution functions f^e and f^i, this means that the electron distribution function is much broader than the ion distribution function.

The maximum of f^e is located at $\mathbf{u}_e$, while that of f^i is at $\mathbf{u}_i$. For the difference velocity, the following holds

$$\boxed{|\mathbf{u}_e - \mathbf{u}_i| \ll v_{te}}\,, \tag{4.347}$$

and we also assume

$$\boxed{\frac{T_i}{T_e} \ll \frac{m_i}{m_e}} \tag{4.348}$$

Let us now consider the Landau tensor (4.236), which depends on the difference velocity $\mathbf{v}_1 - \mathbf{v}_2$. If we choose the velocity $\mathbf{u}_e$, as the reference, then we can expand with respect to $\mathbf{v}_2 - \mathbf{u}_e$, since in the collision term f^i appears as a weighting function, which decays very rapidly for larger values of $|\mathbf{v}_2 - \mathbf{u}_e|$. With

$$\begin{aligned} G_{rs}(\mathbf{v}_1 - \mathbf{u}_e - (\mathbf{v}_2 - \mathbf{u}_e)) &\approx G_{rs}(\mathbf{v}_1 - \mathbf{u}_e) - (v_{2\nu} - u_{e\nu})\partial v_{1\nu} G_{rs}(\mathbf{v}_1 - \mathbf{u}_e) \\ &+ \frac{1}{2}(v_{2\nu} - u_{e\nu})(v_{2\mu} - u_{e\mu})\partial v_{1\nu}\partial v_{1\mu} G_{rs}(\mathbf{v}_1 - \mathbf{u}_e) - \ldots \end{aligned} \tag{4.349}$$

we obtain approximately for the integral

$$\begin{aligned} &\int d^3v_2 G_{rs}(\mathbf{v}_1 - \mathbf{v}_2)\left(\partial_{v_{1s}} - \frac{m_e}{m_i}\partial_{v_{2s}}\right) f^e(\mathbf{v}_1) f^i(\mathbf{v}_2) \\ &\approx n_i \left\{G_{rs}(\mathbf{v}_1) + (u_{e\nu} - u_{i\nu})\left[\partial_{v_{1\nu}} G_{rs}(\mathbf{v}_1)\right]\right. \\ &+ \left.\frac{1}{2}\alpha_{nm}\left[\partial_{v_{1n}}\partial_{v_{1m}} G_{rs}(\mathbf{v}_1)\right]\right\} \partial_{v_{1s}} f^e(\mathbf{v}_1). \end{aligned} \tag{4.350}$$

Thus, a Lorentz form of the collision term for electrons with ions results. This collision term does not depend very sensitively on the form of the ion distribution function; only the first moments n_i, $\mathbf{u}_i$, T_i, π^i_{nm} enter, if we take into account the definitions

$$G_{rs}(\mathbf{v}_1) := \frac{|\mathbf{v}_1 - \mathbf{u}_e|^2\,\delta_{rs} - (v_{1r} - u_{er})(v_{1s} - u_{es})}{|\mathbf{v}_1 - \mathbf{u}_e|^3}, \tag{4.351}$$

$$\alpha_{nm} := k_B \frac{T_i}{m_i}\delta_{nm} + \frac{1}{n_i m_i}\pi^i_{nm} + (u_{en} - u_{in})(u_{em} - u_{im}) \tag{4.352}$$

Here, π^i_{nm} is the dissipative pressure tensor. In fact, the first two terms on the right-hand side of (4.352) are small compared to the third, so that the overall form of the collision term does not become very complicated.

The situation is different if we want to simplify the collisions of ions with electrons. Although the Landau tensor can again be expanded, now the velocity deviations, over which no integration is performed, are small. In the integrand, the electron distribution function remains together with functions of the electron velocity (caused by the Landau tensor), which do not lead to simple moments. The entire (simplified) collision term for ion-electron collisions thus depends more strongly on the details of the electron distribution function than, in the reverse case of electron-ion collisions, the ion distribution function does.

4.7 Kinetics for Strongly Magnetized Plasmas

To conclude the chapter on kinetic equations for a plasma, we will address another important special case. Of course, there are a multitude of interesting cases that could still be discussed, but a comprehensive overview would far exceed the scope of this book. We will address the issue of non-ideal plasmas at the end of the book. Here, as an example, we will focus only on the case of a strongly magnetized classical plasma, in which the motion of individual particles is very well described by the drift approximation. In this case, it is natural to seek a description in which the guiding centers appear as quasi-particles.

The idea behind the following considerations is to simplify the kinetic description of a magnetized plasma (for $\rho_i/L \ll 1$). If we consider the particles, we can decompose their motion into rapid gyrations (with a small gyroradius ρ) and a drift in the direction perpendicular to the magnetic field. The motion parallel to the magnetic field is largely unimpeded, so we retain $v_\parallel$ as an independent variable.

From a simplified perspective, the exact particle positions are replaced by the gyrocenters, the perpendicular velocities are identified as drift velocities, and $\mathbf{r}, v_\parallel, t$ remain as independent variables. As we will see in the next subsection, such a strategy leads to the **drift-kinetic** equation.

However, we should keep in mind that by replacing the particle positions with their gyrocenters, spatial scales on the order of the gyroradius can no longer be resolved. Note that we have a large scale L (on the order of the minor toroidal radius or the density gradient scale). Rapid variations can occur for small, but not necessarily linear, deviations from the stationary state.

If the wavelengths of the rapid variations are on the order of the ion Larmor radius ρ_i, electric fields also vary on the fast scale, and the particles experience different field strengths during their gyro-motion. Taking this effect into account is the goal of the so-called **gyrokinetic** description, which will be discussed in the second subsection.

However, regarding the latter, we should mention that the full nonlinear theory causes several problems; see [55, 56] and the references therein. The main reason is that when decomposing f into

$$f = f_{slow} + \Delta f_{fast} \tag{4.353}$$

$\Delta \ll 1$ is required for magnetized plasmas, since otherwise the plasma would be effectively demagnetized for $\Delta \sim \mathcal{O}(1)$ and $\lambda_{fast} \sim \rho_i$. In the following, we present only a brief overview of the general procedure. For further details, we refer to the literature [55, 56] and the references therein.

Drift-Kinetic Equation

Let us first proceed in a purely *heuristic* manner, and introduce for the quasi-particles a position vector $\mathbf{q}$, the magnetic moment μ, and the (with respect to the external magnetic field) parallel velocity component $v_\parallel$ as coordinates. The gyrophase does not appear; we imagine an averaging over the rapid gyration has been performed.

In order to write down a kinetic equation for

$$f = f_D(v_\parallel, \mu, \mathbf{q}; t) \tag{4.354}$$

(in the Vlasov approximation), we again start from Liouville's theorem and write

$$\boxed{\partial_t f_D + \nabla_\perp \cdot (\mathbf{v}_D f_D) + v_\parallel \partial_z f_D + \partial_{v_\parallel}\left(\frac{F_\parallel}{m} f_D\right) = 0}\,, \tag{4.355}$$

where $\mathbf{v}_D$ is the drift velocity of the particles (electrons or ions) and $F_\parallel$ is the component of the driving force parallel to the magnetic field. The terms $\mathbf{v}_D$ and $F_\parallel$ were discussed in detail in previous chapters and do not require a renewed derivation here. For time-dependent fields, we must in particular also take into account the polarization drift for ions. Due to the different masses, the drift velocities for electrons and ions are different. In the drift-kinetic equation for ions we use

$$\mathbf{v}_{Di} = \frac{\mathbf{E} \times \mathbf{B}}{B^2} + \frac{1}{|\Omega_i| B} \frac{d\mathbf{E}}{dt} + \frac{v_\parallel^2}{|\Omega_i|} (\nabla \times \mathbf{b})_\perp + \frac{\mu_i}{eB} \mathbf{b} \times \nabla_\perp B + v_\parallel \frac{\mathbf{B}_\perp}{B}, \tag{4.356}$$

while for electrons

$$\mathbf{v}_{De} = \frac{\mathbf{E} \times \mathbf{B}}{B^2} - \frac{v_\parallel^2}{|\Omega_e|} (\nabla \times \mathbf{b})_\perp - \frac{\mu_e}{eB} \mathbf{b} \times \nabla_\perp B + v_\parallel \frac{\mathbf{B}_\perp}{B} \tag{4.357}$$

holds approximately.

Example 4.16 (Formulation for Electrons with Other Variables)
As just mentioned, a kinetic equation for magnetized plasmas should be much simpler than, for example, the Landau-Fokker-Planck equation, since it suppresses details on the gyroradius scale. The guiding centers (gyration centers) have a velocity $\mathbf{v}_{gc} = \hat{b} v_\parallel + \mathbf{v}_d$, with the drift velocities $\mathbf{v}_d = \mathbf{v}_E + \frac{1}{\Omega} \hat{b} \times \left(\frac{\mu}{m} \nabla B + v_\parallel^2 \,\check{}\, \right)$. The latter consist of the $E \times B$ drift $\mathbf{v}_E$ and the curvature drift ($\check{} = \hat{b} \cdot \nabla \hat{b}$). Note that to zeroth order and with the scaling

$$\Delta \equiv 0 \quad , \quad \frac{v_E}{v_{th}} \sim \delta \quad , \quad \partial_t \sim \delta \, \Omega \tag{4.358}$$

the electron polarization drift is of higher order. μ is the magnitude of the magnetic moment

$$\mu = \frac{m v_\perp^2}{2B} \quad , \quad \frac{d\mu}{dt} = \mathcal{O}(\delta) \; ; \tag{4.359}$$

μ is an adiabatic invariant.

The total energy of a gyration center is the sum of kinetic and potential energy,

$$U = \frac{m v_\parallel^2}{2} + \mu \, B + e \, \phi \, . \tag{4.360}$$

Note that the last two terms represent the potential energy of a gyration center. In calculating the temporal change of the energy, we make the ansatz

$$\frac{d}{dt} \left[\frac{m}{2} \, v_\parallel^2 \right] \approx \mathbf{v}_{gc} \cdot [e\mathbf{E} - \nabla(\mu \, B)] \tag{4.361}$$

for the change in kinetic energy. The second term on the right-hand side accounts for the mirror force $\mathbf{F}_m = -\nabla(\mu \, B)$. One finds

$$\begin{aligned} \frac{dU}{dt} &\equiv \left(\frac{\partial}{\partial t} + \mathbf{v}_{gc} \cdot \nabla \right) U \approx e \, \frac{d\phi}{dt} + \mu \, \frac{dB}{dt} + \mathbf{v}_{gc} \cdot [e\mathbf{E} - \nabla(\mu \, B)] \\ &= e \, \frac{d\phi}{dt} + \mu \, \frac{dB}{dt} - e \, \mathbf{v}_{gc} \cdot \frac{\partial \mathbf{A}}{\partial t} \, . \end{aligned} \tag{4.362}$$

Without carrying out the exact calculation (see the remarks below), we expect that the distribution function averaged over the gyrophase $\bar{f} = \bar{f}(\mathbf{r}, U, t)$ obeys the drift-kinetic equation:

$$\boxed{\frac{\partial \bar{f}}{\partial t} + \mathbf{v}_{gc} \cdot \nabla \bar{f} + \frac{dU}{dt} \frac{\partial \bar{f}}{\partial U} = 0}. \tag{4.363}$$

In this form of the drift-kinetic equation, we still need to substitute the expression for dU/dt. ■

Example 4.17 (Drift Instability)

Before we turn to the theoretical justification of (4.355), let us first demonstrate the enormous advantage of this drift-kinetic description. We calculate the conductivity of a plasma consisting of particles with charge e and mass m with a spatially inhomogeneous density distribution $n_0(\mathbf{r})$. Furthermore, we assume purely electrostatic perturbations that occur in the plane perpendicular to the density gradient, which is assumed to be linear (in the $q_x = x$ direction). In a system with straight magnetic field lines, $\mathbf{v}_D$ holds:

$$\mathbf{v}_D = \frac{\mathbf{E} \times \mathbf{B}}{B^2} + \frac{1}{\Omega B} \frac{d\mathbf{E}}{dt}. \tag{4.364}$$

If we insert

$$f_D = f_0 + f_1 \tag{4.365}$$

and

$$\mathbf{E} \approx -\nabla \varphi_1 = -i(k_\perp \hat{y} + k_\parallel \hat{z})\varphi_1 \tag{4.366}$$

into (4.355), where $F_\parallel = eE_\parallel$ is, then after linearization we obtain

$$f_1 = \left[\frac{\frac{e}{m} i k_\parallel \frac{\partial f_0}{\partial v_\parallel} + \frac{ik_\perp}{B} \frac{\partial f_0}{\partial x}}{i(k_\parallel v_\parallel - \omega)} - \frac{m}{eB^2} k_\perp^2 f_0 \right] \varphi_1 . \tag{4.367}$$

The last term on the right-hand side of (4.367) originates from the polarization drift

$$\mathbf{v}_p = \frac{1}{\Omega B} \frac{d\mathbf{E}}{dt}, \tag{4.368}$$

where

$$\frac{d}{dt} \to -i(\omega - k_\parallel v_\parallel) \tag{4.369}$$

can be substituted in a Fourier transformation. Typically, f_0 is approximated by a Maxwellian distribution in $v_\parallel$ with x-dependent density, so that after appropriate multiplication by e and integration of (4.367) using the definition

$$\partial_t \int e f_1 dv_\parallel d\mu \equiv e\partial_t n_1 = -\nabla \cdot \mathbf{j}_1 \equiv \sigma \nabla^2 \varphi_1 \tag{4.370}$$

$$\boxed{\sigma = \frac{-i\varepsilon_0 \omega \omega_p^2}{k^2 v_t^2}\left[1 + \frac{\omega - \omega^*}{\sqrt{2}k_\parallel v_t} Z\left(\frac{\omega}{\sqrt{2}k_\parallel v_t}\right)\right] - i\varepsilon_0 \omega \frac{\omega_p^2}{\Omega^2}\frac{k_\perp^2}{k^2}} \tag{4.371}$$

results. Here,

$$\omega^* = -\frac{k_\perp \kappa_n v_t^2}{\Omega} \tag{4.372}$$

is the drift frequency with

$$\kappa_n = -\frac{\partial \ln f_0}{\partial x} . \tag{4.373}$$

Compared to a homogeneous plasma, a Doppler shift occurs, where the magnitude of the shift by $\omega^* \equiv k_\perp v_d$ depends on the strength of the effective diamagnetic drift v_d. The frequency ω^* belongs to the so-called drift mode; however, its physical interpretation requires a somewhat more detailed calculation.

First of all, from (4.371), one sees that Re $\sigma < 0$ becomes for $\omega < \omega^*$. As we will see later, this signifies the onset of an instability, in this case the drift instability (also called the universal instability). If we restrict ourselves to the case $\beta < m_e/m_i$, in order to continue working electrostatically, and require that the wavelengths are larger than the mean gyroradii (for both electrons and ions), then in σ the contribution of the polarization drift is negligible. We can therefore use

$$\sigma \approx -\frac{i\varepsilon_0 \omega \omega_p^2}{k^2 v_t^2}\left[1 + \frac{\omega - \omega^*}{\sqrt{2}k_\parallel v_t} Z\left(\frac{\omega}{\sqrt{2}k_\parallel v_t}\right)\right] \tag{4.374}$$

instead of (4.371). Furthermore, we set the range of the relevant phase velocity to

$$\boxed{v_{te} \gg \frac{\omega}{k_\parallel} \gg v_{ti}} \tag{4.375}$$

This means that we evaluate (4.374) for electrons and ions in different approximations. A short calculation then leads (via the quasineutrality condition $\sigma_i + \sigma_e \approx 0$) to the dispersion relation

$$\varepsilon \approx \frac{1}{k^2 \lambda_{De}^2} + \frac{\omega_{pi}^2 k_\perp^2}{\Omega_i^2 k^2} - \frac{\omega_{pi}^2 k_\parallel^2}{\omega^2 k^2} + \frac{\omega_{pe}^2 k_\perp}{|\Omega_e| \omega k^2}\frac{\partial}{\partial x} \ln n_0$$

$$+\,i\frac{\pi e}{\varepsilon_0 k_\parallel k^2 B}\left(k_\perp\frac{\partial}{\partial x}-|\Omega_e|k_\parallel\frac{\partial}{\partial v_\parallel}\right)f_{oe}\left(\frac{\omega}{k_\parallel}\right)\approx 0. \tag{4.376}$$

The real part of the dispersion relation yields

$$\omega^2\left(1+\frac{c_s^2k_\perp^2}{\Omega_i^2}\right)-\omega k_\perp v_{de}-c_s^2k_\parallel^2=0\,, \tag{4.377}$$

with the diamagnetic drift

$$v_{de}=\frac{\omega_{pe}^2\lambda_{De}^2}{|\Omega_e|}\kappa_n\,. \tag{4.378}$$

If one introduces the parameter

$$\delta=1+\frac{c_s^2k_\perp^2}{\Omega_i^2} \tag{4.379}$$

one finds the solutions

$$\boxed{\omega=\frac{1}{2\delta}\left[k_\perp v_{de}\pm\sqrt{(k_\perp v_{de})^2+4\delta c_s^2k_\parallel^2}\,\right]}\,. \tag{4.380}$$

For fixed $k_\perp$, (4.380) describes two branches $\omega=\omega(k_\parallel)$: a modified ion-acoustic branch $\omega\approx k_\parallel c_s/\sqrt{\delta}$ and the (electrostatic) drift waves $\omega\approx\omega^*/\delta$.

The growth rate γ of the drift instability can be most easily calculated in the regime where the imaginary part of the dielectric constant ε is small. In this case, we evaluate

$$\gamma\approx-\left.\frac{\mathrm{Im}\ \varepsilon}{\partial\varepsilon/\partial\omega}\right|_{\omega_0} \tag{4.381}$$

We will be brief here, as a more detailed evaluation will be discussed in a later section. For now, we simply quote the result

$$\boxed{\gamma\approx\sqrt{\pi\tilde{\beta}}\frac{1}{k_\parallel}\left(\frac{k_\parallel v_{de}}{\delta}\right)^2\left(1-\frac{1}{\delta}\right)\exp\left[-\tilde{\beta}\left(\frac{\omega}{k_\parallel}\right)^2\right]}\,, \tag{4.382}$$

with $\tilde{\beta}=m_e/2k_BT_e$. ■

As has already been stated several times, a satisfactory derivation of the drift-kinetic equation from the Landau-Fokker-Planck equation requires a more systematic approach. In addition to the transformation to new variables [from **r**, **v** to **Y** (gyrocenter), U, μ, and φ (gyrophase)] the collision term must also be reconsidered. The basic idea is to rewrite the Landau-Fokker-Planck equation in the new variables and to recognize that the variation of the distribution function with respect to the gyrophase φ is rapid. This behavior suggests the use of a multiple-timescale formalism.

Furthermore, each part of the distribution function is split into a gyrophase-averaged and a rapidly oscillating part. We will do this shortly (for a systematic derivation see Balescu [57]), in order to arrive at

$$\boxed{\frac{\partial \bar{f}}{\partial t} + \left(v_\parallel \hat{b} + \mathbf{v}_d\right) \cdot \nabla_Y \bar{f} - e v_\parallel \, \hat{b} \cdot \frac{\partial \mathbf{A}}{\partial t} \frac{\partial \bar{f}}{\partial U} \approx \overline{K_0\{f_0, f_1\}} - \overline{K_1\{f_0, f_0\}}} \tag{4.383}$$

The right-hand side will represent the mean linear collision term. Its evaluation is by no means trivial. Note that the spatial derivative is now taken with respect to the gyration center **Y**. In addition, we have written the drift-kinetic equation for a species with electric charge e (including sign). In general, we obtain drift-kinetic equations for each species separately.

Let us now begin the systematic derivation of (4.383). In doing so, we follow the approach proposed by Balescu. The kinetic equation in the variables $\mathbf{q}$, $v_\parallel$, $v_\perp$ and φ can, due to (2.109)–(2.112), be written in the abbreviated form

$$\begin{aligned} &\frac{\partial f^\alpha}{\partial t} + \left[v_\parallel \mathbf{b} + v_\perp \mathbf{n}_1\right] \cdot \frac{\partial f^\alpha}{\partial \mathbf{q}} \\ &+ \left[v_\perp \mathbf{n}_2 \cdot \mathbf{D} + \frac{q_\alpha}{m_\alpha} \mathbf{b} \cdot \mathbf{E}\right] \frac{\partial f^\alpha}{\partial v_\parallel} + \left[-v_\parallel \mathbf{n}_2 \cdot \mathbf{D} + \frac{q_\alpha}{m_\alpha} \mathbf{n}_1 \cdot \mathbf{E}\right] \frac{\partial f^\alpha}{\partial v_\perp} \\ &+ \left[\frac{1}{\varepsilon} \Omega_\alpha + \mathbf{b} \cdot \mathbf{D} - \frac{v_\parallel}{v_\perp} \mathbf{n}_1 \cdot \mathbf{D} - \frac{q_\alpha}{m_\alpha v_\perp} \mathbf{n}_2 \cdot \mathbf{E}\right] \frac{\partial f^\alpha}{\partial \varphi} = K^\alpha \end{aligned} \tag{4.384}$$

Here, $K^\alpha = K^\alpha(f, f)$ denotes the collision term, which we do not wish to specify in detail here (or, in the Vlasov approximation, even neglect entirely). Furthermore, it should be noted that we have indicated the gyrofrequency Ω_α as "large" by the prefactor $\frac{1}{\varepsilon}$. Here, ε is the small (dimensionless) expansion parameter (and not the dielectric constant or anything similar).

The variables of the single-particle distribution function of species α,

$$f^\alpha = f^\alpha\left(\mathbf{q}, v_\parallel, v_\perp, \varphi; t\right), \tag{4.385}$$

have been discussed in detail in the drift approximation. At this point, a reminder is in order. Due to the Hamiltonian structure of the kinetic equation (apart from the collision term), the transformation to the new (pseudocanonical) variables can be written down immediately using the corresponding fundamental Lie brackets.

To introduce the drift approximation into kinetic theory, we perform all those pseudo-canonical transformations that systematically remove the phase φ from the equations of motion, as described earlier. When considering the possible sets of variables for the drift-kinetic equation, it also becomes apparent that even more (possibly adiabatic) constants of motion can be introduced than have been discussed so far. In addition to the magnetic moment M, the energy $\mathcal{E}$ is also suitable. With the parallel velocity component $v_{\parallel}^{\alpha}$ and the drift velocity $\mathbf{v}_D^{\alpha}$ we can write for

$$f^{\alpha} = f^{\alpha}(\mathbf{q}, \mathcal{E}, M, \phi; t) \tag{4.386}$$

the following:

$$\begin{aligned} &\frac{\partial f^{\alpha}}{\partial t} + \left[v_{\parallel}^{\alpha}\mathbf{b} + \varepsilon \mathbf{v}_D^{\alpha}\right] \cdot \frac{\partial f^{\alpha}}{\partial \mathbf{q}} + q_{\alpha} v_{\parallel}^{\alpha} E_{\parallel} \frac{\partial f^{\alpha}}{\partial \mathcal{E}} \\ &+ \left[\frac{1}{\varepsilon}\Omega_{\alpha} + v_{\parallel}^{\alpha}\mathbf{b} \cdot \mathbf{R} - \frac{1}{2} v_{\parallel}^{\alpha}\mathbf{b} \cdot (\nabla \times \mathbf{b})\right] \frac{\partial f^{\alpha}}{\partial \phi} = K^{\alpha} , \end{aligned} \tag{4.387}$$

where $v_{\parallel}^{\alpha}$ is expressed by $\mathcal{E}, M, \phi$ and $\mathbf{q}$ via

$$v_{\parallel}^{\alpha} = \left\{\frac{2}{m_{\alpha}}\left[\mathcal{E} - q_{\alpha} V(\mathbf{q}) - MB(\mathbf{q})\right]\right\}^{1/2} \tag{4.388}$$

must be expressed. The expansion parameter ε was retained in (4.387) in order to keep track of the orders of magnitude for a later expansion. It is also noteworthy about (4.387) that, apart from the (adiabatic) constancy of $\mathcal{E}$ and M, no further assumptions were made, so in particular the phase ϕ still appears explicitly. Although (4.387) already bears some resemblance to (4.355), essential steps are still required to arrive at the drift-kinetic description. In particular, the dependence on the gyrophase ϕ must—up to a certain order in ε—no longer appear explicitly on the left-hand side. We can obtain the drift-kinetic equation after expanding f^{α} in powers of ε,

$$\boxed{f^{\alpha} = f_0^{\alpha} + \varepsilon f_1^{\alpha} + \varepsilon^2 f_2^{\alpha} + \ldots} \tag{4.389}$$

As is well known, such a perturbation series can lead to convergence difficulties (in the long-time behavior) unless one ensures the elimination of secular terms by means of a multiple time scale formalism. The standard procedure uses the formally independent variables

$$t_{-1} = \varepsilon^{-1} t, \quad t_0 = t, \quad t_1 = \varepsilon t, \ldots \tag{4.390}$$

as arguments in the terms $f_0^{\alpha}, f_1^{\alpha}, f_2^{\alpha}$, etc. Thus, in (4.387),$\partial/\partial t$ is replaced by

$$\boxed{\frac{\partial}{\partial t} \to \frac{1}{\varepsilon}\frac{\partial}{\partial t_{-1}} + \frac{\partial}{\partial t_0} + \varepsilon \frac{\partial}{\partial t_1} + \varepsilon^2 \frac{\partial}{\partial t_2} + \ldots} \tag{4.391}$$

The dependence on the time t_{-1} is meant to express the dependence on the fast gyration; t_0 characterizes the relaxation time, and t_1 can, for example, be thought of as the hydrodynamic time. All functions f_ν^α in (4.389) are to be understood as written with the various time variables $t_{-1}, t_0, t_1, \ldots$. We then insert the ansatz (4.389) into (4.387) and systematically collect the various orders in ε. In the orders ε^{-1} and ε^0 we immediately obtain

$$\frac{\partial f_0^\alpha}{\partial t_{-1}} = -\Omega_\alpha \frac{\partial f_0^\alpha}{\partial \phi}, \tag{4.392}$$

$$\begin{aligned}&\frac{\partial f_1^\alpha}{\partial t_{-1}} + \frac{\partial f_0^\alpha}{\partial t_0} + v_\parallel^\alpha \mathbf{b}\cdot\frac{\partial f_0^\alpha}{\partial \mathbf{q}} + \left[v_\parallel^\alpha \mathbf{b}\cdot\mathbf{R} - \frac{1}{2}v_\parallel^\alpha \mathbf{b}\cdot(\nabla\times\mathbf{b})\right]\frac{\partial f_0^\alpha}{\partial \phi}\\ &= K_0^\alpha - \Omega_\alpha \frac{\partial f_1^\alpha}{\partial \phi},\end{aligned} \tag{4.393}$$

whereas at order ε^1

$$\begin{aligned}&\frac{\partial f_2^\alpha}{\partial t_{-1}} + \frac{\partial f_1^\alpha}{\partial t_0} + \frac{\partial f_0^\alpha}{\partial t_1} + v_\parallel^\alpha \mathbf{b}\cdot\frac{\partial f_1^\alpha}{\partial \mathbf{q}} + \mathbf{v}_D^\alpha\cdot\frac{\partial f_0^\alpha}{\partial \mathbf{q}}\\ &+ \left[v_\parallel^\alpha \mathbf{b}\cdot\mathbf{R} - \frac{1}{2}v_\parallel^\alpha \mathbf{b}\cdot(\nabla\times\mathbf{b})\right]\frac{\partial f_1^\alpha}{\partial \phi} + \Omega_\alpha^{(1)}\frac{\partial f_0^\alpha}{\partial \phi} + q_\alpha v_\parallel^\alpha E_\parallel \frac{\partial f_0^\alpha}{\partial \mathcal{E}}\\ &= K_0^\alpha\{f_0, f_1\} + K_1^\alpha\{f_0, f_0\} - \Omega_\alpha \frac{\partial f_2^\alpha}{\partial \phi}\end{aligned} \tag{4.394}$$

results. Here, $\Omega_\alpha^{(1)}$ denotes the terms of first order in ε, that appear in the equation of motion for ϕ. The contributions of the collision term K to first order in ε are written here only formally. To arrive at tractable equations, we follow Balescu and decompose all contributions to the distribution function (f_ν^α) into a part averaged over the gyrophase ϕ) (denoted by $\overline{f}_\nu^\alpha$) and an oscillatory part $\tilde{f}_\nu^\alpha$. From (4.392) it immediately follows that $\overline{f}_0^\alpha$ does not depend on t_{-1}, while the original Eq. (4.392) for the oscillatory part $\tilde{f}_0^\alpha$ possesses the obvious solution

$$\tilde{f}_0^\alpha(\mathbf{q}, \mathcal{E}, M, \phi; t_{-1}, t_0, t_1, \ldots) = \tilde{f}_0^\alpha(\mathbf{q}, \mathcal{E}, M, \phi - \Omega_\alpha t_{-1};\ t_{-1} = 0, t_0, t_1, \ldots) \tag{4.395}$$

For physical reasons, we set this solution equal to zero. The justification for this approach arises from the result for the mean (perpendicular) velocity component (to order ε^0), which we obtain when we average the strongly oscillating (perpendicular) velocity component with f_0^α. Since then obviously only $\tilde{f}_0^\alpha$ contributes, we obtain for $\tilde{f}_0^\alpha \neq 0$ a drift already in zeroth order (ε^0), which contradicts the microscopic theory of drift velocities. Thus, to first order we obtain $f_0^\alpha = \overline{f}_0^\alpha(q, \mathcal{E}, M; t_0, t_1, \ldots)$, where the functional form of $\overline{f}_0^\alpha$ has not yet been determined.

Let us now turn to Eq. (4.393). For the oscillating component, we find

$$\left(\frac{\partial}{\partial t_{-1}} + \Omega_\alpha \frac{\partial}{\partial \phi}\right)\tilde{f}_1^\alpha = \tilde{K}_0^\alpha\left\{\overline{f}_0, \overline{f}_0\right\}, \tag{4.396}$$

while the averaged distribution function follows from

$$-\frac{\partial \overline{f}_1^\alpha}{\partial t_{-1}} = \left(\frac{\partial}{\partial t_0} + v_\parallel^\alpha \mathbf{b} \cdot \frac{\partial}{\partial \mathbf{q}}\right)\overline{f}_0^\alpha - \overline{K}_0^\alpha\left\{\overline{f}_0, \overline{f}_0\right\} \tag{4.397}$$

In the last Eq. (4.397), the right-hand side does not depend on t_{-1}, so the time integration becomes trivial. However, to avoid a divergence such as t_{-1} at $\overline{f}_1^\alpha$, we must set the right-hand side equal to zero, which yields the determining equation

$$\left(\frac{\partial}{\partial t_0} + v_\parallel^\alpha \mathbf{b} \cdot \frac{\partial}{\partial \mathbf{q}}\right)\overline{f}_0^\alpha = \overline{K}_0^\alpha\left\{\overline{f_0}, \overline{f_0}\right\} \tag{4.398}$$

for $\overline{f}_0^\alpha = \overline{f}_0^\alpha(q, \mathcal{E}, M; t_0, \ldots)$. Ultimately, from (4.394), after averaging, we obtain

$$\begin{aligned}-\frac{\partial \overline{f}_2^\alpha}{\partial t_{-1}} = \frac{\partial \overline{f}_1^\alpha}{\partial t_0} + \frac{\partial \overline{f}_0^\alpha}{\partial t_1} + v_\parallel^\alpha \mathbf{b} \cdot \frac{\partial \overline{f}_1^\alpha}{\partial \mathbf{q}} + q_\alpha v_\parallel^\alpha E_\parallel \frac{\partial \overline{f}_0^\alpha}{\partial \mathcal{E}} \\ + \mathbf{v}_D^\alpha \cdot \frac{\partial \overline{f}_0^\alpha}{\partial \mathbf{q}} - \overline{K_0^\alpha\{f_0, f_1\}} - \overline{K_1^\alpha\{f_0, f_0\}}.\end{aligned} \tag{4.399}$$

Without additional assumptions for the collision term, we cannot proceed further at this point. We assume that—even—the collision contributions on the right-hand side of (4.399) also do not depend on t_{-1}. Thus, to again avoid divergences in time t_{-1}, we must set

$$\boxed{\begin{aligned}&\frac{\partial \overline{f}_1^\alpha}{\partial t_0} + \frac{\partial \overline{f}_0^\alpha}{\partial t_1} + v_\parallel^\alpha \mathbf{b} \cdot \frac{\partial \overline{f}_1^\alpha}{\partial \mathbf{q}} + \mathbf{v}_D^\alpha \cdot \frac{\partial \overline{f}_0^\alpha}{\partial \mathbf{q}} + q_\alpha v_\parallel^\alpha E_\parallel \frac{\partial \overline{f}_0^\alpha}{\partial \mathcal{E}} \\ &= \overline{K_0^\alpha\{f_0, f_1\}} + \overline{K_1^\alpha\{f_0, f_0\}}\end{aligned}} \tag{4.400}$$

We will not further discuss the form of the collision terms. Eq. (4.400) is the sought-after drift-kinetic equation. It is formulated in different variables than our original equation (4.355). Nevertheless, it is evident that, for example, for the problem of drift waves discussed at the beginning of this chapter (without collision terms), it leads to the same result. To verify this, one must first discuss the possible solutions for $\overline{f}_0^\alpha$ [from (4.398)]. If one assumes a Maxwellian-like solution and takes (4.388) into account, Fourier transforms in t_0 and $\mathbf{q}$ lead to (4.367).

The preceding discussion was intended to show that, using the methods of theoretical physics (here: pseudo-canonical transformations and Lie formalism), a rigorous derivation of the drift-kinetic equation can be given, which does not have to rely on sometimes opaque averaging procedures. Furthermore, it also becomes clear that with pseudo-canonical transformations and Lie brackets, we have a formalism at hand that enables us to investigate far more complicated situations. We are thinking here of curved and sheared magnetic field lines, which frequently occur in fusion devices. A detailed investigation of plasma transport in such configurations must be based on a well-founded theoretical description, for which the examples in this chapter can only serve as starting points.

Gyrokinetic Approximation

So far, we have assumed that the fields do not change rapidly (on the scale of the gyroradius). However, it is possible for instabilities to develop with $k_\perp \rho_i \sim \mathcal{O}(1)$. For demonstration purposes, let us first assume that the perturbations are electrostatic, with the electric field $\mathbf{E}_1$. During its gyration, the particle will experience different electric field strengths, leading to an additional averaged $E \times B$ velocity:

$$\langle \mathbf{v}_1 \rangle_{es} \approx \frac{1}{B_0} \langle \mathbf{E}_1(\mathbf{r}) \rangle \times \hat{b} \,. \tag{4.401}$$

We denote the current position of a particle by $\mathbf{r} = \mathbf{Y} + \rho$, where

$$\rho = -\frac{v_\perp}{\Omega}\, \mathbf{n}_2 = -\frac{v_\perp}{\Omega}\, (\cos\, \varphi\, \hat{e}_1 - \sin\, \varphi\, \hat{e}_2) \,. \tag{4.402}$$

Let us now discuss a (linear) mode

$$\mathbf{E}_1(\mathbf{r}) = \mathbf{E}_1\, e^{i\mathbf{k}_\perp \cdot \rho} \equiv \mathbf{E}_1(\mathbf{Y})\, e^{i\mathbf{k}_\perp \cdot (\mathbf{Y}+\rho)} \,, \tag{4.403}$$

where different arguments denote different amplitudes.

After averaging, we find

$$\langle \mathbf{E}_1(\mathbf{r}) \rangle \equiv \mathbf{E}_1 \int \frac{d\varphi}{2\pi}\, e^{i\frac{v_\perp}{\Omega}(k_2 \sin\, \varphi - k_1 \cos\, \varphi)} = \mathbf{E}_1\, J_0\Big(\frac{v_\perp}{\Omega}\, k_\perp\Big) \equiv -i\mathbf{k}_\perp\, \phi_A\, J_0(\rho k_\perp) \tag{4.404}$$

with $k_1 = k_\perp \cos\psi$ and $k_2 = k_\perp \sin\psi$. Then, in the collisionless approximation for electrostatic perturbations, we obtain the linearized gyrokinetic equation

$$\frac{\partial \bar{f}_1}{\partial t} + \left(v_\parallel \hat{b} + \mathbf{v}_d\right) \cdot \nabla_Y \bar{f}_1 - e v_\parallel \, \hat{b} \cdot \frac{\partial \mathbf{A}}{\partial t} \frac{\partial \bar{f}_1}{\partial U} \approx i \, \frac{1}{B_0} \, J_0(\rho k_\perp) \phi_A \, (\mathbf{k} \times \hat{b}) \cdot \nabla_Y \bar{f}_0 \tag{4.405}$$

Note that $\bar{f} = \bar{f}_0 + \bar{f}_1$, where $\bar{f}_1$ is the (small) response to the (electrostatic) rapid field perturbation $\mathbf{E}_1(\mathbf{r})$.

In the case of (additional) rapid magnetic perturbations $\mathbf{B}_1(\mathbf{r})$, further first-order terms arise. The solution to the linearized equation of motion

$$m \, \frac{d\mathbf{v}}{dt} = e[\mathbf{v}_1 \times \mathbf{B}_0 + \mathbf{v} \times \mathbf{B}_1] \tag{4.406}$$

is straightforward if we recall the derivation of the $E \times B$ drift:

$$\mathbf{v}_1 \approx -\frac{1}{B_0} \, \hat{b} \times (\mathbf{v} \times \mathbf{B}_1) = -\mathbf{v} \, \frac{\hat{b} \cdot \mathbf{B}_1}{B_0} + \mathbf{B}_1 \, \frac{\hat{b} \cdot \mathbf{v}}{B_0} = -\mathbf{v}_\perp \, \frac{B_{1\parallel}}{B_0} + \mathbf{B}_{1\perp} \, \frac{v_\parallel}{B_0} \, . \tag{4.407}$$

Averaging, we then additionally obtain

$$\begin{aligned} \langle \mathbf{v}_1 \rangle_{em} &= -\frac{1}{B_0} \left\langle B_{1\parallel} \, \mathbf{v}_\perp \right\rangle + \frac{v_\parallel}{B_0} \, \langle \mathbf{B}_{1\perp} \rangle \\ &\approx -i \, \frac{v_\perp}{B_0} \, B_{A\parallel} \, J_1(\rho k_\perp) \, \hat{k}_\perp \times \hat{b} + i \, \frac{v_\parallel}{B_0} \, \mathbf{k}_\perp \times \hat{b} \, A_{A\parallel} \, J_0(\rho k_\perp) \, . \end{aligned} \tag{4.408}$$

Since the field will also be time-dependent $\sim e^{-i\omega t}$, the energy is no longer constant, and we must take into account

$$\begin{aligned} \left\langle \frac{dU}{dt} \right\rangle &= -i \, \omega \, e \left[\langle \phi_1 \rangle - \frac{1}{c} \langle \mathbf{v} \cdot \mathbf{A}_1 \rangle \right] \\ &\approx -i \, \omega \, e \left[J_0(k_\perp \rho) \phi_A - i \, \frac{v_\perp}{c} \, J_1(k_\perp \rho)(\hat{k}_\perp \times \hat{b}) \cdot \mathbf{A}_{A\perp} - \frac{v_\parallel}{c} \, J_0(k_\perp \rho) A_{A\parallel} \right] \end{aligned} \tag{4.409}$$

In addition, the terms

$$-\langle \mathbf{v}_1 \rangle_{em} \cdot \nabla_Y \bar{f}_0 - \langle \mathbf{v}_1 \rangle_{es} \cdot \nabla_Y \bar{f}_0 - \left\langle \frac{dU}{dt} \right\rangle \frac{\partial \bar{f}_0}{\partial U} \tag{4.410}$$

then appear on the right-hand side of the linearized gyrokinetic equation. These terms can be combined, which ultimately leads to

$$\boxed{\begin{aligned}&\frac{\partial \bar{f}_1}{\partial t}+\left(v_{\parallel}\,\hat{b}+\mathbf{v}_d\right)\cdot\nabla_Y\,\bar{f}_1-\frac{e}{c}\,v_{\parallel}\,\hat{b}\cdot\frac{\partial \mathbf{A}}{\partial t}\,\frac{\partial \bar{f}_1}{\partial U}\\&\approx i\left[J_0(k_{\perp}\rho)\left\{\phi_A-\frac{v_{\parallel}}{c}\,A_{A\parallel}\right\}+\frac{v_{\perp}}{ck_{\perp}}\,J_1(k_{\perp}\rho)B_{A\parallel}\right]\\&\quad\times\left\{e\,\omega\,\frac{\partial \bar{f}_0}{\partial U}+\frac{c}{B_0}\left(\mathbf{k}_{\perp}\times\hat{b}\right)\cdot\nabla_Y\,\bar{f}_0\right\}.\end{aligned}} \tag{4.411}$$

This is the (collisionless) linearized gyrokinetic equation for rapid perturbations (on the scale of the gyration). As already mentioned, the linearization breaks down when large perturbations occur; in that case, we must return to the (non-reduced) description using the Landau-Fokker-Planck or Balescu-Lenard-Guernsey equations.

Macroscopic Description

5

Abstract

In this chapter, we move on to the so-called macroscopic description (also called fluid description) of plasmas, which contains less information than the kinetic formulation. Knowledge of a distribution function f provides us with statistical information in the $\mathbf{r}, \mathbf{p}$ phase space at any given time t. (Here, we do not address the difference between descriptions using $\mathbf{p}$ or $\mathbf{v}$, as previously mentioned.) If we are not interested in the details in velocity space, we can integrate over the velocities to obtain equations for the so-called moments of the distribution function. Of course, a finite number of moments mathematically contains less information than the distribution function itself. We should mention that the physical consequences of the reduced, macroscopic description also depend on the time and spatial variables. We assume that the moments vary (slowly) on the so-called hydrodynamic time and length scales, which generally differ from the characteristic microscopic variations.

We will develop the macroscopic theory for plasmas in the following sections. First, we define the moments. When we attempt to derive the equations for the moments, we encounter a hierarchy problem, similar to the difficulties in deriving a (closed) kinetic equation. To illustrate the steps necessary to obtain a closed system of equations for the first three moments, we will, for pedagogical reasons, first summarize the steps known from the theory of neutral gases. In doing so, we will not present all the details of the calculations, as these will appear later when we develop the plasma counterpart.

K.-H. Spatschek, *Theoretical Plasma Physics*,
https://doi.org/10.1007/978-3-662-72828-4_5

5.1 Moments and Their Determining Equations

In this section, we begin with the definition of relevant plasma-dynamic moments and the associated moment equations. The hierarchy problem is noted; however, it is not yet resolved.

The velocity averaging of a quantity (...) is performed by integrating over velocity space according to the following prescription:

$$\boxed{\langle \cdots \rangle_\alpha = \frac{\int d^3v \cdots f^\alpha}{\int d^3v f}} \, . \tag{5.1}$$

We can introduce plasma-dynamic (in the case of neutral fluids, hydrodynamic) variables such as particle density, mean velocity, and temperature. For species α we define the particle density

$$n_\alpha(\mathbf{r}, t) = \int d^3v f^\alpha \tag{5.2}$$

(note the "normalization" of the single-particle distribution function), the mean velocity

$$\mathbf{u} = \langle \mathbf{v} \rangle \tag{5.3}$$

and the temperature (again measured in eV; the Boltzmann constant k_B is omitted)

$$T_\alpha = \frac{1}{3} m_\alpha \langle |\mathbf{v} - \mathbf{u}_\alpha|^2 \rangle \, , \tag{5.4}$$

to mention only the lowest moments.

To derive the dynamic equations for the moments, we start with the kinetic equation in general form

$$\boxed{\frac{\partial f^\alpha}{\partial t} + \mathbf{v} \cdot \frac{\partial f^\alpha}{\partial \mathbf{r}} + \frac{1}{m_\alpha} \mathbf{F}_\alpha \cdot \frac{\partial f^\alpha}{\partial \mathbf{v}} = K^\alpha \{ f^\alpha(t) \}} \, , \tag{5.5}$$

where $\mathbf{F}$ is the force and K^α is the collision term. In deriving the equations for the moments, we take into account that $\mathbf{r}$, $\mathbf{v}$ and t are independent. Furthermore, $f^\alpha \to 0$ holds if $v \to \infty$.

We begin with the zeroth moment n_α. Its dynamic equation is obtained immediately by integrating both sides of (5.5) over velocity space. The independence of the variables as well as Gauss's theorem lead to

$$\boxed{\frac{\partial n_\alpha}{\partial t} + \nabla \cdot (n_\alpha \mathbf{u}_\alpha) = 0} . \tag{5.6}$$

Next, we multiply Eq. (5.5) by $\mathbf{v}$ and integrate over velocity space,

$$\int d^3v\mathbf{v}\left[\frac{\partial f^\alpha}{\partial t} + \mathbf{v} \cdot \frac{\partial f^\alpha}{\partial \mathbf{r}} + \frac{1}{m_\alpha}\mathbf{F}_\alpha \cdot \frac{\partial f^\alpha}{\partial \mathbf{v}}\right] = \int d^3v\mathbf{v}K^\alpha\left\{f^\alpha(t)\right\} . \tag{5.7}$$

Before proceeding with the evaluation of the second integral, we introduce the variable

$$\mathbf{v}' = \mathbf{v} - \mathbf{u}_\alpha \tag{5.8}$$

Since $\mathbf{u} \equiv \mathbf{u}(\mathbf{r}, t)$ is position-dependent, $\mathbf{v}'$ will also be position-dependent. Therefore, in the second term on the right-hand side, we can move the operator $\partial/\partial\mathbf{r}$ should be prioritized first (since it does not act on $\mathbf{v}$, but rather on $\mathbf{v}'$). Furthermore, we assume that the force depends only on $\mathbf{r}$, or, in the more general case, takes the Lorentz form,

$$\mathbf{F}_\alpha = q_\alpha[\mathbf{E}(\mathbf{r}, t) + \mathbf{v} \times \mathbf{B}(\mathbf{r}, t)] . \tag{5.9}$$

Thus, the third term becomes

$$\int d^3v\mathbf{v}q_\alpha[\mathbf{E}(\mathbf{r}, t) + \mathbf{v} \times \mathbf{B}(\mathbf{r}, t)] \cdot \frac{\partial f^\alpha}{\partial \mathbf{v}} = n_\alpha q_\alpha[\mathbf{E}(\mathbf{r}, t) + \mathbf{u}_\alpha \times \mathbf{B}(\mathbf{r}, t)] . \tag{5.10}$$

We abbreviate the collision term as

$$\boxed{\int d^3v\mathbf{v}K^\alpha\left\{f^\alpha(t)\right\} = \frac{1}{m_\alpha}\mathbf{R}_{\alpha\,\alpha'}} \tag{5.11}$$

where

$$\mathbf{R}_{\alpha\,\alpha'} = -\nu_{\alpha\,\alpha'}\, m_\alpha n_\alpha(\mathbf{u}_\alpha - \mathbf{u}_{\alpha'}) \,, \quad \alpha\,, \alpha' = e,\, i \tag{5.12}$$

is written. Taking all contributions into account, we find

$$\boxed{\begin{aligned} &\frac{\partial(n_\alpha \mathbf{u}_\alpha)}{\partial t} + \frac{\partial}{\partial \mathbf{r}} \cdot (n_\alpha \mathbf{u}_\alpha \mathbf{u}_\alpha) \\ &= n_\alpha \frac{q_\alpha}{m_\alpha}[\mathbf{E}(\mathbf{r}, t) + \mathbf{u}_\alpha \times \mathbf{B}(\mathbf{r}, t)] - \frac{1}{m_\alpha}\frac{\partial}{\partial \mathbf{r}} \cdot \overleftrightarrow{\mathbf{P}}_\alpha + \frac{1}{m_\alpha}\mathbf{R}_{\alpha\,\alpha'} . \end{aligned}} \tag{5.13}$$

We have introduced the pressure tensor

$$\overleftrightarrow{P}_\alpha \equiv \overleftrightarrow{\mathbf{P}}_\alpha = m_\alpha \int d^3v'\mathbf{v}'\mathbf{v}'f^\alpha \,, \quad P_{\alpha\,ij} = m_\alpha \int d^3v'\, v_i'\, v_j' f^\alpha \tag{5.14}$$

Without the velocity shift by $\mathbf{u}_\alpha$ we would have the stress tensor

$$\overleftrightarrow{\mathcal{P}}_\alpha = m_\alpha \int d^3v \mathbf{v}\mathbf{v} f^\alpha , \quad \mathcal{P}_{\alpha ij} = m_\alpha \int d^3v \, v_i \, v_j f^\alpha . \tag{5.15}$$

The following holds

$$\overleftrightarrow{\mathcal{P}}_\alpha = \overleftrightarrow{P}_\alpha + m_\alpha \mathbf{u}_\alpha \mathbf{u}_\alpha . \tag{5.16}$$

Furthermore, we have defined

$$\frac{\partial}{\partial \mathbf{r}} \cdot \overleftrightarrow{P}_\alpha \equiv \nabla \cdot \overleftrightarrow{P}_\alpha , \quad \left[\nabla \cdot \overleftrightarrow{P}_\alpha\right]_j = m_\alpha \sum_{i=1}^{3} \frac{\partial}{\partial x_i} \int d^3v \, v_i' \, v_j' f^\alpha . \tag{5.17}$$

Thus, we write

$$\left[\frac{\partial}{\partial \mathbf{r}} \cdot (n_\alpha \mathbf{u}_\alpha \mathbf{u}_\alpha)\right]_j \equiv [\nabla \cdot (n_\alpha \mathbf{u}_\alpha \mathbf{u}_\alpha)]_j = \sum_{i=1}^{3} \frac{\partial}{\partial x_i} \left(n_\alpha u_{\alpha\, i} u_{\alpha\, j}\right) . \tag{5.18}$$

The pressure tensor $\overleftrightarrow{\mathbf{P}}_\alpha$ as a higher moment is so far unknown. For an isotropic distribution function in $\mathbf{v}'$ the diagonal elements are equal to $n_\alpha k_B T_\alpha$ (for a Maxwell distribution), and the off-diagonal elements vanish. If an isotropic distribution is assumed in the following, we introduce the scalar pressure,

$$p_\alpha = \frac{m_\alpha}{3} \int d^3v \mathbf{v}' \cdot \mathbf{v}' f^\alpha . \tag{5.19}$$

Altogether, we write

$$\overleftrightarrow{P}_\alpha = p_\alpha \overleftrightarrow{I} + \overleftrightarrow{\pi}_\alpha \equiv p_\alpha I + \overleftrightarrow{\pi}_\alpha , \tag{5.20}$$

when no preferred direction occurs; $\overleftrightarrow{\pi}_\alpha$ is called the dissipative part of the pressure tensor.

Furthermore, using

$$\frac{\partial (n_\alpha \mathbf{u}_\alpha)}{\partial t} + \frac{\partial}{\partial \mathbf{r}} \cdot (n_\alpha \mathbf{u}_\alpha \mathbf{u}_\alpha) = \mathbf{u}_\alpha \frac{\partial n_\alpha}{\partial t} + \mathbf{u}_\alpha \nabla \cdot (n_\alpha \mathbf{u}_\alpha) + n_\alpha \frac{\partial \mathbf{u}_\alpha}{\partial t} + n_\alpha (\mathbf{u}_\alpha \cdot \nabla) \mathbf{u}_\alpha \tag{5.21}$$

as well as the particle density continuity equation, we arrive at

$$\boxed{\frac{\partial \mathbf{u}_\alpha}{\partial t} + (\mathbf{u}_\alpha \cdot \nabla) \mathbf{u}_\alpha = \frac{q_\alpha}{m_\alpha} [\mathbf{E} + \mathbf{u}_\alpha \times \mathbf{B}] - \frac{1}{m_\alpha n_\alpha} \nabla P_\alpha + \frac{1}{m_\alpha n_\alpha} \mathbf{R}_{\alpha \alpha'}} . \tag{5.22}$$

Next, we multiply (5.5) by $\frac{1}{2} m_\alpha v^2$ and integrate over velocity space,

$$\frac{m_\alpha}{2}\int d^3v\, v^2\left[\frac{\partial f^\alpha}{\partial t}+\mathbf{v}\cdot\frac{\partial f^\alpha}{\partial \mathbf{r}}+\frac{1}{m_\alpha}\mathbf{F}_\alpha\cdot\frac{\partial f^\alpha}{\partial \mathbf{v}}\right]=\frac{m_\alpha}{2}\int d^3v\, v^2 K^\alpha\{f^\alpha(t)\}\,. \tag{5.23}$$

We again use the transformation (5.8) and evaluate term by term. The first term on the left-hand side immediately yields

$$\frac{m_\alpha}{2}\int d^3v\, v^2\frac{\partial f^\alpha}{\partial t}=\frac{\partial}{\partial t}\left(\frac{3}{2}p_\alpha+\frac{m_\alpha}{2}n_\alpha u_\alpha^2\right)\,. \tag{5.24}$$

The second term can be written as

$$\frac{m_\alpha}{2}\nabla\cdot\int d^3v\, v^2\mathbf{v}f^\alpha=\nabla\cdot\left(\mathbf{Q}_\alpha+\overleftrightarrow{P}_\alpha\cdot\mathbf{u}_\alpha+\frac{3}{2}p_\alpha\mathbf{u}_\alpha+\frac{m_\alpha}{2}n_\alpha u_\alpha^2\mathbf{u}_\alpha\right) \tag{5.25}$$

where the heat flux density

$$\boxed{\mathbf{q}_\alpha\equiv\frac{m_\alpha}{2}\int d^3v'\, v'^2\mathbf{v}'f^\alpha} \tag{5.26}$$

has been defined. The latter is related to the energy flux density by

$$\mathbf{Q}_\alpha\equiv\frac{m_\alpha}{2}\int d^3v\, v^2\mathbf{v}f^\alpha \tag{5.27}$$

via

$$\mathbf{Q}_\alpha=\mathbf{q}_\alpha+\overleftrightarrow{P}_\alpha\cdot\mathbf{u}_\alpha+\frac{3}{2}p_\alpha\mathbf{u}_\alpha+\frac{1}{2}m_\alpha n_\alpha u_\alpha^2\mathbf{u}_\alpha\,. \tag{5.28}$$

The third term, which contains the force, is straightforward to determine. First, we rearrange,

$$\frac{q_\alpha}{2}\int d^3v\, v^2[\mathbf{E}+\mathbf{v}\times\mathbf{B}]\cdot\frac{\partial f^\alpha}{\partial \mathbf{v}}=-\frac{q_\alpha}{2}\int d^3v\, v^2\frac{\partial}{\partial \mathbf{v}}\cdot[\mathbf{E}+\mathbf{v}\times\mathbf{B}]f^\alpha\,, \tag{5.29}$$

to find after partial integration

$$\frac{q_\alpha}{2}\int d^3v\, v^2[\mathbf{E}+\mathbf{v}\times\mathbf{B}]\cdot\frac{\partial f^\alpha}{\partial \mathbf{v}}=-q_\alpha n_\alpha\mathbf{E}\cdot\mathbf{u}_\alpha\,. \tag{5.30}$$

The contribution of the collision term is again abbreviated, now in the form

$$\frac{m_\alpha}{2}\int d^3v\, v^2K^\alpha\{f^\alpha(t)\}\equiv-\left.\frac{dW}{dt}\right|_\alpha\,. \tag{5.31}$$

Altogether, this yields

$$\frac{\partial}{\partial t}\left(\frac{3}{2}P_\alpha + \underbrace{\frac{m_\alpha}{2}n_\alpha u_\alpha^2}_{*}\right) + \nabla\cdot\left(\mathbf{q}_\alpha + \frac{5}{2}p_\alpha\mathbf{u}_\alpha + \underbrace{\frac{m_\alpha}{2}n_\alpha u_\alpha^2\mathbf{u}_\alpha}_{**}\right) = q_\alpha n_\alpha \mathbf{E}\cdot\mathbf{u}_\alpha - \left.\frac{dW}{dt}\right|_\alpha . \tag{5.32}$$

The terms marked with $*$ and $**$ can be further simplified by using the continuity equation (5.6) and the momentum equation (5.22). We obtain

$$* = n_\alpha\left(\frac{\partial}{\partial t} + \mathbf{u}_\alpha\cdot\nabla\right)\frac{m_\alpha u_\alpha^2}{2} . \tag{5.33}$$

The right-hand side can be further simplified by using an additional identity. For this, we multiply (5.22) by $\mathbf{u}_\alpha$. We also substitute

$$\mathbf{u}_\alpha\cdot\nabla\mathbf{u}_\alpha = \nabla\left(\frac{u_\alpha^2}{2}\right) - \mathbf{u}_\alpha\times\nabla\times\mathbf{u}_\alpha \tag{5.34}$$

into

$$\mathbf{u}_\alpha\cdot\left\{\frac{\partial\mathbf{u}_\alpha}{\partial t} + (\mathbf{u}_\alpha\cdot\nabla)\mathbf{u}_\alpha\right\} = \mathbf{u}_\alpha\cdot\left\{\frac{q_\alpha}{m_\alpha}[\mathbf{E}+\mathbf{u}_\alpha\times\mathbf{B}] - \frac{1}{m_\alpha n_\alpha}\nabla p_\alpha + \frac{1}{m_\alpha n_\alpha}\mathbf{R}_{\alpha\alpha'}\right\} \tag{5.35}$$

to obtain

$$** = \mathbf{u}_\alpha\cdot\{q_\alpha n_\alpha\mathbf{E} - \nabla p_\alpha + \mathbf{R}_{\alpha\alpha'}\} . \tag{5.36}$$

Thus, the energy equation can be written as

$$\boxed{\left(\frac{\partial}{\partial t} + \mathbf{u}_\alpha\cdot\nabla\right)\frac{3p_\alpha}{2} + \frac{5}{2}p_\alpha\nabla\cdot\mathbf{u}_\alpha = -\nabla\cdot\mathbf{q}_\alpha - \mathbf{R}_{\alpha\alpha'}\cdot\mathbf{u}_\alpha - \left.\frac{dW}{dt}\right|_\alpha} . \tag{5.37}$$

We have made the following substitution:

$$\frac{3}{2}p_\alpha\nabla\cdot\mathbf{u}_\alpha + \overleftrightarrow{P}_\alpha : \nabla\mathbf{u}_\alpha = \frac{5}{2}p_\alpha\nabla\cdot\mathbf{u}_\alpha . \tag{5.38}$$

5.2 Truncation of the Hierarchy in the Boltzmann Case

In this section, we briefly review how, in the classical and for decades well-known case of neutral gases, one arrives at a macroscopic description with fluid equations. We will later use this knowledge to develop a transport theory for plasmas.

The statistical description of neutral particles in the classical regime is known as the famous Boltzmann equation.

$$\frac{\partial f_1}{\partial t} + \mathbf{v}_1 \cdot \frac{\partial f_1}{\partial \mathbf{r}_1} + \frac{\mathbf{F}}{m} \cdot \frac{\partial f_1}{\partial \mathbf{v}_1} = \int d^3 v_2 \int d\Omega \sigma(\Omega) |\mathbf{v}_2 - \mathbf{v}_1| \big(f_2' f_1' - f_2 f_1 \big) \,. \tag{5.39}$$

The Boltzmann equation is a closed integro-differential equation for the single-particle distribution function f, where here and in the following—whenever appropriate—the index refers to the variables, i.e., $f_1 = f(\mathbf{r}_1, \mathbf{v}_1, t)$. The left-hand side of Eq. (5.39) is the free streaming term, while the right-hand side represents the binary collision term. From the Boltzmann equation, one obtains the hydrodynamic equations. The purpose of kinetic plasma theory is to derive the corresponding plasma dynamic equation(s) for a system of charged particles. We now want to show how the hydrodynamic description is linked to kinetic theory using the example of the Boltzmann equation.

Hierarchical Form of the Moment Equations

By direct integration, as already discussed in the previous section, we obtain the generally valid equations

$$\boxed{\frac{\partial n}{\partial t} + \nabla \cdot (n\mathbf{u}) = 0 \,,} \tag{5.40}$$

$$\boxed{\left(\frac{\partial}{\partial t} + \mathbf{u} \cdot \nabla \right) \mathbf{u} = \frac{1}{m} \mathbf{F} - \frac{1}{mn} \nabla \cdot \overleftrightarrow{P} \,,} \tag{5.41}$$

$$\boxed{\left(\frac{\partial}{\partial t} + \mathbf{u} \cdot \nabla \right) T = -\frac{2}{3n} \nabla \cdot \mathbf{q} - \frac{2}{3mn} \overleftrightarrow{P} : \overleftrightarrow{\Lambda} \,.} \tag{5.42}$$

Various moments appear. The components of the pressure tensor $\overleftrightarrow{P}$ are

$$P_{ij} = mn \langle (v_i - u_i)(v_j - u_j) \rangle \,. \tag{5.43}$$

In addition, the heat flux appears

$$\mathbf{q} = \frac{1}{2} mn \langle (\mathbf{v} - \mathbf{u}) |\mathbf{v} - \mathbf{u}|^2 \rangle \,. \tag{5.44}$$

Finally, we have the abbreviation

$$\Lambda_{ij} = \frac{1}{2}m\left(\frac{\partial u_i}{\partial x_j} + \frac{\partial u_j}{\partial x_i}\right) \tag{5.45}$$

as components of a tensor $\overleftrightarrow{\Lambda}$.

It is very important to emphasize that we have not yet obtained the hydrodynamic equations for density, mean velocity, and temperature, since the higher moments are (still) unknown. The calculation of, for example, the pressure tensor and the heat flux is the highly nontrivial task of transport theory. The latter makes quite effective use of kinetic theory, as will now be illustrated with our introductory example of a neutral fluid.

Truncation of the Hierarchy, Transport Coefficients, and Euler Equations
We abbreviate the underlying kinetic equation in the form

$$\boxed{\frac{\partial f}{\partial t} + Df = \frac{1}{\varepsilon}J(f|f)} \tag{5.46}$$

Here, a very important assumption has been introduced. We assume that the (neutral) system is collision-dominated by introducing the (large, but only symbolic) factor $1/\varepsilon$ in front of the collision term. For an equation of this type, the method of multiple scale analysis is the appropriate solution method. In the theory of neutral fluids, it is known as the Chapman-Enskog method. We expand

$$f = f^{(0)} + \varepsilon f^{(1)} + \varepsilon^2 f^{(2)} + \cdots, \quad f^{(\nu)} = f^{(\nu)}(\mathbf{r}, \mathbf{v}; t_0, t_1, t_2, \ldots), \tag{5.47}$$

$$\frac{\partial}{\partial t} \to \frac{\partial}{\partial t_0} + \varepsilon\frac{\partial}{\partial t_1} + \varepsilon^2\frac{\partial}{\partial t_2} + \cdots, \tag{5.48}$$

and obtain to lowest order

$$\boxed{J^{(0)}\left(f^{(0)}|f^{(0)}\right) = 0}\,. \tag{5.49}$$

Its solution is the famous Maxwell distribution

$$f^{(0)}(\mathbf{r}, \mathbf{v}; t) = n\left(\frac{m}{2\pi T}\right)^{3/2} \exp\left[-\frac{m}{2T}(\mathbf{v} - \mathbf{u})^2\right]; \tag{5.50}$$

the hydrodynamic variables appear as parameters. (It is important to note that the parameters appearing in the lowest-order distribution function are by definition the exact hydrodynamic quantities.) With the help of the Maxwell distribution, we can evaluate the higher moments.

By substituting the results into the general equations for the first three moments, we obtain the famous Euler equations (where sometimes only the second equation is referred to as the Euler equation)

$$\boxed{\frac{\partial n}{\partial t} + \nabla \cdot (n\mathbf{u}) = 0\,,} \tag{5.51}$$

$$\boxed{\left(\frac{\partial}{\partial t} + \mathbf{u} \cdot \nabla\right)\mathbf{u} = \frac{\mathbf{F}}{m} - \frac{1}{mn}\nabla p\,,} \tag{5.52}$$

$$\boxed{\left(\frac{\partial}{\partial t} + \mathbf{u} \cdot \nabla\right)T = -\frac{2}{3}T\nabla \cdot \mathbf{u}\,.} \tag{5.53}$$

The factor 3/2 comes from the specific heat capacity at constant volume.

$$c_V = \frac{3}{2}\,, \tag{5.54}$$

The three equations, consisting of the mass conservation equation, the momentum balance, and the temperature equation, are the hydrodynamic equations for dissipation-free fluids. Usually, the entire system is referred to as the Euler equations. Viscosity does not appear.

In the present approximation, the temperature equation and the continuity equation can be combined into the equation of state:

$$\frac{dn}{dt} - c_V \frac{n}{T}\frac{dT}{dt} = 0 \tag{5.55}$$

leads to

$$\frac{d}{dt}\left(nT^{-c_V}\right) = 0\,. \tag{5.56}$$

With $p = nT$ we find

$$pn^{-5/3} = const\,. \tag{5.57}$$

Example 5.1 (Sound wave dispersion from the Euler equations)
By linearizing the Euler equations around a state with constant density (n_0) and constant temperature (T_0), and under the assumption that the zeroth-order flow velocity ($\mathbf{u}_0 = 0$) vanishes, we obtain the linearized Euler equations

$$\frac{\partial \delta n}{\partial t} = -n_0 \nabla \cdot \delta \mathbf{u} \,, \quad m n_0 \frac{\partial \delta \mathbf{u}}{\partial t} = -\nabla \delta p \,, \quad \frac{\delta n}{n_0} = \frac{3}{2} \frac{\delta T}{T_0} \,. \tag{5.58}$$

The first two equations lead to

$$\frac{\partial^2 \delta n}{\partial t^2} = \frac{1}{m} \nabla^2 \delta p \,. \tag{5.59}$$

With $\delta p = n_0 \delta T + T_0 \delta n$ we immediately find

$$\frac{\partial^2 \delta p}{\partial t^2} = \frac{5 T_0}{3m} \nabla^2 \delta p \,. \tag{5.60}$$

Thus, we obtain waves (longitudinal oscillations) with the phase velocity

$$\boxed{c_s = \sqrt{\frac{5 T_0}{3m}}} \,. \tag{5.61}$$

These are the well-known sound waves. ■

Navier-Stokes equations in the next approximation
A better description is achieved when higher-order corrections are included. The equation for $f^{(1)}$ is

$$\boxed{\left(\frac{\partial}{\partial t_0} + D \right) f^{(0)} = J^{(1)} \left(f^{(0)} | f^{(1)} \right)} \,, \tag{5.62}$$

with the constraints

$$\int d^3 v f^{(1)} \begin{pmatrix} 1 \\ \mathbf{v} \\ v^2 \end{pmatrix} = 0 \,. \tag{5.63}$$

The solution is a rather demanding task. We write in abbreviated form (with summation over repeated indices)

$$f^{(1)} = -\left[\frac{1}{T} \frac{\partial T}{\partial x_i} g_i \mathcal{F}(g) + \frac{1}{T} \Lambda_{ij} \left(g_i g_j - \frac{1}{3} \delta_{ij} g^2 \right) \mathcal{G}(g) \right] f^{(0)} \,, \tag{5.64}$$

where $\mathbf{g} := \mathbf{v} - \mathbf{u}$. The functions $\mathcal{F}$ and $\mathcal{G}$ can, for example, be found by series solutions. We will not discuss the details of this calculation here. Instead, in the next section, we present a simplified solution based on the Krook collision term.

Taking into account first order, the fundamental equations are the continuity equation, the Navier-Stokes equation, and the heat conduction equation:

$$\frac{\partial n}{\partial t} + \nabla \cdot (n\mathbf{u}) = 0 \,, \tag{5.65}$$

$$\left(\frac{\partial}{\partial t} + \mathbf{u} \cdot \nabla\right)\mathbf{u} = \frac{\mathbf{F}}{m} - \frac{1}{mn}\nabla\left(p - \frac{\mu}{3}\nabla \cdot \mathbf{u}\right) + \frac{\mu}{mn}\nabla^2\mathbf{u} \,, \tag{5.66}$$

$$\left(\frac{\partial}{\partial t} + \mathbf{u} \cdot \nabla\right)T = -\frac{2}{3}(\nabla \cdot \mathbf{u})T + \frac{2K}{3nm}\nabla^2 T \,. \tag{5.67}$$

In this approximation, the pressure tensor and the heat flux are

$$P_{ij} = p\delta_{ij} - \frac{2\mu}{m}\left(\Lambda_{ij} - \frac{m}{3}\delta_{ij}\nabla \cdot \mathbf{u}\right) , \tag{5.68}$$

$$\mathbf{q} = -K\,\nabla T \,, \tag{5.69}$$

where viscosity and thermal conductivity are given by

$$\mu = \frac{m^2}{15T}\int d^3g\, g^4 f^{(0)}\mathcal{G}(g) \,, \tag{5.70}$$

$$K = \frac{m}{6T}\int d^3g\, g^4 f^{(0)}\mathcal{F}(g) \,, \tag{5.71}$$

Example 5.2 (Krook Collision Term)
Now we simplify the collision term of the Boltzmann equation

$$\left(\frac{\partial f}{\partial t}\right)_{coll} \equiv \int d^3 v_2 \int d\Omega\sigma(\Omega)|\mathbf{v}_2 - \mathbf{v}_1|\left(f_2' f_1' - f_2 f_1\right) , \tag{5.72}$$

for the calculation of $f^{(1)}$. With

$$\delta f_\nu \equiv f_\nu - f_\nu^{(0)} \,, \tag{5.73}$$

where the index ν characterizes the variables $\mathbf{r}$, $\mathbf{v}_\nu$, t, we want to write the linearized collision term as

$$\left(\frac{\partial f}{\partial t}\right)_{coll} \approx \int d^3 v_2 \int d\Omega\sigma(\Omega)|\mathbf{v}_2 - \mathbf{v}_1|\left(f_2^{(0)'}\delta f_1' - f_2^{(0)}\delta f_1 + f_1^{(0)'}\delta f_2' - f_1^{(0)}\delta f_2\right) \tag{5.74}$$

Since we expect each term on the right-hand side to be of similar order, we select the second term for estimation:

$$-\delta f_1 \int d^3 v_2 \int d\Omega \sigma(\Omega)|\mathbf{v}_2 - \mathbf{v}_1| f_2^{(0)} \equiv -\frac{\delta f_1}{\tau} . \tag{5.75}$$

This suggests the following approximation:

$$\boxed{\left(\frac{\partial f}{\partial t}\right)_{coll} \approx -\frac{f - f^{(0)}}{\tau}} , \tag{5.76}$$

which is known as the Krook model. ■

Simplified solution with a Krook collision term

The zeroth order $f^{(0)}$ is a well-known Maxwellian expression, which—besides the explicit velocity dependence—depends on $\mathbf{r}$ and t via n, T, and $\mathbf{u}$.

The first-order distribution function now follows directly from

$$\left(\frac{\partial}{\partial t} + \mathbf{v} \cdot \nabla + \frac{\mathbf{F}}{m}\right) f^{(0)} \approx \frac{f^{(1)}}{\tau} . \tag{5.77}$$

In calculating the left-hand side, we note the following derivatives:

$$\frac{\partial f^{(0)}}{\partial n} = \frac{f^{(0)}}{n} , \quad \frac{\partial f^{(0)}}{\partial T} = \frac{1}{T}\left(\frac{m}{2T} g^2 - \frac{3}{2}\right) f^{(0)} , \tag{5.78}$$

$$\frac{\partial f^{(0)}}{\partial u_i} = \frac{m}{T} g_i f^{(0)} , \quad \frac{\partial f^{(0)}}{\partial v_i} = -\frac{m}{T} g_i f^{(0)} . \tag{5.79}$$

With these derivatives, the first-order solution is

$$f^{(1)} \approx -\tau f^{(0)} \left[\frac{1}{n}\hat{D}n + \frac{1}{T}\left(\frac{m}{2T} g^2 - \frac{3}{2}\right)\hat{D}T + \frac{m}{T}\sum_j g_j \hat{D}u_j - \frac{1}{T}\mathbf{F}\cdot\mathbf{g}\right] , \tag{5.80}$$

with

$$\hat{D} \equiv \frac{\partial}{\partial t} + \mathbf{v} \cdot \nabla . \tag{5.81}$$

The following holds

$$\hat{D}n = -n\nabla \cdot \mathbf{u} + \mathbf{g} \cdot \nabla n , \quad \hat{D}T = -\frac{2}{3} T \nabla \cdot \mathbf{u} + \mathbf{g} \cdot \nabla T , \tag{5.82}$$

$$\hat{D}u_j = \frac{1}{nm}\frac{\partial p}{\partial x_j} + \frac{F_j}{m} + \sum_i g_i \frac{\partial u_j}{\partial x_i} , \quad p = nT . \tag{5.83}$$

Thus, we can approximately write

$$\boxed{f^{(1)} \approx -\tau\left[\frac{1}{T}\sum_i \frac{\partial T}{\partial x_i} g_i\left(\frac{m}{2T}g^2 - \frac{5}{2}\right) + \frac{1}{T}\sum_{i,j}\Lambda_{ij}\left(g_i g_j - \frac{1}{3}\delta_{ij}g^2\right)\right]f^{(0)}} \,. \tag{5.84}$$

The previously introduced functions $\mathcal{F}$ and $\mathcal{G}$ now take the following form:

$$\mathcal{F} \approx \tau\left(\frac{m}{2T}g^2 - \frac{5}{2}\right), \quad \mathcal{G} \approx \tau \,. \tag{5.85}$$

Next, we calculate the heat flux density $\mathbf{q}$. The first-order distribution function $f^{(1)}$ contributes to $\mathbf{q}$ in the form $-K\nabla T$, where

$$K = \frac{m^2\tau}{6T}\int d^3g\, g^4\left(\frac{m}{2T}g^2 - \frac{5}{2}\right)f^{(0)} = \frac{5}{2}\tau T n \,. \tag{5.86}$$

In a similar way, we calculate the pressure tensor $\overleftrightarrow{P}$ to obtain the representation:

$$P_{ij} = p\delta_{ij} + \pi_{ij} \,, \quad \pi_{ij} = -\frac{\tau m}{T}\sum_{k,l}\Lambda_{kl}\int d^3g\, g_i g_j (\underbrace{g_k g_l}_{(1)} - \frac{1}{3}\delta_{kl}g^2)f^{(0)} \,. \tag{5.87}$$

In the detailed determination of the components, we find that π_{ij} is a symmetric tensor with vanishing trace. Its first part, originating from the underlined terms (1), will be proportional to Λ_{ij},

$$\pi_{ij}^{(1)} \sim \Lambda_{ij} \,. \tag{5.88}$$

The previously defined tensor Λ_{ij} is also symmetric, but has a non-vanishing trace

$$\sum_i \Lambda_{ii} = m\sum_i \frac{\partial u_i}{\partial x_i} = m\nabla\cdot\mathbf{u} \,. \tag{5.89}$$

This leads to the ansatz

$$\pi_{ij} \sim \left(\Lambda_{ij} - \frac{m}{3}\delta_{ij}\nabla\cdot\mathbf{u}\right) . \tag{5.90}$$

To determine the proportionality factor, it is sufficient to calculate one component, e.g.

$$\begin{aligned}\pi_{12} &= -\frac{\tau m}{T}\sum_{k,l}\Lambda_{kl}\int d^3g\, g_1 g_2 (g_k g_l - \frac{1}{3}\delta_{kl}g^2)f^{(0)} = -2\frac{\tau m}{T}\Lambda_{12}\int d^3g\, g_1^2 g_2^2 f^{(0)} \\ &= -2\frac{\tau n T}{m}\Lambda_{12} \equiv -2\frac{\mu}{m}\Lambda_{12} \,. \end{aligned} \tag{5.91}$$

From this it then follows

$$\boxed{\pi_{ij} = p\delta_{ij} - 2\frac{\mu}{m}\left(\Lambda_{ij} - \frac{m}{3}\delta_{ij}\nabla \cdot \mathbf{u}\right)} \,. \tag{5.92}$$

The dissipative part of the pressure tensor contains the viscosity.

$$\mu = \tau n T \,. \tag{5.93}$$

For the ratio, it follows

$$\frac{K}{\mu} = \frac{5}{2} = \frac{5}{3}c_V \,. \tag{5.94}$$

5.3 General Outlook and Initial Evaluations

In this section, we take a look ahead at the macroscopic description in plasmas. Simple models and their consequences are presented. We conclude with an initial introduction to the Braginskii equations.
Fundamentally, the plasma dynamic equations are derived in a manner similar to that for neutral gases, where the starting point is the Boltzmann equation. In plasmas, however, the long-range Coulomb forces lead to significant differences. We outline the procedure, starting from the Landau-Fokker-Planck equation. This forms the basis of classical plasma transport theory. Neoclassical transport theory is based on the drift-kinetic equation, which, however, is not treated here. For applications in fusion plasma physics, we refer, for example, to [15, 57, 58].

We begin by recalling that transport theory is based on kinetic equations. The information is reduced by deriving equations for the moments of the distribution function(s). After defining the hydrodynamic variables, we can immediately derive exact—though not closed—equations by integrating the kinetic equation. For the plasma dynamic variables, we calculate

$$\boxed{n_\alpha(\mathbf{r}, t) = \int d^3v f^\alpha(\mathbf{v}, \mathbf{r}, t) \,,} \tag{5.95}$$

$$\boxed{n_\alpha(\mathbf{r}, t)\mathbf{u}_\alpha(\mathbf{r}, t) = \int d^3v \mathbf{v} f^\alpha(\mathbf{v}, \mathbf{r}, t) \,,} \tag{5.96}$$

$$\boxed{n_\alpha T_\alpha = \frac{1}{3}m_\alpha \int d^3v |\mathbf{v} - \mathbf{u}^\alpha|^2 f^\alpha(\mathbf{v}, \mathbf{r}, t) \,.} \tag{5.97}$$

The scalar pressure is $p_\alpha = n_\alpha T_\alpha$ when the temperature is measured in eV.

Direct integration of the Landau-Fokker-Planck equation yields

$$\partial_t n_\alpha + \nabla \cdot \left(n_\alpha \mathbf{u}^\alpha\right) = 0\ , \tag{5.98}$$

$$\begin{aligned}\partial_t\left(m_\alpha n_\alpha u_r^\alpha\right) + \nabla_m\left(m_\alpha n_\alpha u_r^\alpha u_m^\alpha + \delta_{rm} n_\alpha T_\alpha + \Pi_{rm}^\alpha\right)\\ - e_\alpha n_\alpha\left(E_r + \varepsilon_{rmn} u_m^\alpha B_n\right) = R_r^\alpha\ ,\end{aligned} \tag{5.99}$$

$$n_\alpha \partial_t T_\alpha = -n_\alpha\left(\mathbf{u}^\alpha \cdot \nabla\right)T_\alpha - \frac{2}{3} n_\alpha T_\alpha \nabla \cdot \mathbf{u}^\alpha - \frac{2}{3}\Pi_{mn}^\alpha \nabla_m u_n^\alpha - \frac{2}{3}\nabla_m q_m^\alpha + \frac{2}{3} Q^\alpha\ , \tag{5.100}$$

where

$$R_r^\alpha \equiv m_\alpha \int K^\alpha v_r d^3 v\ \ ,\ \ Q^\alpha \equiv \frac{1}{2} m_\alpha \int K^\alpha |\mathbf{v} - \mathbf{u}^\alpha|^2 d^3 v\ . \tag{5.101}$$

Similarly, we compute the dissipative part of the pressure tensor

$$\begin{aligned}\Pi_{rs}^\alpha(\mathbf{r}, t) = m_\alpha \int d^3 v [v_r - u_r^\alpha][v_s - u_s^\alpha] f^\alpha(\mathbf{v}, \mathbf{r}, t)\\ - \frac{1}{3} m_\alpha \delta_{rs} \int d^3 v |\mathbf{v} - \mathbf{u}^\alpha|^2 f^\alpha(\mathbf{v}, \mathbf{r}, t)\end{aligned} \tag{5.102}$$

and the heat flux density

$$q_r^\alpha(\mathbf{r}, t) = \frac{1}{2} m_\alpha \int d^3 v [v_r - u_r^\alpha] |\mathbf{v} - \mathbf{u}^\alpha|^2 f^\alpha(\mathbf{v}, \mathbf{r}, t)\ . \tag{5.103}$$

Both are not yet explicitly known.

The dynamic equations for the mass density ρ, the charge density σ, the mean mass motion $\mathbf{u}$ and the electric current density $\mathbf{j}$ are obtained from these by sums or differences.

Note that all these equations still contain the exact distribution function(s) $f^\alpha(\mathbf{v}, \mathbf{r}, t)$. For further simplification, only phenomena occurring on a hydrodynamic scale, which is much slower than the collision time(s), are considered. Then it can be assumed that the distribution function(s) $f^\alpha(\mathbf{v}, \mathbf{r}, t)$ are close to a Maxwell distribution

$$f_0^\alpha = n_\alpha \left(\frac{m_\alpha}{2\pi T_\alpha}\right)^{3/2} \exp\left[-\frac{m_\alpha |\mathbf{v} - \mathbf{u}^\alpha|^2}{2T_\alpha}\right], \tag{5.104}$$

with varying (thus still exact) expressions n_α, $\mathbf{u}^\alpha$ and T_α.

The small deviations χ^α, where

$$\boxed{f^{\alpha} = f_0^{\alpha}\left[1 + \chi^{\alpha}\right]}, \tag{5.105}$$

are expanded in a particular basis, e.g., Hermite polynomials. With specific truncation procedures (which are discussed in the chapter on linear transport theory), the transport equations can be calculated to a desired degree of accuracy. This yields a closed set of so-called two-fluid equations. For further details, see, for example, the excellent book by Balescu [44, 57].

At this point, we must make an important remark. When we search for (iterated) approximate solutions to the kinetic equation(s) within transport theory, we use a Maxwellian distribution as the dominant part. For a Maxwellian distribution, the collision term vanishes, which is a prerequisite for the zeroth order in the collision-dominated regime. Later, when we consider wave-kinetic equations, we also take into account vanishing collision terms at the lowest order, but in the opposite limiting case of negligible particle collisions.

A Simple Two-Fluid Model

So far, we have introduced the first three moment equations for plasmas in the two-fluid approximation. However, they have a drawback, namely that they do not form a closed system of equations. Further "closures" are necessary. Transport theory will provide the systematic procedure for this.

For an initial insight into standardized problems of plasma dynamics, however, simple models can be helpful. In the following, we will provide a simple two-fluid model using ad hoc arguments. We consider a situation with (approximately) an isotropic Maxwellian distribution and apply the assumption of a scalar pressure. As a result, we expect a model similar to the Euler equations.

First, we consider energy conservation and ignore the collisional contributions that describe energy exchange between the components. This may be appropriate for so-called collisionless plasmas. Nevertheless, the unknown heat flux contribution remains. We expect that, as the thermal conductivity becomes large, the term involving the heat flux density will dominate, unless the temperature becomes spatially uniform. For phenomena with phase velocities much lower than the thermal velocity, heat transfer will equalize the temperatures, and we will postulate constant temperatures. In other words, for

$$\boxed{\frac{v_{phase}}{v_{th\,\alpha}} \ll 1 \;\Rightarrow\; T_\alpha = const \text{ (isothermal)}}. \tag{5.106}$$

The opposite limit is the adiabatic approximation, when the heat flux is neglected:

$$\boxed{\frac{v_{phase}}{v_{th\,\alpha}} \gg 1 \Rightarrow \left(\frac{\partial}{\partial t} + \mathbf{u}_\alpha \cdot \nabla\right)\frac{Np_\alpha}{2} + \frac{N+2}{2}p_\alpha \nabla \cdot \mathbf{u}_\alpha \approx 0 \quad \text{(adiabatic)}} \,. \tag{5.107}$$

The adiabatic equation of state can be specified after introducing

$$\gamma = \frac{N+2}{N}\,, \quad N = 1, 2, 3 \text{ (dimension of the system)} \tag{5.108}$$

Further, we use

$$\left(\frac{\partial}{\partial t} + \mathbf{u}_\alpha \cdot \nabla\right)p_\alpha + \gamma p_\alpha \nabla \cdot \mathbf{u}_\alpha = \frac{dp_\alpha}{dt} - \gamma p_\alpha \frac{1}{n_\alpha}\frac{dn_\alpha}{dt}\,. \tag{5.109}$$

Integration leads to

$$p_\alpha \sim n_\alpha^\gamma\,, \quad p_\alpha = \frac{p_{\alpha 0}}{n_{\alpha 0}^\gamma} n_\alpha^\gamma \quad \text{(adiabatic)}\,. \tag{5.110}$$

In the reverse case, we have

$$p_\alpha = n_\alpha T_\alpha\,, \quad T_\alpha = const \quad \text{(isothermal)}\,. \tag{5.111}$$

In a collisionless, unmagnetized, isotropic plasma, we can use as the simplest two-fluid equations

$$\boxed{\frac{dn_\alpha}{dt} + \nabla \cdot (n_\alpha \mathbf{u}_\alpha) = 0\,,} \tag{5.112}$$

$$\boxed{m_\alpha n_\alpha \left(\frac{d\mathbf{u}_\alpha}{dt} + \mathbf{u}_\alpha \cdot \nabla \mathbf{u}_\alpha\right) = e_\alpha n_\alpha \mathbf{E} - \nabla p_\alpha\,,} \tag{5.113}$$

where we close the system using one of the above pressure equations $p_\alpha = p_\alpha(n_\alpha)$. Of course, the Maxwell equations for the electric and magnetic fields should also be used in addition.

Example 5.3 (Electron Plasma Oscillations)

We now present a brief derivation of electron plasma oscillations (Langmuir oscillations), which we have already discussed within the framework of the kinetic description. From kinetic theory, we already know that the real frequency is approximately given by:

$$\omega^2 \approx \omega_{pe}^2 + 3k^2 v_{the}^2\,. \tag{5.114}$$

If we now seek a re-derivation within the framework of the two-fluid model, we use the fact that the oscillations are high-frequency, $\omega \geq \omega_{pe}$, and in the long-wavelength limit the phase velocity becomes very large, $\omega/k \gg v_{the}$ for $k \to 0$. As a consequence, we can use an adiabatic equation of state for the electrons with

$$p_e \sim n_e^{\gamma} , \quad \text{with } \gamma = 3 \text{ for } N = 1 . \tag{5.115}$$

The high-frequency nature of electron plasma oscillations allows us to treat the ions as an immobile background. By linearizing the simple two-fluid equations around $n_{e0} = n_{i0} \equiv n_0 = \text{const}$ and $\mathbf{u}_{e0} = 0$ we obtain for $\delta n_e \equiv \delta n$ and $\delta \mathbf{u}_e \equiv \delta \mathbf{u}$

$$\frac{\partial \delta n}{\partial t} = -\nabla \cdot (n_0 \delta \mathbf{u}) , \tag{5.116}$$

$$m_e n_0 \frac{\partial \delta \mathbf{u}}{\partial t} = -3T_e \nabla \delta n - e n_0 \delta \mathbf{E} , \tag{5.117}$$

$$\nabla \cdot \delta \mathbf{E} = -\frac{1}{\varepsilon_0} e \delta n . \tag{5.118}$$

In the one-dimensional model, we consider a Fourier mode $\delta n \sim \delta u_x \sim \delta E_x \sim e^{ikx - i\omega t}$ and in the following omit the index x. The linear homogeneous system of equations for the Fourier amplitudes is

$$-\omega \delta n = -k n_0 \delta u , \quad -m n_0 i\omega \delta u = -3T_e ik \delta n - e n_0 \delta E , \quad ik \delta E = -\frac{1}{\varepsilon_0} e \delta n . \tag{5.119}$$

There are nontrivial solutions under the condition

$$\boxed{\omega^2 = \omega_{pe}^2 + 3\frac{T_e}{m_e} k^2} . \tag{5.120}$$

This condition is easily found, for example, by eliminating δn and δE from the above equations. The dispersion relation is the expected result, which we have already derived from kinetic theory. ■

Example 5.4 (Ion-Acoustic Waves)
In an unmagnetized plasma, there also exists a second solution branch, which is generally referred to as the acoustic branch. In an electron-ion plasma, it is known as the ion-acoustic oscillation. For its derivation, we can assume the case $T_e \gg T_i$, which leads to

$$v_{thi} \ll \frac{\omega}{k} \ll v_{the} \tag{5.121}$$

In the end, we will find that this is a consistent assumption. Obviously, due to the low frequency under consideration, we now have to take the ion dynamics into account in the equations of motion. The electron inertia can be neglected, since the frequencies are well below the electron plasma frequency. The electrons are approximately Boltzmann distributed.

After linearization and Fourier transformation of the simple two-fluid model with adiabatic ions and isothermal electrons in a one-dimensional description, we obtain

$$-\omega\delta n_i = -kn_0\delta u_i\ , \quad -m_i n_0 i\omega\delta u_i = en_0\delta E\ , \tag{5.122}$$

$$0 \approx -T_e ik\delta n_e - en_0\delta E\ , \quad ik\delta E = \frac{1}{\varepsilon_0}e(\delta n_i - \delta n_e)\ , \tag{5.123}$$

From the last equation, the electron oscillations follow as

$$\delta n_e = \delta n_i - \frac{ik\varepsilon_0}{e}\delta E\ . \tag{5.124}$$

Using this in the electron momentum balance, we find the electric field oscillations in relation to the ion oscillations,

$$\delta E = -\frac{ikT_e}{en_o\left(1 + k^2\lambda_{De}^2\right)}\delta n_i\ . \tag{5.125}$$

Using the ion momentum balance, we express the ion velocity in terms of the ion density oscillations,

$$- m_i n_0\omega\delta u_i = -\frac{kT_e}{\left(1 + k^2\lambda_{De}^2\right)}\delta n_i\ . \tag{5.126}$$

Substituting these results into the ion continuity equation, a brief rearrangement for non-vanishing ion oscillations yields

$$\boxed{\omega^2 \approx \frac{k^2 c_s^2}{1 + k^2\lambda_{De}^2}}\ , \tag{5.127}$$

where we have defined the ion sound speed

$$c_s = \sqrt{\frac{T_e}{m_i}}\ . \tag{5.128}$$

Obviously, for $T_e \gg T_i$ the initial assumption holds. ■

Example 5.5 (Electromagnetic Waves)

In an unmagnetized plasma, electromagnetic waves can propagate, provided that $\omega > \omega_{pe}$. To show this, we couple the Maxwell equations for a purely electromagnetic wave,

$$\nabla \times \mathbf{E} = -\frac{\partial \mathbf{B}}{\partial t}\ , \quad \nabla \times \mathbf{B} = \mu_0\mathbf{j} + \varepsilon_0\mu_0\frac{\partial \mathbf{E}}{\partial t}\ , \tag{5.129}$$

to the motion of the electrons in the plasma. By combining the Maxwell equations, we immediately obtain

$$\nabla \times (\nabla \times \mathbf{E}) \equiv \nabla(\nabla \cdot \mathbf{E}) - \nabla^2 \mathbf{E} = -\mu_0 \frac{\partial \mathbf{j}}{\partial t} - \varepsilon_0 \mu_0 \frac{\partial^2 \mathbf{E}}{\partial t^2} \,. \tag{5.130}$$

The electric current density $\mathbf{j}$ is calculated from

$$\mathbf{j} \equiv \delta \mathbf{j} \approx \sum_\alpha n_0 e_\alpha \delta \mathbf{u}_\alpha \,. \tag{5.131}$$

By linearizing the equations of motion for $p_\alpha \approx 0$ we obtain

$$m_\alpha \frac{\partial \delta \mathbf{u}_\alpha}{\partial t} \approx e_\alpha \delta \mathbf{E} \,, \tag{5.132}$$

or

$$\frac{\partial \mathbf{j}}{\partial t} \approx \underbrace{\sum_\alpha \frac{e_\alpha^2 n_0}{m_\alpha}}_{\equiv \varepsilon_0 \omega_p^2} \delta \mathbf{E} \,. \tag{5.133}$$

The wave equation then becomes

$$\nabla(\nabla \cdot \mathbf{E}) - \nabla^2 \mathbf{E} = -\varepsilon_0 \mu_0 \mathbf{E} - \varepsilon_0 \mu_0 \frac{\partial^2 \mathbf{E}}{\partial t^2} \,. \tag{5.134}$$

After Fourier transformation, we obtain with $c^2 = 1/\varepsilon_0 \mu_0$

$$\mathbf{k}(\mathbf{k} \cdot \mathbf{E}) - k^2 \mathbf{E} = \frac{\omega^2 - \omega_p^2}{c^2} \mathbf{E} \,. \tag{5.135}$$

Transverse electromagnetic waves have $\mathbf{k} \perp \mathbf{E}$, so that

$$\boxed{\omega^2 \approx \omega_p^2 + k^2 c^2} \tag{5.136}$$

appears as the electromagnetic dispersion relation. In the non-relativistic limit $c \gg v_{the}$ the zero-pressure assumption is justified.

On the other hand, the zero-pressure approximation for electrostatic oscillations with $\mathbf{k} \| \mathbf{E}$ $\omega^2 \approx \omega_p^2$ yields results in agreement with the previous findings. ■

Drift Model

Let us now briefly discuss the truncation procedure for drift waves under the assumption $u_E \sim \delta v_{th}$. The assumed distribution function deviates from a Maxwellian distribution

$$f_M(\mathbf{v}) = \frac{n\, m^{3/2}}{(2\pi T)^{3/2}}\, e^{-\frac{m v^2}{2T}} \tag{5.137}$$

in two terms:

$$f \approx f_M(\mathbf{v})\left[1 + 2\frac{u_\parallel v_\parallel}{v_{th}^2}\right] - \rho \cdot \nabla f_M + \mathcal{O}(\delta^2)\,. \tag{5.138}$$

The first yields the parallel velocity through

$$\int d^3v\, f_M(\mathbf{v})\, \frac{m\, v_\parallel\, u_\parallel}{T}\, v_\parallel = n\, u_\parallel\,, \tag{5.139}$$

while the second leads to the drift velocities:

$$\int d^3v(-\rho \cdot \nabla f_M)\mathbf{v}_\perp = \frac{p}{m\Omega}\mathbf{b} \times \left(\nabla \ln p + \frac{e\nabla\phi}{T}\right) \equiv n\, \mathbf{u}_\perp\,. \tag{5.140}$$

The other moments can also be evaluated, although the algebra is quite tedious. Therefore, we do not present any details here. In summary, by applying the drift ordering in all equations, we ultimately arrive, in the quasineutral case for $P = p_e + p_i$, $p_e \approx p_i \approx p$ and $T_e \approx T_i \approx T$, at the following (drift) model [58]

$$\boxed{\frac{dn}{dt} + n\, \nabla \cdot \mathbf{u} = 0\,,} \tag{5.141}$$

$$\boxed{\frac{dP}{dt} - \mathbf{v}_{pi} \cdot \nabla P + \frac{5}{3}\, P\, \nabla \cdot \left(\mathbf{v}_{E\times B} + \mathbf{b}\, u_\parallel\right) = 0\,,} \tag{5.142}$$

$$\boxed{m_i\, n\left[\frac{d\, \mathbf{v}_{E\times B}}{dt} + \frac{d}{dt}\, (\mathbf{b}\, u_\parallel) - \mathbf{v}_{pi} \cdot \nabla\, (\mathbf{b}\, u_\parallel)\right] + \nabla P - \mathbf{j} \times \mathbf{B} = 0\,,} \tag{5.143}$$

$$\boxed{\mathbf{E} + \mathbf{u} \times \mathbf{B} + \frac{1}{en}\left(\frac{1}{2}\nabla P - \mathbf{j} \times \mathbf{B}\right) - \eta\left[\mathbf{j} - \frac{3}{4}\frac{n}{B}\mathbf{b} \times \nabla\frac{P}{n}\right] = 0\,.} \tag{5.144}$$

The equations

$$\nabla \times \mathbf{B} \approx \mu_0 \mathbf{j}\,, \quad \nabla \times \mathbf{E} = -\frac{\partial \mathbf{B}}{\partial t}\,, \tag{5.145}$$

$$\frac{d}{dt} = \partial_t + \mathbf{u} \cdot \nabla\,, \quad \mathbf{v}_{E\times B} = \frac{\mathbf{E} \times \mathbf{B}}{B^2} \tag{5.146}$$

must be supplemented. The polarization drift of the ions is

$$\mathbf{v}_{pi} = \frac{1}{2\Omega_i \, m_i \, n} \, \mathbf{b} \times \nabla P \, . \tag{5.147}$$

We have again simplified the problem by moving from the detailed description in configuration and velocity space (via the distribution functions f^α) to a reduced description in configuration space only (using a finite number of moments n, $\mathbf{u}$, P, $\mathbf{B}, \ldots$). If we are not interested in detailed wave-particle resonances and are focusing on slow (hydrodynamic) processes, such a reduction is advisable. Nevertheless, the equations of motion still contain important information, e.g., about nonlinear processes, collective effects, etc. [59–62]. To demonstrate this, let us recall as an example the electrostatic drift waves.

Example 5.6 (Electrostatic Drift Waves)
Within the drift model, it is straightforward to derive the dispersion of low-frequency drift waves that occur in inhomogeneous plasmas. Let us discuss here the electrostatic limiting case with $\mathbf{E} = -\nabla\phi$ in a constant magnetic field (index 0: $\mathbf{B} = \mathbf{B}_0$), so that $\nabla_\parallel = \mathbf{b}_0 \cdot \nabla$ holds. Finally, we will discuss the non-resistive case $\eta \approx 0$ and the ion polarization drift v_{pi} be considered small. Based on $\nabla \cdot \mathbf{v}_{E\times B} = 0$, $\nabla \cdot \mathbf{v}_{pi} = -\mathbf{v}_{pi} \cdot \nabla \ln n$ and $\nabla \cdot (\mathbf{b}\, u_\parallel) = \mathbf{b}_0 \cdot \nabla u_\parallel = \nabla_\parallel u_\parallel$, the fundamental Eqs. (5.141)–(5.144) can be significantly simplified. Eq. (5.141) becomes

$$\frac{\partial n}{\partial t} + (\mathbf{u} - \mathbf{v}_{pi}) \cdot \nabla n \approx -n \nabla_\parallel u_\parallel \, , \tag{5.148}$$

The parallel component of (5.143) becomes

$$m_i n \left(\frac{\partial}{\partial t} + (\mathbf{u} - \mathbf{v}_{pi}) \cdot \nabla \right) u_\parallel + \nabla_\parallel u_\parallel \approx 0 \, . \tag{5.149}$$

Again, the parallel component, now of (5.144), is used to find

$$-\nabla_\parallel \phi + \frac{1}{2en} \nabla_\parallel P \approx \eta j_\parallel \approx 0 \, . \tag{5.150}$$

In the linearization within the so-called slab approximation, the zeroth-order quantities are given the index 0, and the first order is denoted by the index 1. We use $u_{\parallel 0} = 0$ and, for example,

$$\nabla n_0 = \hat{x} n_0' \, , \quad \nabla n_1 = i\mathbf{k} n_1 \, , \quad \frac{\partial n_1}{\partial t} = -i\omega n_1 \, . \tag{5.151}$$

We further introduce $\mathbf{k}_\perp = (\mathbf{b} \times \hat{x}) \cdot \mathbf{k} = k_y$ and

$$\omega_* \equiv -k_y \frac{T}{eB_0} \frac{P_0'}{P_0} \, , \quad \omega_E \equiv k_y \frac{\phi_0'}{B_0} \tag{5.152}$$

This leads, for Eqs. (5.148) and (5.150), to

$$-i(\omega - \omega_E)\frac{P_1}{P_0} + i\omega_* \frac{e\phi_1}{T} \approx -i\frac{5}{3}k_\parallel u_{\parallel 1} \,, \tag{5.153}$$

$$-i(\omega - \omega_E)u_{\parallel 1} \approx -ik_\parallel \frac{P_1}{m_i n_0} \,, \tag{5.154}$$

$$\frac{P_1}{P_0} \approx \frac{e\phi_1}{T} \,. \tag{5.155}$$

The solvability condition for nontrivial solutions of the linear homogeneous system is

$$1 - \frac{5}{3}\frac{2k_\parallel^2 T}{m_i(\omega - \omega_E)^2} = \frac{\omega_*}{\omega - \omega_E} \,, \tag{5.156}$$

with the approximate solution

$$\boxed{\omega \approx \omega_E + \omega_* \left[1 + \frac{5}{3}\frac{2k_\parallel^2 T}{m_i \omega_*^2}\right]} \,. \tag{5.157}$$

We observe a Doppler shift (ω_E). Furthermore, for small $k_\parallel$, the characteristic drift wave frequency is ω_*. ■

Braginskii Equations

So far, we have considered simple fluid models that correspond, for example, to the Euler equations for neutral fluids. The derivation of the general two-fluid model is the core task of transport theory. The systematic procedure is outlined in a separate chapter on linear transport theory.

Here, as a summary, we present the result first developed by Braginskii [63]. For electrons, the balance equations for particle density, momentum, and pressure are

$$\frac{\mathrm{d}n_e}{\mathrm{d}t} + \nabla \cdot (n_e \mathbf{u}_e) = 0 \,, \tag{5.158}$$

$$m_e n_e \left(\frac{\mathrm{d}\mathbf{u}_e}{\mathrm{d}t} + \mathbf{u}_e \cdot \nabla \mathbf{u}_e\right) + \nabla p_e + \nabla \cdot \overleftrightarrow{\pi}_e + e n_e(\mathbf{E} + \mathbf{u}_e \times \mathbf{B}) = \mathbf{F} \,, \tag{5.159}$$

$$\frac{3}{2}\left(\frac{\mathrm{d}p_e}{\mathrm{d}t} + \mathbf{u}_e \cdot \nabla p_e\right) + \frac{5}{2}p_e \nabla \cdot \mathbf{u}_e + \overleftrightarrow{\pi}_e : \nabla \mathbf{u}_e + \nabla \cdot \mathbf{q}_e = W_e \,. \tag{5.160}$$

A similar set of equations arises for the ions ($Z = 1$):

$$\frac{\mathrm{d}n_i}{\mathrm{d}t} + \nabla \cdot (n_i \mathbf{u}_i) = 0 \,, \tag{5.161}$$

$$m_i n_i \left(\frac{\mathrm{d}\mathbf{u}_i}{\mathrm{d}t} + \mathbf{u}_i \cdot \nabla \mathbf{u}_i \right) + \nabla p_i + \nabla \cdot \overleftrightarrow{\pi}_i - e n_i (\mathbf{E} + \mathbf{u}_i \times \mathbf{B}) = -\mathbf{F} \,, \tag{5.162}$$

$$\frac{3}{2}\left(\frac{\mathrm{d}p_i}{\mathrm{d}t} + \mathbf{u}_i \cdot \nabla p_i \right) + \frac{5}{2} p_i \nabla \cdot \mathbf{u}_i + \overleftrightarrow{\pi}_i : \nabla \mathbf{u}_i + \nabla \cdot \mathbf{q}_i = W_i \,. \tag{5.163}$$

We should have $n_e \approx n_i \equiv n$ for a fully ionized electron-proton plasma. The important point is that the transport coefficients have been calculated.

For an unmagnetized plasma with $\Omega_e \tau_e$, $\Omega_i \tau_i \ll 1$ we have [63]

$$\mathbf{F} = \frac{ne\mathbf{j}}{\sigma_\parallel} - 0{,}71 n \nabla T_e \,, \quad W_i = \frac{3m_e}{m_i} \frac{n(T_e - T_i)}{\tau_e} \,, \quad W_e = -W_i + \frac{\mathbf{j} \cdot \mathbf{F}}{ne} \,, \tag{5.164}$$

$$\mathbf{j} = -ne(\mathbf{u}_e - \mathbf{u}_i) \,, \quad \sigma_\parallel = 1{,}96 \frac{ne^2 \tau_e}{m_e} \,, \tag{5.165}$$

$$\tau_e = \frac{6\sqrt{2}\pi^{3/2} \varepsilon_0^2 \sqrt{m_e} T_e^{3/2}}{\ln \Lambda n e^4} \,, \quad \tau_i = \frac{12 \pi^{3/2} \varepsilon_0^2 \sqrt{m_i} T_i^{3/2}}{\ln \Lambda n e^4} \,, \tag{5.166}$$

$$\mathbf{q}_e = -\kappa_\parallel^e \nabla T_e - 0{,}71 \frac{T_e}{e} \mathbf{j} \,, \, \mathbf{q}_i = -\kappa_\parallel^i \nabla T_i \,, \tag{5.167}$$

$$\kappa_\parallel^e = 3{,}2 \frac{n \tau_e T_e}{m_e} \,, \quad \kappa_\parallel^i = 3{,}9 \frac{n \tau_i T_i}{m_i} \,, \tag{5.168}$$

$$(\pi_e)_{\alpha\beta} = -\eta_0^e \left(\frac{\mathrm{d}u_\alpha}{\mathrm{d}x_\beta} + \frac{\mathrm{d}u_\beta}{\mathrm{d}x_\alpha} - \frac{2}{3} \nabla \cdot \mathbf{u}_e \delta_{\alpha\beta} \right) , \tag{5.169}$$

$$(\pi_i)_{\alpha\beta} = -\eta_0^i \left(\frac{\mathrm{d}u_\alpha}{\mathrm{d}x_\beta} + \frac{\mathrm{d}u_\beta}{\mathrm{d}x_\alpha} - \frac{2}{3} \nabla \cdot \mathbf{u}_i \delta_{\alpha\beta} \right) , \tag{5.170}$$

$$\eta_0^e = 0{,}73 n \tau_e T_e \,, \quad \eta_0^i = 0{,}96 n \tau_i T_i \,. \tag{5.171}$$

Note that $\ln \Lambda$ depends on the type of collisions and typically has a magnitude of 13 in a fusion plasma.

For a magnetized plasma with $\Omega_e \tau_e$, $\Omega_i \tau_i \geq 1$ the corresponding values are also known [63]; for example, they can be found in [33].

5.4 Transport Theory

Transport theory is a central topic in theoretical plasma physics. We distinguish between linear transport theory (described in the first part of this chapter for straight magnetic field lines) and nonlinear transport, which becomes necessary because linear transport theory fails in the presence of strong fluctuations (as discussed in the second part of this chapter for so-called stochastic transport). Magnetic field topologies, as realized for example in tokamaks and stellarators, require a transport theory—even in the linear regime—that goes beyond the approximation to straight magnetic field lines considered here. In presenting only the fundamental ideas of transport theory, we omit a discussion of neoclassical theory, which is a subject in its own right; see, for example, [57].

In the previous chapter, we introduced the fluid approach, which provides a closed description once some *ad hoc* assumptions about the distribution function(s) are made. Linear transport theory offers us a systematic method for determining transport coefficients by considering the linear kinetic plasma response. The starting point of a linear transport theory is thus a kinetic equation. Here, we will use the Landau-Fokker-Planck equation; the goal is to derive the coefficients summarized in the section on the Braginskii equations. Since the evaluation is quite involved and requires consideration of many details, we will only be able to present an overview here. More practically oriented directions, such as neoclassical transport, which has been developed from the drift-kinetic equation for tokamak- and stellarator-like magnetic field configurations, also cannot be addressed here due to the aforementioned space limitations.

In principle, in transport theory we distinguish between hydrodynamic scales (time scale T_H and length scale L_H, on which variations are slow and weak) and collisional scales (which are fast).

A kinetic equation generally contains a convective term ($\mathbf{v} \cdot \nabla f$), which is mainly responsible for the nonlinear terms in the equations of motion, as well as collision terms K^α. Moments vary on the hydrodynamic scales T_H and L_H, while the collision term is approximately described by

$$\boxed{K^{\alpha\beta} \sim \frac{1}{\tau_{R\alpha\beta}} f^\alpha}, \tag{5.172}$$

where $\tau_{R\alpha\beta}$ is the relaxation time due to collisions. As already discussed, we should distinguish between fast relaxations within a component and slow exchanges between components. For order-of-magnitude estimates, we can use the relation $T_H \sim L_H/v_{th}$ between the hydrodynamic scales. Thus, linear plasma transport theory begins with a zeroth-order approach that satisfies the fast exchanges within the components, and then uses a kinetic equation to determine the (slow and weak) linear response to external forces.

Moments in Linear Transport Theory

Based on the scheme just outlined, we first examine the contributions of K^{ee} and K^{ii}. They are responsible for bringing the local distribution function close to an equilibrium-like form. To zeroth order,

$$K^{\alpha\alpha} \approx 0 \qquad \text{for}\,\alpha = e, i \tag{5.173}$$

is the starting point. The equations are satisfied by Maxwell distributions

$$f_0^\alpha = n_\alpha \left(\frac{m_\alpha}{2\pi k_B T_\alpha} \right)^{3/2} \exp\left[-\frac{m_\alpha |\mathbf{v} - \mathbf{u}^\alpha|^2}{2k_B T_\alpha} \right] , \tag{5.174}$$

which still contain fields (moments or plasma dynamic variables). It is easy to verify that

$$\frac{1}{m_\alpha} \left(\frac{\partial}{\partial v1\nu} - \frac{\partial}{\partial v2\nu} \right) f_0^\alpha(\mathbf{v}_1) f_0^\alpha(\mathbf{v}_2) = -\frac{1}{k_B T_\alpha} g_\nu f_0^\alpha(\mathbf{v}_1) f_0^\alpha(\mathbf{v}_2) \, . \tag{5.175}$$

Furthermore, the collision terms vanish because of

$$G_{\mu\nu} g_\nu = \frac{g^2 \delta_{\mu\nu} - g_\mu g_\nu}{g^3} g_\nu = 0 \, . \tag{5.176}$$

Here, the plasma dynamic variables n_α, T_α and $\mathbf{u}^\alpha$ appear. The general forms of the dynamic equations for these have already been presented. We assume that the temporal and spatial variations of the plasma dynamic variables occur on hydrodynamic scales,

$$n_\alpha = n_\alpha(\mathbf{x}, t) \, , \quad T_\alpha = T_\alpha(\mathbf{x}, t) \, , \quad \mathbf{u}^\alpha = \mathbf{u}^\alpha(\mathbf{x}, t) \, . \tag{5.177}$$

By writing the general solution of a kinetic equation as

$$\boxed{f^\alpha = f_0^\alpha [1 + \chi^\alpha]} \tag{5.178}$$

we introduce the deviation $\chi^\alpha = \chi^\alpha(\mathbf{x}, \mathbf{v}, t)$ as the deviation from the local Maxwell distribution f_0^α. In the following, when we determine χ^α, we require that the moments of f^α exactly match the moments that appear in f_0^α. In other words,

$$\boxed{\int d^3 v f^\alpha = \int d^3 v f_0^\alpha = n_\alpha(\mathbf{x}, t) \, ,} \tag{5.179}$$

$$\boxed{\int d^3v\, \mathbf{v} f^\alpha = \int d^3v\, \mathbf{v} f_0^\alpha = n_\alpha(\mathbf{x}, t)\mathbf{u}^\alpha(\mathbf{x}, t)\ ,} \tag{5.180}$$

$$\boxed{\frac{1}{3}m_\alpha \int d^3v |\mathbf{v} - \mathbf{u}^\alpha|^2 f^\alpha = \frac{1}{3}m_\alpha \int d^3v |\mathbf{v} - \mathbf{u}^\alpha|^2 f_0^\alpha = n_\alpha(\mathbf{x}, t) k_B T_\alpha(\mathbf{x}, t)\ .} \tag{5.181}$$

Before we discuss the consequences for $f_0^\alpha \chi^\alpha$ and χ^α, we introduce the relative velocity $\mathbf{v} - \mathbf{u}^\alpha$, which we normalize with the thermal velocity $v_{th\alpha} = (k_B T_\alpha)^{1/2}/m_\alpha^{1/2}$ for each component α. In the following, we omit the index α as long as no confusion is to be expected. Then, the definition[1]

$$\boxed{\mathbf{c} = \mathbf{c}^\alpha = \left(\frac{m_\alpha}{k_B T_\alpha}\right)^{1/2} (\mathbf{v} - \mathbf{u}^\alpha)}\ . \tag{5.182}$$

Similarly, instead of f_0^α we introduce the normalized function ϕ_0^α, namely by

$$f_0^\alpha = n_\alpha \left(\frac{m_\alpha}{k_B T_\alpha}\right)^{1/2} \phi_0^\alpha\ , \quad \phi_0 = \phi_0^\alpha = \frac{1}{(2\pi)^{3/2}} e^{-\frac{1}{2}c^2}\ . \tag{5.183}$$

Again, not all indices are written out explicitly, as long as no confusion is to be expected.

We will expand χ^α in terms of Gauss-Hermite polynomials. The latter are defined by

$$\tilde{H}^{(m)}_{r_1 \dots r_m}(\mathbf{c}) = (-1)^m\, e^{c^2/2} \frac{\partial}{\partial c_{r1}} \cdots \frac{\partial}{\partial c_{rm}}\, e^{-c^2/2}\ . \tag{5.184}$$

The upper index denotes the order $m = 0, 1, \dots$; the lower indices indicate which m components are used in the derivatives with respect to the velocity components $r_1, \dots, r_m$.

The Hermite polynomials are suitable because we have a Gaussian (Maxwellian) distribution as the weight function. The expansion is formally written as

$$\boxed{\chi^\alpha(\mathbf{x}, \mathbf{c}, t) = \sum_{m=0}^{\infty} \frac{1}{m!} \tilde{h}^{\alpha(m)}_{r_1 \dots r_m}(\mathbf{x}, t) \tilde{H}^{(m)}_{r_1 \dots r_m}(\mathbf{c})} \tag{5.185}$$

with the expansion coefficients

[1] Please do not confuse with the speed of light in vacuum!

$$\boxed{\tilde{h}^{\alpha(m)}_{r_1\ldots r_m}(\mathbf{x},t) = \int d^3c\,\phi_0(c)\chi^\alpha(\mathbf{x},\mathbf{c},t)\tilde{H}^{(m)}_{r_1\ldots r_m}(\mathbf{c})} \,. \tag{5.186}$$

The Einstein convention for summation over repeated velocity indices $r_\nu = x, y, z$ has been applied. By contracting the indices, we obtain, for example,

$$\hat{H}^{(2n)}(\mathbf{c}) = \sum_{s_1,\ldots,s_n} \tilde{H}^{(2n)}_{s_1 s_1 s_2 s_2 \ldots s_n s_n}(\mathbf{c}) \,, \tag{5.187}$$

$$\hat{H}^{(2n+1)}_r(\mathbf{c}) = \sum_{s_1,\ldots,s_n} \tilde{H}^{(2n+1)}_{r s_1 s_1 \ldots s_n s_n}(\mathbf{c}) \,, \tag{5.188}$$

$$\hat{H}^{(2n)}_{r_1 r_2}(\mathbf{c}) = \sum_{s_1,\ldots,s_{n-1}} \tilde{H}_{r_1 r_2 s_1 s_1 \ldots s_{n-1} s_{n-1}}(\mathbf{c}) - \frac{1}{3}\delta_{r_1 r_2}\hat{H}^{(2n)}(\mathbf{c}) \tag{5.189}$$

and so on. In the so-called irreducible representation, we use

$$H^{(0)} = 1 \,, \quad H^{(2)} = \frac{1}{\sqrt{6}}(c^2 - 3) \,, \quad H^{(4)} = \frac{1}{2\sqrt{30}}(c^4 - 10c^2 + 15) \,, \ldots \tag{5.190}$$

$$H^{(1)}_r = c_r \,, \quad H^{(3)}_r = \frac{1}{\sqrt{10}}c_r(c^2 - 5) \,, \quad H^{(5)}_r = \frac{1}{2\sqrt{70}}c_r(c^4 - 14c^2 + 35) \,, \ldots \tag{5.191}$$

$$H^{(2)}_{rs} = \frac{1}{\sqrt{2}}\left(c_r c_s - \frac{1}{3}\delta_{rs}c^2\right) \,, \quad H^{(4)}_{rs} = \frac{1}{2\sqrt{7}}\left(c_r c_s - \frac{1}{3}\delta_{rs}c^2\right)(c^2 - 7) \,, \ldots \tag{5.192}$$

$$H^{(3)}_{rsp} = c_r c_s c_p - \frac{1}{5}c^2(c_r\delta_{sp} + c_s\delta_{rp} + c_p\delta_{rs}) \,, \ldots \tag{5.193}$$

and so on. These are, respectively, scalar, vector, and tensor quantities. Since every symmetric tensor of rank ≥ 2 (here $\tilde{H}^{(m)}_{r_1\ldots r_m}$) can be uniquely decomposed into irreducible components, we can express $\tilde{H}^{(m)}_{r_1\ldots r_m}$ in terms of the quantities just defined $H^{(\nu+\mu)}_{r_1 r_2 \ldots r_\nu}$, $\nu = 0, 1, 2, \ldots, \mu = 0, 2, 4$, for example,

$$\tilde{H}^{(2)}_{rs} = \sqrt{2}\left(H^{(2)}_{rs} + \frac{1}{\sqrt{3}}H^{(2)}\delta_{rs}\right) . \tag{5.194}$$

This leads to the expansion

$$\begin{aligned} \chi^\alpha &= \sum_{m=0}^{\infty} h^{\alpha(2m)}(\mathbf{x},t)H^{(2m)}(\mathbf{c}) + \sum_{m=0}^{\infty} h^{\alpha(2m+1)}_r(\mathbf{x},t)H^{(2m+1)}_r(\mathbf{c}) \\ &+ \sum_{m=1}^{\infty} h^{\alpha(2m)}_{rs}(\mathbf{x},t)H^{(2m)}_{rs}(\mathbf{c}) + \ldots \end{aligned} \tag{5.195}$$

The coefficients h satisfy

$$h^{\alpha(m)}_{r_1\ldots r_n} = \int d^3c\,\phi_0(c)\chi^\alpha(\mathbf{x},\mathbf{c},t)H^{(m)}_{r_1\ldots r_n}(\mathbf{c})\,. \tag{5.196}$$

Due to the constraints

$$\int d^3c\,\phi_0\chi^\alpha = \int d^3c\,\phi_0 c_r\,\chi^\alpha = \int d^3c\phi_0\,c^2\chi^\alpha = 0 \tag{5.197}$$

it follows

$$h^{\alpha(0)} = h^{\alpha(1)}_r = h^{\alpha(2)} = 0\,. \tag{5.198}$$

Since

$$\int d^3c\,1\,H^{(2)}_{rs}\phi_0 = 0\,, \tag{5.199}$$

the dissipative part of the pressure tensor π^α_{rs} can be expressed as

$$\boxed{\pi^\alpha_{rs} = \sqrt{2}\,n_\alpha\,k_BT_\alpha\,h^{\alpha(2)}_{rs}}\,. \tag{5.200}$$

Similarly, for the heat flux we obtain

$$\boxed{q^\alpha_r = \left(\frac{5}{2}\right)^{1/2} m_\alpha\left(\frac{k_BT_\alpha}{m_\alpha}\right)^{3/2} n_\alpha\,h^{\alpha(3)}_r}\,. \tag{5.201}$$

So far, the calculation is exact. However, the coefficients on the right-hand side are still undetermined. To proceed, we truncate the expansion and assume that the vectorial and traceless polynomials of second order are sufficient to describe weakly anisotropic situations. In other words, we will use only the polynomials $H^{(\ldots)}$, $H^{(\ldots)}_r$ and $H^{(\ldots)}_{rs}$ as basis functions. Then the finite expansion is

$$\chi^\alpha(\mathbf{x},\mathbf{c},t) \approx \sum_r c_r B^\alpha_r(c;\mathbf{x},t) + \sum_{r,s}(c_rc_s - \frac{1}{3}c^2\delta_{rs})C^\alpha_{rs}(c;\mathbf{x},t)\,. \tag{5.202}$$

The coefficients B and C still include the summation over m.

For practical reasons, we must also truncate this summation. Within the *13-moment approximation* we use only

$$c_rB^\alpha_r \approx h^{\alpha(3)}_r\,H^{(3)}_r\,,\quad (c_rc_s - \frac{1}{3}c^2\delta_{rs})C^\alpha_{rs} \approx h^{\alpha(2)}_{rs}\,H^{(2)}_{rs}\,. \tag{5.203}$$

The next better approximation is the *21-moment approximation;* it uses

$$c_r B_r^\alpha \approx h_r^{\alpha(3)} H_r^{(3)} + h_r^{\alpha(5)} H_r^{(5)}\,, \quad (c_r c_s - \frac{1}{3} c^2 \delta_{rs}) C_{rs}^\alpha \approx h_{rs}^{\alpha(2)} H_{rs}^{\alpha(2)} + h_{rs}^{\alpha(4)} H_{rs}^{(4)}\,. \tag{5.204}$$

The degree of accuracy can be further improved, for example, by the *29-moment approximation*

$$c_r B_r^\alpha \approx h_r^{\alpha(3)} H_r^{(3)} + h_r^{\alpha(5)} H_r^{(5)} + h_r^{\alpha(7)} H_r^{(7)}\,, \tag{5.205}$$

$$(c_r c_s - \frac{1}{3} c^2 \delta_{rs}) C_{rs}^\alpha \approx h_{rs}^{\alpha(2)} H_{rs}^{\alpha(2)} + h_{rs}^{\alpha(4)} H_{rs}^{\alpha(4)} + h_{rs}^{\alpha(6)} H_{rs}^{\alpha(6)}\,, \tag{5.206}$$

which contains eight more moments.

How do we count the moments? The 13-moment approximation consists of the five plasma dynamic variables n_α, u_r^α and T_α, the three components $h_r^{\alpha(3)}$ and the five independent entries of the symmetric and traceless tensor $h_{rs}^{\alpha(2)}$. Obviously, the multiplicity due to the different species α is not counted.

The next approximation additionally contains three vector and five tensor components. The sequence of approximations would thus comprise 13, 21, 29, 37, … moments. Increasing the number of variables by eight leads to a better approximation; the performance of the approximation can be easily checked. The differences between the 13-moment approximation and the 21-moment approximation are generally significant, while the 29-moment approximation usually does not improve much further.

In summary, we perform two steps. Deviations from anisotropy are considered only through the vectorial and rank-2 tensor components. The contributions of these two non-scalar parts are calculated successively in the 13-, 21-, or 29-moment approximation. For example, in addition to the plasma dynamic variables n_e, n_i, T_e, T_i, $\mathbf{u}^e$ and $\mathbf{u}^i$, in the 13-moment approximation we additionally have $h_r^{\alpha(3)}$ and $h_{rs}^{\alpha(2)}$. In the 21-moment approximation, new elements are added: $h_r^{\alpha(3)}$, $h_r^{\alpha(5)}$, $h_{rs}^{\alpha(2)}$, $h_{rs}^{\alpha(4)}$. In the 29-moment approximation, $h_r^{\alpha(3)}$, $h_r^{\alpha(5)}$, $h_r^{\alpha(7)}$, $h_{rs}^{\alpha(2)}$, $h_{rs}^{\alpha(4)}$ and $h_{rs}^{\alpha(6)}$ are added. In all cases, we have $h^{\alpha(0)} = h^{\alpha(2)} = h_r^{\alpha(1)} = 0$.

Some of the moments are privileged [44]. When we transition to the hydrodynamic variables ρ, $\sigma \approx 0$, $\mathbf{u}$, T_e, T_i and $\mathbf{j}$, we introduce $j_r \equiv e n_e (T_e/m_e)^{1/2} h_r^{(1)}$ (new definition of $h_r^{(1)}$) and find $\pi_{rs}^\alpha \sim h_{rs}^{\alpha(2)}$ and $q_r^\alpha \sim h_r^{\alpha(3)}$. These are the privileged moments in comparison to $h_r^{\alpha(5)}$, $h_{rs}^{\alpha(4)}$, … The reason for this designation will soon become apparent.

We now turn to the 21-moment approximation in the quasineutral case $\sigma \approx 0$. The displacement current is neglected in Maxwell's equations, so that $\nabla \cdot \mathbf{j} \approx 0$ and $\partial_t \sigma \approx 0$

follow. We begin with

$$\partial_t \rho + \nabla \cdot (\rho \mathbf{u}) = 0 \,, \tag{5.207}$$

$$\begin{aligned}\partial_t(\rho u_r) =& - \nabla_s(\rho u_r u_s) - \frac{Zk_B}{m_i}\nabla_r\Big[\rho(T_e + \frac{1}{Z}T_i)\Big] \\ & - \sqrt{2}\frac{Zk_B}{m_i}\nabla_s\Big[\rho\Big(T_e\, h^{e(2)}_{rs} + \frac{1}{Z}T_i\, h^{i(2)}_{rs}\Big)\Big] + e\,\frac{Z\rho}{m_i}\Big(\frac{k_B T_e}{m_e}\Big)^{1/2} \varepsilon_{rsm}\, h^{(1)}_s\, B_m \,,\end{aligned} \tag{5.208}$$

$$\begin{aligned}\partial_t T_e =& - \mathbf{u}\cdot\nabla T_e - \frac{2}{3}T_e \nabla\cdot\mathbf{u} + \Big(\frac{k_B T_e}{m_e}\Big)^{1/2} h^{(1)}_r \nabla_r T_e + \frac{2}{3}T_e\nabla_r\Big[\Big(\frac{k_B T_e}{m_e}\Big)^{1/2} h^{(1)}_r\Big] \\ & - \frac{2\sqrt{2}}{3}T_e\, h^{e(2)}_{rs}\nabla_r u_s + \frac{2\sqrt{2}}{3}T_e\, h^{e(2)}_{rs}\nabla_r\Big[\Big(\frac{k_B T_e}{m_e}\Big)^{1/2} h^{(1)}_s\Big] \\ & - \frac{\sqrt{10}}{3}\frac{1}{\rho}\nabla_r\Big[\rho T_e\Big(\frac{k_B T_e}{m_e}\Big)^{1/2} h^{e(3)}_r\Big] - \frac{1}{Zk_B}Q^{(2)} - \frac{2}{3}T_e\, h^{(1)}_n\, Q^{(1)}_n \,,\end{aligned} \tag{5.209}$$

$$\begin{aligned}\partial_t T_i =& - \mathbf{u}\cdot\nabla T_i - \frac{2}{3}T_i\nabla\cdot\mathbf{u} - \frac{2\sqrt{2}}{3}T_i\, h^{i(2)}_{rs}\nabla_r u_s \\ & - \frac{\sqrt{10}}{3}\frac{1}{\rho}\nabla_r\Big[\rho\, T_i\Big(\frac{k_B T_i}{m_i}\Big)^{1/2} h^{i(3)}_r\Big] + \frac{1}{k_B}Q^{(2)} \,.\end{aligned} \tag{5.210}$$

Here,

$$Q^{(2)} = \frac{2}{3}\frac{m_i}{\rho}Q^{ie} \,, \quad Q^{(1)}_r = -\frac{m_i}{Z\rho m_e}\Big(\frac{m_e}{k_B T_e}\Big)^{1/2} R^{ei}_r \,. \tag{5.211}$$

In addition, we have the dynamic equations for the coefficients, namely,

$$\begin{aligned}\partial_t h^{(1)}_r =& \Big(\frac{m_e}{k_B T_e}\Big)^{1/2}\Big[\frac{e}{m_e}E_r + \frac{1}{m_e n_e}\nabla_r(n_e k_B T_e) + \frac{e}{m_e}\varepsilon_{rmn}\, u_m\, B_n\Big] \\ & - \frac{e}{m_e}\varepsilon_{rmn}\, h^{(1)}_m\, B_n + Q^{(1)}_r + L^{(1)}_r + C^{(1)}_r + N^{(1)}_r,\end{aligned} \tag{5.212}$$

where

$$L^{(1)}_r = \sqrt{2}\Big(\frac{k_B T_e}{m_e}\Big)^{1/2}\frac{1}{\rho k_B T_e}\nabla_s\Big[\rho\Big(k_B T_e\, h^{e(2)}_{rs} - Z\mu\, k_B T_i\, h^{i(2)}_{rs}\Big)\Big], \tag{5.213}$$

$$\begin{aligned}C^{(1)}_r =& - \mathbf{u}\cdot\nabla h^{(1)}_r - h^{(1)}_m\nabla_m u_r + \frac{1}{3}h^{(1)}_r\nabla\cdot\mathbf{u} \\ & + \frac{\sqrt{2}}{3}h^{(1)}_r\, h^{e(2)}_{mn}\nabla_m\Big[u_n - \Big(\frac{k_B T_e}{m_e}\Big)^{1/2} h^{(1)}_n\Big],\end{aligned} \tag{5.214}$$

$$N_r^{(1)} = \left(\frac{k_B T_e}{m_e}\right)^{1/2} \left[\frac{\sqrt{10}}{6} h_r^{(1)} \frac{1}{\rho\,(k_B T_e)^{3/2}} \nabla_m \left(\rho\,(k_B T_e)^{3/2}\, h_m^{e(3)} \right) \right.$$
$$\left. + h_m^{(1)} \nabla_m h_r^{(1)} - \frac{1}{3} h_r^{(1)} \frac{1}{(k_B T_e)^{1/2}} \nabla_m \left((k_B T_e)^{1/2} h_m^{(1)} \right) \right] - \frac{1}{3} h_r^{(1)} h_n^{(1)} Q_n^{(1)} \,. \quad (5.215)$$

The remaining non-privileged moment equations are

$$\partial_t h_r^{\alpha(3)} = - \left(\frac{5}{2}\right)^{1/2} \left(\frac{k_B T_\alpha}{m_\alpha}\right)^{1/2} \frac{1}{T_\alpha} \nabla_r T_\alpha$$
$$+ \frac{e_\alpha}{m_\alpha} \varepsilon_{rmn}\, h_m^{\alpha(3)} B_n + Q_r^{\alpha(3)} + L_r^{\alpha(3)} + M_r^{(3)} + C_r^{\alpha(3)} + N_r^{\alpha(3)} \,, \quad (5.216)$$

$$\partial_t h_r^{\alpha(5)} = \frac{e_\alpha}{m_\alpha c} \varepsilon_{rmn}\, h_m^{\alpha(5)} B_n + Q_r^{\alpha(5)} + L_r^{\alpha(5)} + M_r^{(5)} + C_r^{\alpha(5)} + N_r^{\alpha(5)} \,, \quad (5.217)$$

$$\partial_t h_{rs}^{\alpha(2)} = - \frac{1}{\sqrt{2}} \tau_{rs|pq} \nabla_p u_q + \frac{2 e_\alpha}{m_\alpha c} \tau_{rs|pq} \varepsilon_{pmn}\, h_{qm}^{\alpha(2)} B_n$$
$$+ Q_{rs}^{\alpha(2)} + L_{rs}^{\alpha(2)} + M_{rs}^{\alpha(2)} + C_{rs}^{\alpha(2)} + N_{rs}^{\alpha(2)} \,, \quad (5.218)$$

$$\partial_t h_{rs}^{\alpha(4)} = \frac{2 e_\alpha}{m_\alpha c} \tau_{rs|pq} \varepsilon_{pmn}\, h_{qm}^{\alpha(4)} B_n + Q_{rs}^{\alpha(4)} + L_{rs}^{\alpha(4)} + M_{rs}^{\alpha(4)} + C_{rs}^{\alpha(4)} + N_{rs}^{\alpha(4)} \,. \quad (5.219)$$

The operator $\tau_{rs|pq}$ is a symmetrization operator,

$$\tau_{rs|pq} = \frac{1}{2} \left(\delta_{rp}\delta_{sq} + \delta_{rq}\delta_{sp} - \frac{1}{3}\delta_{rs}\delta_{pq} \right) ; \quad (5.220)$$

the collision contributions are contained in

$$Q_{r_1 r_2 \ldots}^{\alpha(m)} = \frac{1}{n_\alpha} \int d^3 v\, H_{r_1 r_2 \ldots}^{(m)} \left[\left(\frac{m_\alpha}{k_B T_\alpha}\right)^{1/2} (\mathbf{v} - \mathbf{u}^\alpha) \right] K^\alpha \,. \quad (5.221)$$

Furthermore, we see

$$Q_r^{(1)}, Q^{(2)} \sim \mathcal{O}\left(\frac{m_e}{m_i}\right) , \quad (5.222)$$

and we will neglect these contributions to lowest order. Moreover, since we are developing a linear transport theory, nonlinear terms such as $-\frac{2}{3} T_e\, h_n^{(1)}\, Q_n^{(1)}$ or $N_r^{(1)}$ will initially be ignored. This implies that the dynamical equations for ρ, $\mathbf{u}$, T_e and T_i do not explicitly contain collisional terms (ρ, $\mathbf{u}$, T_e and T_i are quasi-conserved hydrodynamic moments).

The dynamic equations for ρ, $\mathbf{u}$, T_e and T_i form a *first group*—the so-called *hydrodynamic* plasma equations. In addition to ρ, $\mathbf{u}$, T_e and T_i, these equations contain a *second group* of variables, namely $h_r^{(1)}$, $h_{rs}^{\alpha(2)}$, $h_r^{\alpha(3)}$, which are referred to as *privileged non-hydrodynamic* moments.
The dynamic equations for $j_r \sim h_r^{(1)}$ (generalized Ohm's law) and $q_r^\alpha \sim h_r^{\alpha(3)}$ (heat conduction equation) as well as for the dissipative part of the pressure tensor $\pi_{rs}^\alpha \sim h_{rs}^{\alpha(2)}$ form the *second group* of equations. Compared to the rest, the second group is privileged because it contains source terms. The latter are gradients of ρ, $\mathbf{u}$, T_e, T_i as well as the electrodynamic fields. In more general terminology, they are referred to as thermodynamic forces. As we shall see, the source terms in the equations for $h^{(1)}$, $h_{rs}^{\alpha(2)}$ and $h_r^{\alpha(3)}$ are the reason for the functional dependence of the privileged non-hydrodynamic variables on the hydrodynamic variables, which suggests a meaningful truncation of the hierarchy.

Before we go into further detail, we want to qualify the terms Q, L, M, C and N. We begin with the collision contributions $Q^{(2)}$, $Q_r^{(1)}$, ... , $Q_{r_1 r_2}^{\alpha(m)}$ .... They are clearly defined in terms of the distribution functions f^α, which appear in the collision term K^α of the kinetic equation. If we substitute the expansion of the distribution function in terms of Hermite polynomials there, we are faced with a straightforward but algebraically overloaded problem, which we will not present here in full detail. The final results [44] begin with

$$Q^{(2)} = -\frac{2Z}{\tau_e}\frac{m_e}{m_i} k_B (T_i - T_e) \,, \tag{5.223}$$

which is of order m_e/m_i because

$$\tau_e = \frac{3(4\pi\varepsilon_0)^2}{4\sqrt{2\pi}} \frac{m_e^{1/2} T_e^{3/2}}{n_i Z^2 e^4 \ln \Lambda_B} \,, \tag{5.224}$$

at

$$\ln \Lambda_B = 4\pi\varepsilon_0 \frac{3(T_e + T_i)\lambda_D}{2Ze^2} \,, \quad \lambda_D = \left[\frac{Ze^2(n_e T_e + n_i T_i)}{\varepsilon_0 T_e T_i (1+Z)}\right]^{-12} . \tag{5.225}$$

The other terms are written in standard notation [44] by introducing $Q_r^{e(1)} = Q_r^{(1)}$ and $h_r^{e(1)} = h_r^{(1)}$. Within a linear approximation, it follows that

$$Q_r^{e(2n+1)} = \sum_{m=0}^{3} C_{2n+1,2m+1}\, h_r^{e(2m+1)}\,, \quad Q_r^{i(2n+1)} = \sum_{m=1}^{3} \overline{C}_{2n+1,2m+1} h_r^{i(2m+1)}\,, \tag{5.226}$$

$$Q_{rs}^{e(2n)} = \sum_{m=1}^{3} C_{2n,2m} h_{rs}^{e(2m)}\,, \quad Q_{rs}^{i(2n)} = \sum_{m=1}^{3} \overline{C}_{2n,2m} h_{rs}^{i(2m)}\,, \tag{5.227}$$

for $n = 0, 1, 2, 3$. Here, $C_{\nu,\mu}$ and $\overline{C}_{\nu,\mu}$ [44] are given as products of numerical constants with τ_e^{-1} (for electrons) or τ_i^{-1} (for ions). Note that

$$\tau_i = \left(\frac{m_i}{m_e}\right)^{1/2} Z^{-2} \left(\frac{T_i}{T_e}\right)^{3/2} \tau_e\,. \tag{5.228}$$

The *hierarchical* terms L contain higher moments. Later, we will examine the terms

$$L_{rs}^{\alpha(2)} = -\frac{2}{\sqrt{5}} \tau_{rs|pq} \left(\frac{k_B T_\alpha}{m_\alpha}\right)^{1/2} \frac{1}{n_\alpha T_\alpha^{3/2}} \nabla_p \left(n_\alpha\, T_\alpha^{3/2}\, h_q^{\alpha(3)}\right), \tag{5.229}$$

$$L_r^{\alpha(3)} = -\left(\frac{14}{5}\frac{k_B T_\alpha}{m_\alpha}\right)^{1/2} \frac{1}{n_\alpha T_\alpha^2} \nabla_m \left(n_\alpha\, T_\alpha^2\, h_{rm}^{\alpha(4)}\right) \tag{5.230}$$

in more detail. The terms M contain derivatives of lower moments. In particular, we will later examine the terms

$$M_{rs}^{\alpha(2)} = \begin{cases} \sqrt{2}\tau_{rs|pq}\left(\frac{k_B T_e}{m_e}\right)^{1/2} \nabla_p h_q^{(1)} & \text{for } \alpha = e, \\ 0 & \text{for } \alpha = i, \end{cases} \tag{5.231}$$

and

$$M_r^{\alpha(3)} = -\frac{2}{\sqrt{5}} \left(\frac{k_B T_\alpha}{m_\alpha}\right)^{1/2} T_\alpha^{-7/2} \nabla_m \left(T_\alpha^{7/2}\, h_{rm}^{\alpha(2)}\right) \tag{5.232}$$

in more detail. The *convective* terms C contain moments of the same or lower order as well as $\mathbf{u}^\alpha$ and its derivatives. In the following discussion, in particular

$$C_{rs}^{\alpha(2)} = -\mathbf{u}^\alpha \cdot \nabla\, h_{rs}^{\alpha(2)} + \frac{2}{3} h_{rs}^{\alpha(2)} \nabla \cdot \mathbf{u}^\alpha - 2\tau_{rs|pq}\, h_{pm}^{\alpha(2)} \nabla_m u_q^\alpha + \frac{2\sqrt{2}}{3} h_{rs}^{\alpha(2)} h_{mn}^{\alpha(2)} \nabla_n\, u_m^\alpha \tag{5.233}$$

and

$$C_r^{\alpha(3)} = -\mathbf{u}^\alpha \cdot \nabla\, h_r^{\alpha(3)} - \frac{7}{5} h_m^{\alpha(3)} \nabla_m\, u_r^\alpha - \frac{2}{5} h_m^{\alpha(3)} \nabla_r\, u_m^\alpha + \frac{3}{5} h_r^{\alpha(3)} \nabla \cdot \mathbf{u}^\alpha + \sqrt{2} h_r^{\alpha(3)} h_{ns}^{\alpha(2)} \nabla_n\, u_s^\alpha \tag{5.234}$$

are relevant. As previously noted, the nonlinear terms N are ignored.

The Hydrodynamic Regime in (Linear) Transport Theory

We now begin with a reduction of the fundamental equations. We note that the hydrodynamic equations for ρ, $\mathbf{u}$, T_e and T_i contain no collision terms, provided that $Q^{(2)}$ for $T_i \approx T_e$ and $m_e \ll m_i$ is neglected. The remainder of the equations contains two characteristic time scales: the collisional ones, $\tau_{coll} = \max(\tau_e, \tau_i)$, where usually $\tau_{coll} \approx \tau_i$ holds, and the hydrodynamic time $\tau_H \equiv T_H$. The latter enters the equations for the privileged non-hydrodynamic moments via the thermodynamic forces (e.g., gradients of the hydrodynamic variables). We define

$$\frac{1}{\tau_{T_\alpha}} = v_{th\,\alpha}|\nabla \ln T_\alpha| \, , \quad \frac{1}{\tau_\rho} = v_{th\,e}|\nabla \ln \rho| \, , \quad \frac{1}{\tau_u} = |\nabla u| \, , \quad \frac{1}{\tau_E} = \frac{e}{m_e v_{th\,e}} E \, . \tag{5.235}$$

The hydrodynamic time τ_H is then of the order of magnitude

$$\boxed{\tau_H \sim \min(\tau_\rho, \tau_{T_\alpha}, \tau_u, \tau_E)} \, . \tag{5.236}$$

The ratio of τ_{coll} and τ_H forms the smallness parameter

$$\boxed{\lambda_H = \tau_{coll}/\tau_H \ll 1} \, , \tag{5.237}$$

which is used in linear transport theory.

To analyze how collisional effects contribute to the evolution of the hydrodynamic variables, we formally write, e.g.,

$$\partial_t h^{(1)} = \frac{\tilde{H}}{\tau_H} - \frac{\tilde{B} h^{(1)}}{\tau_B} + \frac{\tilde{Q}^{(1)}}{\tau_{coll}} + L^{(1)} + C^{(1)} + N^{(1)} \, . \tag{5.238}$$

The term $\tilde{H}$ stands for the source term of the hydrodynamic variables, $\tilde{B}$ is the contribution from the magnetic fields, $\tilde{Q}^{(1)}$ is the collisional contribution, $L^{(1)}$ the hierarchical, $C^{(1)}$ the convective and $N^{(1)}$ the nonlinear contribution (if considered at all). A new time, τ_B, which captures the gyration in the external magnetic field, appears. It can be estimated by $\tau_B \approx |\Omega_e^{-1}|$ or $\tau_B \approx \Omega_i^{-1}$, depending on the species considered. In principle, we could classify τ_B in relation to τ_{coll}. We refrain from doing so and leave the term $\tilde{B}$ without any additional ordering (let us say, of order unity). The balance of the $\tilde{Q}^{(1)}$ term with the $\tilde{H}$ term suggests the scaling

$$h^{(1)} \sim \mathcal{O}(\lambda_H) \, , \tag{5.239}$$

since $\tilde{H}$ in contrast to $\tilde{Q}$ contains no contribution from $h^{(1)}$. Up to first order in λ_H we obtain

$$\partial_t h^{(1)} + \frac{\tilde{B}h^{(1)}}{\tau_B} \approx \frac{\tilde{H}}{\tau_H} + \frac{\tilde{Q}^{(1)}}{\tau_{coll}} . \tag{5.240}$$

The terms $L^{(1)}$, $C^{(1)}$ and $N^{(1)}$ are of higher order (as is evident from (5.238) after multiplication by $\tau_{coll} : h^{(1)} \sim \lambda_H$, $N^{(1)}\tau_{coll} \sim \lambda_H^2$, etc.).

The equations for the non-hydrodynamic moments in the 21-moment approximation are as follows:

$$\begin{aligned}\partial_t h_r^{(1)} - \Omega_e\, \varepsilon_{rmn} h_m^{(1)} b_n = &-\frac{1}{\tau_e}\Big[c_{11} h_r^{(1)} + c_{13} h_r^{e(3)} + c_{15} h_r^{e(5)}\Big] \\ &+ \left(\frac{m_e}{k_B T_e}\right)^{1/2}\left[\frac{e}{m_e} E_r + \frac{1}{m_e \rho}\nabla_r(\rho k_B T_e) + \frac{e}{m_e c}\, \varepsilon_{rmn} u_m\, B_n\right] ,\end{aligned} \tag{5.241}$$

$$\begin{aligned}\partial_t h_r^{e(3)} - \Omega_e\, \varepsilon_{rmn} h_m^{e(3)} b_n = &-\frac{1}{\tau_e}\Big[c_{31} h_r^{(1)} + c_{33} h_r^{e(3)} + c_{35} h_r^{e(5)}\Big] \\ &- \left(\frac{5}{2}\right)^{1/2}\left(\frac{k_B T_e}{m_e}\right)^{1/2}\frac{1}{T_e}\nabla_r T_e ,\end{aligned} \tag{5.242}$$

$$\partial_t h_r^{e(5)} - \Omega_e\, \varepsilon_{rmn} h_m^{e(5)} b_n = -\frac{1}{\tau_e}\Big[c_{51} h_r^{(1)} + c_{53} h_r^{e(3)} + c_{55} h_r^{e(5)}\Big], \tag{5.243}$$

$$\begin{aligned}&\partial_t h_{rs}^{e(2)} - \Omega_e\Big(\varepsilon_{rmn} h_{sm}^{e(2)} + \varepsilon_{smn}\, h_{rm}^{e(2)}\Big) b_n \\ &\quad = -\frac{1}{\tau_e}\Big[c_{22} h_{rs}^{e(2)} + c_{24} h_{rs}^{e(4)}\Big] - \sqrt{2}\tau_{rs|pq}\nabla_p u_q ,\end{aligned} \tag{5.244}$$

$$\partial_t h_{rs}^{e(4)} - \Omega_e\Big(\varepsilon_{rmn} h_{sm}^{e(4)} + \varepsilon_{smn} h_{rm}^{e(4)}\Big) b_n = -\frac{1}{\tau_e}\Big[c_{42} h_{rs}^{e(2)} + c_{44} h_{rs}^{e(4)}\Big], \tag{5.245}$$

$$\partial_t h_r^{i(3)} - \Omega_i\, \varepsilon_{rmn} h_m^{i(3)} b_n = -\frac{1}{\tau_i}\Big[\overline{c}_{33} h_r^{i(3)} + \overline{c}_{35} h_r^{i(5)}\Big] - \left(\frac{5}{2}\right)^{1/2}\left(\frac{k_B T_i}{m_i}\right)^{1/2}\frac{1}{T_i}\nabla_r T_i , \tag{5.246}$$

$$\partial_t h_r^{i(5)} - \Omega_i\, \varepsilon_{rmn} h_m^{i(5)} b_n = -\frac{1}{\tau_i}\Big[\overline{c}_{53} h_r^{i(3)} + \overline{c}_{55} h_r^{i(5)}\Big], \tag{5.247}$$

$$\partial_t h_{rs}^{i(2)} - \Omega_i\Big(\varepsilon_{rmn} h_{sm}^{i(2)} + \varepsilon_{smn} h_{rm}^{i(2)}\Big) b_n = -\frac{1}{\tau_i}\Big[\overline{c}_{22} h_{rs}^{i(2)} + \overline{c}_{24} h_{rs}^{i(4)}\Big] - \sqrt{2}\tau_{rs|pq}\nabla_p u_q , \tag{5.248}$$

$$\partial_t h_{rs}^{i(4)} - \Omega_i\Big(\varepsilon_{rmn} h_{sm}^{i(4)} + \varepsilon_{smn} h_{rm}^{i(4)}\Big) b_n = -\frac{1}{\tau_i}\Big[\overline{c}_{42} h_{rs}^{i(2)} + \overline{c}_{44} h_{rs}^{i(4)}\Big] . \tag{5.249}$$

We have abbreviated

$$b_n = \frac{B_n}{B} , \quad C_{\ldots} = -\frac{1}{\tau_e} c_{\ldots} , \quad \overline{C}_{\ldots} = -\frac{1}{\tau_i}\overline{c}_{\ldots} . \tag{5.250}$$

The gyrofrequencies include a sign, i.e., $\Omega_e = -(eB/m_e)$ and $\Omega_i = (ZeB/m_i)$. We have not listed all known constants, for example $c_{11} = 1$, $c_{13} = 3/\sqrt{10}$, $\bar{c}_{33} = 2\sqrt{2}/5$, etc.

The vectorial moment equations decouple from the tensorial ones. Next, we will simplify by using the previously mentioned ordering $\mu = m_e/m_i \ll 1$, $\sigma \approx 0$, $\tau_{coll}/\tau_H \ll 1$, $h^{\alpha}_{\ldots} \sim \lambda_H$, etc.

It would be practically impossible and, fortunately, physically meaningless if transport theory required the initial value problem of the aforementioned differential equations to be solved exactly.

The physical reason for this is quite simple. If the non-hydrodynamic variables depend on time only functionally through the hydrodynamic variables, the structure of the equations can be reduced as follows:

$$\tau_{coll}\,\partial_t h^{\alpha(m)}_{\ldots} \sim \frac{\tau_{coll}}{\tau_H}\lambda_H = \lambda_H^2 \ll \lambda_H \;, \tag{5.251}$$

which means that we can effectively solve algebraic equations after neglecting the left-hand sides:

$$\partial_t h^{\alpha(m)}_{\ldots} \approx 0 \;. \tag{5.252}$$

Example 5.7 (Initial values in the solution of the moment equations)
The mathematical argument is as follows. We summarize the previously mentioned structure of the equations (now with a single equation and not for a system of equations; the latter can be discussed with similar arguments and conclusions) in the form

$$\partial_t y = -\frac{|c|}{\tau_{coll}}\, y + q \;, \tag{5.253}$$

where q is the source term containing the thermodynamic forces. The solution of the homogeneous part of the equation is

$$y = \alpha \exp\left(-\frac{|c|t}{\tau_{coll}}\right) . \tag{5.254}$$

Variation of constants leads to

$$y(t) = y(0)\ \exp\left(-\frac{|c|t}{\tau_{coll}}\right) + \int_0^t d\tau\, q(\tau) e^{-|c|(t-\tau)/\tau_{coll}} \;. \tag{5.255}$$

Since $|c| > 0$ is of order one, the influence of the initial values disappears after $t \gg \tau_{coll}$. The temporal variation of the source term $q(t)$ occurs on the scale $\tau_H \gg \tau_{coll}$. Therefore, we can approximate

$$\int_0^t d\tau\, q(\tau)e^{-|c|(t-\tau)/\tau_{coll}} = \int_0^t d\sigma\, q(t-\sigma)e^{-|c|\sigma/\tau_{coll}} \approx \int_0^\infty d\sigma\, q(t-\sigma)e^{-|c|\sigma/\tau_{coll}}\,, \tag{5.256}$$

for $t \gg \tau_{coll}$. We expand q for $t \gg \tau_{coll}$,

$$q(t-\sigma) \approx q(t) + \mathcal{O}\left(\frac{\tau_{coll}}{\tau_H}\right), \tag{5.257}$$

and obtain the result

$$y(t) \approx \frac{\tau_{coll}}{|c|} q(t) \qquad \text{for } t \gg \tau_{coll}\,. \tag{5.258}$$

This approximate solution follows for (5.253), if we approximate (5.253) by an algebraic equation, which is formally obtained by setting $\partial_t y \approx 0$. ■

In summary, in the so-called hydrodynamic regime

$$\tau_{coll} \ll t \sim \mathcal{O}(\tau_H) \tag{5.259}$$

the equations for the non-hydrodynamic variables become simple, i.e., (linear) algebraic equations that can be solved without difficulty. Substituting the solutions into the equations for the hydrodynamic variables leads to the (linear) transport equations.

Summary of the (linear) Transport Equations

We summarize the results of the transport calculations outlined in the two previous paragraphs. We begin with the equations for the hydrodynamic variables and assume that $\sigma \approx 0$. Then it follows

$$\partial_t \rho + \nabla \cdot (\rho \mathbf{u}) = 0\,, \tag{5.260}$$

$$\begin{aligned}\rho \partial_t u_r = &- \rho u_s \nabla_s u_r - \frac{Z}{m_i} \nabla_r \left[\rho\left(k_B T_e + \frac{1}{Z} k_B T_i\right)\right] \\ &- \sqrt{2}\frac{Z}{m_i} \nabla_s \left[\rho\left(k_B T_e h_{rs}^{e(2)} + \frac{1}{Z} k_B T_i h_{rs}^{i(2)}\right)\right]\end{aligned}$$

$$+ e\,\frac{Z\rho}{m_i}\left(\frac{k_B T_e}{m_e}\right)^{1/2} \varepsilon_{rsm} h_s^{(1)} B_m\,, \tag{5.261}$$

$$\begin{aligned}
\partial_t T_e =& -\mathbf{u}\cdot\nabla T_e - \frac{2}{3}T_e\nabla\cdot\mathbf{u} \\
&+ \left(\frac{k_B T_e}{m_e}\right)^{1/2} h_r^{(1)}\nabla_r T_e + \frac{2}{3}T_e\nabla_r\left[\left(\frac{k_B T_e}{m_e}\right)^{1/2} h_r^{(1)}\right] \\
&- \frac{2\sqrt{2}}{3}T_e h_{rs}^{e(2)}\nabla_r u_s + \frac{2\sqrt{2}}{3}T_e h_{rs}^{e(2)}\nabla_r\left[\left(\frac{k_B T_e}{m_e}\right)^{1/2} h_s^{(1)}\right] \\
&- \frac{\sqrt{10}}{3}\,\frac{1}{\rho}\nabla_r\left[\rho T_e\left(\frac{k_B T_e}{m_e}\right)^{1/2} h_r^{e(3)}\right],
\end{aligned} \tag{5.262}$$

$$\begin{aligned}
\partial_t T_i =& -\mathbf{u}\cdot\nabla T_i - \frac{2}{3}T_i\nabla\cdot\mathbf{u} - \frac{2\sqrt{2}}{3}T_i\, h_{rs}^{i(2)}\nabla_r u_s \\
&- \frac{\sqrt{10}}{3}\,\frac{1}{\rho}\nabla_r\left[\rho T_i\left(\frac{k_B T_i}{m_i}\right)^{1/2} h_r^{i(3)}\right].
\end{aligned} \tag{5.263}$$

They contain the privileged non-hydrodynamic coefficients $h_r^{(1)}$, $h_{rs}^{(2)}$ and $h_r^{\alpha(3)}$, which are related to the electric current density $\mathbf{j}$, the dissipative part of the pressure tensor π_{rs}^{a}, and the heat flux $\mathbf{q}^{\alpha}$ by the following relation:

$$\begin{aligned}
j_r &= en_e\left(\frac{k_B T_e}{m_e}\right)^{1/2} h_r^{(1)}\,, \quad \pi_{rs}^{\alpha} = \sqrt{2}n_\alpha\, k_B T_\alpha h_{rs}^{\alpha(2)}\,, \\
q_r^{\alpha} &= \left(\frac{5}{2}\right)^{1/2} m_\alpha\left(\frac{k_B T_\alpha}{m_\alpha}\right)^{3/2} n_\alpha\, h_r^{\alpha(3)}\,.
\end{aligned} \tag{5.264}$$

The particle densities are approximately $n_e \approx Z\rho/m_i$ and $n_i \approx \rho/m_i$. The quantities $\mathbf{j}$, π_{rs}^{e}, q_r^{e}, π_{rs}^{i} and q_r^{i} can be expressed as functions of the thermodynamic forces. The numerical factors change depending on the degree of approximation. We present the result for the electric current density in the following form:

$$\begin{aligned}
\mathbf{j} =& \mathbf{b}\left\{\sigma_\parallel\hat{\mathbf{E}}\cdot\mathbf{b} + \alpha_\parallel(-\nabla k_B T_e)\cdot\mathbf{b}\right\} + \mathbf{b}\times\left\{\sigma_\wedge\hat{\mathbf{E}} + \alpha_\wedge(-\nabla k_B T_e)\right\} \\
&+ \mathbf{b}\times\left\{\sigma_\perp(\hat{\mathbf{E}}\times\mathbf{b}) + \alpha_\perp(-\nabla k_B T_e)\times\mathbf{b}\right\}.
\end{aligned} \tag{5.265}$$

Here,

$$\hat{\mathbf{E}} = \mathbf{E} + (\mathbf{u}\times\mathbf{B}) + \frac{1}{e\rho}\nabla(\rho k_B T_e)\,. \tag{5.266}$$

The coefficients are

$$\sigma_A = \frac{e^2 Z\rho}{m_e m_i}\tau_e\,\tilde{\sigma}_A\,, \quad \alpha_A = \left(\frac{5}{2}\right)^{1/2}\frac{eZ\rho}{m_e m_i}\tau_e\,\tilde{\alpha}_A\,, \tag{5.267}$$

with the index A, which characterizes the direction with respect to the external magnetic field: $A = \|$, i.e., parallel to $\mathbf{b}$; $A = \wedge$, i.e., along $\mathbf{b}\times\nabla\cdots$; $A = \perp$, i.e., along $\mathbf{b}\times[(\nabla\cdots)\times\mathbf{b}]$. Numerical values of $\tilde{\sigma}_A, \tilde{\alpha}_A, \tilde{\kappa}_A^e$ can be calculated (see below).

The heat fluxes are

$$\begin{aligned}\mathbf{q}^{e} &= \mathbf{b}\Big\{\alpha_\| k_B T_e(\hat{\mathbf{E}}\cdot\mathbf{b}) + \kappa_\|^e(-\nabla k_B T_e)\cdot\mathbf{b}\Big\} + \mathbf{b}\times\left\{\alpha_\wedge k_B T_e\hat{\mathbf{E}} + \kappa_\wedge^e(-\nabla k_B T_e)\right\} \\ &\quad + \mathbf{b}\times\left\{\alpha_\perp k_B T_e(\hat{\mathbf{E}}\times\mathbf{b}) + \kappa_\perp^e(-\nabla k_B T_e)\times\mathbf{b}\right\}\,,\end{aligned} \tag{5.268}$$

$$\mathbf{q}^{i} = -\kappa_\|^i\mathbf{b}(\mathbf{b}\cdot\nabla k_B T_i) - \kappa_\wedge^i(\mathbf{b}\times\nabla k_B T_i) - \kappa_\perp^i\{\mathbf{b}\times[(\nabla k_B T_i)\times\mathbf{b}]\}\,, \tag{5.269}$$

with the additional coefficients

$$\kappa_A^e = \frac{5}{2}\frac{Z\rho k_B T_e}{m_e m_i}\tau_e\,\tilde{\kappa}_A^e\,, \quad \kappa_A^i = \frac{5}{2}\frac{\rho k_B T_i}{m_i^2}\tau_i\,\tilde{\kappa}_A^i\,. \tag{5.270}$$

With

$$\nu_{rs} = 2\tau_{rs|pq}\,\nabla_p u_q\,, \tag{5.271}$$

one can write the components of the dissipative part of the pressure tensor as follows:

$$\pi_{zz}^\alpha = -\eta_\|^\alpha\,\nu_{zz}\,, \tag{5.272}$$

$$\pi_{xz}^\alpha = -\eta_2^\alpha\,\nu_{xz} + \eta_1^\alpha\,\nu_{yz}\,, \tag{5.273}$$

$$\pi_{yz}^\alpha = -\eta_1^\alpha\,\nu_{xz} - \eta_2^\alpha\,\nu_{yz}\,, \tag{5.274}$$

$$\pi_{xx}^\alpha = -\frac{1}{2}\left(\eta_\|^\alpha + \eta_4^\alpha\right)\nu_{xx} - \frac{1}{2}\left(\eta_\|^\alpha - \eta_4^\alpha\right)\nu_{yy} + \eta_3^\alpha\nu_{xy}\,, \tag{5.275}$$

$$\pi_{yy}^\alpha = -\frac{1}{2}\left(\eta_\|^\alpha - \eta_4^\alpha\right)\nu_{xx} - \frac{1}{2}\left(\eta_\|^\alpha + \eta_4^\alpha\right)\nu_{yy} - \eta_3^\alpha\,\nu_{xy}\,, \tag{5.276}$$

$$\pi_{xy}^\alpha = -\frac{1}{2}\eta_3^\alpha\,\nu_{xx} + \frac{1}{2}\eta_3^\alpha\,\nu_{yy} - \eta_4^\alpha\,\nu_{xy}\,, \tag{5.277}$$

where

$$\eta_A^e = \frac{Z\rho}{m_i}k_B T_e\tau_e\,\tilde{\eta}_A^e\,, \quad \eta_A^i = \frac{\rho}{m_i}k_B T_i\tau_i\,\tilde{\eta}_A^i\,, \quad A = \|, 1, 2, 3, 4\,. \tag{5.278}$$

Traditionally, the following quantities are referred to as transport coefficients for the components $\alpha = e, i$: three coefficients for the electrical conductivity σ (or resistance η), three coefficients for the thermal conductivity κ^α, three thermoelectric coefficients α, and five viscosity coefficients η^α. The dimensionless quantities $\tilde{\sigma}$, $\tilde{\alpha}$, $\tilde{\kappa}^e$, $\tilde{\kappa}^i$, $\tilde{\eta}^e$, $\tilde{\eta}^i$ depend on Z and $\Omega_\alpha \tau_\alpha$ (except for $\|$).

In the following, we summarize the characteristic values for $Z = 1$. We begin with $A = \|$. The following values have been obtained (up to the 29-moment expansion):

$$\tilde{\sigma}_\| \approx 1{,}9\,, \quad \tilde{\alpha}_\| \approx -0{,}8\,, \quad \tilde{\kappa}^e_\| \approx 1{,}6\,,$$
$$\tilde{\kappa}^i_\| \approx 2{,}2\,, \quad \tilde{\eta}^e_\| \approx 0{,}7\,, \quad \tilde{\eta}^i_\| \approx 1{,}3\,. \tag{5.279}$$

For the perpendicular coefficients ($\wedge, \perp$), a nontrivial behavior is observed with variations in $\Omega_\alpha \tau_\alpha$. $\tilde{\sigma}_\perp$ decreases monotonically and has the characteristic value 0.4 at $|\Omega_e|\tau_e = 1$. $\tilde{\sigma}_\wedge$ exhibits a maximum at $|\Omega_e|\tau_e \approx 0{,}3$ and a value of 0.7 at $|\Omega_e|\tau_e = 1.\tilde{\alpha}_\perp$ increases monotonically and reaches the value 0.02 at $|\Omega_e|\tau_e = 1$. $\tilde{\alpha}_\wedge$ has a minimum at $|\Omega_e|\tau_e \approx 0{,}3$ and the value -0.3 at $|\Omega_e|\tau_e = 1$. $\tilde{\kappa}^e_\perp$ decreases monotonically and has the value 0.3 at $|\Omega_e|\tau_e = 1$. $\tilde{\kappa}^e_\wedge$ shows a maximum at $|\Omega_e|\tau_e \approx 0{,}3$ and the value 0.5 at $|\Omega_e|\tau_e = 1$. $\tilde{\eta}^e_2$ decreases monotonically and has the value 0.2 at $|\Omega_e|\tau_e = 1$. A similar behavior is observed for $\tilde{\eta}^e_4$ with a value of 0.4 at $|\Omega_e|\tau_e = 1$. $\tilde{\eta}^e_1$ shows a maximum at $|\Omega_e|\tau_e \approx 0{,}3$ and has the value 0.3 at $|\Omega_e|\tau_e = 1$. The maximum of $\tilde{\eta}^e_3$ is less pronounced; a characteristic value is 0.3 at $|\Omega_e|\tau_e = 1$.

The behavior of $\tilde{\kappa}^i$ and $\tilde{\eta}^i$ for ions is similar to that for electrons; characteristic values at $\Omega_i \tau_i = 1$ are $\tilde{\kappa}^i_\perp \approx 0{,}4$, $\tilde{\kappa}^i_\wedge \approx -0{,}7$, $\tilde{\eta}^i_2 \approx 0{,}2$, $\tilde{\eta}^i_4 \approx 0{,}4$, $\tilde{\eta}^i_1 \approx -0{,}4$ and $\tilde{\eta}^i_3 \approx -0{,}6$.

For strong magnetic fields ($|\Omega_\alpha|\tau_\alpha \gg 1$) the following formulas are asymptotically correct:

$$\tilde{\sigma}_\perp \approx \frac{1}{(\Omega_e \tau_e)^2}\,, \qquad \tilde{\sigma}_\wedge \approx -\frac{1}{\Omega_e \tau_e}\,, \tag{5.280}$$

$$\tilde{\alpha}_\perp \approx \frac{3/\sqrt{10}}{(\Omega_e \tau_e)^2}\,, \qquad \tilde{\alpha}_\wedge \approx 0, \tag{5.281}$$

$$\tilde{\kappa}^\alpha_\perp \approx \frac{c^\alpha_{33}}{(\Omega_\alpha \tau_\alpha)^2}\,, \qquad \tilde{\kappa}^\alpha_\wedge \approx -\frac{1}{\Omega_\alpha \tau_\alpha}, \tag{5.282}$$

$$\tilde{\eta}^\alpha_2 \approx \frac{c^\alpha_{22}}{4(\Omega_\alpha \tau_\alpha)^2}\,, \qquad \tilde{\eta}^\alpha_1 \approx -\frac{1}{\Omega_\alpha \tau_\alpha}\,, \tag{5.283}$$

$$\tilde{\eta}^\alpha_4 \approx \frac{c^\alpha_{22}}{(\Omega_\alpha \tau_\alpha)^2}\,, \qquad \tilde{\eta}^\alpha_3 \approx -\frac{1}{2\Omega_\alpha \tau_\alpha}\,. \tag{5.284}$$

Corrections are of the order of $(\Omega_\alpha \tau_\alpha)^{-3}$. Also note the following abbreviations

$$c_{22}^\alpha = \begin{cases} \frac{3}{5}\left(2 + \sqrt{2}\frac{1}{Z}\right) & \text{for } \alpha = e \,, \\ \frac{3\sqrt{2}}{5} & \text{for } \alpha = i \,, \end{cases} \tag{5.285}$$

$$c_{33}^\alpha = \begin{cases} \frac{1}{10}\left(13 + 4\sqrt{2}\frac{1}{Z}\right) & \text{for } \alpha = e \,, \\ \frac{2\sqrt{2}}{5} & \text{for } \alpha = i \,. \end{cases} \tag{5.286}$$

For further details, we refer to the more specialized literature, e.g., [33, 44].

Finally, we discuss transport in the absence of an external magnetic field, i.e., $\mathbf{B} = 0$. In the case $\mathbf{B} = 0$ the system is isotropic, and the transport coefficients are scalars. We have

$$\mathbf{j} = \sigma_\parallel \hat{\mathbf{E}} + \alpha_\parallel (-\nabla k_B T_e) \,, \tag{5.287}$$

$$\mathbf{q}^{\,e} = \alpha_\parallel k_B T_e \hat{\mathbf{E}} + \kappa_\parallel^e (-\nabla k_B T_e) \,, \tag{5.288}$$

$$\mathbf{q}^{\,i} = \kappa_\parallel^i (-\nabla k_B T_i) \,, \tag{5.289}$$

$$\pi_{rs}^\alpha = -\eta_\parallel^\alpha \, \nu_{rs} \,, \tag{5.290}$$

with the abbreviation

$$\hat{\mathbf{E}} = \mathbf{E} + \frac{1}{e n_e} \nabla p_e \,. \tag{5.291}$$

Formally, these equations agree with the previous ones for $\sigma_\perp = \sigma_\parallel$, $\sigma_\wedge = 0$, $\eta_2^\alpha = \eta_4^\alpha = \eta_\parallel^\alpha$, $\eta_1^\alpha = \eta_3^\alpha = 0$. Obviously, a temperature gradient can cause an electric current (thermoelectric current), and an electric field will induce a heat flow. Heat conduction is mainly carried by electrons,

$$\boxed{\kappa_\parallel^i \ll \kappa_\parallel^e} \,, \tag{5.292}$$

when $T_e \geq T_i$. On the other hand, momentum transport, and thus viscosity, is dominated by the ions, as reflected in

$$\boxed{\eta_\parallel^i \gg \eta_\parallel^e} \,. \tag{5.293}$$

We conclude this section by mentioning another transport coefficient, namely diffusion.

Example 5.8 (Diffusion Coefficient)
The simplest estimate of a diffusion coefficient D in a stationary, non-magnetized plasma can be made using the random-walk formula:

$$\boxed{D \equiv \chi_{\parallel} = \frac{(\Delta x)^2}{2\Delta t} \approx \nu \lambda_{mfp}^2 \approx \frac{v_{th}^2}{\nu}}, \tag{5.294}$$

where λ_{mfp} is the mean free path, v_{th} the thermal velocity, and ν the collision frequency. Here, we do not specify the particle species explicitly, which could easily be done by introducing appropriate indices (e.g., e for electrons and i for ions). It is obvious how this approach can be generalized to transport perpendicular to an external magnetic field. The characteristic step size is the (thermal) Larmor radius ρ_L, which leads to the following:

$$D \equiv \chi_{\perp} = \frac{(\Delta x)^2}{2\Delta t} \approx \nu \rho_L^2 \approx v_{th}^2 \frac{\nu}{\Omega_L^2}, \tag{5.295}$$

where $\Omega_L = \frac{eB}{m}$ is the cyclotron frequency. Clearly, from this one can predict (where the temperature T is measured in eV):

$$\boxed{\chi_{\perp} = \frac{\nu m T}{e^2 B^2} \sim \frac{T}{B^2}}, \tag{5.296}$$

as has also been done for other transport coefficients in the limit of strong magnetic fields. However, in most fusion devices, the magnitude and scaling of transport do not agree with the B^{-2} prediction. The ambipolar diffusion coefficient can exhibit an anomalous factor of 10^2. In addition, the electronic thermal conductivity often deviates by a factor of 10 to 1000 from theoretical predictions, and the ionic thermal conductivity can differ by a factor of 10. ■

5.5 Simple Diffusion Models for Transport

After the "intense," mathematically oriented elaborations of linear transport theory, we now "recover" with simple, physically motivated models. Despite their simplicity, they have considerable potential for application.

Collision-dominated models

Example 5.9 (Photon Transport in the Sun)
We now examine how quickly photons travel from the interior of the Sun to its surface. Instead of a more precise calculation, we will be content here with a rough estimate. Using the Thomson cross section for the scattering of photons by electrons, we estimate the mean free path,

$$\lambda = \frac{1}{n\sigma_T} \approx 1{,}5 \times 10^{24}\ \text{cm}^{-2}\ \frac{1}{n}\ . \tag{5.297}$$

The particle density n in the Sun is, to order of magnitude,

$$n \leq \frac{\rho_c}{\bar{m}} \approx \frac{1{,}5 \times 10^2\ \text{g cm}^{-3}}{m_{\text{Proton}}} \approx 10^{26}\ \text{cm}^{-3}\ . \tag{5.298}$$

From this it follows directly

$$\lambda \geq 1{,}5 \times 10^{-2}\ \text{cm} \tag{5.299}$$

or

$$\lambda \sim O(\text{ mm})\ . \tag{5.300}$$

How long does it take a photon to travel from the center to the periphery of the Sun? We calculate the corresponding time using a diffusion model (see Fig. 5.1).

If a photon diffuses (due to scattering) in a medium, it covers the distance ν between the $(\nu + 1)$ -th and the $\mathbf{l}_\nu$ -th scattering process. The total ("zigzag") path is

$$\mathbf{L} = \sum_{\nu=1}^{N} \mathbf{l}_\nu\ . \tag{5.301}$$

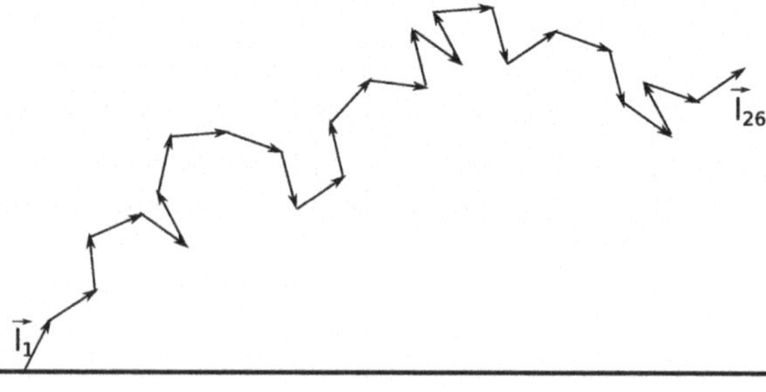

Fig. 5.1 Sketch of a "random walk" for calculating photon diffusion. The individual path steps are labeled with $\mathbf{l}_\nu$

The square $(\mathbf{L})^2$ is averaged, and we assume that the changes in direction occur randomly:

$$\boxed{\langle(\mathbf{L})^2\rangle \equiv D^2 = \left\langle \sum_{\nu,\mu} \mathbf{l}_\nu \cdot \mathbf{l}_\mu \right\rangle \approx \sum_{\nu=1}^{N} l_\nu^2 \approx N\lambda^2} \,. \tag{5.302}$$

D^2 is the mean square displacement after N scatterings, where for simplicity we have estimated the mean distance between two scatterings by the mean free path λ. If now

$$D \approx R_\odot \tag{5.303}$$

is to be reached, then for

$$N \approx \frac{R_\odot^2}{\lambda^2} \tag{5.304}$$

scatterings, the average flight time

$$t_{\text{diff}} \approx \frac{\lambda}{c} N \approx \frac{R_\odot^2}{\lambda c} \,, \tag{5.305}$$

is required to reach the surface. Compared to the undisturbed flight time

$$t_{\text{direct}} \approx \frac{R_\odot}{c} \approx 2\ \text{s} \tag{5.306}$$

this is much longer, specifically by the factor

$$\boxed{\frac{R_\odot}{\lambda} \approx 10^{12}} \,, \tag{5.307}$$

A photon takes more than 60,000 years to travel from the interior of the Sun to its edge. ■

Example 5.10 (Heat Transport by Electrons and Ions)
We now want to calculate classical heat transport by particles that interact with each other (collide) in a simpler way. For this, we use the following approach: The temporal change of the local internal energy density u is given by a classical heat flux $\mathbf{j}$:

$$\frac{\partial u}{\partial t} + \nabla \cdot \mathbf{j} = 0 \,. \tag{5.308}$$

The heat flux is driven (in the simplest case) by a temperature gradient. In a one-dimensional Cartesian model, we consider a flux through a surface (perpendicular to the x-axis) at x. Particles reach or leave the surface with different energy values from an average distance, which can be estimated by the mean free path l. For simplicity, in the balance we assume that one sixth of the particles make a jump by l in the positive x-direction (see Fig. 5.2) [or

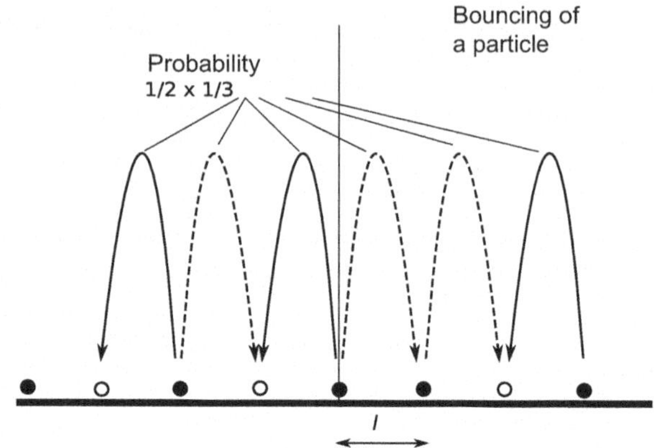

Fig. 5.2 Illustration of a three-dimensional diffusive transport, when the "hopping" is shown only along the x-axis

vice versa] and carry their energy with them. Then it follows

$$\begin{aligned} j_x(x) &\approx \frac{1}{6} \ \text{v} \ u(x-l) - \frac{1}{6} \ \text{v} \ u(x+l) \approx -\frac{1}{3} \ \text{v} \ l \frac{du}{dx} \\ &\approx -\frac{1}{3} \ \text{v} \ l \, C \frac{dT}{dx} \approx -K \frac{dT}{dx} \ , \end{aligned} \tag{5.309}$$

where $C = du/dT$ is the heat capacity and

$$\boxed{K = \frac{1}{3} \ \text{v} \ l \, C} \tag{5.310}$$

is the thermal conductivity. If we insert characteristic values for the electrons,

$$\text{v} \approx \sqrt{\frac{3kT_e}{m_e}} \quad , \quad u_e \approx \frac{3}{2} n_e \, k \, T_e \quad , \quad C_e \approx \frac{3}{2} n_e \, k \ , \tag{5.311}$$

and calculate

$$l \approx \frac{1}{n_i \, \sigma} \tag{5.312}$$

for energetically dominant electron-ion collisions, we are still missing the cross section. We estimate the latter by

$$\sigma \approx \pi r^2 \quad , \quad \frac{Ze^2}{4\pi \varepsilon_0 r} \approx k \, T \tag{5.313}$$

With this, the classical formula for the thermal conductivity by electrons follows

$$\boxed{K_e = \frac{kn_e}{2\pi n_i}\sqrt{\frac{3kT}{m_e}}\left(\frac{4\pi\varepsilon_0 kT}{Ze^2}\right)^2} . \tag{5.314}$$

A comparison shows that the ionic thermal conductivity is much smaller,

$$\boxed{K_i = \frac{1}{Z^2}\sqrt{\frac{m_e}{m_i}}K_e \ll K_e} . \tag{5.315}$$

These are—order of magnitude—the correct values for collision-dominated heat transport by electrons and ions, respectively. ■

Example 5.11 (Heat Transport by Photons)
When electrons and photons are in equilibrium, the temperature can also change through photon transport. We now transfer the results of the simple model for electrons or ions to photons. In doing so, we expect that the equation

$$\mathbf{F} = -\frac{c}{3\kappa\rho}\,\nabla(aT^4) \equiv -\frac{4}{3}\,aT^3\,lc\,\nabla T \tag{5.316}$$

results, and thus the thermal conductivity due to photons is given by

$$K_P = \frac{4}{3}\,a\,l\,c\,T^3 \tag{5.317}$$

In the context of transferring the results of our simple diffusion model to photons (index P), we set

$$\mathrm{v} = c \quad , \quad u_P = a\,T^4 \quad , \quad C_P = 4\,a\,T^3 . \tag{5.318}$$

Thus, the thermal conductivity is then

$$K_P \approx \frac{4}{3}\,c\,l_P\,a\,T^3 \approx \frac{4acT^3}{3\kappa\rho} . \tag{5.319}$$

If we calculate $l_P = 1/\sigma n_e$ with the Thomson cross section

$$\sigma_T = \frac{8\pi}{3}\left[\frac{e^2}{4\pi\varepsilon_0 m_e c^2}\right]^2 \tag{5.320}$$

then it follows that

$$\boxed{K_P = \frac{1}{2\pi}\,\frac{acT^3}{n_e}\left[\frac{4\pi\varepsilon_0 m_e c^2}{e^2}\right]^2} . \tag{5.321}$$

■

Example 5.12 (Heat Transport in the Sun)
Now we can compare the thermal conductivity due to photons with that for electrons. One possibility is to use the pressures

$$P_P = \frac{1}{3} u = \frac{1}{3} a T^4 \tag{5.322}$$

and

$$P_e = n_e k T \tag{5.323}$$

and to find

$$\frac{K_P}{K_e} \approx \sqrt{3} Z^2 \frac{P_P}{P_e} \left[\frac{m_e c^2}{kT} \right]^{5/2} \tag{5.324}$$

Now, if we take typical parameters for the solar interior, $T = 6 \times 10^6$ K, $\rho = 1{,}4 \times 10^3$ kg m^{-3}, it follows that

$$kT = 10^{-3} m_e c^2 , \tag{5.325}$$

$$P_P = 3 \times 10^{11} \text{ Pa} , \tag{5.326}$$

$$P_e = 7 \times 10^{13} \text{ Pa} \tag{5.327}$$

and thus

$$K_P \approx 2 \times 10^5 K_e . \tag{5.328}$$

The thermal conductivity in the solar interior is therefore determined by the photons. ■

The models can be extended with additional processes. However, instead of a quantitative improvement of collision-dominated transport, we now turn to a qualitatively new form of transport.

Convective Transport

In addition to classical (collision-dominated) transport (including photon transport), there is another important transport mechanism: convection.

Convection sets in when "classical" transport can no longer maintain a certain temperature gradient. For example, consider the boiling of water. The task is to determine the critical temperature gradient above which (i.e., for larger temperature differences) convection begins. One can derive a fairly precise instability criterion for the onset of convective cells in the Rayleigh-Bénard instability. In this section, however, we will forgo a detailed calculation of this.

Rayleigh-Bénard instability occurs in a fluid heated from below (see Fig. 5.3). In the following simple (adiabatic) model calculation, we present a physically intuitive derivation of the instability criterion, allowing for variations in the mass density ρ. However, the criterion remains qualitative and is therefore quantitatively incomplete.

The inhomogeneous system is characterized by the boundary conditions for temperature T, pressure P, and mass density ρ at

$$\text{height } z : T, \quad P, \quad \rho \,, \tag{5.329}$$

$$\text{height } z + \Delta z : T + \Delta T, \; P + \Delta P, \; \rho + \Delta \rho \,, \tag{5.330}$$

ΔT, ΔP and $\Delta \rho$ are not independent. From $P = nkT = \rho kT/\bar{m}$ it follows that

$$\Delta P = \frac{P}{\rho}\Delta \rho + \frac{P}{T}\Delta T \,. \tag{5.331}$$

We further assume ΔT, ΔP and $\Delta \rho < 0$. We now perform a virtual displacement of a "bubble," assuming during the displacement by δz that the pressure "instantaneously" adjusts to the surroundings. If we denote the changes during the displacement by $\delta z = \Delta z > 0$ with the symbol δ, then for the pressure the following should hold:

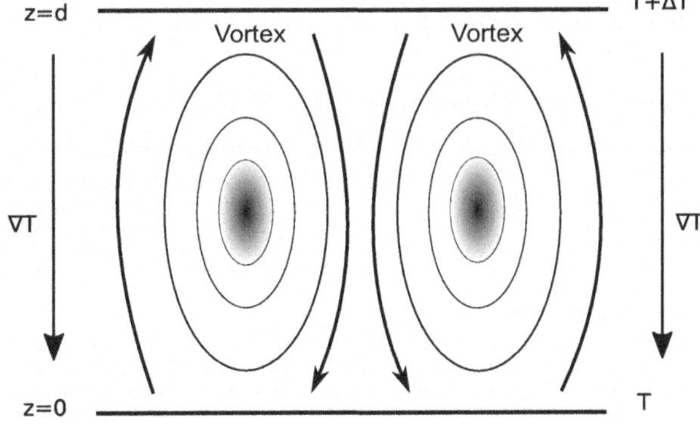

Fig. 5.3 Illustration of the formation of convective cells in the Rayleigh-Bénard experiment. The temperature is higher at the bottom, so $\Delta T < 0$

$$\delta P = \Delta P \, . \tag{5.332}$$

No heat exchange is to be possible (so quickly), so that an adiabatic change occurs. With $P \sim \rho^{\gamma}$ it then follows that

$$\frac{\delta P}{P} = \gamma \frac{\delta \rho}{\rho} \, . \tag{5.333}$$

According to Archimedes' principle, the "bubble" will continue to rise if

$$\delta \rho < \Delta \rho \tag{5.334}$$

is satisfied, i.e.,

$$\delta \rho = \frac{1}{\gamma} \frac{\rho}{P} \Delta P < \Delta \rho = \rho \left(\frac{\Delta P}{P} - \frac{\Delta T}{T} \right) \tag{5.335}$$

or

$$\frac{\Delta T}{T} < \frac{\gamma - 1}{\gamma} \frac{\Delta P}{P} \, . \tag{5.336}$$

Since $\gamma \geq 0$ and $\Delta T, \Delta P < 0$, we can also write the condition in the form

$$\left| \frac{dT}{dz} \right| > \frac{\gamma - 1}{\gamma} \frac{T}{P} \left| \frac{dP}{dz} \right| \tag{5.337}$$

Thus, if we know the pressure gradient and have a temperature gradient that exceeds the above critical value (right-hand side), convection will occur. We can also rewrite the above condition in the form

$$\boxed{\frac{d \ln P}{d \ln T} < \frac{\gamma}{\gamma - 1}} \tag{5.338}$$

For $\gamma = \frac{5}{3}$ the right-hand side is 2.5.

Example 5.13 (Convection in the Sun)
If we combine the equation for hydrostatic equilibrium

$$\frac{dP}{dr} = -\frac{Gm(r)\rho(r)}{r^2} \tag{5.339}$$

with the equation

$$\frac{dT}{dr} = -\frac{3\rho(r)\kappa}{4acT^3} \frac{L}{4\pi r^2} \tag{5.340}$$

for a gradient in heat conduction by photons, we obtain the value for a critical luminosity,

$$\frac{3\kappa}{4acT^3}\frac{L_c(r)}{4\pi r^2} = \frac{\gamma - 1}{\gamma}\frac{T}{P}\frac{Gm(r)}{r^2} \ . \tag{5.341}$$

With

$$P_P = \frac{1}{3}aT^4 \tag{5.342}$$

we can transform to

$$\left[\frac{L(r)}{m(r)}\right]_c = \frac{\gamma - 1}{\gamma}\frac{16\pi Gc}{\kappa}\frac{P_P}{P} \tag{5.343}$$

For the variation of luminosity in the energy-generating region, we set

$$\frac{dL}{dr} = 4\pi r^2 \varepsilon(r) \tag{5.344}$$

where $\varepsilon(r)4\pi r^2 dr$ is the power in a spherical shell. With the numerical values $\gamma = \frac{5}{3}$, $P = 1{,}7 \times 10^{16}$ Pa , $T = 13{,}7 \times 10^6$ K , $\kappa = 0{,}138\ \text{m}^2\ \text{kg}^{-1}$ we find the critical value

$$\boxed{\left[\frac{L(r)}{m(r)}\right]_c \approx 1{,}5 \times 10^{-3}\ \text{W kg}^{-1}} \ . \tag{5.345}$$

At the center of the Sun, we have a value $L/m \approx 1{,}35 \times 10^{-3}\ \text{W kg}^{-1}$, so no convection is expected in the core. However, convection does occur in the outer regions.

If we further analyze this example, we gain an increasingly differentiated picture of the structure of a star. Using the Sun as an example, we illustrate this in Fig. 5.4. The innermost core, with very high temperature and very high mass density, meets the criterion for thermonuclear burning. Fusion provides energy that, on the one hand, enables continuous burning in the core, but on the other hand, also reaches the surface. The transport of energy and heat is accomplished by two fundamentally different mechanisms.

Near the core, energy transport by photons is dominant. A corresponding solar model even predicts that this region extends up to about $0{,}74\,R_\odot$. Beyond this, up to just below the surface, convection is the dominant transport mechanism. Note the enormous temperature difference between the core and the surface.

Here, the surface still refers to the transition region between solid and gaseous states. Beyond the actual solar radius $R_\odot$ there are still the photosphere, chromosphere, and corona, which are particularly relevant for the radiation we observe. Enormous temperature variations are observed there. ■

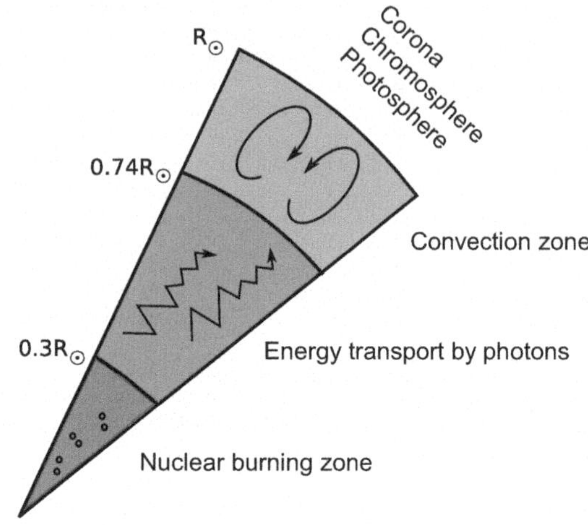

Fig. 5.4 Schematic structure of our Sun with the characteristic model zones

5.6 MHD Model

So far, we have described the plasma components (electrons, ions, ...) separately using the corresponding moments (electron density, ion density, ...) and derived the (macroscopic) transport equations for them. Now we ask whether we can go a step further and describe the plasma as a whole using quantities such as (total) mass density, (local) plasma velocity, and so on. As we will show, this is indeed possible for certain questions (and with appropriate restrictions). The result is the magnetohydrodynamic (MHD) model, which we will now introduce.

So far, we have defined plasma dynamic moments for each species. We recall the particle density

$$n_\alpha(\mathbf{r}, t) = \int d^3v f^\alpha(\mathbf{r}, \mathbf{v}; t) \tag{5.346}$$

of species $\alpha = e, i$. The particle current density of the individual components

$$\boldsymbol{\Gamma}_\alpha(\mathbf{r}, t) \equiv n_\alpha(\mathbf{r}, t)\mathbf{u}_\alpha(\mathbf{r}, t) = \int d^3v\, \mathbf{v} f^\alpha(\mathbf{r}, \mathbf{v}; t) \tag{5.347}$$

contains the mean velocity $\mathbf{u}_\alpha$. The energy density is initially defined with the total velocity in the form

$$n_\alpha \mathcal{E}_\alpha = \frac{1}{2} m_\alpha \int d^3v\, v^2 f^\alpha(\mathbf{r}, \mathbf{v}; t)\ . \tag{5.348}$$

Taking into account the mean velocity of component α, the thermal energy density is defined with the (random) relative velocity, so that

$$n_\alpha \varepsilon_\alpha = \frac{1}{2} m_\alpha \int d^3v |\mathbf{v} - \mathbf{u}_\alpha|^2 f^\alpha(\mathbf{r}, \mathbf{v}; t) \,. \tag{5.349}$$

Obviously, we have

$$n_\alpha \mathcal{E}_\alpha = \frac{1}{2} m_\alpha n_\alpha |\mathbf{u}_\alpha|^2 + n_\alpha \varepsilon_\alpha \,. \tag{5.350}$$

In thermodynamic equilibrium and for $\Lambda \gg 1$ we expect

$$p_\alpha = \frac{2}{3} n_\alpha \varepsilon_\alpha \tag{5.351}$$

for an ideal gas. This suggests the definition

$$p_\alpha(\mathbf{r}, t) = \frac{1}{3} m_\alpha \int d^3v |\mathbf{v} - \mathbf{u}_\alpha|^2 f^\alpha(\mathbf{r}, \mathbf{v}; t) \tag{5.352}$$

for the scalar pressure. We define the pressure tensor via

$$\overleftrightarrow{P}_\alpha = m_\alpha \int d^3v (\mathbf{v} - \mathbf{u}_\alpha)(\mathbf{v} - \mathbf{u}_\alpha) f^\alpha(\mathbf{r}, \mathbf{v}; t) \,. \tag{5.353}$$

In light of the equilibrium relation

$$\varepsilon_\alpha = \frac{3}{2} k_B T_\alpha \tag{5.354}$$

we define the temperature T_α by

$$n_\alpha k_B T_\alpha(\mathbf{r}, t) = \frac{1}{3} m_\alpha \int d^3v \; |\mathbf{v} - \mathbf{u}_\alpha|^2 f^\alpha(\mathbf{r}, \mathbf{v}; t) \,. \tag{5.355}$$

The variables $n_\alpha, \mathbf{u}_\alpha, T_\alpha$ (for $\alpha = e, i$) are the plasma dynamic variables in a two-fluid description of a plasma.

However, by suitable addition and subtraction, we can introduce new "variables" such as the mass density

$$\rho(\mathbf{r}, t) = \sum_\alpha m_\alpha n_\alpha \,, \tag{5.356}$$

the charge density

$$\Sigma(\mathbf{r}, t) = \sum_{\alpha} e_\alpha n_\alpha \,, \tag{5.357}$$

the total momentum density and thus the center-of-mass velocity **u**,

$$\rho\mathbf{u}(\mathbf{r}, t) = \sum_{\alpha} m_\alpha n_\alpha \mathbf{u}_\alpha \,, \tag{5.358}$$

the electric current density

$$\mathbf{j}(\mathbf{r}, t) = \sum_{\alpha} e_\alpha n_\alpha \mathbf{u}_\alpha \,, \tag{5.359}$$

and the energy density

$$\rho\mathcal{E} = \sum_{\alpha} n_\alpha \mathcal{E}_\alpha = \frac{1}{2}\rho u^2 + \rho\varepsilon \tag{5.360}$$

can be introduced.

Note that, in general, ε does not coincide with $\sum_\alpha \varepsilon_\alpha$. We retain the two temperatures T_e and T_i, assuming rapid thermalization within each component, but only a small energy exchange between the components. The set of hydrodynamic variables $\rho, \mathbf{u}, T_e, T_i$ plus electrodynamic variables Σ and $\mathbf{j}$ is, of course, equivalent to the complete set of plasma dynamic variables $n_\alpha, \mathbf{u}_\alpha, T_\alpha (\alpha = e, i)$. The latter can be obtained from the hydrodynamic plus electrodynamic variables $\rho, \mathbf{u}, T_e, T_i, \Sigma, \mathbf{j}$. If we set $e_i = Ze$ and $e_e = -e$, then

$$\boxed{n_e = \frac{Ze\rho - m_i\Sigma}{e(m_i + Zm_e)} \,, \quad n_i = \frac{e\rho + m_e\Sigma}{e(m_i + Zm_e)} \,,} \tag{5.361}$$

$$\boxed{\mathbf{u}_e = \frac{Ze\rho\mathbf{u} - m_i\mathbf{j}}{Ze\rho - m_i\Sigma} \,, \quad \mathbf{u}_i = \frac{\rho e\mathbf{u} + m_e\mathbf{j}}{e\rho + m_e\Sigma} \,.} \tag{5.362}$$

Thus, we can readily translate the equations for $n_\alpha, \mathbf{u}_\alpha, T_\alpha (\alpha = e, i)$ into those for $\rho, \mathbf{u}, T_e, T_i, \Sigma, \mathbf{j}$ without any simplification resulting. However, in certain parameter regimes, some of the latter variables are more important than others, and significant simplification may become possible.

We will first introduce the widely used MHD (magnetohydrodynamics) simplification in an ad hoc manner. We assume that the model should be valid on large spatial scales L. Large here means in comparison to the Debye length and the (ion) Larmor radius. Then,

$$\boxed{\Sigma \approx 0}$$

is used. Furthermore, let us assume that the temperatures are comparable, $T_e \approx T_i \sim T$. The single-fluid description further assumes that electrons and ions respond more or less together. The time scales τ should be long, e.g., compared to the inverse of the ion gyrofrequency. To better understand the synchronized response of electrons and ions, let us consider the momentum balance of species α, which can be reformulated as

$$n_\alpha \mathbf{u}_{\alpha\perp} = \frac{1}{m_\alpha \Omega_\alpha} \mathbf{b} \times \left[\partial_t (m_\alpha\, n_\alpha\, \mathbf{u}_\alpha) + \nabla \cdot \overleftrightarrow{\mathcal{P}}_\alpha - e_\alpha n_\alpha \mathbf{E} - \mathbf{R}_\alpha \right] , \tag{5.363}$$

where $\overleftrightarrow{\mathcal{P}}_\alpha$ is the stress tensor and $\mathbf{b}$ is the unit vector of the magnetic field. Electrons and ions respond together when, on the right-hand side, the $E \times B$ velocity

$$\mathbf{u}_\perp \approx \frac{\mathbf{E} \times \mathbf{B}}{B^2} \equiv \mathbf{u}_{E\times B} \tag{5.364}$$

dominates. The diamagnetic drift should be smaller than the $E \times B$ drift. Therefore, large electric fields (perpendicular to the magnetic field) should be allowed. The resulting $E \times B$ velocity $\mathbf{u}_{E\times B}$ can be of the order of the thermal velocities. The corresponding MHD approximation thus allows for the possibility of rapid and very violent motion. As a result, the electromagnetic contribution generally dominates over the electrostatic one.

Example 5.14 (Comparison of the electrostatic and electromagnetic contributions)
This can be seen as follows. With the electric field $\mathbf{E} = -\nabla\phi - \partial_t \mathbf{A}$, we can easily estimate the various contributions under the assumption

$$\frac{e\phi}{T_e} \sim \mathcal{O}(1) , \quad A \sim \mathcal{O}(BL) , \quad \frac{L}{\tau} \sim \mathcal{O}(v_{th}) \tag{5.365}$$

for $v_{th} \sim v_{the} \sim v_{thi}$. To this end, we also introduce

$$\delta_\alpha \equiv \frac{\rho_\alpha}{L} \ll 1 , \quad \rho_\alpha = \frac{v_{th\alpha}}{|\Omega_\alpha|} . \tag{5.366}$$

The classifications

$$\frac{|\frac{e_\alpha}{m_\alpha} \nabla\phi|}{|\Omega_\alpha|} \sim \frac{T_\alpha}{L m_\alpha |\Omega_\alpha|} \sim \frac{\rho_\alpha}{L} v_{th\alpha} \sim \delta_\alpha\, v_{th\alpha} , \quad \frac{|\frac{e_\alpha}{m_\alpha} \partial_t \mathbf{A}|}{|\Omega_\alpha|} \sim \frac{L}{\tau} \sim v_{th} \tag{5.367}$$

follow. ■

Note: $A \sim \mathcal{O}(BL)$ and $E \sim |\partial\mathbf{A}/\partial t| \sim v_{th} B$ are essential. The truncation of the moment hierarchy is then performed using Maxwellian distributions (index M) for each component,

$$f^\alpha \approx f_M^\alpha(\mathbf{v} - \mathbf{u}) + \mathcal{O}(\delta_\alpha) , \tag{5.368}$$

where $\mathbf{u} \approx \mathbf{u}_{E\times B} + \mathbf{b}u_\parallel$ is. So far, the parallel flux $u_\parallel$ has not yet been specified [29]. The assumption of Maxwellian distributions implies that the pressure tensor is isotropic and the heat flux becomes negligible. In the following, we will show that these assumptions lead to the following system:

$$\boxed{\frac{d\rho}{dt} + \rho\nabla\cdot\mathbf{u} = 0\,,} \tag{5.369}$$

$$\boxed{\rho\frac{d\mathbf{u}}{dt} - \mathbf{j}\times\mathbf{B} + \nabla p = 0\,,} \tag{5.370}$$

$$\boxed{\mathbf{E} + \mathbf{u}\times\mathbf{B} = \eta\mathbf{j}\,,} \tag{5.371}$$

$$\boxed{\nabla\times\mathbf{E} + \partial_t\mathbf{B} = 0\,,} \tag{5.372}$$

$$\boxed{\nabla\times\mathbf{B} - \mu_0\mathbf{j} = 0\,,} \tag{5.373}$$

$$\boxed{\frac{dp}{dt} + \frac{5}{3}p\nabla\cdot\mathbf{u} = 0\,.} \tag{5.374}$$

Here,

$$\frac{d}{dt} = \frac{\partial}{\partial t} + \mathbf{u}\cdot\nabla\,, \tag{5.375}$$

and p is the total pressure.

In ideal MHD, resistivity is neglected, i.e., $\eta \approx 0$. Note that two equations can immediately be used to eliminate $\mathbf{E}$ and $\mathbf{j}$. This leaves eight scalar equations for eight components of $\rho, p, \mathbf{u}, \mathbf{B}$.

Systematic Derivation

To systematically "derive" the MHD equations [64], we begin with the kinetic equation, for example, the Landau-Fokker-Planck equation without particle sources and sinks. Binary collisions are taken into account. We will pay attention to the corresponding conservation laws during (elastic) collisions. We abbreviate the kinetic equation as:

$$\boxed{\partial_t f^\alpha + \mathbf{v}\cdot\nabla f^\alpha + \frac{e_\alpha}{m_\alpha}(\mathbf{E}+\mathbf{v}\times\mathbf{B})\cdot\partial_{\mathbf{v}} f^\alpha = K^\alpha \equiv \sum_\beta K^{\alpha\beta}}\,, \tag{5.376}$$

with $f^\alpha \equiv f^\alpha(\mathbf{r}, \mathbf{v}; t)$. We multiply by powers of the velocity $\mathbf{v}$, sum over the species α, and finally integrate over velocity space. Of course, the resulting moment equations must be supplemented by Maxwell's equations,

$$\nabla\cdot\mathbf{E} = \frac{1}{\varepsilon_0}\sum_\alpha e_\alpha \int d^3v f^\alpha\,, \tag{5.377}$$

$$\nabla\times\mathbf{E} = -\frac{\partial\mathbf{B}}{\partial t}\,, \tag{5.378}$$

$$\nabla\cdot\mathbf{B} = 0\,, \tag{5.379}$$

$$\nabla\times\mathbf{B} = \mu_0\varepsilon_0\frac{\partial\mathbf{E}}{\partial t} + \mu_0\sum_\alpha e_\alpha\int d^3v\,\mathbf{v} f^\alpha\,. \tag{5.380}$$

For the zeroth moment, the calculation is straightforward. Taking into account the conservation properties of the collision term, we immediately obtain

$$\frac{\partial\rho}{\partial t} + \nabla\cdot(\rho\mathbf{u}) = 0\,, \tag{5.381}$$

Next, we consider

$$\int d^3v\sum_\alpha m_\alpha\mathbf{v}\left[\partial_t f^\alpha + \mathbf{v}\cdot\nabla f^\alpha + \frac{e_\alpha}{m_\alpha}(\mathbf{E}+\mathbf{v}\times\mathbf{B})\cdot\partial_{\mathbf{v}} f^\alpha\right] = \sum_{\alpha\beta}\int d^3v\mathbf{v}K^{\alpha\beta} = 0\,. \tag{5.382}$$

Before we introduce the random velocity $\mathbf{v}' = \mathbf{v} - \mathbf{u}$ relative to the mean mass flow velocity $\mathbf{u} = \mathbf{u}(\mathbf{r}, t)$, we make use of the independence of $t, \mathbf{r}, \mathbf{v}$. We introduce the pressure tensor:

$$\boxed{P_{rs} = \sum_\alpha m_\alpha\int d^3v(v_r - u_r)(v_s - u_s)f^\alpha(\mathbf{r}, \mathbf{v}; t)}\,, \tag{5.383}$$

which in general does not coincide with the sum over the pressure tensors of the components,

$$P_{rs} \neq \sum_\alpha P_{rs}^\alpha \tag{5.384}$$

The total scalar pressure follows from

$$p = \frac{1}{3} \operatorname{Tr} \overleftrightarrow{P} , \tag{5.385}$$

so that $P_{rs} = p\delta_{rs} + \pi_{rs}$, if the situation is isotropic to first order. The dissipative part of the pressure tensor,

$$\begin{aligned} \pi_{rs} = &\sum_{\alpha} m_\alpha \int d^3v (v_r - u_r)(v_s - u_s) f^\alpha(\mathbf{r}, \mathbf{v}; t) \\ &- \frac{1}{3} \sum_{\alpha} m_\alpha \int d^3v |\mathbf{v} - \mathbf{u}|^2 f^\alpha(\mathbf{r}, \mathbf{v}; t)\, \delta_{rs} \end{aligned} \tag{5.386}$$

is assumed to be of higher order in the MHD approximation.

In a similar way, we define the heat flux vector

$$\boxed{\mathbf{q}(\mathbf{r}, t) = \frac{1}{2} \sum_{\alpha} m_\alpha \int d^3v\, (\mathbf{v} - \mathbf{u}) |\mathbf{v} - \mathbf{u}|^2 f^\alpha(\mathbf{r}, \mathbf{v}; t)} . \tag{5.387}$$

A short calculation leads to

$$\frac{\mathrm{d}\mathbf{u}}{\mathrm{d}t} + \nabla \cdot (\rho \mathbf{u}\mathbf{u}) = \left[\sum_{\alpha} n_\alpha e_\alpha \right] \mathbf{E} + \mathbf{j} \times \mathbf{B} - \nabla \cdot \overleftrightarrow{P} . \tag{5.388}$$

First, it is assumed that the plasma is essentially neutral due to $\Sigma(\mathbf{r}, t) = \sum_\alpha e_\alpha n_\alpha \approx 0$. Note that we will apply MHD to large-scale phenomena. Second, similar to the two-fluid equations, we can use (5.381) to rearrange the terms, resulting in

$$\rho \frac{d\mathbf{u}}{dt} - \mathbf{j} \times \mathbf{B} + \nabla p = 0 . \tag{5.389}$$

A simple form of Ohm's law for the current density $\mathbf{j}$ is obtained from the electron momentum equation,

$$m_e \left(\frac{\mathrm{d}\mathbf{u}_e}{\mathrm{d}t} + \mathbf{u}_e \cdot \nabla \mathbf{u}_e \right) \approx -e(\mathbf{E} + \mathbf{u}_e \times \mathbf{B}) - \frac{1}{n_e} \nabla p_e - \nu_{ei} m_e (\mathbf{u}_e - \mathbf{u}_i) . \tag{5.390}$$

Within the MHD scaling, we can neglect the terms on the left-hand side. They are small compared to the magnetic force term on the right-hand side:

$$\left| \frac{\mathrm{d}\mathbf{u}_e}{\mathrm{d}t} + \mathbf{u}_e \cdot \nabla \mathbf{u}_e \right| \sim \frac{v_{the}}{\tau} \ll v_{the} |\Omega_e| \sim \frac{e}{m_e} |\mathbf{u}_e \times \mathbf{B}| . \tag{5.391}$$

Furthermore, we substitute

$$\mathbf{u}_e = \mathbf{u}_i - \frac{1}{en_e} \mathbf{j} \approx \mathbf{u} - \frac{1}{en_e} \mathbf{j} \tag{5.392}$$

and write (5.390) as

$$\mathbf{E} + \mathbf{u} \times \mathbf{B} - \frac{1}{en_e}\mathbf{j} \times \mathbf{B} + \frac{1}{en_e}\nabla p_e \approx \eta\mathbf{j} , \tag{5.393}$$

with the resistivity

$$\eta = \frac{m_e \nu_{ei}}{e^2 n_e} . \tag{5.394}$$

Example 5.15 (Neglecting the Hall Term)
Except in the so-called Hall-MHD model, the Hall term $\frac{1}{en_e}\mathbf{j} \times \mathbf{B}$ is normally omitted. Arguments for this assumption can be found by comparing the Hall term either with the resistive term (on the right-hand side) or with the pressure term. In resistive MHD, the ratio of the Hall term to the resistive term is of the order of

$$\frac{|\Omega_e|}{\nu_{ei}} \ll 1 . \tag{5.395}$$

Otherwise, if (5.389) leads to $j \sim \rho v_{th}/(\tau B)$, we compare the Hall term with $\mathbf{u} \times \mathbf{B}$ to obtain the ratio

$$\frac{1}{|\Omega_i|\tau} \ll 1 \tag{5.396}$$

within the framework of MHD scaling. Therefore, the Hall term is negligible when the pressure term in the MHD equation of motion (5.389) is small. In this case, we write Ohm's law as

$$\mathbf{E} + \mathbf{u} \times \mathbf{B} = \eta\mathbf{j} . \tag{5.397}$$

Let us compare in (5.389) the pressure term with the magnetic force term. We find

$$\frac{|\nabla p|}{|\mathbf{j} \times \mathbf{B}|} \sim \frac{p}{\frac{B^2}{2\mu_0}} \equiv \beta . \tag{5.398}$$

The ratio of plasma pressure to magnetic pressure is known as the plasma β. For $\beta \ll 1$, the Hall term will also be small. If we add Maxwell's induction equation (Faraday's law)

$$\nabla \times \mathbf{E} + \partial_t \mathbf{B} = 0 , \tag{5.399}$$

and supplement it with Ampère's law

$$\nabla \times \mathbf{B} - \mu_0 \mathbf{j} = 0 , \tag{5.400}$$

for sufficiently slow processes (phase velocities much less than the speed of light), we neglect the displacement current. ■

We are almost finished with the derivation of the closed set of MHD equations. For the pressure, we can use the adiabatic law:

$$\frac{dp}{dt} + \frac{5}{3}p\nabla \cdot \mathbf{u} \equiv \frac{dp}{dt} + \frac{5}{3}\frac{p}{\rho}\frac{d\rho}{dt} = 0 . \tag{5.401}$$

Here we recall

$$\frac{d}{dt} = \frac{\partial}{\partial t} + \mathbf{u} \cdot \nabla \tag{5.402}$$

as the total derivative in the one-fluid system. Finally, we can assume that the diagonal terms in the pressure tensor dominate, so that the MHD equations (5.369)–(5.374) follow.

Example 5.16 (Equations of State)
We should note that instead of (5.374), which expresses the adiabatic equation of state,

$$\frac{d}{dt}\left(\frac{p}{\rho^{5/3}}\right) = 0 , \tag{5.403}$$

other closure conditions can also be used. The isothermal equation of state is

$$\frac{d}{dt}\left(\frac{p}{\rho}\right) = 0 . \tag{5.404}$$

In some situations, incompressibility may be the better choice. If

$$\nabla \cdot \mathbf{u} = 0 \tag{5.405}$$

applies, Eq. (5.374) leads to

$$\frac{d}{dt}p = 0 . \tag{5.406}$$

■

We conclude with some additional remarks. The continuity equation for the charge density

$$\partial_t \Sigma + \nabla(\mathbf{j}) = 0 \tag{5.407}$$

is consistent with the derived set of MHD equations. The neglect of the dissipative part of the pressure tensor is related to the fact that

$$\pi_{rs} \sim \mathcal{O}(\delta) . \tag{5.408}$$

More detailed models are sometimes used, for example, the so-called electron MHD.

The literature on MHD is extensive. Important problems fill many books. We refer to [65, 66] and the references therein for a comprehensive overview of MHD.

6 Waves and Instabilities in Vlasov Systems

Abstract

Starting from a time-independent (stationary) state of a plasma, one can investigate how the plasma behaves dynamically under (small) perturbations. Essentially, within the framework of the Vlasov description, we have already begun addressing this problem when we derived the plasma dispersion function and discussed the consequences for plasma oscillations in the form of Landau damping. In this section, when we revisit the question of stability, we intend to do so from a more general perspective, that is, not always assuming a Maxwellian distribution as the unperturbed state. We will retain the linear approximation and therefore will not address the issue of nonlinear stability here. Furthermore, we will also base our discussion on the Vlasov model.

6.1 Fundamentals

Stability theory is a rather well-developed mathematical theory—however, primarily in the area of ordinary differential equations. For partial differential equations, the principles are clearly formulated, but almost every practical evaluation brings new difficulties. This is one reason why, in this section, we do not take the rigorous standpoint that stability should only be demonstrated with Lyapunov functionals and instability only on the basis of variational principles. Even with these restrictions, there are still enough methods available that allow conclusions about stability or instability. Here, we will particularly highlight normal mode analysis for instability and statements about stability based on conserved quantities.

K.-H. Spatschek, *Theoretical Plasma Physics*,
https://doi.org/10.1007/978-3-662-72828-4_6

Let us begin, however, with a somewhat more detailed analysis of the plasma dispersion equation, which we have already derived. There, in (4.90), with an undisturbed distribution function g,

$$\boxed{k^2 = G\left(\frac{\omega}{k}\right) \equiv \omega_{pe}^2 \int \frac{dg}{du} \frac{du}{u - \frac{\omega}{k}}}, \tag{6.1}$$

where the Landau prescription for the integration path must be chosen. The relation (6.1) is relevant for a single-component system with a smeared-out ion background. Since we have used $\omega = ip$, the question arises for a statement about instability as to whether solutions with

$$\text{Im } \omega > 0 \tag{6.2}$$

exist. If so, we can infer exponentially growing modes. Under the condition (6.2), the integration path in (6.1) can be chosen along the real u-axis (from $-\infty$ to $+\infty$)). On the real u_p-axis, with the phase velocity

$$u_p = \frac{\omega}{k} = \frac{\text{Re } \omega}{k}, \tag{6.3}$$

the following holds:

$$\boxed{G(u_p + i0) = \omega_{pe}^2 \left\{ \mathcal{P} \int_{-\infty}^{+\infty} \frac{g'(u)}{u - u_p} du + i\pi g'(u_p) \right\}}. \tag{6.4}$$

The idea for the further approach is as follows. Eqs. (6.1) and (6.4) can be viewed as a mapping of the upper u_p half-plane onto a surface in the (complex) G-plane. This surface is bounded by the curve (6.4), with the upper half-plane being mapped into the interior. This statement is justified by the fact that $G(R + i0)$ is bounded and continuous, and $G(u_p)$ is a holomorphic function in the upper u_p-plane. The directed curve $G(u_p)$ starts and ends, due to $G(\pm\infty) = 0$, at $G = 0$, when u_p varies from $-\infty$ to $+\infty$. If G_0 is any point that is not on $G(R + i0)$ lies, then the curve (6.4) winds (counterclockwise) around G_0 as many times as $G(u_p)$ takes the value G_0 in the upper u_p-plane.

Instabilities (i.e., solutions with $\text{Im } \omega > 0$) are possible if the directed curve (6.4) crosses the positive real G-axis, since then solutions of (6.1) exist in the upper ω-half-plane. When crossing the real G-axis (from below to above), $g'(u_p)$ must of course change its sign.

Therefore, following Penrose, a stability criterion can be formulated as follows: The two conditions are necessary and sufficient for (exponential) instability:
(a) The distribution function g has at least one minimum.
(b) At this minimum,

$$\mathcal{P}\int_{-\infty}^{+\infty}\frac{g'(u)}{u-u_p}du>0\,. \tag{6.5}$$

That the extremum must be a minimum follows from the transition of the boundary curve (6.4) [in the G-plane] from the lower G-half-plane to the upper (so that positive real values lie to its left). As u_p increases, $g'(u_p)$ is initially negative and then positive.

The Penrose criterion allows one to see from a rather simple representation whether unstable modes exist. Before that, however, we rewrite the rather unwieldy condition equation (6.5). After partial integration, we find

$$\begin{aligned}
&\mathcal{P}\int_{-\infty}^{+\infty}\frac{g'(u)}{u-u_p}du\\
&=\lim_{\varepsilon\to 0}\left[\left\{\int_{-\infty}^{u_p-\varepsilon}+\int_{u_p+\varepsilon}^{\infty}\right\}\left\{\frac{g(u)}{(u-u_p)^2}du\right\}-\frac{g(u_p-\varepsilon)}{\varepsilon}-\frac{g(u_p+\varepsilon)}{\varepsilon}\right]\\
&=\mathcal{P}\int_{-\infty}^{+\infty}\frac{g(u)-g(u_p)}{(u-u_p)^2}du+\lim_{\varepsilon\to 0}\left[\frac{g(u_p)-g(u_p-\varepsilon)}{\varepsilon}-\frac{g(u_p+\varepsilon)-g(u_p)}{\varepsilon}\right]\\
&=\int_{-\infty}^{+\infty}\frac{g(u)-g(u_p)}{(u-u_p)^2}du
\end{aligned} \tag{6.6}$$

because of

$$g(u_p)\lim_{\varepsilon\to 0}\left\{\int_{-\infty}^{u_p-\varepsilon}+\int_{u_p+\varepsilon}^{\infty}\right\}\left\{\frac{du}{(u-u_p)^2}\right\}=2\lim_{\varepsilon\to 0}\frac{g(u_p)}{\varepsilon}\,. \tag{6.7}$$

In the result (6.6), we can omit the principal value sign $\mathcal{P}$, since at the point of interest $u_p=u_p^*$ the function g is supposed to have a minimum. The criterion (6.5), evaluated at the minimum point u_p^* of $g(u_p)$, thus reads

$$\boxed{\int_{-\infty}^{+\infty}\frac{g(u)-g(u_p^*)}{(u-u_p^*)^2}du>0}\,. \tag{6.8}$$

We now give two examples.

Example 6.1 (Evaluation for a Maxwell Distribution)
Let us first assume a Maxwell distribution for g

$$g(u) = \frac{1}{(2\pi)^{1/2} v_t} \exp[-u^2/2v_t^2] \tag{6.9}$$

which leads to

$$G(u_p + i0) = -\frac{\omega_{pe}^2 \, v_t^{-3}}{(2\pi)^{1/2}} \left\{ \mathcal{P} \int_{-\infty}^{+\infty} \frac{u \, \exp(-u^2/2v_t^2)}{u - u_p} du + i\pi \, u_p \exp\left[-\frac{u_p^2}{2v_t^2} \right] \right\} \tag{6.10}$$

Obviously, it is not possible to construct an intersection with the positive real G-axis. The image of (6.10) in the complex G-plane is shown in Fig. 6.1. ■

Example 6.2 (Evaluation for Two Superposed Dirac Distributions)
The direct generalization of this example leads us to the next case of two superposed Maxwell distributions whose maxima do not coincide. The overall distribution function can then possess a minimum, and the Penrose criterion allows for instability. Since in the general case an analysis can only be performed numerically, we restrict ourselves here to a limiting case that is still analytically tractable. We let the thermal width of the Maxwell distributions shrink to zero, i.e., we transition to Dirac delta functions,

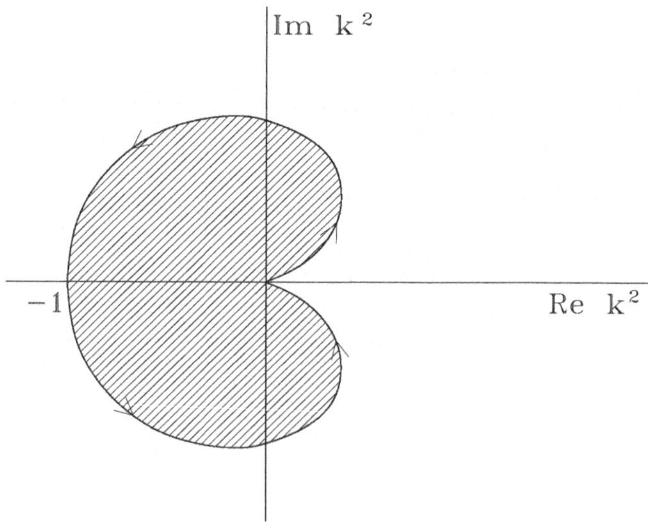

Fig. 6.1 Mapping of the upper u_p-plane into the k^2-plane (inner dashed region) via $G(R + i0)$ for a Maxwell distribution

$$g \approx [\delta(u - U) + \delta(u + U)] \,. \tag{6.11}$$

Due to symmetry, it is of course clear that a minimum occurs at

$$u = 0 \tag{6.12}$$

at this minimum, $g(0) = 0$. In the discussion of (6.8), the integral

$$\int_{-\infty}^{+\infty} \frac{g(u)}{u^2} du = \frac{2}{U^2} > 0 \tag{6.13}$$

must be evaluated. The result allows us to clearly conclude instability. [In Fig. 6.2, the curve (6.4) is plotted for two Maxwell distributions that lead to a minimum. One can clearly see the intersection with the positive G-axis, which indicates instability.] ■

Example 6.3 (Direct Solution for Two Dirac Distributions)
In the limiting case (6.11), the general dispersion relation can also be evaluated directly. We have already given the result in (4.71), although there ν may only be set to 1 and 2. Landau damping, which can become quite significant when choosing Maxwellian distributions with finite thermal widths, does not play a decisive role here, and we can directly evaluate

$$\omega_{pe}^2 \left[\frac{1}{(kU - \omega)^2} + \frac{1}{(kU + \omega)^2} \right] = 1 \tag{6.14}$$

For

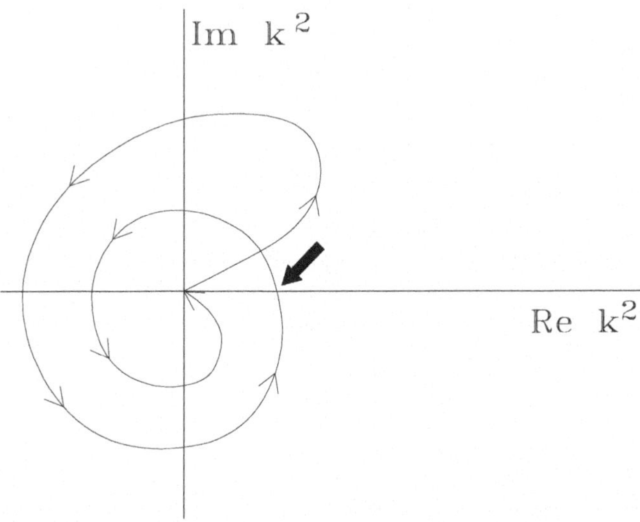

Fig. 6.2 $G(u_p + i0)$ in the complex k^2-plane for an unstable distribution function with two maxima

$$|k| < \sqrt{2}\,\frac{\omega_{pe}}{U} \tag{6.15}$$

there are only two purely real solutions to Eq. (6.14), which is of fourth order. In addition, there exist two complex solutions, which correspond to exponentially growing and exponentially damped solutions, respectively. The instability, also known as the two-stream instability, has a maximum growth rate

$$\gamma = \omega_{pe}/2\,. \tag{6.16}$$

(Finite thermal widths, of course, change the result!) We will return to this point later, when we discuss the normal mode analysis of the linearized Vlasov equation in context. ■

At this point, let us bring another question to the forefront, namely, how one might—admittedly in the simple case of spatially homogeneous ground states—approach the (complementary) stability problem. To this end, we attempt to find an argument that excludes temporally growing modes.

The linearization of the Vlasov equation starts from a specific stationary state, which we here define as an isotropic distribution in velocity space. No external electric field is present, whereas a constant external magnetic field $\mathbf{B}_0$ is allowed. The (in principle arbitrary) function

$$g = g\left(\frac{m_e}{2}v^2\right) \tag{6.17}$$

is then an exact solution of the stationary Vlasov equation to zeroth order. When linearizing the full Vlasov equation around the stationary state g, i.e.,

$$\boxed{f = g + \delta f}\,, \tag{6.18}$$

electric and magnetic fields $\mathbf{E}_1$ and $\mathbf{B}_1$ appear to first order. However, it should be noted that due to our choice of the stationary initial state g

$$\frac{q_e}{m_e}\mathbf{v} \times \mathbf{B}_1 \cdot \partial_{\mathbf{v}} g = 0 \tag{6.19}$$

holds. The linearized Vlasov equation then reads

$$\boxed{\partial_t \delta f + \mathbf{v} \cdot \partial_{\mathbf{r}}\, \delta f + \frac{q_e}{m_e}\mathbf{E}_1 \cdot \partial_{\mathbf{v}}\, g + \frac{q_e}{m_e}(\mathbf{v} \times \mathbf{B}_0) \cdot \partial_{\mathbf{v}}\, \delta f = 0}\,. \tag{6.20}$$

If we multiply both sides of this equation by δf and divide by

$$g' = \frac{d}{d\varepsilon} g(\varepsilon) , \tag{6.21}$$

with $\varepsilon = \frac{1}{2} m_e v^2$, then, after integrating over velocities and a brief calculation,

$$\partial_t \int d^3 v \frac{(\delta f)^2}{g'} + 2 q_e \int d^3 v \mathbf{E}_1 \cdot \mathbf{v} \, \delta f = 0 . \tag{6.22}$$

follows. Due to the assumption of an isotropic distribution function g, the magnetic field contribution vanishes. A corresponding calculation can be carried out in a (multi-component) plasma for each individual component $\alpha (= e, i)$. Summing the corresponding contributions (6.22) leads to

$$\partial_t \sum_\alpha \int \frac{(\delta f_\alpha)^2}{g'_\alpha} d^3 v + 2 \mathbf{E}_1 \cdot \mathbf{j}_1 = 0 \tag{6.23}$$

with the electric current density $\mathbf{j}_1$. Based on Maxwell's equations

$$\mu_0 \mathbf{j}_1 = \nabla \times \mathbf{B}_1 - \mu_0 \varepsilon_0 \partial_t \mathbf{E}_1 , \tag{6.24}$$

$$\partial_t \mathbf{B}_1 = - \nabla \times \mathbf{E}_1 \tag{6.25}$$

and the relation

$$\int \mathbf{B}_1 \cdot (\nabla \times E_1) d^3 r = \int \mathbf{E}_1 \cdot (\nabla \times \mathbf{B}_1) d^3 r \tag{6.26}$$

it finally follows, after integration over spatial coordinates,

$$\partial_t \left[\sum_\alpha \int d^3 v \, d^3 r \, \frac{(\delta f_\alpha)^2}{g'_\alpha} - \varepsilon_0 \int d^3 r \, (\mathbf{E}_1^2 + c^2 \mathbf{B}_1^2) \right] = 0 , \quad c^2 = \frac{1}{\varepsilon_0 \mu_0} . \tag{6.27}$$

The quantity in square brackets on the left-hand side is thus a conserved quantity. For

$$\boxed{g'_\alpha < 0} \tag{6.28}$$

it is negative definite. From

$$\sum_\alpha \int d^3 v \, d^3 r \frac{(\delta f_\alpha)^2}{(-g'_\alpha)} + \varepsilon_0 \int d^3 r \, (\mathbf{E}_1^2 + c^2 \mathbf{B}_1^2) = const \tag{6.29}$$

it then follows that, under the condition (6.28), there can be no exponentially growing (unstable) modes $\sim \exp(\gamma t)$. This is consistent with the statement of the Penrose criterion, according to which distribution functions without a minimum ("single-humped distributions") are stable.

The argument just presented can also be motivated thermodynamically. From equilibrium thermodynamics, we know the stability criterion that requires a minimum of the free energy in the stable state.

If we strictly apply the approaches of equilibrium thermodynamics to our system, we must assume Maxwell distributions for g_α and set $\mathbf{E}_0$ as well as $\mathbf{B}_0$ equal to zero. The internal energy is given by

$$U = \sum_\alpha \int d^3v\, d^3r\, \frac{m_\alpha}{2} v^2 f_\alpha + \frac{\varepsilon_0}{2} \int d^3r\, (\mathbf{E}^2 + c^2\mathbf{B}^2) \tag{6.30}$$

while the entropy S can be written as

$$S = -k_B \sum_\alpha \int d^3v\, d^3r\, f_\alpha \;\ln f_\alpha \tag{6.31}$$

If we vary the free energy

$$\boxed{F = U - TS}\,, \tag{6.32}$$

we find from the equilibrium condition ($\delta F = 0$) to first order

$$f_{\alpha 0} = g_\alpha, \tag{6.33}$$

i.e., the Maxwell distributions, and to second order

$$\delta^2 F = \sum_\alpha \int d^3v\, d^3r\, k_B T_\alpha \frac{(\delta f_\alpha)^2}{g_\alpha} + \frac{\varepsilon_0}{2} \int d^3r[(\delta\mathbf{E})^2 + c^2(\delta\mathbf{B})^2]\,. \tag{6.34}$$

Without doubt, the Maxwell distribution is stable in a system without external fields! In order to investigate more general distributions g_α, we must generalize the expression for the entropy (6.31). We choose

$$\boxed{S = -k_B \sum_\alpha \int d^3v\, d^3r\, F_\alpha(f_\alpha)} \tag{6.35}$$

with a functional F_α. still to be determined. From the first variation of the free energy F it follows

$$F'_\alpha(g_\alpha) = -\frac{1}{2} m_\alpha v^2\,. \tag{6.36}$$

Here, the prime (′) denotes differentiation with respect to the argument. In order for (6.36) to be invertible and the solution g_α to be localized for large v, the following must hold:

$$F''_\alpha > 0 \tag{6.37}$$

The approach (6.36) restricts us to isotropic distribution functions. If we assume the solution $g_\alpha = g_\alpha(+\frac{1}{2}m_\alpha v^2)$ to be known, then from (6.36) by differentiation with respect to the argument $\frac{1}{2}m_\alpha v^2$

$$F''_\alpha g'_\alpha = -1\,, \tag{6.38}$$

and the second variation of the free energy can be written as

$$\boxed{\delta^2 F = \sum_\alpha \int d^3v\, d^3r \frac{1}{2}\frac{(\delta f_\alpha)^2}{(-g'_\alpha)} + \frac{\varepsilon_0}{2}\int d^3r\big[(\delta\mathbf{E})^2 + c^2(\delta\mathbf{B})^2\big] \geq 0} \tag{6.39}$$

We see again that for distribution functions without a minimum, no unstable modes can exist.

The considerations outlined so far should give us an idea of how, in principle, necessary and sufficient (linear) stability criteria can be established. However, they also show that it is not easy to actually find necessary and sufficient conditions that are seamlessly connected in all relevant cases. We will examine this aspect in more detail for spatially homogeneous Vlasov systems.

In the spatially homogeneous case without an external magnetic field $\mathbf{B}_0$, any (mathematically and physically well-defined) function of the velocity $\mathbf{v}$, d. h. $g = g(\mathbf{v})$, can initially be a stationary solution. In the presence of an external magnetic field $\mathbf{B}_0 = const$, this class is restricted to arbitrary functions of $v_\parallel$ ($\parallel$ meaning parallel to $\mathbf{B}_0$) and $v_\perp^2$ ($\perp$ meaning perpendicular to $\mathbf{B}_0$), but the variety is still enormous.

The proofs of stability, as indicated above in an example, apply only to isotropic distribution functions, i.e. $g = g(v^2)$. We derived the Penrose criterion for systems without external fields, which are subject only to electrostatic perturbations depending on $x\delta f_\alpha(\mathbf{v}, x, t)$. We now want to address this situation from the perspective of stability. For this, we start from

$$\boxed{\partial_t \delta f_\alpha + \mathbf{v}\cdot\partial_\mathbf{r}\,\delta f_\alpha + \frac{q_\alpha}{m_\alpha}\mathbf{E}_1\cdot\partial_\mathbf{v}\,g_\alpha = 0} \tag{6.40}$$

and integrate in real space over y and z as well as in velocity space over v_y and v_z. Note that in (6.40) the Lorentz term is absent due to the electrostatic approximation. (This term would also be absent if we abandoned the electrostatic approximation, but instead assumed isotropic stationary ground states $g_\alpha(v^2)$.) With

$$\delta\tilde{f}_\alpha = \int dv_y\, dv_z\, dy\, dz\, \delta f_\alpha \ , \tag{6.41}$$

$$\tilde{E}_1 = \int dy\, dz\, E_{1x} \ , \tag{6.42}$$

$$\tilde{g}_\alpha = \int dv_y\, dv_z\, g_\alpha \tag{6.43}$$

we obtain

$$\partial_t \delta\tilde{f}_\alpha + v_x \partial_x \delta\tilde{f}_\alpha + \frac{q_\alpha}{m_\alpha}\tilde{E}_1 \frac{\partial}{\partial v_x}\tilde{g}_\alpha = 0 \ . \tag{6.44}$$

The now customary transformation—multiplication by $m_\alpha v_x \delta\tilde{f}_\alpha$, division by $d\tilde{g}_\alpha/dv_x$, and integration over the x and v_x coordinates—finally yields

$$\sum_\alpha \int\int dxdv_x \left(-m_\alpha v_x \left(\frac{d\tilde{g}_\alpha}{dv_x} \right)^{-1} (\delta\tilde{f}_\alpha)^2 \right) + \varepsilon_0 \int dx\, \tilde{E}_1^2 = const \ , \tag{6.45}$$

so that, with a definite sign of the first term on the left-hand side of (6.45), we obtain agreement with the Penrose criterion.

A remark is interesting in this context. Up to Eq. (6.44)—as already noted—the above calculation also applies to non-electrostatic, but only x-dependent perturbations, provided the ground state g_α is isotropic in $\mathbf{v}$. Then, the simple transformation

$$\begin{aligned} \tilde{g}_\alpha &= 2\pi \int_0^\infty dv_\perp v_\perp\, g_\alpha \Big[\frac{m_\alpha}{2}(v_x^2 + v_\perp^2) \Big] \\ &= \frac{4\pi}{m_\alpha} \int_{\varepsilon_x}^\infty d\varepsilon_\perp\, g_\alpha(\varepsilon_\perp) \end{aligned} \tag{6.46}$$

for

$$\varepsilon_x = \frac{m_\alpha}{2} v_x^2 \ , \qquad \varepsilon_\perp = \frac{m_\alpha}{2} v_\perp^2 \ , \tag{6.47}$$

shows that

$$g_\alpha'(\varepsilon_x) < 0 \tag{6.48}$$

holds. By a calculation similar to that in (6.27), but here with

$$\hat{e}_x \cdot [\nabla \times \mathbf{B}_1(x,t)] = 0 \ , \tag{6.49}$$

we now obtain

$$\sum_\alpha \int dxdv_x \frac{(\delta f_\alpha)^2}{(-g_\alpha')} + \varepsilon_0 \int dx\, E_1^2 = const. \tag{6.50}$$

The stability now follows immediately—because of (6.48) without further restrictions!

As soon as—even in spatially homogeneous initial systems—the distribution function deviates from the previous assumptions, we know so far—apart from the Penrose criterion—very little. We aim to partially address this deficit in the following sections.

6.2 Dispersion in Homogeneous Vlasov Systems

In this section, we investigate the stability behavior in Vlasov systems, placing particular emphasis on those systems not covered by the previous considerations. We start from a homogeneous plasma in a homogeneous magnetic field $\mathbf{B}_0 = B_0\hat{z}$ and with vanishing external electric field to zeroth order ($\mathbf{E}_0 = 0$). Various modes are discussed.

The stationary Vlasov equation allows in zeroth order, i.e., for g_α arbitrary functions of the parallel velocity v_z and the magnitude of the perpendicular (to $\mathbf{B}_0$) velocity component $v_\perp$, provided no external electric field is present.

We write

$$g_\alpha = g_\alpha(v_\perp^2, v_z) \ . \tag{6.51}$$

After linearization by means of

$$f_\alpha = g_\alpha + \delta f_\alpha \ , \tag{6.52}$$

a Fourier transformation of the linearized equation in space, and a restriction to

$$\text{normal modes} \ \sim \ \exp(-i\omega t + i\mathbf{k}\cdot\mathbf{r}) \tag{6.53}$$

we obtain

$$\boxed{i(\mathbf{k}\cdot\mathbf{v} - \omega)\delta f_\alpha + (\mathbf{v}\times\boldsymbol{\Omega}_\alpha)\cdot\partial_\mathbf{v}\,\delta f_\alpha = -\frac{q_\alpha}{m_\alpha}(\mathbf{E}_1 + \mathbf{v}\times\mathbf{B}_1)\cdot\partial_\mathbf{v}\,g_\alpha} \ . \tag{6.54}$$

Here we have introduced

$$\boldsymbol{\Omega}_\alpha = \frac{q_\alpha\mathbf{B}_0}{m_\alpha} = \Omega_\alpha\hat{z} \tag{6.55}$$

Without loss of generality, we can set

$$\mathbf{k} = k_x\hat{x} + k_z\hat{z} \tag{6.56}$$

and

$$v_x = v_\perp \cos\varphi \,, \qquad v_y = v_\perp \sin\varphi \tag{6.57}$$

A simple transformation leads to

$$(\mathbf{v} \times \boldsymbol{\Omega}_\alpha) \cdot \partial_\mathbf{v}\, \delta f_\alpha = -\Omega_\alpha \frac{\partial\, \delta f_\alpha}{\partial\varphi} \,, \tag{6.58}$$

so that the homogeneous part of (6.54) can be written as

$$i(\omega - k_x v_\perp \cos\varphi - k_z v_z)\delta f_\alpha^h + \Omega_\alpha \frac{\partial}{\partial\varphi}\delta f_\alpha^h = 0 \tag{6.59}$$

The solution to this is

$$\delta f_\alpha^h = h(v_\perp, v_z) e^{ik_x v_\perp \sin\varphi/\Omega_\alpha}\, e^{-i(\omega - k_z v_z)\varphi/\Omega_\alpha} \,; \tag{6.60}$$

and it can be used, via the method of "variation of constants $h(v_\perp, v_z)$", to compute the general solution of (6.54). If we abbreviate the right-hand side of (6.54) by $I_\alpha(v_\perp \cos\varphi, v_\perp \sin\varphi, v_z)$, it follows that

$$\boxed{\begin{aligned}\delta f_\alpha = & \frac{1}{\Omega_\alpha}\int^{\varphi} d\varphi' I_\alpha(v_\perp \cos\varphi', v_\perp \sin\varphi', v_z) e^{i(k_x v_\perp/\Omega_\alpha)(\sin\varphi - \sin\varphi')} \\ & \times e^{-i(\omega - k_z v_z)(\varphi - \varphi')/\Omega_\alpha} \,.\end{aligned}} \tag{6.61}$$

In the case of (exponential) instability, we can assume $\operatorname{Im}\omega > 0$. Furthermore, we are free to formally use $\varphi' = \infty$ as the lower limit in the particular solution (6.61). With the substitution

$$\tau := (\varphi - \varphi')/\Omega_\alpha \tag{6.62}$$

we obtain from (6.61)

$$\begin{aligned}\delta f_\alpha = & -\int_{-\infty}^{0} d\tau\; I_\alpha[v_\perp \cos(\varphi - \Omega_\alpha\tau),\; v_\perp \sin(\varphi - \Omega_\alpha\tau),\; v_z] \\ & \times e^{i(k_x v_\perp/\Omega_\alpha)[\sin\varphi - \sin(\varphi - \Omega_\alpha\tau)]} e^{i(k_z v_z - \omega)\tau} \,.\end{aligned} \tag{6.63}$$

For further evaluation, let us examine the expression

$$\begin{aligned}
I_\alpha(v_x, v_y, v_z) &= -\frac{q_\alpha}{m_\alpha}(\mathbf{E}_1 + \mathbf{v}\times\mathbf{B}_1)\cdot\partial_{\mathbf{v}} g_\alpha \\
&= -\frac{q_\alpha}{m_\alpha}\left\{\frac{1}{v_\perp}\left[(\mathbf{E}_1\cdot\mathbf{v}_\perp) + \frac{1}{\omega}(\mathbf{k}\cdot\mathbf{v}_\perp)(\mathbf{E}_1\cdot\mathbf{v}) - \frac{1}{\omega}(\mathbf{v}\cdot\mathbf{k})(\mathbf{E}_1\cdot\mathbf{v}_\perp)\right]\frac{\partial g_\alpha}{\partial v_\perp}\right. \\
&\quad \left. + \left[E_{1z} + \frac{1}{\omega}k_z(\mathbf{E}_1\cdot\mathbf{v}) - \frac{1}{\omega}E_{1z}(\mathbf{v}\cdot\mathbf{k})\right]\frac{\partial g_\alpha}{\partial v_z}\right\} \\
&= -\frac{q_\alpha}{m_\alpha}\left\{\frac{1}{v_\perp}\mathbf{E}_1\cdot\mathbf{v}_\perp\left(1 - \frac{k_z v_z}{\omega}\right)\frac{\partial g_\alpha}{\partial v_\perp} + \mathbf{E}_1\cdot\mathbf{v}_\perp\frac{k_z}{\omega}\frac{\partial g_\alpha}{\partial v_z}\right. \\
&\quad \left. + E_{1z}\left[\left(1 - \frac{\mathbf{k}_\perp\cdot\mathbf{v}_\perp}{\omega}\right)\frac{\partial g_\alpha}{\partial v_z} + \frac{1}{\omega}\mathbf{k}_\perp\cdot\mathbf{v}_\perp\frac{v_z}{v_\perp}\frac{\partial g_\alpha}{\partial v_\perp}\right]\right\}
\end{aligned} \tag{6.64}$$

in more detail. In this transformation, the Maxwell equation $\mathbf{B}_1 = \frac{1}{\omega}\mathbf{k}\times\mathbf{E}_1$ played a crucial role. If we substitute (6.64) into the integrand on the right-hand side of (6.63), we must take into account that for the components (6.57) of $\mathbf{v}_\perp$ the shift $\varphi \to \varphi - \Omega_\alpha\tau$ must be performed. In addition, we can use the relation

$$e^{i\kappa[\sin\varphi - \sin(\varphi-\chi)]} = \sum_{m,n=-\infty}^{+\infty} J_m(\kappa)J_n(\kappa)e^{i(m-n)\varphi + in\chi} \tag{6.65}$$

to simplify integrations. From (6.63)–(6.65), a closed-form expression for δf_α can be obtained. Using the definition

$$\mathbf{j}_1 = \sum_\alpha q_\alpha \int \mathbf{v}\,\delta f_\alpha\, d^3v \tag{6.66}$$

the linear current then follows. If we substitute this again into the electrodynamic result

$$\mathbf{k}\times(\mathbf{k}\times\mathbf{E}_1) + \mu_0\varepsilon_0\omega^2\mathbf{E}_1 + i\omega\mu_0\mathbf{j}_1 = 0 \tag{6.67}$$

we obtain the homogeneous system of equations

$$\boxed{\mathcal{D}\cdot\mathbf{E}_1 = 0}\,. \tag{6.68}$$

General Dispersion Relation

Nontrivial solutions for $\mathbf{E}_1$ exist provided that

$$\boxed{\det\ \mathcal{D} = 0} \tag{6.69}$$

is satisfied. The relation (6.69) represents the dispersion relation. The calculation of the individual elements of $\mathcal{D}$ is admittedly tedious, but can be carried out without any fundamental difficulties. The result is given, for example, in the handbook article by R.C. Davidson [67]:

$$\begin{aligned} D_{xx}(\mathbf{k},\omega) =& 1-\frac{c^2k_z^2}{\omega^2}+\sum_\alpha\frac{\omega_{p\alpha}^2}{\omega}\sum_{n=-\infty}^{\infty}\int d^3v\, v_\perp\frac{(n^2/b_\alpha^2)J_n^2(b_\alpha)}{(\omega-n\Omega_\alpha-k_zv_z)}\\ &\times\left[\frac{\partial F_\alpha}{\partial v_\perp}-\frac{k_zv_z}{\omega}\left(\frac{\partial F_\alpha}{\partial v_\perp}-\frac{v_\perp}{v_z}\frac{\partial F_\alpha}{\partial v_z}\right)\right], \end{aligned} \tag{6.70}$$

$$\begin{aligned} D_{xy}(\mathbf{k},\omega) =& i\sum_\alpha\frac{\omega_{p\alpha}^2}{\omega}\sum_{n=-\infty}^{\infty}\int d^3v\, v_\perp\frac{(n/b_\alpha)J_n(b_\alpha)J_n'(b_\alpha)}{(\omega-n\Omega_\alpha-k_zv_z)}\\ &\times\left[\frac{\partial F_\alpha}{\partial v_\perp}-\frac{k_zv_z}{\omega}\left(\frac{\partial F_\alpha}{\partial v_\perp}-\frac{v_\perp}{v_z}\frac{\partial F_\alpha}{\partial v_z}\right)\right]\\ =& -D_{yx}(\mathbf{k},\omega)\,, \end{aligned} \tag{6.71}$$

$$\begin{aligned} D_{xz}(\mathbf{k},\omega) =& \frac{c^2k_zk_\perp}{\omega^2}+\sum_\alpha\frac{\omega_{p\alpha}^2}{\omega}\sum_{n=-\infty}^{\infty}\int d^3v\, v_\perp\frac{(n/b_\alpha)J_n^2(b_\alpha)}{(\omega-n\Omega_\alpha-k_zv_z)}\\ &\times\left[\frac{\partial F_\alpha}{\partial v_z}+\frac{n\Omega_\alpha}{\omega}\left(\frac{v_z}{v_\perp}\frac{\partial F_\alpha}{\partial v_\perp}-\frac{\partial F_\alpha}{\partial v_z}\right)\right], \end{aligned} \tag{6.72}$$

$$\begin{aligned} D_{yy}(\mathbf{k},\omega) =& 1-\frac{c^2(k_z^2+k_\perp^2)}{\omega^2}+\sum_\alpha\frac{\omega_{p\alpha}^2}{\omega}\sum_{n=-\infty}^{\infty}\int d^3v\, v_\perp\frac{\left[J_n'(b_\alpha)\right]^2}{(\omega-n\Omega_\alpha-k_zv_z)}\\ &\times\left[\frac{\partial F_\alpha}{\partial v_\perp}-\frac{k_zv_z}{\omega}\left(\frac{\partial F_\alpha}{\partial v_\perp}-\frac{v_\perp}{v_z}\frac{\partial F_\alpha}{\partial v_z}\right)\right], \end{aligned} \tag{6.73}$$

$$\begin{aligned} D_{yz}(\mathbf{k},\omega) =& -i\sum_\alpha\frac{\omega_{p\alpha}^2}{\omega}\sum_{n=-\infty}^{\infty}\int d^3v\, v_\perp\frac{J_n(b_\alpha)J_n'(b_\alpha)}{(\omega-n\Omega_\alpha-k_zv_z)}\\ &\times\left[\frac{\partial F_\alpha}{\partial v_z}+\frac{n\Omega_\alpha}{\omega}\left(\frac{v_z}{v_\perp}\frac{\partial F_\alpha}{\partial v_\perp}-\frac{\partial F_\alpha}{\partial v_z}\right)\right], \end{aligned} \tag{6.74}$$

$$\begin{aligned} D_{zx}(\mathbf{k},\omega) =& \frac{c^2k_zk_\perp}{\omega^2}+\sum_\alpha\frac{\omega_{p\alpha}^2}{\omega}\sum_{n=-\infty}^{\infty}\int d^3v\, v_z\frac{(n/b_\alpha)J_n^2(b_\alpha)}{(\omega-n\Omega_\alpha-k_zv_z)}\\ &\times\left[\frac{\partial F_\alpha}{\partial v_\perp}-\frac{k_zv_z}{\omega}\left(\frac{\partial F_\alpha}{\partial v_\perp}-\frac{v_\perp}{v_z}\frac{\partial F_\alpha}{\partial v_z}\right)\right], \end{aligned} \tag{6.75}$$

$$D_{zy}(\mathbf{k},\omega) = i\sum_\alpha\frac{\omega_{p\alpha}^2}{\omega}\sum_{n=-\infty}^{\infty}\int d^3v\, v_z\frac{J_n(b_\alpha)J_n'(b_\alpha)}{(\omega-n\Omega_\alpha-k_zv_z)}$$

$$\times \left[\frac{\partial F_\alpha}{\partial v_\perp} - \frac{k_z v_z}{\omega} \left(\frac{\partial F_\alpha}{\partial v_\perp} - \frac{v_\perp}{v_z} \frac{\partial F_\alpha}{\partial v_z} \right) \right], \tag{6.76}$$

$$D_{zz}(\mathbf{k}, \omega) = 1 - \frac{c^2 k_\perp^2}{\omega^2} + \sum_\alpha \frac{\omega_{p\alpha}^2}{\omega} \sum_{n=-\infty}^{\infty} \int d^3v \, v_z \frac{J_n^2(b_\alpha)}{(\omega - n\Omega_\alpha - k_z v_z)}$$
$$\times \left[\frac{\partial F_\alpha}{\partial v_z} + \frac{n\Omega_\alpha}{\omega} \left(\frac{v_z}{v_\perp} \frac{\partial F_\alpha}{\partial v_\perp} - \frac{\partial F_\alpha}{\partial v_z} \right) \right]. \tag{6.77}$$

Here, we have factored out the density dependence of the distribution functions via

$$g_\alpha = n_\alpha F_\alpha \tag{6.78}$$

defined

$$b_\alpha = k_\perp v_\perp / \Omega_\alpha \tag{6.79}$$

and set $k_x = k_\perp$. The prime ($'$) on the Bessel functions J_n denotes differentiation with respect to the argument. The dispersion relation (6.69) contains a wealth of information, which we cannot discuss here in full detail. Some typical examples will be examined in more detail below.

Simplification for Propagation in the Direction of $\mathbf{B}_0$

If the wave propagation occurs predominantly in the direction of $\mathbf{B}_0$, then we can let $k_\perp \to 0$ go to zero and immediately obtain

$$D_{xz} = D_{zx} = D_{yz} = D_{zy} = 0 \, . \tag{6.80}$$

The remaining elements simplify to

$$D_{xx} = D_{yy}$$
$$= 1 - \frac{c^2 k_z^2}{\omega^2} + \frac{1}{4} \sum_\alpha \frac{\omega_{p\alpha}^2}{\omega} \int d^3v \, v_\perp \left[\frac{\partial F_\alpha}{\partial v_\perp} - \frac{k_z v_z}{\omega} \left(\frac{\partial F_\alpha}{\partial v_\perp} - \frac{v_\perp}{v_z} \frac{\partial F_\alpha}{\partial v_z} \right) \right]$$
$$\times \left[\frac{1}{\omega - \Omega_\alpha - k_z v_z} + \frac{1}{\omega + \Omega_\alpha - k_z v_z} \right], \tag{6.81}$$

$$D_{xy} = -D_{yx}$$
$$= i \sum_\alpha \frac{\omega_{p\alpha}^2}{\omega} \int d^3v \, v_\perp \left[\frac{\partial F_\alpha}{\partial v_\perp} - \frac{k_z v_z}{\omega} \left(\frac{\partial F_\alpha}{\partial v_\perp} - \frac{v_\perp}{v_z} \frac{\partial F_\alpha}{\partial v_z} \right) \right]$$
$$\times \left[\frac{1}{\omega - \Omega_\alpha - k_z v_z} - \frac{1}{\omega + \Omega_\alpha - k_z v_z} \right], \tag{6.82}$$

$$D_{zz} = 1 + \sum_\alpha \frac{\omega_{p\alpha}^2}{\omega} \int d^3v \frac{v_z\, \partial F_\alpha / \partial v_z}{\omega - k_z v_z} . \tag{6.83}$$

The last expression has already appeared in the plasma dispersion function (in plasmas without magnetic fields).

Altogether, in addition to the electrostatic branch

$$D_{zz} = 0 \tag{6.84}$$

we also obtain the electromagnetic branch

$$D_{xx} D_{yy} + D_{xy}^2 = 0 , \tag{6.85}$$

which we can also write in the form

$$(D_{xx} + iD_{xy})(D_{xx} - iD_{xy}) \equiv D^+ D^- = 0 \tag{6.86}$$

From this we immediately find the dispersion equation

$$D^\pm(k_z, \omega) = 1 - \frac{c^2 k_z^2}{\omega^2} + \sum_\alpha \frac{\omega_{p\alpha}^2}{\omega} \int d^3v \frac{v_\perp}{2} \left[\frac{\partial F_\alpha}{\partial v_\perp} - \frac{k_z v_z}{\omega} \left(\frac{\partial F_\alpha}{\partial v_\perp} - \frac{v_\perp}{v_z} \frac{\partial F_\alpha}{\partial v_z} \right) \right] \frac{1}{\omega \pm \Omega_\alpha - k_z v_z} = 0 . \tag{6.87}$$

We will return to the discussion of the various solutions to these equations later.

Simplification for Propagation Perpendicular to $\mathbf{B}_0$

Propagation perpendicular to $\mathbf{B}_0$,, i.e., the limiting case $k_z \to 0$,, is treated under the additional assumption that there is no particle flux parallel to $\mathbf{B}_0$ present,

$$\int_{-\infty}^{+\infty} dv_z v_z F_\alpha(v_\perp, v_z) = 0 . \tag{6.88}$$

Then the elements

$$D_{xz} = D_{zx} = D_{yz} = D_{zy} = 0 , \tag{6.89}$$

vanish, and the remaining expressions are

$$D_{xx} = 1 + \sum_\alpha \frac{\omega_{p\alpha}^2}{\omega} \sum_{n=-\infty}^{+\infty} \int d^3v\, v_\perp \frac{n^2 J_n^2(b_\alpha)/b_\alpha^2}{\omega - n\Omega_\alpha} \frac{\partial F_\alpha}{\partial v_\perp}\,, \tag{6.90}$$

$$D_{xy} = -D_{yx} = i \sum_\alpha \frac{\omega_{p\alpha}^2}{\omega} \sum_{n=-\infty}^{+\infty} d^3v\, v_\perp \frac{n J_n(b_\alpha) J_n'(b_\alpha)/b_\alpha}{\omega - n\Omega_\alpha} \frac{\partial F_\alpha}{\partial v_\perp}\,, \tag{6.91}$$

$$D_{yy} = 1 - \frac{c^2 k_\perp^2}{\omega^2} + \sum_\alpha \frac{\omega_{p\alpha}^2}{\omega} \sum_{n=-\infty}^{+\infty} \int d^3v\, v_\perp \frac{[J_n'(b_\alpha)]^2}{\omega - n\Omega_\alpha} \frac{\partial F_\alpha}{\partial v_\perp}\,, \tag{6.92}$$

$$D_{zz} = 1 - \frac{c^2 k_\perp^2}{\omega^2} - \sum_\alpha \frac{\omega_{p\alpha}^2}{\omega^2} + \sum_\alpha \frac{\omega_{p\alpha}^2}{\omega^2} \sum_{n=-\infty}^{+\infty} \int d^3v \frac{n\Omega_\alpha J_n^2(b_\alpha)}{\omega - n\Omega_\alpha} \frac{v_z^2}{v_\perp} \frac{\partial F_\alpha}{\partial v_\perp}\,. \tag{6.93}$$

Again, the dispersion relation splits into two branches, namely

$$D_{zz} = 0 \tag{6.94}$$

as well as

$$D_{xx} D_{yy} + D_{xy}^2 = 0\,, \tag{6.95}$$

where the first contains the ordinary mode with transverse electromagnetic polarization, and the second the extraordinary mode with mixed polarization.

Two further limiting cases of the general dispersion function are of interest.

In the case of purely *electrostatic* perturbations, we obtain the dispersion relation from the equation

$$\begin{aligned} D_{elst}(\mathbf{k}, \omega) =& 1 + \sum_\alpha \frac{\omega_{p\alpha}^2}{k^2} \sum_{n=-\infty}^{+\infty} \int d^3v \frac{J_n^2(b_\alpha)}{\omega - n\Omega_\alpha - k_z v_z} \\ & \times \left[k_z \frac{\partial F_\alpha}{\partial v_z} + \frac{n\Omega_\alpha}{v_\perp} \frac{\partial F_\alpha}{\partial v_\perp} \right] = 0\,. \end{aligned} \tag{6.96}$$

The electrostatic limiting case follows from the general dispersion relation in the limit $\omega^2 \ll k^2 c^2$ or $\omega_{p\alpha}^2 \ll k^2 c^2$.

Simplification for the Zero Magnetic Field Case

Finally, we discuss the case where *no external field* $\mathbf{B}_0$ is present. If the distribution function F_α in v_x and v_y is isotropic, one obtains

$$D_{xx} = D_{yy} = 1 - \frac{c^2 k_z^2}{\omega^2} + \sum_\alpha \frac{\omega_{p\alpha}^2}{\omega} \int d^3v \frac{v_\perp^2/2}{\omega - k_z v_z} \left[\left(1 - \frac{k_z v_z}{\omega}\right) \frac{\partial F_\alpha}{\partial v_\perp^2} + \frac{k_z}{\omega} \frac{\partial F_\alpha}{\partial v_z} \right], \tag{6.97}$$

whereas otherwise

$$D_{xx} = 1 - \frac{c^2 k_z^2}{\omega^2} + \sum_\alpha \frac{\omega_{p\alpha}^2}{\omega} \int d^3v \frac{v_x}{\omega - k_z v_z} \left[\left(1 - \frac{k_z v_z}{\omega}\right) \frac{\partial F_\alpha}{\partial v_x} + \frac{k_z v_x}{\omega} \frac{\partial F_\alpha}{\partial v_z} \right] \tag{6.98}$$

and

$$D_{yy} = 1 - \frac{c^2 k_z^2}{\omega^2} + \sum_\alpha \frac{\omega_{p\alpha}^2}{\omega} \int d^3v \frac{v_y}{\omega - k_z v_z} \left[\left(1 - \frac{k_z v_z}{\omega}\right) \frac{\partial F_\alpha}{\partial v_y} + \frac{k_z v_y}{\omega} \frac{\partial F_\alpha}{\partial v_z} \right] \tag{6.99}$$

apply. The off-diagonal elements vanish, e.g.

$$D_{xy} = D_{yx} = 0 \,, \quad \text{etc. ;} \tag{6.100}$$

D_{zz} has the same form (6.83) as for propagation parallel to $\mathbf{B}_0$.

Before we later turn to the unstable solutions in homogeneous Vlasov plasmas, we will summarize the characteristic oscillations in stable systems.

Solutions for $\mathbf{B}_0 = 0$

Example 6.4 (Electrostatic Oscillations)

In the field-free case $B_0 = 0$, the *electrostatic* relation $D_{zz} = 0$ yields the high-frequency Langmuir oscillations

$$\boxed{\omega_r^2 \approx \omega_{pe}^2 (1 + 3k^2 \lambda_{De}^2)} \tag{6.101}$$

for the real part ω_r of ω for $\omega_r/k \gg (k_B T_e/m_e)^{1/2}$. The imaginary part ω_i is determined by Landau damping,

$$\omega_i \approx -\left(\frac{\pi}{8}\right)^{1/2} \frac{\omega_{pe}}{k^3\lambda_{De}^3} \exp\left(-\frac{1}{2k^2\lambda_{De}^2} - \frac{3}{2}\right). \tag{6.102}$$

For $T_e \gg T_i$ and

$$(k_B T_i/m_i)^{1/2} \ll \omega_r/k \ll (k_B T_e/m_e)^{1/2} \tag{6.103}$$

ion-acoustic modes arise with

$$\boxed{\omega_r^2 \approx \frac{k^2 c_s^2}{1 + k^2\lambda_{De}^2}} \tag{6.104}$$

and electron Landau damping

$$\omega_i \approx -\left(\frac{\pi}{8}\right)^{1/2} \left(\frac{m_e}{m_i}\right)^{1/2} (1 + k^2\lambda_{De}^2)^{-3/2} |\omega_r| \tag{6.105}$$

appear. ■

Example 6.5 (Electromagnetic Waves)
The *electromagnetic* modes follow, for example, from (6.98), i.e.

$$\begin{aligned} D_{xx}(k_z, \omega) =& 1 - \frac{c^2 k_z^2}{\omega^2} - \sum_\alpha \frac{\omega_{p\alpha}^2}{\omega^2} \\ &+ \sum_\alpha \frac{\omega_{p\alpha}^2}{\omega^2} \int d^3v \frac{v_x^2}{\omega - k_z v_z} k_z \frac{\partial F_\alpha}{\partial v_z} = 0 \,. \end{aligned} \tag{6.106}$$

If we introduce $\omega_p^2 = \sum_\alpha \omega_{p\alpha}^2$ and

$$F = \sum_\alpha \frac{\omega_{p\alpha}^2}{\omega_p^2} \int dv_x \, dv_y \frac{v_x^2}{W^2} F_\alpha(\mathbf{v}) \tag{6.107}$$

where W is the mean transverse velocity, then

$$\omega^2 = \omega_p^2 + k_z^2 c^2 + \omega_p^2 \, \widehat{G}(\omega/k_z) \,. \tag{6.108}$$

Here,

$$\widehat{G}(u) = W^2 \int_{-\infty}^{+\infty} dv \frac{F'(v)}{v - u} = W^2 \int_{-\infty}^{+\infty} dv \frac{F(v)}{(v - u)^2} \,. \tag{6.109}$$

Because $\omega/k_z \geq c$, we can neglect the last term on the right-hand side of (6.108) and obtain

$$\boxed{\omega^2 \approx \omega_{pe}^2 + c^2k_z^2} \,. \tag{6.110}$$

■

Solutions for $\mathbf{B}_0 \neq 0$

Example 6.6 (Ordinary Mode)
For $\mathbf{B}_0 = B_0\hat{z} \neq 0$ and $k_z = 0$ (propagation perpendicular to an external magnetic field), it follows from (6.94) that an electromagnetic wave propagates whose $\mathbf{E}$ vector is parallel to $\mathbf{B}_0$, while the propagation vector $\mathbf{k} = k_x\hat{x}$ is perpendicular to $\mathbf{B}_0$. The solution of (6.94) is generally quite complicated. However, one can look for cases where only a single term in the sum over n contributes. For wavelengths much larger than the gyroradii, we obtain—except for $n = 0$—contributions only when ω is close to $n\Omega_\alpha$. A similar statement holds for $k^2c^2 \gg \omega_{pe}^2$. For $n = 0$ one obtains

$$\omega^2 \approx c^2k^2 + \omega_{pe}^2 \tag{6.111}$$

and for $n \neq 0$

$$\omega \approx n\Omega_\alpha \,. \tag{6.112}$$

■

Example 6.7 (Bernstein Mode)
In contrast to the ordinary mode (6.111), in the dispersion relation (6.95) the electron dynamics are strongly influenced by the magnetic field. If $\omega_{pe}^2 \ll k^2c^2$ holds, we can neglect D_{xy}, and the two branches $D_{xx} \approx 0$ and $D_{yy} \approx 0$ represent approximate solutions. It is clear that the first of the two, due to $\mathbf{k} = k_x\hat{x}$, describes almost electrostatic oscillations and can therefore also be approximated by (6.96). Both expressions lead to

$$k^2 + \sum_\alpha \sum_{n=1}^{\infty} \frac{\omega_{p\alpha}^2 n^2 \Omega_\alpha^2}{\omega^2 - n^2\Omega_\alpha^2} \int J_n^2(b_\alpha) \frac{\partial F_\alpha}{\partial v_\perp^2} d^3v = 0 \,. \tag{6.113}$$

This equation is known as the dispersion relation for Bernstein modes. Bernstein waves propagate in frequency ranges that lie between harmonics of the cyclotron frequencies Ω_α. To evaluate the integral in (6.113) for Maxwellian distributions, the following formula is used:

$$\int_0^\infty \exp(-a^2x^2)\, x\, J_n(px)J_n(qx)dx = \frac{1}{2a^2} \exp\left(-\frac{p^2+q^2}{4a^2}\right) I_n\left(\frac{pq}{2a^2}\right) ; \tag{6.114}$$

I_n is the modified Bessel function of order n,

$$I_n(z) = e^{-i\pi n/2} J_n\left(e^{i\pi/2} z\right) . \tag{6.115}$$

If one then expands for small k values, one finds for the first solution

$$\omega_H \approx \sqrt{\omega_{pe}^2 + \Omega_e^2} , \tag{6.116}$$

i.e., the upper electron hybrid frequency, while all others are located near the harmonics of the cyclotron frequency (for electrons and ions, respectively), e.g.

$$\omega_n \approx n\Omega_e \qquad \text{for } n \geq 2 . \tag{6.117}$$

If $\omega_H > 2|\Omega_e|$ holds (at high plasma densities), the lowest frequency appears somewhat below twice the cyclotron frequency, and all others are located near (above or below, depending on the plasma density) the higher harmonics of the cyclotron frequency. ■

Example 6.8 (Cyclotron Waves)
The so-called electromagnetic cyclotron waves or extraordinary modes follow from (6.92) and $D_{yy} \approx 0$, i.e.

$$k^2c^2 - \omega^2 - 2\pi\omega \sum_\alpha \sum_n \omega_{p\alpha}^2 \int dv_z dv_\perp^2 \frac{v_\perp^2 \left[J_n'(b_\alpha)\right]^2}{\omega - n\Omega_\alpha} \frac{\partial F_\alpha}{\partial v_\perp^2} = 0 . \tag{6.118}$$

The transverse modes ($\mathbf{k} = k_x\hat{x}$, $\mathbf{E} \approx E_y\hat{y}$) for $n = \pm 1$ satisfy

$$k^2c^2 - \omega^2 + \sum_\alpha \frac{\omega^2\omega_{p\alpha}^2}{\omega^2 - \Omega_\alpha^2} \approx 0 , \tag{6.119}$$

i.e., in the low-frequency limit ($\omega < \Omega_i$)

$$\omega^2 = \frac{k^2v_A^2}{1 + v_A^2/c^2} . \tag{6.120}$$

These are then magnetosonic waves with the Alfvén velocity

$$v_A = \frac{B_0}{\sqrt{\mu_0 n_{e0} m_i}} . \tag{6.121}$$

At higher frequencies ($\Omega_i < \omega < |\Omega_e| < \omega_{pe}$) we have

$$\omega \approx (\Omega_i|\Omega_e|)^{1/2} . \tag{6.122}$$

■

For wave propagation parallel to the external magnetic field $\mathbf{B}_0$, i.e., for $k_\perp = 0$, (6.84) applies. It describes Langmuir waves or ion-acoustic modes, which can propagate along the magnetic field, unaffected by B_0,.

Example 6.9 (Whistler Mode)
The kinetic dispersion equations $D^{\pm} = 0$, identical to

$$\omega^2 = k^2c^2 + 2\pi\omega \sum_\alpha \omega_{p\alpha}^2 \int dv_\perp dv_z v_\perp^3 (k_z v_z - \omega \pm \Omega_\alpha)^{-1} \times \left[\frac{\partial F_\alpha}{\partial v_\perp^2} \left(1 - \frac{k_z v_z}{\omega}\right) + \frac{k_z v_z}{\omega} \frac{\partial F_\alpha}{\partial v_z^2} \right], \tag{6.123}$$

reduce in the cold plasma approximation $[F_\alpha \sim \delta(v_z)]$ to the hydrodynamic form

$$\frac{c^2 k_z^2}{\omega^2} = \varepsilon_\pm \, , \tag{6.124}$$

with

$$\varepsilon_+ = \varepsilon_1 + \varepsilon_2 \, , \tag{6.125}$$

$$\varepsilon_- = \varepsilon_1 - \varepsilon_2 \, , \tag{6.126}$$

$$\varepsilon_1 = 1 + \frac{\omega_{pe}^2}{\Omega_e^2 - \omega^2} + \frac{\omega_{pi}^2}{\Omega_i^2 - \omega^2} \, , \tag{6.127}$$

$$\varepsilon_2 = -\frac{\Omega_e}{\omega} \frac{\omega_{pe}^2}{\Omega_e^2 - \omega^2} - \frac{\Omega_i}{\omega} \frac{\omega_{pi}^2}{\Omega_i^2 - \omega^2} \, . \tag{6.128}$$

For high frequencies, i.e., $\omega \gg \Omega_i$, we find the two solutions

$$k = \frac{\omega}{c} \left(1 - \frac{\omega_{pe}^2}{\omega(\omega \mp |\Omega_e|)}\right)^{1/2} \, . \tag{6.129}$$

These correspond to circularly polarized electromagnetic waves (more precisely: the upper sign corresponds to right-circularly polarized waves, in which the sense of rotation coincides with the cyclotron motion of the electrons; the lower sign corresponds to left-circularly polarized waves). For the right-circularly polarized wave (upper sign), we observe the cyclotron resonance $\omega \approx |\Omega_e|$, at which kinetic effects become significant.

For frequencies in the range

$$\Omega_i < \omega \ll |\Omega_e| \tag{6.130}$$

only the propagation of a right-circularly polarized wave is possible, with

$$k \approx \frac{\omega_{pe}}{c}\left(\frac{\omega}{|\Omega_e|}\right)^{1/2} . \tag{6.131}$$

The waves with this dispersion are called whistlers. In the case of broadband excitation, high-frequency components travel faster than low-frequency ones, so that disturbances in the ionosphere can be heard as a characteristic whistling tone. ■

Example 6.10 (Alfvén Waves)

Eq. (6.131) is only approximately correct, since for low-frequency phenomena we cannot neglect ion dynamics. More precisely, we have

$$k^2 = \frac{\omega^2}{c^2}\left[1 - \frac{\omega_{pe}^2 + \omega_{pi}^2}{(\omega \pm \Omega_i)(\omega \mp |\Omega_e|)}\right] , \tag{6.132}$$

and from this we see that for $\omega \ll \Omega_i$ the new branch

$$k^2 \approx \frac{\omega^2}{c^2}\left[1 + \frac{n_{e0}(m_e + m_i)}{\varepsilon_0 B_0^2}\right] \tag{6.133}$$

appears. With the Alfvén velocity (6.120) we have

$$\frac{\omega}{k} = \frac{v_A}{(1 + v_A^2/c^2)^{1/2}} . \tag{6.134}$$

This velocity does not depend on the sense of rotation. Differences only appear at order ω/Ω_i. ■

Finally, a word on damping. For electrostatic waves propagating along $\mathbf{B}_0$, we can apply the previous results for electron and ion Landau damping. Waves with $k_z = 0$ (exactly) are not damped, since the waves do not propagate synchronously with the particle motion along B_0. For $k_z \neq 0$, in addition to Landau damping, a new phenomenon can occur: cyclotron damping. Assuming $\omega_i \ll \omega_r$, we can carry out calculations similar to those for Landau damping in the electrostatic case. The only difference is that ω is replaced by $\omega \mp \Omega_\alpha$, and the damping depends on the number of particles with a velocity

$$v_z = \frac{\omega_r \mp \Omega_\alpha}{k_z} \tag{6.135}$$

.

Example 6.11 (Damping of Alfvén Waves)
As examples, we first choose the Alfvén waves (6.133). Assuming an isotropic Maxwellian distribution, we then obtain

$$\omega_i \approx -\frac{\omega_{pi}^2}{|k_z|}\frac{1}{1+c^2/v_A^2}\left(\frac{\pi m_i}{8k_BT_i}\right)^{1/2}\exp\left(-\frac{B_0^2}{2\mu_0 n_{eo}k_BT_i}\frac{\Omega_i^2}{\omega_r^2}\right). \tag{6.136}$$

An analogous calculation yields, for whistler waves in the frequency range $\Omega_i \ll \omega \ll |\Omega_e|$

$$\omega_i \approx -\frac{\omega_{pe}^2}{|k_z|}\left(\frac{m_e}{2k_BT_e}\right)^{1/2}\frac{1}{1+k^2c^2/\omega_r^2}\exp\left(-\frac{\Omega_e^2}{k_z^2}\frac{m_e}{2k_BT_e}\right), \tag{6.137}$$

with similar limitations as in previous approximate calculations of damping rates. ■

6.3 Instabilities in Homogeneous Vlasov Systems

We have already gained an initial insight into the possible unstable behavior of a plasma in the previous sections. There, we discussed the Penrose criterion for electrostatic modes at $B_0 = 0$. (Based on the preceding considerations, we know that these statements also apply for $B_0 \neq 0$ for $k_\perp = 0$.) We can easily extend the calculations to the electromagnetic case (6.106) at $B_0 = 0$. For example, it is easy to see that completely isotropic distribution functions F_α do not allow for any instability.
In this section, we will turn to the concrete analysis of certain unstable configurations and introduce some of the best-known plasma instabilities.

We will organize the following discussion according to the sources of free energy.

Plasma Without Magnetic Field with Flow

We begin with *magnetically field-free* homogeneous plasmas and initially use the *electrostatic* approximation. If we assume as the zeroth-order distribution function a superposition of two Maxwellian distributions, we can quickly arrive at concrete results. The superposition can exist within a single component ("bump on tail") or represent the sum of the distribution functions of different components α. For example, if we choose the normalized one-dimensional distribution function

$$\boxed{F(v) = \frac{1}{\sqrt{\pi}}\left[\frac{1-\varepsilon}{v_T}e^{-v^2/v_T^2} + \frac{\varepsilon}{v_B}e^{-(v-v_D)^2/v_B^2}\right]} \tag{6.138}$$

with coefficients ε, v_B, v_T and v_D. still to be determined. In the normalized quantities

$$\xi_1 = \frac{\omega}{kv_T} \tag{6.139}$$

and

$$\xi_2 = \left(\frac{\omega}{k} - v_D\right)/v_B \tag{6.140}$$

the dispersion relation reads

$$1 - \frac{\omega_p^2}{k^2}\left[\frac{1-\varepsilon}{v_T^2}Z'(\xi_1) + \frac{\varepsilon}{v_B^2}Z'(\xi_2)\right] = 0\,. \tag{6.141}$$

Here, $\omega_p^2 = \omega_{p\alpha}^2 + \omega_{p\beta}^2$ has been introduced, with $\omega_{p\alpha}^2 = n_\alpha e^2/\varepsilon_0 m_\alpha$ and $n_\alpha = (1-\varepsilon)\, n_{ges}$; $\omega_{p\beta}^2$ defined accordingly.

Two meaningful and analytically tractable limiting cases in which the Z-function can be expanded are

$$|\xi_1| \gg 1 \quad \text{and} \quad |\xi_2| \ll 1\,, \tag{6.142}$$

and

$$|\xi_1| \gg 1 \quad \text{and} \quad |\xi_2| \gg 1\,. \tag{6.143}$$

Let us first consider the first case. Eq. (6.141) yields approximately

$$\begin{aligned} 1 + 2\frac{\omega_p^2}{k^2}\Bigg\{&\frac{1-\varepsilon}{v_T^2}\left[-\frac{1}{2}\xi_1^{-2} - \frac{3}{4}\xi_1^{-4} - \dots\right] \\ &+ \frac{\varepsilon}{v_B^2}\left[1 - 2\xi_2^2 + i\sqrt{\pi}\,\xi_2 + \dots\right]\Bigg\} = 0\,. \end{aligned} \tag{6.144}$$

Now we interpret F as a superposition of the ion distribution function with a shifted Maxwellian for electrons. Since for this application we set

$$\varepsilon = \frac{\omega_{pe}^2}{\omega_p^2} \approx 1\,, \qquad 1 - \varepsilon = \frac{\omega_{pi}^2}{\omega_p^2} \approx \frac{m_e}{m_i} \tag{6.145}$$

and accordingly

$$v_T = \sqrt{2}\,v_{ti}\,, \qquad v_B = \sqrt{2}\,v_{te} \tag{6.146}$$

we have considered only imaginary contributions (for electron Landau damping) in the electron term. Without a drift v_D in the electron component, the present calculation would be identical to the evaluation for ion-acoustic oscillations. With v_D, a so-called *ion-acoustic instability* can occur. To calculate it, we introduce

$$u = \frac{\omega}{k} \tag{6.147}$$

as an abbreviation and obtain from (6.144) to lowest order

$$0 \approx 1 - \frac{\omega_{pi}^2}{u^2 k^2} + \frac{\omega_{pe}^2}{v_{te}^2 k^2} \tag{6.148}$$

with the well-known solution

$$u^2 \approx u_0^2 = \frac{c_s^2}{1 + k^2 \lambda_{De}^2} . \tag{6.149}$$

Corrections to u_0 follow to next order from

$$0 \approx \frac{2\omega_{pi}^2}{u_0^3 k^2} u_1 + i\sqrt{\frac{\pi}{2}} \frac{\omega_{pe}^2}{v_{te}^3 k^2}(u_0 - v_D) \tag{6.150}$$

with the result

$$u_1 = -i\sqrt{\frac{\pi}{8}} \frac{m_i}{m_e} \frac{u_0^3}{v_{te}^3} \frac{u_0 - v_D}{u_0} u_0 . \tag{6.151}$$

It is clear that for $v_D > u_0$ the condition Im $\omega > 0$ for instability is fulfilled. However, the relation $|\xi_2| \approx |c_s - v_D|/v_{te} \ll 1$ must be correct in order of magnitude.

Another interpretation of F is possible for beam instability ("bump on tail instability"). We consider an electron plasma with a smeared-out ion background and describe by ε the fraction of electrons that, with velocity v_D, represents a beam of width v_B. To zeroth order in ε, it then follows from (6.144)

$$1 \approx \frac{\omega_{pe}^2}{k^2}\left(\frac{1}{u^2} + 3\frac{v_{te}^2}{u^4}\right) \tag{6.152}$$

with the approximate solution $\omega \approx \omega_{pe}$ or

$$u^2 \approx u_0^2 = \frac{\omega_{pe}^2}{k^2}(1 + 3k^2\lambda_{De}^2) . \tag{6.153}$$

For $k\lambda_{De} \ll 1$ the condition $\xi_1 \gg 1$ is well satisfied.

To first order in ε we have

$$0 \approx -\varepsilon\left[-\frac{1}{2u_0^2} - \frac{3v_{te}^2}{2u_0^4}\right] + \frac{u_1}{u_0^3} + 6u_1\frac{v_{te}^2}{u_0^5} + i\sqrt{\pi}\,\varepsilon\frac{u_0 - v_D}{v_B^3}\,, \tag{6.154}$$

from which we can infer

$$\mathrm{Im}\ \omega = -\sqrt{\pi}\ k\varepsilon\left(\frac{u_0}{v_B}\right)^3\frac{u_0 - v_D}{1 + 6\frac{v_{te}^2}{u_0^2}}\,. \tag{6.155}$$

For $v_D > u_0$ instability occurs. This formula is only valid for $|\xi_2| \ll 1$, i.e.

$$\left|\frac{\frac{\omega_{pe}}{k} - v_D}{v_B}\right| \ll 1\,. \tag{6.156}$$

In the last example of this class of instabilities, we focus on section (6.3.6). We discuss the *Buneman instability* ("electron-ion two-stream instability") by making the same identifications as in (6.145) and (6.146), but now in the region

$$v_{ti} \ll u \ll v_D\,, \tag{6.157}$$

$$v_{te} \ll v_D \tag{6.158}$$

to evaluate.

The dispersion relation (6.141) is then approximately

$$0 \approx 1 + \frac{2}{k^2}\left\{\frac{\omega_{pi}^2}{v_{ti}^2}\left[-\frac{1}{2}\left(\frac{v_{ti}}{u}\right)^2 - \frac{3}{2}\left(\frac{v_{ti}}{u}\right)^4 + \ldots\right]\right.$$
$$\left. + \frac{\omega_{pe}^2}{v_{te}^2}\left[-\frac{1}{2}\,\frac{v_{te}^2}{(u - v_D)^2} - \frac{3}{2}\,\frac{v_{te}^4}{(u - v_D)^4} + \ldots\right]\right\}. \tag{6.159}$$

From the lowest-order contributions,

$$k^2 \approx \frac{\omega_{pi}^2}{u^2} + \frac{\omega_{pe}^2}{(u - v_D)^2}\,, \tag{6.160}$$

we determine the stability behavior. Without explicitly solving the quartic equation, we assume the k^2 value

$$k^2 = \frac{\omega_{pe}^2}{v_D^2} \tag{6.161}$$

in advance. As a more detailed analysis shows, this approximately corresponds to the value at the maximum growth rate. From (6.160) we then obtain

$$\frac{m_e}{m_i} \approx \frac{u^2}{v_D^2}\left[1 - \left(1 - \frac{u}{v_D}\right)^{-2}\right] . \tag{6.162}$$

If we define

$$x := \frac{-u}{v_D} \ll 1 , \tag{6.163}$$

then from (6.162) we obtain

$$x^3 \approx \frac{m_e}{2m_i} \tag{6.164}$$

or

$$x \approx \left(-\frac{1}{2} - i\frac{1}{2}\sqrt{3}\right)\left(\frac{m_e}{2m_i}\right)^{1/3} . \tag{6.165}$$

Thus, the growth rate, just like the real frequency, is proportional to $(m_e/m_i)^{1/3}$.

Plasma without Magnetic Field with Anisotropy

All instabilities considered so far had their origin (i.e., the source of free energy) in the plasma flow. It is to be expected that anisotropies can also cause plasma instabilities. Here, we will not discuss how anisotropies (e.g., in temperature) are maintained in the plasma, but will instead turn directly to some examples.

An anisotropic distribution function (of the component α) can, for example, be chosen in the form

$$\boxed{\begin{aligned} F_\alpha =& \pi^{-3/2}(w_{\alpha\perp}^2 w_{\alpha\|})^{-1} \exp\left[-\frac{v_z^2}{w_{\alpha\|}^2} - \frac{v_y^2}{w_{\alpha\perp}^2}\right] \\ & \times \frac{1}{2}\left\{\exp\left[-\frac{(v_x + v_\alpha)^2}{w_{\alpha\perp}^2}\right] + \exp\left[-\frac{(v_x - v_\alpha)^2}{w_{\alpha\perp}^2}\right]\right\} \end{aligned}} \tag{6.166}$$

Here, we have set $w^2_{\alpha\parallel,\perp} = 2(k_B T_{\alpha\parallel,\perp}/m_\alpha)$. The dispersion relation for electromagnetic perturbations is then given by ($B_0 = 0, k = k_\parallel$)

$$-\omega^2 + k^2c^2 + \omega_p^2 - \sum_\alpha \omega_{p\alpha}^2 \frac{w^2_{\alpha\perp} + 2v^2_\alpha}{w^2_{\alpha\parallel}}[1 + \xi_\alpha Z(\xi_\alpha)] = 0 \,. \qquad (6.167)$$

This dispersion equation allows for two types of solutions: stable waves with a large phase velocity ($> c$) and unstable, purely growing perturbations. The latter are obtained when we set $\omega = i\gamma$ and assume the real part of ω to be zero. Since then the argument ξ_α is also purely imaginary and

$$\xi_\alpha Z(\xi_\alpha) < 0 \qquad (6.168)$$

applies, we immediately find the instability range for purely growing perturbations as

$$0 < k^2 < k_0^2 = \sum_\alpha \frac{\omega^2_{p\alpha}}{c^2}\left(\frac{w^2_{\alpha\perp} + 2v^2_\alpha}{w^2_{\alpha\parallel}} - 1\right) . \qquad (6.169)$$

The instability condition is

$$\sum_\alpha \frac{\omega^2_{p\alpha}}{c^2}\left(\frac{w^2_{\alpha\perp} + 2v^2_\alpha}{w^2_{\alpha\parallel}} - 1\right) > 0 \,. \qquad (6.170)$$

If we assume a plasma with equal charge magnitude for ions and electrons, the condition (6.170) can be written as

$$\frac{k_B T_{e\perp} + m_e v_e^2}{k_B T_{e\parallel}} + \frac{m_e}{m_i}\frac{k_B T_{i\perp} + m_i v_i^2}{k_B T_{i\parallel}} > 1 + \frac{m_e}{m_i} \,. \qquad (6.171)$$

A further specification for the *Weibel instability* is achieved by neglecting

$$v_e = v_i = 0 \qquad (6.172)$$

and assuming isotropically distributed ions ($T_{i\perp} = T_{i\parallel}$).

Then the instability region reduces to

$$k^2 < \frac{\omega^2_{pe}}{c^2}\left(\frac{T_{e\perp}}{T_{e\parallel}} - 1\right) \qquad (6.173)$$

with the condition $T_{e\perp} > T_{e\parallel}$. The growth rate is determined from the dispersion relation (6.167). Assuming

$$\varepsilon^2 = |T_{e\perp} - T_{e\parallel}|/T_{e\parallel} \ll 1 \tag{6.174}$$

it follows that

$$\gamma^2 + k^2c^2 - \omega_{pe}^2\varepsilon^2 - \omega_{pe}^2 \frac{T_{e\perp}}{T_{e\parallel}} \xi_e Z(\xi_e) - \omega_{pi}^2\, \xi_i Z(\xi_i) = 0\,. \tag{6.175}$$

Since $\xi_i = i\gamma/(\sqrt{2}\, kv_{ti})$ holds and $|\xi_i| \gg 1$ can be assumed, we obtain for $|\xi_e| \ll 1$ and $\varepsilon^2\omega_{pe}^2 \gg \omega_{pi}^2$

$$k^2c^2 - \omega_{pe}^2\varepsilon^2 + \sqrt{\pi}\, \frac{\gamma}{kw_{e\parallel}} \omega_{pe}^2 \frac{T_{e\perp}}{T_{e\parallel}} \approx 0 \tag{6.176}$$

with the approximate solution

$$\gamma \approx \frac{1}{\sqrt{\pi}} \frac{T_{e\parallel}}{T_{e\perp}} \left(\frac{k_0^2}{k^2} - 1\right) \frac{c^2k^2}{\omega_{pe}^2} kw_{e\parallel}\,. \tag{6.177}$$

The maximum growth rate (for $k^2 = k_0^2/2$) is

$$\gamma_{max} \approx \frac{1}{2} k_0 v_{te} \left(\frac{T_{e\parallel}}{T_{e\perp}}\right)^{1/2} \left|\frac{T_{e\perp}}{T_{e\parallel}} - 1\right|^{1/2}. \tag{6.178}$$

Anisotropic Plasma with External Magnetic Field

After these examples in plasma situations without magnetic fields, we now turn to typical instabilities in magnetized plasmas. We will no longer consider the situations **k** parallel to $B_0\hat{z}$ and $D_{zz}(k, \omega) = 0$, since they do not provide anything new compared to the field-free case ($B_0 = 0$). Let us proceed directly to (6.87). Here, for $k_\perp = 0$

$$D^\pm(k, \omega) = 1 - \frac{c^2k^2}{\omega^2} \tag{6.179}$$

$$-\sum_\alpha \frac{\omega_{p\alpha}^2}{\omega^2} \int d^3v \left\{ \left[(\omega - k_z v_z) F_\alpha - k_z \frac{v_x^2 + v_y^2}{2} \frac{\partial F_\alpha}{\partial v_z} \right] (\omega \pm \Omega_\alpha - kv_z)^{-1} \right\} = 0\,.$$

From this expression it is clear that the detailed dependence of the distribution function F_α on $v_\perp$ is not at all decisive for the instability calculation. We therefore write

$$\boxed{F_\alpha = \pi^{-1/2} w_{\alpha\parallel}^{-1} \exp\left(-v_z^2/w_{\alpha\parallel}^2\right) G_\alpha(v_\perp^2)} \tag{6.180}$$

with the normalized distribution function G_α,

$$\int d^2 v_\perp G_\alpha(v_\perp^2) = 1 \,. \tag{6.181}$$

The effective "perpendicular" temperature is related to $w_{\alpha\perp}$ via

$$k_B T_{\alpha\perp} = \frac{m_\alpha}{2} w_{\alpha\perp}^2 = \frac{m_\alpha}{2} \int d^2 v_\perp v_\perp^2 G_\alpha(v_\perp^2) \tag{6.182}$$

If we insert (6.180) into (6.179), we obtain (note $k = k_z$)

$$\begin{aligned} 0 =& 1 - \frac{k^2c^2}{\omega^2} + \sum_\alpha \frac{\omega_{p\alpha}^2}{\omega^2} \left\{ \frac{\omega}{k w_{\alpha\parallel}} Z(\xi_\alpha^\pm) \right. \\ & \left. - \left(1 - \frac{T_{\alpha\perp}}{T_{\alpha\parallel}}\right) [1 + \xi_\alpha^\pm Z(\xi_\alpha^\pm)] \right\} , \end{aligned} \tag{6.183}$$

where

$$\xi^\pm = (\omega \pm \Omega_\alpha)/k w_{\alpha\parallel} \equiv \omega_\alpha^\pm / k w_{\alpha\parallel} \tag{6.184}$$

was defined.

Depending on the external conditions ($T_{\alpha\parallel} > T_{\alpha\perp}$ or $T_{\alpha\parallel} < T_{\alpha\perp}$) and the considered wavenumber range, there are various instabilities that are described by (6.183). Let us begin with $|\xi_\alpha| \gg 1$ for the so-called firehose instability.

We then approximate (6.183) as

$$0 \approx 1 - \frac{k^2c^2}{\omega^2} - \sum_\alpha \frac{\omega_{p\alpha}^2}{\omega \omega_\alpha^\pm} + \sum_\alpha \frac{\omega_{p\alpha}^2}{\omega^2} \left(1 - \frac{T_{\alpha\perp}}{T_{\alpha\parallel}}\right) \frac{k^2 T_{\alpha\parallel}}{(\omega_\alpha^\pm)^2 \, m_\alpha} \,. \tag{6.185}$$

Assuming that $|\omega| \ll |\Omega_\alpha|$ holds for all α, we can further simplify, especially if we use the relations

$$\sum_\alpha \frac{\omega_{p\alpha}^2}{\Omega_\alpha} = 0 \,, \tag{6.186}$$

$$\sum_\alpha \frac{\omega_{p\alpha}^2}{\Omega_\alpha^2} = \frac{c^2}{v_A^2} \tag{6.187}$$

and

$$\sum_\alpha \omega_{p\alpha}^2 k_B (T_{\alpha\parallel} - T_{\alpha\perp}) / m_\alpha \Omega_\alpha^2 = \frac{1}{2} c^2 (\beta_\parallel - \beta_\perp) \tag{6.188}$$

Here, β is defined as the ratio of kinetic pressure to magnetic pressure,

$$\beta \equiv 8\pi \sum_{\alpha} n_\alpha k_B T_\alpha / B_0^2 \equiv 8\pi p / B_0^2 \,.$$

We then see that (6.185) can be written in the form

$$0 \approx 1 - \frac{c^2k^2}{\omega^2} + \frac{c^2}{v_A^2} + \frac{c^2k^2}{\omega^2}\frac{1}{2}(\beta_\parallel - \beta_\perp) \tag{6.189}$$

A solution with $\omega = i\gamma$, is obtained as

$$\gamma^2 = \frac{1}{2}\frac{k^2v_A^2}{1 + v_A^2/c^2}(\beta_\parallel - \beta_\perp - 2) \,; \tag{6.190}$$

which indicates that for

$$\beta_\parallel > \beta_\perp + 2 \tag{6.191}$$

a purely growing instability is possible. The apparent divergence for large k is not physical; for larger k values, we must take into account higher-order terms in the expansion of the Z function with respect to ξ_i!

While the derivation of the instability just found is based on the same assumptions as the derivation of Alfvén waves (for $\beta_\parallel = \beta_\perp$ we have simply taken the Alfvén frequency as a real ω solution), we can also look for instabilities in the range of the electron cyclotron and ion cyclotron frequencies. In both frequency ranges, instabilities exist in the presence of temperature anisotropies. We will not go into the details here, but refer to the specialized literature.

We conclude our survey of instabilities in Vlasov systems with a discussion of modes that can propagate perpendicular to an external magnetic field. Among the well-known cases (electromagnetic instability of the ordinary mode, electrostatic instabilities due to a deficit in a certain angular range, etc.), we will focus on the so-called "ordinary-mode electromagnetic instability."

The dispersion relation (6.94) is then under consideration, i.e.,

$$1 - \frac{c^2k^2}{\omega^2} - \sum_{\alpha}\frac{\omega_{p\alpha}^2}{\omega^2} + \sum_{\alpha}\frac{\omega_{p\alpha}^2}{\omega^2}\sum_{n=-\infty}^{+\infty}\frac{n\Omega_\alpha}{\omega - n\Omega_\alpha}\int d^3v\, J_n^2(b_\alpha)\frac{v_z^2}{v_\perp}\frac{\partial F_\alpha}{\partial v_\perp} = 0 \tag{6.192}$$

with $b_\alpha = kv_\perp/\Omega_\alpha$. In the present case, we have set $\mathbf{k} = k\hat{x}$; the magnetic field is in the z direction, $\mathbf{B}_0 = B_0\hat{z}$. For F_α we introduce an anisotropic distribution function, e.g.,

$$F_\alpha(v_\perp^2, v_z) = \pi^{-3/2} w_{\alpha\|}^{-1}\, w_{\alpha\perp}^{-2}\, \exp(-v_\perp^2/w_{\alpha\perp}^2)\exp(-v_z^2/w_{\alpha\|}^2)\,, \tag{6.193}$$

and use

$$\int_0^\infty dx x e^{-a^2x^2}\, J_n(px)J_n(qx) = \frac{1}{2a^2}\exp\left[-\frac{p^2+q^2}{4a^2}\right] I_n\left(\frac{pq}{2a^2}\right)\,, \tag{6.194}$$

then (6.192) can be written as

$$\omega^2 = k^2c^2 + \sum_\alpha \omega_{p\alpha}^2 \left\{1 + \sum_{n=-\infty}^{+\infty} \frac{T_{\alpha\|}}{T_{\alpha\perp}}\, \frac{n\Omega_\alpha}{\omega - n\Omega_\alpha} \exp(-\lambda_\alpha) I_n(\lambda_\alpha)\right\}\,. \tag{6.195}$$

The parameter λ_α is given by

$$\lambda_\alpha = \frac{k^2 k_B T_{\alpha\perp}}{m_\alpha \Omega_\alpha^2} = \frac{1}{2}\,\frac{c^2k^2}{\omega_{p\alpha}^2}\beta_\perp\,, \tag{6.196}$$

and I_n is the modified Bessel function of the first kind and order n. As in the case of a plasma without a magnetic field, in addition to the purely real high-frequency solution, there is a purely growing mode with Re $\omega = 0$. If we use the following transformations

$$I_n(x) = I_{-n}(x)\,, \tag{6.197}$$

$$\frac{n^2\Omega_\alpha^2}{\gamma^2 + n^2\Omega_\alpha^2} = 1 - \frac{\gamma^2}{\gamma^2 + n^2\Omega_\alpha^2}\,, \tag{6.198}$$

$$\sum_n I_n(\lambda_\alpha) = e^{\lambda_\alpha}\,, \tag{6.199}$$

where we have set $\omega = i\gamma$ and assume the imaginary part γ to be small compared to Ω_α, then the dispersion relation (6.195) can be written in the form

$$\gamma^2 L(k, \gamma^2) = R(k) \tag{6.200}$$

with

$$L(k, \gamma^2) = 1 + \sum_\alpha \sum_{n\neq 0} \frac{T_{\alpha\|}}{T_{\alpha\perp}}\, \frac{\omega_{p\alpha}^2}{\gamma^2 + n^2\Omega_\alpha^2} \exp(-\lambda_\alpha) I_n(\lambda_\alpha)\,, \tag{6.201}$$

$$R(k) = \sum_\alpha \omega_{p\alpha}^2 \left[\frac{T_{\alpha\|}}{T_{\alpha\perp}} - 1 - \frac{T_{\alpha\|}}{T_{\alpha\perp}} \exp(-\lambda_\alpha) I_0(\lambda_\alpha)\right] - c^2k^2\,. \tag{6.202}$$

Because $I_n(x) > 0$ holds, $L(k, \gamma^2) > 0$, applies, and instability requires

$$R(k) > 0\,. \tag{6.203}$$

This condition, for given plasma parameters, defines the unstable k range. In the unstable region,

$$\gamma^2 \geq R(k)/L(k,0) \ . \tag{6.204}$$

applies. If one chooses $T_i = T_{i\|} = T_{i\perp}$ and lets m_e/m_i approach zero, the expressions for L and R simplify considerably, e.g.

$$R(k) \approx \omega_{pe}^2 \frac{T_{e\|}}{T_{e\perp}} \left[1 - \frac{T_{e\perp}}{T_{e\|}} - G(\lambda_e) \right] , \tag{6.205}$$

where

$$G(\lambda_e) = \frac{2}{\beta_{e\|}} \lambda_e + \exp(-\lambda_e) \, I_0(\lambda_e) \tag{6.206}$$

is. $\beta_{e\|}$ is defined by

$$\beta_{e\|} = 8\pi n_{eo} k_B T_{e\|} / B_0^2 \tag{6.207}$$

.

If one analyzes the determining equation (6.200) for γ^2 in the $T_{e\perp}/T_{e\|}$, $\beta_{e\|}$ parameter space, one finds instability for small $T_{e\perp}/T_{e\|}$ (≤ 1) and large $\beta_{e\|}$ (≥ 2). With this, we conclude our demonstration of plasma instabilities in spatially homogeneous systems.

6.4 Stationary Inhomogeneous Vlasov Systems

At the beginning of the chapter, we already addressed questions of existence and stability in Vlasov systems—but with the essential restriction to spatially homogeneous configurations. Here, we wish to lift this restriction and move on to statements concerning spatially inhomogeneous systems. Naturally, the algebra then becomes even more complicated, and accordingly, we will now limit ourselves to some fundamental and typical statements and examples. Calculations of instability require knowledge of the sources of free energy and an exact description of the initial situation. We will therefore first discuss under which conditions stationary, spatially inhomogeneous initial configurations exist.

Stationary solutions of the Vlasov equation must satisfy

$$\boxed{\left[\mathbf{v}\cdot\nabla+\frac{e_\alpha}{m_\alpha}(\mathbf{E}+\mathbf{v}\times\mathbf{B})\cdot\partial_\mathbf{v}\right]g_\alpha(\mathbf{q},\mathbf{v})=0} \tag{6.208}$$

Physically, it should be noted that, in general, the equilibrium states (sometimes also called meta-equilibria) only need to be quasi-stationary on a timescale shorter than the mean collision time.

In general, we can first state that the (time-dependent) Vlasov equation is solved by $f_\alpha(\mathbf{q},\mathbf{v};t)=f_\alpha(c_1,c_2,\ldots)$ if $c_1,c_2,\ldots$ are constants of particle motion [see (6.209) and (6.210)]. The reasoning is simple. We can, in fact, write the Vlasov equation in abbreviated form as $df_\alpha/dt\ =\ \sum_i(\partial f_\alpha/\partial c_i)(dc_i/dt)\ =\ 0$. If the constants of motion c_i do not depend explicitly on time, $c_i\ =\ c_i(\mathbf{q},\mathbf{v})$, we obtain from $f_\alpha\ =\ g_\alpha(c_1,c_2,\ldots)$ solutions of (6.208). In other words: If we use the formal solution of the Vlasov equation by integrating along the trajectories, we are led to the differential equations

$$\dot{\mathbf{q}}'=\mathbf{v}', \tag{6.209}$$

$$\dot{\mathbf{v}}'=\frac{e_\alpha}{m_\alpha}\left[\mathbf{E}+\mathbf{v}'\times\mathbf{B}\right] \tag{6.210}$$

From these, we determine the orbit $\mathbf{q}'(t')$ and $\mathbf{v}'(t')$, that coincide at time $t=t'$ with $\mathbf{q}$ and $\mathbf{v}$. If $a=a(\mathbf{q}',\mathbf{v}')$ and $b=b(\mathbf{q}',\mathbf{v}')$ are constants of motion of (6.209) and (6.210), then

$$g_\alpha=g_\alpha[a(\mathbf{q}',\mathbf{v}'),b(\mathbf{q}',\mathbf{v}')] \tag{6.211}$$

fulfills the stationary Vlasov equation (6.208). From the Hamiltonian function corresponding to (6.209) and (6.210)

$$H=\frac{1}{2m_\alpha}(\mathbf{p}'-e_\alpha\mathbf{A})^2+q_\alpha\phi \tag{6.212}$$

and the corresponding Hamiltonian theory, we know that in the case of time-independent fields, H itself is a conserved quantity. If one of the spatial components q_k' is a cyclic coordinate, then the corresponding generalized momentum component p_k' is also a conserved quantity.

The description becomes quite trivial when no external fields are present ($E_0 = B_0 = 0$), since in this case g_α can be an arbitrary function of the three velocity components v_x', v_y' and v_z'.

The situation is different, for example, in the case $E_0 = 0$ and $\mathbf{B}_0(\mathbf{q})\hat{z}$. It can be easily shown that

$$W_\alpha = \frac{1}{2}m_\alpha\left(v_x'^2 + v_y'^2 + v_z'^2\right), \tag{6.213}$$

$$p'_{\alpha\|} = m_\alpha v'_z , \tag{6.214}$$

$$L_z = (\mathbf{q}' \times \mathbf{p}') \cdot \hat{z} , \tag{6.215}$$

$$\xi_y = v'_y + \frac{e_\alpha}{m_\alpha}\int B_0(\mathbf{q}')\, dq'_x , \tag{6.216}$$

$$\xi_x = v'_x - \frac{e_\alpha}{m_\alpha}\int B_0(\mathbf{q}')\, dq'_y \tag{6.217}$$

are constants of motion (6.209) and (6.210), and g_α can accordingly be constructed from them. In addition to these constants of motion, which can be used to characterize the equilibrium distribution functions, there are others, for example,

$$H_\alpha = \int G_\alpha(g_\alpha)\, d^3q\, d^3v , \tag{6.218}$$

where G_α can be any bounded functional of g_α with a finite derivative. We have previously written down the entropy, see (6.31).

Specifically, in the following we consider stationary states in which the external fields do not explicitly depend on time and one coordinate is cyclic. For the (external) scalar electric potential ϕ^0 and the (external) vector potential $\mathbf{A}^0$ we assume

$$\frac{\partial}{\partial q_k}\phi^0 = \frac{\partial}{\partial q_k}\mathbf{A}^0 = 0 \quad \text{for } k = x \text{ or } y \tag{6.219}$$

.

Then the stationary distribution function g_α can be written as

$$\boxed{g_\alpha = g_\alpha(\varepsilon_\alpha^0, p_{\alpha k}^0)} \tag{6.220}$$

with

$$\varepsilon_\alpha^0 = \frac{m_\alpha}{2}v^2 + e_\alpha\phi^0 \tag{6.221}$$

and

$$p^0_{\alpha k} = m_\alpha v_k + e_\alpha A^0_k \tag{6.222}$$

By integrating over velocity space, we obtain

$$\rho^0 = \sum_\alpha \int d^3v \; e_\alpha \; g_\alpha(\varepsilon^0_\alpha, \; p^0_{\alpha k}) \tag{6.223}$$

and

$$j^0_k = \sum_\alpha \int d^3v \; e_\alpha v_k \; g_\alpha(\varepsilon^0_\alpha, \; p^0_{\alpha k}) \; . \tag{6.224}$$

Thus, the charge density ρ^0 and the current density $\mathbf{j}^0$ are functionally dependent on ϕ^0 and A^0_k, i.e.

$$\rho^0 = \rho^0(\phi^0, A^0_k) \; , \tag{6.225}$$

$$j^0_k = j^0_k(\phi^0, A^0_k) \; . \tag{6.226}$$

Both obey the differential equations

$$\nabla^2 \phi^0 = -\frac{1}{\varepsilon_0} \rho^0 \; , \tag{6.227}$$

$$\nabla^2 A^0_k = \mu_0 j^0_k \; . \tag{6.228}$$

Example 6.12 (Generating Function)
For this equilibrium, there exists a generating function

$$W = W(\phi^0, A^0_k) \; , \tag{6.229}$$

such that

$$\frac{\partial W}{\partial \phi^0} = -\rho^0 \tag{6.230}$$

and

$$\frac{\partial W}{\partial A^0_k} = j^0_k \tag{6.231}$$

hold. The proof is straightforward, since we can show that

$$\frac{\partial \rho^0}{\partial A_k^0} = -\frac{\partial j_k^0}{\partial \phi^0} \tag{6.232}$$

is true. ■

So much for the existence of equilibria (or more precisely: stationary solutions) in inhomogeneous Vlasov systems.

A sufficient condition for stability can also be formulated for such systems. Lyapunov functionals play an outstanding role in this context.

We use the stationary distribution (6.220) and the time-dependent distribution function $f_\alpha = g_\alpha + \delta f_\alpha$, where, for the sake of clarity, we allow only x as the relevant spatial variable, assume y to be cyclic ($k = y$) and suppress z (that is, physically speaking, we consider spatially two-dimensional arrangements). Note that $\partial_y = 0$ is also to apply to the time-dependent, perturbed states. We construct a Lyapunov functional from the constants

$$E = \sum_\alpha \int d^3q\, d^3v \frac{m_\alpha}{2} v^2 f_\alpha + \frac{\varepsilon_0}{2} \int d^3q\, (\mathbf{E}^2 + c^2 \mathbf{B}^2)\,, \quad c^2 = \frac{1}{\varepsilon_0 \mu_0}\,, \tag{6.233}$$

and

$$H = \sum_\alpha \int d^3q\, d^3v\, G_\alpha(f_\alpha, p_{\alpha y}) \tag{6.234}$$

where G_α will be specified later.

We now show that

$$\boxed{L := E + H} \tag{6.235}$$

can be used as a sufficient stability criterion. More precisely: We show that the stationary state g_α is stable if L possesses a minimum for it.

The first variation δL of L must vanish. We find, due to (6.235) and $f_\alpha = g_\alpha + \delta f_\alpha$

$$\begin{aligned}\delta L = &\sum_\alpha \int d^3q\, d^3v \left\{ \frac{\partial G_\alpha}{\partial g_\alpha} \delta f_\alpha + \frac{m_\alpha}{2} v^2 \delta f_\alpha + \frac{\partial G_\alpha}{\partial p_{\alpha y}}\, e_\alpha \delta A_y \right\} \\ &+ \varepsilon_0 \int d^3q \{\mathbf{E}^0 \cdot \delta\mathbf{E} + c^2 \mathbf{B}^0 \cdot \delta\mathbf{B}\}\,.\end{aligned} \tag{6.236}$$

The coefficients appearing here to first order are calculated using the zeroth-order values $[\phi^0(x), \mathbf{A}^0(x), \mathbf{E}^0, \varepsilon_\alpha^0, p_{\alpha y}^0]$. In the last integral on the right-hand side of (6.236), partial integration is appropriate. Furthermore, we can use

$$m_\alpha \frac{\partial}{\partial p_{\alpha y}^0} G_\alpha(g_\alpha,\, p_{\alpha y}^0) = \frac{\partial}{\partial v_y} G_\alpha\left(g_\alpha,\, m_\alpha v_y + q_\alpha A_y^0\right) - \frac{\partial g_\alpha}{\partial v_y} \frac{\partial}{\partial g_\alpha} G_\alpha(g_\alpha,\, p_{\alpha y}^0) \tag{6.237}$$

to write (6.236) as

$$\begin{aligned} \delta L &= \sum_\alpha \int d^3q\, d^3v \left\{ \left[\frac{\partial G_\alpha}{\partial g_\alpha} + \frac{m_\alpha}{2} v^2 \right] \delta f_\alpha - \frac{e_\alpha}{m_\alpha} \delta A_y \frac{\partial G_\alpha}{\partial g_\alpha} \frac{\partial g_\alpha}{\partial v_y} \right\} \\ &\quad + \varepsilon_0 \int d^3q \{ \phi^0\, \nabla \cdot \delta \mathbf{E} + c^2 \delta \mathbf{A} \cdot (\nabla \times \mathbf{B}_0) \} \\ &= \sum_\alpha \int d^3q\, d^3v \left\{ \left[\frac{\partial G_\alpha}{\partial g_\alpha} + \varepsilon_\alpha^0 \right] \delta f_\alpha - \frac{e_\alpha}{m_\alpha} \delta A_y \frac{\partial G_\alpha}{\partial g_\alpha} \frac{\partial g_\alpha}{\partial v_y} \right\} \\ &\quad + \int d^3q\, \delta \mathbf{A} \cdot \mathbf{j}^0 \end{aligned} \tag{6.238}$$

A simple rearrangement leads to

$$\begin{aligned} \delta L = \sum_\alpha \int d^3q\, d^3v \Bigg\{ & \left[\frac{\partial G_\alpha}{\partial g_\alpha} + \varepsilon_\alpha^0 \right] \delta f_\alpha \\ & + e_\alpha \delta A_y \left[\frac{1}{m_\alpha} \frac{\partial}{\partial v_y} \left(\frac{\partial G_\alpha}{\partial g_\alpha} \right) + v_y \right] g_\alpha \Bigg\} . \end{aligned} \tag{6.239}$$

This equation is used to determine G_α with the aim of achieving $\delta L = 0$. If we set

$$\boxed{\frac{\partial G_\alpha}{\partial g_\alpha} = -\varepsilon_\alpha^0} \tag{6.240}$$

then not only does the coefficient of δf_α, vanish in (6.239), but, due to (6.221), so does the term proportional to δA_y.

Let $\tilde{g}_\alpha$ be the inverse function of g_α with respect to $-\varepsilon_\alpha^0$, i.e.

$$-\varepsilon_\alpha^0 = \tilde{g}_\alpha(g_\alpha,\, p_{\alpha y}^0) , \tag{6.241}$$

then we can use this to solve (6.240). We have

$$G_\alpha = G_\alpha(g_\alpha,\, p_{\alpha y}^0) = \int_0^{g_\alpha} d\xi\; \tilde{g}_\alpha(\xi, p_{\alpha y}^0) . \tag{6.242}$$

Everything now depends on the sign of the second variation $\delta^2 L$. From (6.235) it follows that

$$\boxed{\begin{aligned}\delta^2 L =& \frac{1}{2}\sum_\alpha \int d^3q\, d^3v \left\{ \frac{\partial^2 G_\alpha}{\partial g_\alpha^2}(\delta f_\alpha)^2 + 2e_\alpha \frac{\partial^2 G_\alpha}{\partial g_\alpha \partial p_{\alpha y}}\, \delta f_\alpha\, \delta A_y \right. \\ & \left. + e_\alpha^2 \frac{\partial^2 G_\alpha}{\partial p_{\alpha y}^2}\left(\delta A_y\right)^2 \right\} + \frac{\varepsilon_0}{2}\int d^3q \left[(\delta \mathbf{E})^2 + c^2 (\delta \mathbf{B})^2\right].\end{aligned}} \tag{6.243}$$

If we use the expressions

$$\delta \mathbf{E} = -\nabla \phi - \frac{\partial}{\partial t}\delta \mathbf{A}\,, \tag{6.244}$$

$$(\delta \mathbf{B})^2 = \left(\delta B_y\right)^2 + \left(\nabla \delta A_y\right)^2\,, \tag{6.245}$$

with the gauge

$$\nabla \cdot \delta \mathbf{A} = 0\,, \tag{6.246}$$

then from (6.243) it follows that

$$\begin{aligned}\delta^2 L =& \frac{1}{2}\sum_\alpha \int d^3q\, d^3v \left\{ \frac{\partial^2 G_\alpha}{\partial g_\alpha^2}\left[\delta f_\alpha + \frac{(\partial^2 G_\alpha / \partial g_\alpha \partial p_{\alpha y})}{(\partial^2 G_\alpha / \partial g_\alpha^2)}\, e_\alpha \delta A_y\right]^2 \right. \\ & \left. + \left[\frac{\partial^2 G_\alpha}{\partial p_{\alpha y}^2} - \frac{(\partial^2 G_\alpha / \partial g_\alpha \partial p_{\alpha y})^2}{(\partial^2 G_\alpha / \partial g_\alpha^2)}\right]\left(e_\alpha \delta A_y\right)^2 \right\} \\ & + \frac{\varepsilon_0}{2}\int d^3q \left\{ (\nabla \delta \phi)^2 + c^2 (\nabla \delta A_y)^2 + \left(\frac{\partial}{\partial t}\delta \mathbf{A}\right)^2 + c^2 (\delta B_y)^2 \right\}.\end{aligned} \tag{6.247}$$

From (6.240) and (6.241) it follows that

$$\frac{\partial^2 G_\alpha}{\partial g_\alpha^2} = \frac{\partial \tilde{g}_\alpha}{\partial g_\alpha} = -\left(\frac{\partial g_\alpha}{\partial \varepsilon^0}\right)^{-1}. \tag{6.248}$$

If we also differentiate both sides of (6.240) with respect to $p_{\alpha y}^0$, that is,

$$0 = \frac{\partial^2 G_\alpha}{\partial g_\alpha^2}\frac{\partial g_\alpha}{\partial p_{\alpha y}^0} + \frac{\partial^2 G_\alpha}{\partial g_\alpha \partial p_{\alpha y}} \tag{6.249}$$

we obtain

$$\frac{\partial^2 G_\alpha}{\partial g_\alpha \partial p_{\alpha y}} = -\frac{1}{e_\alpha}\frac{\partial^2 G_\alpha}{\partial g_\alpha^2}\frac{\partial g_\alpha}{\partial A_y^0}\,. \tag{6.250}$$

Furthermore, we have

$$\sum_\alpha e_\alpha \int d^3v \frac{\partial G_\alpha}{\partial p_{\alpha y}} = -\sum_\alpha \int d^3v \frac{e_\alpha}{m_\alpha} \frac{\partial G_\alpha}{\partial g_\alpha} \frac{\partial g_\alpha}{\partial v_y} = -j_y^0 \,. \tag{6.251}$$

This further implies

$$\frac{\partial j_y^0}{\partial A_y^0} = -\sum_\alpha \int d^3v \left[e_\alpha^2 \frac{\partial^2 G_\alpha}{\partial p_{\alpha y}^2} + e_\alpha \frac{\partial^2 G_\alpha}{\partial g_\alpha \partial p_{\alpha y}} \frac{\partial g_\alpha}{\partial A_y^0} \right] \tag{6.252}$$

or

$$\frac{\partial j_y^0}{\partial A_y^0} = -\sum_\alpha \int d^3v \, e_\alpha^2 \left[\frac{\partial^2 G_\alpha}{\partial p_{\alpha y}^2} - \frac{(\partial^2 G_\alpha / \partial g_\alpha \partial p_{\alpha y})^2}{(\partial^2 G_\alpha / \partial g_\alpha^2)} \right] . \tag{6.253}$$

If we insert all this into (6.247), we obtain $\delta^2 L$ in a form that is accessible to stability arguments:

$$\boxed{\begin{aligned} \delta^2 L =& \frac{1}{2} \sum_\alpha \int d^3q \, d^3v \left\{ -\left(\frac{\partial g_\alpha}{\partial \varepsilon_\alpha^0} \right)^{-1} \left(\delta f_\alpha - \frac{\partial g_\alpha}{\partial A_y^0} \delta A_y \right)^2 \right\} \\ &+ \frac{\varepsilon_0}{2} \int d^3q \left\{ (\nabla \delta\phi)^2 + c^2 (\nabla \delta A_y)^2 - \frac{1}{\varepsilon_0} \frac{\partial j_y^0}{\partial A_y^0} (\delta A_y)^2 \right. \\ &\left. + (\partial_t \delta \mathbf{A})^2 + c^2 (\delta B_y)^2 \right\} . \end{aligned}} \tag{6.254}$$

Example 6.13 (Monotonic Distribution Function)
As an example, we choose monotonic distribution functions g_α, for which

$$\frac{\partial g_\alpha}{\partial \varepsilon_\alpha^0} < 0 \tag{6.255}$$

holds. We then see that the discussion of the sign of $\delta^2 L$ requires the solution of a Schrödinger eigenvalue problem

$$-\nabla^2 \delta A_y - \mu_0 \frac{\partial j_y^0}{\partial A_y^0} \delta A_y = \lambda \, \delta A_y \tag{6.256}$$

The "potential" depends on $\partial j_y^0 / \partial A_y^0$. The eigenvalue condition $\lambda \geq 0$ represents a sufficient stability criterion. Of course, we cannot expect it to be necessary as well. Evaluating the stability criterion requires specifying the stationary (equilibrium) state. ∎

Further considerations would have to address the question of necessity (in a more general formulation) as well as the problem, so far omitted, of the relevant neighboring states g_α. The second issue is closely related to the question of structural stability, more precisely: the stability of an invariant set. However, a corresponding detailed analysis would go beyond the scope of this discussion.

6.5 Instabilities in Inhomogeneous Vlasov Systems

As already demonstrated for homogeneous Vlasov systems, instability can be inferred within the framework of a normal mode analysis. In inhomogeneous Vlasov systems, a dispersion relation $\det \mathcal{D}(\mathbf{k}, \omega) = 0$ can also be derived, which, however, represents only an approximation in local coordinates. The better, though much more difficult, analysis uses integral equations, which we will not discuss here. For the presentation, we will also restrict ourselves to simple cases.

Similar to the previous section, let x be the only relevant spatial coordinate (of the stationary state), and let an external magnetic field have the form $\mathbf{B}_0 = B_0(x)\hat{z}$, so that the vector potential can be taken as $\mathbf{A}_0 = A_0(x)\hat{y}$. Furthermore, we exclude an external electric field, $\phi^0 \equiv \varphi_0 = 0$. In addition, we will consider here only perturbations in the *electrostatic* approximation ($\mathbf{E}_1 = -\nabla\delta\varphi$).

We therefore start with the linearized Vlasov equation in the form

$$\boxed{\partial_t \, \delta f_\alpha + \mathbf{v} \cdot \partial_{\mathbf{q}} \, \delta f_\alpha + \frac{e_\alpha}{m_\alpha}(\mathbf{v} \times \mathbf{B}_0) \cdot \partial_{\mathbf{v}} \, \delta f_\alpha = \frac{e_\alpha}{m_\alpha} \nabla \delta\varphi \cdot \partial_{\mathbf{v}} \, g_\alpha} \, , \tag{6.257}$$

where the linearization is performed around a state g_α that can depend on $\varepsilon_{\alpha\perp} = \frac{1}{2} m_\alpha \left(v_x^2 + v_y^2\right)$, $p_{\alpha y} = m_\alpha v_y + e_\alpha A_0(x)$ and v_z, i.e.

$$g_\alpha = g_\alpha(\varepsilon_{\alpha\perp}, p_{\alpha y}, v_z) \, . \tag{6.258}$$

The spatial dependence thus appears via $p_{\alpha y}$ in the distribution function. The solution of (6.257) is obtained by integrating along the trajectories

$$\frac{d\mathbf{q}'}{dt'} = \mathbf{v}' \, , \tag{6.259}$$

$$\frac{d\mathbf{v}'}{dt'} = \frac{e_\alpha}{m_\alpha} \, \mathbf{v}' \times \mathbf{B}_0(\mathbf{q}') \tag{6.260}$$

with the initial conditions $\mathbf{q}'(t' = t) = \mathbf{q}$ and $\mathbf{v}'(t' = t) = \mathbf{v}$. We obtain

$$\delta f_\alpha = \frac{e_\alpha}{m_\alpha} \int_{-\infty}^{t} dt' \, \nabla' \delta\varphi(\mathbf{q}', \mathbf{v}', t') \cdot \partial_{\mathbf{v}'} g_\alpha \, . \tag{6.261}$$

It is important to note here that the initial conditions contain the "variables" **q** and **v**. Since x is the only relevant spatial coordinate, it is natural to perform a Fourier transform in y and z, as well as a Fourier (or Laplace) transform in time. This procedure is equivalent to the so-called normal mode analysis, in which

$$\delta f_\alpha(x, y, z, \mathbf{v}, t) = \delta f_\alpha(x, k_y, k_z, \mathbf{v}, t)\ \exp\left[ik_y y + ik_z z - i\omega t\right] \tag{6.262}$$

is assumed. Instead of (6.261), we therefore obtain for the normal mode

$$\boxed{\delta f_\alpha = \frac{e_\alpha}{m_\alpha}\int_{-\infty}^{t} dt'\ \nabla'\delta\varphi \cdot \partial_{\mathbf{v}'} g_\alpha\ e^{ik_y(y'-y)+ik_z(z'-z)-i\omega(t'-t)}}\ . \tag{6.263}$$

The further calculation is based on the assumption that the spatial dependence in the distribution function g_α is only weak. Since the x-variation enters via the magnetic field $B_0(x)$, a weak x-dependence of the magnetic field is thus assumed. We can therefore use $\widehat{B}$ with $B_0(x) \approx \widehat{B} = const$ near the position x as a suitable normalization constant.

The trajectory equations (6.259) and (6.260) have the exact solutions

$$v_z' = v_z\ , \quad z' - z = v_z(t' - t)\ , \quad v_\perp^2 \equiv v_x^2 + v_y^2 = v_x'^2 + v_y'^2 \tag{6.264}$$

as well as the approximate solutions

$$\boxed{v_x' = v_\perp \cos(\Omega_\alpha \tau - \chi)\ ,} \tag{6.265}$$

$$\boxed{v_y' = -v_\perp \sin(\Omega_\alpha \tau - \chi)\ ,} \tag{6.266}$$

$$\boxed{x' = x + v_\perp[\sin(\Omega_\alpha \tau - \chi) + \sin\chi]/\,\Omega_\alpha\ ,} \tag{6.267}$$

$$\boxed{y' = y + v_\perp[\cos(\Omega_\alpha \tau - \chi) - \cos\chi]/\,\Omega_\alpha\ .} \tag{6.268}$$

Here, $\tau = t' - t$, and it should be noted that the index α for the particle species has not been used consistently everywhere.

We express the weak spatial dependence of the distribution function g_α by means of a smallness parameter δ_α, with which we scale the dependence of the distribution function g_α on $p_{\alpha y}$. If we suppress the variables $\varepsilon_{\alpha\perp}$ and v_z, which are not of interest at the moment, we write

$$g_\alpha = g_\alpha\left[\ldots, \delta_\alpha\left(\frac{v_y}{\widehat{\Omega}_\alpha L} + \frac{A_0}{\widehat{B}L}\right)\right] \tag{6.269}$$

with $\widehat{\Omega}_\alpha = e_\alpha \widehat{B}/m_\alpha$ and a characteristic inhomogeneity length L. For small δ_α we have

$$g_\alpha \approx g_\alpha^{(0)}(\dots) + \delta_\alpha \left(\frac{v_y}{\widehat{\Omega}_\alpha L} + \frac{A_0}{\widehat{B}L} \right) g_\alpha^{(1)}(\dots) . \tag{6.270}$$

Since $g_\alpha^{(0)}$ and $g_\alpha^{(1)}$ are isotropic in v_y and v_x and the plasma as a whole is to be electrically neutral, we can set

$$\delta_i = \delta_e \equiv \delta \tag{6.271}$$

.$A_0(x)$ follows from the corresponding Maxwell equation

$$\partial_x^2 A_0 = \delta \sum_\alpha \int d^3v \, \frac{\mu_0 e_\alpha}{\widehat{\Omega}_\alpha L} \, v_y^2 \, g_\alpha^{(1)} . \tag{6.272}$$

We can estimate the right-hand side with

$$\beta = \frac{2\mu_0 p}{\widehat{B}^2} \tag{6.273}$$

as $\delta\beta\widehat{B}/L$. This x-independent constant allows us to write the solution of (6.271) in the form

$$A_0(x) = \frac{1}{2L}\delta\beta\widehat{B}x^2 + \widehat{B}x \tag{6.274}$$

We see that $\beta \ll 1$ is a meaningful condition for the approximation method. We can now evaluate the anisotropic part of g_α according to (6.270) and obtain in total

$$g_\alpha = g_\alpha^{(0)}(\dots) + \delta \left(\frac{v_y}{\widehat{\Omega}_j L} + \frac{x}{L} + \frac{1}{2}\delta\beta \frac{x^2}{L^2} \right) g_\alpha^{(1)}(\dots) . \tag{6.275}$$

For the evaluation of (6.263), we make the further assumption

$$|\partial_x \delta\varphi| \ll |k_y \delta\varphi| , |k_z \delta\varphi| . \tag{6.276}$$

With this, all meaningful approximations and tools are assembled with which we can solve the linearized Vlasov equation. We start with

$$\begin{aligned}\delta f_\alpha =& i\frac{e_\alpha}{m_\alpha}\delta\varphi \int_{-\infty}^{0} d\tau \left[k_y \frac{\partial g_\alpha}{\partial v_y'} + k_z \frac{\partial g_\alpha}{\partial v_z'} \right] \exp\left\{ i\frac{v_\perp}{\Omega_\alpha}[\cos(\Omega_\alpha \tau - \chi) - \cos\chi] k_y \right\} \\ & \times \exp\{i(k_z v_z - \omega)\tau\} .\end{aligned} \tag{6.277}$$

In this expression, we must take into account

$$\frac{1}{m_\alpha}\frac{\partial g_\alpha}{\partial v'_y} = v'_y \frac{\partial g_\alpha}{\partial \varepsilon_{\alpha\perp}} + \frac{\partial g_\alpha}{\partial p_{\alpha y}} = -v_\perp \sin(\Omega_\alpha \tau - \chi)\frac{\partial g_\alpha}{\partial \varepsilon_{\alpha\perp}} + \frac{\partial g_\alpha}{\partial p_{\alpha y}} \tag{6.278}$$

and

$$\frac{\partial g_\alpha}{\partial v'_z} = \frac{\partial g_\alpha}{\partial v_z} \tag{6.279}$$

Furthermore, we use the formula

$$\boxed{e^{i\xi \sin\delta} = \sum_{n=-\infty}^{\infty} J_n(\xi) e^{in\delta}}\,, \tag{6.280}$$

to, with

$$\xi_\alpha = k_y v_\perp / \Omega_\alpha \tag{6.281}$$

after some calculation, obtain

$$\begin{aligned} \delta f_\alpha =& \frac{e_\alpha}{m_\alpha}\delta\varphi \sum_{n=-\infty}^{\infty}\left[n\Omega_\alpha m_\alpha \frac{\partial g_\alpha}{\partial \varepsilon_{\alpha\perp}} + m_\alpha \frac{\partial g_\alpha}{\partial p_{\alpha y}} k_y + \frac{\partial g_\alpha}{\partial v_z} k_z\right] \\ &\times \frac{\exp\left\{in\left(\frac{\pi}{2} - \chi\right) - i\xi_\alpha \cos\chi\right\}}{k_z v_z - \omega + n\Omega_\alpha} J_n(\xi_\alpha) \end{aligned} \tag{6.282}$$

This last expression can also be written differently, namely as

$$\begin{aligned} \delta f_\alpha &= \frac{e_\alpha}{m_\alpha}\delta\varphi \sum_{n=-\infty}^{\infty}\left[m_\alpha \frac{\partial g_\alpha}{\partial \varepsilon_{\alpha\perp}} J_n(\xi_\alpha) - G_\alpha \frac{J_n(\xi_\alpha)}{\omega - k_z v_z - n\Omega_\alpha}\right] e^{in\left(\frac{\pi}{2}-\chi\right) - i\xi_\alpha\cos\chi} \\ &= \frac{e_\alpha}{m_\alpha}\delta\varphi \left\{ m_\alpha \frac{\partial g_\alpha}{\partial \varepsilon_{\alpha\perp}} - G_\alpha \sum_{n=-\infty}^{+\infty} \frac{J_n(\xi_\alpha)}{\omega - k_z v_z - n\Omega_\alpha} e^{in\left(\frac{\pi}{2}-\chi\right) - i\xi_\alpha\cos\chi} \right\}, \end{aligned} \tag{6.283}$$

where

$$G_\alpha := \omega m_\alpha \frac{\partial g_\alpha}{\partial \varepsilon_{\alpha\perp}} + k_z\left(\frac{\partial g_\alpha}{\partial v_z} - v_z m_\alpha \frac{\partial g_\alpha}{\partial \varepsilon_{\alpha\perp}}\right) + k_y m_\alpha \frac{\partial g_\alpha}{\partial p_{\alpha y}} \tag{6.284}$$

is. In the last transformation, we used

$$e^{i\xi_\alpha \cos\chi} = \sum_{n=-\infty}^{\infty} J_n(\xi_\alpha) e^{in\left(\frac{\pi}{2}-\chi\right)} \tag{6.285}$$

Assuming that the function G_α depends only weakly on $p_{\alpha y}$, we can approximate

$$\begin{aligned} G_\alpha &\approx G_\alpha\left(\varepsilon_{\alpha\perp}, v_z, \frac{e_\alpha}{m_\alpha}A_0\right) + m_\alpha v_y \frac{\partial G_\alpha}{\partial p_{\alpha y}} \\ &= \widehat{G}_\alpha + m_\alpha v_y \frac{\partial \widehat{G}_\alpha}{\partial p_{\alpha y}} = \widehat{G}_\alpha + m_\alpha v_\perp \sin\chi \frac{\partial \widehat{G}_\alpha}{\partial p_{\alpha y}} \end{aligned} \tag{6.286}$$

Now, δf_α can be averaged over χ, with the result

$$\boxed{\frac{1}{2\pi}\int_0^{2\pi} d\chi\, \delta f_\alpha = \frac{e_\alpha}{m_\alpha}\delta\varphi\left\{m_\alpha \frac{\partial \hat{g}_\alpha}{\partial \varepsilon_{\alpha\perp}} - \sum_{n=-\infty}^{+\infty} \frac{J_n^2(\xi_\alpha)}{\omega - k_z v_z - n\Omega_\alpha} G_{1\alpha}\right\}}\,, \tag{6.287}$$

where

$$\hat{g}_\alpha = g_\alpha\left(\varepsilon_{\alpha\perp}, v_z, \frac{e_\alpha}{m_\alpha}A_0\right) \tag{6.288}$$

and

$$G_{1\alpha} = \widehat{G}_\alpha + \frac{m_\alpha \Omega_\alpha}{k_y} n \frac{\partial \widehat{G}_\alpha}{\partial p_{\alpha y}} \approx \widehat{G}_\alpha + \frac{n}{k_y}\frac{\partial \widehat{G}_\alpha}{\partial x} \tag{6.289}$$

were used. In deriving this result, (6.285) was also employed; likewise, for the Bessel functions,

$$J_{n-1}(\zeta) + J_{n+1}(\zeta) = \frac{2n}{\zeta} J_n(\zeta) \tag{6.290}$$

was used. The right-hand side of the relation (6.289) is understandable to lowest order from the formulas for $p_{\alpha y}$ and (6.274).

The result (6.287) directly enables us, via the Poisson equation, to write down the electrostatic dispersion relation in inhomogeneous plasmas; we have

$$1 - \sum_\alpha \frac{4\pi e_\alpha^2}{m_\alpha k^2}\int d^3v \left\{\frac{\partial \hat{g}_\alpha}{\partial \varepsilon_\perp} - \sum_{n=-\infty}^{\infty} \frac{J_n^2(\xi_\alpha)}{\omega - n\Omega_\alpha - k_z v_z} G_{1\alpha}\right\} = 0\,. \tag{6.291}$$

To simplify the notation, we have introduced $\varepsilon_\perp = \varepsilon_{\alpha\perp}/m_\alpha$. In the dispersion relation (6.291) we set $\hat{g}_\alpha = \hat{g}_\alpha(\varepsilon, e_\alpha A_0/m_\alpha)$ with $A_0 \approx \widehat{B}_0 x$, so that for weak x variations,

$$\hat{g} \approx \hat{g}(\varepsilon, 0) + x\partial_x \hat{g}(\varepsilon, 0) \tag{6.292}$$

follows. Here, $\varepsilon = \varepsilon_\perp + v_z^2/2$; applies; we have made the simplification that the dependence is not separately on $\varepsilon_\perp$ and v_z, but only on ε. As a result, in (6.284)

$$k_z\left(\frac{\partial g_\alpha}{\partial v_z} - v_z\frac{\partial g_\alpha}{\partial \varepsilon_\perp}\right) = 0 \tag{6.293}$$

results. If we then combine the definitions (6.284) and (6.292), it follows that

$$\widehat{G}_\alpha = \omega\frac{\partial}{\partial \varepsilon}\left(\hat{g}_\alpha + x\hat{g}'_\alpha\right) + \frac{k_y}{\Omega_\alpha}\hat{g}'_\alpha\ ; \tag{6.294}$$

the prime (′) denotes differentiation with respect to x. The function $G_{1\alpha}$ on the right-hand side of (6.291) follows from the definition (6.289) as

$$\begin{aligned} G_{1\alpha} &\approx \omega\frac{\partial}{\partial \varepsilon}\left(\hat{g}_\alpha + x\hat{g}'_\alpha\right) + \frac{k_y}{\Omega_\alpha}\hat{g}'_\alpha + \frac{n}{k_y}\omega\frac{\partial}{\partial \varepsilon}\hat{g}'_\alpha \\ &\approx \omega\frac{\partial}{\partial \varepsilon}\hat{g}_\alpha + \frac{k_y}{\Omega_\alpha}\hat{g}'_\alpha + \frac{n}{k_y}\omega\frac{\partial}{\partial \varepsilon}\hat{g}'_\alpha\ . \end{aligned} \tag{6.295}$$

In the last step, we have crucially made use of the so-called local approximation by setting $x \approx 0$.

In concrete applications, we further specify $\hat{g}_\alpha$, for example in the form

$$\boxed{\hat{g}_\alpha = n_\alpha(x)\left[\frac{2\pi k_B T_\alpha}{m_\alpha}\right]^{-\frac{3}{2}}\exp\left[-\frac{m_\alpha v^2}{2k_B T_\alpha}\right]}, \tag{6.296}$$

so that

$$\frac{\partial}{\partial \varepsilon}\hat{g}_\alpha = -\frac{m_\alpha}{k_B T_\alpha}\hat{g}_\alpha \tag{6.297}$$

and

$$\hat{g}'_\alpha = \frac{\partial \hat{g}_\alpha}{\partial x} = n'_\alpha\frac{\hat{g}_\alpha}{n_\alpha(x)} + T'_\alpha\frac{\partial \hat{g}_\alpha}{\partial T_\alpha} \tag{6.298}$$

hold. For the sake of clarity, in the following presentation we set $T'_\alpha = 0$. According to (6.295),$G_{1\alpha}$ results as

$$\boxed{G_{1\alpha} \approx -\frac{\omega m_\alpha}{k_B T_\alpha}\left[1 + \left(\frac{n}{k_y} - \frac{k_y k_B T_\alpha}{\Omega_\alpha m_\alpha \omega}\right)\kappa\right]\hat{g}_\alpha}, \tag{6.299}$$

where

$$\boxed{\kappa = \partial_x\left[\ln n_\alpha(x)\right]} \tag{6.300}$$

applies. Furthermore, we introduce the parameter

$$z_\alpha := \frac{k_y^2 k_B T_\alpha}{m_\alpha \Omega_\alpha^2} = k_y^2 \rho_\alpha^2 \tag{6.301}$$

and, in the following, restrict ourselves to low-frequency modes. Even if $z_\alpha < 1$ is assumed, let

$$\frac{\Omega_\alpha z_\alpha}{\omega} \gg 1 \,. \tag{6.302}$$

This assumption implies that on the right-hand side of (6.299), the term proportional to n can be neglected. If we furthermore exploit

$$\int_0^\infty d\xi \; \xi e^{-a^2\xi^2} J_n^2(p\xi) = \frac{1}{2a^2}\, e^{-p^2/2a^2}\, I_n\left(\frac{p^2}{2a^2}\right) \tag{6.303}$$

we can further simplify the dispersion relation (6.291) [for a Maxwell distribution]. The result is

$$\boxed{1 + \sum_\alpha \frac{1}{k^2\lambda_{D\alpha}^2}\left\{1 + \frac{\omega - \omega_\alpha^*}{\sqrt{2}k_z v_{t\alpha}} \sum_{n=-\infty}^{\infty} Z\left(\frac{\omega - n\Omega_\alpha}{\sqrt{2}k_z v_{t\alpha}}\right) I_n(z_\alpha)\, e^{-z_\alpha}\right\} = 0} \tag{6.304}$$

with

$$\omega_\alpha^* = k_y \rho_\alpha^2 \kappa \Omega_\alpha \,. \tag{6.305}$$

We now evaluate the dispersion relation (6.304) for two important cases. Further examples can be found in the specialized literature.

Example 6.14 (Drift Instability)

The so-called low-frequency drift instability occurs in the range

$$\omega \ll \Omega_i \ll |\Omega_e| \,, \tag{6.306}$$

$$k_z v_{t\alpha} \ll |\Omega_\alpha| \,, \qquad \text{for} \quad \alpha = e, i \,, \tag{6.307}$$

$$k_z v_{ti} \ll \omega \ll k_z v_{te} \,, \tag{6.308}$$

$$k^2 \lambda_{De}^2 \ll z_i \,, \tag{6.309}$$

$$z_e \ll 1 \,, \tag{6.310}$$

$$z_i < 1 . \tag{6.311}$$

In this case, the plasma dispersion function Z can be evaluated asymptotically by expansion. To lowest order, one obtains for the real part of the dispersion relation

$$\omega^2\left(1+\frac{c_s^2k_y^2}{\Omega_i^2}\right) - \omega\omega_e^* - c_s^2k_z^2 \approx 0 . \tag{6.312}$$

Here, in addition to the drift branch, the ion sound is also included; the solution of (6.312) can be written in the form

$$\omega \approx \frac{1}{2\delta}\left\{k_y v_{de} \pm \left[(k_y v_{de})^2 + 4\delta c_s^2 k_z^2\right]^{1/2}\right\} \tag{6.313}$$

where

$$\delta = 1 + \frac{c_s^2 k_y^2}{\Omega_i^2} \tag{6.314}$$

and

$$v_{de} = \omega_e^*/k_y \tag{6.315}$$

have been introduced. $\omega_e^* = k_y v_{de}$ is referred to as the drift frequency.

For small growth rates γ we can use the formulaγ to calculate

$$\boxed{\gamma \approx -\frac{\text{Im }\ \varepsilon(\omega_0, k)}{\partial\varepsilon/\partial\omega|_{\omega_0}}} \tag{6.316}$$

where ω_0 satisfies the relation Re $\varepsilon(\omega_0, k) = 0$. We have

$$\begin{aligned}\text{Im }\ \varepsilon \approx & \frac{4\pi^2 ec}{k_z k^2 B_0} n_{e0}(x)\left(\frac{m_e}{2\pi k_B T_e}\right)^{1/2} \exp\left[-\frac{m_e\omega_0^2}{2k_B T_e k_z^2}\right] \\ & \times\left(k_y\kappa - \frac{m_e\omega_0\Omega_e}{k_B T_e}\right) .\end{aligned} \tag{6.317}$$

It is noteworthy that the two terms in the last bracket have opposite signs due to $\omega_0 \approx \omega_e^*$. From the real part of the dispersion relation, one obtains

$$\left.\frac{\partial\varepsilon}{\partial\omega}\right|_{\omega_0} \approx \left(\frac{\delta}{k\lambda_{De}}\right)^2 \frac{1}{k_y v_{de}} , \tag{6.318}$$

so that ultimately for $k^2 v_{de}^2 \gg \delta c_s^2\, k_z^2$ the growth rate is given by

$$\boxed{\gamma \approx (\pi \tilde{\beta})^{1/2}\, k_z^{-1} \left(\frac{k_y v_{de}}{\delta}\right)^2 \left(1 - \frac{1}{\delta}\right) \exp\left[-\tilde{\beta}\left(\frac{\omega}{k_z}\right)^2\right]} \qquad (6.319)$$

Here, $\tilde{\beta} = m_e/2k_B T_e$ was set. A more precise evaluation must refer to the complete Z function. It turns out that the maximum growth rate for $k_y \rho_s$ occurs on the order of 1 and is, in terms of magnitude, quite comparable to the frequency ω_0. ■

Example 15. (Drift Cyclotron Instability)
As a second example, we consider the so-called drift cyclotron instability in the range

$$\omega \sim \ell\, \Omega_i \sim \omega_i^* \ , \quad |\omega - \ell\, \Omega_i\, | \gg k_z v_{ti} \ , \qquad (6.320)$$

$$\omega \gg k_z v_{te} \ , \quad z_i \gg 1 \ , \quad z_e < 1 \qquad (6.321)$$

The susceptibilities ($\varepsilon = 1 + \chi_e + \chi_i$) are then approximately

$$\chi_e \approx -(k_y \lambda_{Di})^{-2}\, (\omega_i^*/\omega) + \left(\omega_{pe}^2/\,\Omega_e^2\right) , \qquad (6.322)$$

$$\chi_i \approx (k_y \lambda_{Di})^{-2} \left\{ 1 + \frac{\omega - \omega_i^*}{\omega - \ell \Omega_i} I_l(z_i) e^{-z_i} \right\} . \qquad (6.323)$$

A corresponding analysis as in the case of the drift instability yields the growth rate

$$\gamma \approx \frac{\ell^{3/2} \Omega_i}{(2\pi)^{1/4}} \left(\frac{m_e}{m_i} + \frac{\Omega_i}{\omega_{pi}}\right)^{1/2} \left(\frac{\kappa}{\rho_i}\right)^{1/2} . \qquad (6.324)$$

For further details, see, for example, the handbook article by Mikhailovskii [68]. ■

Wave-Plasma Interaction 7

Abstract

The linear approximation that we have used so far in the study of collective effects can easily break down. Finite and non-negligible amplitudes are generally unavoidable, and often even desirable, in experimentally generated waves. Moreover, initially small wave amplitudes can grow as a result of an instability and quickly leave the linear regime. Various physical processes then become significant. Clearly, a description in terms of individual Fourier modes is generally no longer adequate, since couplings (convolution integrals) and nonlinear interactions occur. In addition, particles can become truly trapped in pronounced wave troughs ("particle trapping"). Furthermore, a large-amplitude wave alters key plasma parameters, such as local particle density, temperature, and so on. All these phenomena require extensive analysis. In this chapter, we focus on some important effects in coherent wave phenomena, where phase information is tracked deterministically. Later, in the context of laser-plasma interaction, we will refer back to parts of this chapter.

7.1 Maxwell Fluid Model

In this section, we compile fundamental basic equations. These are models that describe the coupling of a plasma to waves. Electromagnetic waves are described by Maxwell's equations, in which charge densities and electric currents appear as sources. The latter can be captured by plasma dynamic equations. However, a fluid description ignores typical kinetic effects such as Landau damping and resonances.

K.-H. Spatschek, *Theoretical Plasma Physics*,
https://doi.org/10.1007/978-3-662-72828-4_7

Relativistic Maxwell Electron Fluid Model

In the study of high-frequency plasma phenomena, we consider the motion of electrons but ignore the motion of ions (immobile ion approximation).

From Maxwell's equations in SI units, we obtain for the vector potential $\mathbf{A}$ and the scalar potential ϕ in the Coulomb gauge $\nabla \cdot \mathbf{A} = 0$

$$-\nabla^2 \mathbf{A} + \frac{1}{c^2}\frac{\partial^2 \mathbf{A}}{\partial t^2} = -\frac{1}{c^2}\frac{\partial \nabla\phi}{\partial t} + \mu_0 \mathbf{j}\,, \tag{7.1}$$

where we have used: $\mathbf{B} = \nabla \times \mathbf{A}$, $\mathbf{E} = -\nabla\phi - \frac{\partial \mathbf{A}}{\partial t}$, $\nabla \times \nabla \times \mathbf{a} = \nabla(\nabla \cdot \mathbf{a}) - \nabla^2 \mathbf{a}$, and $\varepsilon_0 \mu_0 = c^{-2}$.

For immobile ions, the electric current density is

$$\mathbf{j} \approx -e n_e \mathbf{v}_e\,; \tag{7.2}$$

which follows from the electron velocity $\mathbf{v}_e$ and the particle density n_e. The wave equation is coupled to the continuity equation for the electron density

$$\frac{\partial n_e}{\partial t} + \nabla \cdot (n_e \mathbf{v}_e) = 0 \tag{7.3}$$

and the electron momentum balance

$$\left(\frac{\partial}{\partial t} + \mathbf{v}_e \cdot \nabla\right)\mathbf{p}_e = -e\left[-\nabla\phi - \frac{\partial \mathbf{A}}{\partial t} + \mathbf{v}_e \times (\nabla \times \mathbf{A})\right] \tag{7.4}$$

which describes the nonlinear response of the medium. The pressure term has been neglected. Due to the relativistic mass factor, for

$$\mathbf{v}_e = \frac{\mathbf{p}_e}{m_e\, \gamma_e}\,, \tag{7.5}$$

where m_e is the rest mass of the electron and γ_e is the relativistic factor

$$\gamma_e = \frac{1}{\sqrt{1 - \left(\frac{\mathbf{v}_e}{c}\right)^2}} = \sqrt{1 + \left(\frac{\mathbf{p}_e}{m_e\, c}\right)^2} \tag{7.6}$$

the momentum balance contains nonlinear terms.

With a few simple transformations, the momentum balance can be written in the following form:

$$\frac{\partial}{\partial t}(\mathbf{p}_e - e\mathbf{A}) = e\,\nabla\phi - m_e c^2\,\nabla\gamma_e + \frac{1}{m_e \gamma_e}\mathbf{p}_e \times [\nabla \times (\mathbf{p}_e - e\mathbf{A})]\,, \tag{7.7}$$

where we have made use of the vector identity

$$\mathbf{a} \times (\nabla \times \mathbf{a}) = \frac{1}{2}\nabla a^2 - \mathbf{a} \cdot \nabla \mathbf{a} \tag{7.8}$$

To normalize the equations, we use a constant density n_0 (for example, the constant ion density, which should be identical to the zeroth-order electron density), the length $L = c\omega_{pe}^{-1}$ and the time $T = \omega_{pe}^{-1}$, where

$$\omega_{pe} \equiv \sqrt{\frac{n_0 e^2}{\varepsilon_0 m_e}} \tag{7.9}$$

is the electron plasma frequency. The scalar potential is measured in units of $m_e c^2/e$, while the vector potential is measured in units of $m_e c/e$. Velocities are measured in units of c, and momenta are normalized with $m_e c$. From now on, we omit the index e for electrons.

Using this normalization, the Maxwell-electron fluid model for a plasma with immobile ions can be written in the following form:

$$\boxed{\frac{\partial^2 \mathbf{A}}{\partial t^2} - \nabla^2 \mathbf{A} + \frac{\partial \nabla \phi}{\partial t} = -n\frac{\mathbf{p}}{\gamma}\,,} \tag{7.10}$$

$$\boxed{\nabla^2 \phi = n - 1\,,} \tag{7.11}$$

$$\boxed{\frac{\partial n}{\partial t} + \nabla \cdot \left(n\frac{\mathbf{p}}{\gamma}\right) = 0\,,} \tag{7.12}$$

$$\boxed{\frac{\partial}{\partial t}(\mathbf{p} - \mathbf{A}) - \frac{\mathbf{p}}{\gamma} \times \nabla \times (\mathbf{p} - \mathbf{A}) = \nabla(\phi - \gamma)\,,} \tag{7.13}$$

where $\gamma = \sqrt{1 + \mathbf{p}^2}$ is the relativistic factor. We use the Coulomb gauge $\nabla \cdot \mathbf{A} = 0$.

Relativistic Maxwell Two-Fluid Model

It is quite straightforward to generalize the Maxwell-electron fluid model to a two-component plasma with mobile ions. For many applications, we can assume that the plasma is cold (with "vanishing" ion and electron temperatures). However, in the following, we will include simple scalar pressure terms (at finite temperatures) for later use in stability considerations. The forms of the temperature (and pressure) terms used here are only valid in weakly relativistic ($A \ll 1$) low-temperature regimes ($T_e \ll 1$, which means $k_B T_e \ll m_e c^2$). Furthermore, the isothermal equation of state (with constant temperatures) is only appropriate when the phase velocities of the phenomena are smaller than the thermal velocities. Otherwise, an adiabatic approximation may be suitable. Dimensionless quantities are used in the same form as before, with the species now denoted by α, and $\varepsilon = m_e/m_i$ is introduced as a small parameter. In the Coulomb gauge, the Maxwell equations for the vector and scalar potentials $\mathbf{A}$ and ϕ, as well as the hydrodynamic equations (particle and momentum balance) for the densities n_α and the canonical momentum P_α of the electrons and ions, each written in dimensionless form as

$$\boxed{\nabla^2 \mathbf{A} - \frac{\partial^2 \mathbf{A}}{\partial t^2} - \frac{\partial \nabla \phi}{\partial t} = n_e \mathbf{v}_e - n_i \mathbf{v}_i \ ,} \tag{7.14}$$

$$\boxed{\nabla^2 \phi = n_e - n_i \ ,} \tag{7.15}$$

$$\boxed{\frac{\partial n_\alpha}{\partial t} + \nabla \cdot (n_\alpha \mathbf{v}_\alpha) = 0 \ ,} \tag{7.16}$$

$$\boxed{\frac{\partial \mathbf{P}_e}{\partial t} - \mathbf{v}_e \times \nabla \times \mathbf{P}_e = \nabla(\phi - \gamma_e) - T_e \nabla \ln n_e \ ,} \tag{7.17}$$

$$\boxed{\frac{\partial \mathbf{P}_i}{\partial t} - \mathbf{v}_i \times \nabla \times \mathbf{P}_i = \nabla(-\phi - \gamma_i/\varepsilon) - T_i \nabla \ln n_i \ ,} \tag{7.18}$$

where $\mathbf{P}_\alpha$ and γ_α are related to the kinetic momentum $\mathbf{p}_\alpha$ by $\mathbf{P}_e = \mathbf{p}_e - \mathbf{A}$, $\mathbf{P}_i = \mathbf{p}_i + \mathbf{A}$, $\gamma_e = \sqrt{1 + \mathbf{p}_e^2}$, $\gamma_i = \sqrt{1 + \varepsilon^2 \mathbf{p}_i^2}$, where $\mathbf{v}_e = \mathbf{p}_e/\gamma_e$ and $\mathbf{v}_i = \varepsilon \mathbf{p}_i/\gamma_i$ are the (dimensionless) fluid velocities. The factor $\varepsilon = m_e/m_i$ appears due to the normalization $\mathbf{p}_e/m_e c \rightarrow \mathbf{p}_e$ and $\mathbf{p}_i/m_e c \rightarrow \mathbf{p}_i$. As already mentioned, we will neglect the pressure terms (by setting $T_e = T_i = 0$) in most applications.

One-Dimensional Propagation in the *x*-Direction

We now write the fundamental equations for immobile ions under the assumption that the wave propagates in the x-direction, so that all variables depend only on a single spatial coordinate, i.e.

$$\mathbf{A} = \mathbf{A}(x, t) \ , \quad n = n(x, t) \ , \quad \phi = \phi(x, t) \ . \tag{7.19}$$

The following holds $p \equiv p_x$. In the Coulomb gauge, the purely transverse nature of the waves ($A_x = 0,\ \mathbf{A} = \mathbf{A}_\perp$) becomes evident. The dimensionless form of the wave Eq. (7.10) for $\mathbf{A}_\perp$ is now

$$\frac{\partial^2}{\partial x^2}\mathbf{A}_\perp - \frac{\partial^2}{\partial t^2}\mathbf{A}_\perp = n\,\frac{\mathbf{p}_\perp}{\gamma}\,. \tag{7.20}$$

The longitudinal component of the wave equation simplifies to

$$\frac{\partial^2 \phi}{\partial t \partial x} + n\frac{p}{\gamma} = 0\,. \tag{7.21}$$

The perpendicular component of the electron momentum balance

$$\frac{\partial}{\partial t}(\mathbf{p}_\perp - \mathbf{A}_\perp) + \left(\frac{p}{\gamma}\right)\frac{\partial(\mathbf{p}_\perp - \mathbf{A}_\perp)}{\partial x} = 0 \tag{7.22}$$

has the particular solution

$$\boxed{\mathbf{p}_\perp = \mathbf{A}_\perp}\,. \tag{7.23}$$

Assuming the initial values are chosen accordingly, the last relation simplifies the longitudinal electron momentum balance

$$\frac{\partial p}{\partial t} = \mathbf{p}_\perp \cdot \frac{\partial(\mathbf{p}_\perp - \mathbf{A}_\perp)}{\partial x} + \frac{\partial(\phi - \gamma)}{\partial x} = \frac{\partial(\phi - \gamma)}{\partial x}\,. \tag{7.24}$$

This leads to the fundamental system of equations for one-dimensional fully relativistic wave propagation in a plasma with immobile ions:

$$\frac{\partial^2}{\partial x^2}\mathbf{A}_\perp - \frac{\partial^2}{\partial t^2}\mathbf{A}_\perp = n\,\frac{\mathbf{A}_\perp}{\gamma}\,, \tag{7.25}$$

$$\frac{\partial^2 \phi}{\partial t \partial x} + n\frac{p}{\gamma} = 0\,, \tag{7.26}$$

$$\frac{\partial^2 \phi}{\partial x^2} = n - 1\,, \tag{7.27}$$

$$\frac{\partial n}{\partial t} + \frac{\partial}{\partial x}\left(\frac{np}{\gamma}\right) = 0\,, \tag{7.28}$$

$$\frac{\partial p}{\partial t} = \frac{\partial(\phi - \gamma)}{\partial x} . \tag{7.29}$$

Eqs. (7.27) and (7.28) lead to

$$\frac{\partial}{\partial x}\left[\frac{\partial^2 \phi}{\partial t \partial x} + n\frac{p}{\gamma}\right] = 0 . \tag{7.30}$$

In principle, integrating (7.30) yields an arbitrary constant on the right-hand side. However, in view of (7.26), the constant should be zero.

Assuming *linearly polarized* waves ($\mathbf{A}_\perp = A\hat{y}$), with $E = -\partial\phi/\partial x$, as well as neglecting the ion response and assuming $T_e = T_i = 0$, these equations agree with the model [69].

$$\boxed{\frac{\partial E}{\partial t} = \frac{n_e p_e}{\gamma_e} , \qquad \frac{\partial E_y}{\partial t} + \frac{\partial B_z}{\partial x} = \frac{n_e A}{\gamma_e} ,} \tag{7.31}$$

$$\boxed{\frac{\partial B_z}{\partial t} + \frac{\partial E_y}{\partial x} = 0 , \qquad \frac{\partial A}{\partial t} = -E_y ,} \tag{7.32}$$

$$\boxed{\frac{\partial n_e}{\partial t} = -\frac{\partial}{\partial x}\left(\frac{n_e p_e}{\gamma_e}\right) , \qquad \frac{\partial p_e}{\partial t} = -E - \frac{\partial \gamma_e}{\partial x} ,} \tag{7.33}$$

$$\boxed{\gamma = \sqrt{1 + p_e^2 + A^2} .} \tag{7.34}$$

In addition to discussing the interplay between backward and forward Raman scattering, the modulation of broad light pulses, "down-cascading" in the frequency spectrum, photon condensation, and the splitting of the original laser beam, this model is also suitable for investigating slow 1D solitons on the electron time scale.

The one-dimensional relativistic Maxwell two-fluid model can be treated in a similar manner as the previously discussed Maxwell electron fluid model. We abbreviate $\hat{x} \cdot \mathbf{p}_\alpha \equiv p_\alpha$. For a laser pulse that starts in vacuum, we obtain $\mathbf{P}_{\perp\alpha} = 0$. The system of equations

$$\boxed{\frac{\partial^2 \mathbf{A}_\perp}{\partial x^2} - \frac{\partial^2 \mathbf{A}_\perp}{\partial t^2} = n_e \frac{\mathbf{A}_\perp}{\gamma_e} + \varepsilon n_i \frac{\mathbf{A}_\perp}{\gamma_i} ,} \tag{7.35}$$

$$\boxed{\frac{\partial^2 \phi}{\partial t \partial x} + \frac{n_e p_e}{\gamma_e} - \varepsilon \frac{n_i p_i}{\gamma_i} = 0 ,} \tag{7.36}$$

$$\boxed{\frac{\partial^2 \phi}{\partial x^2} = n_e - n_i ,} \tag{7.37}$$

$$\boxed{\frac{\partial n_e}{\partial t} + \frac{\partial}{\partial x}\left(\frac{n_e p_e}{\gamma_e}\right) = 0\,,} \tag{7.38}$$

$$\boxed{\frac{\partial n_i}{\partial t} + \varepsilon\frac{\partial}{\partial x}\left(\frac{n_i p_i}{\gamma_i}\right) = 0\,,} \tag{7.39}$$

$$\boxed{\frac{\partial p_e}{\partial t} = \frac{\partial(\phi - \gamma_e)}{\partial x} - T_e\frac{\partial \ln n_e}{\partial x}\,,} \tag{7.40}$$

$$\boxed{\frac{\partial p_i}{\partial t} = \frac{\partial(-\phi - \gamma_i/\varepsilon)}{\partial x} - T_i\frac{\partial \ln n_i}{\partial x}} \tag{7.41}$$

forms a closed system that generalizes the previous single-fluid model. Whether we can use an isothermal equation of state (as assumed here) depends on the phase velocity in relation to the thermal velocity.

The Weakly Relativistic Limit

It is not always appropriate (and in fact, in vacuum it is even impossible) to use the inverse plasma frequency as the unit of time. It makes sense to use the inverse laser frequency ω_0^{-1} as a new unit of time. Formally, we can define it via a density, the so-called critical density n_c, as follows:

$$\omega_0 \equiv \sqrt{\frac{n_c e^2}{\varepsilon_0 m_e}}\,. \tag{7.42}$$

In the following, we will use the inverse laser frequency for time normalization and $c\omega_0^{-1}$ for length normalization. With these new units, the Maxwell-fluid Eqs. (7.10)–(7.13) in full 3D spatial dimensions become

$$\frac{\partial^2 \mathbf{A}}{\partial t^2} - \nabla^2\mathbf{A} + \frac{\partial \nabla\phi}{\partial t} = -\frac{n_0}{n_c} n\frac{\mathbf{p}}{\gamma}\,, \tag{7.43}$$

$$\nabla^2\phi = -\frac{n_0}{n_c}(1-n)\,, \tag{7.44}$$

$$\frac{\partial n}{\partial t} + \nabla\cdot(n\frac{\mathbf{p}}{\gamma}) = 0\,, \tag{7.45}$$

$$\frac{\partial}{\partial t}(\mathbf{p}-\mathbf{A}) - \frac{\mathbf{p}}{\gamma}\times\nabla\times(\mathbf{p}-\mathbf{A}) = \nabla(\phi-\gamma)\,, \tag{7.46}$$

where we omit the index e for the electron density and have used the relativistic factor $\gamma = \sqrt{1+\mathbf{p}^2}$. As before, we have applied the Coulomb gauge condition $\nabla \cdot \mathbf{A} = 0$. The vector potential $\mathbf{A}$ is measured in the unit $e/m_e c$, while for the dimensionless electrostatic potential ϕ we use the unit $e/m_e c^2$. The unit of density is n_0. In principle, this can be arbitrary, but in the presence of plasma we choose the ion background density, which in general differs from the critical density n_c. For the propagation of the laser in an underdense plasma, $n_0/n_c < 1$ applies. The momentum $\mathbf{p} \equiv \mathbf{p}_e$ is measured in $m_e c$, where m_e, e, and c are the electron mass, the elementary charge, and the speed of light, respectively. Sometimes, for the Laplace operator ∇^2 we also use the notation Δ.

We emphasize once again that, compared to the previous section, the unit of time has been changed from ω_{pe}^{-1} to ω_0^{-1}. This has the advantage of facilitating comparison with the vacuum case $n_0 = 0$.

The momentum balance (7.46) can be further simplified by choosing an initial gauge condition (which, of course, should not restrict the general validity).

Example 7.1 (Decomposition with Projection Operators)
We define the projection operators Π_c and Π_g, so that any vector field $\mathbf{u}$ is decomposed into $\mathbf{u} = \mathbf{v} + \mathbf{w}$, with the following properties:

$$\Pi_c \mathbf{u} = \mathbf{v} \equiv \mathbf{u}_c \quad \nabla \times \mathbf{v} = 0\,, \quad \text{but generally} \quad \nabla \cdot \mathbf{v} \neq 0\,, \tag{7.47}$$

$$\Pi_g \mathbf{u} = \mathbf{w} \equiv \mathbf{u}_g\,, \quad \nabla \cdot \mathbf{w} = 0\,, \quad \text{but generally} \quad \nabla \times \mathbf{w} \neq 0\,, \tag{7.48}$$

where

$$\boxed{\Pi_c + \Pi_g = \mathbf{1}}\,. \tag{7.49}$$

Obviously, $\mathbf{v}$ is a gradient field, and $\mathbf{w}$ is a rotational field. The operators can be represented as follows:

$$\Pi_c = \nabla \Delta^{-1} \nabla \cdot \tag{7.50}$$

$$\Pi_g = 1 - \nabla \Delta^{-1} \nabla \cdot\,. \tag{7.51}$$

The application of the projection operators to the momentum balance (7.46) allows the equation to be split into a divergence-free and a curl-free part. The equation

$$\frac{\partial}{\partial t}(\mathbf{p}_g - \mathbf{A}) - \Pi_g \left[\frac{\mathbf{p}}{\gamma} \times \{\nabla \times (\mathbf{p}_g - \mathbf{A})\} \right] = 0 \tag{7.52}$$

describes the convective transport of the divergence-free part of the canonical momentum $\mathbf{P}_{\text{can}} = \mathbf{p} - \mathbf{A}$. This implies that, for the initial condition $\mathbf{p}_g = \mathbf{A}$, the canonical momentum remains curl-free for all times, i.e.,

$$\mathbf{P}_{\text{can}} = \mathbf{p}_g + \mathbf{p}_c - \mathbf{A} = \mathbf{p}_c \,. \tag{7.53}$$

The initial condition simplifies the curl-free part,

$$\frac{\partial}{\partial t}\mathbf{p}_c = \nabla(\phi - \gamma) \,. \tag{7.54}$$

Since $\nabla \times \mathbf{p}_c = 0$ is, $\mathbf{p}_c$ can be written with a Clebsch potential: $\mathbf{p}_c = \nabla\psi$. The integration of (7.54) then leads to

$$\frac{\partial\psi}{\partial t} = \phi - \gamma + 1 \,. \tag{7.55}$$

The application of the decomposition with Π_g and Π_c to the wave Eq. (7.43) for $\mathbf{A}$ yields, under the condition $\mathbf{p}_g = \mathbf{A}$, for the divergence-free part

$$\frac{\partial^2 \mathbf{A}}{\partial t^2} - \nabla^2\mathbf{A} = -\frac{n_0}{n_c}(1 - \nabla\Delta^{-1}\nabla\cdot)\left\{\frac{n}{\gamma}(\mathbf{A} + \nabla\psi)\right\} \tag{7.56}$$

and for the curl-free part

$$\frac{\partial}{\partial t}\nabla\phi = -\frac{n_0}{n_c}\nabla\Delta^{-1}\nabla\cdot\left\{\frac{n}{\gamma}(\mathbf{A} + \nabla\psi)\right\} . \tag{7.57}$$

Simplifications of the vector operations on the right-hand side lead to

$$\frac{\partial^2 \mathbf{A}}{\partial t^2} - \nabla^2\mathbf{A} = -\frac{n_0}{n_c}\left[\frac{n}{\gamma}\mathbf{A} - \Delta^{-1}\left\{\nabla\left(\mathbf{A}\cdot\nabla\frac{n}{\gamma}\right) + \nabla\times\left[\left(\nabla\frac{n}{\gamma}\right)\times(\nabla\psi)\right]\right\}\right] , \tag{7.58}$$

$$\frac{\partial\nabla\phi}{\partial t} = -\frac{n_0}{n_c}\left[\frac{n}{\gamma}\nabla\psi + \Delta^{-1}\left\{\nabla\left(\mathbf{A}\cdot\nabla\frac{n}{\gamma}\right) + \nabla\times\left[\left(\nabla\frac{n}{\gamma}\right)\times(\nabla\psi)\right]\right\}\right] . \tag{7.59}$$

The equations for $\mathbf{A}$ and ϕ are coupled to n and ψ. The evolution of the latter is given by (7.55) in the form

$$\frac{\partial\psi}{\partial t} = \phi - \gamma + 1 = \phi - \sqrt{1 + [\mathbf{A} + \nabla\psi]^2} + 1 \,. \tag{7.60}$$

The density n can be obtained from the Poisson Eq. (7.44),

$$n = 1 + \left(\frac{n_0}{n_c}\right)^{-1}\nabla^2\phi \,. \tag{7.61}$$

In the following, we will use the Maxwell fluid equations in the reformulated form just presented to derive simpler models. ■

Let us consider the weakly relativistic limiting case. For small amplitudes, we scale the variables. First, we introduce various smallness parameters. We scale the perpendicular variation [with respect to the parallel changes in the propagation direction x (scaling parameter α)] and introduce the smallness parameters ε, μ, β, ρ and δ for the amplitudes of the physical variables (note that ε is a general smallness parameter and is no longer fixed to m_e/m_i):

$$\mathbf{A}(\mathbf{r},t) = \varepsilon \left\{ \mathbf{A}_\perp(x, \alpha \mathbf{r}_\perp, t) + \mu\, \mathbf{e}_x A_\parallel(x, \alpha \mathbf{r}_\perp, t) \right\} , \tag{7.62}$$

$$n(\mathbf{r},t) = 1 + \beta n_e^1(x, \alpha \mathbf{r}_\perp, t) , \tag{7.63}$$

$$\phi(\mathbf{r},t) = \rho\, \phi^1(x, \alpha \mathbf{r}_\perp, t) , \tag{7.64}$$

$$\psi(\mathbf{r},t) = \delta\, \psi^1(x, \alpha \mathbf{r}_\perp, t) , \tag{7.65}$$

$$\gamma(\mathbf{r},t) = \sqrt{1 + \left(\varepsilon \mathbf{A}_\perp + \varepsilon\mu\, \mathbf{e}_x A_\parallel + \delta \nabla \psi^1 \right)^2} . \tag{7.66}$$

The various smallness parameters are not completely independent of each other:

$$\mu \sim \alpha \sim \varepsilon \ll 1 \quad \text{und} \quad \delta \sim \rho \sim \beta \sim \varepsilon^2 \ll 1. \tag{7.67}$$

Example 7.2 (Relation between Smallness Parameters)
The relations are justified as follows. Due to the Coulomb gauge, we have

$$\nabla \cdot \mathbf{A} = \varepsilon \{ \alpha\, \nabla_\perp \cdot \mathbf{A}_\perp + \mu\, \partial_x A_\parallel \} = 0 \;\Rightarrow\; \mu \sim \alpha \ , \tag{7.68}$$

except when $A_\parallel \equiv 0$. The Laplace Eq. (7.44) for ϕ leads to

$$\rho\, \nabla^2 \phi^1 = \frac{n_0}{n_c}(n-1) = \frac{n_0}{n_c}\, \beta\, n_e^1 \;\Rightarrow\; \rho \sim \beta . \tag{7.69}$$

The reduced momentum balance (7.55) yields

$$\delta\, \partial_t \psi^1 = \rho\, \phi^1 - (\gamma - 1) = \rho\, \phi^1 + \mathcal{O}(\varepsilon^2) + \cdots \;\Rightarrow\; \delta \sim \rho \sim \varepsilon^2 . \tag{7.70}$$

A combination of the Poisson Eq. (7.69) with the continuity Eq. (7.45) leads to

$$\rho\, \nabla \cdot \frac{\partial}{\partial t} \nabla \phi^1 = \frac{n_0}{n_c} \beta\, \frac{\partial}{\partial t} n_e^1 = -\frac{n_0}{n_c}\, \delta\, \nabla^2 \psi^1 + \cdots \;\Rightarrow\; \rho \sim \delta , \tag{7.71}$$

which is consistent with $\delta \sim \rho \sim \beta \sim \varepsilon^2$. The scaling is then also compatible with the wave equation for **A**, i.e.,

$$\varepsilon\left\{\frac{\partial^2}{\partial t^2}\mathbf{A} - \frac{\partial^2}{\partial x^2}\mathbf{A} - \alpha^2\Delta_\perp\mathbf{A}\right\} = -\varepsilon\left\{\frac{n_0}{n_c}(1+\beta\, n_e^1)\left[1-\frac{\varepsilon^2}{2}\mathbf{A}^2\right]\mathbf{A}\right\} + \dots . \tag{7.72}$$

■

Based on the arguments just presented for a consistent scaling, we now introduce a smallness parameter ε and scale the physical quantities as follows:

$$\boxed{\mathbf{A}(\mathbf{r},t) = \varepsilon\,\{\mathbf{A}_\perp(x,\varepsilon\mathbf{r}_\perp,t) + \varepsilon\,\mathbf{e}_x A_\parallel(x,\varepsilon\mathbf{r}_\perp,t)\}\ ,} \tag{7.73}$$

$$\boxed{n(\mathbf{r},t) = 1 + \varepsilon^2\delta n(x,\varepsilon\mathbf{r}_\perp,t)\ ,} \tag{7.74}$$

$$\boxed{\phi(\mathbf{r},t) = \varepsilon^2\,\delta\phi(x,\varepsilon\mathbf{r}_\perp,t)\ ,} \tag{7.75}$$

$$\boxed{\psi(\mathbf{r},t) = \varepsilon^2\,\delta\psi(x,\varepsilon\mathbf{r}_\perp,t)\ ,} \tag{7.76}$$

$$\boxed{\gamma(\mathbf{r},t) = \sqrt{1+\left(\varepsilon\mathbf{A}_\perp + \varepsilon^2\,\mathbf{e}_x A_\parallel + \varepsilon^2\nabla\delta\psi\right)^2} \approx 1 + \frac{\varepsilon^2}{2}\mathbf{A}_\perp^2 + \mathcal{O}(\varepsilon^3)\ .} \tag{7.77}$$

Because

$$\nabla\frac{n}{\gamma} = \varepsilon^2\left(\nabla\delta n - n_e^0\nabla\frac{\mathbf{A}_\perp^2}{2}\right) + \mathcal{O}(\varepsilon^3) \tag{7.78}$$

and

$$\mathbf{A}\cdot\nabla = \varepsilon^2\mathbf{A}_\perp\cdot\nabla_\perp + \varepsilon^2 A_\parallel\,\partial_x \tag{7.79}$$

it follows

$$\nabla\left(\mathbf{A}\cdot\nabla\frac{n}{\gamma}\right) = \mathcal{O}(\varepsilon^4) \tag{7.80}$$

and

$$\nabla\times\left[\left(\nabla\frac{n}{\gamma}\right)\times(\nabla\psi)\right] = \mathcal{O}(\varepsilon^5)\ . \tag{7.81}$$

The inverse Laplace operator does not change the order of the dominant terms, since

$$\Delta^{-1} \hat{=} \mathcal{F}^{-1} \frac{1}{k_\parallel^2 + \varepsilon^2 k_\perp^2} \approx \mathcal{F}^{-1} \frac{1}{k_\parallel^2} \left(1 - \varepsilon^2 \frac{k_\perp^2}{k_\parallel^2}\right) \hat{=} \left(\frac{\partial^2}{\partial x^2}\right)^{-1} + \mathcal{O}(\varepsilon^2) , \tag{7.82}$$

where $\mathcal{F}^{-1}$ is the inverse Fourier transform.

To obtain consistent equations, we include all terms up to order ε^3 and neglect terms of order ε^4 and higher.

From the continuity Eq. (7.45), we obtain, using $\mathbf{p} = \mathbf{A} + \nabla \psi$ after a short calculation

$$\varepsilon^2 \frac{\partial^2 \delta n}{\partial t^2} \approx -\varepsilon^2 \frac{\partial}{\partial t} \nabla^2 \delta \psi + \mathcal{O}(\varepsilon^4) . \tag{7.83}$$

On the other hand, by applying the Laplace operator to (7.55) in conjunction with the Poisson Eq. (7.61) we obtain

$$\varepsilon^2 \frac{\partial}{\partial t} \nabla^2 \delta \psi = \frac{n_0}{n_c} \varepsilon^2 \delta n - \frac{\varepsilon^2}{2} \nabla^2 \mathbf{A}_\perp^2 + \mathcal{O}(\varepsilon^4) . \tag{7.84}$$

By combining these, we obtain to lowest order

$$\frac{\partial^2 \delta n}{\partial t^2} + \frac{n_0}{n_c} \delta n \approx \frac{1}{2} \nabla^2 \mathbf{A}_\perp^2 . \tag{7.85}$$

Together with the (perpendicular component of the) wave equation

$$\frac{\partial^2 \mathbf{A}_\perp}{\partial t^2} - \nabla^2 \mathbf{A}_\perp \approx -\frac{n_0}{n_c} \left\{1 - \frac{1}{2} \mathbf{A}_\perp^2 + \delta n\right\} \mathbf{A}_\perp \tag{7.86}$$

we obtain a closed system of two equations. Instead of a series expansion, we can retain the full γ factor, keeping in mind at the end the valid accuracy of the results. For the initial condition $A_\parallel = 0$ the parallel component will remain approximately zero during the temporal evolution.

Some explanations are necessary to fully understand (7.86). Due to the zeroth-order dispersion relation, we find to lowest order

$$\frac{\partial^2 \mathbf{A}_\perp}{\partial t^2} - \frac{\partial^2 \mathbf{A}_\perp}{\partial x^2} + \frac{n_0}{n_c} \mathbf{A}_\perp = \varepsilon \, 0 + \mathcal{O}(\varepsilon^3) , \tag{7.87}$$

which shows that the system is indeed correctly closed at the intended order.

The weakly relativistic model for a *linearly* polarized electromagnetic wave with $\mathbf{A} \approx \mathbf{A}_\perp = a \hat{\mathbf{e}}_z$ is

$$\frac{\partial^2 a}{\partial t^2} = \nabla^2 a - \frac{n_0}{n_c}\frac{(1+\delta n)}{\gamma}\,a\,, \tag{7.88}$$

$$\frac{\partial^2 \delta n}{\partial t^2} = -\frac{n_0}{n_c}\delta n + \nabla^2 \gamma\,, \tag{7.89}$$

where γ is the relativistic factor,

$$\gamma \approx 1 + \frac{1}{2}a^2 \quad \text{and} \quad \gamma^{-1} \approx 1 - \frac{1}{2}a^2\,. \tag{7.90}$$

The consistency of the model requires the use of the aforementioned approximate expressions and a consistent truncation at orders ε^3 and ε^2.

For circular polarization, it is possible to write for the complex-valued scalar field a

$$a(\mathbf{r},t) = A_y(\mathbf{r},t) + iA_z(\mathbf{r},t) \quad \text{with} \quad \mathbf{A}^2 = |a|^2\,. \tag{7.91}$$

Then follows

$$\frac{\partial^2 a}{\partial t^2} = \nabla^2 a - \frac{n_0}{n_c}\frac{(1+\delta n)}{\gamma}\,a\,, \tag{7.92}$$

$$\frac{\partial^2 \delta n}{\partial t^2} = -\frac{n_0}{n_c}\delta n + \nabla^2 \gamma \tag{7.93}$$

where the relativistic factor takes the form

$$\gamma \approx 1 + \frac{1}{2}|a|^2 \quad \text{and} \quad \gamma^{-1} \approx 1 - \frac{1}{2}|a|^2\,. \tag{7.94}$$

Note the difference in the relativistic factor compared to the model for linear polarization.

In summary, the spatial variations of the γ factor cause density modulations and drive plasma oscillations. The ion motion is neglected for rapid propagation. The parameter n_0/n_c (ion background density to critical density) should be less than 1 for wave propagation in an underdense medium and greater than 1/4 to avoid Raman instability. Note that in vacuum $n_0 \equiv 0$ and therefore $\delta n \equiv n_0 \delta n \equiv 0$.

Weakly Relativistic 1D Maxwell Two-Fluid Model

Let us briefly reconsider Eq. (7.87). For a physical understanding, we will recapitulate the key steps of its derivation. One starts with the wave equation in Coulomb gauge

$$\nabla^2 \mathbf{A}_\perp - \mu_0 \varepsilon_0 \frac{\partial^2 \mathbf{A}_\perp}{\partial t^2} = -\mu_0 \mathbf{j}_\perp \tag{7.95}$$

and the approximations

$$\mathbf{j}_\perp \approx -en_e \mathbf{v}_{e\perp} = -\frac{en_e}{m_e \gamma} \mathbf{p}_{e\perp} \approx -\frac{e^2 n_e}{m_e \gamma} \mathbf{A}_\perp \,, \tag{7.96}$$

where the last relation follows from

$$\frac{d\mathbf{p}_{e\perp}}{dt} \approx e \frac{\partial \mathbf{A}_\perp}{\partial t} \leftrightarrow \mathbf{p}_{e\perp} \approx e\mathbf{A}_\perp \tag{7.97}$$

Since $\gamma \equiv \gamma_e \approx \sqrt{1 + e^2 \mathbf{A}_\perp^2}$, the evolution of the γ factor leads to the previously derived form (7.87). Now we introduce

$$\mathbf{A}_\perp \approx \frac{1}{e} \mathbf{p}_{e\perp} \approx \frac{m_e}{e} \mathbf{v}_{e\perp} \left(1 + \frac{1}{2c^2} \mathbf{v}_{e\perp}^2 \right) \tag{7.98}$$

and $\mathbf{v}_\perp \equiv \mathbf{v}_{e\perp}$ to obtain in 1D

$$c^2 \frac{\partial^2 \mathbf{v}_\perp}{\partial x^2} - \frac{\partial^2 \mathbf{v}_\perp}{\partial t^2} - \omega_{pe}^2 \mathbf{v}_\perp \approx -\frac{1}{2} \frac{\partial^2 \mathbf{v}_\perp^3}{\partial x^2} + \frac{1}{2c^2} \frac{\partial^2 \mathbf{v}_\perp^3}{\partial t^2} + \frac{\omega_{pe}^2}{n_0} \delta n_e \, \mathbf{v}_\perp \,, \tag{7.99}$$

where $\mathbf{v}_\perp^3 = (\mathbf{v}_\perp \cdot \mathbf{v}_\perp) \mathbf{v}_\perp$. This form was first derived by Gorbunov and Kirsanov [70].

The other Eq. (7.85) describes the *high-frequency* electron response. Recall that we derived it solely from the electron dynamics with immobile ions. A simplified derivation for the one-dimensional case begins with the electron continuity equation

$$\frac{\partial n_e}{\partial t} + \frac{\partial}{\partial x} \left(\frac{n_e p_{ex}}{m_e \gamma} \right) = 0 \tag{7.100}$$

and the momentum balance in the x direction

$$\frac{\partial p_{ex}}{\partial t} + v_{ex} \frac{\partial}{\partial x} p_{ex} = -eE_x - e[\mathbf{v}_e \times \mathbf{B}]_x \,. \tag{7.101}$$

With the approximate solution $\mathbf{p}_{e\perp} \approx e\mathbf{A}_{e\perp}$ the last term on the right-hand side becomes

$$- e[\mathbf{v}_e \times \mathbf{B}]_x \approx -\mathbf{v}_{e\perp} \cdot \frac{\partial}{\partial x} \mathbf{p}_{e\perp} \,. \tag{7.102}$$

Altogether, we obtain

$$\frac{\partial p_{ex}}{\partial t} \approx -eE_x - m_e c^2 \frac{\partial \gamma}{\partial x} . \tag{7.103}$$

Combining with the derivative of (7.100), we obtain in the weakly relativistic case

$$\boxed{\frac{\partial^2 \delta n_e}{\partial t^2} + \frac{\omega_{pe}^2}{\gamma} \delta n_e \approx \frac{n_0 c^2}{\gamma} \frac{\partial^2 \gamma}{\partial x^2}} . \tag{7.104}$$

If one proceeds in the same way as discussed previously, (7.85) follows immediately.

The question remains as to when a low-frequency ion response should be included. It is clear that when transforming from the laboratory frame with x, t to a moving reference frame with $\tau = t$, $\xi = x - Vt$ the partial time derivative at constant ξ becomes $\partial/\partial\tau$, and the time derivative in the laboratory frame $\partial/\partial t$ (at constant x) transforms as follows:

$$\frac{\partial}{\partial t} \rightarrow \frac{\partial}{\partial \tau} - V \frac{\partial}{\partial \xi} . \tag{7.105}$$

For example, when we consider standing waves or pulses in a moving reference frame, for

$$\left| V \frac{\partial}{\partial \xi} \right| \sim \frac{V}{L} \gg \omega_{pi} \tag{7.106}$$

the response of the ions can be neglected. This condition is fulfilled for rapidly moving and sufficiently narrow pulses (characteristic length L). If the condition is not met, ion dynamics play a role for $\tau > \omega_{pi}^{-1}$.

The low-frequency ion response (e.g., non-relativistic) is obtained from

$$\frac{\partial n_i}{\partial t} + \nabla \cdot (n_i \mathbf{v}_i) = 0 , \tag{7.107}$$

$$\frac{\partial \mathbf{v}_i}{\partial t} + \mathbf{v}_i \cdot \nabla \mathbf{v}_i = -\frac{e}{m_i} \nabla \phi - \frac{T_i}{m_i} \nabla \ln n_i , \tag{7.108}$$

when we include the ion pressure term with an isothermal equation of state (see below). For low ion temperatures (where $T_i \approx 0$ is set) and small ion velocities (due to the large ion mass), we obtain

$$\begin{aligned} \frac{\partial^2 n_i}{\partial t^2} + \nabla \cdot \frac{\partial}{\partial t}(n_i \mathbf{v}_i) &\approx \frac{\partial^2 n_i}{\partial t^2} + \nabla \cdot \left\{ \mathbf{v}_i \nabla \cdot (n_i \mathbf{v}_i) - n_i \mathbf{v}_i \cdot \nabla \mathbf{v}_i - \frac{e}{m_i} n_i \nabla \phi \right\} \\ &\approx \frac{\partial^2 n_i}{\partial t^2} - \frac{e}{m_i} n_0 \nabla^2 \phi \approx 0 . \end{aligned} \tag{7.109}$$

For low-frequency phenomena, the inertia term of the electrons in the electron momentum balance can be neglected,

$$\frac{\partial}{\partial t}\mathbf{p}_e + \underbrace{m_e c^2 \nabla\tilde{\gamma} - \mathbf{v}_e \times \nabla \times \mathbf{p}_e}_{\mathbf{v}_e\cdot\nabla\mathbf{p}_e} \approx e\nabla\phi + e\frac{\partial \mathbf{A}}{\partial t} - e\mathbf{v}_e \times \nabla \times \mathbf{A} - T_e \nabla \ln n_e \,. \quad (7.110)$$

With $\mathbf{p}_e \approx e\mathbf{A}_\perp$ we find

$$m_e c^2 \nabla\tilde{\gamma} \approx e\nabla\phi - T_e \nabla \ln n_e \,, \quad (7.111)$$

where $\tilde{\gamma} = \sqrt{1+\mathbf{a}^2}$ and $\mathbf{a} = e\mathbf{A}_\perp/m_e c$ are. Quasineutrality $n_e \approx n_i$ should hold for low-frequency responses, so that a combination of (7.111) with (7.109) leads to

$$\boxed{\frac{\partial^2 \delta n}{\partial t^2} - c_s^2 \nabla^2 \delta n \approx \frac{m_e}{m_i} c^2 \frac{n_0}{2} \nabla^2 \mathbf{a}^2 \Big|_{\mathrm{lf}}} \quad (7.112)$$

with the ion sound speed $c_s = \sqrt{T_e/m_i}$. We have indicated by the subscript "lf" that the ions respond only to the low-frequency (lf) part of the forces.

Example 7.3 (Adiabatic Changes)

When investigating phase velocities that are greater than the thermal electron velocity, the adiabatic law $P_e \sim n_e^\gamma$ should be used instead of the kinetic electron pressure P_e. It should be noted that γ here is the adiabatic exponent and must not be confused with the relativistic γ factor. This means that in (7.110) we should replace

$$\underbrace{-T_e \nabla \ln n_e}_{\text{isothermal}} \longrightarrow \underbrace{-\gamma n_e^{\gamma-2} \frac{P_{e0}}{n_0^\gamma} \nabla n_e}_{\text{adiabatic}} \,. \quad (7.113)$$

Then, instead of (7.111), we obtain

$$m_e c^2 \nabla\tilde{\gamma} \approx e\nabla\phi - \gamma T_0 \frac{1}{n_0} \nabla n_e \,, \quad (7.114)$$

with the result

$$\boxed{\frac{\partial^2 \delta n}{\partial t^2} - c_s^2 \gamma \nabla^2 \delta n \approx \frac{m_e}{m_i} c^2 \frac{n_0}{2} \nabla^2 \mathbf{a}^2 \Big|_{\mathrm{lf}}} \quad (7.115)$$

for $|V/L| \leq \omega_{pi}$. ■

Both Eq. (7.112) and (7.115) imply that, on the slow time scale, even weakly relativistic laser amplitudes

$$|\mathbf{a}| \sim \frac{v_{the}}{c} \tag{7.116}$$

can create density cavities that empty a certain spatial region. If we approximate with $c \approx 3\times 10^{10}$ cm/s and $v_{the} \approx 4.19\times 10^{7}\sqrt{T_e[\text{eV}]}$ cm/s, we see that temperatures on the order of 50 keV are required to avoid complete evacuation at $|\mathbf{a}| \approx 0.3$.

If we replace $-\omega_{pe}A \to E$ and use the isothermal equation of state, Eq. (7.112) will change to

$$\boxed{\frac{\partial^2 \delta n}{\partial t^2} - c_s^2 \nabla^2 \delta n \approx \varepsilon_0 \frac{\nabla^2 E^2}{2m_i}} \tag{7.117}$$

This equation was first introduced by Zakharov in a model that became known as the Zakharov equations [71–75].

Finally, we note that the equation describing the nonlinear density response within a two-field description can take different forms depending on the physics under consideration.

Nonrelativistic Limit

The formulation chosen in the previous subsection might lead us to believe that there is no effect on the density when relativistic effects are absent ($\gamma \equiv 1$). However, we should be cautious when taking the limit $\gamma \to 1$. The correct procedure, for example for linear polarization, is

$$\frac{\partial^2 a}{\partial t^2} = \nabla^2 a - \frac{n_0}{n_c}\frac{(1+\delta n)}{\gamma}\, a \;\to\; \nabla^2 a - \frac{n_0}{n_c}(1+\delta n)\, a\ , \tag{7.118}$$

$$\frac{\partial^2 \delta n}{\partial t^2} = -\frac{n_0}{n_c}\delta n + \nabla^2 \gamma \,\hat{=}\, -\frac{n_0}{n_c}\delta n + \frac{1}{2}\nabla \cdot \frac{\nabla a^2}{\gamma} \;\to\; -\frac{n_0}{n_c}\delta n + \frac{1}{2}\nabla^2 a^2\ . \tag{7.119}$$

In the last equation, we first showed that within the present scaling, the effect of the electromagnetic wave on the density perturbation arises from the *relativistic ponderomotive force* ($\sim -\frac{1}{2\gamma}\nabla a^2$), which correctly reduces to its well-known nonrelativistic form. In this sense, the mathematically correct term $\nabla^2 \gamma$ (up to order a^2) points to a somewhat misleading origin.

Single-Field Models

The coupled, weakly relativistic equations sometimes allow a further reduction to single-field models. In the case of circular polarization (7.92) and (7.93), the driving term in the density equation is low-frequency, and the system could be reduced to a standard form of the cubic nonlinear Schrödinger equation. The situation is different for linear polarization (7.88) and (7.89), where both low-frequency responses and higher harmonics occur.

Circular Polarization

For low-frequency responses, the coupled weakly relativistic Eqs. (7.92) and (7.93) for circularly polarized waves must be supplemented by the ion density dynamics, provided the pulse duration is longer than ω_{pi}^{-1}. On the other hand, for short pulses, the purely electronic response can be used.

For the nearly stationary electronic response, we obtain (using ω_{pe}^{-1} for time normalization)

$$\frac{\partial^2 a}{\partial t^2} = \frac{\partial^2 a}{\partial x^2} - \left(1 + \delta n - \frac{1}{2}|a|^2\right) a \, , \quad \delta n \approx \frac{1}{2}\frac{\partial^2 |a|^2}{\partial x^2} \, . \tag{7.120}$$

This yields in 1D the single-field model

$$\boxed{\frac{\partial^2 a}{\partial t^2} - \frac{\partial^2 a}{\partial x^2} + a = \frac{1}{2}\left(\frac{\partial^2 |a|^2}{\partial x^2} - |a|^2\right) a} \, , \tag{7.121}$$

which is applicable for $V/L \gg \omega_{pi}$.

Linear Polarization

For linear polarization, we start from (7.88)

$$\frac{\partial^2 a}{\partial t^2} - \frac{\partial^2 a}{\partial x^2} = -\left(1 + \delta n + \frac{1}{2}a^2\right) a \tag{7.122}$$

and (7.89) in the form

$$\frac{\partial^2 \delta n}{\partial t^2} = -\delta n + \frac{1}{2}\frac{\partial^2 a^2}{\partial x^2} \tag{7.123}$$

with ω_{pe}^{-1} for time normalization. This means that we assume short pulses with $|V/L| \gg \omega_{pi}$, so that we can neglect the ion dynamics. For the density, we need to solve an equation for a driven linear oscillator. The general procedure is discussed in the section on the generation of wakefields. Here, we present a less general approach.

Let us consider the modification of the first harmonic of a linear transverse mode by writing

$$a \equiv a^{(0)} = \frac{1}{2}\left[A(x,t)e^{-i\omega_\perp^{(0)}t} + A^*(x,t)e^{+i\omega_\perp^{(0)}t}\right] . \tag{7.124}$$

For the driven density response, we make the ansatz

$$\delta n = N_0 + \frac{1}{2}\left(N_2 e^{-2i\omega_\perp^{(0)}t} + N_2^* e^{2i\omega_\perp^{(0)}t}\right) . \tag{7.125}$$

The frequency of the transverse mode is denoted by $\omega_\perp^{(0)}$; we will return to this definition later.

By substituting (7.125) and (7.124) into (7.123), a short calculation yields

$$N_0 = \frac{1}{4}\frac{\partial^2 |A|^2}{\partial x^2} , \quad N_2 = -\frac{1}{16\omega_\perp^{(0)\,2} - 4}\frac{\partial^2 A^2}{\partial x^2} . \tag{7.126}$$

We substitute (7.124) and (7.125) together with (7.126) into (7.122), and after some calculations, for the coefficient proportional to $\exp[-i\omega_\perp^{(0)}t]$, we obtain

$$\boxed{\begin{aligned} &\frac{1}{2}\frac{\partial^2 A}{\partial t^2} - i\omega_\perp^{(0)}\frac{\partial A}{\partial t} + \frac{1}{2}(1-\omega_\perp^{(0)\,2})A - \frac{1}{2}\frac{\partial^2 A}{\partial x^2} + \frac{1}{8}\frac{\partial^2 |A|^2}{\partial x^2}A \\ &- \frac{1}{64\omega_\perp^{(0)\,2} - 16}\frac{\partial^2 A^2}{\partial x^2}A^* - \frac{3}{16}|A|^2 A = 0 . \end{aligned}} \tag{7.127}$$

In principle, since a low-frequency density response occurs, the ion dynamics should also be included. However, for short pulses with a pulse duration shorter than the inverse of the ion oscillation frequencies, the ions can be considered immobile.

7.2 Relativistic Parametric Instabilities

Nonlinear effects generally prevent complete analytical solutions of the Maxwell-fluid systems. Therefore, a precise stability analysis of relativistic wave models usually has to be carried out numerically, unless one linearizes around a simple stationary solution. The latter are, for example, constant waves, often referred to as pump waves. The instabilities thus found can be regarded as generalizations of known nonrelativistic instabilities for plane wave solutions. Of course, relativistic effects also give rise to new types of instabilities, such as the relativistic modulation instability. Waves in plasmas are the source of many types of instabilities [24, 76, 77]. Electrostatic decay and modulation instabilities occur. When electromagnetic modes are mainly involved, the two-plasmon decay instability as well as the

Raman and Brillouin scattering instabilities develop. Kinetic limits, in which quasi-modes are involved, are referred to as Compton scattering. From the multitude of instabilities, we will discuss the Raman and Brillouin instabilities in the following.

Stimulated Raman and Brillouin Scattering in the Classical Regime

The physics of the Raman instability is relatively easy to understand [24]. When a light wave propagates in a plasma, it can generate density waves in the direction of propagation. The oscillating electrons produce a transverse current, which can generate a scattered light wave, provided the wave numbers and frequencies are properly matched. The scattered light wave interferes with the incident light wave and produces a ponderomotive pressure, which acts mainly on the electrons and affects the generated density waves. Through this feedback loop, an instability can arise.

We begin with the wave equation for the vector potential $\mathbf{A}$ in the Coulomb gauge $\nabla \cdot \mathbf{A} = 0$:

$$\boxed{\left(\frac{1}{c^2}\frac{\partial^2}{\partial t^2} - \nabla^2\right)\mathbf{A} = \mu_0 \mathbf{j} - \mu_0 \varepsilon_0 \frac{\partial}{\partial t}\nabla\phi} \ . \tag{7.128}$$

We split the current density in the wave equation into a transverse part $\mathbf{j}_\perp$ (associated with the light wave) and a longitudinal part $\mathbf{j}_\parallel$. The longitudinal part arises from the space charge (electron density) oscillations. The continuity equation for the space charge and the Poisson equation lead to

$$\nabla \cdot \left(\frac{\partial}{\partial t}\nabla\phi - \frac{1}{\varepsilon_0}\mathbf{j}\right) \approx \nabla \cdot \left(\frac{\partial}{\partial t}\nabla\phi - \frac{1}{\varepsilon_0}\mathbf{j}_\parallel\right) = 0 \ , \tag{7.129}$$

since to lowest order $\nabla \cdot \mathbf{j}_\perp = 0$. Equation (7.128) then reduces to

$$\left(\frac{1}{c^2}\frac{\partial^2}{\partial t^2} - \nabla^2\right)\mathbf{A} = \mu_0 \mathbf{j}_\perp \approx -\mu_0 e n_e \mathbf{v}_{e\perp} \approx -\frac{e^2}{c^2 \varepsilon_0 m_e} n_e \mathbf{A} \ , \tag{7.130}$$

where, as in previous models, the nonrelativistic electron momentum has been replaced by $\mathbf{p}_{e\perp} = m_e \mathbf{v}_{e\perp} \approx e\mathbf{A} \hat{=} e\mathbf{A}_\perp$. We can use this equation for the scattered electromagnetic wave after substituting $n_e = n_0 + \delta n$ and $\mathbf{A} = \mathbf{A}_0 + \delta\mathbf{A}$, to obtain to lowest order for $\delta\mathbf{A}$:

$$\left(\frac{\partial^2}{\partial t^2} - c^2\nabla^2 + \omega_{pe}^2\right)\delta\mathbf{A} = -\frac{e^2}{\varepsilon_0 m_e}\delta n \mathbf{A}_0 \ . \tag{7.131}$$

The density response follows from the conservation laws for electron density and momentum,

$$\frac{\partial n_e}{\partial t} + \nabla \cdot (n_e \mathbf{v}_e) = 0 \,, \tag{7.132}$$

$$\frac{\partial \mathbf{v}_e}{\partial t} + \mathbf{v}_e \cdot \nabla \mathbf{v}_e = -\frac{e}{m_e}(\mathbf{E} + \mathbf{v}_e \times \mathbf{B}) - \frac{1}{n_e m_e} \nabla p_e \,, \tag{7.133}$$

together with the adiabatic equation of state (for one degree of freedom) $p_e \sim n_e^3$.

If we set $\mathbf{v}_e = \mathbf{v}_{e\|} + e\mathbf{A}/m_e$, the parallel component of (7.133) leads to

$$\frac{\partial \mathbf{v}_{e\|}}{\partial t} = \frac{e}{m_e}\nabla\phi - \frac{1}{2}\nabla\left(\mathbf{v}_{e\|} + \frac{e\mathbf{A}}{m_e}\right)^2 - \frac{1}{n_e m_e}\nabla p_e \,. \tag{7.134}$$

By identifying $\mathbf{v}_{e\|} \equiv \delta\mathbf{v}$ and $\phi = \delta\phi$, we obtain for the fluctuations to lowest order

$$\frac{\partial \delta n}{\partial t} + n_0 \nabla \cdot \delta\mathbf{v} = 0 \,, \tag{7.135}$$

$$\frac{\partial \delta\mathbf{v}}{\partial t} = \frac{e}{m_e}\nabla\delta\phi - \frac{e^2}{m_e^2}\nabla(\mathbf{A}_0 \cdot \delta\mathbf{A}) - \frac{3v_{the}^2}{n_0}\nabla\delta n \,. \tag{7.136}$$

These equations can be combined to:

$$\left(\frac{\partial^2}{\partial t^2} + \omega_{pe}^2 - 3v_{the}^2\nabla^2\right)\delta n = \frac{n_0 e^2}{m_e^2}\nabla^2(\mathbf{A}_0 \cdot \delta\mathbf{A}) \,. \tag{7.137}$$

The Eqs. (7.131) and (7.137) are the coupled equations for the electrostatic (Langmuir) and electromagnetic (scattered) waves. For the dispersion relation of the Raman instability, we set $\mathbf{A}_0 = \mathbf{A}_0 \cos(\mathbf{k}_0 \cdot \mathbf{r} - \omega_0 t)$ [note the change in notation] and perform a Fourier transformation,

$$\left(\omega^2 - c^2k^2 - \omega_{pe}^2\right)\delta\mathbf{A}(\mathbf{k}, \omega) = \frac{e^2}{2\varepsilon_0 m_e}\mathbf{A}_0[\delta n(\mathbf{k} - \mathbf{k}_0, \omega - \omega_0) + \delta n(\mathbf{k} + \mathbf{k}_0, \omega + \omega_0)] \,, \tag{7.138}$$

$$(\omega^2 - \omega_k^2)\delta n(\mathbf{k}, \omega) = \frac{k^2 e^2 n_0}{2m_e^2}\mathbf{A}_0[\delta\mathbf{A}(\mathbf{k} - \mathbf{k}_0, \omega - \omega_0) + \delta\mathbf{A}(\mathbf{k} + \mathbf{k}_0, \omega + \omega_0)] \,, \tag{7.139}$$

with $\omega_k = \sqrt{\omega_{pe}^2 + 3k^2 v_{te}^2}$. We consider the modes $\mathbf{k}, \omega$ with $\Re\omega \approx \omega_{pe}$ (plasma wave) and the sidebands $\mathbf{k} - \mathbf{k}_0, \omega - \omega_0$ and $\mathbf{k} + \mathbf{k}_0, \omega + \omega_0$ Assuming that the harmonics $\mathbf{k} \pm 2\mathbf{k}_0, \omega \pm 2\omega_0$ are nonresonant, we can easily eliminate $\delta\mathbf{A}$ in (7.139) by using (7.138), in order to obtain, for nontrivial solutions, a sixth-order polynomial in ω:

$$\boxed{\omega^2 - \omega_k^2 = \frac{\omega_{pe}^2 k^2 v_0^2}{4}\left[\frac{1}{D_-} + \frac{1}{D_+}\right]} \,. \tag{7.140}$$

Here, $v_0 = eE_0/m_e\omega_0 = eA_0/m_e$, $D_\pm = \omega_\pm^2 - k_\pm^2 c^2 - \omega_{pe}^2$, $\omega_\pm^2 = (\omega \pm \omega_0)^2$ and $k_\pm^2 = (\mathbf{k} \pm \mathbf{k}_0)^2$ are defined.

Scattering can occur in all directions. For backward or sideways scattering, we neglect the nonresonant, upshifted light wave as nonresonant and make the assumption $\omega = \omega_k + \delta\omega$ with $\delta\omega \ll \omega_k$. The maximum growth occurs when the scattered light wave is also resonant ($D_- \approx 0$). In this case, the growth rate $\gamma = -i\delta\omega$ is maximized for backward scattering (which exhibits the greatest growth) to

$$\boxed{\gamma = \frac{kv_0}{4}\sqrt{\frac{\omega_{pe}^2}{\omega_k(\omega_0 - \omega_k)}} \quad \text{at} \quad k = k_0 + \frac{\omega_0}{c}\sqrt{1 - \frac{2\omega_{pe}}{\omega_0}}} \,. \tag{7.141}$$

It is obvious that $n \leq n_{\text{krit}}$ (quarter-critical density) is necessary for backward scattering to occur in the case of Raman instability.

Further details, especially on forward Raman and stimulated Compton scattering, can be found in the specialized literature [24, 76, 78, 79].

Stimulated Raman scattering describes the scattering from a high-frequency electron plasma oscillation (Langmuir wave).
For the Brillouin instability, the density fluctuation δn is a low-frequency oscillation associated with an ion-acoustic wave.

For the (more or less immediate) electron response, we can also use the approach for Brillouin scattering (7.136), but with an isothermal instead of an adiabatic equation of state. Furthermore, the electron inertia is negligible, so that

$$0 \approx \frac{e}{m_e}\nabla\delta\phi - \frac{e^2}{m_e^2}\nabla(\mathbf{A}_0 \cdot \delta\mathbf{A}) - \frac{v_{the}^2}{n_0}\nabla\delta n \,. \tag{7.142}$$

The scalar potential transfers the ponderomotive force to the ions. In the ion equations, we neglect the ion temperature for simplicity; thus, we obtain

$$\frac{\partial n_i}{\partial t} + \nabla \cdot (n_i \mathbf{v}_i) = 0 \,, \tag{7.143}$$

$$\frac{\partial \mathbf{v}_i}{\partial t} + \mathbf{v}_i \cdot \nabla \mathbf{v}_i \approx -\frac{Ze}{m_i}\nabla\phi \,. \tag{7.144}$$

To lowest order, we find for the ion density perturbation

$$\frac{\partial \delta n_i}{\partial t} + n_0 \nabla \cdot \delta \mathbf{v}_i = 0 \,, \tag{7.145}$$

$$\frac{\partial \delta \mathbf{v}_i}{\partial t} = -\frac{Ze}{m_i} \nabla \delta\phi \,. \tag{7.146}$$

From this, we obtain

$$\frac{\partial^2 \delta n_i}{\partial t^2} - \frac{n_0 Ze}{m_i} \nabla^2 \delta\phi = 0 \,. \tag{7.147}$$

For quasi-neutral perturbations, $Z\delta n_i \approx \delta n_e \equiv \delta n$ holds, so that we can eliminate $\delta\phi$ from (7.142), to finally obtain:

$$\left(\frac{\partial^2}{\partial t^2} - c_s^2 \nabla^2 \right) \delta n = \frac{n_0 Z e^2}{m_e m_i} \nabla^2 (\mathbf{A}_0 \cdot \delta \mathbf{A}) \,, \tag{7.148}$$

where $c_s = \sqrt{ZT_e/m_i}$ is the ion sound speed. The equation for δn is coupled to the wave Eq. (7.131),

$$\boxed{\left(\frac{\partial^2}{\partial t^2} - c^2 \nabla^2 + \omega_{pe}^2 \right) \delta \mathbf{A} = -\frac{e^2}{\varepsilon_0 m_e} \delta n \mathbf{A}_0} \,. \tag{7.149}$$

The further analysis is similar to that for Raman scattering. We now obtain the dispersion relation

$$\boxed{\omega^2 - k^2 c_s^2 = \frac{\omega_{pi}^2 k^2 v_0^2}{4} \left[\frac{1}{D_-} + \frac{1}{D_+} \right]} \,, \tag{7.150}$$

where $\omega_{pi} = \sqrt{Zn_0 e^2/\varepsilon_0 m_i}$ is.

Proceeding as before, one arrives at the instability result for backscattering

$$\boxed{\gamma = \frac{1}{2\sqrt{2}} \frac{k_0 v_0 \omega_{pi}}{\sqrt{\omega_0 k_0 c_s}} \quad \text{at} \quad k = 2k_0 - \frac{2\omega_0}{c} \frac{c_s}{c}} \,. \tag{7.151}$$

In summary, the general form of the dispersion relation, which takes into account the entire spectrum of electromagnetic excitations (electrostatic as well as electromagnetic), can be written as [78]

$$
\begin{aligned}
&1+\chi_e(\mathbf{k},\omega)+\chi_i(\mathbf{k},\omega)\\
&=-\chi_e(\mathbf{k},\omega)[1+\chi_i(\mathbf{k},\omega)]\frac{k^2}{4}\left\{\frac{|\mathbf{k}_+\times\mathbf{v}_0|^2}{k_+^2D_+}+\frac{|\mathbf{k}_-\times\mathbf{v}_0|^2}{k_-^2D_-}-\frac{|\mathbf{k}_+\cdot\mathbf{v}_0|^2}{k_+^2\omega_+^2\varepsilon_+}-\frac{|\mathbf{k}_-\cdot\mathbf{v}_0|^2}{k_-^2\omega_-^2\varepsilon_-}\right\}.
\end{aligned}
\tag{7.152}
$$

Here, $\mathbf{v}_0 = e\mathbf{A}_0/m_e = e\mathbf{E}_0/m_e\omega_0$, χ_j denote the electric susceptibilities in the respective frequency ranges, and $\varepsilon_\pm = 1+\chi_\pm$ is the high-frequency relative permittivity. In the case of Raman scattering, for example, we have:

$$
\chi_i(\mathbf{k},\omega)\approx 0\,,\quad \chi_e(\mathbf{k},\omega)\approx -\frac{\omega_{pe}^2}{\omega^2}+\cdots(\text{thermal effects})\,,\quad \varepsilon_\pm = 1-\frac{\omega_{pe}^2}{\omega^2}\,. \tag{7.153}
$$

On the other hand, for Brillouin scattering we have

$$
\chi_i(\mathbf{k},\omega)\approx -\frac{\omega_{pi}^2}{\omega^2}\,,\quad \chi_e(\mathbf{k},\omega)\approx \frac{1}{k^2\lambda_{De}^2}\,,\quad 1+\chi_e+\chi_i\approx\chi_e+\chi_i\,,\quad 1+\chi_i\approx\chi_i\,, \tag{7.154}
$$

and $\varepsilon_\pm$ is as before. The more general formulation (7.152) has the advantage that damping effects can be easily incorporated to determine thresholds for the scattering instabilities [76].

Scattering Instabilities in the Relativistic Regime

Style3 In the weakly relativistic regime, we should replace the rest mass with the relativistic mass, which contains the relativistic factor γ. Apart from this—rather trivial—substitution, an additional term appears in the wave equation, which results from a derivative of the γ factor. In the case of circular polarization in the weakly relativistic regime (as long as the ions behave non-relativistically), it is straightforward to generalize the non-relativistic results. We should replace the high-frequency wave Eq. (7.131) by

$$
\left(\frac{\partial^2}{\partial t^2}-c^2\nabla^2+\omega_{pe}^2\right)\delta\mathbf{A} = -\frac{e^2}{\varepsilon_0 m_e}\delta n\mathbf{A}_0+\frac{e^2 n_0}{\varepsilon_0 m_e}\mathbf{A}_0\frac{e^2}{m_e^2c^2}(\mathbf{A}_0\cdot\delta\mathbf{A}) \tag{7.155}
$$

. The second term on the right-hand side arises from the variation of the relativistic mass. The plasma responses can be followed using the relativistic fluid equations (using the momentum $\mathbf{p}_e$ instead of the velocity $\mathbf{v}_e$) to obtain results similar to the non-relativistic case (of course, in the end, the rest mass is replaced by the relativistic mass).

In the case of Raman scattering, one can consult references [80–82], where the following dispersion relation was obtained:

$$\omega^2 - \frac{\omega_{pe}^2}{\gamma_e} = \frac{\omega_{pe}^2 k^2 v_0^2}{4\gamma_e^3} \underbrace{\left(\frac{k^2c^2 - \omega^2 + \frac{\omega_{pe}^2}{\gamma_e}}{k^2c^2} \right)}_{\approx 1 \text{ for } c \to \infty} \left[\frac{1}{\tilde{D}_-} + \frac{1}{\tilde{D}_+} \right], \tag{7.156}$$

with m_e as the electron rest mass, $v_0 = eE_0/m_e\omega_0 = eA_0/m_e$, $\tilde{D}_\pm = \omega_\pm^2 - k_\pm^2 c^2 - \frac{\omega_{pe}^2}{\gamma_e}$, $\omega_\pm^2 = (\omega \pm \omega_0)^2$, $k_\pm^2 = (\mathbf{k} \pm \mathbf{k}_0)^2$ and $\gamma_e = \sqrt{1 + \frac{v_0^2}{2c^2}}$. Thermal effects have been neglected.

Note that this dispersion relation reduces to (7.140), i.e.

$$\omega^2 - \omega_k^2 = \frac{\omega_{pe}^2 k^2 v_0^2}{4} \left[\frac{1}{D_-} + \frac{1}{D_+} \right], \tag{7.157}$$

when $\gamma_e \to 1$ and the nonrelativistic limit for $c \to \infty$ is assumed. The temporal dependence of the relativistic factor is only "simple" for circular polarization. Therefore, the presented dispersion relations are only valid for circular polarization. Otherwise, the relativistic factor should, for example, be expanded for small amplitudes and appropriately incorporated into the Fourier matching. The amplitude dependence of the dispersion relation in the relativistic regime leads to interesting conclusions regarding the possibility of realistic wave propagation [82]. These conclusions help to interpret the numerically obtained growth rates.

The scheme for relativistic effects on Raman scattering can (for circular polarization) be transferred to Brillouin scattering. In this case, the relativistic electron mass must be used in the dispersion relation (as long as the ions can be treated nonrelativistically). For Brillouin scattering in the (weakly) relativistic regime, one thus obtains

$$\omega^2 - k^2 c_s^2 = \frac{\omega_{pi}^2 k^2 v_0^2}{4\gamma_e^2} \underbrace{\left(\frac{k^2c^2 - \frac{m_i}{\gamma_e m_e}(\omega^2 - k^2 c_s^2)}{k^2c^2} \right)}_{\approx 1 \text{ for } c \to \infty} \left[\frac{1}{\tilde{D}_-} + \frac{1}{\tilde{D}_+} \right], \tag{7.158}$$

where $\omega_{pi} = \sqrt{Zn_0 e^2/\varepsilon_0 m_i}$ and $c_s = \sqrt{T_e/m_i}$ are. The remaining notations are the same as in the case of Raman scattering, in particular, m_e and m_i are the rest mass of the electron and ion, respectively. In the nonrelativistic limit $\gamma_e \to 1$ and $c \to \infty$ this dispersion relation reduces to (7.150), i.e.

$$\omega^2 - k^2 c_s^2 = \frac{\omega_{pi}^2 k^2 v_0^2}{4} \left[\frac{1}{D_-} + \frac{1}{D_+} \right]. \tag{7.159}$$

7.3 Three-Wave Interaction

In this section, we generalize the perspective of parametric instabilities by allowing for a back-action of the unstable modes on the pump wave. We also move away from plane waves and consider pulses of finite width. This leads us to a three-wave interaction model, which finds numerous applications in optics and plasma physics.

Raman Scattering with Pump Wave Depletion

We begin with a Raman model of three-wave interaction. The high-frequency pump transfers energy to a lower-frequency pulse and a Langmuir wave. The following three-wave model is well known [24]:

$$\boxed{\left(\frac{\partial^2}{\partial t^2} - c^2\nabla^2 + \omega_{pe}^2\right)\mathbf{A}_0 = -\omega_{pe}^2\frac{n}{n_0}\mathbf{A}_1,} \tag{7.160}$$

$$\boxed{\left(\frac{\partial^2}{\partial t^2} - c^2\nabla^2 + \omega_{pe}^2\right)\mathbf{A}_1 = -\omega_{pe}^2\frac{n}{n_0}\mathbf{A}_0,} \tag{7.161}$$

$$\boxed{\left(\frac{\partial^2}{\partial t^2} - 3v_{the}^2\nabla^2 + \omega_{pe}^2\right)\frac{n}{n_0} = \frac{e^2}{c^2 m_e^2}\nabla^2(\mathbf{A}_0 \cdot \mathbf{A}_1)\,.} \tag{7.162}$$

The equations are formulated for the vector potentials of the pump wave (index 0) and the scattered wave (index 1), as well as for the low-frequency density perturbations (n). In the following, we consider a spatially one-dimensional and linearly polarized case and introduce a

$$\mathbf{A}_0 = \hat{y}\frac{m_e c^2}{2e}a_0(x,t)e^{i(k_0 x - \omega_0 t)} + c.c.\,, \tag{7.163}$$

$$\mathbf{A}_1 = \hat{y}\frac{m_e c^2}{2e}a_1(x,t)e^{i(-k_1 x - \omega_1 t)} + c.c.\,, \tag{7.164}$$

$$\frac{n}{n_0} = i\frac{ck_2}{2\omega_{pe}}\sqrt{\frac{\omega_0}{\omega_{pe}}}a_2(x,t)e^{i(k_2 x - \omega_2 t)} + c.c.\,. \tag{7.165}$$

For backscattering, the following notation was used: $\omega_s \equiv \omega_1 = \omega_0 - \omega \equiv \omega_0 - \omega_2$, $\mathbf{k}_s = \mathbf{k}_0 - \mathbf{k}$, $\mathbf{k} \equiv k_2\hat{x}$, $\mathbf{k}_s \equiv -k_1\hat{x}$ and $\mathbf{k}_0 \equiv k_0\hat{x}$. This means that $k_0 = -k_1 + k_2$ holds.

Assuming a slowly varying envelope approximation, we suppose that the amplitudes change slowly compared to the phases. Substituting the above expressions into the wave

Eqs. (7.160)–(7.162) and neglecting the second derivatives of the amplitudes, we obtain

$$\left(\frac{\partial}{\partial t}+\frac{c^2k_0}{\omega_0}\frac{\partial}{\partial x}\right)a_0 \approx \frac{ck_2}{4}\sqrt{\frac{\omega_{pe}}{\omega_0}}a_1a_2\,, \tag{7.166}$$

$$\left(\frac{\partial}{\partial t}-\frac{c^2k_1}{\omega_1}\frac{\partial}{\partial x}\right)a_1 \approx -\frac{ck_2}{4}\sqrt{\frac{\omega_{pe}\omega_0}{\omega_1^2}}a_1a_2^*\,, \tag{7.167}$$

$$\left(\frac{\partial}{\partial t}+i\delta\omega\right)a_2 \approx -\frac{ck_2}{4}\frac{\omega_{pe}}{\omega_0}\sqrt{\frac{\omega_{pe}}{\omega_0}}a_0a_1^*\,. \tag{7.168}$$

Here, $\delta\omega \approx \omega_L - \omega_2$ applies, with $\omega_L = \sqrt{\omega_{pe}^2 + 3k_2^2 v_{the}^2}$, so that $\omega_L^2 - \omega_2^2 \approx 2\delta\omega\omega_L \approx 2\delta\omega\omega_2$. In the resonant case, we assume that $\delta\omega \approx 0$.

For backscattering with $k_2 \approx 2k_0$ and in the case $\omega_0 \approx \omega_1 \gg \omega_{pe}$ further simplification is straightforward. But even in the slightly subcritical case, we can simplify by changing the notation to

$$\frac{a_0}{\sqrt{\omega_0\omega_1}} \equiv \tilde{a}_0\,, \quad \frac{a_1}{\omega_0} \equiv \tilde{a}_1\,, \quad \frac{a_2}{\sqrt{\omega_0\omega_1}} \equiv \tilde{a}_2\,. \tag{7.169}$$

We introduce the group velocities c_{pump} and c_{seed} of the pump wave and a "seed wave" (hereafter also referred to as the seed pulse). Then we can write:

$$\left(\frac{\partial}{\partial t}+c_{\text{pump}}\frac{\partial}{\partial x}\right)\tilde{a}_0 = \frac{\sqrt{\omega_0\omega_{pe}}}{2}\tilde{a}_1\tilde{a}_2\,, \tag{7.170}$$

$$\left(\frac{\partial}{\partial t}-c_{\text{seed}}\frac{\partial}{\partial x}\right)\tilde{a}_1 = -\frac{\sqrt{\omega_0\omega_{pe}}}{2}\tilde{a}_0\tilde{a}_2^*\,, \tag{7.171}$$

$$\left(\frac{\partial}{\partial t}+i\delta\omega\right)\tilde{a}_2 = -\frac{\sqrt{\omega_0\omega_{pe}}}{2}\tilde{a}_0\tilde{a}_1^*\,. \tag{7.172}$$

If time is measured in units of $2/\sqrt{\omega_0\omega_{pe}}$ and the spatial coordinate x in $2c_{\text{pump}}/\sqrt{\omega_0\omega_{pe}}$, we obtain for

$$\tilde{a}_0 \equiv E_p\,, \quad \tilde{a}_1 \equiv E_s\,, \quad \tilde{a}_2 \equiv -N^*\,, \quad \delta\omega \equiv 0 \tag{7.173}$$

the three-wave interaction equations in standard form

$$\left(\frac{\partial}{\partial t}+\frac{\partial}{\partial x}\right)E_p = -N^*E_s\,, \tag{7.174}$$

$$\left(\frac{\partial}{\partial t} - v_{sp}\frac{\partial}{\partial x}\right)E_s = E_p N \ , \tag{7.175}$$

$$\frac{\partial N}{\partial t} = E_s E_p^* \ . \tag{7.176}$$

We have introduced the ratio $v_{sp} = \frac{c_{\text{seed}}}{c_{\text{pump}}} = \frac{\sqrt{1-\omega_{pe}^2/\omega_1^2}}{\sqrt{1-\omega_{pe}^2/\omega_0^2}} \neq 1$ of the corresponding group velocities for cases in which the frequency of the Langmuir wave is not small compared to the pump frequency. In the following demonstrations, the ratio of plasma density to critical density $n/n_c = 0.04$ is chosen, which means that $\omega_0/\omega_{pe} = 5$ holds. Furthermore, we assume $\omega_1/\omega_0 = 0.79$.

It should be noted that the equation for the "seed pulse" is sometimes generalized to

$$\left(\frac{\partial}{\partial t} - v_{sp}\frac{\partial}{\partial x}\right)E_s = E_p N - i\kappa\frac{\partial^2 E_s}{\partial t^2} + i\chi|E_s|^2 E_s \ . \tag{7.177}$$

The equation contains two additional effects [83] compared to the simpler form. First, group velocity dispersion may play a role. This can be taken into account by an additional term that includes the group velocity dispersion term κ.

Second, in the relativistic regime, the nonlinearities due to relativistic mass variation, which lead to self-interaction, should be taken into account. All these effects could be included in current simulations of the three-wave model, but in all demonstrations we will choose $\kappa = \chi = 0$ for comparison.

The Linear Regime

First, for the small-signal Raman amplification of pulses, we treat the pump wave as constant, $E_p = const$, and "ignore" in (7.174)–(7.176) the first equation for the pump wave E_p. Since this is a "classical" linear initial value problem, the remaining two equations can, for example, be solved by means of integral transformations.

It should be noted that solutions to the linear initial value problem have been presented in the form of Green's functions [84–86].

After performing a Fourier transformation in space and a Laplace transformation in time, we obtain

$$\boxed{\left(p - ik - \frac{|E_p|^2}{p}\right)E_s(k,p) = E_s(k,t=0)} \ . \tag{7.178}$$

The temporal evolution of a Fourier mode of the pulse is described by $E_s(k,t) = A_k(t)E_s(k,t=0)$. Clearly, the coefficient $A_k(t)$ is equivalent to $\exp(\gamma_R t)$, which appears in the parametric amplification of plane waves. For demonstration, we use Gaussian initial pulses

$$E_s(x,t=0) \approx a e^{-x^2/D^2} . \tag{7.179}$$

It is straightforward to calculate $A_k(t)$. For example, for $\kappa = \chi = 0$ and $v_{sp} = 1$

$$\begin{aligned} A_k(t) =& e^{\frac{1}{2}ikt}\left[\cosh\left(\frac{1}{2}|E_p|t\sqrt{4-\frac{k^2}{|E_p|^2}}\right)\right. \\ &\left. + i\frac{k}{|E_p|}\frac{\sinh\left(\frac{1}{2}|E_p|t\sqrt{4-\frac{k^2}{|E_p|^2}}\right)}{\sqrt{4-\frac{k^2}{|E_p|^2}}}\right] . \end{aligned} \tag{7.180}$$

The next step is the inverse Fourier transformation, which can also be carried out straightforwardly. For the inverse transformation, we multiply by e^{ikx}. This factor can be combined with an exponential factor that appears in $A_k(t)$, i.e., $e^{\frac{1}{2}ikt}e^{ikx} = e^{ik(x+\frac{1}{2}t)} \equiv e^{ikz}$ with $z = x + \frac{1}{2}t$. We then obtain

$$E_s(z,t) = \frac{1}{\sqrt{2\pi}}\int_{-\infty}^{\infty} \tilde{E}_s(k,t)e^{ikz}dk , \tag{7.181}$$

with

$$\boxed{\tilde{E}_s(k,t) = \cosh\left(\frac{1}{2}t\sqrt{4-k^2}\right) + ik\frac{\sinh\left(\frac{1}{2}t\sqrt{4-k^2}\right)}{\sqrt{4-k^2}}} . \tag{7.182}$$

We note that for $|k| \geq 2$ the following holds: $\sqrt{4-k^2} = i\sqrt{k^2-4}$ as well as $\cosh(ix) = \cos(x)$ and $\sinh(ix) = i\sin(x)$. This leads to the result that (for the initially symmetric Gaussian pulse shapes assumed here) only real values of $E_s(z,t)$ appear after the inverse transformation.

From the introduction of the coordinate z, we can conclude that the amplified seed propagates to the left with a velocity of $\frac{1}{2}$. Within the z coordinate system, the leading edge propagates with the (additional) approximate velocity of $\frac{1}{2}$. This is in full agreement with results from the works [87, 88].

Let us now consider two qualitatively different initial seed configurations.

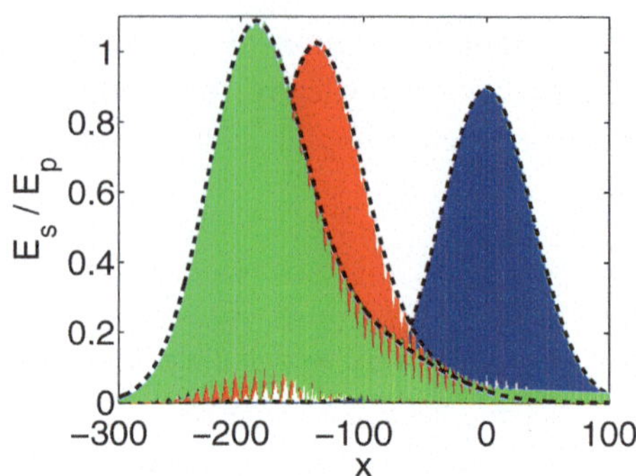

Fig. 7.1 Growth of a broad seed pulse with $D = 52.5$ in the linear regime, as found from the analytical solution of the linear Raman equations (for $\chi = \kappa = 0$) [black solid line] and from the Vlasov code, for $E_p = 0.005$ at the times $t = 0$, $t = 141$ and $t = 192$

First, at $t = 0$, we start with a relatively broad seed pulse with $a = 0.9$ and $D = 52.5$. The constant pump wave is $E_p = 0.005$. The temporal evolution of the seed is shown in Fig. 7.1. After an initial (slight) restructuring, the pulse moves to the left, broadens, and is amplified.

The analytically predicted behavior is fully reproduced for short times when the 1D version of the PDE2D code is used to solve (7.174)–(7.176). Furthermore, the Vlasov simulations shown in Fig. 7.1 in the linear regime agree with these results.

Next, we reduce the width of the initial seed by changing D to $D = 5.25$. As shown in Fig. 7.2, the behavior is completely different compared to the broad seed situation in Fig. 7.1.

The narrow initial seed pulse at $t = 0$ is initially less amplified, but generates a broad secondary pulse. The width of the latter can reach a similar (or even greater) extent as the initially broad pulses used in the first case. The broad secondary pulse is then amplified. However, it takes much longer times to reach a similar amplitude amplification as in the case shown in Fig. 7.1.

Thus, for effective Raman seed amplification in the *linear* regime, there is a lower limit for the width of the initial ($t = 0$) seed pulse.

The essential differences between Figs. 7.1 and 7.2 can be understood with the help of (7.180). Note that the effective initial growth time is $T \equiv |E_p|t$ and instead of k the ratio $K \equiv k/|E_p|$ appears. When these variables are chosen, $A_K(T)$ is a universal function. The

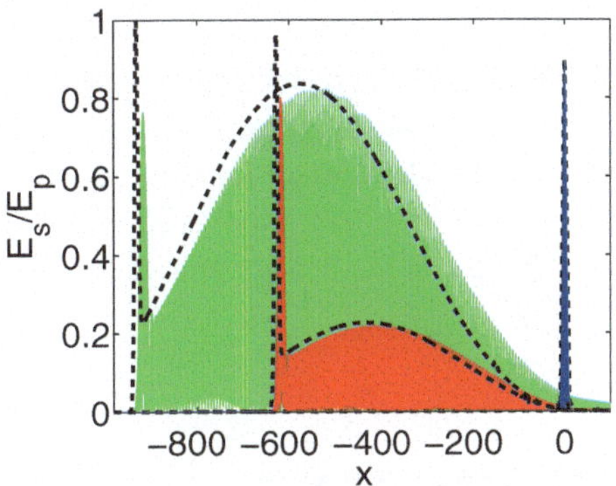

Fig. 7.2 Same as in Fig. 7.1, but for a narrow seed pulse with $D = 5.25$ and at the times $t = 0$, $t = 627$ and $t = 930$

logarithm of its real value, divided by t, can be interpreted as the effective growth rate $\gamma(K)$ in k-space.

The Fourier transform of an initial Gaussian seed is $E_s(K, t = 0) = \frac{aD}{\sqrt{2}} \exp[-D^2|E_p|^2K^2/4]$. This shows that the width $W_s \sim 2/D|E_p|$ must be compared with the (fixed) width W_γ of $\gamma(K)$. Such a comparison for narrow and broad initial seed pulses is shown in Fig. 7.3. The broad initial pulse in Fig. 7.3(a) is directly amplified, while a narrow initial seed first generates a pulse of width $\sim 1/W_\gamma$ (in x-space), which should subsequently continue to grow. This explains the restructuring during the "start-up period" reported in Ref. [89].

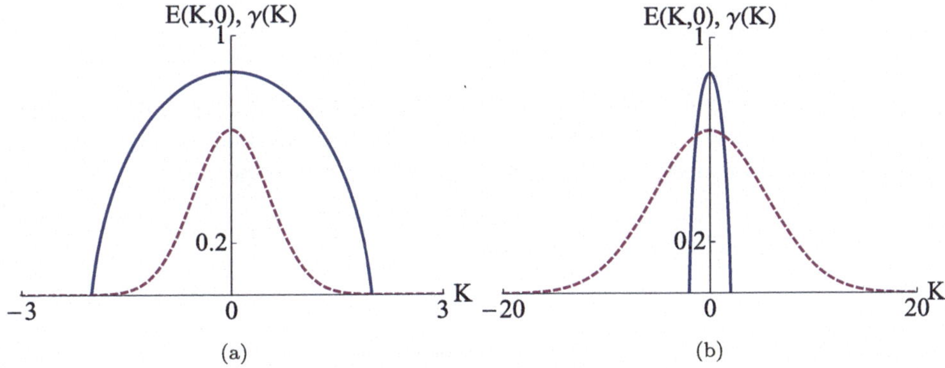

Fig. 7.3 Comparison of the Fourier transform $E_s(K, t = 0)$ (dashed line) with the effective (universal) growth rate $\gamma(K)$ (solid line) for (**a**) a broad pulse with $D = 52.5$ and (**b**) a narrow pulse with $D = 5.25$. We have chosen $E_p = 0.05$. The Fourier distributions were normalized in amplitude

The interesting result of the linear regime (with nearly exponential growth) is therefore the following: An initially very narrow seed pulse can evolve into a broad pulse, while a broad initial seed generally remains broad. The difference between narrow and broad initial seed pulses depends on the ratio of W_s to W_γ. The latter depends on the density and the pump strength.

Self-Similar Solution

We now summarize the procedure for obtaining a family of self-similar solutions for the amplification of a Raman seed pulse. The self-similar solution for the three-wave interaction model of Raman (7.174)–(7.176) is well known [86, 87, 90]. Here, we present the main steps of the derivation and add a generalization that can be applied to other types of three-wave interactions, such as Brillouin seed pulse amplification [91]. The general formalism is compared with the "standard" formalism for π-pulses in Raman backscattering.

For simplicity, we set $v_{sp} = 1$ in (7.174)–(7.176) and introduce the time coordinate $\zeta = t + x$ as well as the spatial coordinate $\tau = -x$. In this coordinate system, the fundamental equations for $a_0 \equiv E_p$, $a_1 \equiv E_s$ and $a_2 \equiv -N^*$

$$\left(2\frac{\partial}{\partial\zeta} - \frac{\partial}{\partial\tau}\right)a_0 = a_1 a_2 \,, \tag{7.183}$$

$$\frac{\partial}{\partial\tau}a_1 = -a_0 a_2^* \,, \tag{7.184}$$

$$\frac{\partial}{\partial\zeta}a_2 = -a_0 a_1^* \,. \tag{7.185}$$

During an advanced nonlinear phase, in which the amplification length of the pumped pulse is much greater than the pulse duration, in the first equation the derivative with respect to τ can be neglected compared to the derivative with respect to ζ, yielding

$$\boxed{2\frac{\partial}{\partial\zeta}a_0 = a_1 a_2 \,,} \tag{7.186}$$

$$\boxed{\frac{\partial}{\partial\tau}a_1 = -a_0 a_2^* \,,} \tag{7.187}$$

$$\boxed{\frac{\partial}{\partial\zeta}a_2 = -a_0 a_1^* \,.} \tag{7.188}$$

Before we make use of some specific properties of the system, we proceed in a rather general way. We introduce the self-similar variable

$$\boxed{\xi = 2\zeta^{\alpha}\tau^{\beta}} \tag{7.189}$$

, which (with the new variables ξ and τ) leads to the following substitutions:

$$\left.\frac{\partial}{\partial\tau}\right|_{\zeta} \to \left.\frac{\partial}{\partial\tau}\right|_{\xi} + \beta\tau^{-1}\xi\left.\frac{\partial}{\partial\xi}\right|_{\tau} , \tag{7.190}$$

$$\left.\frac{\partial}{\partial\zeta}\right|_{\tau} \to 2^{1/\alpha}\alpha\xi^{(\alpha-1)/\alpha}\tau^{\beta/\alpha}\left.\frac{\partial}{\partial\xi}\right|_{\tau} . \tag{7.191}$$

Furthermore, we make the ansatz for self-similar solutions

$$\boxed{a_0 = \tau^{\gamma}A_0(\xi) , \quad a_1 = \tau^{\delta}A_1(\xi) , \quad a_2 = \tau^{\lambda}B(\xi)} . \tag{7.192}$$

The parameters that have so far remained free, α, β, γ, δ and λ are determined such that all explicit τ-dependencies disappear. A short calculation leads to the linear equations

$$\frac{\beta}{\alpha} = \delta + \lambda , \tag{7.193}$$

$$\delta - 1 = \gamma + \lambda , \tag{7.194}$$

$$\lambda + \frac{\beta}{\alpha} = \gamma + \delta . \tag{7.195}$$

For $\gamma = 0$ one obtains

$$\beta = \alpha , \quad \delta = 1 , \quad \lambda = 0 . \tag{7.196}$$

Without loss of generality, we can choose

$$\beta = \alpha = \frac{1}{2} \tag{7.197}$$

which leads to

$$\boxed{4\xi^{-1}\frac{dA_0}{d\xi} = A_1 B ,} \tag{7.198}$$

$$\boxed{A_1 + \frac{1}{2}\xi\frac{dA_1}{d\xi} = -A_0 B^* ,} \tag{7.199}$$

$$\boxed{2\xi^{-1}\frac{dB}{d\xi} = -A_0 A_1^* ,} \tag{7.200}$$

for the self-similar functions $A_0(\xi)$, $A_1(\xi)$ and $B(\xi)$. For small ξ we expand:

$$A_0 = 1 + a_{01}\xi + a_{02}\xi^2 + \cdots , \tag{7.201}$$

$$A_1 = \varepsilon + a_{11}\xi + a_{12}\xi^2 + \cdots , \tag{7.202}$$

$$B = b_0 + b_1\xi + b_2\xi^2 + \cdots . \tag{7.203}$$

Substituting into (7.198)–(7.200) shows that a one-dimensional family of solutions exists with

$$A_1(\xi = 0) = \varepsilon , \quad \varepsilon \text{ is a free parameter} . \tag{7.204}$$

Since $b_0 = -\varepsilon$ should hold, Fig. 7.4 shows, for two values of the parameter ε, the solutions of the system (7.198)–(7.200).

Example 7.4 (Transformation to the Sine-Gordon Equation)
A "standardized procedure" for determining the self-similar Raman solution starts directly with (7.186)–(7.188) and the ansatz

$$a_0(\zeta, \tau) = \cos\left(\frac{u(\zeta, \tau)}{2}\right) , \quad a_2 = \sqrt{2}\sin\left(\frac{u(\zeta, \tau)}{2}\right) , \tag{7.205}$$

which takes into account the conservation of $2a_0^2 + a_2^2$. It is then obvious that

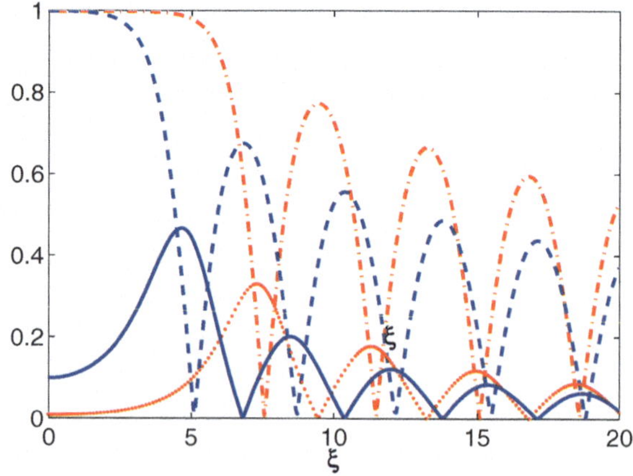

Fig. 7.4 Two examples of the self-similar solutions $A_0(\xi)$ and $A_1(\xi)$. The blue curves are for $\varepsilon = 0.1$: A_0 (dashed line) and A_1 (solid line). The red curves are for $\varepsilon = 0.01$: A_0 (dash-dotted line) and A_1 (dotted line)

$$a_1 = -\frac{1}{\sqrt{2}}\frac{\partial u}{\partial \zeta} \,. \tag{7.206}$$

Subsequently, (7.187) leads to the Sine-Gordon equation (we use this designation, as it is also common in German)

$$\boxed{\frac{\partial^2 u(\zeta,\tau)}{\partial\zeta\,\partial\tau} = \sin u(\zeta,\tau)}\,, \tag{7.207}$$

which can be further analyzed. For the Sine-Gordon equation, it is known that in certain regions a self-similar solution acts as an attractor. We do not repeat the analysis of self-similarity here and merely note that the variables

$$\xi = 2\sqrt{\zeta\tau}\,, \quad u(\zeta,\tau) \to U(\xi) \tag{7.208}$$

are introduced. With

$$\frac{\partial}{\partial\zeta} = 2\frac{\tau}{\xi}\frac{\partial}{\partial\xi}\,, \quad \frac{\partial}{\partial\tau} = 2\frac{\zeta}{\xi}\frac{\partial}{\partial\xi} \tag{7.209}$$

and

$$\frac{\partial^2 U}{\partial\tau\,\partial\zeta} = \frac{\partial}{\partial\tau}\left(\frac{\sqrt{\tau}}{\sqrt{\zeta}}\frac{\partial U}{\partial\xi}\right) = \frac{1}{\xi}\frac{\partial U}{\partial\xi} + \frac{\sqrt{\tau}}{\sqrt{\zeta}}2\frac{\zeta}{\xi}\frac{\partial^2 U}{\partial\xi^2} \tag{7.210}$$

we ultimately arrive at the ordinary differential equation

$$\boxed{\frac{d^2 U}{d\xi^2} + \frac{1}{\xi}\frac{dU}{d\xi} = \sin U} \tag{7.211}$$

for $U = U(\xi)$.

We can directly "translate" this procedure to the present formulation with the three-wave equations. In the case of the three-wave Raman equations, Eqs. (7.198) and (7.200) satisfy the conservation law

$$\frac{d}{d\xi}(2A_0^2 + B^2) = 0\,, \tag{7.212}$$

which allows A_0 and B to be rewritten:

$$A_0(\xi) = \cos\left(\frac{U(\xi)}{2}\right), \quad B(\xi) = \sqrt{2}\sin\left(\frac{U(\xi)}{2}\right). \tag{7.213}$$

Then (7.198) or (7.200) leads to

$$A_1(\xi) = -\frac{\sqrt{2}}{\xi}\frac{dU(\xi)}{d\xi} \,. \tag{7.214}$$

Ultimately, (7.199) leads to

$$\frac{d^2U(\xi)}{d\xi^2} + \frac{1}{\xi}\frac{dU(\xi)}{d\xi} = \sin U(\xi) , \tag{7.215}$$

which coincides with (7.211). ■

Example 7.5 (A One-Parameter Family)
We have shown that the free initial value ε, as defined in (7.204), generates a one-parameter family of self-similar solutions. We now discuss the relationship to the physically relevant variable a_1. The function $U(\xi)$ plays a central role in the comparison. For large ξ we have $U(\xi) = -\pi + \delta U$ with $\delta U \sim J_0(\xi)$, while for small ξ we have $U(\xi) \sim I_0(\xi)$. One factor remains undetermined. We will fix it in the form

$$U(\xi) \sim -\sqrt{2}\varepsilon I_0(\xi) \sim -\sqrt{2}\varepsilon \quad \text{for } \xi \to 0 , \tag{7.216}$$

since then

$$\frac{dU}{d\xi} \sim -\sqrt{2}\varepsilon I_1(\xi) \sim -\frac{\varepsilon}{\sqrt{2}}\xi \quad \text{for } \xi \to 0 . \tag{7.217}$$

Equation (7.214) is consistent with (7.204).

On the other hand, Eqs. (7.216) and (7.206) immediately yield

$$u(\zeta, \tau = 0) = -\sqrt{2}\varepsilon = -\sqrt{2}\int a_1(\zeta, \tau = 0)\, d\zeta , \tag{7.218}$$

which for ε means

$$\varepsilon = \int a_1(\zeta, \tau = 0)\, d\zeta . \tag{7.219}$$

For a "narrow" initial seed, we obtain

$$a_1(x = 0, t) = \varepsilon\delta(t) = a_1(\zeta, \tau = 0) = \varepsilon\delta(\zeta) . \tag{7.220}$$

Thus, the members of the family distinguished by the parameter ε belong to different regions of the initial seed pulses. ■

Brillouin Scattering with Pump Wave Reaction

Here we briefly summarize the three-wave interaction model used for Brillouin scattering. The basic three-wave interaction model is well known [24]. Due to recent applications of laser-plasma interaction for the generation of high-intensity laser beams, we focus on Brillouin scattering in the strongly coupled (sc) regime.

For a standard analysis of three-wave interaction, the fundamental equations can be found in Ref. [24] [see, e.g., Eq. (8.1) and (8.8) of [24]]. They are usually formulated for the vector potentials of the pump wave (index 0) and the scattered wave (index 1):

$$\boxed{\left(\frac{\partial^2}{\partial t^2} - c^2\nabla^2 + \omega_{pe}^2\right)\mathbf{A}_0 = -\omega_{pe}^2 \frac{n}{n_0}\mathbf{A}_1,} \tag{7.221}$$

$$\boxed{\left(\frac{\partial^2}{\partial t^2} - c^2\nabla^2 + \omega_{pe}^2\right)\mathbf{A}_1 = -\omega_{pe}^2 \frac{n}{n_0}\mathbf{A}_0,} \tag{7.222}$$

$$\boxed{\left(\frac{\partial^2}{\partial t^2} - c_s^2\nabla^2\right)\frac{n}{n_0} = \frac{Ze^2}{c^2 m_e m_i}\nabla^2(\mathbf{A}_0 \cdot \mathbf{A}_1)\,.} \tag{7.223}$$

The high-frequency waves are coupled to low-frequency density perturbations n. Here, Z is the ion charge state, $c_s = \sqrt{\frac{ZT_e}{m_i}}$ is the ion sound speed, and n_0 is the unperturbed electron (and ion) density. In the following, we consider a spatially one-dimensional and linearly polarized situation.

Next, we introduce electric fields $\mathbf{E} = -\frac{\partial \mathbf{A}}{\partial t}$ and apply a slowly varying envelope approximation for the electromagnetic waves. The phase factors are introduced by $E_0 \sim \exp[i(k_0 x - \omega_0 t)]$, $E_1 \sim \exp[i(-k_1 x - \omega_1 t)]$, and $\frac{n}{n_0} \sim \frac{\tilde{n}}{n_0}\exp(ik_2 x)$. The propagation of the pump wave is assumed to be in the positive x-direction, while the backscattered wave propagates in the negative x-direction; wave vectors are denoted by k, and frequencies by ω.

For comparison with other recent publications [91] on Brillouin seed amplification, we first change the notation $E_0 \to E_p^*$ and $E_1 \to E_s^*$. We then define $\frac{\tilde{n}}{n_0} = -\frac{\delta n_p^*}{n_0} \equiv -\frac{\delta n^*}{n_0}$. Andreev et al. [91] used $\delta n_p = \delta n_s^* \equiv \delta n$. We modify this slightly by introducing $\delta\tilde{n} \equiv \frac{\delta n^*}{n_e}$ and the equivalent form

$$\left(\frac{\partial}{\partial t} + v_g \frac{\partial}{\partial x}\right)E_p = -i\frac{\omega_{pe}^2}{2\omega_0}\delta\tilde{n}^* E_s\,, \tag{7.224}$$

$$\left(\frac{\partial}{\partial t} - v_g \frac{\partial}{\partial x}\right)E_s = -i\frac{\omega_{pe}^2}{2\omega_0}\delta\tilde{n} E_p\,, \tag{7.225}$$

$$\left(\frac{\partial^2}{\partial t^2} - c_s^2 \frac{\partial^2}{\partial x^2}\right)\delta\tilde{n} = -\frac{Ze^2}{c^2 m_e m_i} E_p^* E_s \tag{7.226}$$

for the envelopes E_p of the pump wave and E_s of the backscattered wave. Here, v_g is the group velocity of the high-frequency waves. Note that no exact frequency resonance with a low-frequency ion-acoustic wave was assumed, but rather $\omega_1 \approx \omega_0$ is used, since we are scattering from low-frequency oscillations. The density equation remains almost unchanged.

Next, we introduce the units t_{unit}, $x_{\text{unit}} \equiv v_g t_{\text{unit}}$, $E_{\text{unit}} \equiv |E_p^{\text{typical}}|$ and $\delta\tilde{n}_{\text{unit}}$ for time, space, electric field, and density perturbations. The new variables are

$$\frac{t}{t_{\text{unit}}} \to t\,, \quad \frac{x}{x_{\text{unit}}} \to x\,, \quad \frac{E}{E_{\text{unit}}} \to E\,, \quad \frac{\delta\tilde{n}}{\delta\tilde{n}_{\text{unit}}} \to N\,. \tag{7.227}$$

Postulating (in order to obtain simple equations as shown below)

$$\frac{\omega_{pe}^2}{2\omega_0} t_{\text{unit}} \delta\tilde{n}_{\text{unit}} = 1\,, \quad \frac{t_{\text{unit}}^2}{\delta\tilde{n}_{\text{unit}}} \frac{Ze^2}{c^2 m_e m_i} |E_p^{\text{typical}}|^2 = 1\,, \tag{7.228}$$

we find the units

$$\frac{1}{t_{\text{unit}}} = \sqrt[3]{\frac{\omega_0 \omega_{pi}^2}{2} \frac{v_{osc}^2}{c^2}} \sim \gamma_B^{sc}\,, \quad \delta\tilde{n}_{\text{unit}} = \frac{2\omega_0}{t_{\text{unit}} \omega_{pe}^2}\,. \tag{7.229}$$

For $|E_p^{\text{typical}}|$ we use the initial pump amplitude. As a side note, according to Kruer [24], we can also introduce $E_{\text{unit}} \equiv |E_p^{\text{typical}}| \equiv k_0 A_p^{\text{Kruer}}$ and $v_{osc} \equiv a\,c = \frac{eA_p^{\text{Kruer}}}{m_e c}$.

Using these units, the final dimensionless three-wave interaction equations are

$$\boxed{\left(\frac{\partial}{\partial t} + \frac{\partial}{\partial x}\right) E_p = -iN^* E_s\,,} \tag{7.230}$$

$$\boxed{\left(\frac{\partial}{\partial t} - \frac{\partial}{\partial x}\right) E_s = -iN E_p\,,} \tag{7.231}$$

$$\boxed{\left(\frac{\partial^2}{\partial t^2} - \frac{c_s^2}{v_g^2} \frac{\partial^2}{\partial x^2}\right) N = -E_s E_p^*\,.} \tag{7.232}$$

A brief comparison with the three-wave interaction equations already discussed for Raman scattering:

$$\left(\frac{\partial}{\partial t} + \frac{\partial}{\partial x}\right) E_p = -N^* E_s\,, \tag{7.233}$$

$$\left(\frac{\partial}{\partial t} - \frac{\partial}{\partial x}\right) E_s = E_p N \,, \tag{7.234}$$

$$\frac{\partial N}{\partial t} = E_s E_p^* \,. \tag{7.235}$$

The main difference between (7.230)–(7.232) for Brillouin and (7.233)–(7.235) for Raman is the absence of the (slowly varying) envelope approximation in (7.232). Outside the strongly coupled case, when the Stokes component becomes resonant, we could replace (7.232) by

$$i\frac{\partial N}{\partial t} = -E_s E_p^* \,. \tag{7.236}$$

Mathematically, the systems (7.230)–(7.231) and (7.236) on the one hand, and (7.233)–(7.235) on the other, are equivalent. This becomes apparent when one transitions from Raman to Brillouin by substituting $N \to iN^*$, $E_s \to iE_s^*$, and $E_p \to iE_p^*$.

Physically, of course, there is a significant difference. All nonrelativistic scattering instabilities require underdense plasmas, and in addition, Raman scattering only occurs below the quarter-critical density. This well-known fact follows from the frequency and wavenumber resonance conditions. Furthermore, we point out the different normalizations of the density perturbations:

$$\delta\tilde{n}_{\text{unit}}^{\text{Brillouin}} = \sqrt[3]{\frac{\omega_0 \omega_{pi}^2}{2} \frac{v_{osc}^2}{c^2} \frac{2\omega_0}{\omega_{pe}^2}} \ll \delta\tilde{n}_{\text{unit}}^{\text{Raman}} = \frac{ck_0}{\omega_{pe}} \sqrt{\frac{\omega_0}{\omega_{pe}}} \,. \tag{7.237}$$

Brillouin Amplification and Pulse Shape Changes

Next, we examine the amplification of a seed pulse in the strongly coupled regime. We start from Eqs. (7.230)–(7.232). We consider the pump wave as constant, $E_p = 1$, and "neglect" the first equation for the quantity E_p. Then, in the small-signal case, i.e., when the pump amplitude is considered constant, we obtain simple linear equations. In the strongly coupled (sc) case, the equations (see also [92]) for $c_s^2 \ll v_g^2$ are

$$\boxed{\left(\frac{\partial}{\partial t} - \frac{\partial}{\partial x}\right) E_s = -iN \,,} \tag{7.238}$$

$$\boxed{\frac{\partial^2}{\partial t^2} N = -E_s \,.} \tag{7.239}$$

For an initial seed pulse

$$E_s(x, t = 0) \approx a e^{-x^2/D^2} , \tag{7.240}$$

the Fourier transform is

$$E_s(k, t = 0) \approx \frac{aD}{\sqrt{2}} e^{-D^2k^2/4} . \tag{7.241}$$

The linear initial value problem can be solved by Fourier transformation in space and Laplace transformation in time. If we first perform the Fourier transformation in space and then the Laplace transformation in time, we obtain

$$(p - ik)E_s(k, p) = -iN(k, p) + E_s(k, t = 0) , \tag{7.242}$$

$$p^2 N(k, p) = -E_s(k, p) , \tag{7.243}$$

which can be combined:

$$\left(p - ik - \frac{i}{p^2}\right) E_s(k, p) = E_s(k, t = 0) . \tag{7.244}$$

For the inverse Laplace transformation we use

$$\frac{1}{2\pi i} \int_{\sigma - i\infty}^{\sigma + i\infty} \frac{1}{p - ik - \frac{i}{p^2}} e^{pt} dp = -\sum_{j=1}^{3} \frac{\alpha_j e^{\alpha_j t}}{-3\alpha_j + 2ik} \equiv A_k(t) , \tag{7.245}$$

where α_j are the three roots of $\alpha_j^3 - ik\alpha_j^2 - i = 0$ for $j = 1, 2, 3$. These roots can easily be determined by symbolic programs, such as MATHEMATICA. Even before such programs were available, the transformations were carried out fully analytically, although this required a great deal of algebra [93].

For the time evolution of a Fourier mode we find $E_s(k, t) = A_k(t) E_s(k, t = 0)$. Obviously, the coefficient $A_k(t)$ is equivalent to the factor $\exp(\gamma t)$ that appears in the parametric amplification of plane waves.

The next step, the inverse Fourier transformation, can also be performed using symbolic programs such as MATHEMATICA. The results are shown in Fig. 7.5 for two cases. In each case, the seed pulse is amplified, changes its shape, and propagates to the left. However, its velocity is lower than one might have expected. In addition to amplification, broadening, and propagation, the pulse shape changes. A comparison with the Vlasov simulations and the solutions of the three-wave interaction model, both at early times, shows complete agreement.

In Fig. 7.5 we present two qualitatively different seed configurations at the time $t = 0$.

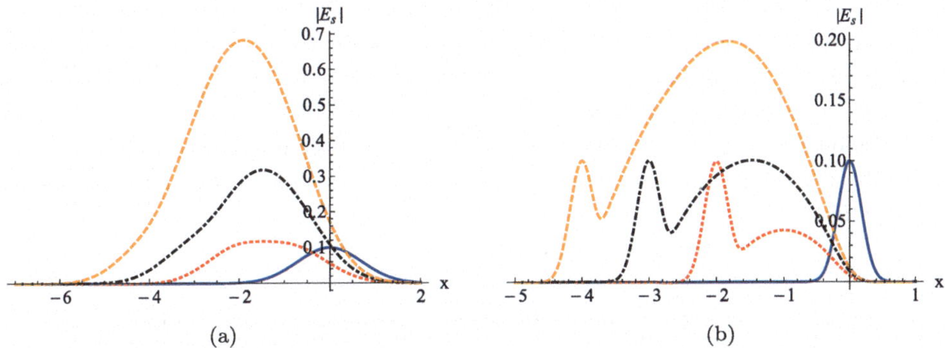

Fig. 7.5 Evolution of the seed pulse, obtained by Fourier-Laplace transformation of the linearized three-wave interaction equations. The initial maximum amplitude of the seed is one tenth of the pump amplitude. The initial seed pulse has the form of a Gaussian function $0.1\exp[-x^2/D^2]$ with (**a**) $D = 1$ and (**b**) $D = 0.25$. In each subfigure, the temporal evolution is shown at the times $t = 0$ (blue solid line), $t = 2$ (red dotted line), $t = 3$ (black dash-dotted line), and $t = 4$ (orange dashed line)

First, in subfigure (a) at $t = 0$ we start with a relatively broad seed pulse with $a = 0.1$ and $D = 1$. The temporal evolution of the seed is shown in the linear regime. After an initial (slight) rearrangement, the pulse moves to the left, spreads out, and is amplified.

The analytically predicted behavior is fully reproduced for short times by the three-wave interaction code and the Vlasov simulations.

In subfigure (b) we reduce the width of the initial seed by changing D to $D = 0.25$. The behavior is completely different compared to the broad seed situation in (a). The narrow initial seed pulse at $t = 0$ is only slightly amplified at first, but mainly generates a broad secondary pulse.

Its width can reach a similar (or even broader) width as the initially broad pulses in (a). The broad secondary pulse is then amplified. It takes a similar amount of time to reach the same amplitude amplification as in case (a). Thus, the *linear* evolution shows that the width of the amplified pulse is generally not identical to the initial width.

The essential differences between cases (a) and (b) can be explained by comparing the bandwidths of the initial seed pulse with the bandwidth of the sc-Brillouin instability. The latter can be determined from (7.245). The function $A_k(t)$ has a certain width in

k-space, which should be compared with the width of $E_s(k, t = 0) = \frac{aD}{\sqrt{2}} \exp[-D^2k^2/4]$. The broad initial pulse in Fig. 7.5(a) has a narrow distribution in k-space and will therefore experience a relatively uniform amplification. On the other hand, a broad distribution in k-space corresponds to the narrow initial seed distribution chosen in Fig. 7.5(b). The amplification factor $A_k(t)$ will filter out a narrow region for amplification. This is similar to the rearrangements during the "startup phase" reported in Ref. [89] for Raman amplification.

In summary, in the linear regime (with nearly exponential growth), an initially very narrow seed pulse can become a broad pulse, while a broad initial seed generally remains broad. The (relative) terminology "narrow" and "broad" (in x-space) arises from comparing the bandwidth of the seed with the bandwidth of the sc-Brillouin growth in the conjugate k-space.

Example 7.6 (Self-similar solution for Brillouin scattering)
The behavior observed in the numerical simulations is interpreted by means of a self-similar solution [91] of the three-wave interaction model. To derive the self-similar solution, we start with Eqs. (7.230)–(7.232). We transform into the reference frame moving with the seed by introducing $\zeta = x + t$ and $\tau = -x$. During an advanced nonlinear phase, in which the amplification length of the pumped pulse is much greater than the pulse duration, the τ derivative can be neglected compared to the ζ derivative. The new equations are

$$\boxed{2\frac{\partial}{\partial \zeta} E_p(\zeta, \tau) = -iN^*(\zeta, \tau) E_s(\zeta, \tau) \,,} \tag{7.246}$$

$$\boxed{\frac{\partial}{\partial \tau} E_s(\zeta, \tau) = -iN(\zeta, \tau) E_p(\zeta, \tau) \,,} \tag{7.247}$$

$$\boxed{\frac{\partial^2}{\partial \zeta^2} N(\zeta, \tau) = -E_s(\zeta, \tau) E_p^*(\zeta, \tau) \,.} \tag{7.248}$$

We now proceed with a self-similar ansatz. The basic idea was published by Andreev et al. [91]. Here, we briefly summarize the procedure, mainly to identify the free parameter ε, which generates a family of self-similar solutions.

With the ansatz

$$\boxed{\xi = \zeta^\alpha \tau^\beta} \tag{7.249}$$

one obtains, after changing from ζ , τ to ξ , τ

$$\left.\frac{\partial}{\partial\tau}\right|_\zeta \to \left.\frac{\partial}{\partial\tau}\right|_\xi + \left.\frac{\partial}{\partial\xi}\frac{\partial\xi}{\partial\tau}\right|_\zeta = \left.\frac{\partial}{\partial\tau}\right|_\xi + \beta\tau^{\beta-1}\zeta^\alpha\frac{\partial}{\partial\xi}\,, \tag{7.250}$$

$$\frac{\partial}{\partial\zeta} \to \tau^\beta\alpha\zeta^{\alpha-1}\frac{\partial}{\partial\xi}\,, \tag{7.251}$$

$$\frac{\partial^2}{\partial\zeta^2} \to \tau^\beta\alpha(\alpha-1)\zeta^{\alpha-2}\frac{\partial}{\partial\xi} + \tau^{2\beta}\alpha^2\zeta^{2\alpha-2}\frac{\partial^2}{\partial\xi^2}\,. \tag{7.252}$$

We then set:

$$\boxed{E_s = \tau^{\delta_r}A_{sr}(\xi) + i\tau^{\delta_i}A_{si}(\xi)\,,} \tag{7.253}$$

$$\boxed{E_p = \tau^{\gamma_r}A_{pr}(\xi) + i\tau^{\gamma_i}A_{pi}(\xi)\,,} \tag{7.254}$$

$$\boxed{N = \tau^{\lambda_r}b_r(\xi) + i\tau^{\lambda_i}b_i(\xi)\,,} \tag{7.255}$$

The parameters δ_r, δ_i, γ_r, γ_i, λ_r, λ_i are still free.

If this ansatz is inserted into Eqs. (7.246)–(7.248), this leads to equations that still contain both ξ and τ. (In expressions that originally contain ζ, we substitute $\zeta^\alpha = \xi\tau^{-\beta}$.)

Next, we postulate that the explicit τ dependence should disappear. Direct calculations yield

$$\lambda_r = \lambda_i\,, \quad \delta_r = \delta_i\,, \quad \gamma_r = \gamma_i\,. \tag{7.256}$$

The conclusion is that we can use a complex notation in which the real and imaginary parts have the same τ dependence. Therefore, we can simplify the self-similar ansatz to the form described in [91]:

$$\boxed{E_s = \tau^\delta A_s(\xi)\,,} \tag{7.257}$$

$$\boxed{E_p = \tau^\gamma A_p(\xi)\,,} \tag{7.258}$$

$$\boxed{N = \tau^\lambda B^*(\xi)\,.} \tag{7.259}$$

This approach leads to five equations, which can essentially be reduced to the following:

$$\frac{\beta}{\alpha} = \lambda + \delta\,, \quad \delta - 1 = \lambda + \gamma\,, \quad \lambda + 2\frac{\beta}{\alpha} = \delta + \gamma\,. \tag{7.260}$$

One can quickly find:

$$\lambda = -\frac{1}{4}\,, \quad \gamma = -\frac{3}{4} + \delta\,, \quad \frac{\beta}{\alpha} = -\frac{1}{4} + \delta\,. \tag{7.261}$$

The values of α and δ are still free. Without loss of generality, one can choose $\alpha = 1$. The pump pulse should not increase with time. For $\delta = \frac{3}{4}$ the maximum amplitude of the pump pulse remains constant, which leads to $\beta = \frac{1}{2}$. In this way, we arrive at the scaling of Andreev et al. [91]

$$\xi = z\sqrt{t}\,, \quad E_s = t^{3/4} A_s(\xi)\,, \quad E_p = A_p(\xi)\,, \quad N = t^{-1/4} B^*(\xi)\,. \tag{7.262}$$

Thus, the ordinary differential equations for the amplitudes become

$$\boxed{2\frac{dA_p}{d\xi} = -iBA_s\,,} \tag{7.263}$$

$$\boxed{\frac{\xi}{2}\frac{dA_s}{d\xi} + \frac{3}{4}A_s = -iB^* A_p\,,} \tag{7.264}$$

$$\boxed{\frac{d^2B}{d\xi^2} = -A_p A_s^*\,.} \tag{7.265}$$

With the choice $A_p(0) = 1$ and $A_s(0) = \varepsilon$ (free parameter), $B(0) = -i\frac{3}{4}\varepsilon$ we perform an expansion for small ξ. Since the density is then proportional to $B^* + B$, the initial value for the self-similar density variable vanishes.

In summary, a one-dimensional family of self-similar solutions (depending on ε) can be constructed from (7.263)–(7.265) with the following initial values.

$$A_p(\xi = 0) = 1\,, \tag{7.266}$$

$$A_s(\xi = 0) = \varepsilon\,, \tag{7.267}$$

$$B(\xi = 0) = -i\frac{3}{4}\varepsilon\,, \tag{7.268}$$

$$\left.\frac{dB}{d\xi}\right|_{\xi=0} = 0\,, \tag{7.269}$$

where ε is a free parameter. Typical solutions are shown in Fig. 7.6.

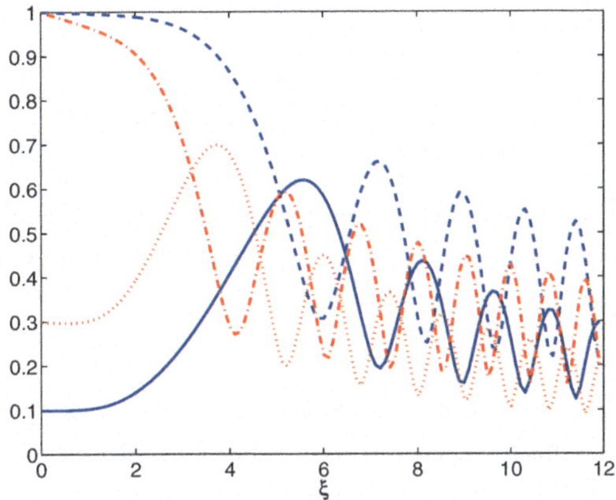

Fig. 7.6 Two self-similar solutions obtained by integrating Eqs. (7.263)–(7.265). Shown are the pump amplitudes A_p (dashed blue line, dashed red-dotted line) and the seed amplitudes A_s (solid blue line, dotted red line) for $\varepsilon = 0.3$ and $\varepsilon = 0.1$ respectively.

In Fig. 7.6 we have chosen two different values for the free parameter ε. For smaller ε, the self-similar solution for the seed shows a lower maximum value. Note that, in the frame of the seed with the "spatial coordinate τ", the widths of the oscillations decrease with time, since $\xi = \zeta\sqrt{\tau}$. The amplitude of the field E_s will grow with $\tau^{3/4}$. ■

7.4 Excitation of Wakefields

The discussion of the Maxwell-fluid models has shown that there is a strong coupling between electromagnetic and electrostatic components. Originally purely transverse laser light (in vacuum) can excite electrostatic oscillations in a plasma. For an intense laser pulse, electrostatic oscillations arise at the maximum of the electromagnetic field. In addition to the expected local behavior, nonlocal phenomena such as the excitation of wakefields (hereafter referred to as wakefields) can also occur.

Excitation of Quasi-Stationary Wakefields

Density reactions are driven by an intense laser field. In the weakly relativistic limit, the excited density is obtained, for example, from Eq. (7.85). In the case of linear polarization, the right-hand side contains higher harmonics and must be treated separately. Let us first consider stationary solutions in the moving reference frame $\xi = x - Vt$ for circular polarization. Using the dimensional form (7.104) in the new variables $\tau = t$ and ξ, we obtain

$$\left(\frac{\partial}{\partial\tau} - V\frac{\partial}{\partial\xi}\right)^2 \delta n_e + \omega_{pe}^2 \delta n_e \approx \frac{n_0 e^2}{2m_e^2} \frac{\partial^2 \mathbf{A}_\perp^2}{\partial\xi^2} . \tag{7.270}$$

Example 7.7 (Inhomogeneous Oscillator Equation)
For $\partial/\partial\tau \equiv 0$, Eq. (7.270) can be abbreviated as a driven oscillator equation:

$$V^2\ddot{y} + \tilde{\omega}^2 y = \ddot{f} , \tag{7.271}$$

where the dot denotes the derivative with respect to ξ, $\tilde{\omega} \equiv \omega_{pe}$ is, and $\ddot{f}$ denotes the driving term on the right-hand side of (7.270). The homogeneous part of the differential equation describes plasma oscillations $y \sim \exp[i(\tilde{\omega}/V)\xi]$. To determine the driven (forced) oscillations, we use the method of variation of parameters. We introduce

$$y(\xi) = Y(\xi)e^{i(\tilde{\omega}/V)\xi} \tag{7.272}$$

into (7.271). A single integration yields

$$V^2\dot{Y} + 2i\tilde{\omega}VY = \int_\infty^\xi d\xi' \ddot{f} e^{-i(\tilde{\omega}/V)\xi'} . \tag{7.273}$$

Rearranging the last term on the right-hand side leads to

$$\int_\infty^\xi d\xi' \ddot{f} e^{-i(\tilde{\omega}/V)\xi'} = \dot{f} e^{-i(\tilde{\omega}/V)\xi} + i\frac{\tilde{\omega}}{V} f e^{-i(\tilde{\omega}/V)\xi} - \frac{\tilde{\omega}^2}{V^2} \int_\infty^\xi d\xi' f e^{-i(\tilde{\omega}/V)\xi'} . \tag{7.274}$$

We have assumed that the integration constants vanish at $\xi = \infty$. If the laser front propagates from left to right along the x-axis, this requires that for finite times, nothing has happened far ahead yet. Next, the notation is changed to

$$Y(\xi) = W(\xi)e^{-2i(\tilde{\omega}/V)\xi} . \tag{7.275}$$

This leads to

$$V^2\dot{W} = \frac{d}{d\xi}\left[f e^{i(\tilde{\omega}/V)\xi}\right] - \frac{\tilde{\omega}^2}{V^2} e^{2i(\tilde{\omega}/V)\xi} \int_\infty^\xi d\xi' f e^{-i(\tilde{\omega}/V)\xi'} . \tag{7.276}$$

The integration and rearrangement of the last term (partial integration) leads to

$$W(\xi) = \frac{1}{V^2} f e^{i(\tilde{\omega}/V)\xi} + \frac{\tilde{\omega}}{V^3} f e^{i(\tilde{\omega}/V)\xi} \int_\infty^\xi d\xi' f \sin[k_p(\xi' - \xi)] , \tag{7.277}$$

where

$$k_p = \frac{\tilde{\omega}}{V} \equiv \frac{\omega_{pe}}{V} . \tag{7.278}$$

■

For *circular* polarization we use

$$a = \frac{e}{m_e c}(A_y + iA_z)\ , \quad \frac{e^2}{m_e^2 c^2}\mathbf{A}_\perp^2 = |a|^2\ , \quad f = \frac{n_0 e^2}{2m_e^2}\mathbf{A}_\perp^2 = \frac{n_0 c^2}{2}|a|^2\ ; \tag{7.279}$$

this leads to

$$\frac{\delta n_e}{n_0} = \frac{c^2}{2V^2}\left\{|a(\xi)|^2 - k_p \int_\infty^\xi d\xi' |a(\xi')|^2 \sin[k_p(\xi - \xi')]\right\}\ . \tag{7.280}$$

In this case, we have only a low-frequency driving force, for which the assumption $a = a(\xi)$ is reasonable. The above solution can be used to excite wakefields. Various limiting cases illustrate the physical content of the solution.

First, for narrow laser pulses, e.g., in the simple approximation (width $l \to 0$)

$$|a(\xi')|^2 = |\tilde{a}|^2 \delta(\xi')\tilde{L}\ , \tag{7.281}$$

with $V \approx v_{gr}$ as the group velocity of the laser, we obtain for finite wavenumbers k_p the constant envelope wakefield

$$\frac{\delta n_e}{n_0} = -\frac{c^2 k_p}{2V^2}|\tilde{a}|^2 \tilde{L} \sin(k_p x - \omega_{pe} t) \approx -\frac{c^2 k_p \tilde{L}}{2v_{gr}^2}|\tilde{a}|^2 \sin(k_p x - \omega_{pe} t)\ . \tag{7.282}$$

In general, wakefield generation vanishes when $k_p l \gg 1$ holds, where l denotes the width of the driver. Significant contributions occur for

$$k_p l \sim \mathcal{O}(1)\ . \tag{7.283}$$

On the other hand, in the limiting case $k_p \to 0$, and for finite V,

$$\frac{\delta n_e}{n_0} = \frac{c^2}{2V^2}|a|^2\ , \tag{7.284}$$

i.e., a density hump instead of the density depression caused by the ponderomotive force.

Finally, for $V \approx \frac{c^2 k}{\omega_{pe}} \to 0$, ω_{pe} finite and $k_p \to \infty$, we can obtain from (7.280) the ponderomotive density depression caused by a moving laser pulse

$$\frac{\delta n_e}{n_0} = \frac{c^2}{2V^2 k_p^2} \frac{d^2 |a|^2}{d\xi^2} \tag{7.285}$$

[94, 95].

The argument is as follows. A first partial integration in (7.280) yields

$$\frac{\delta n_e}{n_0} = \frac{c^2}{2V^2} \int_\infty^\xi d\xi' \, \frac{d|a(\xi')|^2}{d\xi'} \cos[k_p(\xi - \xi')] \, . \tag{7.286}$$

If we perform a second one, we obtain

$$\frac{\delta n_e}{n_0} = \frac{c^2}{2V^2} \frac{1}{k_p} \int_\infty^\xi d\xi' \, \frac{d^2|a(\xi')|^2}{d\xi'^2} \sin[k_p(\xi - \xi')] \, . \tag{7.287}$$

A further partial integration yields

$$\begin{aligned} \frac{\delta n_e}{n_0} &= \frac{c^2}{2V^2 k_p^2} \left\{ \frac{d^2|a|^2}{d\xi^2} - \int_\infty^\xi d\xi' \, \frac{d^3|a(\xi')|^2}{d\xi'^3} \cos[k_p(\xi - \xi')] \right\} \\ &\approx \frac{c^2}{2V^2 k_p^2} \frac{d^2|a|^2}{d\xi^2} + \mathcal{O}\left(\frac{1}{k_p}\right) . \end{aligned} \tag{7.288}$$

This result was already obtained earlier in soliton physics.

For *linear* polarization we write

$$\mathbf{A}_\perp(x, t) = A_\perp(x, t)\hat{\mathbf{e}}_z \, , \quad a = \frac{eA_\perp}{m_e c} \tag{7.289}$$

and make the ansatz

$$a(x, t) = \frac{1}{2}\left[\tilde{a}(\xi)e^{-i\omega t + ikx} + \tilde{a}^*(\xi)e^{i\omega t - ikx}\right] , \tag{7.290}$$

which leads to

$$\boxed{a^2 = \underbrace{\frac{1}{2}|\tilde{a}|^2}_{\text{low-frequency}} + \underbrace{\frac{1}{4}\tilde{a}^2 e^{-2i\omega t + 2ikx} + \frac{1}{4}\tilde{a}^{*2} e^{2i\omega t - 2ikx}}_{\text{high-frequency}}} \tag{7.291}$$

.

Since we have different frequencies in the driving term, we decompose the density perturbation δn_e into two parts: $\delta n_e = \delta n_e^{(0)} + \delta n_e^{(2)}$, where the index 0 denotes the low-frequency part and the index 2 indicates that a perturbation of the second harmonic will also occur. From the low-frequency density component $\delta n_e^{(0)}$ we obtain an expression similar to the case of circular polarization, i.e., we can use

$$f \sim \frac{n_0 e^2}{2m_e^2}\mathbf{A}_\perp^2 \sim \frac{n_0 c^2}{4}|\tilde{a}|^2\,, \tag{7.292}$$

and thus

$$\boxed{\frac{\delta n_e^{(0)}}{n_0} = \frac{c^2}{4V^2}\left\{|\tilde{a}(\xi)|^2 - k_p\int_\infty^\xi d\xi'|\tilde{a}(\xi')|^2\sin[k_p(\xi-\xi')]\right\}}\,. \tag{7.293}$$

Example 7.8 (Dominant High-Frequency Component)
For the high-frequency density component $\delta n_e^{(2)}$ we return to Eq. (7.104), which then reads as follows:

$$\boxed{\frac{\partial^2\delta n_e^{(2)}}{\partial t^2} + \omega_{pe}^2\delta n_e^{(2)} \approx \frac{n_0 c^2}{8}\frac{\partial^2}{\partial x^2}\left\{\tilde{a}^2 e^{-2i\omega t+2ikx} + c.c.\right\}}\,. \tag{7.294}$$

We abbreviate (including the *c.c.* part):

$$\ddot{y} + \tilde{\omega}^2 y = \left[h e^{-2i\omega t+2ikx}\right]''\,, \tag{7.295}$$

where the dot denotes a partial time derivative, the prime denotes the spatial derivative, $y = \delta n_e^{(2)}$, $\tilde{\omega} = \omega_{pe}$ and $h = n_0 c^2\tilde{a}^2/8$. Equation (7.295) is again solved using the method of variation of constants. We introduce:

$$y(x,t) = Y(x,t)e^{i\tilde{\omega}t}\,. \tag{7.296}$$

Integration yields

$$\dot{Y} + 2i\tilde{\omega}Y = \int_{-\infty}^t dt'\left[h e^{-2i\omega t'+2ikx}\right]'' e^{-i\tilde{\omega}t'}\,. \tag{7.297}$$

With

$$Y(x,t) = W(x,t)e^{-2i\tilde{\omega}t} \tag{7.298}$$

we can write

$$W = \int_{-\infty}^t dt''\, e^{2i\tilde{\omega}t''}\int_{-\infty}^{t''} dt'\left[h(x,t')e^{-2i\omega t'+2ikx}\right]'' e^{-i\tilde{\omega}t'}\,. \tag{7.299}$$

When evaluating the integrals, we begin with a systematic expansion for weak spatial dependence of h. To lowest order, we approximate

$$\int_{-\infty}^{t''} dt' \Big[h(x,t')e^{-2i\omega t'+2ikx}\Big]'' e^{-i\tilde{\omega}t'} \approx -4k^2\, h e^{2ikx} \int_{-\infty}^{t''} dt' e^{-i(2\omega+\tilde{\omega})t'}$$
$$= -4ik^2\, h e^{2ikx} \frac{e^{-i(2\omega+\tilde{\omega})t''}}{2\omega+\tilde{\omega}} . \tag{7.300}$$

By substituting this into the expression for W, we find

$$W \approx -\frac{4ik^2\, h e^{2ikx}}{2\omega+\tilde{\omega}} \int_{-\infty}^{t} dt''\, e^{-i(2\omega-\tilde{\omega})t''} = \frac{4k^2\, h}{4\omega^2-\tilde{\omega}^2} e^{2ikx} e^{-i(2\omega-\tilde{\omega})t} . \tag{7.301}$$

Thus, to lowest order we obtain

$$\frac{\delta n_e^{(2)}}{n_0} \approx \frac{c^2k^2}{2} \frac{1}{4\omega^2-\tilde{\omega}^2} \tilde{a}^2 e^{-2i\omega t+2ikx} . \tag{7.302}$$

For $\omega \gg \tilde{\omega} \equiv \omega_{pe}$ we can approximate as follows:

$$\frac{1}{4\omega^2-\tilde{\omega}^2} \approx \frac{1}{4\omega^2}\left(1+\frac{\omega_{pe}^2}{4\omega^2}\right) , \tag{7.303}$$

which leads to

$$\boxed{\delta n_e^{(2)} \approx \frac{n_0 c^2 k^2}{8\omega^2} e^{-2i\omega t+2ikx} \tilde{a}^2 \left(1+\frac{\omega_{pe}^2}{4\omega^2}\right) + c.c.} \tag{7.304}$$

when we include the previously suppressed $c.c.$ ■

Example 7.9 (Corrections)

The main approximation in the current evaluation was made in Eq. (7.300). In principle, the bracketed expression in the integrand

$$\Big[h(x,t')e^{-2i\omega t'+2ikx}\Big]'' = \big[h'' + 4ikh' - 4k^2\, h\big] e^{-2i\omega t'+2ikx} \tag{7.305}$$

leads to more contributions than previously considered. The last term in the bracket on the right-hand side of (7.305) was previously taken into account by approximating the integral (after partial integration)

$$\int_{-\infty}^{t} dt'\, h e^{-i(2\omega-\tilde{\omega})t'} = h \int_{-\infty}^{t} dt'\, e^{-i(2\omega-\tilde{\omega})t'} - \int_{-\infty}^{t} dt' \frac{\partial h}{\partial t'} \frac{e^{-i(2\omega-\tilde{\omega})t'}}{-i(2\omega-\tilde{\omega})} \tag{7.306}$$

using only the first contribution. However, within the framework of a systematic expansion with respect to the derivatives of h, the second term can be included in the next order as follows:

$$\int_{-\infty}^{t} dt' \frac{\partial h}{\partial t'} \frac{e^{-i(2\omega-\tilde{\omega})t'}}{-i(2\omega-\tilde{\omega})} \approx V \frac{\partial h}{\partial \xi} \frac{1}{i(2\omega-\tilde{\omega})} \int_{-\infty}^{t} dt'\, e^{-i(2\omega-\tilde{\omega})t'} . \tag{7.307}$$

In addition, the second term in the bracket on the right-hand side of (7.305) should also be taken into account. If we do this, we obtain for $\omega \gg \tilde{\omega} \equiv \omega_{pe}$

$$\delta n_e^{(2)} \approx \frac{n_0 c^2 k}{8\omega^2} e^{-2i\omega t+2ikx} \left[k\tilde{a}^2 \left(1 + \frac{\omega_{pe}^2}{4\omega^2} \right) - i \left(1 - \frac{kV}{\omega} \right) \frac{d\tilde{a}^2}{d\xi} \right] + c.c. \\ + \mathcal{O}\left(\frac{d^2\tilde{a}^2}{d\xi^2} \right). \tag{7.308}$$

Exploiting the definition $V = v_{gr} \equiv \frac{c^2 k}{\omega}$, we can further approximate

$$\left(1 - \frac{kV}{\omega} \right) \approx \frac{\omega_{pe}^2}{\omega^2} . \tag{7.309}$$

The next corrections of order $\frac{d^2\tilde{a}^2}{d\xi^2}$ are not shown. However, they may become important in the limiting case $k \to 0$. In this case, the evaluation yields

$$\left. \frac{\delta n_e^{(2)}}{n_0} \right|_{k\to 0} \approx -\frac{c^2}{32\omega^2} \left(1 + \frac{\omega_{pe}^2}{4\omega^2} \right) \frac{d^2\tilde{a}^2}{d\xi^2} e^{-2i\omega t+2ikx} + c.c. \tag{7.310}$$

We recall the definition $\xi = x - Vt$. ■

When comparing the contributions of the zeroth harmonic with those of the second harmonic, we see roughly that

$$\frac{\delta n_e^{(0)}}{n_0} \sim \frac{c^2}{V^2} \tilde{a}^2 , \qquad \frac{\delta n_e^{(2)}}{n_0} \sim \frac{c^2 k^2}{\omega^2} \tilde{a}^2 , \tag{7.311}$$

so that for $V \approx v_{gr}$ (group velocity) and $v_{ph} = \omega/k$ the ratio

$$\frac{\delta n_e^{(2)}}{\delta n_e^{(0)}} \sim \frac{V^2 k^2}{\omega^2} \sim \frac{v_{gr}^2}{v_{ph}^2} \sim \frac{v_{gr}^4}{c^4} \ll 1 \tag{7.312}$$

is negligible in many applications.

In summary, we first note that the phase velocity of the excited wakefield coincides with the group velocity v_{gr} of the laser pulse.

If we determine the laser frequency ω_L from equation $\omega_L^2 = \omega_p^2 + k^2 c^2$, we find

$$v_{gr} = \frac{kc^2}{\omega_L} \hat{=} v_{phase}^{wake} \equiv v_{ph} \ . \tag{7.313}$$

By substituting the expression for the laser frequency and then replacing the wavenumber using the laser frequency formula, we obtain

$$v_{gr} = c\sqrt{1 - \frac{\omega_p^2}{\omega_L^2}} \ . \tag{7.314}$$

For particle acceleration, we can consider particle velocities close to the phase velocity of the wave (particles surfing on the wave) and introduce

$$\beta_p = \frac{v_{ph}}{c} = \frac{v_{gr}}{c} = \sqrt{1 - \frac{\omega_p^2}{\omega_L^2}} \ . \tag{7.315}$$

For the particles, the gamma factor is

$$\gamma_p = \frac{1}{\sqrt{1 - \beta_p^2}} = \frac{\omega_L}{\omega_p} = \sqrt{\frac{n_c}{n_0}} \ ; \tag{7.316}$$

it is completely determined by the electron density n_0 and the critical density n_c, at which the laser frequency matches the plasma frequency. The plasma should be strongly undercritical ($n_0 < n_c$) in order to achieve high velocities.

Strongly Relativistic Nonlinear Wakefields

So far, we have calculated the amplitude of the excited density oscillations in the weakly relativistic regime. Let us now move on to the strongly relativistic case. Since the excited wakefield is mainly high-frequency and electrostatic, we begin with an *electrostatic* Maxwell electron fluid model:

$$\boxed{\frac{\partial n}{\partial t} + \frac{\partial}{\partial x}(nv) = 0 \ ,} \tag{7.317}$$

$$\boxed{\left(\frac{\partial}{\partial t} + v\frac{\partial}{\partial x}\right)(\gamma m v) = -eE \ ,} \tag{7.318}$$

$$\boxed{\frac{\partial E}{\partial x} = \frac{1}{\varepsilon_0} e(n_0 - n) \ .} \tag{7.319}$$

We have omitted the index e (for electrons). In addition, we have neglected temperature effects in the model for 1D propagation (cold plasma). We use

$$\gamma = \frac{1}{\sqrt{1-\beta^2}} \,, \quad \beta = \frac{v}{c} \,, \tag{7.320}$$

together with the assumptions

$$n(x,t) = n(\tau) \,, \quad v(x,t) = v(\tau) \,, \quad E(x,t) = E(\tau) \,, \tag{7.321}$$

at

$$\tau = \omega_p\left(t - \frac{x}{V}\right), \quad V \approx v_{gr} \,. \tag{7.322}$$

We normalize

$$\frac{n}{n_0} \to n \,, \quad \frac{v}{c} \to v \mathrel{\hat{=}} \beta \,, \quad \frac{eE}{m\omega_p c} \to E \tag{7.323}$$

and obtain from (7.317)–(7.319)

$$\boxed{n\left(1 - \frac{\beta}{\beta_p}\right) = const \equiv 1 \,,} \tag{7.324}$$

$$\boxed{\left(1 - \frac{\beta}{\beta_p}\right)\frac{d(\beta\gamma)}{d\tau} = -E \,,} \tag{7.325}$$

$$\boxed{\frac{d\,E}{d\tau} = \frac{\beta}{1 - \frac{\beta}{\beta_p}} \,.} \tag{7.326}$$

Combining the last two equations leads to

$$E\frac{d\,E}{d\tau} \equiv \frac{d}{d\tau}\left(\frac{E^2}{2}\right) = -\beta\frac{d\beta\gamma}{d\tau} \,. \tag{7.327}$$

With

$$-\frac{d\gamma}{d\tau} = -\frac{\beta\gamma}{1-\beta^2}\frac{d\beta}{d\tau} \tag{7.328}$$

and

$$-\beta\frac{d(\beta\gamma)}{d\tau} = -\frac{\beta}{1-\beta^2}\gamma\frac{d\beta}{d\tau} \tag{7.329}$$

we find

$$\frac{d}{d\tau}\left(\frac{E^2}{2}\right) = -\frac{d\gamma}{d\tau} \,. \tag{7.330}$$

That is, we have found a conserved quantity, which we can express in the form:

$$\boxed{E = E(\gamma) = \sqrt{2(\gamma_m - \gamma)}}\ . \tag{7.331}$$

Here, γ_m is the integration constant. We have $E = 0$ at $\gamma = \gamma_m$. The constant γ_m corresponds to the maximum fluid velocity v_m with $\beta_m = v_m/c$ and

$$\gamma_m = \frac{1}{\sqrt{1-\beta_m^2}}\ . \tag{7.332}$$

The maximum electric field

$$E_m = \sqrt{2(\gamma_m - 1)} \tag{7.333}$$

occurs at locations where $v = 0$ and $\gamma = 1$. From Eqs. (7.325) and (7.331) we obtain

$$\boxed{\pm d\tau = \frac{\left(1 - \frac{\beta}{\beta_p}\right) d(\beta\gamma)}{\sqrt{2(\gamma_m - \gamma)}}}\ . \tag{7.334}$$

Example 7.10 (Limiting Cases)
For comparison with the weakly relativistic case

$$\beta \ll 1\ , \quad \gamma \approx 1\ , \quad \sqrt{2(\gamma_m - \gamma)} \approx \sqrt{\beta_m^2 - \beta^2} \tag{7.335}$$

we approximate

$$\frac{\left(1 - \frac{\beta}{\beta_p}\right) d(\beta\gamma)}{\sqrt{2(\gamma_m - \gamma)}} \approx \frac{\left(1 - \frac{\beta}{\beta_p}\right) d(\beta/\beta_m)}{\sqrt{(1 - (\beta/\beta_m)^2}}\ . \tag{7.336}$$

By using this relation and introducing the abbreviation $x = \beta/\beta_m$ we can rewrite (7.334) as

$$\pm\, d\tau \approx \frac{dx}{\sqrt{1-x^2}} - \frac{\beta_m}{\beta_p} \frac{x dx}{\sqrt{1-x^2}}\ , \tag{7.337}$$

which leads to

$$\pm\,(\tau - \tau_0) \approx \arcsin\left(\frac{\beta}{\beta_m}\right) + \frac{\beta_m}{\beta_p}\sqrt{1 - \left(\frac{\beta}{\beta_m}\right)^2} \tag{7.338}$$

. For small wave amplitudes $\beta_m \ll \beta_p$ Eq. (7.338) can be inverted, yielding the linear wave result

$$\beta \approx \beta_m \sin(\tau - \tau_0)\ , \tag{7.339}$$

$$n \approx \frac{1}{1 - \beta/\beta_p} + \frac{\beta_m}{\beta_p} \sin(\tau - \tau_0) , \tag{7.340}$$

$$E \approx \sqrt{\beta_m^2 - \beta^2} \approx \beta_m \cos(\tau - \tau_0) . \tag{7.341}$$

■

The extrema of $\beta(\tau)$, $n(\tau)$ and $E(\tau)$ do not shift in τ when β_m is increased, whereas the zeros of $\beta(\tau)$ and $n(\tau)$ as well as the extrema of $E(\tau)$ are shifted in such a way that, in the nonlinear high-amplitude regime, velocity and density develop sharp wave crests with broad troughs in between. The electric field takes on a sawtooth shape. In the half-waves with negative $E(\tau)$, electrons can be accelerated. This effect is utilized for plasma wave accelerators.

7.5 Wave Breaking at Large Wave Amplitudes

Wave breaking is an important and fundamental process in nonlinear wave dynamics. It becomes particularly relevant when electron acceleration is excited in electrostatic plasma oscillations whose phase velocities are close to the speed of light. For rapid electron acceleration, stability criteria for localized (inhomogeneous) and relativistic nonlinear electrostatic fields must be known. Wave breaking criteria define the stability of excited wakefields. Wave breaking can also lead to an efficient injection mechanism. Therefore, general criteria for wave breaking are urgently needed both for a better understanding of fundamental plasma processes and for applications in accelerator physics.

A skillful injection of particles into the wakefield of a short laser pulse [96] can result in their being trapped in the electrostatic field. The particles can then be accelerated to high energies, since the phase velocity of the wakefield oscillations is close to the speed of light. The generation of wakefields [70] has become a very interesting phenomenon in short-pulse laser physics, with great potential for practical applications. Laser-plasma accelerators have been proposed as the next generation of compact accelerators, as they can sustain large electric fields [97–99].

In conventional accelerators, the accelerating fields are limited to energy gains of a few MeV m^{-1} due to material damage at the walls. In contrast, laser-plasma accelerators can sustain fields of 100 GeV m^{-1} and more [97]. However, beam quality, including large energy spreads, remains the main challenge [100–106].

In addition to destructive phenomena, the maximum acceleration rate is determined by the maximum amplitude of the oscillation that the plasma can sustain over sufficiently long periods. Wave breaking is a well-known limitation. It generally occurs when the amplitude of the wave reaches the point where the wave crest overturns. This picture is taken from surface water waves. For other waves, such as electromagnetic waves, definitions of breaking also exist [107]. These have been applied to waves with slowly varying envelopes. A new, strongly nonlinear regime arises for laser pulses that are shorter than $\lambda_p = 2\pi c/\omega_{pe}$ (where ω_{pe} is the electron plasma frequency) and whose relativistic intensities are high enough to break the plasma wave already during the first oscillation [108, 109].

There are many highly interesting studies in which wave breaking and injection mechanisms in plasma waves have been discussed. The maximum amplitude of relativistic plasma oscillations in a thermal plasma was determined using a combination of a one-dimensional waterbag model and a fluid model [110]. Conditions for the onset of wave breaking have been specified for both leading and trailing pulses [111]. The trapping and acceleration of a test electron in a nonlinear plasma wave was analyzed in one dimension using Hamiltonian dynamics [112]. Transverse wave breaking [113], electron injection into plasma fields by colliding laser pulses [114], as well as injection mechanisms due to nonlinear wakefield wave breaking [115] have been discussed. The structure and dynamics of the wakefield (wake field) in a plasma column [116–118] have been investigated analytically and numerically, explaining, for example, channel formation. Recently, the advantages and disadvantages of various particle injection mechanisms into wakefields have been theoretically studied [119–124]. The fundamental acceleration mechanisms have been generalized to positrons, photons, and ions [122, 125–127].

The analysis of wave breaking dates back to classical works by Dawson [128] as well as Davidson and Schram [129]. It was shown that planar oscillations in a uniform plasma are stable below a critical amplitude. For larger amplitudes, a mixing process sets in already during the first oscillation. Later, a one-dimensional, nonlinear, relativistic second-order differential equation for the electron fluid velocity was derived in Ref. [130] in Lagrangian coordinates. This formulation enables the analysis of the dynamics up to wave breaking. The Lagrangian analysis, combined with suitable numerical integration of the second-order differential equation, provided further insights into the wave breaking process.

Nonrelativistic Wave Breaking

> The following Lagrangian analysis is based on Dawson's presentation [128], according to which longitudinal wave breaking in a cold one-dimensional (1D) plasma occurs when elements of the electron fluid that initially had different positions overtake each other. From the literature [131] we cite that this overtaking occurs in both nonrelativistic and relativistic plasmas when the maximum fluid velocity equals the phase velocity of the plasma wave [128, 132].

The classical, nonrelativistic analysis of wave breaking begins with the simple electron fluid equations

$$\boxed{\frac{\partial n_e}{\partial t} + \frac{\partial}{\partial x} n_e v_e = 0 \,,} \tag{7.342}$$

$$\boxed{\frac{\partial v_e}{\partial +} v_e \frac{\partial v_e}{\partial x} = -\frac{e}{m_e} E \,,} \tag{7.343}$$

$$\boxed{\frac{\partial E}{\partial x} = \frac{e}{\varepsilon_0}(n_o - n_e) \,.} \tag{7.344}$$

Thermal effects, caused by a pressure gradient in (7.343), have been neglected (see below). Simple stationary solutions in a reference frame moving with u relative to the laboratory system can be found in the form $n_e = n_e(\xi)$, $v_e = v_e(\xi)$, $\phi = \phi(\xi)$ for $\xi = x - ut$. Eqs (7.342) and (7.343) then read

$$\frac{\partial}{\partial \xi}[n_e(v_e - u)] = 0, \tag{7.345}$$

$$\frac{\partial}{\partial \xi}\left[\frac{m_e}{2}(v_e - u)^2 - e\phi\right] = 0. \tag{7.346}$$

Solutions can be given directly (note $v_e = 0$ and $n_e = n_0$ for $\phi = 0$) :

$$n_e(v_e - u) = -n_0 u, \tag{7.347}$$

$$(v_e - u) = -\left[\frac{2e\phi}{m_e} + u^2\right]^{1/2} ; \tag{7.348}$$

they are subsequently used in (7.344),

$$\boxed{\frac{d^2\phi}{d\xi^2} = -\frac{1}{\varepsilon_0} en_0 \left[1 - u\left(\frac{2e\phi}{m_e} + u^2 \right)^{-1/2} \right]} . \tag{7.349}$$

After multiplication by the factor $d\phi/d\xi$ we can integrate. We determine the integration constant that arises in such a way that for $d^2\phi/d\xi^2 = 0$ the electric field $E = -d\phi/d\xi$ is extremal and equal to E_m, i.e.,

$$\frac{1}{2}\left(\frac{d\phi}{d\xi}\right)^2 + 4\pi n_0 m_e \left[\frac{e\phi}{m_e} - u\left(\frac{2e\phi}{m_e} + u^2 \right)^{1/2} \right] = \frac{E_m^2}{2} - 4\pi n_0 m_e u^2 . \tag{7.350}$$

If we write this last equation in the form

$$\boxed{\frac{1}{2}\left(\frac{d\phi}{d\xi}\right)^2 + V(\phi) = 0} , \tag{7.351}$$

then the analogy to the motion of a Newtonian particle in a potential becomes obvious [ϕ corresponds to position, ξ to time, $\frac{1}{2}\left(\frac{d\phi}{d\xi}\right)^2$ is the kinetic energy, and $V(\phi)$ the potential energy]. We have

$$V(\phi) = 4\pi n_0 m_e u^2 \left[\frac{e\phi}{m_e u^2} + 1 - \frac{E_m^2}{8\pi n_0 m_e u^2} - \left(\frac{2e\phi}{m_e u^2} + 1 \right)^{1/2} \right] . \tag{7.352}$$

Note that we have chosen the minus sign on the right-hand side of (7.348), which then allows solutions in (7.351) and (7.352) in the limit $e\phi/m_e u^2 \ll 1$. For a detailed discussion, we introduce the dimensionless variables

$$\psi := \frac{2e\phi}{m_e u^2} , \quad \theta := \frac{E_m^2}{4\pi n_0 m_e u^2} , \tag{7.353}$$

$$\zeta := (2\xi/u)(4\pi n_0 e^2/m_e)^{1/2} \tag{7.354}$$

in order to write (7.351) as

$$\left(\frac{d\psi}{d\zeta}\right)^2 + \psi + 2 - \theta - 2(1+\psi)^{1/2} = 0 \tag{7.355}$$

The potential $V(\psi;\theta)$ is shown in Fig. 7.7.

We observe that for $0 \leq \theta \ll 1$ there are nearly symmetric periodic solutions with $\psi_{min} \approx -2\sqrt{\theta}$ and $\psi_{max} \approx +2\sqrt{\theta}$. Small θ correspond to small amplitudes of the electric field E. With increasing $\theta \to 1$, the oscillations become increasingly larger and more asymmetric. For the "field" $d\psi/d\zeta$ we observe a pronounced steepening of the periodic

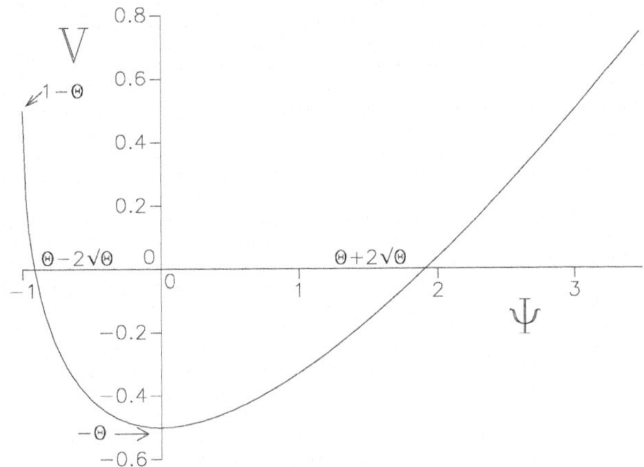

Fig. 7.7 Sketch of the potential (7.352) after introducing the dimensionless variables (7.353) and (7.354) for $\theta = 0.5$

motion. This is because for $\psi = -1$ the slope of the potential $V(\psi)$ is infinite, and with $\theta \to 1$ the left zero $\theta - 2\sqrt{\theta}$ approaches $\psi = -1$. Beyond $\theta = 1$ the condition $\psi + 2 - \theta > 0$ at $\psi = \theta - 2\sqrt{\theta}$ is no longer satisfied, i.e., the "zero" $\psi = \theta - 2\sqrt{\theta}$ corresponds to the negative root in (7.355), and no periodic solutions exist anymore. At the critical value $\theta = 1$, for $u = \omega/k \approx \omega_{pe}/k$,

$$\boxed{\frac{e^2 E_m^2 k^2}{m_e^2 \omega_{pe}^4} \approx 1,} \tag{7.356}$$

i.e., the mean amplitude of the electron oscillations $(eE_m/m_e\omega_{pe}^2)$ becomes comparable to the wavelength $(\sim k^{-1})$. The phenomenon of wave steepening and wave breaking thus becomes understandable by recognizing that the wavelength becomes a function of the wave amplitude. As a result, the phase velocity also becomes amplitude-dependent, and the wave crest increasingly overtakes the wave trough.

This behavior can be seen more precisely mathematically if one switches to the so-called Lagrangian coordinates ξ and τ (in contrast to the Eulerian coordinates x and t in the laboratory frame).

In contrast to the Eulerian coordinates x and t, the Lagrangian coordinates ξ and τ measure the position of each fluid element with respect to its initial position. We define

$$\boxed{\tau = t \quad , \quad \xi = x - \int_0^\tau d\tau' v_e(\xi, \tau')} \, . \tag{7.357}$$

The chain rule leads to

$$\frac{\partial}{\partial \xi} = \frac{\partial x}{\partial \xi}\frac{\partial}{\partial x} \,, \quad \frac{\partial}{\partial \tau} = \frac{\partial t}{\partial \tau}\frac{\partial}{\partial t} + \frac{\partial x}{\partial \tau}\frac{\partial}{\partial x} \,. \tag{7.358}$$

From this, one immediately finds

$$\frac{\partial}{\partial x} = \left[1 + \int_0^\tau d\tau' \frac{\partial v_e(\xi, \tau')}{\partial \xi}\right]^{-1} \frac{\partial}{\partial \xi} \,, \tag{7.359}$$

$$\frac{\partial}{\partial t} = \frac{\partial}{\partial \tau} - v_e(\xi, \tau)\left[1 + \int_0^\tau d\tau' \frac{\partial v_e(\xi, \tau')}{\partial \xi}\right]^{-1} \frac{\partial}{\partial \xi} \,. \tag{7.360}$$

Transforming (7.342) into Lagrangian coordinates yields

$$\frac{\partial}{\partial \tau}\left\{n_e\left[1 + \int_0^\tau d\tau' \frac{\partial v_e(\xi, \tau')}{\partial \xi}\right]\right\} = 0 \,. \tag{7.361}$$

Similarly for (7.343):

$$\frac{\partial v_e(\xi, \tau)}{\partial \tau} = -\frac{e}{m_e} E(\xi, \tau) \,. \tag{7.362}$$

Finally, we take the τ-derivative of (7.344),

$$\frac{\partial}{\partial \tau}\frac{\partial E}{\partial \xi} = \frac{1}{\varepsilon_0} e \frac{\partial}{\partial \tau}\left\{\left[1 + \int_0^\tau d\tau' \frac{\partial v_e(\xi, \tau')}{\partial \xi}\right](n_0 - n_e)\right\} = \frac{e n_0}{\varepsilon_0}\frac{\partial v_e(\xi, \tau)}{\partial \xi} \,, \tag{7.363}$$

to obtain a simple oscillator equation:

$$\boxed{\frac{\partial^2 v_e(\xi, \tau)}{\partial \tau^2} + \omega_{pe}^2 v_e(\xi, \tau) = 0} \,. \tag{7.364}$$

At first glance, this is rather surprising. Starting from a nonlinear system, we end up with a linear equation. Unfortunately, we must apply the nonlinear transformation (7.358) to return to the Eulerian coordinate x.

The solution of (7.364) can be written as follows:

$$v_e(\xi, \tau) = V_e(\xi)\cos(\omega_{pe}\tau) + \omega_{pe} X_e(\xi)\sin(\omega_{pe}\tau) \,. \tag{7.365}$$

For

$$v_e(\xi, \tau = 0) = V_e(\xi) \equiv 0 \tag{7.366}$$

the momentum balance (7.362) yields

$$E(\xi,\tau) = -\frac{m_e}{e}\omega_{pe}^2 X_e(\xi)\cos(\omega_{pe}\tau)\ . \tag{7.367}$$

Next, from the Poisson Eq. (7.363) at time $\tau = 0$

$$\frac{\partial X_e(\xi)}{\partial \xi} = \frac{n_e(\xi,\tau=0)}{n_0} - 1\ . \tag{7.368}$$

Now, the density continuity Eq. (7.361) can be rewritten as

$$n_e(\xi,\tau) = \frac{n_e(\xi,0)}{\left\{1+\frac{\partial X_e(\xi)}{\partial \xi}\left[1-\cos(\omega_{pe}\tau)\right]\right\}}\ . \tag{7.369}$$

If the free function $X_e(\xi)$ is assumed in the form

$$X_e(\xi) = \frac{\Delta}{k}\sin(k\xi)\ , \tag{7.370}$$

it contains the parameters Δ and k. Their meaning becomes clear from the corresponding initial density distribution:

$$n_e(\xi,0) = n_0[1+\Delta\cos(k\xi)]\ . \tag{7.371}$$

In Lagrange coordinates, v_e, E, and n_e are simple harmonic functions of τ. The back transformation to the laboratory (Euler) coordinates is given by

$$t=\tau,\quad x=\xi+\frac{2\Delta}{k}\sin^2(\omega_{pe}\tau/2)\sin(k\xi)\ . \tag{7.372}$$

Everything is linear for $\Delta \to 0$; otherwise, it is nonlinear. For $|\Delta| > \frac{1}{2}$ the transformation is not unique, which corresponds to wave breaking.

Example 7.11 (Explicit Form of E(x,t))
Let us take an explicit look at the solution $E(x,t)$. We expand

$$\sin[k\xi(x,t)] = \sum_{n=1}^{\infty} a_n(t)\sin(nkx)\ , \tag{7.373}$$

where the coefficients are:

$$a_n(t) = \frac{k}{\pi}\int_0^{2\pi/k} dx\,\sin(nkx)\sin[k\xi(x,t)]\ . \tag{7.374}$$

When evaluating the integral on the right-hand side, we use $dx = d\xi\ [1 + \alpha\cos(k\xi)]$ with $\alpha = 2\Delta\sin^2(\omega_{pe}\tau/2) \equiv \alpha(t)$. In addition to the addition theorems for sine and cosine functions, the representation of the Bessel functions J_n of order n

$$J_n(z) = \frac{1}{\pi}\int_0^{\pi} \cos[z\sin\theta - n\theta]d\theta \tag{7.375}$$

is used. Recall

$$J_{-n}(z) = (-1)^n J_n(z) \tag{7.376}$$

and

$$J_{\nu-1}(z) + J_{\nu+1}(z) = \frac{2\nu}{z} J_\nu(z)\ . \tag{7.377}$$

A short calculation of the coefficients yields

$$a_n(t) = (-1)^{n+1}\frac{2}{n\alpha(t)} J_n[n\alpha(t)]\ . \tag{7.378}$$

We can then express the electric field as follows:

$$E(x,t) = -\frac{m_e}{e}\frac{\omega_{pe}^2}{k}\sum_{n=1}^{\infty}(-1)^{n+1}\frac{1}{n\sin^2(\omega_{pe}t/2)} J_n[2n\Delta\sin^2(\omega_{pe}t/2)] \times \sin(nkx)\cos(\omega_{pe}t)\ . \tag{7.379}$$

This is not a purely harmonic function. Higher harmonics in k appear, and the time dependence is also quite complicated. Of course, this description only makes sense as long as $|\Delta| < \frac{1}{2}$. ■

Relativistic Formulation

Dawson [128] showed that wave breaking (within the first period) can occur, provided that the distance traveled by the fluid with the maximum fluid velocity V in the time T, which is approximately the inverse plasma frequency, exceeds the inhomogeneity length L (which is approximately the inverse wavenumber k). In this scenario, the wave breaking time T is limited to the first period. For a given k, there exists an amplitude threshold A of the wave ($V \sim A$). If we are below this threshold, no wave breaking should occur [128].

However, the relativistic nonlinearity produces a nonlinear frequency shift that can become spatially dependent. This gives rise to a nonlinear inhomogeneity length L, which depends on the amplitude of the oscillation and thus on A. The condition $VT > L$ can now be fulfilled without a threshold. For small amplitudes A, the breaking time T can become large and for $A \to 0$ scale like $T \sim 1/A^3$.

We begin with a description using a cold fluid and Lagrangian coordinates in one spatial dimension. The fluid element was located at x_0 at time $t = 0$ and is located at x_L at time t, i.e., the Lagrangian coordinate is $x_L = x_L(x_0, t)$ with $x_L(x_0, t = 0) = x_0$. Furthermore, we define the Lagrangian momentum of the electrons $p_L = p_L(x_0, t) = p(x_L, t)$, the Lagrangian electrostatic field $E_L = E_L(x_0, t) = E(x_L, t)$ and the electron density $n_L = n_L(x_0, t) = n(x_L, t)$. The ion density N is assumed to be constant.

The Maxwell equations (we consider electrostatic fields and neglect the magnetic fields) yield

$$\frac{\partial E_L}{\partial t} = \frac{1}{\varepsilon_0} e n_L v_L \,. \tag{7.380}$$

The total time derivative of E_L is

$$\frac{dE_L}{dt} = \frac{\partial E}{\partial x_L}\frac{dx_L}{dt} + \frac{\partial E}{\partial t} \,. \tag{7.381}$$

From the Poisson equation

$$\frac{\partial E}{\partial x_L} = \frac{1}{\varepsilon_0}(N - n_L)e \tag{7.382}$$

as well as from the definition

$$\boxed{\frac{dx_L}{dt} = v_L = \frac{p_L}{m\gamma}} \tag{7.383}$$

we obtain, together with (7.380)

$$\frac{dE_L}{dt} = \frac{1}{\varepsilon_0} N e v_L \,. \tag{7.384}$$

Example 7.12 (Constant Ion Density)
We first consider the case of a homogeneous ion density N. After integration, we obtain

$$E_L = \frac{1}{\varepsilon_0} N e x_L + E_0(x_0) - \frac{1}{\varepsilon_0} N e x_0 \,, \tag{7.385}$$

where we have defined $E_0(x_0) = E_L(x_0, t = 0)$. Eq. (7.385) is a fundamental relation that we will use in the following. The (longitudinal) momentum balance reads

$$\boxed{\frac{dp_L}{dt} = -eE_L = -\frac{1}{\varepsilon_0}Ne^2 x_L - eE_0(x_0) + \frac{1}{\varepsilon_0}Ne^2 x_0} \ . \qquad (7.386)$$

Together with Eq. (7.383), these form the fundamental equations of motion for constant ion density. Here,

$$\gamma = \sqrt{1 + \frac{p_L^2}{m^2c^2}} \ . \qquad (7.387)$$

■

Example 7.13 (Inhomogeneous Ion Density)
If the ion density distribution is not homogeneous (but the ion dynamics are still neglected), then $N_L = N_L(x_0, t) = N(x_L)$ holds. The generalization of Eq. (7.384) is

$$\frac{dE_L}{dt} = \frac{1}{\varepsilon_0} N_L e \frac{dx_L}{dt} \ . \qquad (7.388)$$

If we introduce the function $Y_i(\xi)$ via

$$N(x_L) = \left.\frac{dY_i}{d\xi}\right|_{\xi = x_L} \qquad (7.389)$$

we find

$$E_L = \frac{1}{\varepsilon_0} e[Y_i(x_L) - Y_i(x_0)] + E_0(x_0) \ , \qquad (7.390)$$

which replaces the previous Eq. 7.385). Now Eq. (7.386) can be generalized to

$$\boxed{\frac{dp_L}{dt} = -\frac{1}{\varepsilon_0} e^2[Y_i(x_L) - Y_i(x_0)] - eE_0(x_0)} \ , \qquad (7.391)$$

which, together with (7.383), forms the basic system of equations for inhomogeneous plasmas.
■

Relativistic wave breaking for N=const

We now illustrate wave breaking for N = const; the conclusions and calculations are similar for inhomogeneous situations $N(x)$. Using Poisson's equation, we obtain

$$\frac{\partial E_L}{\partial x_0} = \frac{\partial E}{\partial x_L}\frac{\partial x_L}{\partial x_0} = \frac{1}{\varepsilon_0}(N - n_L)e\frac{\partial x_L}{\partial x_0} \ . \qquad (7.392)$$

On the other hand, we have

$$\frac{\partial E_L}{\partial x_0} = \frac{\partial}{\partial x_0}\left(\frac{1}{\varepsilon_0}Nex_L + E_0(x_0) - \frac{1}{\varepsilon_0}Nex_0\right) . \tag{7.393}$$

A short calculation leads to

$$\frac{\partial E_L}{\partial x_0} = \frac{1}{\varepsilon_0}e\left(\frac{\partial x_L}{\partial x_0}N - n_0\right) , \tag{7.394}$$

where we have used

$$\frac{\partial E_0}{\partial x_0} = \frac{1}{\varepsilon_0}e(N - n_0) \tag{7.395}$$

with $n_0 = n_L(x_0, t = 0)$. Comparing the right-hand sides of (7.392) and (7.394), it follows that

$$n_L = \frac{n_0}{\frac{\partial x_L}{\partial x_0}} . \tag{7.396}$$

The condition

$$\boxed{\frac{\partial x_L}{\partial x_0} = 0} \tag{7.397}$$

defines wave breaking. The relation (7.396) shows that "infinite" density corresponds to wave breaking.

For $N = \text{const}$ we insert the following definition into (7.386):

$$y_L := \frac{1}{\varepsilon_0}Ne^2x_L + eE_0 - \frac{1}{\varepsilon_0}e^2x_0N . \tag{7.398}$$

We can then write the fundamental equations of motion as

$$\frac{dy_L}{dt} = \frac{Ne^2}{\varepsilon_0 m}\frac{p_L}{\gamma} , \tag{7.399}$$

$$\frac{dp_L}{dt} = -y_L . \tag{7.400}$$

Due to the appearance of γ, this is an equation for a nonlinear oscillator. Instead of ω_{pe0}, a frequency that depends on the amplitude generally appears.

Example 7.14 (Nonlinear Oscillator)
From the equation of the nonlinear oscillator

$$\frac{d^2p}{dt^2} = -\frac{p}{\sqrt{1 + p^2}} \tag{7.401}$$

we calculate the nonlinear frequency. Momentum and time are now normalized by mc and ω_{pe}^{-1}, respectively. The amplitude-dependent frequency can be determined using perturbation theory. First, we expand the square root for small amplitudes in order to obtain for the oscillator (with linear frequency $\omega = 1$)

$$\frac{d^2p}{dt^2} + \omega^2 p = -(1-\omega^2)p - \frac{1}{2}p^3. \tag{7.402}$$

Multiplying both sides by $\cos(\omega t)$ and integrating over t from $-\pi/\omega$ to $+\pi/\omega$, as well as the approximation $p(t) \approx \tilde{A}\cos(\omega t)$ within the integrals, lead to the approximate result for the frequency at small amplitudes:

$$\omega^2 \approx 1 - \frac{3}{8}\tilde{A}^2 \tag{7.403}$$

This is the result that will be used in Eq. (7.424). It is represented by the crosses in Fig. 7.8. We can even obtain a better approximation, valid for larger amplitudes, if we do not expand the square root. By following the same steps as before, we arrive at

$$\omega^2 \approx \frac{1}{\pi}\int_{-\pi}^{\pi} d\tau \frac{\cos^2(\tau)}{\sqrt{1+\tilde{A}^2\cos^2(\tau)}} \tag{7.404}$$

This result is also shown in Fig. 7.8 by the lightly dashed curve. In comparison with the exact numerical result, we find excellent agreement up to quite large amplitudes. ■

For inhomogeneous $N = N(x)$, an approximate calculation is possible, e.g., when $|x_L - x_0| \ll 1$. Then we start with the equations

$$\frac{dp_L}{dt} = -\frac{1}{\varepsilon_0}e^2 N(x_0)[x_L - x_0] - eE_0(x_0) \equiv -\tilde{y}_L\,, \tag{7.405}$$

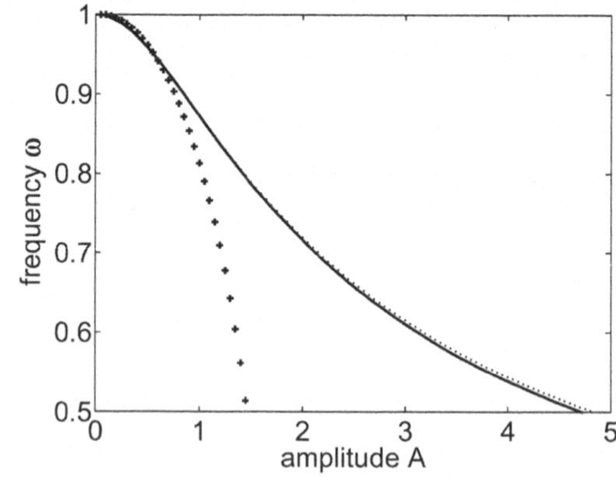

Fig. 7.8 Amplitude dependence of the frequency of the nonlinear oscillator (7.401). The exact result (represented by the solid line) is compared with the approximation for small amplitudes (7.403) (crosses) and (7.404) (dashed curve) [133]

$$\frac{d\tilde{y}_L}{dt} = \frac{e^2 N(x_0)}{\varepsilon_0 m \gamma} p_L . \tag{7.406}$$

Now the frequency depends not only on amplitude but also on position. The solutions for $\tilde{y}_L$ and $x_L - x_0$ are straightforward to obtain.

Dynamic ions with $N = N(x, t)$ can only be treated numerically.

Non-relativistic limit for $\gamma = 1$ *and N=const*

For the non-relativistic limit, we start with a harmonic field at $t = 0$

$$E_0 = A(x_0) \sin(kx_0 + \varphi) \tag{7.407}$$

and the momentum

$$p_0 = -\sqrt{\frac{\varepsilon_0 m}{N}} A(x_0) \cos(kx_0 + \varphi) , \tag{7.408}$$

where φ is an arbitrary phase. We can write the wave solution as follows:

$$E_L = A(x_0) \sin(kx_0 - \omega_{pe0} t + \varphi) . \tag{7.409}$$

The reason for this is that in this case Eqs. (7.399) and (7.400) simplify to

$$\frac{d^2 E_L}{dt^2} = -\omega_{pe0}^2 E_L . \tag{7.410}$$

Then the fundamental relation (7.385) leads to

$$x_L = x_0 + \frac{\varepsilon_0}{Ne} A(x_0) \left[\sin(kx_0 - \omega_{pe0} t + \varphi) - \sin(kx_0 + \varphi) \right] , \tag{7.411}$$

and for constant amplitudes $A(x_0) = A$ we obtain

$$\frac{\partial x_L}{\partial x_0} = 1 + \frac{\varepsilon_0 k}{Ne} A \left[\cos(kx_0 - \omega_{pe0} t + \varphi) - \cos(kx_0 + \varphi) \right] . \tag{7.412}$$

The necessary condition for wave breaking becomes

$$\boxed{\frac{\varepsilon_0 k}{Ne} A \geq \frac{1}{2} ,} , \tag{7.413}$$

which corresponds to the previous result [128] (note the factor of 2).

Wave breaking occurs when the change in maximum fluid velocity is greater than the phase velocity, i.e.

$$\Delta v_L|_{max} = \frac{2\varepsilon_0 A\omega_{pe0}}{Ne} > \frac{\omega_{pe0}}{k} \,. \tag{7.414}$$

Dawson [128] obtained the result $\varepsilon_0 A/Ne \geq 1/k$ for specific initial conditions.

The following should be said about the famous result (7.413). First, we can use (7.413) to determine the onset of wave breaking as a function of, for example, the amplitude A. By introducing $T = \omega_{pe0}^{-1}$, $V = \Delta v_L|_{max}$, and $L = k^{-1}$, we obtain from

$$VT \approx L \tag{7.415}$$

the threshold amplitude for fixed k. If, for wakefields, we approximately set $\frac{\omega_{pe0}}{k} \approx c$ and use the normalized maximum field amplitude $a = \frac{\varepsilon_0 A\omega_{pe0}}{Nec}$ (normalized with mc), the criterion (7.413) reads:

$$a \geq \frac{1}{2} \,. \tag{7.416}$$

According to (7.413), small amplitudes therefore do not lead to wave breaking. On the other hand, (7.416) for wakefields requires a relativistic treatment.

In the homogeneous, nonrelativistic limit (not necessarily for wakefields), the threshold prediction (7.413) for wave breaking in finite time can be compared with numerical simulations, and an excellent agreement is found [133].

Relativistic calculation for $\gamma = 1$ and N=const

In the fully relativistic description, instead of (7.411), we obtain a nonlinear oscillator solution, which we abbreviate in the form:

$$x_L = x_0 + \frac{\varepsilon_0}{Ne} A \sin(kx_0)[F(\omega t, x_0) - 1] \,, \tag{7.417}$$

where ω is the nonlinear frequency, which in the weakly relativistic limit corresponds to $\omega \approx \omega_{pe0}$. The function F remains finite and is 2π-periodic, i.e., $F(y \equiv \omega t + 2\pi, x_0) = F(y \equiv \omega t, x_0)$ holds. If we start with $E_0 = A\sin(kx_0)$, then differentiating now leads to

$$\frac{\partial x_L}{\partial x_0} = 1 + \frac{\varepsilon_0}{Ne}\left\{\frac{\partial E_0}{\partial x_0}\big[F(y \equiv \omega t, x_0) - 1\big] + E_0 \frac{\partial F(y \equiv \omega t, x_0)}{\partial y}\frac{\partial \omega}{\partial x_0} t + E_0 \frac{\partial F(y \equiv \omega t, x_0)}{\partial x_0}\right\} . \tag{7.418}$$

The right-hand side contains, in addition to the Dawson criterion (7.413), a second cause for wave breaking. In any case, for $t \to \infty$ the term growing linearly in t will dominate. For a given

$$\frac{\partial \omega}{\partial x_0} \neq 0 \tag{7.419}$$

we can always find a series of intervals such that

$$\frac{\varepsilon_0}{Ne} E_0 \frac{\partial F(y \equiv \omega t, x_0)}{\partial y} \frac{\partial \omega}{\partial x_0} < 0 \tag{7.420}$$

and wave breaking can occur without a threshold.

This changes the statement from [131] that "for nonrelativistic and relativistic plasmas, this overtaking occurs when the maximum fluid velocity equals the phase velocity." The breaking considered here may not necessarily occur during the first oscillation, but only later, i.e., after many electron plasma periods. In general, (7.420) allows for an estimate of the time until breaking. We would like to emphasize that the time for wave breaking now follows from

$$VT \approx L \tag{7.421}$$

[see (7.415), which was used there for the amplitude threshold], but now the inhomogeneity length

$$L \approx \left| \left[\frac{\partial \ln \omega}{\partial x_0} \right]^{-1} \right| \tag{7.422}$$

should be used. Quantitative predictions require precise knowledge of the nonlinear oscillation.

Obviously, the frequency $x_0$becomes position-dependent when the plasma is inhomogeneous. Likewise, nonlinearity can introduce a spatially dependent frequency. We demonstrate this analytically by, for example, solving the (extended) equation

$$\frac{d^2 p_L}{dt^2} = -\frac{\omega_{pe0}^2}{\gamma} p_L \approx -\omega_{pe0}^2 \left(1 - \frac{1}{2} \frac{p_L^2}{m^2 c^2} \right) p_l \tag{7.423}$$

for small amplitudes. If we assume $p_0 \sim \tilde{A}$, we obtain approximately

$$\omega^2 \approx \omega_{pe0}^2 \left(1 - \frac{3}{8} \frac{\tilde{A}^2}{m^2 c^2} \right) . \tag{7.424}$$

For example, with $\tilde{A} = \tilde{A}(x_0) = A(x_0) \cos(k x_0)$ we can determine the explicit spatial dependence of the frequency.

If we apply this to the wakefield (ponderomotive field) excited by a pulsed laser beam, we find that the initially excited wakefield exhibits a spatially dependent frequency. Consequently, the wakefield will break (sooner or later, depending on the strength of the excited field), even if (7.413) is not satisfied. For $\gamma \to 1$ the time to breaking approaches infinity. The scaling is

$$T \sim a^{-3} \tag{7.425}$$

in the small amplitude limit. Here, $a = \frac{e\tilde{A}}{mc} \hat{=} \frac{E_0 e}{\omega_{pe} mc}$. The result (7.413) can be understood as the (nonrelativistic) prediction for the breaking of (general) electrostatic oscillations in finite time. If we choose relativistic wave fields, numerical simulations always show wave breaking in agreement with the analytical predictions.

Finite Lengths

Dawson [128] considered a nonrelativistic field of infinite length. To investigate the influence of a finite field length, one can start with an electric field at $t = 0$ in the following form:

$$E_L = A_\infty e^{-x_0^2/\sigma^2} \cos(kx_0) \sin(\omega_{pe0} t) \ . \tag{7.426}$$

The fundamental result (7.385) leads to

$$x_L = x_0 + \frac{\varepsilon_0 A_\infty}{Ne} \cos(kx_0) \sin(\omega_{pe0} t) e^{-x_0^2/\sigma^2} \tag{7.427}$$

and yields after differentiation

$$\frac{\partial x_L}{\partial x_0} = 1 - \frac{\varepsilon_0 A_\infty}{Ne} \left(k \sin(kx_0) + 2\frac{x_0}{\sigma^2} \cos(kx_0) \right) \sin(\omega_{pe0} t) e^{-x_0^2/\sigma^2} \ . \tag{7.428}$$

Therefore, we expect wave breaking for

$$\frac{\varepsilon_0 A_\infty}{Ne} \left(k \sin(kx_0) + 2\frac{x_0}{\sigma^2} \cos(kx_0) \right) e^{-x_0^2/\sigma^2} \geq 1 \ . \tag{7.429}$$

For $\sigma \to \infty$ we obtain Dawson's criterion [128] for breaking, namely $\frac{\varepsilon_0 A_\infty k}{Ne} \geq 1$. In the limiting case $\sigma \to 0$ it is expected that the field will break even for arbitrarily small field amplitudes. Between these two limits, it is most practical to evaluate the inequality (7.429) numerically. We define

$$\Phi = \frac{1}{\max\limits_{x_0} \tilde{f}} \ , \quad \text{with} \quad \tilde{f} \equiv \tilde{f}(x_0; k, \sigma) = \left(\sin(kx_0) + 2\frac{x_0}{k\sigma^2} \cos(kx_0) \right) e^{-x_0^2/\sigma^2} \ . \tag{7.430}$$

In Fig. 7.9, Φ is shown as a function of σ for $\frac{kc}{\omega_{pe}} = 2$. The breaking criterion is

$$\boxed{\frac{\varepsilon_0 A_\infty k}{Ne} \geq \Phi} \ , \tag{7.431}$$

where, for simplicity, we consider k as a parameter. For very wide fields, i.e., large σ, the criterion resembles Dawson's result [128]. As σ decreases, some fields with amplitudes A_∞ that satisfy the breaking condition for infinite length will no longer break in the

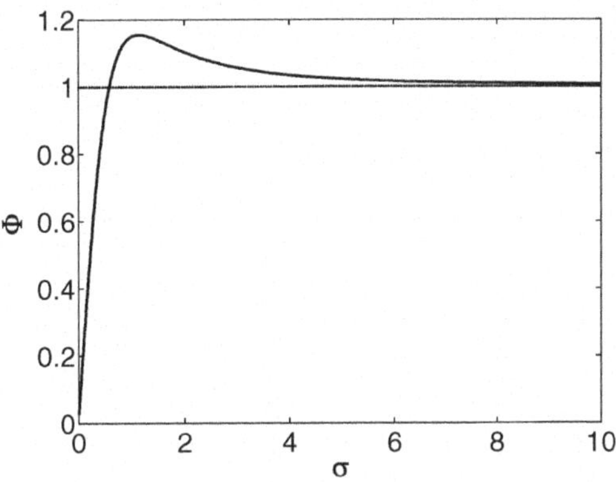

Fig. 7.9 Plot of ϕ as a function of σ for $\frac{kc}{\omega_{pe}} = 2$. Fields with $\frac{\varepsilon_0 A_\infty k}{Ne} \geq \Phi$ should break [133]

nonrelativistic limit. For very small σ, even in the nonrelativistic case, fields with very small amplitudes can be subject to breaking.

Nonlinear Oscillations in Plasmas

8

Abstract

In this chapter, we present general methods for modeling waves in plasmas with larger amplitudes, which we then refer to as nonlinear. In principle, we would have to do this separately for each individual wave type. However, we illustrate common methods using only a few types (ion-acoustic oscillations, Langmuir oscillations, and drift vortices), with the expectation that once the fundamental procedure has been demonstrated for some modes, transferring and generalizing it to other modes should not be too difficult. Central to this, of course, is whether the nonlinear oscillatory states thus found are also observable, i.e., stable from a mathematical standpoint. Among other things, we present exact solutions using the inverse scattering method. While most of the chapter focuses on coherent states, we address turbulent states at the end. However, this is done without any claim to a comprehensive presentation of the important but also extremely challenging current approaches to turbulence.

8.1 Ion-Acoustic Solitons and the KdV Equation

Some nonlinear wave configurations are called solitons or solitary waves. In the context of solitons, plasma physics played a pioneering role for the entire field of physics. In plasma physics, ion-acoustic waves were studied in great physical detail; these studies led to a crucial insight into the behavior of nonlinear waves with dispersion. And, in particular, to a new mathematical method for finding exact time-dependent solutions. In the following, we present the hydrodynamic model for ion-acoustic nonlinear waves and discuss its implications.

K.-H. Spatschek, *Theoretical Plasma Physics*,
https://doi.org/10.1007/978-3-662-72828-4_8

KdV Equation

Let us assume that the ion temperature T_i is much smaller than the electron temperature T_e. Furthermore, we restrict ourselves to low-frequency phenomena ($\omega \ll \omega_{pe}$) and can therefore, to a good approximation, assume a Boltzmann distribution for the electrons. (For obvious reasons, such a model is often referred to as the "$m_e \to 0$ approximation." Note that the Boltzmann distribution follows from the electron momentum balance in the limit of negligible electron inertia, if the nonlinearity due to the convective derivative is neglected.) For simplicity, the isothermal equation of state is also assumed

$$p_e = n_e k_B T_e \tag{8.1}$$

for $T_e = const$. The low-frequency dynamics are essentially determined by the ion motion. The latter follows from the momentum balance (one-dimensional)

$$\partial_t v_i + v_i \partial_x v_i = -\frac{e}{m_i}\,\partial_x \phi, \tag{8.2}$$

where the ambipolar potential ϕ is related via the Poisson equation

$$\partial_x^2 \phi = \frac{1}{\varepsilon_0} e[n_e - n_i] \tag{8.3}$$

to a connection between the ion and electron densities, n_i and n_e, respectively. The ion density follows from the particle balance

$$\partial_t n_i + \partial_x (n_i v_i) = 0, \tag{8.4}$$

while the electron density

$$n_e = n_0 \exp(e\phi / k_B T_e) \tag{8.5}$$

is distributed according to Boltzmann. Eqs. (8.2)–(8.5) constitute a closed system for n_i, v_i and ϕ. At this point, it is advisable to introduce dimensionless variables by means of

$$\boxed{x/\lambda_{De} \to x, \quad t\omega_{pi} \to t, \quad e\phi/k_B T_e \to \phi, \quad n_i/n_0 \to n, \quad v_i/c_s \to v} \tag{8.6}$$

where the characteristic length scale is the electron Debye length λ_{De},, the characteristic time is the inverse ion plasma frequency ω_{pi}^{-1}, and the characteristic velocity is the ion sound speed $c_s = \lambda_{De}\omega_{pi}$. The spatially homogeneous mean density is n_0. The basic equations in the new variables are

$$\partial_t n + \partial_x (nv) = 0, \tag{8.7}$$

$$\partial_t v + v\partial_x v = -\partial_x \phi, \tag{8.8}$$

$$\partial_x^2 \phi = e^\phi - n. \tag{8.9}$$

The dispersion relation

$$\boxed{\omega = k/(1+k^2)^{1/2} \approx k} \tag{8.10}$$

follows directly after linearization and Fourier transformation. (For quantities with dimensions, $\omega \approx kc_s$ holds. Since the speed of sound is much smaller than the thermal velocity v_{te} of the electrons, the Boltzmann distribution for electrons is justified in retrospect.)

Within the framework of a weakly nonlinear theory, we assume that the deviations from the stationary and spatially homogeneous values $n = 1,\ \phi = 0$ and $v = 0$ are small. We take this into account by introducing a smallness parameter ε (to be examined in more detail later) and expand

$$n = 1 + \varepsilon n^{(1)} + \varepsilon^2 n^{(2)} + \dots\ , \tag{8.11}$$

$$\phi = \varepsilon \phi^{(1)} + \varepsilon^2 \phi^{(2)} + \dots\ , \tag{8.12}$$

$$v = \varepsilon v^{(1)} + \varepsilon^2 v^{(2)} + \dots\ . \tag{8.13}$$

In addition, we must scale the spatial and temporal dependencies. We do this in the form

$$\xi = \varepsilon^{1/2}(x - t), \tag{8.14}$$

$$\tau = \varepsilon^{3/2}\, t. \tag{8.15}$$

This last step still needs to be motivated. One can take a rather formal standpoint and introduce new smallness parameters ξ and τ for the new spatial and temporal variables, respectively, as well as new smallness parameters ε and μ. A ε-μ relation can then be found on the basis of nontrivial solvability conditions. Here, however, we take a different approach—one often used in practice—by first seeking stationary localized solutions of (8.7)–(8.9).

Stationary Solution
With the Mach number M we write

$$\partial_t = -M\,\partial_x, \tag{8.16}$$

and find directly from (8.7)

$$n = M/(M - v). \tag{8.17}$$

Here, we have directly satisfied the boundary condition $n \to 1,\ v \to 0,\ \phi \to 0$ for $x \to \pm\infty$. At the same time, (8.8) yields

$$(M - v)^2 = M^2 - 2\phi. \tag{8.18}$$

We can now, by virtue of the last two relations, express v and n in terms of ϕ. If we then multiply both sides of (8.9) by $\partial_x\phi$ and integrate, we obtain

$$\boxed{\frac{1}{2}(\partial_x\phi)^2 = -V(\phi) \equiv e^{\phi} + M(M^2 - 2\phi)^{1/2} - (M^2 + 1)}\,. \tag{8.19}$$

A sketch of the "potential" V as a function of ϕ can be found in Fig. 8.1. Owing to the analogy with the motion of a Newtonian particle in the "potential" V, it is easy to see that a localized solution ($\phi \to 0$ for $x \to \pm\infty$) exists, whose maximum amplitude is ϕ_{max}.

At the moment, we are only interested in the weakly nonlinear case, in which we can expand the e-function,

$$e^{\phi} \approx 1 + \phi + \frac{1}{2}\phi^2 + \frac{1}{6}\phi^3 + \dots\ . \tag{8.20}$$

Furthermore, we expect only a small deviation from the linear propagation velocity ($\omega/k \approx 1$) and therefore set

$$0 < \delta M \equiv M - 1 \ll 1. \tag{8.21}$$

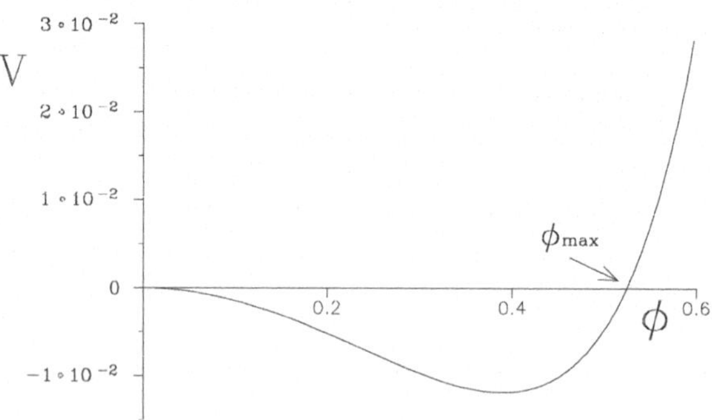

Fig. 8.1 Potential $V(\phi)$ as a function of ϕ for $M = 1.2$

Because

$$(M^2 - 2\phi)^{1/2} = [1 + 2\delta M + (\delta M)^2 - 2\phi]^{1/2}$$
$$\approx 1 + \delta M - \phi + \frac{1}{2}(\delta M)^2 - \frac{1}{2}(\delta M - \phi)^2$$
$$- \frac{1}{2}(\delta M - \phi)(\delta M)^2 + \frac{1}{2}(\delta M - \phi)^2 - \ldots \tag{8.22}$$

we obtain, taking into account the lowest orders in δM and ϕ

$$V \approx -\frac{2}{3}\phi^2(3\delta M - \phi). \tag{8.23}$$

The equation

$$(\partial_x \phi)^2 = \frac{2}{3}\phi^2(3\delta M - \phi) \tag{8.24}$$

can be integrated immediately; we find the solitary wave solution

$$\boxed{\phi = 3\delta M \ \mathrm{sech}^2\left[\left(\frac{1}{2}\delta M\right)^{1/2}(x - Mt)\right]}\,. \tag{8.25}$$

Since

$$(\delta M)^{1/2}(x - Mt) = (\delta M)^{1/2}(x - t) - (\delta M)^{3/2}\, t \tag{8.26}$$

holds, the scaling (8.14) and (8.15) become immediately evident. Moreover, with $\varepsilon \sim \delta M$ we also immediately obtain an intuitive meaning for the expansion parameter ε.

KdV equation

After this interim consideration, we can proceed with the scaling (8.11)–(8.15). Substituting into (8.7)–(8.9), we obtain equations that must be satisfied separately at different orders of ε. Taking into account

$$\partial_x = \varepsilon^{1/2}\partial_\xi, \tag{8.27}$$

$$\partial_t = \varepsilon^{3/2}\partial_\tau - \varepsilon^{1/2}\partial_\xi \tag{8.28}$$

we obtain, in the lowest orders of ε and $\varepsilon^{3/2}$

$$\phi^{(1)} = n^{(1)}, \tag{8.29}$$

$$\partial_\xi n^{(1)} = \partial_\xi v^{(1)}. \tag{8.30}$$

The boundary conditions for localized solutions only allow

$$n^{(1)} = \phi^{(1)} = v^{(1)} \tag{8.31}$$

The next orders in ε yield

$$\partial_\xi^2 \phi^{(1)} = \phi^{(2)} - \frac{1}{2}\left[\partial_\xi \phi^{(1)}\right]^2 - n^{(2)}, \tag{8.32}$$

$$-\partial_\xi n^{(2)} + \partial_\tau n^{(1)} + \partial_\xi \left[n^{(1)} v^{(1)}\right] + \partial_\xi v^{(2)} = 0, \tag{8.33}$$

$$-\partial_\xi v^{(2)} + \partial_\tau v^{(1)} + v^{(1)}\, \partial_\xi v^{(1)} = -\partial_\xi \phi^{(2)}. \tag{8.34}$$

Of course, we can immediately use (8.31) here. In the next steps, we solve (8.32) for $n^{(2)}$ and substitute into (8.34). If we also substitute $\partial_\xi v^{(2)}$ from (8.33), then, "fortunately," $\partial_\xi \phi^{(2)}$ drops out.

We are left with a closed equation for $n^{(1)}$,

$$\partial_\tau n^{(1)} + n^{(1)} \partial_\xi n^{(1)} + \frac{1}{2} \partial_\xi^3 n^{(1)} = 0. \tag{8.35}$$

Equation (8.35) is the Korteweg–de Vries (KdV) equation, whose simplest stationary (single-soliton) solution

$$n^{(1)}(\xi - c\tau) = 3c \operatorname{sech}^2\left[\left(\frac{1}{2}c\right)^{1/2} (\xi - c\tau)\right] \text{ with a constant } c \tag{8.36}$$

is.

The agreement with (8.25) was to be expected. Due to the scaling, both of the amplitude and the spatial dependence, the KdV equation is a one-dimensional model for weakly nonlinear and weakly dispersive ($k^2 \ll 1$) ion-acoustic waves in unmagnetized plasmas.

At this point, a brief digression is permitted, in which we exemplify the essential phenomena hidden within the KdV equation. For this purpose, we compare the KdV equation (8.35), now written in a somewhat clearer notation

$$u_t + uu_x + \kappa u_{xxx} = 0 \tag{8.37}$$

with the simplified equations

$$u_t + cu_x = 0, \tag{8.38}$$

$$u_t + uu_x = 0, \tag{8.39}$$

$$u_t + cu_x + \kappa u_{xxx} = 0. \tag{8.40}$$

Here, κ and c are constants.

Example 8.1 Comparison with (8.38)
The first simplification (8.38), in which the nonlinear and dispersive terms are neglected, is trivial to solve. The general solution

$$u = f(x - ct) \tag{8.41}$$

can be written with any differentiable function f, which is determined by the initial condition. For example, if we require

$$u(x, 0) = \begin{cases} a^2 - x^2 & \text{for } |x| < a, \\ 0 & \text{for } |x| \geq a, \end{cases} \tag{8.42}$$

then at time t we have

$$u(x, t) = \begin{cases} a^2 - (x - ct)^2 & \text{for } |x - ct| < a, \\ 0 & \text{for } |x - ct| \geq a. \end{cases} \tag{8.43}$$

This means that an initial pulse propagates through space without changing its shape. ■

Example 8.2 Comparison with (8.39)
In the model (8.39), in contrast to (8.38), nonlinearity is permitted. The general solution

$$u = f(x - ut) \tag{8.44}$$

shows that u is preserved along the characteristic $dx/dt = u$. But in contrast to the previous example (with $dx/dt = c$), the characteristics are not parallel to each other. If we again start from the initial distribution (8.42), we now obtain

$$u(x, t) = \begin{cases} a^2 - (x - ut)^2 & \text{for } |x - ut| < a, \\ 0 & \text{for } |x - ut| \geq a. \end{cases} \tag{8.45}$$

This equation represents a quadratic equation for u, whose solutions

$$u(x,t) = \frac{1}{2t^2}\left[(2xt-1) \pm (1-4xt+4a^2t^2)^{1/2}\right] \quad (8.46)$$

are valid for $|x - ut| < a$. In (8.46), we choose the positive sign for $t \to 0$ and $|x| < a$. Now, if we look at the first derivative u_x, we have

$$u_x(x,t) = \frac{1}{2t^2}\left\{2t - \frac{2t}{[1-4xt+4a^2t^2]^{1/2}}\right\}, \quad (8.47)$$

and specifically

$$u_x(a,t) = \frac{1}{2t^2}\left[2t - \frac{2t}{1-2at}\right]. \quad (8.48)$$

Recalling $u_x(a,0) = -2a$, we see that at the time $T = 1/(2a)$ the sign of the derivative changes. We have already identified this phenomenon as wave breaking. For $t > T$, both signs in (8.46) become relevant. ■

Example 8.3 Comparison with (8.40)
Ultimately, (8.40) differs from the equation just discussed in that the nonlinearity is absent, but the dispersive term is taken into account. The dispersive terms become relevant when the wave fronts steepen. The solution of the linear wave equation (8.40) with dispersion shows a dispersive broadening. We recall the solution method using Fourier transformation. With $\omega(k) = ck - \kappa k^3$ we can write the solution of (8.40) as

$$u(x,t) = \int_{-\infty}^{\infty} A(k)\exp\{i[kx - \omega(k)t]\}dk \quad (8.49)$$

where the function $A(k)$ is determined by the initial distribution. Evaluating the integral on the right-hand side, for example using the method of steepest descent, shows the well-known dispersive broadening. ■

Multiple spatial dimensions

We now ask the question of which model equation is valid for ion-acoustic modes in multiple spatial dimensions or in magnetized plasmas.

We address the questions simultaneously, as they are methodologically closely related. To do so, we return to the fundamental equations (8.7)–(8.9) and add the Lorentz force to the right-hand side of (8.8). In the variables already chosen to be dimensionless, we then have, instead of (8.8):

$$\partial_t \mathbf{v} + \mathbf{v} \cdot \nabla \mathbf{v} + \nabla \phi + \Omega \hat{z} \times \mathbf{v} = 0, \tag{8.50}$$

where

$$\Omega = \Omega_i / \omega_{pi} = eB/(m_i\, \omega_{pi}) \tag{8.51}$$

applies for a magnetic field in the z-direction. In the three-dimensional generalization, furthermore,

$$\partial_t n + \nabla \cdot (n\mathbf{v}) = 0 \tag{8.52}$$

and

$$\nabla^2 \phi = e^{\phi} - n \tag{8.53}$$

are to be used. Let us first consider once again the case without a magnetic field ($\Omega = 0$). Of course, the linear dispersion relation is then again given by (8.10), but now

$$k = (k_x^2 + k_y^2 + k_z^2)^{1/2} \equiv (k_\perp^2 + k_z^2)^{1/2} \tag{8.54}$$

applies; $\hat{z}$ denotes a preferred direction—chosen arbitrarily in the absence of a magnetic field. For small k and weak transverse dependencies ($k_x^2,\ k_y^2 \ll k_z^2$), the linear dispersion relation reduces to

$$\begin{aligned} \omega &\approx k\left(1 - \frac{1}{2}k^2\right) \approx k_z\left(1 + \frac{1}{2}\frac{k_\perp^2}{k_z^2}\right)\left[1 - \frac{1}{2}k_z^2\left(1 + \frac{k_\perp^2}{k_z^2}\right)\right] \\ &\approx k_z - \frac{1}{2}k_z^2 + \frac{1}{2}\frac{k_\perp^2}{k_z}. \end{aligned} \tag{8.55}$$

These preliminary considerations now allow us to more purposefully tackle the formulation of a multidimensional model equation. With regard to the quantities n, ϕ, v_z [$\widehat{=}\ v$ (one-dimensional)], z [$\widehat{=}\ x$ (one-dimensional)], and t, we can adopt the scalings (8.11)–(8.15). In addition, we must scale $\mathbf{v}_\perp, x$ and y.

The assumption of a weak transverse dependence, which, however, must not be so weak that the corresponding terms do not appear at all in the relevant orders, leads us additionally [note $\xi = \varepsilon^{1/2}(z - t)$]

$$\boxed{\zeta = \varepsilon x \quad \text{and} \quad \eta = \varepsilon y} \tag{8.56}$$

as well as

$$\boxed{\mathbf{v}_\perp = \varepsilon^{3/2}\,\mathbf{v}_\perp^{(1)} + \varepsilon^{5/2}\mathbf{v}_\perp^{(2)} + \ldots} \tag{8.57}$$

to be assumed.

Up to order $\varepsilon^{3/2}$ one obtains

$$n^{(1)} = \phi^{(1)} = v_z^{(1)} \tag{8.58}$$

similarly to the one-dimensional case. At order ε^2 we obtain

$$\partial_\xi v_y^{(1)} = \partial_\eta \phi^{(1)}, \tag{8.59}$$

$$\partial_\xi v_x^{(1)} = \partial_\zeta \phi^{(1)}, \tag{8.60}$$

$$n^{(2)} = \phi^{(2)} - \partial_\xi^2 \phi^{(1)} + \frac{1}{2}\left[\phi^{(1)}\right]^2. \tag{8.61}$$

Ultimately, we must proceed up to order $\varepsilon^{5/2}$. The corresponding equations are

$$\begin{aligned} -\partial_\xi n^{(2)} + \partial_\tau n^{(1)} + \partial_\xi\left[n^{(1)} v_z^{(1)}\right] + \partial_\xi v_z^{(2)} \\ + \partial_\zeta v_x^{(1)} + \partial_\eta v_y^{(1)} = 0, \end{aligned} \tag{8.62}$$

$$-\partial_\xi v_z^{(2)} + \partial_\tau v_z^{(1)} + v_z^{(1)} \partial_\xi v_z^{(1)} = -\partial_\xi \phi^{(2)}. \tag{8.63}$$

If we differentiate (8.61) with respect to ξ and substitute the result for $\partial_\xi n^{(2)}$, as well as $\partial_\xi v_z^{(2)}$ from (8.63) into (8.62), after a short calculation we obtain

$$\boxed{\partial_\tau n^{(1)} + n^{(1)} \partial_\xi n^{(1)} + \frac{1}{2}\partial_\xi^3 n^{(1)} + \frac{1}{2}\int^\xi d\xi' (\partial_\zeta^2 + \partial_\eta^2) n^{(1)}(\xi') = 0}\,. \tag{8.64}$$

This equation is called the Kadomtsev–Petviashvili equation. If we compare with (8.35), we notice an additional (transverse) dispersive term, which is suggested by the right-hand side of (8.55).

Let us now turn to $\Omega \neq 0$. A straightforward calculation yields the linear dispersion relation

$$\omega^2 = \frac{1}{2}\left(\Omega^2 + \frac{k^2}{1+k^2}\right) \pm \left[\frac{1}{4}\left(\Omega^2 + \frac{k^2}{1+k^2}\right)^2 - \frac{k_z^2}{1+k^2}\Omega^2\right]^{1/2}. \tag{8.65}$$

Let us first consider parameter values $\Omega \sim \mathcal{O}(1)$, where for $k \ll 1$ we also require $\omega \ll \Omega$. Without specifying an order-of-magnitude relationship between k_z and $k_\perp$ (the z-direction is singled out by $\mathbf{B}$), we have

$$\boxed{\omega \approx k_z\left[1 - \frac{1}{2}k_z^2 - \frac{1}{2}(1+\Omega^{-2})k_\perp^2\right]} . \tag{8.66}$$

To derive the nonlinear equation corresponding to the mode (8.66), we use the scaling

$$\xi = \varepsilon^{1/2}(z-t), \quad \zeta = \varepsilon^{1/2}x, \quad \eta = \varepsilon^{1/2}y, \quad \tau = \varepsilon^{3/2}t, \tag{8.67}$$

$$n = 1 + \varepsilon n^{(1)} + \varepsilon^2 n^{(2)} + \dots , \tag{8.68}$$

$$v_z = \varepsilon v_z^{(1)} + \varepsilon^2 v_z^{(2)} + \dots , \tag{8.69}$$

$$\mathbf{v}_\perp = \varepsilon^{3/2}\,\mathbf{v}_\perp^{(1)} + \varepsilon^2\,\mathbf{v}_\perp^{(2)} + \dots , \tag{8.70}$$

$$\phi = \varepsilon\phi^{(1)} + \varepsilon^2\phi^{(2)} + \dots . \tag{8.71}$$

Note that, compared to the previous scaling, x and y scale as $\varepsilon^{1/2}$ (in order to properly capture the $E \times B$ drift), and $\mathbf{v}_\perp$ is ordered in steps of $\varepsilon^{1/2}$ (to properly describe the polarization drift).

Up to order $\varepsilon^{3/2}$ one obtains

$$n^{(1)} = v_z^{(1)} = \phi^{(1)} \tag{8.72}$$

together with the $E \times B$-drift

$$\Omega v_y^{(1)} = \partial_\zeta\phi^{(1)}, \tag{8.73}$$

$$\Omega v_x^{(1)} = -\partial_\eta\phi^{(1)}. \tag{8.74}$$

The next order ε^2 yields

$$\partial_\xi v_y^{(1)} = \Omega v_x^{(2)}, \tag{8.75}$$

$$-\partial_\xi v_x^{(1)} = \Omega v_y^{(2)}, \tag{8.76}$$

$$n^{(2)} = \phi^{(2)} - \left(\partial_\xi^2 + \partial_\zeta^2 + \partial_\eta^2\right)\phi^{(1)} + \frac{1}{2}\left[\phi^{(1)}\right]^2. \tag{8.77}$$

Finally, at order $\varepsilon^{5/2}$ one obtains

$$-\partial_\xi n^{(2)} + \partial_\tau n^{(1)} + \partial_\xi \left[n^{(1)} v_z^{(1)}\right] + \partial_\xi v_z^{(2)} + \partial_\zeta v_x^{(2)} + \partial_\eta v_y^{(2)} = 0, \tag{8.78}$$

$$-\partial_\xi v_z^{(2)} + \partial_\tau v_z^{(1)} + v_z^{(1)} \partial_\xi v_z^{(1)} = -\partial_\xi \phi^{(2)}. \tag{8.79}$$

The combination of all these equations leads to the Zakharov-Kuznetsov equation

$$\boxed{2\partial_\tau n^{(1)} + \partial_\xi^3 n^{(1)} + 2n^{(1)} \partial_\xi n^{(1)} + (1 + \Omega^{-2}) \partial_\xi (\partial_\zeta^2 + \partial_\eta^2) n^{(1)} = 0}\,. \tag{8.80}$$

Obviously, this equation is a good approximation for nonlinear ion-acoustic waves in strong magnetic fields. In the absence of external fields, the Kadomtsev-Petviashvili equation (8.64) applies.

The transition between the two model equations can also be discussed by scaling Ω itself. If we set Ω to be of order $\varepsilon^{1/2}$ and otherwise use the scaling of the Kadomtsev-Petviashvili equation, then a calculation—similar to those we have already performed several times above—yields the result

$$\boxed{2\partial_\tau n^{(1)} + \partial_\xi^3 n^{(1)} + 2n^{(1)} \partial_\xi n^{(1)} + \left(\partial_\xi^2 + \Omega^2\right)^{-1} \partial_\xi \left(\partial_\zeta^2 + \partial_\eta^2\right) n^{(1)} = 0}\,. \tag{8.81}$$

An equivalent form is

$$2\partial_\tau n^{(1)} + \partial_\xi^3 n^{(1)} + 2n^{(1)} \partial_\xi n^{(1)} + \int^\xi d\xi' \cos[\Omega(\xi - \xi')](\partial_\zeta^2 + \partial_\eta^2) n^{(1)}(\xi') = 0. \tag{8.82}$$

It is immediately apparent that for $\Omega = 0$ the Kadomtsev-Petviashvili equation results. In contrast, for $\Omega^2 \gg \partial_\xi^2$ the equation reduces to a form corresponding to the Zakharov-Kuznetsov equation.

Electrostatic Approximation

Finally, we briefly discuss the electrostatic approximation, which has been adopted in all previous models. The question of the validity of the electrostatic approximation is relevant because we have considered weakly nonlinear corrections, but have neglected the electromagnetic components due to a non-vanishing vector potential.

We answer the question in the simplest—though not entirely complete—way by discussing the contributions to the linear dispersion relation for $\omega/\Omega_i \ll 1$. For simplification, we neglect compression effects here by writing for the vector potential $\mathbf{A} \approx A(x, y, t)\hat{z}$. Compression effects, in turn, will be proportional to β (the ratio of hydrodynamic pressure to magnetic pressure), whereby it will become apparent in the following that $\beta \ll m_e/m_i$ must hold anyway.
In the drift approximation ($\omega/\Omega_i \ll 1$) we have

$$\mathbf{v}_{i\perp} \approx \frac{m_i}{eB_0^2}\frac{\partial \mathbf{E}_\perp}{\partial t} + \frac{1}{B_0^2}\mathbf{E} \times \mathbf{B}_0, \tag{8.83}$$

where the index 0 denotes the external fields. The momentum balance in the z-direction,

$$\partial_t v_{iz} \approx -\frac{e}{m_i}(\partial_z \phi + \partial_t A), \tag{8.84}$$

then yields directly, together with (8.83), in the continuity equation for the ion density

$$\partial_t^2 n_i - \frac{n_0}{B_0\Omega_i}\partial_t^2 \nabla_\perp^2 \phi + n_0 \partial_z\left[-\frac{e}{m_i}(\partial_z\phi + \partial_t A)\right] = 0. \tag{8.85}$$

For $\omega/k_z \ll v_{te}$ the electron momentum balance reads

$$0 \approx \frac{e}{m_e}\partial_z\phi - \frac{k_B T_e}{m_e}\partial_z n_e + \frac{e}{m_e}\partial_t A. \tag{8.86}$$

Since we are not currently interested in higher dispersive effects, we assume quasineutrality $n_i \approx n_e$. From the equation

$$\nabla \times \nabla \times A\hat{z} \approx \mu_0 \mathbf{j}, \tag{8.87}$$

i.e., without displacement current for low-frequency processes, it then follows after some simple rearrangements

$$\partial_z \nabla_\perp^2 A \approx -\frac{1}{v_A^2}\partial_t \nabla^2 \phi. \tag{8.88}$$

Here, $v_A = B_0/(\mu_0 n_0 m_i)^{1/2}$ is the Alfvén velocity. From this we can estimate

$$\boxed{\left|\frac{\partial A}{\partial t}\right| \approx \frac{m_i}{m_e}\beta|\nabla\phi|} \tag{8.89}$$

This means that for $\beta \ll m_e/m_i$ electromagnetic contributions in the linear dispersion relation can be neglected. A more precise calculation yields the dispersion relation

$$\left(\omega^2 - \frac{k_z^2 c_s^2}{1 + k_\perp^2 \rho_s^2}\right)\left[\omega^2 - k_z^2 v_A^2 (1 + k_\perp^2 \rho_s^2)\right] = \frac{\omega^2}{\omega_{pi}^2} \frac{k_z^2 k_\perp^2 c_s^2 c^2 \beta}{1 + k_\perp^2 \rho_s^2}, \quad \beta = \frac{p}{\frac{B^2}{2\mu_0}}, \tag{8.90}$$

and from this, the electrostatic branch again follows in the limit $\beta \to 0$.

When we compare with the (additional) weak nonlinearities, we can specify the neglect of electromagnetic corrections to $\beta \leq \ \varepsilon m_e / m_i$ more precisely.

Integrability of the KdV Equation

Around 1965, Zabusky and Kruskal numerically solved the temporal initial value problem of the KdV equation for spatially periodic boundary conditions. From a sinusoidal initial distribution, localized pulses developed that appeared quite stable and whose velocities turned out to be proportional to the amplitude. After collisions, the pulses emerged practically unchanged in form. At that time, the term *soliton* was coined to characterize the particle-like behavior of the pulses.

In 1967, Gardner, Greene, Kruskal, and Miura presented the proof of the integrability of the KdV equation using the inverse scattering transform (IST). The IST can be regarded as a generalized Fourier transform. It is a generalization because the Fourier transform (or Laplace transform) can only be used for linear partial differential equations to solve the initial value problem by integration.

A linear partial differential equation [for $u(x,t)$] with the dispersion relation $\omega = \omega(k)$ allows the following procedure. First, from the initial distribution $u(x, 0)$, the spectrum

$$u(k, 0) = \frac{1}{\sqrt{2\pi}} \int_{-\infty}^{+\infty} u(x, 0) e^{ikx} dx \tag{8.91}$$

at time $t = 0$ is determined. Then, using the linear dispersion relation, the spectrum at time t is calculated,

$$u(k, t) = u(k, 0) e^{i\omega(k)t}. \tag{8.92}$$

and finally, the solution $u(x,t)$ can be obtained via the inverse Fourier transform

$$u(x, t) = \frac{1}{\sqrt{2\pi}} \int_{-\infty}^{+\infty} u(k, t) e^{-ikx}\, dk \tag{8.93}$$

The IST for the KdV equation now proceeds according to a formally similar prescription. First, a Schrödinger scattering problem is solved for the potential $u = u(x, 0)$, i.e., the scattering data at time $t = 0$ (reflection and transmission coefficients, discrete eigenvalues) are determined. Subsequently, the scattering data at time t can be obtained via ordinary differential equations. Finally, the "potential" $u(x,t)$, i.e., the solution at time t, is reconstructed from the scattering data at time t.

Example 8.4 (Modified KdV Equation)
Before we discuss the method in detail, let us briefly shed light on the historical background of this initially quite surprising approach. In his search for constants of motion, Miura found that the modified KdV equation

$$\boxed{v_t + 6v^2 v_x + v_{xxx} = 0} \tag{8.94}$$

is simply related to the KdV equation

$$q_t + 6qq_x + q_{xxx} = 0 \tag{8.95}$$

The transformation

$$q = v^2 - iv_x \tag{8.96}$$

yields

$$q_t + 6qq_x + q_{xxx} = (2v - i\partial_x)[v_t + 6v^2 v_x + v_{xxx}]. \tag{8.97}$$

Equation (8.96) can be interpreted as a Riccati equation for v. A Riccati equation can be linearized using the variable ϕ, where

$$v = v(x, t) = -i\frac{\phi_x}{\phi} \tag{8.98}$$

is. Then, from (8.96), a linear Schrödinger equation [for energy eigenvalue zero] results. Taking into account the Galilean invariance of the KdV equation (with $q \to q + \lambda$), the solution of $\phi = \phi(x; t)$ remains for

$$\phi_{xx} + [\lambda + q(x, t)]\phi = 0. \tag{8.99}$$

It was natural to ask how the eigenvalue λ and the eigenfunction ϕ evolve in time when $q = q(x, t)$ represents a time-dependent solution of the KdV equation (8.95). ■

We now turn to the exact proof. In contrast to (8.95), we use the variable $u = -q$, i.e.

$$\boxed{u_t - 6uu_x + u_{xxx} = 0}\,. \tag{8.100}$$

At time t, $u(x, 0)$ is given; the solution $u(x,t)$ is sought for $t > 0$ and $-\infty < x < +\infty$. Of course, $u(x, 0)$ cannot be completely arbitrary, but must satisfy certain integrability conditions, e.g.

$$\int_{-\infty}^{+\infty} \left| \frac{d^n u}{dx^n} \right|^2 dx < \infty \quad \text{for} \quad n = 0, 1, 2, 3, 4, \tag{8.101}$$

$$\int_{-\infty}^{+\infty} (1 + |x|)|u|dx < \infty. \tag{8.102}$$

With the requirement that u vanishes sufficiently rapidly for $x \to \pm\infty$, we can use the asymptotic form of the scattering problem (8.99), or of

$$\boxed{L\psi \equiv [-\partial_x^2 + u(x, t)]\psi = \lambda\psi}\,, \tag{8.103}$$

with

$$\psi_{xx} \simeq -\lambda\psi \quad \text{for} \quad x \to \pm\infty \tag{8.104}$$

as an ansatz. ψ is then, in the asymptotic region, representable as a combination of $\exp(\pm i\sqrt{\lambda}x)$.

We must distinguish two cases: $\lambda < 0$ for bound states and $\lambda > 0$ in the continuum.

Let us begin with the bound states and define

$$\kappa = \sqrt{-\lambda} > 0. \tag{8.105}$$

If

$$\psi(x) \simeq \alpha e^{\kappa x} \quad \text{for} \quad x \to -\infty \tag{8.106}$$

is used as an ansatz, then in general

$$\psi(x) \simeq \beta e^{\kappa x} + \gamma e^{-\kappa x} \quad \text{for} \quad x \to +\infty \tag{8.107}$$

will hold. The constants β and γ are proportional to α and depend on κ and the form of the potential u. Only for special values of κ (discrete eigenvalues λ) will ψ remain limited for $x \to +\infty$. If κ^2 is greater than $-u$ (for a negative "potential" u) for all x, it follows from (8.103) that $\psi_{xx}/\psi > 0$ for all x, and ψ cannot vanish on both sides. Therefore, it must be

$$\boxed{0 < \kappa^2 < -u_{min}} \tag{8.108}$$

so that the boundary condition $\psi < \infty$ at $x \to \pm\infty$ can be satisfied. Furthermore, we know that for $u_{min} < 0$ there exists a finite number $p > 0$ of discrete eigenvalues λ_n with square-integrable eigenfunctions ψ_n, for which

$$u_{min} < \lambda_1 < \lambda_2 < \ldots < \lambda_p < 0 \tag{8.109}$$

holds. ψ_n has $n-1$ (a finite number of) zeros. For the normalization of the eigenfunctions, we choose

$$\psi_n \simeq e^{\kappa_n x} \quad \text{for} \quad x \to -\infty, \tag{8.110}$$

from which

$$\boxed{\psi_n \simeq b_n(t)e^{-\kappa_n x} \quad \text{for} \quad x \to +\infty} \tag{8.111}$$

follows.

We now turn to the continuous eigenvalue spectrum. For $\lambda > 0$ the asymptotic behavior is harmonic (sine or cosine functions) and boundedness is automatically satisfied. Therefore, for all $\lambda > 0$ there exist bounded but not square-integrable eigenfunctions. Every solution can be interpreted as a superposition of an incident wave [e.g., from the right $(x = +\infty)$], a reflected, and a transmitted wave. If we set $k = \sqrt{\lambda} > 0$, this means

$$\boxed{\psi \simeq \begin{cases} e^{-ikx} + R(k,t)e^{ikx} & \text{for} \quad x \to +\infty, \\ T(k,t)e^{-ikx} & \text{for} \quad x \to -\infty \end{cases}} \tag{8.112}$$

with the transmission coefficient T and the reflection coefficient R.

The set consisting of κ_n, $b_n (n = 1, \ldots, p)$, $R(k,t)$ and $T(k,t)$ is referred to as the scattering data.
We now present the result, due to Lax, that λ is independent of t if u satisfies the KdV equation.

To this end, we substitute from (8.103)

$$u = \frac{\psi_{xx}}{\psi} + \lambda \tag{8.113}$$

into the KdV equation (8.100). We have

$$u_t = \frac{\psi_{xxt}}{\psi} - \frac{\psi_{xx}\psi_t}{\psi^2} + \lambda_t, \tag{8.114}$$

$$u_x = \frac{\psi_{xxx}}{\psi} - \frac{\psi_{xx}\psi_x}{\psi^2} \tag{8.115}$$

and so on. Ultimately, from (8.100) it follows

$$\lambda_t\psi^2 + [\psi Q_x - \psi_x Q]_x = 0 \tag{8.116}$$

with

$$Q = \psi_t + \psi_{xxx} - 3(u+\lambda)\psi_x, \tag{8.117}$$

where u is to be replaced by (8.113). If ψ vanishes and is square integrable for $x \to \pm\infty$, then it follows that

$$\boxed{\lambda_t = 0}\,. \tag{8.118}$$

Furthermore, from (8.116) it follows that $\psi Q_x - \psi_x Q$ can only depend on time. We have

$$\psi Q_{xx} - Q\psi_{xx} = 0, \tag{8.119}$$

$$Q_{xx} - \frac{\psi_{xx}}{\psi}Q = Q_{xx} + (\lambda - u)Q = 0. \tag{8.120}$$

Therefore, Q itself satisfies the Schrödinger equation. Thus, a solution is $Q = c\psi$. Since the Schrödinger equation is second order, there is a second solution, which, however, is not bounded. Using the method of variation of constants, $Q = X\psi$, it follows from (8.120) and (8.103)

$$\frac{X_{xx}}{X_x} + \frac{2\psi_x}{\psi} = 0. \tag{8.121}$$

The solution is

$$X_x = \frac{D}{\psi^2} \tag{8.122}$$

or

$$X = D\int \frac{dx}{\psi^2} + C. \tag{8.123}$$

For Q we thus have, in addition to $Q = c\psi$, the second independent form

$$Q = \psi \int^x \frac{dx}{\psi^2}. \tag{8.124}$$

The lower limit in the integral on the right-hand side of (8.124) is chosen so small ($\to -\infty$), that the asymptotic expansion of ψ is already good. Now we know that Q [defined in (8.117)] must be a linear combination of the two independent solutions; we call the coefficients C and D:

$$\psi_t + \psi_{xxx} - 3(u + \lambda)\psi_x = C\psi + D\psi \int \frac{dx}{\psi^2}. \tag{8.125}$$

For the discrete eigenvalues $\lambda = \lambda_n \psi = \psi_n$ vanishes for $x \to \pm\infty$. Therefore,

$$D = D_n \equiv 0 \tag{8.126}$$

must hold. The normalization (8.110) of the bound states ($x \to -\infty$) together with $\psi_{nt} = 0$ for $x \to -\infty$, leads to

$$C = C_n = 4(-\lambda_n)^{3/2}. \tag{8.127}$$

If we use all this in (8.125), it follows that

$$\boxed{\psi_{nt} + \psi_{nxxx} - 3\frac{\psi_{nxx}\psi_{nx}}{\psi_n} - 6\lambda_n\psi_{nx} = 4(-\lambda_n)^{3/2}\psi_n}. \tag{8.128}$$

We now discuss this equation in the limit $x \to +\infty$. We obtain the time evolution of the coefficient b_n via

$$\begin{aligned} b_{nt} &= \left[\kappa_n^3 - 3\kappa_n^3 - 6\lambda_n\kappa_n + 4(-\lambda_n)^{3/2}\right]b_n \\ &= 8(-\lambda_n)^{3/2} b_n \end{aligned} \tag{8.129}$$

with the solution

$$\boxed{b_n(t) = b_n(0)\exp\left[8(-\lambda_n)^{3/2}t\right]}. \tag{8.130}$$

For the continuous eigenvalues $\lambda = k^2$ we can again start from (8.125), but now we make use of the boundary conditions (8.112). For $x \to -\infty$ we obtain

$$T_t + 4ik^3T = CT + \frac{D}{T}\int^x dx\, e^{2ikx}. \tag{8.131}$$

Since all terms except the last are independent of x,$D = 0$ must be, and for the time evolution of T it follows that

$$T_t - (C - 4ik^3)T = 0. \tag{8.132}$$

For $x \to +\infty$ the boundary condition (8.112) in (8.125)

$$(R_t - 4ik^3R - CR)e^{ikx} + (4ik^3 - C)e^{-ikx} = 0 \tag{8.133}$$

fixes. This, of course

$$C = 4ik^3, \tag{8.134}$$

$$R_t = 8ik^3R \tag{8.135}$$

as a consequence.

Thus, (8.132) yields $T_t = 0$ with the solution

$$\boxed{T(k,t) = T(k,0)}\,. \tag{8.136}$$

Together with

$$\boxed{R(k,t) = R(k,0)\exp(8ik^3t)\,, \qquad \lambda_n(t) = \lambda_n(0)} \tag{8.137}$$

we have determined the time dependencies of all scattering data.

Example 8.5 (Normalization)
Before we proceed further with the IST, let us briefly discuss the normalization integrals for the bound states. If we define

$$c_n(t) = \left[\int_{-\infty}^{+\infty} \psi_n^2 dx\right]^{-1}, \tag{8.138}$$

then it follows that

$$\begin{aligned}\frac{d}{dt}(c_n^{-1}) &= 2\int_{-\infty}^{+\infty} \psi_n\psi_{nt}dx \\ &= -2\int_{-\infty}^{+\infty} \psi_n\psi_{nxxx}dx + 6\int_{-\infty}^{+\infty} \psi_{nxx}\psi_{nx} \\ &+ 12\lambda_n\int_{-\infty}^{+\infty} \psi_n\psi_{nx}dx + 8(-\lambda_n)^{3/2}\int_{-\infty}^{+\infty} \psi_n\psi_n dx \\ &= 8(-\lambda_n)^{3/2}c_n^{-1}\end{aligned} \tag{8.139}$$

with the solution

$$\boxed{c_n(t) = c_n(0)\exp[-8(-\lambda_n)^{3/2}t]}\,. \tag{8.140}$$

■

A second remark is also appropriate. We have not assumed that the scattering data κ_n, b_n, R and T as defined by us represent the minimal set of data that uniquely characterize the potential. In fact, the transmission coefficient can be obtained from the remaining data. One, albeit not sufficient, relation for this is the relation known from quantum mechanics

$$\boxed{1 = |R|^2 + |T|^2} \tag{8.141}$$

Therefore, it should not surprise us that not all scattering data are needed to reconstruct the potential.

The problem of determining a potential distribution from scattering experiments was solved some time ago (particularly with regard to questions in atomic and nuclear physics) by physicists and mathematicians. The essential ideas go back to Gel'fand, Levitan, and Marchenko as well as Kay and Moses. Here, we only summarize the results.
The *linear* integral equation

$$K(x, y, t) + B(x + y, t) + \int_x^{\infty} K(x, z, t)B(y + z, t)dz = 0 \tag{8.142}$$

for $y > x$ is referred to as the Gel'fand-Levitan-Marchenko equation for K.

Here, the kernel B is given by

$$B(x + y, t) = \sum_{n=1}^{p} c_n(t)e^{-\kappa_n(x+y)} + \frac{1}{2\pi}\int_{-\infty}^{+\infty} R(k, t)e^{ik(x+y)}dk \tag{8.143}$$

The solution of the KdV equation (in the language of atomic and nuclear physicists: the potential) then follows from

$$\boxed{u(x, t) = -2\partial_x K(x, x, t)}\,, \tag{8.144}$$

where in the solution $K(x,y,t)$ from (8.142) we first set $y = x$ and then differentiate with respect to x. It would go beyond the scope of this plasma physics course to derive the relations (8.142)–(8.144) and to discuss the solutions of the integral equation, especially since this is a major topic in any lecture on nonlinear waves. Here, just a brief remark.

The solitons arise from the existence of the discrete spectrum. Each discrete eigenvalue corresponds to a soliton which—considered in isolation—has a height, a width, and a velocity that are proportional to $-\lambda_n$, $\sqrt{-\lambda_n}$ or $-\lambda_n$, respectively. The so-called radiation corresponds to the continuous spectrum.

Lax formalism

We now place the solution method just presented within the Lax formalism. We have shown that the KdV equation (8.100) can be interpreted as the integrability condition for two *linear* equations. The first equation,

$$(-L + \lambda)\psi \equiv \psi_{xx} + [\lambda - u(x, t)]\psi = 0, \tag{8.145}$$

is identical to (8.103). The second follows for $D = 0$ from (8.125),

$$\psi_t = -4\psi_{xxx} + 3u\psi_x + 3(u\psi)_x + C\psi, \tag{8.146}$$

where, due to (8.145), we have replaced $\lambda\psi_x$ by

$$\lambda\psi_x = -\psi_{xxx} + (u\psi)_x \tag{8.147}$$

Lax formulated this result in a more general way, recognizing that the KdV equation with the commutator [,] can also be written in the form

$$\boxed{L_t = AL - LA \equiv [A, L]} \tag{8.148}$$

where the operators L and A (the so-called Lax pair) are defined by

$$L = -\partial_x^2 + u \tag{8.149}$$

and

$$A = -4\partial_x^3 + 3(u\partial_x + \partial_x u) + C \tag{8.150}$$

Of course, the correspondence with the operators appearing in (8.145) and (8.146) is not coincidental.

An advantage of the formulation (8.148) is that, for an entire class of flows $u(x,t)$, one can demonstrate the time-independence of the spectrum of L $[L\psi = \lambda\psi]$. The following calculation steps, starting from (8.148), illustrate this:

$$L_t\psi = A\lambda\psi - LA\psi, \tag{8.151}$$

$$\partial_t(L\psi) - L\psi_t = \lambda A\psi - LA\psi, \tag{8.152}$$

$$\lambda_t\psi + \lambda\psi_t - L\psi_t = (\lambda - L)A\psi, \tag{8.153}$$

$$\lambda_t\psi = (\lambda - L)(A\psi - \psi_t). \tag{8.154}$$

If the time dependence of the eigenfunctions is given by

$$\psi_t = A\psi \tag{8.155}$$

[cf. (8.146)], it follows that $\lambda_t = 0$. We can also interpret (8.154) differently. After multiplying by ψ^* and subsequent integration, we obtain for the discrete states

$$\lambda_t \int \psi^*\psi dx = 0, \tag{8.156}$$

provided L is self-adjoint. Then the time evolution of the eigenstates ψ is given by the linear differential equation (8.155), and the solution via the IST is possible.

Conservation Laws

An integrable equation should possess as many conserved quantities as it has degrees of freedom. In the case of the KdV equation, we therefore expect infinitely many constants of motion. The IST provides a constructive method for determining them. However, we present here a simplified computational procedure.

If in the KdV equation (8.100) we replace u by

$$u = -w - \varepsilon w_x + \varepsilon^2 w^2, \tag{8.157}$$

i.e., a Riccati equation for w, we obtain

$$\begin{aligned}
-(u_t - 6uu_x + u_{xxx}) &= w_t + \varepsilon w_{xt} - 2\varepsilon^2 ww_t \\
&+ 6(w + \varepsilon w_x - \varepsilon^2 w^2)(w_x + \varepsilon w_{xx} - 2\varepsilon^2 ww_x) \\
&+ w_{xxx} + \varepsilon w_{xxxx} - 2\varepsilon^2 (ww_x)_{xx} \\
&= (1 + \varepsilon\partial_x - 2\varepsilon^2 w)[w_t + 6(w - \varepsilon^2 w^2)w_x + w_{xxx}].
\end{aligned} \tag{8.158}$$

Here, ε is a real parameter. From (8.158) we can conclude that the KdV equation is satisfied by $u(x,t)$ if w solves the so-called Gardner equation

$$w_t + 6(w - \varepsilon^2 w^2)w_x + w_{xxx} = 0 \tag{8.159}$$

This equation has the form $\partial_t w + \partial_x\{\ldots\} = 0$,, i.e., the form of a local conservation law. Now we attempt to find a solution to (8.157) using a power series ansatz (for $\varepsilon \to 0$),

$$w = \sum_{n=0}^{\infty} \varepsilon^n w_n(u). \tag{8.160}$$

Comparing coefficients in (8.157) leads to

$$w_0 = -u\ , \tag{8.161}$$

$$w_1 = -w_{0x} = u_x\ , \tag{8.162}$$

$$w_2 = -w_{1x} + w_0^2 = -u_{xx} + u^2 \tag{8.163}$$

and so on. Substituting this into (8.159), we successively obtain

$$u_t + (-3u^2 + u_{xx})_x = 0, \tag{8.164}$$

$$(-u_x)_t + (6uu_x - u_{xxx})_x = 0, \tag{8.165}$$

$$(u_{xx} - u^2)_t + (4u^3 - 8uu_{xx} - 5u_x^2 + u_{xxxx})_x = 0 \tag{8.166}$$

and so on. For localized functions u, the equations of even order in ε yield the constants of motion

$$\boxed{\int_{-\infty}^{+\infty} u\, dx = const\ ,} \tag{8.167}$$

$$\boxed{\int_{-\infty}^{+\infty} u^2 dx = const\ ,} \tag{8.168}$$

$$\boxed{\int_{-\infty}^{+\infty} \left(u^3 + \frac{1}{2}u_x^2\right) dx = const} \tag{8.169}$$

and so on. This procedure works in all orders in ε and yields all conservation laws. For reasons of space, we omit a more detailed justification.

In conclusion, let us once again highlight the physically highly interesting result regarding the dynamics of solitons. Solitons are stable nonlinear wave solutions that even survive the interaction process without changing their shape. Within the framework of a one-dimensional KdV equation, this surprising phenomenon [with the IST] can be mathematically rigorously justified. Solitons therefore present themselves as building blocks of a nonlinear dynamics.

8.2 Langmuir Oscillations and the NLS Equation

In a fully ionized and unmagnetized plasma, in addition to ion-acoustic waves, there are Langmuir oscillations, whose characteristic frequency is the electron plasma frequency ω_{pe}. In this section, we discuss nonlinear models for these nearly electrostatic and—in contrast to ion-acoustic waves—high-frequency phenomena. Here, a nonlinear coupling with the low-frequency plasma dynamics occurs. Characteristic model equations are the Zakharov equations, which in the simplest case reduce to the cubic nonlinear Schrödinger equation. We present their integrability.

Model Equations

The starting equations are Maxwell's equations together with the density continuity equations for electrons and ions, as well as the electron and ion momentum balances. In the latter, equations of state $p_{i,e} \sim n_{i,e}^{\gamma_{i,e}}$ are used. For adiabatic changes of state, the exponent $\gamma_{i,e}$ is given by the ratio of specific heats [$\gamma_{i,e} = (2+N)/N$, where N is the number of degrees of freedom]. For isothermal changes of state, $\gamma_{i,e} = 1$ applies.

Since the coupling wave types differ greatly in frequency, all quantities are split into high-frequency (h) and low-frequency (s) components as follows:

$$\boxed{n_{i,e} = n_0 + n_{i,e}^s + n_{i,e}^h ,} \tag{8.170}$$

$$\boxed{\mathbf{v}_{i,e} = \mathbf{v}_{i,e}^s + \mathbf{v}_{i,e}^h ,} \tag{8.171}$$

$$\boxed{\mathbf{E} = \mathbf{E}^s + \mathbf{E}^h .} \tag{8.172}$$

Since we are considering an unmagnetized plasma here, for the magnetic field $\mathbf{B}$ we only take into account the high-frequency contributions $\mathbf{B}^h$. Furthermore, for each high-frequency component, we separate out the rapid time variation, e.g.

$$n^h_{i,e} = \frac{1}{2}\Big[\tilde{n}_{i,e}(\mathbf{r}, t)e^{-i\omega t} + c.c.\Big], \tag{8.173}$$

where the remaining amplitudes $\tilde{n}_{i,e}$, $\tilde{\mathbf{v}}_{i,e}$, $\tilde{\mathbf{E}}$, $\tilde{\mathbf{B}}$ are assumed to depend only weakly on time. If one then performs an averaging $\langle \ldots \rangle$ over the high-frequency oscillations in the original equations, one obtains

$$\nabla \cdot \mathbf{E}^s = \frac{1}{\varepsilon_0} e(n^s_i - n^s_e), \tag{8.174}$$

$$\nabla \times \mathbf{E}^s = 0, \tag{8.175}$$

$$\mathbf{j}^s + \varepsilon_0 \frac{\partial \mathbf{E}^s}{\partial t} = 0, \tag{8.176}$$

$$\frac{\partial n^s_e}{\partial t} + \nabla \cdot \Big[\big(n_0 + n^s_e\big)\mathbf{v}^s_e + \big\langle n^h_e \mathbf{v}^h_e \big\rangle\Big] = 0, \tag{8.177}$$

$$\frac{\partial n^s_i}{\partial t} + \nabla \cdot \Big[\big(n_0 + n^s_i\big)\mathbf{v}^s_i + \big\langle n^h_i \mathbf{v}^h_i \big\rangle\Big] = 0, \tag{8.178}$$

$$\begin{aligned} \frac{\partial \mathbf{v}^s_e}{\partial t} + \mathbf{v}^s_e \cdot \nabla \mathbf{v}^s_e = {} & -\frac{e}{m_e}\Big\{\mathbf{E}^s + \big\langle \mathbf{v}^h_e \times \mathbf{B}^h \big\rangle\Big\} - \big\langle \mathbf{v}^h_e \cdot \nabla \mathbf{v}^h_e \big\rangle \\ & - \frac{\gamma_e k_B}{m_e} \big\langle T_e \nabla \ln\big(n_0 + n^s_e\big)\big\rangle, \end{aligned} \tag{8.179}$$

$$\begin{aligned} \frac{\partial \mathbf{v}^s_i}{\partial t} + \mathbf{v}^s_i \cdot \nabla \mathbf{v}^s_i = {} & \frac{e}{m_i}\Big\{\mathbf{E}^s + \big\langle \mathbf{v}^h_i \times \mathbf{B}^h \big\rangle\Big\} - \big\langle \mathbf{v}^h_i \cdot \nabla \mathbf{v}^h_i \big\rangle \\ & - \frac{\gamma_i k_B}{m_i} \big\langle T_i \nabla \ \ln\,(n_0 + n^s_i)\big\rangle. \end{aligned} \tag{8.180}$$

From (8.175) it follows that the low-frequency electric field can be represented by a potential ϕ^s in the form

$$\boxed{\mathbf{E}^s = -\nabla \phi^s} \tag{8.181}$$

is representable.

Completely analogously, the equations for the high-frequency components follow. If one neglects the higher harmonics in these, one obtains

$$\nabla \cdot \mathbf{E}^h = 4\pi e(n^h_i - n^h_e)\ , \tag{8.182}$$

$$\nabla \times \mathbf{E}^h = -\frac{\partial \mathbf{B}^h}{\partial t} , \tag{8.183}$$

$$\nabla \times \mathbf{B}^h = \mu_0 \mathbf{j}^h + \mu_0 \varepsilon_0 \frac{\partial \mathbf{E}^h}{\partial t} , \tag{8.184}$$

$$\frac{\partial n_e^h}{\partial t} + \nabla \cdot \Big((n_0 + n_e^s) \mathbf{v}_e^h + n_e^h \mathbf{v}_e^s \Big) = 0 , \tag{8.185}$$

$$\frac{\partial n_i^h}{\partial t} + \nabla \cdot \Big((n_0 + n_i^s) \mathbf{v}_i^h + n_i^h \mathbf{v}_i^s \Big) = 0 , \tag{8.186}$$

$$\frac{\partial \mathbf{v}_e^h}{\partial t} + \mathbf{v}_e^h \cdot \nabla \mathbf{v}_e^s + \mathbf{v}_e^s \cdot \nabla \mathbf{v}_e^h + \frac{e}{m_e} \mathbf{E}^h + \frac{e}{m_e} \mathbf{v}_e^s \times \mathbf{B}^h + \frac{\gamma_e k_B}{m_e} \left(\frac{T_e \nabla n_e}{n_e} \right)^h = 0, \tag{8.187}$$

$$\frac{\partial \mathbf{v}_i^h}{\partial t} + \mathbf{v}_i^h \cdot \nabla \mathbf{v}_i^s + \mathbf{v}_i^s \cdot \nabla \mathbf{v}_i^h - \frac{e}{m_i} \mathbf{E}^h - \frac{e}{m_i} \mathbf{v}_i^s \times \mathbf{B}^h + \frac{\gamma_i k_B}{m_i} \left(\frac{T_i \nabla n_i}{n_i} \right)^h = 0 . \tag{8.188}$$

For not too large amplitudes, the low-frequency (directed) velocities of the electrons and ions are always small compared to the thermal velocity of the electrons,

$$v_{i,e}^s \ll v_{te}, \tag{8.189}$$

while the phase velocities v_{ph} are always greater than v_{te}. We can therefore safely neglect terms of the order of $v_{i,e}^s / v_{ph}$ compared to unity.

These are, in the high-frequency momentum balance, the convective terms and the force density arising from the high-frequency magnetic field; specifically, this means

$$|\mathbf{v}^s \cdot \nabla \mathbf{v}^h| \leq \mathcal{O}\left(\frac{v^s}{v_{ph}} \right) \left| \partial \mathbf{v}^h / \partial t \right| , \tag{8.190}$$

$$|\mathbf{v}^h \cdot \nabla \mathbf{v}^s| \leq \mathcal{O}\left(\frac{v^s}{v_{ph}} \right) |\partial \mathbf{v}^h / \partial t| , \tag{8.191}$$

$$\left| \frac{e}{m_{i,e}} \mathbf{v}^s \times \mathbf{B}^h \right| \leq \mathcal{O}\left(\frac{v^s}{v_{ph}} \right) \left| \frac{e}{m_{i,e}} \mathbf{E}^h \right|, \tag{8.192}$$

where here and in the following we assume that all characteristic lengths are greater than the electron Debye length.

If the term

$$\left|\nabla\left(n_{e,i}^{h}\mathbf{v}_{e,i}^{s}\right)\right| \leq \mathcal{O}\left(\frac{v_{e,i}^{s}}{v_{ph}}\right)\left|\partial n_{e,i}^{h}/\partial t\right| \tag{8.193}$$

is neglected in the continuity equation, it follows that

$$\nabla \cdot \mathbf{E}^{h} = \frac{1}{\varepsilon_0} e\left(n_i^h - n_e^h\right), \tag{8.194}$$

$$\nabla \times \mathbf{E}^{h} = -\frac{\partial \mathbf{B}^{h}}{\partial t}, \tag{8.195}$$

$$\nabla \times \mathbf{B}^{h} = \mu_0 \mathbf{j}^{h} + \mu_0 \varepsilon_0 \frac{\partial \mathbf{E}^{h}}{\partial t}, \tag{8.196}$$

$$\nabla \cdot \mathbf{B}^{h} = 0 \tag{8.197}$$

$$\frac{\partial n_e^h}{\partial t} + \nabla \cdot \left[(n_0 + n_e^s)\mathbf{v}_e^h\right] = 0, \tag{8.198}$$

$$\frac{\partial n_i^h}{\partial t} + \nabla \cdot \left[(n_0 + n_i^s)\mathbf{v}_i^h\right] = 0, \tag{8.199}$$

$$\frac{\partial \mathbf{v}_e^h}{\partial t} + \frac{e}{m_e}\mathbf{E}^h + \frac{\gamma_e k_B}{m_e}\left(\frac{T_e \nabla n_e}{n_e}\right)^h = 0, \tag{8.200}$$

$$\frac{\partial \mathbf{v}_i^h}{\partial t} - \frac{e}{m_i}\mathbf{E}^h + \frac{\gamma_i k_B}{m_i}\left(\frac{T_i \nabla n_i}{n_i}\right)^h = 0. \tag{8.201}$$

From (8.198) and (8.199) one can deduce

$$n_e^h \sim \mathcal{O}\left(\frac{n_0 + n_e^s}{v_{ph}} v_e^h\right), \tag{8.202}$$

$$n_i^h \sim \mathcal{O}\left(\frac{n_0 + n_i^s}{v_{ph}} v_i^h\right) \tag{8.203}$$

With this, the order of magnitude of the pressure terms in the momentum balances can be specified as follows:

$$\frac{\gamma_e k_B}{m_e}\left(T_e \frac{\nabla n_e}{n_e}\right)^h \sim \mathcal{O}\left(\frac{v_{te}^2}{v_{ph}^2}\right)\frac{\partial v_e^h}{\partial t}, \tag{8.204}$$

$$\frac{\gamma_i k_B}{m_i}\left(T_i \frac{\nabla n_i}{n_i}\right)^h \sim \mathcal{O}\left(\frac{v_{ti}^2}{v_{ph}^2}\right)\frac{\partial v_i^h}{\partial t}. \tag{8.205}$$

In the following, we will consider only modes whose temperature fluctuations can be neglected. Furthermore, from the momentum balances (8.200) and (8.201), it can be seen that both the high-frequency ion velocity and the ion density are smaller by a factor of m_e/m_i than the corresponding quantities for the electrons.

In the further calculation, we will disregard terms of the order of m_e/m_i and terms that are smaller than v_{te}^2/v_{ph}^2.

Under these assumptions, the relevant high-frequency equations reduce to

$$\nabla \cdot \mathbf{E}^h = -\frac{1}{\varepsilon_0} e n_e^h \,, \tag{8.206}$$

$$\nabla \times \mathbf{E}^h = -\frac{\partial \mathbf{B}^h}{\partial t} \,, \tag{8.207}$$

$$\frac{\partial^2 \mathbf{E}^h}{\partial t^2} + c^2 \nabla \times \nabla \times \mathbf{E}^h = -\frac{1}{\varepsilon_0} \frac{\partial \mathbf{j}^h}{\partial t} \,, \tag{8.208}$$

$$\frac{\partial \mathbf{v}_e^h}{\partial t} + \frac{e}{m_e} \mathbf{E}^h + \frac{\gamma_e k_B T_e}{m_e} \frac{\nabla n_e^h}{(n_0 + n_e^s)} = 0. \tag{8.209}$$

The time derivative of the electric current density is

$$\frac{\partial \mathbf{j}^h}{\partial t} = \frac{\partial}{\partial t} [e(n_i \mathbf{v}_i - n_e \mathbf{v}_e)]^h \tag{8.210}$$

or

$$\frac{\partial \mathbf{j}^h}{\partial t} = e \frac{\partial}{\partial t} \Big[(n_0 + n_i^s) \mathbf{v}_i^h + n_i^h \mathbf{v}_i^s - (n_0 + n_e^s) \mathbf{v}_e^h - n_e^h \mathbf{v}_e^s \Big]. \tag{8.211}$$

The terms $(n_0 + n_i^s)\mathbf{v}_i^h$, $n_i^h \mathbf{v}_i^s$ can be neglected, since they are smaller by a factor of m_e/m_i than the corresponding terms for the electrons. If we further neglect the term

$$n_e^h v_e^s \le \mathcal{O}\left(\frac{v_e^s}{v_{ph}} \right) (n_0 + n_e^s) v_e^h \tag{8.212}$$

and take into account in the time derivative that $\omega^h \gg \omega^s$ is valid, it follows that

$$\frac{\partial \mathbf{j}^h}{\partial t} \approx -e(n_0 + n_e^s) \frac{\partial \mathbf{v}_e^h}{\partial t} . \tag{8.213}$$

If we insert this expression into Eq. (8.208) for the high-frequency electric field and take into account Eqs. (8.206) and (8.209), we obtain

$$\boxed{\frac{\partial^2 \mathbf{E}^h}{\partial t^2} + c^2 \nabla \times \nabla \times \mathbf{E}^h + \omega_{pe}^2 \left(1 + \frac{n_e^s}{n_0}\right) \mathbf{E}^h - 3 v_{te}^2 \nabla \nabla \cdot \mathbf{E}^h = 0}\,, \tag{8.214}$$

where, for high-frequency variations, the adiabatic equation of state is appropriate.

To complete (8.214), we need the equations for the low-frequency components, from which the unknown low-frequency density n_e^s is to be determined. After obvious simplifications, they are

$$\nabla \cdot \mathbf{E}^s = \frac{1}{\varepsilon_0} e(n_i^s - n_e^s)\,, \tag{8.215}$$

$$\frac{\partial \mathbf{v}_i^s}{\partial t} + \mathbf{v}_i^s \cdot \nabla \mathbf{v}_i^s = -\frac{e}{m_i} \nabla \phi^s - \frac{\gamma_i k_B T_i}{m_i} \nabla \ \ln\,(n_0 + n_i^s)\,, \tag{8.216}$$

$$\frac{\partial n_i^s}{\partial t} + \nabla \cdot \left[(n_0 + n_i^s) \mathbf{v}_i^s\right] = 0\,, \tag{8.217}$$

$$\frac{\gamma_e k_B T_e}{m_e} \nabla \ \ln\ (n_0 + n_e^s) = \frac{e}{m_e} \nabla \phi^s - \mathbf{F}_e. \tag{8.218}$$

In the continuity equation (8.217), the term $\langle n_i^h \mathbf{v}_i^h \rangle$ and in the electron momentum balance (8.218) the inertia term $\frac{\partial \mathbf{v}_e^s}{\partial t} + \mathbf{v}_e^s \cdot \nabla \mathbf{v}_e^s$ have been neglected, since these terms are, in terms of order of magnitude, smaller by a factor of m_e/m_i than the remaining terms. (This can easily be shown, for example, by solving for the ambipolar potential ϕ^s.)

The interaction between the high-frequency wave and the plasma is described by the ponderomotive force

$$\mathbf{F}_e = \langle \mathbf{v}_e^h \cdot \nabla \mathbf{v}_e^h \rangle + \frac{e}{m_e} \langle \mathbf{v}_e^h \times \mathbf{B}^h \rangle \tag{8.219}$$

Since the ponderomotive force acting on the ions is smaller by a factor of m_e/m_i than that acting on the electrons, we have not included this ion contribution in (8.216). Using Eqs. (8.207) and (8.209) to lowest order $\left(\frac{v_{te}^2}{v_{ph}^2}\right) \ll 1$,

$$\frac{\partial \mathbf{v}_e^h}{\partial t} \approx -\frac{e}{m_e} \mathbf{E}^h, \tag{8.220}$$

as well as the splitting (8.173), we obtain

$$\mathbf{B}^h \sim -\frac{1}{\omega} \nabla \times \mathbf{E}^h\,, \tag{8.221}$$

$$\mathbf{v}_e^h \sim -\frac{e}{\omega m_e}\mathbf{E}^h \,. \tag{8.222}$$

This yields the ponderomotive force

$$\mathbf{F}_e \approx \frac{\omega_{pe}^2}{4\pi n_0 m_e \omega^2}\Big\langle \mathbf{E}^h \cdot \nabla \mathbf{E}^h + \mathbf{E}^h \times \nabla \times \mathbf{E}^h \Big\rangle. \tag{8.223}$$

The expressions (8.221)–(8.223) are written somewhat loosely in that we must split into harmonics $\exp(-i\omega t)$ and $\exp(i\omega t)$ [with $\partial_t = -i\omega$ and $\partial_t = +i\omega$ respectively] and subsequently, when forming products, we may only collect the low-frequency contributions $\exp(-i\omega t + i\omega t)$. The following manipulations are to be understood in this sense. The expression (8.223) can, using the vector identity $\nabla(\mathbf{a}\cdot\mathbf{a}) = 2[\mathbf{a}\cdot\nabla\mathbf{a} + \mathbf{a}\times(\nabla\times\mathbf{a})]$ and with $\langle \mathbf{E}^h \cdot \mathbf{E}^h \rangle = \frac{1}{2}|\tilde{\mathbf{E}}\cdot\tilde{\mathbf{E}}^*|$ be simplified to

$$\boxed{\mathbf{F}_e = \frac{\omega_{pe}^2}{16\pi\omega^2}\,\frac{\nabla|\tilde{\mathbf{E}}\cdot\tilde{\mathbf{E}}^*|}{n_0 m_e}} \tag{8.224}$$

This is used below in (8.218). Accordingly, we rewrite (8.214) in the new notation and set $\omega \approx \omega_{pe}$; then it follows

$$-2i\omega_{pe}\frac{\partial\tilde{\mathbf{E}}}{\partial t} + c^2\nabla\times\nabla\times\tilde{\mathbf{E}} + \omega_{pe}^2\frac{n_e^s}{n_0}\tilde{\mathbf{E}} - 3v_{te}^2\nabla\nabla\cdot\tilde{\mathbf{E}} = 0. \tag{8.225}$$

Here it has been taken into account that $\tilde{\mathbf{E}}$ is only weakly time-dependent. Therefore,

$$\frac{\partial^2\tilde{\mathbf{E}}}{\partial t^2} \ll 2\omega_{pe}\frac{\partial\tilde{\mathbf{E}}}{\partial t} \tag{8.226}$$

is neglected.

We now introduce dimensionless quantities: the length $\sqrt{3}\,\lambda_{De}$, the time $\sqrt{3}/\omega_{pi}$, the potential $k_B T_e/e$, the density n_0, the electric field $(16\pi n_0 k_B T_e)^{1/2}$ and the velocity $c_s = (k_B T_e/m_i)^{1/2}$ are defined as units. This yields the system of equations

$$\boxed{i\varepsilon\frac{\partial\mathbf{E}}{\partial t} + \nabla^2\mathbf{E} - Q\nabla\times\nabla\times\mathbf{E} - (n_e - 1)\mathbf{E} = 0 \,,} \tag{8.227}$$

$$\boxed{\frac{\partial n_i}{\partial t} + \nabla(n_i\mathbf{v}_i) = 0 \,,} \tag{8.228}$$

$$\boxed{\frac{\partial \mathbf{v}_i}{\partial t} + \mathbf{v}_i \cdot \nabla \mathbf{v}_i = -\nabla \phi - \frac{T_i}{T_e} \nabla \ \ln \ n_i \ ,} \tag{8.229}$$

$$\boxed{\frac{\nabla n_e}{n_e} = \nabla \phi - \nabla |\mathbf{E}|^2 \ ,} \tag{8.230}$$

$$\boxed{\frac{1}{3} \nabla^2 \phi = n_e - n_i \ .} \tag{8.231}$$

The index s has been omitted; in addition, we have simplified the notation: $n_{i,e} \mathrel{\widehat{=}} 1 + n^s_{i,e}/n_0$ and $\mathbf{E} \mathrel{\widehat{=}} \tilde{\mathbf{E}}$. Furthermore, $\varepsilon = 2(m_e/3m_i)^{1/2}$ and $Q = (m_e c^2/k_B T_e) - 1$.

So far, no statements have been made about the low-frequency state changes of electrons and ions. For the electrons, the relation

$$\frac{\omega^s}{k^s v_{te}} \approx M \frac{c_s}{v_{te}} \ll 1 \tag{8.232}$$

can be given, where M is the Mach number. (M is the velocity of the solitons measured in c_s.) From this, one can see that the processes for the electrons proceed so slowly that they can always be described as isothermal. On the other hand, for the ions the equation

$$\frac{\omega^s}{k^s v_{ti}} = M \frac{c_s}{v_{ti}} = M \left(\frac{T_e}{T_i} \right)^{1/2} . \tag{8.233}$$

applies. Because of the generally large factor $(T_e/T_i)^{1/2}$ we must distinguish two cases: For

$$M < \left(\frac{T_i}{T_e} \right)^{1/2} \tag{8.234}$$

the ions can be described as isothermal. If the ions must be described adiabatically, the pressure term is generally small. In this case,

$$M > \left(\frac{T_i}{T_e} \right)^{1/2} , \tag{8.235}$$

applies and the approximation "$T_i = 0$" is good.

In the past, many theories for Langmuir solitons have been proposed. In this section, we discuss small-amplitude models. For this purpose, we start from the fundamental equations (8.227)–(8.231), which, in the one-dimensional case, reduce to

$$i\varepsilon\frac{\partial E}{\partial t}+\frac{\partial^2 E}{\partial x^2}-(n_e-1)E=0\ , \tag{8.236}$$

$$\frac{\partial n_i}{\partial t}+\frac{\partial}{\partial x}(n_i v_i)=0\ , \tag{8.237}$$

$$\frac{\partial v_i}{\partial t}+v_i\frac{\partial v_i}{\partial x}=-\frac{\partial \phi}{\partial x}\ , \tag{8.238}$$

$$\frac{\partial n_e}{\partial x}=n_e\left[\frac{\partial \phi}{\partial x}-\frac{\partial |E|^2}{\partial x}\right]\ , \tag{8.239}$$

$$\frac{1}{3}\frac{\partial^2 \phi}{\partial x^2}=n_e-n_i \tag{8.240}$$

Here it is assumed that the ions can be described adiabatically.

NLS Equation

There are various well-known model equations that we can now derive within the framework of different scalings. The order parameter of the scaling is ε. For

$$n_e=1+\varepsilon^2 n_e^{(1)}+\varepsilon^3 n_e^{(2)}+\dots\ , \tag{8.241}$$

$$n_i=1+\varepsilon^2 n_i^{(1)}+\varepsilon^3 n_i^{(2)}+\dots\ , \tag{8.242}$$

$$v_i=\varepsilon^2 v_i^{(1)}+\varepsilon^3 v_i^{(2)}+\dots\ , \tag{8.243}$$

$$\phi=\varepsilon^2 \phi^{(1)}+\varepsilon^3 \phi^{(2)}+\dots\ , \tag{8.244}$$

$$E=\varepsilon E^{(1)}+\varepsilon^2 E^{(2)}+\dots\ , \tag{8.245}$$

$$\xi=\varepsilon(x-Mt)\ , \tag{8.246}$$

$$\tau=\varepsilon^2 t \tag{8.247}$$

it follows after a short calculation that the cubic nonlinear Schrödinger equation

$$\boxed{\begin{aligned}&i\varepsilon\partial_\tau E^{(1)}-iM\partial_\xi E^{(1)}+\partial_\xi^2 E^{(1)}-n_e^{(1)}E^{(1)}=0\ ,\\ &n_e^{(1)}=-|E^{(1)}|^2/(1-M^2)\ .\end{aligned}} \tag{8.248}$$

The fact that the smallness parameter ε still appears explicitly in the first term is a relic of the historical development. The following transformations also show how this “blemish” can be remedied.

The stationary solutions can be written in the form

$$E^{(1)}=E_s\exp\left\{i\left[\left(\frac{\eta^2}{\varepsilon}+\frac{M^2\varepsilon}{4}\right)\tau+\frac{M\varepsilon}{2}\xi\right]\right\} \tag{8.249}$$

and

$$n_e^{(1)} = -E_s^2/(1 - M^2) \tag{8.250}$$

write, where

$$\boxed{E_s = \sqrt{2}\sqrt{1 - M^2}\ \eta\ \text{sech}\,(\eta\xi)} \tag{8.251}$$

applies. This yields the ambipolar potential

$$\phi^{(1)} = -\frac{2\eta^2 M^2}{(1 - M^2)}\ \text{sech}^2(\eta\xi). \tag{8.252}$$

The solutions E_s and $n_e^{(1)}$ are plotted in Figs. 8.2 and 8.3, respectively. The solution E_s has the shape of a bell, and $n_e^{(1)}$ that of an inverted bell.

Zakharov equations
If one changes the scaling (8.246)–(8.247) with respect to x and t in

$$\xi = \varepsilon x, \tag{8.253}$$

$$\tau = \varepsilon t \tag{8.254}$$

one arrives, after a short calculation, at the (one-dimensional) Zakharov equations

$$\boxed{i\partial_t E^{(1)} + \partial_\xi^2 E^{(1)} - n_e^{(1)} E^{(1)} = 0\ ,} \tag{8.255}$$

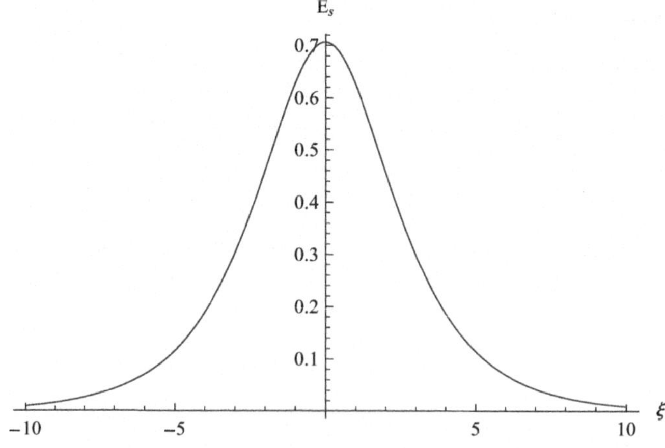

Fig. 8.2 Envelope electric field E_s of a "bell soliton" (8.251) as a function of ξ

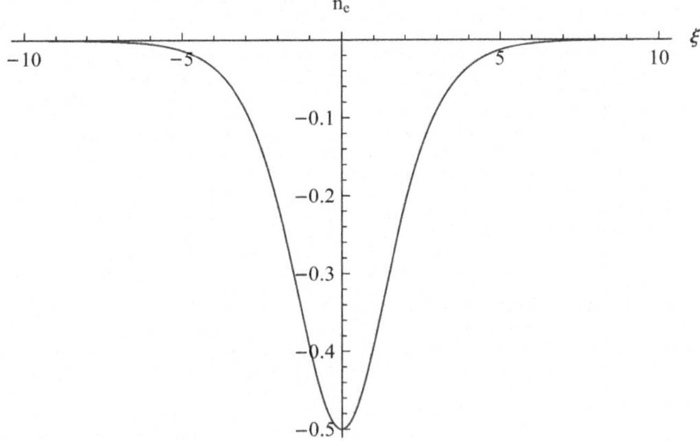

Fig. 8.3 Density depression $n_e = n_e^{(1)}$ for a "bell soliton"

$$\boxed{\partial_\tau^2 n_e^{(1)} - \partial_\xi^2 n_e^{(1)} = \partial_\xi^2 |E^{(1)}|^2 \ ,} \tag{8.256}$$

whose stationary solutions correspond to those of the Schrödinger equation. Only now, the ion dynamics are also included.

KdV-type equations
For velocities near the ion sound speed, that is, for $M \to 1$, it is also possible to derive simple model equations. With the scaling

$$n_e = 1 + \varepsilon n_e^{(1)} + \varepsilon^2 n_e^{(2)} + \dots \ , \tag{8.257}$$

$$n_i = 1 + \varepsilon n_i^{(1)} + \varepsilon^2 n_i^{(2)} + \dots \ , \tag{8.258}$$

$$v_i = \varepsilon v_i^{(1)} + \varepsilon^2 v_i^{(2)} + \dots \tag{8.259}$$

$$\phi = \varepsilon \phi^{(1)} + \varepsilon^2 \phi^{(2)} + \dots \ , \tag{8.260}$$

$$E = \varepsilon E^{(1)} + \varepsilon^2 E^{(2)} + \dots \ , \tag{8.261}$$

$$\xi = \varepsilon^{1/2}(x - t) \ , \tag{8.262}$$

$$\tau = \varepsilon^{3/2}\, t \tag{8.263}$$

it follows to lowest order (ε) $n_e^{(1)} = n_i^{(1)}$. In the next order ($\varepsilon^{3/2}$) we obtain

$$\partial_\xi n_i^{(1)} = \partial_\xi v_i^{(1)}, \tag{8.264}$$

$$\partial_\xi v_i^{(1)} = \partial_\xi \phi^{(1)}, \tag{8.265}$$

$$\partial_\xi n_e^{(1)} = \partial_\xi \phi^{(1)}. \tag{8.266}$$

As a localized solution we obtain

$$n_e^{(1)} = n_i^{(1)} = v_i^{(1)} = \phi^{(1)}. \tag{8.267}$$

At order ε^2 the contributions are (note that we still allow for a faster phase variation)

$$i\varepsilon^{3/2}\partial_\tau E^{(1)} - \varepsilon^{1/2}\partial_\xi E^{(1)} + \partial_\xi^2 E^{(1)} - n_e^{(1)} E^{(1)} = 0, \tag{8.268}$$

$$\frac{1}{3}\partial_\xi^2 \phi^{(1)} = n_e^{(2)} - n_i^{(2)}. \tag{8.269}$$

Only at order $\varepsilon^{5/2}$ does a closed system of equations result:

$$\partial_\tau n_i^{(1)} - \partial_\xi n_i^{(2)} + \partial_\xi (n_i^{(1)} v_i^{(1)}) + \partial_\xi v_i^{(2)} = 0, \tag{8.270}$$

$$\partial_\tau v_i^{(1)} - \partial_\xi v_i^{(2)} + v_i^{(1)} \partial_\xi v_i^{(1)} + \partial_\xi \phi^{(2)} = 0, \tag{8.271}$$

$$\partial_\xi n_e^{(2)} = n_e^{(1)} \partial_\xi \phi^{(1)} + \partial_\xi \phi^{(2)} - \partial_\xi |E^{(1)}|^2. \tag{8.272}$$

By combining Eqs. (8.270)–(8.272) we can obtain

$$2\,\partial_\tau n_e^{(1)} + \partial_\xi \left[n_e^{(1)}\right]^2 + \frac{1}{3}\partial_\xi^3 n_e^{(1)} + \partial_\xi |E^{(1)}|^2 = 0 \tag{8.273}$$

This is a KdV-type equation. Equations (8.268) and (8.273) describe nearly sonic Langmuir solitons.

Let us move to a coordinate system comoving with the Mach number M and write the stationary solution $E^{(1)}$ in the form

$$E^{(1)} = E_s(\xi) \exp\left[i\left(\frac{\eta^2}{\varepsilon^{3/2}} + \frac{M\varepsilon^{1/2}}{4} \right)\tau + \frac{M\varepsilon^{1/2}}{2}\xi \right], \tag{8.274}$$

then the stationary solutions are given by the equations

$$\partial_\xi^2 E_s - \eta^2 E_s - n_e^{(1)} E_s = 0\,, \tag{8.275}$$

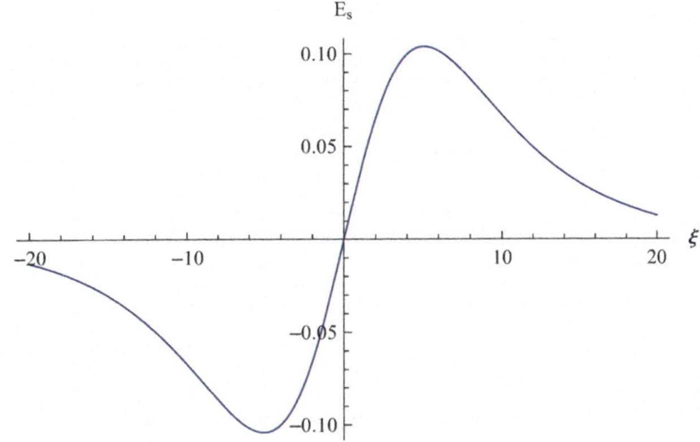

Fig. 8.4 Envelope electric field (8.277) in its spatial distribution

$$\frac{1}{3}\partial_\xi^2 n_e^{(1)} + \left[n_e^{(1)}\right]^2 - 2(M-1)n_e^{(1)} + |E^{(1)}|^2 = 0, \tag{8.276}$$

where $n_e^{(1)}$ is the stationary density. Solutions of (8.275) and (8.276) are

$$E_s = \sqrt{48}\,\eta^2 \operatorname{sech}(\eta\xi)\tanh(\eta\xi), \tag{8.277}$$

$$n_e^{(1)} = -6\eta^2 \operatorname{sech}^2(\eta\xi). \tag{8.278}$$

In addition, the relation

$$\eta^2 = \frac{3}{10}(1-M) \tag{8.279}$$

between η^2 and M must be satisfied. It is immediately apparent that (8.277) is an odd solution, whose form is shown in Fig. 8.4. Figure 8.5 shows the corresponding density distribution.

There are also other one-dimensional models for nonlinear Langmuir oscillations of small amplitude; however, for reasons of space, we will not go into further detail here. Instead, to conclude, a word on the multidimensional formulation.

Spatially multidimensional formulation

We choose the Zakharov model, which, following the one-dimensional considerations that led to (8.255) and (8.256), we can immediately write in the form

$$\boxed{i\partial_\tau \mathbf{E} + \nabla^2\mathbf{E} - Q\nabla\times\nabla\times\mathbf{E} - \delta n\mathbf{E} = 0\,,} \tag{8.280}$$

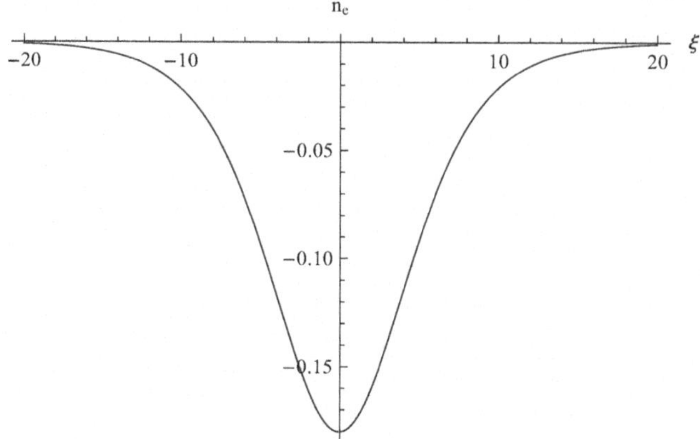

Fig. 8.5 Density depression $n = n_e^{(1)}$ as a function of ξ

$$\boxed{\partial_t^2 \delta n - \nabla^2 \delta n = \nabla^2 |\mathbf{E}|^2} \tag{8.281}$$

here. Since Q is a large parameter in nonrelativistic plasmas, it is convenient to write (8.280) in the form

$$\nabla \times \nabla \times \mathbf{E} = \frac{1}{Q}\left[i\partial_\tau \mathbf{E} + \nabla^2 \mathbf{E} - \delta n \mathbf{E}\right] \tag{8.282}$$

If we expand $\mathbf{E}$ in terms of Q^{-1} with

$$\mathbf{E} = -\nabla\varphi + \frac{1}{Q}\mathbf{E}_1 + \ldots, \tag{8.283}$$

then, to lowest order, we obtain

$$-\nabla \times \nabla \times \mathbf{E}_1 = i\partial_\tau \nabla\varphi + \nabla^2 \nabla\varphi - \delta n \nabla\varphi. \tag{8.284}$$

The solvability condition of this equation for $\mathbf{E}_1$ requires that the divergence of the right-hand side vanishes. That is,

$$\nabla \cdot \left[i\partial_\tau \nabla\varphi + \nabla^2 \nabla\varphi - \delta n \nabla\varphi\right] = 0\,. \tag{8.285}$$

This equation must be solved together with

$$\partial_t^2 \delta n - \nabla^2 \delta n = \nabla^2 |\nabla\varphi|^2 \tag{8.286}$$

In the adiabatic approximation, i.e., neglecting the ion inertia term, we find a generalized cubic nonlinear Schrödinger equation

$$\boxed{\nabla \cdot \left[i\partial_t \nabla\varphi + \nabla^2\nabla\varphi + |\nabla\varphi|^2\nabla\varphi\right] = 0}\,. \tag{8.287}$$

Note the significant difference from the equation often incorrectly used for Langmuir solitons

$$i\partial_\tau\varphi + \nabla^2\varphi + |\varphi|^2\varphi = 0. \tag{8.288}$$

In the next section, we return to the more formal aspects of the Schrödinger equation.

The Cubic Nonlinear Schrödinger Equation

After having seen in the previous section that, in the very simplest case and under very restrictive assumptions, the one-dimensional cubic nonlinear Schrödinger equation arises, we will now discuss its structure in somewhat more detail. We also choose this simple example in the hope that the fundamental results will reappear in more complicated generalizations.

While the KdV equation is applicable for modes with the (linear) dispersion relation $\omega = ak + bk^3$ in the weakly nonlinear case, the Schrödinger equation applies to waves with the (linear) dispersion $\omega = a + bk^2$. The Schrödinger equation describes the envelope $\tilde{E}$ of the field

$$E(x, t) = \tilde{E}(x, t)e^{i(kx-\omega t)} + c.c., \tag{8.289}$$

where the spatial and temporal dependencies of $\tilde{E}$ are assumed to be weak (compared to the phase variation). We now present three simplified derivations of the Schrödinger equation, with the intention of highlighting its universality more clearly.

Example 8.6 (Expansion of a Nonlinear Dispersion Function)
If the linear dispersion relation is written as

$$\varepsilon_L(\omega, k; \alpha_{j0})E(k, \omega) = 0 \tag{8.290}$$

with plasma parameters α_j (e.g., n,T, etc.), one expects a modification of the quantities α_j. for finite amplitudes. The values α_{j0} should be independent of the wave amplitudes. We formally write the nonlinear dispersion relation in the form

$$\boxed{e^{i(kx-\omega t)}\varepsilon_{NL}\left\{\left(\omega + i\frac{\partial}{\partial t}\right), \left(k - i\frac{\partial}{\partial x}\right);\ (\alpha_{j0} + \Delta\alpha_j)\right\}\tilde{E}(x, t) = 0} \tag{8.291}$$

The derivatives with respect to x and t take into account the (weak) modification due to the nonlinear changes in the plasma parameters. Within a Taylor expansion of ε_{NL},

$$\varepsilon_{NL} \approx \varepsilon_L + i\frac{\partial \varepsilon_L}{\partial \omega}\frac{\partial}{\partial t} - i\frac{\partial \varepsilon_L}{\partial k}\frac{\partial}{\partial x} - \frac{1}{2}\frac{\partial^2 \varepsilon_L}{\partial k^2}\frac{\partial^2}{\partial x^2} - \frac{1}{2}\frac{\partial^2 \varepsilon_L}{\partial \omega^2}\frac{\partial^2}{\partial t^2} + \frac{\partial^2 \varepsilon_L}{\partial \omega \partial k}\frac{\partial}{\partial x}\frac{\partial}{\partial t} + \sum_j \Delta\alpha_j \frac{\partial \varepsilon_L}{\partial \alpha_{j0}} + \dots , \tag{8.292}$$

we can now systematically calculate the dynamics of $\tilde{E}$ within the framework of perturbation theory. Since $\varepsilon_L(\omega, k; \alpha_{j0}) = 0$ holds and, by virtue of $\xi = x - v_g t,\ \tau = t$, we can transform into a coordinate system moving with the group velocity

$$v_g = \left.\frac{d\omega}{dk}\right|_{\varepsilon_L=0} = -\frac{\partial \varepsilon_L/\partial k}{\partial \varepsilon_L/\partial \omega} \tag{8.293}$$

it follows for the envelope of the harmonic $\exp\{i(kx - \omega t)\}$ to first order in ∂_τ and to second order in ∂_ξ

$$i\frac{\partial \tilde{E}}{\partial \tau} + \frac{1}{2}\frac{\partial^2 \omega}{\partial k^2}\frac{\partial^2 \tilde{E}}{\partial \xi^2} - \frac{\partial \omega}{\partial |\tilde{E}|^2}|\tilde{E}|^2\tilde{E} = 0. \tag{8.294}$$

We have used

$$\begin{aligned}\frac{\partial^2 \omega}{\partial k^2} &= \left.\frac{\partial}{\partial k}\frac{\partial \omega}{\partial k}\right|_{\varepsilon_L=0} = -\frac{\partial}{\partial k}\left(\frac{\partial \varepsilon_L/\partial k}{\partial \varepsilon_L/\partial \omega}\right)_{\varepsilon_L=0} \\ &= -\frac{\partial^2 \varepsilon_L/\partial k^2}{\partial \varepsilon_L/\partial \omega} - \frac{\partial^2 \varepsilon_L/\partial \omega^2}{\partial \varepsilon_L/\partial \omega}\frac{(\partial \varepsilon_L/\partial k)^2}{(\partial \varepsilon_L/\partial \omega)^2} + 2\frac{\partial^2 \varepsilon_L/\partial \omega \partial k}{\partial \varepsilon_L/\partial \omega}\frac{\partial \varepsilon_L/\partial k}{\partial \varepsilon_L/\partial \omega}\end{aligned} \tag{8.295}$$

and also

$$\sum_j \frac{\partial \varepsilon_L}{\partial \alpha_{j0}}\left(\frac{\partial \varepsilon_L}{\partial \omega}\right)^{-1}\Delta\alpha_j = -\frac{\partial \omega}{\partial |\tilde{E}|^2}|\tilde{E}|^2 \tag{8.296}$$

for $\Delta\alpha_j \sim |\tilde{E}|^2$. The relation (8.296) can be explicitly verified in the simplest case for the ponderomotive force at $n = n_0 - \delta|\tilde{E}|^2$ and $\varepsilon_L(k = 0) = \varepsilon_L = 1 - 4\pi ne^2/(m_e\omega^2)$. Furthermore, if we normalize the time by means of

$$\frac{\partial^2 \omega}{\partial k^2}\tau \to \tau, \tag{8.297}$$

we obtain the cubic nonlinear Schrödinger equation in standard form

$$\boxed{i\frac{\partial \tilde{E}}{\partial \tau} + \frac{1}{2}\frac{\partial^2 \tilde{E}}{\partial \xi^2} + \kappa|\tilde{E}|^2\tilde{E} = 0} \tag{8.298}$$

with

$$\kappa = -\frac{\partial \omega}{\partial |\tilde{E}|^2} / \frac{\partial^2 \omega}{\partial k^2}. \tag{8.299}$$

■

Example 8.7 (Feedback of the Harmonics)
The second demonstration starts from a nonlinear model equation

$$\boxed{u_{tt} - c^2 u_{xx} + \omega_p^2 u = \frac{1}{6}\omega_p^2 u^3}, \tag{8.300}$$

which we treat as representative of more complicated models. With the ansatz

$$\boxed{u = \sum_{\nu=-\infty}^{\infty} \sum_{\mu=|\nu|}^{\infty} \varepsilon^{\mu} u_{\nu\mu}(\xi, \tau) e^{i\nu(kx-\omega t)}}, \tag{8.301}$$

where ω satisfies the linear dispersion relation $\omega^2 = \omega_p^2 + c^2 k^2$, we substitute into (8.300). Here, $\tau = \varepsilon^2 t$ and $\xi = \varepsilon(x - v_g t)$; both denote the weak time and spatial dependence of the envelope. Since u is to be real, it must hold that

$$u_{-\nu\mu} = u_{\nu\mu}^* \tag{8.302}$$

At order ε^1 we obtain $u_{01} = 0$, and the remaining equation for u_{11} is, due to the linear dispersion relation, trivially satisfied. At order ε^2 we can sort according to the different harmonics $\nu = 0, 1, 2$. Let us begin with $\nu = 0$. One immediately sees $u_{02} = 0$. The contributions of the first harmonics ($\nu = 1$) yield

$$(-\omega^2 + c^2 k^2 + \omega_p^2) u_{12} = 2i(c^2 k - \omega v_g) \partial_\xi u_{11}, \tag{8.303}$$

where

$$\partial_t = \varepsilon^2 \partial_\tau - \varepsilon v_g \partial_\xi - i\nu\omega \tag{8.304}$$

and

$$\partial_x = \varepsilon \partial_\xi + i\nu k \tag{8.305}$$

were used. The group velocity is

$$v_g = \frac{d\omega}{dk} = \frac{c^2 k}{\omega}; \tag{8.306}$$

therefore, (8.303) is identically satisfied. For the second harmonic, a similar equation follows, with the crucial difference that the coefficient on the left-hand side is not exactly zero as in (8.303). Therefore, $u_{22} = 0$ must hold.

At order ε^3 we must take into account $\nu = 0, 1, 2, 3$. $\nu = 0$ leads to

$$(v_g^2 - c^2)\partial_\xi^2 u_{01} + \omega_p^2 u_{03} = 0, \tag{8.307}$$

from which

$$u_{03} = 0 \tag{8.308}$$

follows. The contributions of the first harmonics ($\nu = 1$) yield

$$\begin{aligned}(-\omega^2 + c^2k^2 + \omega_p^2)u_{13} - 2i(c^2k - \omega v_g)u_{12} - 2i\omega\partial_\tau u_{11} \\ + v_g^2\partial_\xi^2 u_{11} - c^2\partial_\xi^2 u_{11} = \frac{1}{6}\omega_p^2\, 3u_{11}^2 u_{11}^*.\end{aligned} \tag{8.309}$$

For

$$U = u_{11} \tag{8.310}$$

one thus obtains the closed equation

$$\boxed{\partial_\tau U - \frac{i\omega''}{2}\partial_\xi^2 U - \frac{i\omega_p^2}{4\omega}|U|^2 U = 0}\,, \tag{8.311}$$

where

$$\frac{c^2 - v_g^2}{\omega} = \frac{c^2\omega_p^2}{\omega^3} = \frac{d^2\omega}{dk^2} = \omega'' \tag{8.312}$$

was used. In this way, we have once again obtained the cubic nonlinear Schrödinger equation—albeit in a more systematic manner. [In fact, we have clarified using a simple example how the nonlinear dispersion relation (8.291) can be derived.] ■

Example 8.8 (Multiple-scale formalism)
In the following third approach, we demonstrate that the transition from (8.300) to (8.311) can be achieved using the same method of the multiple-scale formalism that was previously applied in the kinetic description. We only need to require the dominance of the first harmonic, which is expressed in the ansatz

$$\boxed{\begin{aligned}u = \varepsilon u_1(\xi_1, \xi_2, ...; \tau_1, \tau_2, ..) \exp[i(kx - \omega t)] \\ + \varepsilon^2 u_2(x, \xi_1, \xi_2, ...; t, \tau_1, \tau_2, ...) + ... + c.c.\end{aligned}} \tag{8.313}$$

We have various spatial and temporal scales $\xi_1, \xi_2, \ldots; \tau_1, \tau_2; \ldots$ with

$$\partial_x \to \partial_x + \varepsilon \partial_{\xi_1} + \varepsilon^2 \partial_{\xi_2} + \ldots, \tag{8.314}$$

$$\partial_t \to \partial_t + \varepsilon \partial_{\tau_1} + \varepsilon^2 \partial_{\tau_2} + \ldots. \tag{8.315}$$

The scaling is chosen so that at order ε^1 we obtain the linear dispersion relation. At order ε^2 we obtain

$$\left[\partial_t^2 - c^2\partial_x^2 + \omega_p^2\right]u_2 = \left[2i\omega\partial_{\tau_1}u_1 + 2ikc^2\partial_{\xi_1}u_1\right]e^{i(kx-\omega t)}. \tag{8.316}$$

With respect to the variables x and t, the right-hand side is proportional to $\exp\{i(kx - \omega t)\}$. Either u_2 is also a pure contribution to the first harmonic (which we can also take into account directly in u_1), or the solution u_2 grows linearly ("secular growth"), e.g.

$$u_2(x, t, \ldots) \sim \frac{[\ldots]t}{-2i\omega} e^{i(kx-\omega t)}, \tag{8.317}$$

where we have abbreviated the coefficient of the exponential function on the right-hand side of (8.316) by [...]. To avoid such developments that violate the order relations, [...] must be zero. In any case, the right-hand side of (8.316) must vanish, i.e.

$$\partial_{\tau_1}u_1 = -\frac{c^2k}{\omega}\partial_{\xi_1}u_1 = -v_g\partial_{\xi_1}u_1 \tag{8.318}$$

must hold, and without loss of generality we set $u_2 = 0$.

To order ε^3 follows

$$\left[\partial_t^2 - c^2\partial_x^2 + \omega_p^2\right]u_3 = \frac{\omega_p^2}{6}u_1^3\, e^{3i(kx-\omega t)}$$

$$+\left[2i\omega\partial_{\tau_2}u_1 - \partial_{\tau_1}^2u_1 + c^2\partial_{\xi_1}^2\, u_1 + \frac{1}{2}\omega_p^2|u_1|^2u_1\right]e^{i(kx-\omega t)}. \tag{8.319}$$

Now we can carry out the same considerations as in the previous case regarding a contribution $u_3 \sim \exp\{i(kx - \omega t)\}$. Except when

$$2i\omega\partial_{\tau_2}u_1 - \partial_{\tau_1}^2u_1 + c^2\partial_{\xi_1}^2u_1 + \frac{1}{2}\omega_p^2|u_1|^2u_1 = 0 \tag{8.320}$$

applies, there exist growing solutions that contradict the original ansatz (8.313) ("secular growth"). u_3 will therefore only have a contribution $\sim \exp\{3i(kx - \omega t)\}$, and taking into account (8.318), from (8.320) we obtain exactly the form (8.311). ■

With these three examples, we have been able to demonstrate the universality of the cubic nonlinear Schrödinger equation for waves whose linear dispersion is given by $\omega^2 \approx \omega_p^2 + c^2k^2 + \ldots$ in the case of weak nonlinearity and weak dispersion. In the following, we will deal with (simple stationary) solutions of the Schrödinger equation.

We return to the form (8.298) and slightly simplify the notation,

$$i\phi_\tau + \frac{1}{2}\phi_{\xi\xi} + \kappa|\phi|^2\phi = 0 \tag{8.321}$$

with $\kappa = -(\partial\omega/\partial|\phi|^2)/(\partial^2\omega/\partial k^2)$. For the complex envelope ϕ we set

$$\boxed{\phi = \rho^{1/2}e^{i\sigma}} \tag{8.322}$$

this leads in (8.321) to the real and imaginary parts

$$\rho_\tau + (\rho\sigma_\xi)_\xi = 0, \tag{8.323}$$

$$\kappa\rho - \frac{1}{8}\frac{1}{\rho^2}\left(\frac{\partial\rho}{\partial\xi}\right)^2 + \frac{1}{4}\frac{1}{\rho}\frac{\partial^2\rho}{\partial\xi^2} = \frac{\partial\sigma}{\partial\tau} + \frac{1}{2}\left(\frac{\partial\sigma}{\partial\xi}\right)^2. \tag{8.324}$$

A stationary solution is defined by $\rho_\tau = 0$. (A nonlinear wavenumber and frequency shift may still occur via σ.) Integration of (8.323) then yields

$$\rho\sigma_\xi = c(\tau). \tag{8.325}$$

The Eq. (8.324) further determines

$$\sigma_\tau + \frac{1}{2}\sigma_\xi^2 = f(\xi) \tag{8.326}$$

where c and f are still undetermined functions of τ and ξ, respectively.

Example 8.9 (Determination of the constants in (8.325))
We now prove that $c = const$ is the only permissible choice. From (8.326) and (8.325) it follows that

$$\begin{aligned}\sigma_{\tau\tau\xi} &= -\frac{1}{2}\frac{\partial^2}{\partial\xi\partial\tau}\left(\frac{\partial\sigma}{\partial\xi}\right)^2 = -\frac{1}{2}\frac{\partial^2}{\partial\xi\partial\tau}\frac{c^2}{\rho^2} \\ &= \frac{1}{\rho^3}\frac{d\rho}{d\xi}\frac{dc^2}{d\tau} = \frac{1}{\rho}\frac{d^2c}{d\tau^2}.\end{aligned} \tag{8.327}$$

From this, by separation of variables (provided that $c_\tau \neq 0$ holds), we obtain

$$\left(\frac{d^2c}{d\tau^2}\right)\left(\frac{dc^2}{d\tau}\right)^{-1} = \frac{1}{\rho^2}\frac{d\rho}{d\xi}. \tag{8.328}$$

Since the possible solution $\rho^{-2}\rho_\xi = const$ leads to

$$\rho = \frac{1}{\rho_0^{-1} + const(\xi - \xi_0)} \tag{8.329}$$

for which $\int \rho \, d\xi$ diverges, we must require

$$c(\tau) = const = c_1 \tag{8.330}$$

■

The integration of (8.325) then leads to

$$\sigma = \int^{\xi} \frac{c_1}{\rho} d\xi' + A(\tau). \tag{8.331}$$

Since, due to (8.325), σ_ξ is now only a function of ξ, and due to (8.326), σ_τ is only a function of ξ, this means

$$\frac{dA}{d\tau} = \Omega = const. \tag{8.332}$$

Now, if we insert the solution

$$\sigma = \int^{\xi} \frac{c_1}{\rho} d\xi' + \Omega\tau \tag{8.333}$$

into (8.324), we obtain

$$\kappa\rho - \frac{1}{8}\frac{1}{\rho^2}\left(\frac{d\rho}{d\xi}\right)^2 + \frac{1}{4}\frac{1}{\rho}\frac{d^2\rho}{d\xi^2} = \Omega + \frac{1}{2}\left(\frac{c_1}{\rho}\right)^2. \tag{8.334}$$

After multiplication by $8\rho_\xi$ this corresponds to

$$\frac{d}{d\xi}\left[\frac{1}{\rho}\left(\frac{d\rho}{d\xi}\right)^2\right] = -4\kappa\frac{d\rho^2}{d\xi} + 8\Omega\frac{d\rho}{d\xi} - 4c_1^2\frac{d}{d\xi}\frac{1}{\rho}. \tag{8.335}$$

We can integrate this equation, taking into account $\rho_\xi = 0$ for $\xi \to \pm\infty$. Subsequent multiplication by ρ leads to

$$\left(\frac{d\rho}{d\xi}\right)^2 = -4\kappa\rho^3 + 8\Omega\rho^2 + c_2\rho - 4c_1^2$$
$$= -4\kappa(\rho - \rho_\infty)^2(\rho - \rho_s). \tag{8.336}$$

c_2 is the integration constant, and ρ_ξ should vanish only at two points: at the extremum ρ_s and at ρ_∞ for $\xi \to \pm\infty$.

$\kappa > 0$

Let us first consider the case $\kappa > 0$. Since $\rho \geq 0$ and $-4c_1^2 = 4\kappa\rho_\infty^2\rho_s \geq 0$ it follows that $c_1 = \rho_\infty = 0$. This naturally results in $c_2 = 0$ and from (8.336) we obtain

$$\left(\frac{d\rho}{d\xi}\right)^2 = -4\kappa\rho^2(\rho - \rho_s) \tag{8.337}$$

with

$$\rho_s = 2\Omega/\kappa. \tag{8.338}$$

The solution is

$$\boxed{\rho = \rho_s \operatorname{sech}^2\left[(\kappa\rho_s)^{1/2}\xi\right]}. \tag{8.339}$$

For σ it then follows from (8.333)

$$\sigma = \Omega\tau. \tag{8.340}$$

If we take into account $\xi = x - v_g t$, we first see that the amplitudes of the solitary solution (8.339) are independent of (and can be freely chosen with respect to) the velocity. Furthermore, the Schrödinger equation is Galilean invariant, i.e., if $\phi(\xi, \tau)$ is a solution of the (time-dependent) Schrödinger equation, then the same holds for

$$\phi'(\xi, \tau) = \phi(\xi - u\tau, \tau)e^{i(u\xi - \frac{1}{2}u^2\tau)} \tag{8.341}$$

as well.

$\kappa < 0$

As the next case, we consider $\kappa < 0$. Now, $c_1 \neq 0$ is acceptable, and instead of (8.336) we obtain

$$\left(\frac{d\rho}{d\xi}\right)^2 = 4|\kappa|(\rho - \rho_\infty)^2(\rho - \rho_s). \tag{8.342}$$

Solutions are

$$\rho = \rho_\infty\left\{1 - a^2 \operatorname{sech}^2\left[(|\kappa|\rho_\infty)^{1/2}a\xi\right]\right\} \tag{8.343}$$

with

$$a^2 = \frac{\rho_\infty - \rho_s}{\rho_\infty} \leq 1, \tag{8.344}$$

$$c_1^2 = |\kappa|\rho_\infty^3(1 - a^2), \tag{8.345}$$

$$|\kappa|\rho_\infty(3 - a^2) = -2\Omega, \tag{8.346}$$

$$\sigma = \sin^{-1}\left\{a \tanh\left[(|\kappa|\rho_\infty)^{1/2} a\xi\right] / \left[1 - a^2 \operatorname{sech}^2((|\kappa|\rho_\infty)^{1/2} a\xi)\right]\right\} + \Omega\tau. \tag{8.347}$$

In contrast to the envelope soliton (8.339), (8.343) is referred to as an envelope hole. The latter has an additional free parameter a. For $a = 1$ the following holds

$$\boxed{\rho = \rho_\infty \tanh^2\left[(|\kappa|\rho_\infty)^{1/2}\xi\right]}\,. \tag{8.348}$$

Moving solutions can again be generated via a Galilean transformation.

The solutions (8.343) are shown in Fig. 8.6 for various parameter values of a.

Just like the KdV equation, the cubic nonlinear Schrödinger equation is completely integrable. We will return to this in Chap. 9.

Modulation instability

We now turn to modulation instability, as it follows within the framework of a Schrödinger description. To this end, we examine the dynamics of a wave train (with initially constant amplitude).

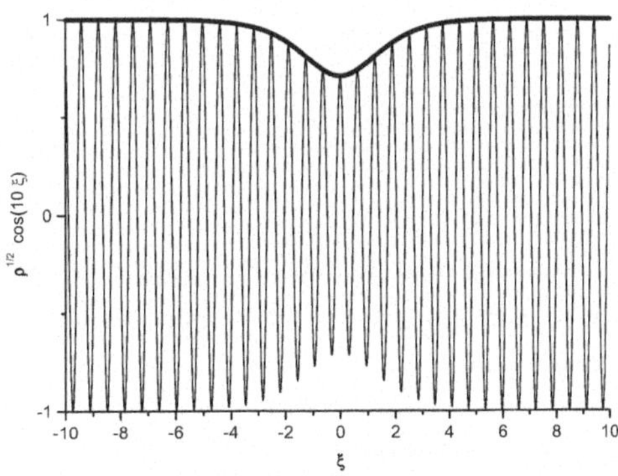

Fig. 8.6 Envelope $\sqrt{\rho}$ according to (8.343), i.e., $\sqrt{1 - a^2\, sech^2}$, together with the rapid phase variation for $a^2 = 0.5$

Due to the boundary conditions $\phi \to \phi_0$ for $\xi \to \pm\infty$ we now start from

$$\boxed{i\phi_\tau + \frac{1}{2}\phi_{\xi\xi} + \kappa(|\phi|^2 - |\phi_0|^2)\phi = 0}\,. \tag{8.349}$$

The simplest way to understand this model is to repeat the derivation of (8.294), this time expanding around the small values $|\tilde{E}_0|^2$ with the deviation $|\tilde{E}|^2 - |\tilde{E}_0|^2$ taken into account. However, it should be emphasized that this argument is not without issues, since in fact $|\tilde{E}_0 + \Delta\tilde{E}|^2$ appears in the dispersion relation. Therefore, we consider (8.349) merely as a simplified demonstration model.
If we linearize using the ansatz

$$\begin{pmatrix} \rho \\ \sigma \end{pmatrix} = \begin{pmatrix} \rho_0 \\ 0 \end{pmatrix} + \mathrm{Re}\begin{pmatrix} \rho_1 \\ \sigma_1 \end{pmatrix} \exp[i(K\xi - \Omega\tau)] \tag{8.350}$$

and start from the coupled equations

$$\rho_\tau + (\rho\sigma_\xi)_\xi = 0, \tag{8.351}$$

$$\kappa(\rho - \rho_0) + \frac{\rho_{\xi\xi}}{4\rho} - \sigma_\tau - \frac{1}{2}(\sigma_\xi)^2 - \frac{1}{8}\frac{\rho_\xi^2}{\rho^2} = 0 \tag{8.352}$$

a short calculation leads to

$$\boxed{\Omega^2 = \frac{1}{4}K^4 - \kappa\rho_0 K^2}\,. \tag{8.353}$$

For $\kappa > 0$ and $K \to 0$ it is immediately apparent that $\Omega^2 < 0$ results. This means instability (for long wavelengths of the perturbations); due to the modulation of the original wave train, it is called modulational instability. The maximum growth rate

$$\gamma_{max} = \kappa\rho_0 \equiv \kappa|\phi_0|^2 \tag{8.354}$$

occurs at

$$K_{max} = (2\kappa\rho_0)^{1/2} \tag{8.355}$$

Modulational instability requires $\kappa > 0$. In this case, the effective potential in (8.349) is attractive. If the quasiparticle density $|\phi|^2$ is increased, the potential depth grows. The attraction then becomes stronger, more quasiparticles are drawn in, and a self-reinforcement sets in.

Integrability of the Nonlinear Schrödinger Equation

The Lax formulation $L_t = [A, L]$ with the Schrödinger operator $L = -\partial_x^2 + u$ brought, as we showed in the previous section, a significant advance in solving the KdV equation. However, for the second major nonlinear wave equation in plasma physics, the cubic nonlinear Schrödinger equation (NLS), no IST could be found for several years. The reason is that the NLS equation is based on a different scattering problem. Here, we do not describe the solution originally presented by Zakharov and Shabat, but rather a generalization due to Ablowitz, Kaup, Newell, and Segur.

Instead of the formulation

$$L\psi = \lambda\psi, \tag{8.356}$$

$$\psi_t = A\psi \tag{8.357}$$

with scalar functions ψ, we now generalize to n-component vectors $\mathbf{v}$ and write, using $n \times n$ matrices $\mathcal{X}$ and $\mathcal{T}$

$$\mathbf{v}_x = \mathcal{X}\mathbf{v}, \tag{8.358}$$

$$\mathbf{v}_t = \mathcal{T}\mathbf{v}. \tag{8.359}$$

The solvability condition $\mathbf{v}_{xt} = \mathbf{v}_{tx}$ results in the compatibility requirement

$$\mathcal{X}_t - \mathcal{T}_x + [\mathcal{X}, \mathcal{T}] = 0. \tag{8.360}$$

Analogous to the Lax formulation, $\mathcal{X}$ should contain a parameter (ζ) that plays the role of the eigenvalue and for which $\zeta_t = 0$ holds. Since—as we now know—the NLS equation can already be solved within a two-component theory, we will restrict ourselves in the following to this simplest case. We therefore make the ansatz for (8.358)

$$v_{1x} = -i\zeta v_1 + q v_2, \tag{8.361}$$

$$v_{2x} = i\zeta v_2 + r v_1, \tag{8.362}$$

which for $r = -1$ and $q = -u$ contains the Schrödinger scattering problem

$$\boxed{v_{2xx} - u v_2 = -\zeta^2 v_2} \tag{8.363}$$

The most general *linear* time evolution (8.359) is explicitly given by

$$\boxed{v_{1t} = Av_1 + Bv_2 \ ,} \tag{8.364}$$

$$\boxed{v_{2t} = Cv_1 + Dv_2} \tag{8.365}$$

with scalar functions A, B, C, and D, which, however, must not depend on $\mathbf{v}$. This linear ansatz guarantees that the generalized scattering data also obey linear equations.
The compatibility condition (8.360) yields

$$A_x = qC - rB, \tag{8.366}$$

$$B_x + 2i\zeta B = q_t - (A - D)q, \tag{8.367}$$

$$C_x - 2i\zeta C = r_t + (A - D)r, \tag{8.368}$$

$$-D_x = qC - rB \ . \tag{8.369}$$

From (8.366) and (8.369) it follows that

$$D = -A + const. \tag{8.370}$$

Without loss of generality, we can set the constant equal to zero. The reason is simple: if we substitute (8.370) into (8.367) and (8.368), the constant can be eliminated by the redefinition $2A' = 2A - const$. Thus, the remaining task is to solve the coupled equations

$$A_x = qC - rB, \tag{8.371}$$

$$B_x + 2i\zeta B = q_t - 2Aq, \tag{8.372}$$

$$C_x - 2i\zeta C = r_t + 2Ar. \tag{8.373}$$

Example 8.10 (Power series approach for *A*,*B* and *C*)
One possible solution method consists of an expansion in powers of ζ. We demonstrate this with the ansatz

$$\boxed{A = A_2\zeta^2 + A_1\zeta + A_0 \ ,} \tag{8.374}$$

$$\boxed{B = B_2\zeta^2 + B_1\zeta + B_0 \ ,} \tag{8.375}$$

$$\boxed{C = C_2\zeta^2 + C_1\zeta + C_0 .} \tag{8.376}$$

Substituting into (8.371)–(8.373) and collecting powers of ζ^3, ζ^2, ζ^1 and ζ^0 we obtain

$$B_2 = C_2 = 0, \quad A_2 = const , \tag{8.377}$$

$$B_1 = iA_2 q , \tag{8.378}$$

$$C_1 = iA_2 r , \tag{8.379}$$

$$A_1 = const = 0 \ (!) , \tag{8.380}$$

$$B_0 = -A_2 q_x/2, \tag{8.381}$$

$$C_0 = A_2 r_x/2, \tag{8.382}$$

$$A_0 = A_2 qr/2 + const = A_2 qr/2 \ (!) . \tag{8.383}$$

Finally, (8.372) and (8.373) yield

$$-\frac{1}{2}A_2 q_{xx} = q_t - A_2 q^2 r, \tag{8.384}$$

$$\frac{1}{2}A_2 r_{xx} = r_t + A_2 q r^2. \tag{8.385}$$

If we now set

$$r = \mp q^* \tag{8.386}$$

here and choose $A_2 = 2i$ to be purely imaginary, we end up with the NLS equation

$$\boxed{iq_t = q_{xx} \pm 2q^2 q^*} . \tag{8.387}$$

With this, the crucial breakthrough for the NLS equation has been achieved. We now "only" need to fully work out the scattering problem. Before we do that, let us take a brief detour that will lead us to other important nonlinear wave equations.

■

Example 8.11 (Generalization with third powers of ζ)
If we extend the approach (8.374)–(8.376) to third powers of ζ, after some calculation we obtain

$$A = A_3\zeta^3 + \frac{1}{2}A_3qr\zeta - \frac{i}{4}A_3(qr_x - q_xr), \tag{8.388}$$

$$B = iA_3q\zeta^2 - \frac{1}{2}A_3q_x\zeta + \frac{i}{2}A_3q^2r - \frac{i}{4}A_3q_{xx}, \tag{8.389}$$

$$C = iA_3r\zeta^2 + \frac{1}{2}A_3r_x\zeta + \frac{i}{2}A_3qr^2 - \frac{i}{4}A_3r_{xx} \tag{8.390}$$

with $A_3 = const$. In (8.372) and (8.373) this leads to

$$q_t + \frac{i}{4}A_3(q_{xxx} - 6qrq_x) = 0, \tag{8.391}$$

$$r_t - \frac{1}{4}A_3(r_{xxx} - 6qrr_x) = 0. \tag{8.392}$$

For $r = -1, A_3 = -4i$ and $q = -u$ the KdV equation arises. The integrable modified KdV equation

$$\boxed{q_t \pm 6q^2q_x + q_{xxx} = 0} \tag{8.393}$$

results for $A_3 = 4i$ and $r = \pm q$. ■

Example 8.12 (Generalization with powers of ζ^{-1})
Ultimately, we can also perform an expansion in powers of ζ^{-1}. For example, if we set

$$A = \frac{a(x,t)}{\zeta}, \tag{8.394}$$

$$B = \frac{b(x,t)}{\zeta}, \tag{8.395}$$

$$C = \frac{c(x,t)}{\zeta} \tag{8.396}$$

then the now well-known calculation yields

$$a_x = \frac{i}{2}(qr)_t, \tag{8.397}$$

$$q_{xt} = -4iaq, \tag{8.398}$$

$$r_{xt} = -4iar. \tag{8.399}$$

With

$$a = \frac{i}{4}\cos u, \tag{8.400}$$

$$b = c = \frac{i}{4}\sin u, \tag{8.401}$$

$$q = -r = -\frac{u_x}{2} \tag{8.402}$$

we have demonstrated the integrability of the sine-Gordon equation

$$\boxed{u_{xt} = \sin u} \tag{8.403}$$

■

Scattering data

But now, back to the nonlinear Schrödinger equation (8.387) and the development of the scattering problem. We assume that q and r vanish sufficiently rapidly for $x \to \pm\infty$. The scattering problem can be formulated for real ζ and two independent (two-component) sets of solutions ϕ and $\overline{\phi}$ [or ψ and $\overline{\psi}$] with the boundary conditions

$$\phi \simeq \begin{pmatrix} 1 \\ 0 \end{pmatrix} e^{-i\zeta x}, \quad \text{for} \quad x \to -\infty, \tag{8.404}$$

$$\overline{\phi} \simeq \begin{pmatrix} 0 \\ -1 \end{pmatrix} e^{i\zeta x}, \quad \text{for} \quad x \to -\infty \tag{8.405}$$

or

$$\psi \simeq \begin{pmatrix} 0 \\ 1 \end{pmatrix} e^{i\zeta x}, \quad \text{for} \quad x \to +\infty, \tag{8.406}$$

$$\overline{\psi} \simeq \begin{pmatrix} 1 \\ 0 \end{pmatrix} e^{-i\zeta x}, \quad \text{for} \quad x \to +\infty \tag{8.407}$$

formulated. The solutions ϕ and $\overline{\phi}$ or ψ and $\overline{\psi}$ constitute the natural solutions of the potential-free problem (8.361) and (8.362).

If

$$u = \begin{pmatrix} u_1 \\ u_2 \end{pmatrix} \quad \text{and} \quad v = \begin{pmatrix} v_1 \\ v_2 \end{pmatrix} \tag{8.408}$$

two solutions of (8.361) and (8.362) are [which we have abbreviated above (and in the following) as ϕ and $\overline{\phi}$ or ψ and $\overline{\psi}$], then the Wronskian determinant is

$$W(u, v) \equiv u_1 v_2 - u_2 v_1. \tag{8.409}$$

It is easy to verify

$$\frac{d}{dx} W(u, v) = 0 \tag{8.410}$$

this, and thus the value of the Wronskian determinant can be determined from the asymptotic region. For ϕ and $\overline{\phi}$ or ψ and $\overline{\psi}$ this

$$W(\phi, \overline{\phi}) = -W(\psi, \overline{\psi}) = -1 \tag{8.411}$$

follows.

The scattering data $a, b, \overline{a}$ and $\overline{b}$ are defined via the linear relationship between the two sets of independent solutions,

$$\phi = a(\zeta)\overline{\psi} + b(\zeta)\psi, \tag{8.412}$$

$$\overline{\phi} = -\overline{a}(\zeta)\psi + \overline{b}(\zeta)\overline{\psi}. \tag{8.413}$$

At this point, we would need to insert a longer section in which we analyze the mathematical properties of the scattering data. For reasons of space, we will forgo a detailed presentation and refer the reader to the specialized literature for further details. Only a few brief remarks will suffice here.

With the help of the integral representations, it can be shown that $\phi_i e^{i\zeta x}$, $\psi_i e^{-i\zeta x}$ ($i = 1, 2$) and $a(\zeta)$ are analytic in the upper complex ζ-half-plane.

The zeros of $a(\zeta)$ [for $\zeta = \zeta_k, k = 1, 2, \ldots$] correspond to the bound states (with discrete eigenvalues). For the Schrödinger case with $r = \pm q^*$ the following holds explicitly

$$\overline{\psi}(x, \zeta) = \begin{pmatrix} \psi_2^*(x, \zeta^*) \\ \pm\psi_1^*(x, \zeta^*) \end{pmatrix}, \tag{8.414}$$

$$\overline{\phi}(x, \zeta) = \begin{pmatrix} \mp\phi_2^*(x, \zeta^*) \\ -\phi_1^*(x, \zeta^*) \end{pmatrix}, \tag{8.415}$$

$$\overline{a}(\zeta) = a^*(\zeta^*), \tag{8.416}$$

$$\overline{b}(\zeta) = \pm b^*(\zeta^*), \tag{8.417}$$

$$\overline{\zeta}_\kappa = \zeta_\kappa^*. \tag{8.418}$$

Inverse Problem

After this brief summary of the most important mathematical properties, we now turn to the inverse problem, i.e., the reconstruction of the potential from the scattering data. If one defines

$$\psi = \begin{pmatrix} 0 \\ 1 \end{pmatrix} e^{i\zeta x} + \int_x^\infty K(x, s) e^{i\zeta s} ds \tag{8.419}$$

and

$$\overline{\psi} = \begin{pmatrix} 1 \\ 0 \end{pmatrix} e^{-i\zeta x} + \int_x^\infty \overline{K}(x, s) e^{-i\zeta s} ds \tag{8.420}$$

one can show that the kernels

$$K = \begin{pmatrix} K_1(x, s) \\ K_2(x, s) \end{pmatrix} \tag{8.421}$$

and $\overline{K}$ are independent of ζ.

The kernels obey the integral equations

$$\boxed{\overline{K}(x, y) + \begin{pmatrix} 0 \\ 1 \end{pmatrix} F(x + y) + \int_x^\infty K(x, s) F(s + y) ds = 0 \,,} \tag{8.422}$$

$$\boxed{K(x, y) - \begin{pmatrix} 1 \\ 0 \end{pmatrix} \overline{F}(x + y) - \int_x^\infty \overline{K}(x, s) \overline{F}(s + y) ds = 0} \tag{8.423}$$

with

$$F(x) := \frac{1}{2\pi} \int_C \frac{b(\zeta)}{a(\zeta)} e^{i\zeta x} d\zeta \tag{8.424}$$

$$= \frac{1}{2\pi} \int_{-\infty}^{+\infty} \frac{b(\xi)}{a(\xi)} e^{i\xi x} d\xi - i \sum_{j=1}^n \Gamma_j \exp(i\zeta_j x), \tag{8.425}$$

$$\overline{F}(x) := \frac{1}{2\pi} \int_{\overline{C}} \frac{\overline{b}(\zeta)}{\overline{a}(\zeta)} e^{-i\zeta x} d\zeta \tag{8.426}$$

$$= \frac{1}{2\pi}\int_{-\infty}^{+\infty}\frac{\overline{b}(\xi)}{\overline{a}(\xi)}e^{-i\xi x}d\xi + i\sum_{j=1}^{\overline{n}}\overline{\Gamma}_j \exp(-i\zeta_j x). \tag{8.427}$$

Some further explanation is required for the last transformations. The path C runs above all zeros of $a(\zeta)$ of $-\infty + i0$ up to $+\infty + i0$. On the other hand, the path $\overline{C}$ lies below all zeros of $\overline{a}(\zeta)$ in the complex ζ-plane. In evaluating the integrals, simple isolated zeros were assumed, and accordingly the residues

$$\Gamma_\kappa = \frac{b(\zeta_k)}{\dot{a}(\zeta_k)} \tag{8.428}$$

and

$$\overline{\Gamma}_\kappa = \frac{\overline{b}(\overline{\zeta}_k)}{\dot{\overline{a}}(\overline{\zeta}_k)} \tag{8.429}$$

were defined. If $r = \pm q^*$ holds, then

$$\overline{F}(x) = \mp F^*(x), \tag{8.430}$$

$$\overline{K}(x, y) = \begin{pmatrix} K_2^*(x, y) \\ \pm K_1^*(x, y) \end{pmatrix}, \tag{8.431}$$

and the coupled integral equations (8.422) and (8.423) reduce, for example, to

$$K_1(x, y) \pm F^*(x + y) \mp \int_x^\infty \int_x^\infty K_1(x, z)F(z + s)F^*(s + y)ds\, dz = 0\,. \tag{8.432}$$

The reward for all this effort is then the solution

$$\boxed{q(x) = -2K_1(x, x)}\,, \tag{8.433}$$

which constructively demonstrates the integrability of the Schrödinger equation.

Of course, for the complete solution we still need the time dependence of the quantities a and b. We find these—in fact, very similarly to the previous chapter for the KdV equation—from the linear time evolution equations (8.364) and (8.365) of the scattering functions (8.361) and (8.362). A short calculation yields

$$a(\zeta_k, t) = a(\zeta_k, 0)\,, \tag{8.434}$$

$$a(\xi, t) = a(\xi, 0), \tag{8.435}$$

$$\partial_t b(\zeta_k, t) = -4i\zeta_k^2 b(\zeta_k, t), \tag{8.436}$$

$$\partial_t b(\xi, t) = -4i\xi^2 b(\xi, t). \tag{8.437}$$

Conservation laws

Finally, we provide a construction rule for the theorem of infinitely many conserved quantities. To this end, we note, due to (8.434) and the boundary conditions (8.404)–(8.407)

$$a(\zeta) = \lim_{x\to\infty} \phi_1 e^{i\zeta x}. \tag{8.438}$$

$a(\zeta)$ is analytic for Im $\zeta > 0$ and approaches 1 for $|\zeta| \to \infty$. Furthermore, $a(\zeta)$ is time-independent [see (8.434) and (8.436)]. In the scattering problem (8.361) and (8.362)

$$\phi_{1x} = -i\zeta\phi_1 + q\phi_2, \tag{8.439}$$

$$\phi_{2x} = i\zeta\phi_2 + r\phi_1 \tag{8.440}$$

we eliminate ϕ_2,

$$q\left(q^{-1}\phi_{1x} + i\zeta q^{-1}\phi_1\right)_x = i\zeta(\phi_{1x} + i\zeta\phi_1) + rq\phi_1. \tag{8.441}$$

If we introduce the new variable h by

$$\phi_1 = \exp[-i\zeta x + h] \tag{8.442}$$

then the following holds

$$2i\zeta h_x = (h_x)^2 - qr + q\left(\frac{h_x}{q}\right)_x. \tag{8.443}$$

Since h vanishes for $|\zeta| \to \infty$ and Im $\zeta > 0$, we can use the expansion

$$h_x = \sum_{n=0}^{\infty} \frac{g_n(x, t)}{(2i\zeta)^{n+1}} \tag{8.444}$$

and in (8.443) collect terms according to powers of $1/\zeta$.

Successively, we obtain

$$g_0 = -qr, \tag{8.445}$$

$$g_1 = -qr_x, \tag{8.446}$$

$$g_{n+1} = q\left(\frac{g_n}{q}\right)_x + \sum_{k=0}^{n-1} g_k\, g_{n-k-1} \quad \text{for} \quad n \geq 1 \tag{8.447}$$

and thus a unique recursion formula for the g_n. Now, from (8.438), (8.442), and (8.444) we deduce

$$\ln a(\zeta) = h(x = +\infty) = \sum_{n=0}^{\infty} \frac{1}{(2i\zeta)^{n+1}} \int_{-\infty}^{+\infty} g_n \, dx. \tag{8.448}$$

Since $\ln\ a(\zeta)$ is time-independent for all ζ with $\mathrm{Im}\ \zeta > 0$, we can infer

$$\int_{-\infty}^{+\infty} g_n dx = C_n = const \tag{8.449}$$

The integrals thus calculated using (8.445)–(8.447) represent the mutually independent conserved quantities of the nonlinear Schrödinger equation.

The soliton solutions of the cubic nonlinear Schrödinger equation exhibit in one spatial dimension the same stability behavior that we already know from the solitons of the KdV equation. Even in the face of strong perturbations of the initial distribution, and indeed even in collisions with other solitons, the NLS solitons prove to be robust and shape-stable. This result of the IST makes Langmuir solitons appear as attractive nonlinear structures in plasma, which can lead to qualitatively new phenomena. In the field of optical fibers and optical data transmission, Schrödinger solitons still play a significant and practically relevant role today.

8.3 Drift Vortices

In this section, we turn to the nonlinear theory of drift waves. We choose the hydrodynamic approach and the drift approximation for the motion perpendicular to **B**. A new type of model equation is derived in the form of the Hasegawa-Mima equation. Its vortex solutions—including their stability—are briefly discussed.

As linear velocities, we use in the drift approximation

$$\boxed{\mathbf{v}_E = \frac{1}{B_0} \hat{z} \times \nabla\varphi = \mathbf{v}_{E\times B}\ ,} \tag{8.450}$$

$$\boxed{\mathbf{v}_{pe} = \frac{1}{B_0|\Omega_e|} \partial_t \nabla_\perp \varphi\ ,} \tag{8.451}$$

$$\boxed{\mathbf{v}_{pi} = -\frac{1}{B_0\Omega_i}\partial_t \nabla_\perp \varphi \,,} \tag{8.452}$$

$$\boxed{\mathbf{v}_{De} = -\frac{v_{te}^2}{|\Omega_e|}\hat{z} \times \nabla \ln n_e \,,} \tag{8.453}$$

$$\boxed{\mathbf{v}_{Di} = \frac{v_{ti}^2}{\Omega_i}\hat{z} \times \nabla \ln n_i .} \tag{8.454}$$

For the polarization drifts $\mathbf{v}_{pi}$ and $\mathbf{v}_{pe}$ it is noticeable that the ion polarization drift $\mathbf{v}_{pi}$ is significantly larger than the electron polarization drift $\mathbf{v}_{pe}$. This is the main reason why, in nonlinear generalizations (when the inertia terms $\mathbf{v} \cdot \nabla \mathbf{v}$ are also taken into account in the drift approximation), only the so-called nonlinear ion polarization drift

$$\boxed{\tilde{\mathbf{v}}_{pi} = \mathbf{v}_{pi} - \frac{1}{\Omega_i}\{\mathbf{v}_E \cdot \nabla_\perp \mathbf{v}_E + \mathbf{v}_E \cdot \nabla_\perp \mathbf{v}_{Di} + \mathbf{v}_{Di} \cdot \nabla_\perp \mathbf{v}_E + \mathbf{v}_{Di} \cdot \nabla_\perp \mathbf{v}_{Di}\} \times \hat{z}} \tag{8.455}$$

becomes significant. Here is a brief plausibility consideration as a review.

Example 8.13 (Nonlinear Ion Polarization Drift)
If we solve an equation of motion of the form

$$\partial_t \mathbf{v} + \mathbf{v} \cdot \nabla \mathbf{v} = \frac{q}{m}(\mathbf{E} + \mathbf{v} \times \mathbf{B}) - \frac{1}{nm}\nabla p \tag{8.456}$$

iteratively, we multiply vectorially with $\mathbf{B}$ from the left, in order to obtain

$$\mathbf{B} \times \partial_t \mathbf{v} + \mathbf{B} \times \mathbf{v} \cdot \nabla \mathbf{v} = \frac{q}{m}(\mathbf{B} \times \mathbf{E}) + \frac{q}{m}B^2\mathbf{v} - \frac{1}{nm}\mathbf{B} \times \nabla p \tag{8.457}$$

If we assign the terms on the left-hand side for $\partial_t \ll \Omega$ and small amplitudes a higher order, we first obtain

$$\mathbf{v}^{(0)} = \frac{1}{B^2}\mathbf{E} \times \mathbf{B} + \frac{1}{qnB^2}\mathbf{B} \times \nabla p \mathrel{\widehat{=}} \mathbf{v}_E + \mathbf{v}_D . \tag{8.458}$$

If we use this zeroth order on the left-hand side of (8.457), it follows that

$$\mathbf{v}^{(1)} = \frac{1}{B\Omega}\Big[\mathbf{B} \times \partial_t \mathbf{v}^{(0)} + \mathbf{B} \times \mathbf{v}^{(0)} \cdot \nabla \mathbf{v}^{(0)}\Big]. \tag{8.459}$$

This expression corresponds to (8.455). ■

At the same time, from this consideration we have learned how we can include a viscous damping term $\mu \nabla^2 \mathbf{v}$ on the right-hand side of (8.456) at first order. In addition to the contributions (8.459), we obtain $-(\mu/B\Omega)\nabla^2 \mathbf{E}_\perp$. Specifically, with regard to the nonlinear ion polarization drift, this means that in the following we will start from

$$\mathbf{v}'_{pi} = \tilde{\mathbf{v}}_{pi} + \frac{\mu_i}{B_0\Omega_i}\nabla_\perp^2\nabla_\perp\varphi \tag{8.460}$$

with

$$\mu_i = 0.3k_BT_i\nu_i/m_i\Omega_i^2 \tag{8.461}$$

for the nonlinear ion polarization drift.

In the closed model to be derived now, we do not restrict ourselves from the outset to the electrostatic approximation, although it will be the main focus in the subsequent analysis.

We begin with the equation for the vorticity ("vorticity equation"), which we obtain from the Poisson equation by differentiating with respect to time,

$$-\partial_t\nabla^2\varphi = -\frac{1}{\varepsilon_0}\nabla_\| j_\| - \frac{1}{\varepsilon_0}\nabla_\perp\cdot\mathbf{j}_\perp. \tag{8.462}$$

Here, $\mathbf{j} = \mathbf{j}_\| + \mathbf{j}_\perp$ is the total current density

$$\mathbf{j} = \mathbf{j}_i + \mathbf{j}_e = en_i\mathbf{v}_i - en_e\mathbf{v}_e \tag{8.463}$$

and

$$\boxed{\nabla_\| := \partial_z + \frac{\mathbf{B}_\perp}{B_0}\cdot\nabla_\perp}. \tag{8.464}$$

The derivative $\nabla_\|$ is thus taken along the total magnetic field $\mathbf{B}$ (where electromagnetic field components in addition to the external magnetic field $B_0\hat{z}$ are taken into account). Equation (8.464) follows from

$$\nabla_\| = \hat{b}(\hat{b}\cdot\nabla) \approx \hat{z}\left(\partial_z + \frac{\mathbf{B}_\perp}{B_0}\cdot\nabla_\perp\right), \tag{8.465}$$

where $\hat{b}$ is the unit vector in the direction of $\mathbf{B}$ (total).

We use this more precise distinction between ∂_z and $\nabla_\|$, because the dynamics along the total field (and not just along $\hat{z}$) fundamentally differs from the dynamics perpendicular to the magnetic field. We must take this, in itself small, difference into account, since later on we may assume the z-variation to be small and therefore cannot generally assume $B_0^{-1}\mathbf{B}_\perp \cdot \nabla_\perp$ to be negligible compared to ∂_z. The situation is different for the perpendicular variation $\nabla_\perp = \mathbf{b}_\perp(\mathbf{b}_\perp \cdot \nabla)$, which we can well approximate by $\hat{x}\partial_x + \hat{y}\partial_y$, since in the following the x and y dependencies are never assumed to be weak.

If we denote by ψ the parallel component of $\mathbf{A}$,, then from the Maxwell equation $\nabla \times \mathbf{B} = \mu_0 \mathbf{j} + \mu_0 \varepsilon_0 \dot{\mathbf{E}}$

$$\mu_0 j_\parallel = -\nabla^2 \psi - \partial_t E_\parallel \tag{8.466}$$

follows. It should be noted that for electromagnetic oscillations, the transverse component $\mathbf{B}_\perp$ is assumed to be essentially different from zero. For weak variations along $\hat{z}$, the approximation $\mathbf{B}_\perp \approx \nabla\psi \times \hat{z}$, holds, and we obtain $\hat{b} \cdot \nabla \times \mathbf{B}_\perp \approx \hat{z} \cdot \nabla \times \mathbf{B}_\perp \approx -\nabla_\perp^2 \psi$. In the drift approximation, we obtain ($e = e_i = -e_e$)

$$\begin{aligned}\mathbf{j}_\perp &= e\mathbf{v}_E(n_i - n_e) - en_i \frac{1}{\Omega_i}[\partial_t + (\mathbf{v}_E + \mathbf{v}_{Di}) \cdot \nabla_\perp](\mathbf{v}_E + \mathbf{v}_{Di}) \times \hat{z} \\ &+ en_i \frac{1}{B_0 \Omega_i} \mu_i \nabla_\perp^2 \nabla_\perp \varphi,\end{aligned} \tag{8.467}$$

where, due to $m_e/m_i \ll 1$, the electron polarization drift has only been taken into account to lowest order. If we substitute the individual drifts into (8.467), we obtain

$$\begin{aligned}\mathbf{j}_\perp &= -\varepsilon_0 \mathbf{v}_E \nabla^2 \varphi - en_i \frac{1}{B_0 \Omega_i}(\partial_t + \mathbf{v}_E \cdot \nabla_\perp)\nabla_\perp \varphi \\ &- en_i \frac{v_{ti}^2}{\Omega_i^2} \frac{1}{L_n B_0} \partial_y \nabla_\perp \varphi + en_i \frac{1}{B_0 \Omega_i} \mu_i \nabla_\perp^2 \nabla_\perp \varphi.\end{aligned} \tag{8.468}$$

Since we will use the approximation $\omega_{pi}^2 \gg \Omega_i^2$ in the following, we neglect the first term on the right-hand side of (8.468) [for example, compared to the third]. For the complete determination of the current $j_\parallel$ we also note that

$$E_\parallel = -(\hat{b} \cdot \nabla)\varphi - \frac{\partial \psi}{\partial t} \tag{8.469}$$

applies. The expressions (8.466) [together with (8.469)] and (8.468) can now be inserted into (8.462). If we also take into account that, again for $\omega_{pi}^2 \gg \Omega_i^2$, the divergence of the ion polarization current dominates over contributions due to deviations from quasineutrality [left-hand side of (8.462)], then from (8.462) we ultimately obtain

$$\boxed{\begin{aligned} 0 \approx & c^2(\hat{b}\cdot\nabla)\nabla_\perp^2\psi + \frac{\omega_{pi}^2}{\Omega_i^2}\left[\partial_t - \frac{1}{B_0}(\nabla_\perp\varphi\times\hat{z})\cdot\nabla_\perp\right]\nabla_\perp^2\varphi \\ & + \frac{\omega_{pi}^2}{\Omega_i^2}\frac{v_{ti}^2}{L_n\Omega_i}\partial_y\nabla_\perp^2\varphi - \frac{\omega_{pi}^2}{\Omega_i^2}\mu_i\nabla_\perp^2\nabla_\perp^2\varphi. \end{aligned}} \tag{8.470}$$

This is the first equation of the generalized drift-vortex model, which links φ and ψ with each other.

We obtain another relation from the momentum balances, which we combine in an obvious way into a generalized Ohm's law,

$$\boxed{\frac{m_e}{e^2}\frac{d}{dt}\left(\frac{j_\parallel - j_{i\parallel}}{n_e}\right) - \frac{1}{en_e}(\hat{b}\cdot\nabla)p_e = E_\parallel - \eta(j_\parallel - j_{i\parallel})} \tag{8.471}$$

with

$$\eta \approx \frac{m_e\nu_{ei}}{n_e e^2} \tag{8.472}$$

and

$$\frac{d}{dt} = \partial_t + \mathbf{v}_E\cdot\nabla_\perp + v_{e\parallel}(\hat{b}\cdot\nabla). \tag{8.473}$$

Taking into account the expression for $j_\parallel$ already given in (8.466), we are now still missing equations for p_e and $j_{i\parallel}$.

For isothermal (because slowly varying) processes, $p_e \sim n_e^\gamma$ holds with $\gamma = 1$. Furthermore, we have

$$\frac{dp_e}{dt} = \frac{p_e}{n_e}\frac{dn_e}{dt} = -p_e\nabla\cdot\mathbf{v}_e\,. \tag{8.474}$$

The relation

$$\frac{dp_e}{dt} = \frac{p_e}{e}\hat{b}\cdot\nabla\frac{j_\parallel - j_{i\parallel}}{n_e} - p_e\nabla_\perp\cdot\mathbf{v}_{e\perp} \tag{8.475}$$

can be easily evaluated if one recalls $\nabla\cdot\mathbf{v}_E = 0$. In addition, we change the notation by writing $p_e = p_{e0}(x) + \delta p_e \to p_{e0} + p_e$; p_e hereafter denotes only the pressure deviations. Thus, (8.475) becomes

$$\boxed{\begin{aligned} \frac{\partial}{\partial t}p_e + \frac{1}{B_0}\hat{z}\times\nabla\varphi\cdot\nabla p_e - \frac{1}{B_0}p_{e0}\frac{1}{L_n}\partial_y\varphi - \frac{(p_e + p_{e0})}{e}(\hat{b}\cdot\nabla)\frac{j_\parallel - j_{i\parallel}}{n_e} \\ - \frac{j_\parallel - j_{i\parallel}}{en_e}(\hat{b}\cdot\nabla)p_e = 0. \end{aligned}} \tag{8.476}$$

Here, $p_{e0} = n_{e0}k_B T_e$ applies. The last remaining equation is obtained from the ion momentum balance along the magnetic field

$$\boxed{\partial_t j_{i\|} + \frac{e^2 n_i}{m_i}(\hat{b}\cdot\nabla)\varphi + \mathbf{v}_E \cdot \nabla_\perp j_{i\|} = 0}\,. \tag{8.477}$$

At this point, it becomes clear that these calculations are based on a specific scaling. The motion along the magnetic field, as well as the variation along the magnetic field, is only included to lowest order. However, we refrain from an explicit presentation of the formal procedure here, so as not to obscure the physical processes.

Equations (8.470), (8.471), (8.476), and (8.477) represent our fundamental equations. We further simplify them by introducing $j_\| - j_{i\|} = -en_e v_{e\|}$, and noting that for isothermal processes, density fluctuations can easily be converted into pressure fluctuations p_e. For the variables p_e, $v_{e\|}$, $j_{i\|}$ and φ we then obtain—summarized once again—the following equations:

$$\frac{\omega_{pi}^2}{\Omega_i^2}\left[\partial_t - \frac{1}{B_0}(\nabla_\perp\varphi \times \hat{z})\cdot\nabla_\perp\right]\nabla_\perp^2\varphi + c^2(\hat{b}\cdot\nabla)\nabla_\perp^2\psi$$
$$+\frac{\omega_{pi}^2}{\Omega_i^2}\frac{v_{ti}^2}{L_n\Omega_i}\partial_y\nabla_\perp^2\varphi - \frac{\omega_{pi}^2}{\Omega_i^2}\mu_i\nabla_\perp^2\nabla_\perp^2\varphi = 0, \tag{8.478}$$

$$-\frac{m_e}{e}\left[\partial_t + \frac{1}{B_0}(\hat{z}\times\nabla\varphi)\cdot\nabla_\perp + v_{e\|}(\hat{b}\cdot\nabla)\right]v_{e\|} - \frac{1}{en_e}(\hat{b}\cdot\nabla)p_e$$
$$+(\hat{b}\cdot\nabla)\varphi + \frac{\partial\psi}{\partial t} = en_e\eta v_{e\|}, \tag{8.479}$$

$$\left[\frac{\partial}{\partial t} + \frac{1}{B_0}(\hat{z}\times\nabla\varphi)\cdot\nabla_\perp\right]p_e - \frac{1}{B_0}k_B T_e n_{e0}\frac{1}{L_n}\partial_y\varphi$$
$$+k_B T_e n_e(\hat{b}\cdot\nabla)v_{e\|} + v_{e\|}(\hat{b}\cdot\nabla)p_e = 0, \tag{8.480}$$

$$\left[\frac{\partial}{\partial t} + \frac{1}{B_0}(\hat{z}\times\nabla\varphi)\cdot\nabla_\perp\right]j_{i\|} + \frac{n_i e^2}{m_i}(\hat{b}\cdot\nabla)\varphi = 0. \tag{8.481}$$

Here, based on (8.466), we must also note the following relation for low-frequency processes:

$$\boxed{\mu_0(j_{i\|} - en_e v_{e\|}) = -\nabla_\perp^2\psi} \tag{8.482}$$

With regard to $\nabla_\|$ we refer to (8.465); we write

$$\nabla_\| = \partial_z + \frac{1}{B_0}(\nabla\psi \times \hat{z})\cdot\nabla_\perp. \tag{8.483}$$

Furthermore, we have in mind a scaling in which all coefficients in the above equations (in particular n_{e0}, $L_n^{-1} = \partial_x \ln n_{e0}(x)$, T_e)) can be regarded as constant.

If we introduce the quantities

$$\phi = \frac{e\varphi}{k_B T_e}, \quad \nabla = \rho_s \nabla_\perp, \quad \nabla_\parallel = \frac{v_A}{\Omega_i} \hat{b} \cdot \nabla, \tag{8.484}$$

$$v_e = v_{e\parallel}/v_A, \quad j_i = j_{i\parallel}/en_0 v_A, \quad \tilde{\mu} = \frac{\mu_i}{c_s \rho_s}, \tag{8.485}$$

$$\frac{e\psi}{k_B T_e} \frac{\Omega_i}{\omega_{pi}} \to \psi, \quad v_A = B_0/(\mu_0 m_i n_0)^{1/2}, \quad \kappa_n = \rho_s/L_n, \tag{8.486}$$

$$\tilde{\eta} = \eta \Omega_e/\mu_0 v_{te}^2, \quad \delta = T_i/T_e, \quad t\Omega_i \to t \tag{8.487}$$

the fundamental equations are

$$\boxed{[\partial_t + \hat{z} \times \nabla\phi \cdot \nabla]\nabla^2\phi + \nabla_\parallel \nabla^2 \psi + \delta\kappa_n \partial_y \nabla^2 \phi - \tilde{\mu} \nabla^2 \nabla^2 \phi = 0\,,} \tag{8.488}$$

$$\boxed{-[\partial_t + \hat{z} \times \nabla\phi \cdot \nabla + v_e \nabla_\parallel] v_e + \frac{v_{te}^2}{v_A^2} \partial_t \psi - \frac{v_{te}^2}{v_A^2} \nabla_\parallel (p_e - \phi) = \frac{v_{te}^2}{v_A^2} \tilde{\eta} v_e,} \tag{8.489}$$

$$\boxed{[\partial_t + \hat{z} \times \nabla\phi \cdot \nabla] p_e - \kappa_n \partial_y \phi + \nabla_\parallel v_e + v_e \nabla_\parallel p_e = 0,} \tag{8.490}$$

$$\boxed{[\partial_t + \hat{z} \times \nabla\phi \cdot \nabla] j_i + \frac{c_s^2}{v_A^2} \nabla_\parallel \phi = 0,} \tag{8.491}$$

$$\boxed{j_i - v_e = -\nabla^2 \psi.} \tag{8.492}$$

Now, $\nabla_\parallel = \partial_z + (\nabla\psi \times \hat{z}) \cdot \nabla$. If we neglect higher orders in the terms with v_e (which, following the previous approximations, is only consistent) and introduce

$$d_t := \partial_t + \hat{z} \times \nabla\phi \cdot \nabla \tag{8.493}$$

as a new operator, then the above fundamental equations become

$$d_t \nabla^2 \phi + \nabla_\parallel \nabla^2 \psi + \delta\kappa_n \partial_y \nabla^2 \phi - \tilde{\mu} \nabla^2 \nabla^2 \phi = 0, \tag{8.494}$$

$$\frac{v_A^2}{v_{te}^2} d_t v_e - \partial_t \psi = \nabla_\parallel (\phi - p_e) - \tilde{\eta} v_e, \tag{8.495}$$

$$d_t p_e - \kappa_n \partial_y \phi + \nabla_\parallel v_e = 0, \tag{8.496}$$

$$\frac{v_A^2}{c_s^2} d_t j_i + \nabla_\| \phi = 0, \tag{8.497}$$

where

$$v_e = \nabla^2 \psi + j_i \tag{8.498}$$

applies.

We now move to the electrostatic limit, where $\psi \to 0$ approaches zero. However, we must be careful, since according to Maxwell's equations, currents give rise to magnetic fields. The electrostatic approximation means that these induced magnetic field components are very small. We formally transition to the electrostatic limit by rescaling the dimensionless quantities,

$$V_e = \frac{v_A}{v_{te}} v_e, \quad J_i = \frac{v_A}{c_s} j_i, \quad \frac{v_{te}}{v_A} \nabla_\| = \partial_z, \tag{8.499}$$

and thus write (8.492) in the form

$$V_e = \frac{v_A}{v_{te}} \nabla^2 \psi + \left(\frac{m_e}{m_i}\right)^{1/2} J_i \tag{8.500}$$

Then we let $\psi \to 0$ and $v_{te}/v_A \to 0$ approach zero, in order to obtain

$$d_t \nabla^2 \phi + \partial_z \left[V_e - \left(\frac{m_e}{m_i}\right)^{1/2} J_i \right] + \delta \kappa_n \partial_y \nabla^2 \phi - \tilde{\mu} \nabla^2 \nabla^2 \phi = 0, \tag{8.501}$$

$$d_t V_e = \partial_z (\phi - p_e) - \tilde{\tilde{\eta}} V_e, \tag{8.502}$$

$$d_t p_e - \kappa_n \partial_y \phi + \partial_z V_e = 0, \tag{8.503}$$

$$d_t J_i + \left(\frac{m_e}{m_i}\right)^{1/2} \partial_z \phi = 0 \tag{8.504}$$

Now, $\tilde{\tilde{\eta}} = (v_{te}^2/v_A^2)\tilde{\eta}$. holds. Furthermore, the rescaling simply means that we measure the (parallel) electron velocity v_e in units of v_{te},, the (parallel) ion velocity in units of c_s, and the characteristic length along the magnetic field in v_{te}/Ω_i. The remaining units remain unchanged.

The dynamics of nonlinear drift waves become more transparent if we measure *all* velocities (i.e., of ions and electrons) in units of c_s and all lengths (perpendicular and parallel to the magnetic field) in units of ρ_s. Then, from (8.501)–(8.504) we obtain

$$\boxed{d_t \nabla^2 \phi + \partial_Z (V - J) + \delta \kappa_n \partial_y \nabla^2 \phi - \tilde{\mu} \nabla^2 \nabla^2 \phi = 0\,,} \tag{8.505}$$

$$\boxed{d_t V = \frac{m_i}{m_e} \partial_Z (\phi - P) - \tilde{\tilde{\eta}} V \ ,} \tag{8.506}$$

$$\boxed{d_t P - \kappa_n \partial_y \phi + \partial_Z V = 0 \ ,} \tag{8.507}$$

$$\boxed{d_t J + \partial_Z \phi = 0 \ ,} \tag{8.508}$$

where, to distinguish from the previous set of equations, we have chosen the new variables $V = (v_{te}/c_s) V_e$, $J = J_i$, $P = p_e$ and $Z = (v_{te}/c_s) z$.

We are now in a position to directly derive well-known model equations as simplifications of the general system of equations (8.505)–(8.508). For the sake of clarity, we will omit the "tildes" over the dissipative coefficients in the following (i.e., $\tilde{\tilde{\eta}} \to \eta$ and $\tilde{\mu} \to \mu$).).

If we set $\partial_Z \equiv 0$, it follows that

$$d_t \nabla^2 \phi + \delta \kappa_n \partial_y \nabla^2 \phi - \mu \nabla^2 \nabla^2 \phi = 0 \tag{8.509}$$

is the equation for the so-called convective cells.
The Hasegawa-Mima equation is obtained by solving (8.506) to leading order in $\phi \approx P$ and then subtracting (8.507) from (8.505):

$$d_t (1 - \nabla^2) \phi - \kappa_n (1 + \delta \nabla^2) \partial_y \phi + \mu \nabla^2 \nabla^2 \phi = 0. \tag{8.510}$$

This formulation contains only damping terms. To capture the instability mechanism due to collisions ("collisional drift instability"), we proceed as follows. To lowest order, we still use $P \approx \phi$; to introduce corrections, but we solve (8.506) approximately and substitute into (8.507):

$$V \approx \frac{1}{\eta} \frac{m_i}{m_e} \partial_z (\phi - P) \approx 0, \tag{8.511}$$

$$\partial_t \phi - \kappa_n \partial_y \phi + \frac{1}{\eta} \frac{m_i}{m_e} \partial_Z^2 (\phi - P) \approx 0. \tag{8.512}$$

From the last equation we obtain

$$P \approx \phi - \frac{\eta m_e}{k_\parallel^2 m_i} (\partial_t - \kappa_n \partial_y) \phi. \tag{8.513}$$

Of course, the transition to (8.513) is only approximate, since a Fourier decomposition can only be meaningfully assumed for the linear relationships. The wavenumber $k_\parallel$ represents an average characteristic z-dependence. [A similar procedure is used for collisionless damping, when $\delta n_e \approx n_{e0}\left(\dfrac{e\phi}{k_B T_e}\right)(1-\delta_k)$ is assumed, where δ_k is obtained from the linearized Vlasov equation.]

If we insert (8.513) into (8.507), then subtract (8.505) to eliminate $\partial_Z V$, and set $J \equiv 0$, i.e., do not consider ion-acoustic modes, we finally obtain

$$\partial_t\left[1-\nabla^2-\frac{\eta m_e}{k_\parallel^2 m_i}(\partial_t-\kappa_n\partial_y)\right]\phi$$

$$-\kappa_n\partial_y\phi-\delta\kappa_n\partial_y\nabla^2\phi+\mu\nabla^2\nabla^2\phi=(z\times\nabla\phi)\cdot\nabla\nabla^2\phi. \tag{8.514}$$

Compared to (8.510), this equation contains the excitation mechanism (via the instability rate) for drift waves, which makes it particularly well suited for self-consistent calculations.

Because of the fundamental importance of the Hasegawa-Mima equation (8.510), which is usually discussed for $\mu=\delta=0$ in the form

$$\boxed{\partial_t(1-\nabla^2)\phi-\kappa_n\partial_y\phi-(\hat{z}\times\nabla\phi)\cdot\nabla\nabla^2\phi=0} \tag{8.515}$$

we will now provide a shortened derivation, which has the advantage over the deduction from the general model in that the essential physical assumptions become more transparent. We then discuss dipole solutions of (8.515).

Example 8.14 (Hasegawa-Mima Equation)

Under the assumptions of quasineutrality, $n_e \approx n_i$, and Boltzmann-distributed electrons,

$$n_e \approx n_0 \exp(e\varphi/k_B T_e), \tag{8.516}$$

we can very quickly obtain the Hasegawa-Mima equation from the particle balance and the drift approximation for ions. With

$$\frac{d}{dt}=\partial_t+\mathbf{v}_i\cdot\nabla \tag{8.517}$$

and a purely two-dimensional model (sometimes also referred to as a $2\frac{1}{2}$-dimensional description, since the Boltzmann distribution of the electrons arises due to their rapid motion in the third dimension along the magnetic field), we write the particle balance for the ions as

$$\frac{d}{dt}\ln n_i+\nabla_\perp\cdot\mathbf{v}_{i\perp}\approx 0. \tag{8.518}$$

The velocity of the ions in the plane perpendicular to the external magnetic field is approximated by

$$\mathbf{v}_{i\perp} \approx \frac{\hat{z} \times \nabla\varphi}{B_0} - \frac{1}{B_0 \Omega_i}\left(\partial_t + \frac{\hat{z} \times \nabla\varphi}{B_0} \cdot \nabla_\perp\right)\nabla_\perp \varphi. \tag{8.519}$$

The combination of the last two equations leads to

$$\partial_t\left(\frac{1}{\rho_s^2} - \nabla_\perp^2\right)\varphi - u\partial_\eta\left[\left(\frac{1}{\rho_s^2} + \frac{\Omega_i \kappa_n}{u}\right)\varphi\right]$$

$$+ u\partial_\eta \nabla^2\varphi - \frac{1}{B_0}(\hat{z} \times \nabla\varphi) \cdot \nabla_\perp \nabla_\perp^2 \varphi + \frac{\kappa_T}{\rho_s^2 B_0}\varphi\partial_\eta\varphi = 0, \tag{8.520}$$

where we have directly switched to a moving coordinate system $\eta = y - ut$. Here, $\kappa_T = \partial_x \ln T_e(x)$ is the contribution of a temperature gradient, which we have not yet taken into account. For $u = \kappa_T = 0$, (8.520) agrees with (the equation written in dimensionless variables) Eq. (8.515). ■

The problem of temperature dependence will not be treated in full here; the following brief remarks should suffice. If $\kappa_T \neq 0$ holds, i.e., $T_e = T_e(x)$, it seems natural to require $\rho_s = \rho_s(x)$ and thus to end up with a rather complicated model equation. At first glance, $\kappa_T \neq 0$ and $\rho_s = const$ may appear inconsistent. However, a systematic analysis based on a multiple-scale formalism explains that this is actually not the case for most practical applications. All calculations with $\kappa_T \neq 0$ and $\rho_s = const$ are valid if the variation of temperature occurs on a length scale that is large compared to ρ_s. A systematic analysis then shows for $T_e = T_e(\varepsilon x)$, that (8.520), including the last term on the left-hand side with $\kappa_T = const \neq 0$ and $\rho_s = const$ correctly describes the variation of the potential on the ρ_s scale. [A similar problem, incidentally, also arises in the case of density inhomogeneity alone. Even then, we may choose $\kappa_n = const \neq 0$ and $n_{e0} = const$ in the coefficients, provided the density varies on a scale that is large compared to ρ_s.]

Equation (8.520) contains two types of nonlinearities: the so-called scalar nonlinearity $\sim \varphi\partial_y\varphi$ and the so-called vector nonlinearity $\sim (\hat{z} \times \nabla\varphi) \cdot \nabla\nabla^2\varphi$. The former dominates for broad structures and leads to monopole vortices, while the latter prevails for variations on the ρ_s scale and is the reason for dipole vortices.

We now conclude by considering solutions of the Hasegawa-Mima equation

$$\partial_t(1 - \nabla^2)\phi - \{\phi, \nabla^2\phi\} + u\frac{\partial}{\partial\eta}\nabla^2\phi - u\left(1 + \frac{\kappa_n}{u}\right)\frac{\partial}{\partial\eta}\phi = 0 \tag{8.521}$$

in a moving coordinate system.

The Poisson bracket is defined as

$$\{a, b\} = \frac{\partial a}{\partial x}\frac{\partial b}{\partial \eta} - \frac{\partial a}{\partial \eta}\frac{\partial b}{\partial x} \tag{8.522}$$

For stationary solutions $\phi = \phi_s$ (in the coordinate system moving with velocity u), we set $\partial_t = 0$.. The stationary form of Eq. (8.521) can also be written in the form

$$\boxed{\hat{z} \times \nabla(\phi_s - ux) \cdot \nabla\big[(1 - \nabla^2)\phi_s + \kappa_n x\big] = 0} \tag{8.523}$$

Obviously, this equation is solved by

$$(1 - \nabla^2)\phi_s + \kappa_n x = F(\phi_s - ux) \tag{8.524}$$

if F is any (but differentiable) function of its argument. Depending on the choice of this function, there may be different solutions. In the following case, analytical solutions can even be given.

For localized solutions (ϕ_s, $\nabla^2\phi_s \to 0$ for $x, \eta \to \pm\infty$), (8.524) yields

$$\kappa_n x \simeq F(-ux), \quad x, \eta \to \pm\infty. \tag{8.525}$$

[Incidentally, this also holds for $\eta \to \pm\infty$ and finite x.] We now require (8.525) for the entire domain $r = (x^2 + \eta^2)^{1/2} > a$, where a is an (arbitrary) radius at which the outer solution is later matched to the inner solution:

$$F(\xi) \equiv F_>(\xi) = -\frac{\kappa_n}{u}\xi \qquad \text{for} \quad r > a. \tag{8.526}$$

In the inner region $r < a$, any other dependence $F_<(\xi)$ can be chosen for F, where $F_<(a) \neq F_>(a)$ may apply. The simplest choice is

$$F(\xi) \equiv F_<(\xi) = d\xi \qquad \text{for} \qquad r < a\,; \tag{8.527}$$

d is a constant yet to be determined. With the choice (8.526) or (8.527), one arrives at the determining equations

$$\nabla^2\phi_s = \begin{cases} \left(1 + \frac{\kappa_n}{u}\right)\phi_s, & r > a, \\ (1 - d)\phi_s + (\kappa_n + ud)x, & r < a. \end{cases} \tag{8.528}$$

Solutions of these equations are localized if

$$\rho^2 := 1 + \frac{\kappa_n}{u} > 0 \tag{8.529}$$

holds. For $r \to 0$ the solution has negative curvature if

$$-\lambda^2 = 1 - d < 0 \tag{8.530}$$

is satisfied. Also, ρ^2 and λ^2 are as κ_n, u and d are constants. In polar coordinates $x = r\cos\theta$, $\eta = r\sin\theta$ the corresponding solution is

$$\phi_s = \begin{cases} \dfrac{u+\kappa_n}{\rho^2} a \dfrac{K_1(\rho r)}{K_1(\rho a)} \cos\theta, & r > a, \\ -\dfrac{u+\kappa_n}{\lambda^2}\left[a\dfrac{J_1(\lambda r)}{J_1(\lambda a)} - \left(1 + \dfrac{\lambda^2}{\rho^2}\right) r \right] \cos\theta, & r < a. \end{cases} \tag{8.531}$$

J_1 and K_1 are the Bessel function and the modified Bessel function of the first order, respectively. The prefactors are already chosen such that ϕ_s and $\nabla^2\phi_s$ are continuous at the point $r = a$. In order for the first derivatives to also satisfy the continuity requirement, the "dispersion relation"

$$-\frac{1}{\lambda a}\frac{J_2(\lambda a)}{J_1(\lambda a)} = \frac{1}{\rho a}\frac{K_2(\rho a)}{K_1(\rho a)} \tag{8.532}$$

must be satisfied. For given values of a, κ_n and u, (8.532) together with (8.529) and (8.530) determines the constant d. Apparently, bands of solutions exist; ground states are defined for

$$j_{1,1} \le \lambda a \le j_{2,1} \tag{8.533}$$

when ρa varies within the range

$$0 \le \rho a \le \infty \tag{8.534}$$

Here, $j_{m,1}$ denotes the first zero of the Bessel function of order mJ_m. Fig. 8.7 shows a 3D representation. Because of their special shape, these solutions are called dipoles or "modons." It should be noted that, by construction, these vortex solutions are twice continuously differentiable. Already the third derivative exhibits a finite jump in the radial direction at the point $r = a$.

The velocities are constrained by (8.529). Obviously, in the (complementary to the phase velocities of linear drift waves $0 > \omega/k > -\kappa_n$) regions

$$u > 0 \tag{8.535}$$

and

$$u < -\kappa_n < 0 \tag{8.536}$$

nonlinear solutions are possible. Note the choice $\kappa_n = \partial_x \ln n_{e0} > 0$.

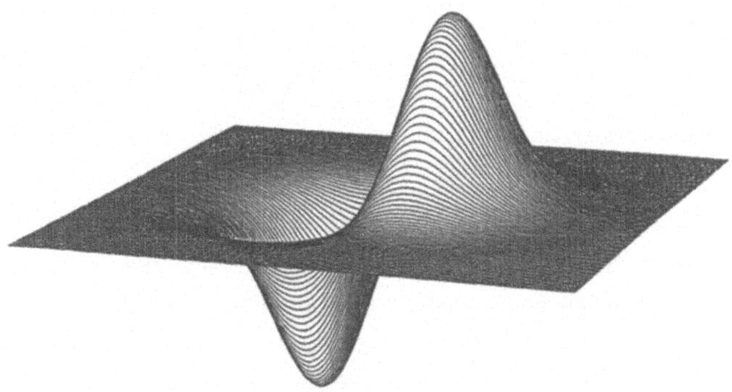

Fig. 8.7 $3D$ plot of a modon as a solution of the Hasegawa-Mima equation

Linear stability analyses lead to the result of stability for $\rho^2 < 1$, i.e., $u < -\kappa_n < 0$. In atmospheric physics, $\kappa_n < 0$ is assumed, so that we have $u < 0$ or $u > -\kappa_n > 0$ as existence domains. The stable solutions with $\rho^2 < 1$ correspond to $u > -\kappa_n > 0$, i.e., westward-propagating dipoles.

Several remarks are appropriate at this point.

The strict coupling between electron density and potential via the Boltzmann distribution is generally quite a good assumption for determining the potential distribution, but is not always appropriate for calculating the density from the potential profile. In particular, when determining plasma transport in the presence of nonlinear drift waves, better models (or iterations of existing models) must be used in order to obtain a finite contribution from drift wave turbulence.

Numerical simulations show both monopole and dipole vortices. For temperature gradients [cf. (8.456)], a scalar nonlinearity appears in the basic equations, which can be held responsible for the existence of monopoles.

The question of stability then becomes important, whereby we must distinguish between two fundamentally different types of stability analysis: first, the so-called Lyapunov stability with respect to perturbations in the initial distributions, and second, the so-called structural stability with respect to structural perturbations in the basic equations themselves. The stability statements mentioned so far apply, for example, to Hasegawa-Mima dipoles under perturbations in the initial distribution.

However, if we add structural perturbations in the form of a scalar nonlinearity (for finite temperature gradients) to the Hasegawa-Mima equation, the dipoles decay, i.e., they are structurally unstable. The monopole solutions that are then possible, however, are stable!

Finally, it should be noted that in atmospheric physics (whether of the Earth or Jupiter) very similar basic equations arise, leading to similar theoretical investigations as in the case of nonlinear drift waves. A long-observed, quite stable vortex is Jupiter's Great Red Spot, which should also have its analogue in terrestrial plasma systems.

8.4 BGK Modes

We conclude this overview of fundamental models of nonlinear waves with a brief discussion of typical kinetic effects and their impact on mode structures. In doing so, we rely on a classic work by Bernstein, Greene, and Kruskal [BGK] on periodic potential distributions in Vlasov-Poisson systems.

BGK modes could represent a special type of "dissipation-free wave solution" in a plasma. It turns out that there are stationary, nonlinear waves in a plasma that can be constructed from a specific distribution function. The BGK solutions describe electrostatic potential waves in which electrons and ions are distributed in such a way that a self-consistent, stationary structure arises. These modes would be remarkable in that they should exhibit no Landau damping and could be stable in the long term.

BGK modes should evade Landau damping because they are stationary, self-consistent, nonlinear solutions in which the velocity distribution adjusts itself so that there is no effective net energy transfer from the wave to the particles. In particular, the trapping of particles in the potential waves should prevent the classical Landau resonance.

A simple self-consistent solution
The procedure is quickly outlined for the simplest case. We restrict ourselves to a spatially one-dimensional model with local variations only in the x-direction. In the single-particle distribution function, we imagine an integration over the v_y- and v_z-dependencies; we call this $v_x = v$. Then, the Vlasov equation for stationary solutions of particle species s reads

$$\boxed{\left(v\partial_x - \frac{q_s}{m_s}\frac{d\varphi}{dx}\partial_v\right)f_s(x, v) = 0}\,. \tag{8.537}$$

The potential φ must satisfy the Poisson equation

$$\boxed{\frac{d^2\varphi}{dx^2} = \frac{1}{\varepsilon_0} e\left[\int_{-\infty}^{\infty} dv\, f_e(x,v) - \int_{-\infty}^{\infty} dv\, f_i(x,v)\right]} \tag{8.538}$$

where we have assumed a two-component electron-ion system ($s = e, i$). Equations (8.537) and (8.538) form a coupled system. However, the solutions of (8.537) can, in general, be written as

$$f_s = f_s\left[v^2 + 2(q_s/m_s)\varphi(x)\right] \tag{8.539}$$

where the functional form is still undetermined. If we choose two solutions for $s = e, i$ with initially arbitrary functions f_e and f_i and substitute them into (8.538), we obtain the integro-differential equation

$$\frac{d^2\varphi}{dx^2} = \frac{1}{\varepsilon_0} e\left\{\int_{-\infty}^{+\infty} dv\, f_e[v^2 - 2e\varphi(x)/m_e] - \int_{-\infty}^{+\infty} dv\, f_i[v^2 + 2e\varphi(x)/m_i]\right\}. \tag{8.540}$$

We have not yet specified the boundary conditions. We now set them to be periodic. [Alternatively, one could also search for soliton-like states.] To make Eq. (8.540) more tractable, we will focus on a further simplification in the following. We take f_e as the distribution of a sharp electron beam,

$$f_e(x,v) = 2n_{e0}v_e\delta\left[v^2 - 2e\varphi(x)/m_e - v_e^2\right], \tag{8.541}$$

where v_e is a constant. For v we further restrict ourselves to positive values only, i.e., when solving for v in the argument of the delta function, we choose the positive root. Another remark, which is of utmost physical importance, deserves special attention. If we

$$v_e^2 > -\min_x\left\{\frac{2e\varphi(x)}{m_e}\right\} \tag{8.542}$$

require this, we exclude particle trapping.

$$v = +\left[v_e^2 + 2e\varphi(x)/m_e\right]^{1/2} \equiv v_{e0} \tag{8.543}$$

is the relevant zero of the argument in the delta function (8.541). Because

$$\delta(v^2 - v_{e0}^2) = \frac{\delta(v - v_{e0})}{2v_{e0}} \tag{8.544}$$

it follows that

$$f_e(x,v) = n_{e0}\frac{v_e}{v_{e0}}\delta(v - v_{e0})\,. \tag{8.545}$$

We proceed similarly for ions:

$$f_i(x, v) = 2n_{i0}v_i\delta\left[v^2 + 2e\varphi(x)/m_i - v_i^2\right], \tag{8.546}$$

$$v_i^2 > \max_x\left\{\frac{2e\varphi(x)}{m_i}\right\}, \tag{8.547}$$

$$v_{i0} \equiv +\left[v_i^2 - 2e\varphi(x)/m_i\right]^{1/2} \tag{8.548}$$

and thus

$$f_i(x, v) = n_{i0}\frac{v_i}{v_{i0}}\delta(v - v_{i0}). \tag{8.549}$$

In the following, we set $n_{i0} = n_{e0} = n_0$. From (8.545) and (8.549) we can directly obtain the density distributions

$$n_e(x) = n_0\frac{v_e}{v_{e0}(x)}, \tag{8.550}$$

$$n_i(x) = n_0\frac{v_i}{v_{i0}(x)} \tag{8.551}$$

The Poisson equation (8.540) becomes

$$\boxed{\frac{d^2\varphi}{dx^2} = \frac{1}{\varepsilon_0}n_0e\left\{\left[1 + \frac{2e\varphi(x)}{m_e v_e^2}\right]^{-1/2} - \left[1 - \frac{2e\varphi(x)}{m_i v_i^2}\right]^{-1/2}\right\}}. \tag{8.552}$$

We introduce the abbreviations $\tau_e = m_e v_e^2,\ \tau_i = m_i v_i^2,\ \phi = e\varphi$ and $X = (n_0 e/\varepsilon_0)^{1/2}x$ in order to write (8.552) in the simplified form

$$\frac{d^2\phi}{dX^2} = \left[1 + \frac{2\phi}{\tau_e}\right]^{-1/2} - \left[1 - \frac{2\phi}{\tau_i}\right]^{-1/2} \tag{8.553}$$

In analogy to the Newtonian motion of a particle in a potential, we define the pseudopotential $V(\phi)$,

$$\frac{d^2\phi}{dX^2} = -\frac{\partial V(\phi)}{\partial\phi} \tag{8.554}$$

with

$$V(\phi) = -\left\{\tau_e\left[1 + \frac{2\phi}{\tau_e}\right]^{1/2} + \tau_i\left[1 - \frac{2\phi}{\tau_i}\right]^{1/2}\right\}. \tag{8.555}$$

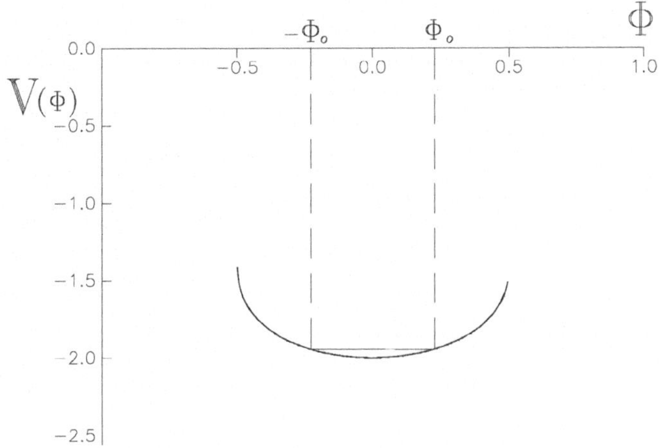

Fig. 8.8 Plot of (8.555) as a function of ϕ

This potential is shown for $\tau_e = \tau_i = 1$ in Fig. 8.8. In this example, particle trapping is only possible for $-\frac{1}{2} \leq \phi \leq \frac{1}{2}$. With a maximum amplitude $\phi_0 < \frac{1}{2}$, oscillations, i.e., periodic solutions $\varphi(x)$, are possible. The amplitude ϕ_0 is determined by the integration constant, which we may refer to as the "pseudoenergy." For small amplitudes, (8.552) becomes

$$\frac{d^2\varphi}{dx^2} + \frac{1}{\varepsilon_0} n_0 e^2 \left(\tau_e^{-1} + \tau_i^{-1}\right)\varphi \approx 0\,, \tag{8.556}$$

and the solution is a harmonic oscillation.

Trapped particles

We now turn to the problem of calculating distribution functions for trapped particles. Figure 8.9 illustrates, for a given potential distribution $\varphi(x)$, regions of trapped ions and electrons. It already becomes apparent that a discussion in terms of energy is more meaningful. With

$$E = \frac{1}{2} m_s v^2 + q_s \varphi \tag{8.557}$$

and

$$dE = m_s v dv = \left[2m_s(E - q_s\varphi)\right]^{1/2} dv \tag{8.558}$$

we have

$$f_s(v)dv = f_s[v(E)]dv = \frac{f_s(E)dE}{[2m_s(E - q_s\varphi)]^{1/2}}\,. \tag{8.559}$$

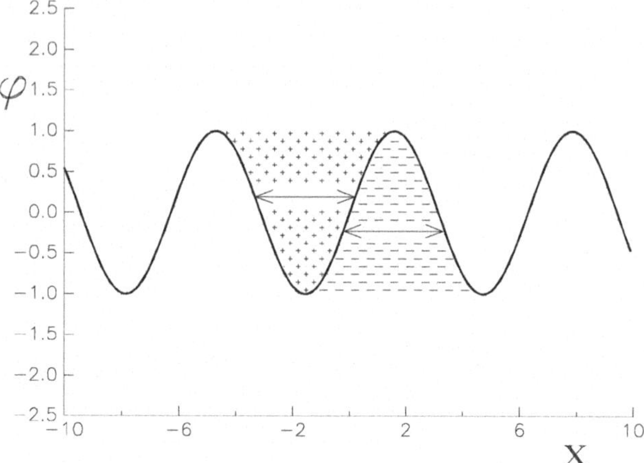

Fig. 8.9 Regions of trapped ions and electrons in a given potential distribution $\varphi(x)$

Equation (8.558) applies for particles with $v > 0$. If we assume an equal number of particles with negative and positive velocity, we can calculate the density using

$$n_s(x) = \int_{-\infty}^{+\infty} dv\, f_s(x, v) = 2\int_0^{\infty} dv\, f_s(x, v)$$
$$= 2\int_{q_s\varphi}^{\infty} dE \frac{f_s(x, E)}{[2m_s(E - q_s\varphi)]^{1/2}} \tag{8.560}$$

The lower limit $q_s\varphi$ arises from the fact that no particle can have an energy less than $q_s\varphi$; the kinetic energy is, after all, positive definite.

Figure 8.9 shows that every ion ($q_i = +e$) with an energy E less than $e\varphi_{max}$ is trapped between neighboring potential maxima. Because of $q_e = -e$ the electrons "see" the potential mirrored, i.e., electrons with an energy $E < -\varphi_{min}$ are trapped between neighboring potential minima. Now let us assume that the entire ion distribution $f_i(E)$ is given, as is the distribution of "free" electrons $f_e(E)$ for $E > -e\varphi_{min}$. Then the Poisson equation requires

$$\frac{d^2\varphi}{dx^2} = -\frac{e}{\varepsilon_0}\int_{e\varphi}^{\infty} dE \frac{2f_i(E)}{[2m_i(E - e\varphi)]^{1/2}} + \frac{e}{\varepsilon_0}\int_{-e\varphi_{min}}^{\infty} dE \frac{2f_e(E)}{[2m_e(E + e\varphi)]^{1/2}}$$
$$+ \frac{e}{\varepsilon_0}\int_{-e\varphi}^{-e\varphi_{min}} dE \frac{2f_e(E)}{[2m_e(E + e\varphi)]^{1/2}}\,. \tag{8.561}$$

The last term on the right-hand side contains $f_e(E)$ for $E < -e\varphi_{min}$,, i.e., the distribution for trapped electrons. We write Eq. (8.561) in the form

$$\int_{-e\varphi}^{-e\varphi_{min}} dE\ 2f_e(E)[2m_e(E + e\varphi)]^{-1/2} = g(e\varphi) \tag{8.562}$$

around, where

$$g(e\varphi) := -\varepsilon_0 N(e\varphi)/e + \int_{e\varphi}^{\infty} dE\ 2f_i(E)[2m_i(E - e\varphi)]^{-1/2}$$
$$- \int_{-e\varphi_{min}}^{\infty} dE\ 2f_e(E)[2m_e(E + e\varphi)]^{-1/2} \tag{8.563}$$

for a given $\varphi(x)$ [and f_i or f_e for $E > -e\varphi_{min}$] is a known inhomogeneity. $g(e\varphi)$ denotes the density of the trapped electrons at the location x; $N(e\varphi)$ is determined by $-d^2\varphi/dx^2$. Equation (8.562) is an integral equation for $f_e(E)$, which can be solved by differentiation. With the variable substitutions

$$E = \tilde{E} - e\varphi_{min}\ , \tag{8.564}$$

$$x = e(\varphi - \varphi_{min})\ , \tag{8.565}$$

$$y = -\tilde{E} \tag{8.566}$$

we arrive at

$$\int_0^x dy\ 2f_e(-y - e\varphi_{min})(2m_e)^{-1/2} \frac{1}{\sqrt{x - y}} = g(x + e\varphi_{min}). \tag{8.567}$$

Example 8.15 (Abelian Integral Equation)
We now have the classical form of an Abelian integral equation (or Volterra integral equation of the first kind), whose kernel becomes infinite for $y = x$. We multiply both sides of (8.567) by $1/\sqrt{\eta - x}$ and integrate from 0 to η with respect to x,

$$\int_0^\eta \frac{1}{\sqrt{\eta - x}} \left[\int_0^x dy \frac{2f_e(-y - e\varphi_{min})}{\sqrt{x - y}} \right] dx = (2m_e)^{1/2} \int_0^\eta \frac{g(x + e\varphi_{min})}{\sqrt{\eta - x}} dx. \tag{8.568}$$

The double integration on the left-hand side is to be carried out such that integration is first performed in the y-direction from 0 to x. The domain of integration of the double integral is therefore the triangle above the diagonal $x = y$ in the y,x plane. If the order of integration is reversed, one first integrates in the x-direction from $x = y$ to $x = \eta$ and then in the y-direction from $y = 0$ to $y = \eta$.. Taking into account

$$\int_y^\eta \frac{dx}{\sqrt{\eta - x}\sqrt{x - y}} = \pi, \tag{8.569}$$

it follows, if we replace η by y,

$$\int_0^y 2f_e(-y' - e\varphi_{min})dy' = \frac{(2m_e)^{1/2}}{\pi} \int_0^y \frac{g(x + e\varphi_{min})}{\sqrt{y - x}} dx. \tag{8.570}$$

On the right-hand side, we integrate by parts, making use of $g(e\varphi_{min}) = 0$. We then differentiate with respect to y and set $E = -y - e\varphi_{min}$, to obtain, with $X = x + E\varphi_{min}$,

$$2f_e(E) = \frac{(2m_e)^{1/2}}{\pi} \int_{e\varphi_{min}}^{-E} dX \frac{dg(X)}{dX} \frac{1}{\sqrt{-E-X}} \tag{8.571}$$

as the solution. ■

This proves that, for a given potential distribution, with known f_i (total) and f_e (for the free electrons), the distribution function of the captured electrons can be determined self-consistently.

If one substitutes (8.563) into (8.571), a few more integrals can be carried out, and the final result is

$$\begin{aligned} f_e(E) &= \frac{(2m_e)^{1/2}}{8\pi^2 e} \int_{e\varphi_{min}}^{-E} \frac{dX}{[-E-X]^{1/2}} \frac{dN(X)}{dX} \\ &+ \frac{1}{\pi} \int_{-e\varphi_{min}}^{\infty} dX \frac{f_e(X)}{X-E} \left[\frac{e\varphi_{min} - E}{X - e\varphi_{min}}\right]^{1/2} \\ &+ \frac{1}{\pi}\left(\frac{m_e}{m_i}\right)^{1/2} \int_{e\varphi_{min}}^{\infty} dX \frac{df_i(X)}{dX} \frac{1}{2} \ln\left\{\frac{(E+X)^2}{[(X-e\varphi_{min})^{1/2} - (-E-e\varphi_{min})^{1/2}]^4}\right\} \end{aligned} \tag{8.572}$$

for $E < -e\varphi_{min}$. Accordingly, one could of course also determine the distribution function for trapped ions.

Finally, we discuss how, using thc method proposed by Bernstein, Greene, and Kruskal, one can in practice construct any potential distribution that possesses a continuous second derivative. We specify an arbitrary distribution, for example the one shown in Fig. 8.10. The x-axis is divided into segments (*AB*,*BC*,*CD*, etc.), with the endpoints characterized by the locations where $d\varphi/dx = 0$. In each segment, φ is monotonic. Now let us assume that for $x \leq B$ we have constructed consistent distribution functions f_e and f_i. In the segment *BC* (since there is supposed to be a potential minimum at *B*), the distribution of *trapped electrons does not* have to match the distribution of trapped electrons in *AB*, whereas the energy dependencies of the distribution functions for *all ions* and the *free electrons* must agree in the segments *BC* and *AB*. The distribution $f_e(E)$ for trapped electrons in *BC* can be obtained from the other data using the procedure described above. If we then proceed to the segment *CD*, the construction process is repeated, except that now all distribution functions except that of the trapped ions can be carried over from *BC*. For the trapped ions in *CD*, we can also apply the construction method for trapped particles described above

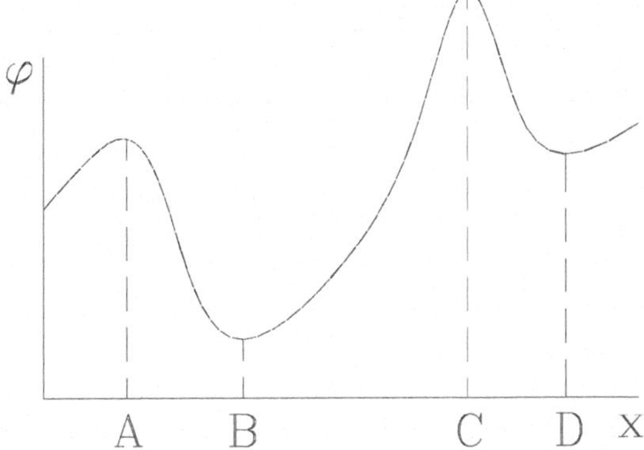

Fig. 8.10 Sketch of a potential distribution with regions of trapped particles

(for f_i instead of f_e). In this way, the distribution functions can be consistently determined for ever larger values of x.

The "message" of this sketch of BGK modes is that plasmas could form coherent, nonlinear wave structures that exist without dissipation and are maintained as equilibria between particle distribution and electric fields. The question of the stability of the BGK modes constructed in this way naturally arises at this point. However, it is difficult to answer and, for the most part, still unresolved. We cannot go into further detail here.

8.5 Collapse in NLS and KdV Systems

In the preceding sections, we have demonstrated the integrability of the cubic nonlinear Schrödinger (NLS) equation and the Korteweg–de Vries equation. However, we must point out a significant limitation of both integrable model equations: Nonlinear Langmuir waves and ion-acoustic modes have only been studied in spatially *one*-dimensional systems. In spatially two- and three-dimensional systems, there is a fundamental difference from the phenomena observed in one dimension. Stable solitary waves can still exist, but in general it is no longer guaranteed that fairly arbitrary initial distributions will evolve into stable soliton-like states. A new phenomenon appears, namely collapse, which we will examine in more detail in this section.

Collapse in NLS Systems

We begin with the multidimensional NLS equation in its simplest form,

$$\boxed{i\partial_t\psi + \nabla^2\psi + |\psi|^p\psi = 0\,,} \tag{8.573}$$

where ∇^2 is to be the Laplace operator in d spatial dimensions. Previously, we have studied the NLS equation for $p = 2$ in the one-dimensional case ($d = 1$). The generalization to the exponent p is to be made here for later similarity considerations. One could even go further and replace $|\psi|^p$ by a function $f(|\psi|^2)$ to be specified later; for much of the following discussion, this would be possible without significant restrictions. However, for the sake of clarity, we will refrain from doing so here.

Conserved Quantities
For a discussion of the solution manifold of (8.573) in d dimensions, it is helpful to determine symmetries and conserved quantities. For this purpose, we make use of Noether's theorem for Lagrangian systems. For (8.573), the action integral is given by

$$S\{\psi\} = \int \mathcal{L}(\psi, \psi_t, \psi_x, \psi_y, \ldots)dt\, d^d r \tag{8.574}$$

with the Lagrangian density

$$\boxed{\mathcal{L} = \frac{i}{2}(\psi^*\psi_t - \psi\psi_t^*) - |\nabla\psi|^2 + \frac{2}{p+2}|\psi|^{p+2}}\;; \tag{8.575}$$

here, $d^d r \equiv d\Gamma$ is the d-dimensional volume element. $\psi = v + iw$ is a complex-valued function; therefore, (8.573) can be regarded as a combination of two real equations. Since in $S\{\psi\} \widehat{=} S\{v, w\}$ the functions v and w with ψ and ψ^* are related via $v = (\psi + \psi^*)/2$ and $w = (\psi - \psi^*)/2i$, we can formally treat ψ and ψ^* as independent variables and compute the variations of the action via

$$\frac{\delta S}{\delta v} = \frac{\delta S}{\delta\psi} + \frac{\delta S}{\delta\psi^*}, \tag{8.576}$$

$$\frac{\delta S}{\delta w} = i\left(\frac{\delta S}{\delta\psi} - \frac{\delta S}{\delta\psi^*}\right) \tag{8.577}$$

calculate. Here, the functional derivative is evaluated via

$$\frac{\delta S}{\delta v} = \frac{\partial\mathcal{L}}{\partial v} - \frac{\partial}{\partial x_j}\frac{\partial\mathcal{L}}{\partial v/\partial x_j} \tag{8.578}$$

with a summation over j, which depends on the dimension d (for $x_0 = t, x_1 = x, x_2 = y, x_3 = z$). This procedure, incidentally, demonstrates the equivalence between the systems $\delta S/\delta v = 0$ and $\delta S/\delta w = 0$ on the one hand, and $\delta S/\delta\psi$ and $\delta S/\delta\psi^*$ on the other.

Example 8.16 (Noether's Theorem)
Noether's theorem states: If the functional $S\{\psi\}$ is invariant under the transformations

$$x_i' = x_i + \sum_k \varepsilon_k \varphi_i^{(k)}(x_0, x_1, \ldots, x_d, \psi, \partial_{x_0}\psi, \partial_{x_1}\psi, \ldots), \tag{8.579}$$

$$\psi' = \psi + \sum_k \varepsilon_k \chi^{(k)}(x_0, x_1, \ldots, x_d, \psi, \partial_{x_0}\psi, \partial_{x_1}\psi, \ldots) \tag{8.580}$$

with infinitesimally small parameters ε_k, then the following holds

$$\sum_{j=0}^{d} \frac{\partial I_j^{(k)}}{\partial x_j} = 0 \tag{8.581}$$

for

$$\begin{aligned} I_j^{(k)} &= \frac{\partial \mathcal{L}}{\partial(\partial\psi/\partial x_j)} \left[\sum_{l=0}^{d} \frac{\partial\psi}{\partial x_l}\varphi_l^{(k)} - \chi^{(k)} \right] \\ &+ \frac{\partial \mathcal{L}}{\partial(\partial\psi^*/\partial x_j)} \left[\sum_{l=0}^{d} \frac{\partial\psi}{\partial x_l}\varphi_l^{(k)} - \chi^{(k)} \right]^* - \mathcal{L}\varphi_j^{(k)}. \end{aligned} \tag{8.582}$$

■

Equation (8.581) has the form of a continuity equation $\partial_t I_0 + \nabla\cdot\mathbf{I} = 0$. If we assume that $I_j^{(k)}$ vanishes at the boundary (e.g., at infinity) for $j = 1, \ldots, d$, then integrating (8.581) yields the conservation law

$$\frac{d}{dt}\int I_0^{(k)} d\Gamma = 0. \tag{8.583}$$

Example 8.17 (Plasmon Conservation)
Since $\mathcal{L}$ [see (8.575)] is gauge invariant, i.e., unchanged under infinitesimal translations in the phase

$$\psi' = \psi e^{i\varepsilon} \approx \psi + \varepsilon i\psi \tag{8.584}$$

we obtain

$$\partial_t \mathcal{N} + \nabla \cdot \mathbf{J} = 0, \tag{8.585}$$

$$\mathcal{N} = |\psi|^2, \tag{8.586}$$

$$\mathbf{J} = i(\psi \nabla \psi^* - \psi^* \nabla \psi), \tag{8.587}$$

that is, plasmon conservation $N = \int \mathcal{N} d\Gamma = const.$ ■

Next, we examine translations in time and space,

$$x_j = x_j + \varepsilon \delta_{ji}. \tag{8.588}$$

Example 8.18 (Conservation of Energy)
For $i = 0$, i.e., shifts in time, energy conservation follows:

$$\partial_t \mathcal{H} + \nabla \cdot \mathbf{Q} = 0, \tag{8.589}$$

$$\mathcal{H} = |\nabla \psi|^2 - \frac{2}{p+2} |\psi|^{p+2}, \tag{8.590}$$

$$\mathbf{Q} = -\psi_t \nabla \psi^* - \psi_t^* \nabla \psi, \tag{8.591}$$

i.e., $H = \int \mathcal{H} d\Gamma = const.$ ■

Example 8.19 (Conservation of Momentum)
For $i = 1, 2$ or 3, i.e., spatial translations, momentum conservation results:

$$\partial_t \mathcal{P} + \nabla \cdot \underline{\underline{\mathcal{W}}} = 0, \tag{8.592}$$

$$\mathcal{P} = \frac{i}{2}(\psi^* \nabla \psi - \psi \nabla \psi^*), \tag{8.593}$$

$$\underline{\underline{\mathcal{W}}} = -2(\nabla \psi)(\nabla \psi^*) - \mathcal{L}\underline{\underline{1}}, \tag{8.594}$$

i.e., $\mathbf{P} = \int \mathcal{P} d\Gamma = const$; $\underline{\underline{1}}$ is the unit tensor. ■

For rotations in three-dimensional space, one can similarly deduce the invariance of angular momentum

$$\mathbf{M} = \int \mathbf{r} \times \mathcal{P} d\Gamma \tag{8.595}$$

It should also be noted that the NLS equation possesses further symmetries (with respect to Galilean transformations and rotations), which, however, do not leave the action integral invariant and therefore do not, according to Noether's theorem, lead to additional conserved quantities.

Example 8.20 (Conformal Invariance)
In the so-called critical case $pd = 4$ there exists the (conformal) invariance

$$t' = \beta - \frac{1}{\alpha^2 t}, \ \mathbf{r}\,' = \frac{\mathbf{r}}{\alpha t}, \tag{8.596}$$

$$\psi' = (\alpha t)^{d/2} \psi \exp(-i|\mathbf{r}|^2/4t), \tag{8.597}$$

where α and β are constants. If we set $\beta = 0$ and insert into

$$H = \int \left[|\nabla' \psi'|^2 - \frac{2}{p+2} |\psi'|^{p+2} \right] d\Gamma' = const \tag{8.598}$$

the transformations (8.596) and (8.597), we obtain

$$\int \left[|\mathbf{r}\psi + 2it\nabla\psi|^2 - \frac{8t^2}{p+2} |\psi|^{p+2} \right] d\Gamma = const\ . \tag{8.599}$$

■

It is interesting to write the transformations (8.596) and (8.597) in a somewhat more general form. With

$$\mathbf{r}\,' = \frac{\mathbf{r}}{a+bt}, \quad t' = \frac{c+\delta t}{a+bt}, \quad a\delta - bc = 1 \tag{8.600}$$

we find, in addition to $\psi(\mathbf{r}, t)$, another exact solution

$$\psi(\mathbf{r}, t) = (a+bt)^{-\frac{d}{2}} \psi(\mathbf{r}\,', t') \exp\left[\frac{ib|\mathbf{r}|^2}{4(a+bt)} \right]. \tag{8.601}$$

[For $a = 0, \ b = \alpha, \ \beta = \delta/\alpha, \ c = -1/\alpha$ the case (8.596) and (8.597) applies.] Now, if we set $a = t_c, \ b = -1$ and $\delta = 0$,, then a stationary solution

$$\psi_s(\mathbf{r}, t) = G(\mathbf{r}; \lambda) e^{i\lambda t} \tag{8.602}$$

is transformed by (8.601) into

$$\psi(\mathbf{r}, t) = (t_c - t)^{-d/2} G\left(\frac{\mathbf{r}}{t_c - t}; \lambda\right) \exp\left[-\frac{i(|\mathbf{r}|^2 - 4\lambda)}{4(t_c - t)}\right], \tag{8.603}$$

which exhibits a singularity after a finite evolution time t_c. Such a solution is called a collapsing solution.

An indication of a collapse can already be (roughly) obtained from the following qualitative arguments. Due to plasmon conservation

$$N = \int |\psi|^2 d\Gamma = const \tag{8.604}$$

a solution such as $\psi \sim L^{-d/2}$ will scale, where L is a typical width scale. The nonlinear term $|\psi|^p \psi$ therefore scales like $L^{-d(p+1)/2}$, while the dispersive term $\nabla^2 \psi \sim L^{-(d/2)-2}$ is. If $pd \geq 4$ holds, a steepening due to nonlinearity can no longer be counteracted by the dispersive contribution.

Zakharov was the first to provide a mathematically rigorous explanation of this phenomenon. The following virial theorem can be derived. We define

$$I := N\langle |\mathbf{r} - \langle \mathbf{r} \rangle|^2 \rangle, \tag{8.605}$$

where the mean value $\langle \ldots \rangle$, is, for example,

$$\langle \mathbf{r} \rangle = N^{-1} \int \mathbf{r}\, |\psi|^2 d\Gamma, \tag{8.606}$$

determined with $|\psi|^2$ as the weighting function. The first time derivative

$$\dot{I} = \frac{d}{dt} \int |\psi|^2 [\mathbf{r}^2 - \langle \mathbf{r} \rangle^2] d\Gamma \tag{8.607}$$

leads via (8.585)–(8.587) to

$$\dot{I} = -\int (\nabla \cdot \mathbf{J})[|\mathbf{r}|^2 - \langle \mathbf{r} \rangle^2] d\Gamma - 2N \langle \mathbf{r} \rangle \cdot \langle \dot{\mathbf{r}} \rangle . \tag{8.608}$$

If we

$$\begin{aligned} \langle \dot{\mathbf{r}} \rangle &= -N^{-1} \int \mathbf{r}(\nabla \cdot \mathbf{J}) d\Gamma = N^{-1} \int \mathbf{J}\, d\Gamma \\ &= -2N^{-1} \mathbf{P} \end{aligned} \tag{8.609}$$

take into account partial integration and (8.592)–(8.594), it follows that

$$\dot{I} = -\int (\nabla \cdot \mathbf{J}) |\mathbf{r}|^2 d\Gamma + 4\mathbf{P} \cdot \langle \mathbf{r} \rangle. \tag{8.610}$$

The second time derivative is calculated with similar arguments as

$$\ddot{I} = -2\int [\nabla \cdot (\nabla \cdot \underline{\underline{\mathcal{W}}})]|\mathbf{r}|^2 d\Gamma - 8N^{-1}|\mathbf{P}|^2. \tag{8.611}$$

The first term on the right-hand side can again be partially integrated,

$$\ddot{I} = -4\int \underline{\underline{1}} : \underline{\underline{\mathcal{W}}}\, d\Gamma - 8N^{-1}|\mathbf{P}|^2, \tag{8.612}$$

and what remains is to transform the "pressure term" $\underline{\underline{\mathcal{W}}}$ in such a way that (8.612) becomes tractable. Via $\mathcal{L}$, $\underline{\underline{\mathcal{W}}}$ still contains the time derivatives ψ_t and ψ_t^*, which we first eliminate using the NLS equation.

A short calculation yields

$$\underline{\underline{\mathcal{W}}} = -2(\nabla\psi)(\nabla\psi^*) + \underline{\underline{1}}\left\{\frac{p}{p+2}|\psi|^{p+2} + \nabla \cdot [\,\mathrm{Re}\,(\psi\nabla\psi^*)]\right\}. \tag{8.613}$$

If we insert this into (8.612), it follows that

$$\ddot{I} = 4\left\{2\int |\nabla\psi|^2 d\Gamma - \frac{pd}{p+2}\int |\psi|^{p+2} d\Gamma - 2N^{-1}|\mathbf{P}|^2\right\}. \tag{8.614}$$

Now, if we use (8.590), (8.614) can be transformed into

$$\boxed{\ddot{I} = 4\left\{2H - 2N^{-1}|\mathbf{P}|^2 + \frac{4-pd}{p+2}\int |\psi|^{p+2} d\Gamma\right\}} \tag{8.615}$$

The first two terms on the right-hand side are constants of motion, while the last term is time-dependent.

Since

$$N^{-1}|\mathbf{P}|^2 \geq 0 \tag{8.616}$$

holds, we can estimate

$$\ddot{I} \leq 8H + 4\frac{4-pd}{p+2}\int |\psi|^{p+2} d\Gamma \tag{8.617}$$

Instead of working with the inequality (8.617), we will operate directly with (8.615), introducing the abbreviations

$$H_0 := H - N^{-1}|\mathbf{P}|^2, \tag{8.618}$$

$$A = 4\frac{4-pd}{p+2}\int |\psi|^{p+2} d\Gamma \tag{8.619}$$

Integrating twice with respect to time yields

$$\boxed{I = 4H_0 t^2 + Bt + C + \int_0^t dt' \int_0^{t'} A(t'')dt''} \tag{8.620}$$

with

$$B = \dot{I}(t=0) = 4 \ \mathrm{Im} \int \mathbf{r} \cdot (\psi_0^* \nabla \psi_0) d\Gamma \ , \tag{8.621}$$

$$C = I(t=0) = \int |\mathbf{r}|^2 |\psi_0|^2 d\Gamma. \tag{8.622}$$

For simplicity, we have set $\langle \mathbf{r} \rangle (t=0) = 0$ and denoted the initial value of $\psi_0 = \psi(\mathbf{r}, t = 0)$ by ψ. It is important to remember that H_0 is a constant of motion and can therefore also be calculated using ψ_0. One final remark concerns (8.616) and (8.618). Since $H_0 \leq H$ holds, we can pass from (8.620) to an inequality $I \leq 4Ht^2 + \ldots$ and formulate the following sufficient criteria for a collapse also using $H\{\psi_0\}$.

If $H_0 < 0$ and $A \leq 0$ hold, then (8.620) states that the mean width of the solution shrinks. Due to $N = const$, the amplitude of ψ must then increase with time. It is intuitively clear that we can describe such solution behavior as collapsing.

The results of a more detailed analysis of (8.620) can be summarized as follows: If $A \leq 0$, that is, $pd \geq 4$, holds and *one* of the conditions

$$H_0 < 0, \tag{8.623}$$

$$H_0 = 0 \quad \text{and} \quad B < 0, \tag{8.624}$$

$$H_0 > 0 \quad \text{and} \quad B \leq -4(H_0 C)^{1/2} \tag{8.625}$$

is satisfied, a singularity will form in finite evolution time.

We now show that for $pd \geq 4$ the solitary wave solutions (8.602) already satisfy $H_0 \leq 0$.

First, the mean momentum $\mathbf{P}$ vanishes,

$$\mathbf{P} = 0 \ , \tag{8.626}$$

and we have $H = H_0$. Next, we multiply both sides of the equation

$$-\lambda G + \nabla^2 G + G^{p+1} = 0 \tag{8.627}$$

for the solitary waves $G = G(\mathbf{r})$ by G and integrate over the (d-dimensional) space;

$$\int (\nabla G)^2 d\Gamma = \int G^{p+2} d\Gamma - \lambda \int G^2 d\Gamma \tag{8.628}$$

is the result. Let us now consider the functional

$$S\{G\} = -[H\{G\} + \lambda N\{G\}] , \tag{8.629}$$

which in the first variation $\delta S = 0$ yields exactly (8.627), for test functions $G_\alpha(\mathbf{r}) = G(\mathbf{r}/\alpha)$ with a real parameter α.

We have

$$-S\{G_\alpha\} = \alpha^{d-2} T\{G\} + \alpha^d [\lambda N\{G\} - I_0\{G\}] \tag{8.630}$$

with

$$T\{\phi\} = \int (\nabla \phi)^2 d\Gamma, \tag{8.631}$$

$$I_0\{\phi\} = \frac{2}{p+2} \int \phi^{p+2} d\Gamma. \tag{8.632}$$

Since G is a solution of $\delta S\{G\} = 0$, it must satisfy

$$-\frac{d}{d\alpha} S\{G_\alpha\}\bigg|_{\alpha=1} = (d-2)T\{G\} + d[\lambda N\{G\} - I_0\{G\}] = 0 \tag{8.633}$$

Together with (8.628), i.e.

$$T\{G\} = \frac{p+2}{2} I_0\{G\} - \lambda N\{G\}, \tag{8.634}$$

we thus have two equations for T, N and I_0, which allow us to express

$$H\{G\} = T\{G\} - I_0\{G\} \tag{8.635}$$

solely in terms of N:

$$H = \frac{4 - pd}{pd - 2(p+2)} \lambda N. \tag{8.636}$$

In the intermediate calculation, we have assumed for localized solutions $\lambda > 0$ and accordingly $p(d-2) < 4$.

Thus, one recognizes a collapse for $d = 1$ and $p > 4$. For $d = 2$ and $p = 2$, states with $\psi \approx G$, collapse, for which $H < 0$ applies [note: in this case $H\{G\} = 0$]. For $p = 2$ and $d = 3$, various numerical calculations have verified the collapse.

The collapse is particularly important from a physical point of view because it represents a novel (nonlinear) dissipation mechanism. The scenario is as follows: Weak turbulence theory predicts a "Bose condensation" of Langmuir wave energy at $k \approx 0$. In this long-wavelength regime, Landau damping is negligible; however, due to modulation instability, localized structures form, which then collapse. When the width of these structures approaches the order of the electron Debye length, effective damping via particle-wave interaction can set in and lead to significant losses.

Collective coordinates

Finally, we present an approximate but very powerful method for determining the critical exponents in the collapse case. It is sometimes referred to as the "method of collective coordinates." If one inserts the trial function

$$\psi(\mathbf{r}, t) \equiv A(t) f[B(t)\mathbf{r}] e^{-i[\phi_0(t) + |\mathbf{r}|^2 \phi_2(t)]} \tag{8.637}$$

into the Lagrangian density (8.575), then after integration over space one obtains

$$\begin{aligned} L = \int \mathcal{L} d\Gamma = \alpha \frac{A^2}{B^d} \dot{\phi}_0 + \beta \frac{A^2}{B^{d+2}} \dot{\phi}_2 - \gamma \frac{A^2}{B^{d-2}} \\ - 4\beta \frac{A^2}{B^{d+2}} \phi_2^2 + \frac{2\delta}{p+2} \frac{A^{p+2}}{B^d}, \end{aligned} \tag{8.638}$$

$$\alpha = \int d\Gamma \, [f(\mathbf{r})]^2, \qquad \beta = \int d\Gamma |\mathbf{r}|^2 [f(\mathbf{r})]^2, \tag{8.639}$$

$$\gamma = \int d\Gamma [\nabla f(\mathbf{r})]^2, \qquad \delta = \int d\Gamma \, [f(\mathbf{r})]^{p+2} . \tag{8.640}$$

The function f must be an element of the Sobolev space $W^{1,2}$, in which the norm

$$\|f\|_{W^{1,2}} \equiv \left[\int |f|^2 d\Gamma + \int |\nabla f|^2 d\Gamma \right]^{1/2} < \infty \tag{8.641}$$

is defined.

The existence of a collapse should manifest itself in $A(t) \to \infty$, $B(t) \to \infty$ for $t \to t_c$. The quantities A, B, ϕ_0, ϕ_2 are called collective coordinates, which are assumed to correctly describe, via (8.637), the relevant degrees of freedom of the infinite-dimensional system (8.573).

The (qualitative) result should not depend significantly on the form of f; for example, for $f(\mathbf{r})$ one chooses a Gaussian function $\sim \exp(-|\mathbf{r}|^2)$. The Lagrangian function $L(A, B, \phi_0, \phi_2, \dot{A}, \dot{B}, \ldots)$ is a function of the generalized coordinates and momenta. The Lagrange equations for this are

$$\dot{C} = 0, \tag{8.642}$$

$$\alpha\dot{\phi}_0 = 2\gamma D^{-1} - \frac{\delta[p(d+2)+4]}{2(p+2)} C^{\frac{p}{2}} D^{-\frac{pd}{4}}, \tag{8.643}$$

$$\dot{D} = -8D\phi_2, \tag{8.644}$$

$$\beta\dot{\phi}_2 = 4\beta\phi_2^2 - \gamma D^{-2} + \frac{\delta p d}{2(p+2)} C^{\frac{p}{2}} D^{-\frac{pd+4}{4}}, \tag{8.645}$$

where we have introduced $C = A^2 B^{-d}$ and $D = B^{-2}$. The two integrals

$$C \equiv C_0 (= const) \tag{8.646}$$

and

$$E \equiv 4\beta D\phi_2^2 + \gamma D^{-1} - \frac{2\delta}{p+2} C_0^{\frac{p}{2}} D^{-\frac{pd}{4}} (= const) \tag{8.647}$$

reflect the conservation of plasmons and energy. An analysis of Eqs. (8.642)–(8.645) first yields the expected result that for $pd < 4$ no collapse occurs. To show this, we solve (8.644) for ϕ_2 and then substitute this ϕ_2 into (8.647). We obtain

$$\frac{\beta}{16}\dot{D}^2 = ED - \gamma + \frac{2\delta C_0^{\frac{p}{2}}}{p+2} D^{-\frac{pd-4}{4}}. \tag{8.648}$$

Since $\beta > 0$ and $\gamma > 0$ are [see (8.639) and (8.640)], in the case $pd < 4$ the collapse condition $D \to 0$ [note $B(t) \to \infty$ and $D = B^{-2}$] for $t \to t_c$ would result in the contradiction $\dot{D}^2 < 0$ for $t \le t_c$.

A further analysis of Eqs. (8.642)–(8.645) leads, after some calculation, to the easily verifiable time dependencies for the collective coordinates C, ϕ_0, D, ϕ_2 and A, B, ϕ_0, ϕ_2, respectively. In the critical case $pd = 4$ there are two different behaviors, depending on the initial conditions. Either

$$A(t) = \frac{C_0^{\frac{1}{2}}\tilde{E}}{4}(t_c - t)^{-\frac{d}{2}} \quad , \quad B(t) = \frac{\tilde{E}}{4}(t_c - t)^{-1}, \tag{8.649}$$

$$\phi_0(t) = \mu_1 (t_c - t)^{-1} \quad , \quad \phi_2(t) = \frac{1}{4}(t_c - t)^{-1} \tag{8.650}$$

for well-defined constants $\tilde{E}$, μ_1 and t_c, or

$$A(t) \sim (t_c - t)^{-\frac{d}{4}} \quad , \quad B(t) \sim (t_c - t)^{-\frac{1}{2}} \, , \tag{8.651}$$

$$\phi_0(t) \sim \ln(t_c - t) \quad , \quad \phi_2(t) \sim (t_c - t)^{-1} \tag{8.652}$$

for $t \to t_c$. In the supercritical case $pd > 4$ the prediction for $t \to t_c$ is unambiguous,

$$A(t) \sim (t_c - t)^{-\frac{2d}{pd+4}} \quad , \quad B(t) \sim (t_c - t)^{-\frac{4}{pd+4}}, \tag{8.653}$$

$$\phi_0(t) \sim (t_c - t)^{-\frac{p(2d-1)-4}{p+4}} \quad , \quad \phi_2(t) \sim (t_c - t)^{-1}. \tag{8.654}$$

The simplest way to verify this is by substitution. Large numerical simulation programs have been developed at various institutes to answer the question of how effective collapse is as a dissipation mechanism. We have already discussed that not all initial states—even in the supercritical case—collapse, which is also reflected in the numerical results. It should be emphasized that no final and generally accepted values exist yet for the critical exponents.

Collapse in KdV Systems

While collapse in NLS systems was already investigated a long time ago, a corresponding discussion for low-frequency modes, described by a KdV-type equation, only gained broader interest later. As a model equation, we use

$$\boxed{q_t + q^p q_x + (\nabla^2 q)_x = 0} \tag{8.655}$$

as a basis [$\nabla^2 = \partial_x^2 + \nabla_\perp^2$, where $\nabla_\perp^2 = 0,\ \partial_y^2$ or $\partial_y^2 + \partial_z^2$ for $d = 1, 2$ or 3]. For (8.655), there exists *no* hierarchical structure of conserved quantities of the form

$$\partial_t \mathcal{N} + \nabla \cdot \mathbf{J} = 0, \tag{8.656}$$

$$-\frac{1}{2}\partial_t \mathbf{J} + \nabla \cdot \underline{\underline{\mathcal{W}}} = 0 \, . \tag{8.657}$$

This may explain why, for a long time, nothing was known about the dynamical behavior of (8.655). The structure (8.656) and (8.657) was essential in deriving a virial theorem for NLS systems. Nevertheless, a (rough) scaling argument also suggests collapse for (8.655) when $pd \geq 4$ is present.

The conserved quantity

$$N_0^2 = \int q^2 d\Gamma \tag{8.658}$$

determines the scaling $q \sim L^{-d/2}$ [note $d\Gamma = d^d r$]. Now, if we compare the dispersive term ($\sim L^{-3}q$) with the nonlinear term ($\sim L^{-1-pd/2}q$), we again see that a steepening for $pd \geq 4$ cannot be prevented by dispersion.

A similar—but mathematically more rigorous—result is provided by a stability analysis for solitary waves. In this section, we restrict ourselves to (8.655), although other generalizations of the KdV equation are also conceivable. The methods discussed here can, for the most part, also be applied to other models.

In addition to N_0^2, the energy

$$H = -\int \left[\frac{1}{(p+1)(p+2)} q^{p+2} - \frac{1}{2} q_x^2 - \frac{1}{2} q_y^2 - \frac{1}{2} q_z^2 \right] d\Gamma \tag{8.659}$$

is also an invariant, where the terms q_y^2 or q_z^2 only appear in H for sufficiently high spatial dimension. Incidentally, it should be noted that (8.655) possesses the Hamiltonian structure

$$\frac{\partial q}{\partial t} = \frac{\partial}{\partial x} \frac{\delta H}{\delta q} \tag{8.660}$$

We begin our discussion with the proof that for $pd < 4$no collapse occurs. In this context, we require for a collapse

$$\int d\Gamma (\nabla q)^2 \to \infty \tag{8.661}$$

for $t \to t_c$. Since the total energy (8.659) is a (finite) conserved quantity, it follows that in the case of collapse, the integral

$$I_{p+2} = \int d\Gamma q^{p+2}. \tag{8.662}$$

also diverges. In the one-dimensional case ($d = 1$) we use, instead of (8.661), the measure

$$N_1^2 = \int d\Gamma (q_x)^2. \tag{8.663}$$

If we succeed in specifying a finite upper bound for N_1^2 that depends only on constants of motion, collapse is excluded. In the one-dimensional case, we write (8.659) as

$$H = -\frac{1}{(p+1)(p+2)} I_{p+2} + \frac{1}{2} N_1^2. \tag{8.664}$$

For further estimation, it is useful to establish relationships between I_{p+2}, N_0^2 and N_1^2. The Schwarz inequality yields

$$I_3 \leq N_0 I_4^{1/2}; \tag{8.665}$$

I_4 can be estimated by

$$I_4 \leq \max_x(q^2) N_0^2 \leq 2 \int_{-\infty}^{+\infty} |q q_x| dx \, N_0^2 \leq 2 N_0^3 N_1 \tag{8.666}$$

This is already sufficient for $p = 2$. Starting from the last two relations, we can proceed successively in the same manner and estimate

$$I_{p+2} \leq 2 N_0 N_1 I_p \tag{8.667}$$

which, for example,

$$I_5 \leq 2\sqrt{2} N_0^{7/2} N_1^{3/2}, \tag{8.668}$$

$$I_6 \leq 4 N_0^4 N_1^2 \tag{8.669}$$

means. If we substitute (8.665) and (8.667) into (8.664), we obtain for $p = 1$

$$\boxed{H \geq -\frac{1}{6}\sqrt{2} N_0^{5/2} N_1^{1/2} + \frac{1}{2} N_1^2 \equiv f_1(N_1)} \,. \tag{8.670}$$

Since H and N_0 are constants (determined by the initial conditions), N_1 cannot grow without bound.

Example 8.21 (Estimates at $d = 1$ for $p = 2, 3, 4$ and 5)
For $p = 2$ a similar estimate follows,

$$H \geq f_2(N_1) \equiv -\frac{1}{6} N_0^{3/2} N_1^{1/2} + \frac{1}{2} N_1 \,, \tag{8.671}$$

while $p = 3$

$$H \geq f_3(N_1) \equiv -\frac{1}{10}\sqrt{2} N_0^{7/4} N_1^{3/4} + \frac{1}{2} N_1 \tag{8.672}$$

yields. The situation is already different for $p = 4$. In

$$H \geq f_4(N_1) \equiv N_1^2\left[\frac{1}{2} - \frac{2}{15}N_0^4\right] \tag{8.673}$$

we cannot assume $N_0^4 < \frac{15}{4}$, since already for the soliton solutions

$$q = (15v)^{1/4}\,\mathrm{sech}^{1/2}(2v^{1/2}x) \tag{8.674}$$

$$N_0^2 = \frac{1}{2}\sqrt{15}\int_{-\infty}^{+\infty} \mathrm{sech}\ z\,dz = \frac{1}{2}\pi\sqrt{15} > \frac{1}{2}\sqrt{15} \tag{8.675}$$

is the case. For $d = 1$ and $p = 4$ a collapse can therefore no longer be ruled out; further estimates, e.g.

$$H \geq f_5(N_1) \equiv -\frac{2}{21}\sqrt{2}N_0^{11/2}N_1^{5/2} + \frac{1}{2}N_1^2 \tag{8.676}$$

for $p = 5$, show that this is no longer possible even for larger p values. ∎

Example 8.22 (Estimates for $d = 2$ or 3)
For the argumentation with $d = 2$ and $d = 3$ we need the estimates

$$\begin{aligned} I_4 &\leq \left[\int dy \max_x(q^2)\right]\left[\int dx \max_y(q^2)\right] \\ &\leq 4\left[\int dxdy|qq_x|\right]\left[\int dxdy|qq_y|\right] \\ &\leq 4N_0^2N_1N_2 \leq 2N_0^2(N_1^2+N_2^2) \end{aligned} \tag{8.677}$$

for $d = 2$, or

$$\begin{aligned} I_4 &\leq 4\int dz\left\{\left[\int dxdyq^2\right]\left[\int dxdy|q_x|^2\right]\right\}^{1/2}\left[\int dxdy|q_y|^2\right]^{1/2} \\ &\leq 4\left[\max_z \int dxdyq^2\right]\int dz\left\{\left[\int dxdy|q_x|^2\right]\left[\int dxdy|q_y|^2\right]\right\} \\ &\leq 8N_0N_3N_2N_1 \leq \frac{8}{\sqrt{6}}N_0(N_1^2+N_2^2+N_3^2)^{3/2} \end{aligned} \tag{8.678}$$

for $d = 3$. ∎

With these results, it is easy to estimate for $p = 1$ and $d = 2$

$$H \geq g_1(X = N_1^2 + N_2^2) \equiv \frac{1}{2}X - \frac{1}{6}\sqrt{2}N_0^2X^{1/2} \tag{8.679}$$

while for $p = 2$ and $d = 2$

$$H \geq g_2(X) \equiv \frac{1}{2}X - \frac{1}{3}N_0^2 X \tag{8.680}$$

already follows. Here, we have defined N_2^2 analogously to (8.663) for the y-derivative. With (8.679) we can rule out a collapse, whereas such a conclusion is no longer possible for $d = 2$ and $p \geq 2$.

Corresponding calculations for $d = 3$ and $p = 1$,

$$\begin{aligned} H &\geq h_1(Y \equiv N_1^2 + N_2^2 + N_3^2) \\ &\equiv \frac{1}{2}Y - \frac{\sqrt{2}}{3 \times 6^{1/4}} N_0^{3/2} Y^{3/4}, \end{aligned} \tag{8.681}$$

or for $d = 3$ and $p = 2$,

$$H \geq h_2(Y) \equiv \frac{1}{2}Y - \frac{4}{3\sqrt{6}} Y^{3/2}, \tag{8.682}$$

prove our assertion that for $pd < 4$ no collapse is possible.

Collective Coordinates

To demonstrate the possibility of collapse for $pd > 4$, we use the method of collective coordinates discussed in the previous section. With

$$q \equiv \phi_x \tag{8.683}$$

for $d = 1$ (and this is the only dimension we will examine in the following demonstration example), the Lagrangian density is

$$\boxed{\mathcal{L} = \phi_t \phi_x - \phi_{xx}^2 + \frac{2}{(p+1)(p+2)} \phi_x^{p+2}}\,. \tag{8.684}$$

If we insert the test functions

$$\begin{aligned} q(x,t) = &A(t)B(t)g\{B(t)[x - C(t)]\} \\ &+ D(t)B(t)u\{B(t)[x - C(t)]\} \end{aligned} \tag{8.685}$$

with an even (g) and an odd (u) component into the Lagrangian density and integrate over space (note: d = 1),

$$L = \int \mathcal{L} dx, \tag{8.686}$$

we obtain a Lagrangian function with the "coordinates" A, B, C, D and the "velocities" $\dot{A}, \dot{B}, \dot{C}, \dot{D}$. The functions g and u should be localized; the assumption of equal width B^{-1} is not necessary, but as will be shown, is entirely sufficient. A short calculation yields

$$\boxed{\begin{aligned} L =& a_1 AD\frac{\dot{B}}{B} - a_2 A^2 B\dot{C} - a_3 D^2 B\dot{C} \\ &+ a_4(\dot{A}D - A\dot{D}) - B^2(a_5 A^2 B + a_6 D^2 B) \\ &+ B^{p/2}\sum_{n=0}^{N} a_{7+n}(A^2 B)^{(p+2)/2-n}\,(D^2 B)^n. \end{aligned}} \tag{8.687}$$

Here,

$$N = \begin{cases} \frac{1}{2}(p+2) \text{ for even } p, \\ \frac{1}{2}(p+1) \text{ for odd } p \end{cases} \tag{8.688}$$

and

$$a_1 = 2\int_{-\infty}^{\infty} yu(y)g(y)dy, \quad a_2 = \int_{-\infty}^{+\infty} g^2(y)dy\,, \tag{8.689}$$

$$a_3 = \int_{-\infty}^{+\infty} u^2(y)dy, \quad a_4 = -\int_{-\infty}^{\infty} g(y)\int_{-\infty}^{y} u(y')dy'dy, \tag{8.690}$$

$$a_5 = \int_{-\infty}^{\infty}\left(\frac{dg}{dy}\right)^2 dy, \quad a_6 = \int_{-\infty}^{\infty}\left(\frac{du}{dy}\right)^2 dy, \tag{8.691}$$

$$a_{7+n} = \frac{2}{(p+1)(p+2)}\binom{p+2}{2n}\int_{-\infty}^{+\infty} g^{p+2-2n}(y)u^{2n}(y)dy. \tag{8.692}$$

If we introduce the new variables

$$X = A^2 B, \tag{8.693}$$

$$K = D^2 B + \frac{a_2}{a_3}X \tag{8.694}$$

the Lagrangian function then follows as

$$\begin{aligned} L = L_0\frac{\dot{K}}{B} + L_1(X,K)\frac{\dot{X}}{B} - a_3 K\dot{C} \\ - B^2 L_2(X,K) + B^{p/2}L_3(X,K), \end{aligned} \tag{8.695}$$

where

$$L_0 = \frac{1}{2}\sigma(a_1 - a_4)X[X(K - \alpha X)]^{-1/2}, \tag{8.696}$$

$$L_1 = \frac{1}{2}\sigma[(a_1 + a_4)K - 2\alpha a_1 X][X(K - \alpha X)]^{-1/2}, \tag{8.697}$$

$$L_2 = (a_5 - \alpha a_6)X + \alpha_6 K\ , \tag{8.698}$$

$$L_3 = \sum_{n=0}^{N} a_{7+n} X^{p/2+1-n}(K - \alpha X)^n \tag{8.699}$$

was defined; σ is an abbreviation for the sign of the square root in L_1, and $\alpha = a_2/a_3$.

The variations of L with respect to C, K, X, and B yield the Euler-Lagrange equations

$$\boxed{\dot{K} = 0,} \tag{8.700}$$

$$\boxed{\begin{aligned} a_3\dot{C} = {} & \left(\frac{\partial L_1(X,K)}{\partial K} - \frac{\partial L_0}{\partial X}\right)\frac{\dot{X}}{B} - B^2\frac{\partial L_2(X,K)}{\partial K} \\ & + B^{p/2}\frac{\partial L_3(X,K)}{\partial K} + L_0\frac{\dot{B}}{B^2}, \end{aligned}} \tag{8.701}$$

$$\boxed{\frac{L_1\dot{B}}{B^2} = B^2\frac{\partial L_2(X,K)}{\partial X} - B^{p/2}\frac{\partial L_3(X,K)}{\partial X},} \tag{8.702}$$

$$\boxed{L_1\frac{\dot{X}}{B^2} = -2BL_2 + \frac{1}{2}pB^{p/2-1}L_3.} \tag{8.703}$$

$K = const$ reflects the constancy of N_0. Specifically, we have

$$N_0^2 = \int_{-\infty}^{+\infty} q^2 dx = a_3 D^2 B + a_2 A^2 B = a_3 K. \tag{8.704}$$

If, furthermore, we multiply both sides of (8.702) by $\dot{X}$ and both sides of (8.703) by $\dot{B}$ and then subtract, we obtain

$$B^2 L_2(X,K) - B^{p/2} L_3(X,K) = const \tag{8.705}$$

$$= \frac{1}{2}\int_{-\infty}^{+\infty} dx\left[q_x^2 - \frac{2}{(p+1)(p+2)}q^{p+2}\right] \equiv H. \tag{8.706}$$

This result, i.e., the conservation of energy, demonstrates the usefulness of our approach (8.685).

Example 8.23 (Solution of the Euler-Lagrange equations for $d = 1$ and $p = 4$)
It is not difficult to solve Eqs. (8.701)–(8.703) numerically. The numerical solution indeed shows the collapse ($B \to \infty$ for $t \to t_c$) for $p \geq 4$. In the following, we attempt to justify this statement analytically. Let us begin with $p = 4$ and an initial distribution with $H = 0$. Then, from (8.705) it follows

$$L_2(X, K) = L_3(X, K); \tag{8.707}$$

we imagine this equation solved for X,

$$X = X_0(K) = const. \tag{8.708}$$

Substituting into (8.702) yields this solution

$$B(t) = \left[\frac{L_1(X_0, K)}{3\partial P(X, K)/\partial X|_{X_0}}\right]^{1/3} (t_c - t)^{-1/3} \tag{8.709}$$

with

$$P(X, K) = L_2(X, K) - L_3(X, K). \tag{8.710}$$

The width B^{-1} thus decreases (collapsing) and the amplitudes

$$AB = X_0^{1/2} B^{1/2} \sim (t_c - t)^{-1/6}, \tag{8.711}$$

$$DB = (K - \alpha X_0)^{1/2} B^{1/2} \sim (t_c - t)^{-1/6} \tag{8.712}$$

diverge to infinity. ■

The case $p = 4$ and $H \neq 0$ can be treated similarly (although the algebraic effort is greater); we then also obtain

$$B(t) \sim (t_c - t)^{-1/3}. \tag{8.713}$$

Example 8.24 (Solution of the Euler-Lagrange equations for $d = 1$ and $p > 4$)
For $p > 4$ we show that consistent and, for $t \to t_c$, singular solutions exist. Under the assumption $B \to \infty$ for $t \to t_c$, (8.705) tells us $L_3(X, K) \to 0$ for $t \to t_c$ and $p > 4$.. The relevant zero of L_3 is $X = X_0$.. For $t \to t_c$ we then write

$$X(t) = X_0 + \beta(t_c - t)^{\gamma}. \tag{8.714}$$

Inserting this into the energy conservation equation (8.705) and balancing the leading divergent terms, we obtain

$$B \sim (t_c - t)^{-2\gamma/(p-4)}. \tag{8.715}$$

The Eq. (8.703) will then generally be consistently satisfied if

$$\gamma = \frac{p-4}{p+2} \tag{8.716}$$

holds. This means

$$B \sim (t_c - t)^{-2/(p+2)} \tag{8.717}$$

for $t \to t_c$ and $p > 4$.. Note that the velocity

$$\dot{C} \sim (t_c - t)^{-p/(p+2)} \tag{8.718}$$

for $t \to t_c$ also diverges. Due to the conservation law (8.705), Eq. (8.702) is automatically satisfied. ■

These analytical considerations are not yet complete in that only the leading divergences are considered. However, they provide a good indication for numerical solutions of Eq. (8.655), which essentially confirm the stated power laws. In Fig. 8.11 we show a numerical solution for $d = 1$ and $p = 4$.

The initial distribution

$$q(x, t = 0) = (2.02 - 2.2x)e^{-x^2} \tag{8.719}$$

shows, in the coordinate system moving with $\dot{C}$, a temporal and spatial evolution that corresponds to the analytical predictions.

In the final stage of the collapse—as with the NLS equation—the model breaks down. Damping terms, higher-order nonlinearities, and additional dispersive terms then also play a decisive role. However, it is beyond doubt that collapse plays a crucial role as a dissipation mechanism and in the formation of shock fronts for certain low-frequency modes. This must be viewed in the context of the criterion $pd \geq 4$ for collapse. Modes described by a modified KdV equation, e.g., ion-acoustic waves in magnetized plasmas with negative ions and Alfvén waves propagating near a critical angle, collapse in spatially multidimensional situations.

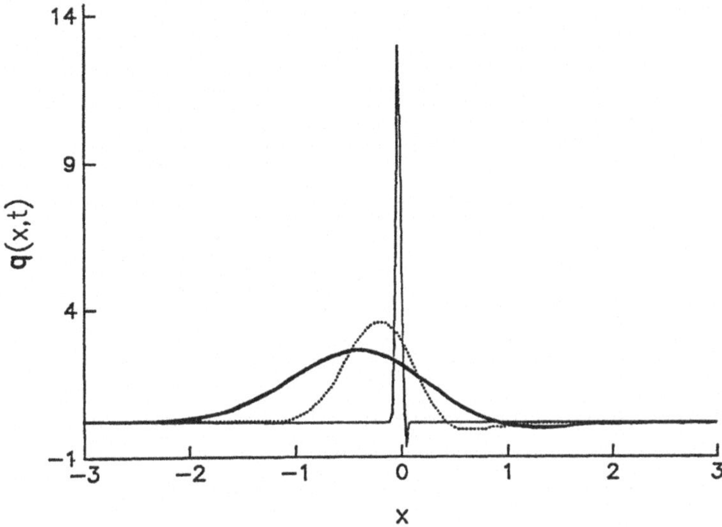

Fig. 8.11 Numerical solution of the KdV equation $q_t + q^4 q_x + q_{xxx} = 0$ in the frame moving with $\dot{C}$. The spatial distributions are shown at times $t = 0$ (thick), $t = 0.176$ (dotted), and $t = 0.2$ (thin)

8.6 Chaos in Schrödinger Systems

So far, we have *not* considered dissipative and external driving terms in the model equations for nonlinear waves in plasmas. We now discuss a simple model in which such terms appear. We investigate (one-dimensional) Langmuir waves with shock damping, which are subject to an external (homogeneous) electric field $E_0 \exp(-i\omega_0 t) + c.c.$. Interestingly, spatially coherent structures can exhibit quite typical temporal dynamics.

We normalize such that all times are measured in ω_{pe}^{-1},, all positions in λ_{De},, velocities in v_{te},, densities in $n_0 = const$, and electric fields in $m_e \omega_{pe} v_{te}/e$. Then, the particle and momentum balances for electrons (the normalized quantities carry no indices) are given by

$$\partial_t n + \partial_x (nv) = 0, \tag{8.720}$$

$$\partial_t v + v \partial_x v = -E - E_0 e^{-i\omega_0 t} - \nu_e v - \partial_x n. \tag{8.721}$$

We calculate the electric field (electrostatically) using

$$\partial_t E = nv, \tag{8.722}$$

where, due to the high frequencies, ($\omega \approx \omega_{pe}$) the ions can be assumed to be an immobile background. External influences and damping are assumed to be small, which we express using the smallness parameter

$$\boxed{\delta^2 = \frac{|\omega_{pe} - \omega_0|}{\omega_{pe}} \ll 1} \tag{8.723}$$

in the form

$$E_0 \sim \delta^3 \quad , \quad \nu_e \sim \delta^2 \tag{8.724}$$

We omit a detailed scaling, as the further procedure is now quite clear. First, we differentiate (8.722) with respect to time and substitute (8.720) and (8.721). Then we separate out the fast time dependence (note: $\omega_{pe} t \widehat{=} t$ in dimensionless times),

$$E = \delta E^{(1)}(x, t) e^{-it}, \tag{8.725}$$

where the amplitude $\delta E^{(1)}$ is still assumed to depend weakly on position and time. We can formally account for this by introducing the new variables $X = \delta x$ and $T = \delta^2 t$. Up to order δ^3 we obtain

$$i\partial_T E^{(1)} - \frac{\Delta n}{2\delta^2} E^{(1)} - \frac{E_0}{2\delta^3} e^{i\omega T} + i\frac{\nu_e}{2\delta^2} E^{(1)} + \frac{1}{2}\partial_X^2 E^{(1)} = 0, \tag{8.726}$$

where

$$\omega = \frac{\omega_{pe} - \omega_0}{\delta^2 \omega_{pe}} \sim \mathcal{O}(1) \tag{8.727}$$

has been set. In an adiabatic approximation, we determine $\Delta n = n - 1$ as

$$\Delta n = \delta^2 n^{(1)} \approx -\delta^2 |E^{(1)}|^2 \tag{8.728}$$

and introduce the new variables

$$q = -\frac{i}{2} E^{(1)}, \quad \gamma = \frac{\nu_e}{2\delta^2}, \quad a = \frac{E_0}{4\delta^3}, \tag{8.729}$$

$$T \to t, \quad \sqrt{2}\, X \to x \tag{8.730}$$

From (8.726) we thus obtain the model equation

$$iq_t + q_{xx} + 2|q|^2 q = -i\gamma q - ia e^{i\omega t}. \tag{8.731}$$

The explicit time dependence can be removed by a transformation $q \to q\exp(i\omega t)$; after this rewriting, the equivalent equation reads

$$\boxed{iq_t + q_{xx} + 2|q|^2 q - \omega q = -i\gamma q - ia}\,. \tag{8.732}$$

Without loss of generality, we can set $\omega = 1$ in numerical calculations, since $q = \omega^{1/2}Q, x = \omega^{-1/2}X,\ t = \omega^{-1}T,\ a = \omega^{3/2}A$ and $\gamma = \omega\Gamma$ to

$$iQ_T + Q_{XX} + 2|Q|^2 Q - Q = -iA - i\Gamma Q \tag{8.733}$$

leads. Against this background, in the following (8.732), we will, in part, use $\omega = 1$.

Analytical Estimates

In contrast to the undisturbed (integrable) NLS equation, we can no longer assume that the solution q vanishes at infinity. Therefore, we first study the boundary conditions that follow from $q_x \to 0$ for $x \to \pm\infty$.

Spatially Homogeneous Distribution

We denote the spatially constant solution by c. It is determined by

$$ic_t + 2|c|^2 c - \omega c = -i\gamma c - ia \tag{8.734}$$

With $c = |c|\exp(i\phi)$ this yields

$$|c|_t = -a\cos(\phi) - \gamma|c|, \tag{8.735}$$

$$\phi_t = 2|c|^2 - \omega + \frac{a}{|c|}\sin(\phi), \tag{8.736}$$

and for stationary solutions we find a cubic equation in $|c|^2$,

$$[(2|c|^2 - \omega)^2 + \gamma^2]|c|^2 = a^2\,. \tag{8.737}$$

Depending on whether the discriminant is positive or negative, this equation has one or three real solutions. For small values of a there is only the one solution

$$c \approx i\frac{a}{\omega}, \tag{8.738}$$

which we will use in the following. A stability analysis further shows that the branch (8.738) is stable.

Example 8.25 (Spatially Independent Stability Analysis)
If we linearize Eqs. (8.735) and (8.736) about the solution (8.738) [hereafter written as $|c_0| \exp(i\phi_0)$] in the form

$$|c| = |c_0| + d, \tag{8.739}$$

$$\phi = \phi_0 + \varphi, \tag{8.740}$$

we obtain

$$d_t = \Gamma d = -\gamma d + (\omega|c_0| - 2|c_0|^3)\varphi, \tag{8.741}$$

$$\varphi_t = \Gamma\varphi = \left(6|c_0| - \frac{\omega}{|c_0|}\right)d - \gamma\varphi. \tag{8.742}$$

The eigenvalues of the stability problem are

$$\begin{aligned}\Gamma &= -\gamma \pm \sqrt{-12|c_0|^4 + 8\omega|c_0|^2 - \omega^2} \\ &= -\gamma \pm \sqrt{\gamma^2 - \frac{\partial a^2}{\partial |c_0|^2}}.\end{aligned} \tag{8.743}$$

Since for (8.738) $\partial|a|^2/\partial|c_0|^2 > 0$ holds, stability follows within the framework of the model (8.734). ■

Example 8.26 (Spatially Dependent Stability Analysis)
However, we must also examine the stability of (8.738) with respect to x-dependent perturbations. For this, we linearize (8.732) in the form

$$q(x,t) = c + \delta q(x,t) \tag{8.744}$$

to obtain

$$i\delta q_t + \delta q_{xx} + 4|c|^2\delta q + 2c^2\delta q^* - \omega\delta q + i\gamma\delta q = 0. \tag{8.745}$$

As a modulation, we make the ansatz:

$$\delta q(x,t) = q_1 e^{i(\Omega t - kx) + \Gamma t} + q_2^* e^{-i(\Omega t - kx) + \Gamma t}. \tag{8.746}$$

This yields

$$\begin{aligned}0 = &\left[(i\Gamma - \Omega - k^2 + 4|c|^2 - \omega + i\gamma)q_1 + 2c^2 q_2\right]e^{i(\Omega t - kx) + \Gamma t} \\ &+ \left[(i\Gamma + \Omega - k^2 + 4|c|^2 - \omega + i\gamma)q_2^* + 2c^2 q_1^*\right]e^{-i(\Omega t - kx) + \Gamma t}.\end{aligned} \tag{8.747}$$

The condition for the existence of nontrivial solutions is that the determinant of this system of equations vanishes, i.e.,

$$(4|c|^4 - k^2 - \omega)^2 - (\Omega - i(\Gamma + \gamma))^2 - 4|c|^4 = 0. \tag{8.748}$$

The growth rate Γ is positive if and only if

$$4|c|^4 - (4|c|^2 - k^2 - \omega)^2 - \gamma^2 > 0 \tag{8.749}$$

holds. This equation can evidently only be satisfied for a k^2 interval of finite width, whose boundaries are

$$\begin{aligned} k_{1,2}^2 &= 4|c|^2 - \omega \pm \sqrt{4|c|^4 - \gamma^2} \\ &= 4|c|^2 - \omega \pm \sqrt{(4|c|^2 - \omega)^2 - \frac{\partial a^2}{\partial |c|^2}} \end{aligned} \tag{8.750}$$

The inequality (8.749) has real solutions k only if at least $k_1^2 \geq 0$ holds. For $\partial a^2/\partial|c|^2 \approx \omega^2 > 0$ and $|c|^2 \ll \omega$ (i.e., small a), (8.750) has no positive solutions k_1^2, and therefore the branch (8.738) [with $|c| \equiv |c_0|$] is also stable with respect to modulations. ■

We conclude this discussion of the spatially homogeneous system (8.734) with two remarks. The solutions of (8.734) cannot exhibit chaotic behavior, since they follow from an autonomous system with two degrees of freedom. The solution (8.738) represents a stable fixed point.

Solitary Fixed Points

The question now arises whether, in addition to the trivial fixed point (8.738), i.e.,

$$c = |c|e^{i\phi_c} \equiv c_0 e^{i\phi_c} \tag{8.751}$$

with $c_0 \approx a/\omega$ and $\phi_c \approx \pi/2$, there exist other nontrivial stable solutions in which the x-dependence plays a decisive role.

To do this, let us first consider the soliton solution of (8.732) for $\gamma = a = 0$. With the free amplitude 2η and the velocity -4ξ it reads

$$q(x,t) = 2\eta e^{-2i\xi x} e^{i[4(\eta^2-\xi^2)t - \omega t - \psi_0 - \pi/2]} \operatorname{sech}[2\eta(x - x_0) + 8\eta\xi t]; \tag{8.752}$$

Without loss of generality, we set $\xi = 0$. Furthermore, we expect a phase coupling with the external driver, which should result in $\omega = 4\eta^2$. If we further abbreviate $\varphi_0 = \psi_0 +$

$\pi/2$, then the particular solution

$$q(x, t) = \sqrt{\omega}\ \text{sech}\ (\sqrt{\omega}x)e^{-i\varphi_0} \tag{8.753}$$

should be in resonance with the external driver. The ansatz (8.753) is still too naive for the full Eq. (8.732), since it becomes completely incorrect for $x \to \pm\infty$. We therefore set

$$\boxed{q = (r + c_0)e^{i\phi_c - i\varphi_0}} \tag{8.754}$$

and expect that r—apart from the phase—is well approximated by (8.753). Substituting this into (8.732), we obtain

$$\boxed{ir_t + r_{xx} + 2|r|^2 r - \omega r = -4|r|^2 c_0 - 2r^2 c_0 - 4r\, c_0^2 - 2c_0^2 r^* - i\gamma r}\ . \tag{8.755}$$

A separation into real and imaginary parts $r = A + iB$ yields

$$\begin{aligned} A_t = &-B_{xx} - 2A^2B - 2B^3 + \omega B \\ &- 2c_0A^2 \sin(\varphi_0) - 6c_0B^2 \sin(\varphi_0) - 4c_0AB \cos(\varphi_0) \\ &- 4c_0^2B - 2c_0^2A \sin(2\varphi_0) + 2c_0^2B \cos(2\varphi_0) - \gamma A, \end{aligned} \tag{8.756}$$

$$\begin{aligned} B_t = &A_{xx} + 2A^3 + 2AB^2 - \omega A \\ &+ 6c_0A^2 \cos(\varphi_0) + 2c_0B^2 \cos(\varphi_0) + 4c_0AB \sin(\varphi_0) \\ &+ 4c_0^2A + 2c_0^2A \cos(2\varphi_0) + 2c_0^2B \sin(2\varphi_0) - \gamma B. \end{aligned} \tag{8.757}$$

Example 8.27 (Stationary Solution Components in Zeroth Order)
We now determine the stationary solutions $A_s + iB_s$ of these equations in an expansion in terms of γ and c_0. To do this, we scale as follows:

$$A_s = A_0 + A_1 + \ldots \quad \text{with} \quad A_i = \mathcal{O}(\gamma^i, a^i), \tag{8.758}$$

$$B_s = B_0 + B_1 + \ldots \quad \text{with} \quad B_i = \mathcal{O}(\gamma^i, a^i). \tag{8.759}$$

The zeroth order of (8.756) and (8.757) is determined from

$$0 = -B_{0xx} - 2A_0^2B_0 - 2B_0^3 + \omega B_0, \tag{8.760}$$

$$0 = A_{0xx} + 2A_0^3 + 2B_0^2A_0 - \omega A_0; \tag{8.761}$$

their solution yields the undisturbed soliton

$$\boxed{A_0 = \sqrt{\omega}\ \text{sech}\,(\sqrt{\omega}x),\ B_0 = 0}\ . \tag{8.762}$$

■

Example 8.28 (Stationary Solution Components to First Order)
With the operators

$$\widehat{H}_+ = -\partial_{xx} + \omega - 2A_0^2, \tag{8.763}$$

$$\widehat{H}_- = -\partial_{xx} + \omega - 6A_0^2 \tag{8.764}$$

the first order takes the form

$$\widehat{H}_- A_1 = 6c_0 A_0^2 \cos(\varphi_0), \tag{8.765}$$

$$\widehat{H}_+ B_1 = 2c_0 A_0^2 \sin(\varphi_0) + \gamma A_0 \tag{8.766}$$

These equations are solvable if the inhomogeneities are orthogonal to the kernels of the operators adjoint to $\widehat{H}_\pm$. With respect to the scalar product

$$\langle f|g\rangle = \int_{-\infty}^{\infty} fg dx \tag{8.767}$$

$\widehat{H}_+$ and $\widehat{H}_-$ are self-adjoint on the space of normalizable functions. The kernels are one-dimensional and determined by

$$\widehat{H}_+ A_0 = 0,\ \widehat{H}_- A_{0x} = 0 \tag{8.768}$$

The solvability condition for the first Eq. (8.765), i.e.,

$$6c_0 \cos(\varphi_0)\langle A_{0x} A_0^2\rangle = 0, \tag{8.769}$$

is trivially satisfied, since $A_{0x}A_0^2$ is an odd function of position. The solution

$$\boxed{A_1 = -\frac{2}{\omega} c_0 \cos(\varphi_0) A_0^2 + \alpha_1 A_{0x}} \tag{8.770}$$

is unique only up to a kernel function of $\widehat{H}_-$, i.e., α_1 remains undetermined at this order. The solvability condition for the second Eq. (8.766) is

$$2c_0 \sin(\varphi_0)\langle A_0^3\rangle + \gamma \langle A_0^2\rangle = 0. \tag{8.771}$$

■

After evaluating the integrals, we obtain

$$\sin(\varphi_0) = -\frac{2\gamma}{\pi\sqrt{\omega}c_0}. \tag{8.772}$$

In particular, this solvability condition can only be fulfilled for

$$\gamma \leq \frac{\pi\sqrt{\omega}c_0}{2} \tag{8.773}$$

In this case, however, there are two essential fixed points, which differ in φ_0 (modulo 2π). We refer to them as phase-coupled solitons.

Stability of the Fixed Points

The first order of perturbation theory calculated up to this point is already sufficient to investigate the stability of the fixed points perturbatively. For this, the Schrödinger equation split into real and imaginary parts (8.756) and (8.757) must be linearized around the respective fixed point $A_s + iB_s$.

With

$$A = A_s + \alpha, \qquad B = B_s + \beta \tag{8.774}$$

it follows that

$$\alpha_t = +H_+\beta - 4(A_s + c_0\cos(\varphi_0))(B_s + c_0\sin(\varphi_0))\alpha - \gamma\alpha, \tag{8.775}$$

$$\beta_t = -H_-\alpha + 4(A_s + c_0\cos(\varphi_0))(B_s + c_0\sin(\varphi_0))\beta - \gamma\beta, \tag{8.776}$$

where the operators $H_\pm$ are defined via

$$H_+ = -\partial_{xx} - 2(A_s + c_0\cos(\varphi_0))^2 - 6(B_s + c_0\sin(\varphi_0))^2 + \omega\ , \tag{8.777}$$

$$H_- = -\partial_{xx} - 6(A_s + c_0\cos(\varphi_0))^2 - 2(B_s + c_0\sin(\varphi_0))^2 + \omega \tag{8.778}$$

Equations (8.775) and (8.776) are solved perturbatively, as already indicated. For this, we expand

$$\alpha = \alpha_0 + \alpha_1 + \ldots, \quad \alpha_j = \mathcal{O}(c_0^{j/2}, \gamma^{j/2}), \tag{8.779}$$

$$\beta = \beta_0 + \beta_1 + \ldots, \quad \beta_j = \mathcal{O}(c_0^{j/2}, \gamma^{j/2}), \tag{8.780}$$

$$\Gamma = \Gamma_0 + \Gamma_1 + \ldots, \quad \Gamma_j = \mathcal{O}(c_0^{j/2}, \gamma^{j/2}), \tag{8.781}$$

where we have assumed the time dependence of α and $\beta \sim e^{\Gamma t}$. To zeroth order we obtain

$$\Gamma_0 \alpha_0 = +\widehat{H}_+ \beta_0, \tag{8.782}$$

$$\Gamma_0 \beta_0 = -\widehat{H}_- \alpha_0. \tag{8.783}$$

This system of differential equations possesses a discrete spectrum, which consists of two modes with eigenvalue $\Gamma_0 = 0$, as well as a continuum with imaginary eigenvalues. The discrete modes are known analytically—we will expand around them. This suffices for a sufficient criterion for instability. The two modes with eigenvalue $\Gamma_0 = 0$ are

$$\alpha_0 = 0, \quad \beta_0 = A_0, \tag{8.784}$$

$$\alpha_0 = A_{0x}, \quad \beta_0 = 0. \tag{8.785}$$

In the following, we examine the first (8.784), which is even in x.

To first order, $(c_0^{1/2}, \gamma^{1/2})$, then holds

$$0 = \widehat{H}_+ \beta_1, \tag{8.786}$$

$$\Gamma_1 A_0 = -\widehat{H}_- \alpha_1 \tag{8.787}$$

with the solutions

$$\beta_1 = A_0, \tag{8.788}$$

$$\alpha_1 = -\Gamma_1 \widehat{H}_-^{-1} A_0. \tag{8.789}$$

The second order yields

$$\Gamma_1 \alpha_1 = \widehat{H}_+ \beta_2 - 4A_0^2 (A_1 + c_0 \cos(\varphi_0)), \tag{8.790}$$

$$\Gamma_2 \beta_0 + \Gamma_1 \beta_1 = -\widehat{H}_- \alpha_2 + 4A_0^2 (B_1 + c_0 \sin(\varphi_0)) - \gamma A_0. \tag{8.791}$$

At this order, nontrivial solvability conditions appear for the first time. The second condition is always satisfied, since the modified soliton is an even function of x in all orders of the perturbation expansion. However, the kernel of $\widehat{H}_-$ is odd, so all integrals vanish.

The solvability condition of (8.790) after substitution is

$$\boxed{\begin{aligned}\Gamma_1^2 &= 4c_0\cos(\varphi_0)\frac{\langle A_0^3\rangle - \frac{2}{\omega}\langle A_0^5\rangle}{\langle A_0|\widehat{H}_-^{-1}A_0\rangle}\\ &= 2\pi\omega^{3/2}c_0\cos(\varphi_0).\end{aligned}} \tag{8.792}$$

The two solutions (8.772) for φ_0 differ in the sign of $\cos\varphi_0$. If $\cos(\varphi_0) > 0$, there is a positive growth rate $\Gamma_1 > 0$, and the corresponding fixed point is unstable. Conversely, (8.792) for $\cos(\varphi_0) < 0$ yields no unstable mode; on the contrary: the third order, which we will not explicitly calculate here, yields a damping decrement.

Numerical Simulations

With this, we conclude the presentation of the analytical part and turn to the numerical solution of (8.732).

It should first be emphasized that the numerical results reflect the existence and stability of the fixed points as described above. In particular, a phase-coupled soliton exists as a stable fixed point within a certain parameter range.

In Fig. 8.12 we demonstrate the spatiotemporal behavior for $a = 0.1$ and $\gamma = 0.11$. To this end, we have solved the partial differential equation (8.732) using the phase-locked soliton (or a distribution close to this resonant soliton) as the initial distribution; for $a \approx 0.1$ and $\gamma = 0.11$ the solution evolves towards the distribution shown in Fig. 8.12 as an attractor.

However, if we increase (while keeping the damping decrement γ fixed) the amplitude a of the driver (e.g., to $a = 0.155$ in Fig. 8.13), the behavior changes qualitatively. The phase-locked soliton becomes unstable (Hopf bifurcation), and a temporally modulated solution becomes the new attractor. Interestingly, the spatial coherence largely persists, which can also be seen, at least approximately, in the spatiotemporal representation Fig. 8.13.

Fig. 8.12 Spatiotemporal behavior of a soliton-like solution of (8.732) for $a = 0.1$ and $\gamma = 0.11$

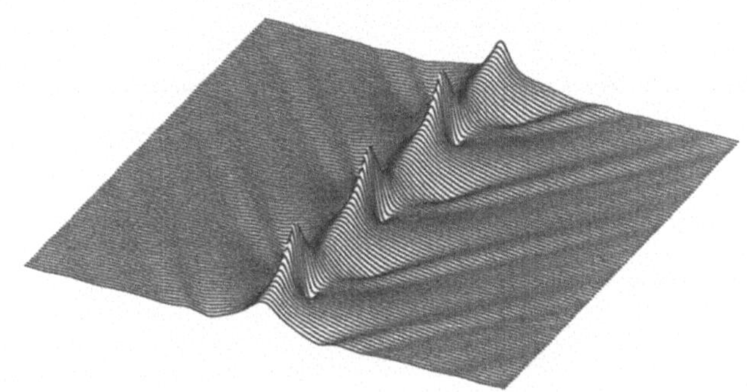

Fig. 8.13 Spatiotemporal behavior of a soliton-like solution of (8.732) for $a = 0.155$ and $\gamma = 0.11$

This raises the question of the temporal behavior in spatial coherence. Numerical analysis shows that the system becomes temporally chaotic (e.g., as seen in frequency spectra). The transition via period doubling à la Feigenbaum is illustrated in Figs. 8.14–8.16. There, in a reduced phase space, the amplitude is plotted against the phase of the solution at $x = 0$. Figure 8.14 illustrates once again the Hopf bifurcation, i.e., the transition from the fixed point (phase-locked soliton) to the limit cycle (spatially coherent solution modulated with *one* frequency). In Fig. 8.15, for the values $a \approx 0.14$ and 0.148, the period doubling in the limit cycle is illustrated, while Fig. 8.16 for $a \approx 0.1504$ and $a \approx 0.16$ presents an 8-cycle and temporal chaos.

Poincaré map

This transition can be precisely analyzed using many diagnostic methods. Here, we discuss only a particularly illustrative evaluation, which leads to the Feigenbaum diagram, Fig. 8.17. In the vicinity of a periodic solution of a differential equation, a Poincaré map can be defined as follows. One selects a hypersurface $\sum$ in phase space that intersects the periodic orbit at a certain point. Trajectories neighboring the periodic orbit also intersect $\sum$. The Poincaré map P now assigns to such an intersection point (n) the next intersection point $(n + 1)$ of the trajectory with $\sum$.

We can numerically define the hypersurface $\sum$ using the condition

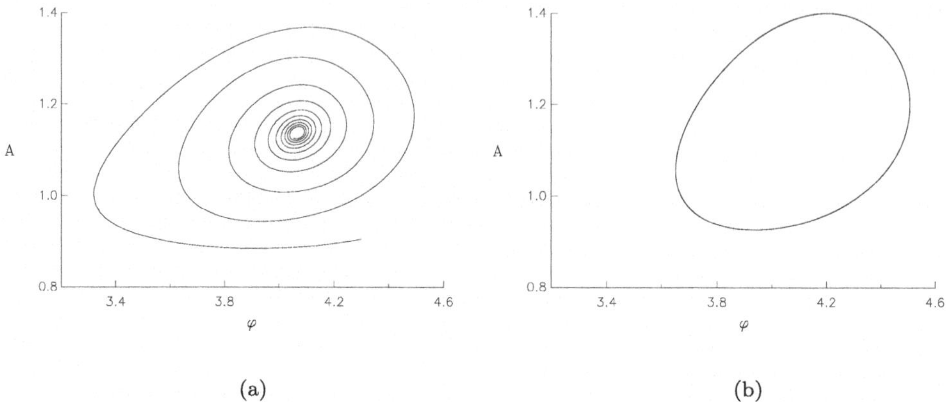

Fig. 8.14 Reduced phase portraits from the numerical solutions of (8.732) for $a = 0.10$ (left) and $a = 0.11$ (right) at $\gamma = 0.11$

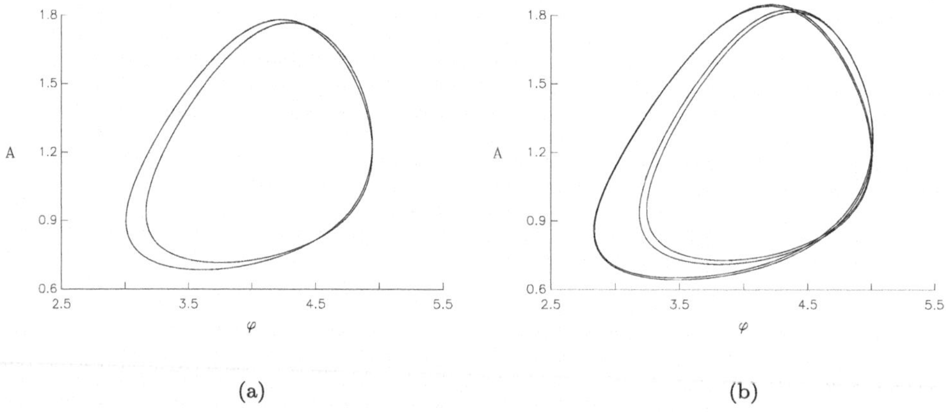

Fig. 8.15 Reduced phase portraits from the numerical solutions of (8.732) for $a = 0.14$ (left) and $a = 0.148$ (right) at $\gamma = 0.11$

$$\frac{\partial}{\partial t}\arg(q(x=0, t_{Pn})) = 0 \quad \text{and} \quad \frac{\partial^2}{\partial t^2}\arg(q(x=0, t_{Pn})) > 0 \tag{8.793}$$

The practical advantage of this choice is that in every periodic orbit, $\arg(q(x = 0, t))$ is necessarily minimal at least once, so $\sum$ is globally defined. Moreover, the phase of the solution can be easily tracked.

The reduction to a map allows for a very compact representation. A limit cycle is a fixed point of the Poincaré map P, a period-n cycle is a so-called n-periodic point of P, i.e., $P^n(q) = q$. If we plot $|q(x = 0, t_P)|$ of these periodic points for various values of the bifurcation parameter, we obtain a bifurcation diagram as shown in Fig. 8.17. There, for

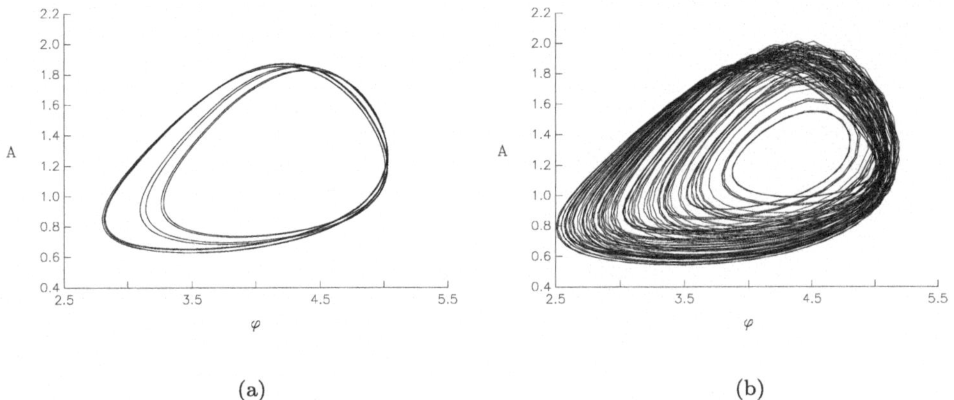

Fig. 8.16 Reduced phase portraits from the numerical solutions of (8.732) for $a = 0.1504$ (left) and $a = 0.16$ (right) at $\gamma = 0.11$

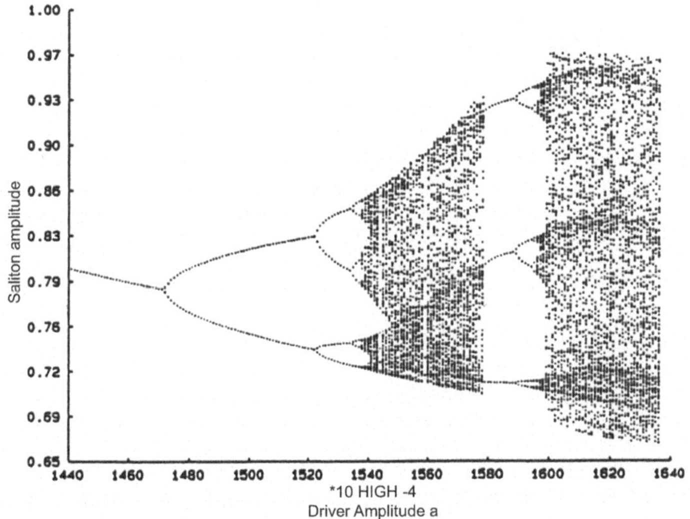

Fig. 8.17 Route to chaos with "solitons" via period doubling

fixed $\gamma = 0.11$, the driver a was varied. The Feigenbaum scaling can be read off from this. The appearance of period-3 windows, etc., also agrees well with results from simple dynamical systems (e.g., in logistic maps).

Collective coordinates

In conclusion, we would like to briefly discuss a theoretical model that explains the phenomena just described quite well. It quickly becomes apparent that a single soliton without further perturbations is not sufficient to determine the relevant collective coordinates. Starting from the two-soliton solution

$$\boxed{q(x,t) = 4ia\frac{e^{-i\phi_1}\eta_1\cosh(2\eta_2 x) + e^{-i\phi_2}\eta_2\cosh(2\eta_1 x)}{a^2\cosh(2(\eta_1-\eta_2)x) + \cosh(2(\eta_1+\eta_2)x) + b\cos(\phi_1-\phi_2)}} \tag{8.794}$$

with

$$a = \frac{\eta_1-\eta_2}{\eta_1+\eta_2}, \qquad b = \frac{4\eta_1\eta_2}{(\eta_1+\eta_2)^2}, \tag{8.795}$$

$$\phi_i = -4\eta_i^2 t + \phi_{i,0}, \ \eta_1 > \eta_2 \geq 0 \tag{8.796}$$

we can obtain for small η_2

$$\begin{aligned} q(x,t) &= 2i\eta_1 e^{-i\phi_1}\,\mathrm{sech}\,(2\eta_1 x) \\ &\quad - 2i\eta_2\Big(e^{-2i\phi_1+i\phi_2}\,\mathrm{sech}^2(2\eta_1 x) - e^{-i\phi_2}\tanh^2(2\eta_1 x)\Big) \end{aligned} \tag{8.797}$$

We interpret the second term on the right-hand side of (8.797) as "$k=0$ radiation."

With

$$\rho = 2i\frac{\pi}{f}\eta_2 e^{i\phi_2}\,, \tag{8.798}$$

$\eta_1 = \eta$ and $\phi_1 = \chi + \pi$ we can write

$$\begin{aligned} r(x,t) &= -i2\eta(t)\,\mathrm{sech}\,(2\eta(t)x)e^{-i\chi(t)} \\ &\quad - \frac{f}{\pi}\Big[\rho(t)e^{-2i\chi(t)}\,\mathrm{sech}^2(2\eta(t)x) + \rho^*(t)\tanh^2(2\eta(t)x)\Big] \end{aligned} \tag{8.799}$$

and, for an ansatz, introduce the collective coordinates η, χ, ρ and ρ^*, where f is a free parameter that [in the always finite domain in practice] represents a normalization constant.

If we insert (8.799) into the Lagrangian density corresponding to (8.755)

$$\begin{aligned} \mathcal{L} = \Big\{&\frac{i}{2}(r_t r^* - r_t^* r) + |r|^4 - |r_x|^2 - \omega|r|^2 \\ &+ 2c_0|r|^2(r^* + r) + c_0^2(r^{*2} + r^2) + 4c_0^2|r|^2\Big\}e^{2\gamma t} \end{aligned} \tag{8.800}$$

and subsequently form

$$L = \int_{-\infty}^{+\infty} \mathcal{L} dx \tag{8.801}$$

we obtain

$$\begin{aligned} L = \Big\{ & 4\eta(\chi_t - \omega) + \frac{16}{3}\eta^3 - 8\pi c_0 \eta^2 \sin(\chi) - 8\pi c_0^2 \eta \cos(2\chi) + 16 c_0^2 \eta \\ & + f \frac{d}{dt}(|\rho| \cos(\chi - \varphi)) + \frac{f}{\pi}|\rho|^2(\varphi_t - \omega) \\ & - 16 \frac{f}{\pi} c_0 \eta |\rho| \cos(2\chi - \varphi) \Big\} e^{2\gamma t}. \end{aligned} \tag{8.802}$$

From this, using $\rho = \rho_r + i\rho_i$, the equations of motion follow

$$\boxed{\begin{aligned} \eta_t = & 2\gamma\eta - 2\pi c_0 \eta^2 \cos(\chi) + 4 c_0^2 \eta \sin(2\chi) \\ & + \gamma \frac{f}{2}(\rho_r \sin(\chi) - \rho_i \cos(\chi)) + 8 \frac{f}{\pi} c_0 \eta [\rho_r \sin(2\chi) - \rho_i \cos(2\chi)], \end{aligned}} \tag{8.803}$$

$$\boxed{\begin{aligned} \chi_t = & \omega - 4\eta^2 + 4\pi c_0 \eta \sin(\chi) + 2 c_0^2 \cos(2\chi) - 4 c_0^2 \\ & + 4 \frac{f}{\pi} c_0 [\rho_r \cos(2\chi) + \rho_i \sin(2\chi)], \end{aligned}} \tag{8.804}$$

$$\boxed{\rho_{rt} = -\omega\rho_i - \gamma\rho_r - \pi\gamma \sin(\chi) - 8\eta c_0 \sin(2\chi),} \tag{8.805}$$

$$\boxed{\rho_{it} = \omega\rho_r - \gamma\rho_i + \pi\gamma \cos(\chi) + 8\eta c_0 \cos(2\chi).} \tag{8.806}$$

It is apparent that these equations pass an initial consistency test. They reproduce the exact time dependence of the parameters of a soliton interacting with "$k = 0$ radiation" for the undamped and undriven case ($c_0 = 0, \gamma = 0$):

$$\eta_t = 0, \quad \chi_t = \omega - 4\eta^2, \quad \rho_t = i\omega\rho. \tag{8.807}$$

Moreover, they reproduce the results of the partial differential equation in a surprisingly accurate manner. Fixed points, Hopf bifurcation, period doubling, and chaos even quantitatively agree with the numerical values from the partial differential equation.

Meanwhile, in Schrödinger systems with other perturbation terms, we have come to know further generic behaviors such as intermittency and the quasiperiodic transition to chaos.

For reasons of space, the example presented here in detail should suffice as an insight into this new and fascinating field.

8.7 Theory of Weak Turbulence

The approaches presented so far in this chapter originate primarily from the field of nonlinear dynamics, which has been steadily developing for decades. With its methodology, the phenomenon of turbulence becomes accessible, and there is much to suggest that the statistical phenomenon of fully developed turbulence should be interpreted as the stochasticity of nonlinearity in the (nonlinear) fundamental equations. For plasmas, theories have existed for many decades that, despite their sometimes rather ad hoc assumptions, show a remarkably good agreement with experimental results. Two essential assumptions are made in the case of weak turbulence in plasmas: The existence of an ensemble of *statistically* fluctuating fields is always assumed (as a consequence of a linear instability), and the ratio of the energy density of the fields to the thermal energy density of the plasma particles is used as a small expansion parameter.

The description and understanding of nonlinear processes is still by no means complete or concluded today—about 35 years after the first edition appeared. It is quite certain that decisive progress can be expected in this area in the coming years. This applies to turbulence research in general and, of course, also to the theory of so-called turbulent plasmas in particular. Even under the assumption of "weak" turbulence, a quantitative theory is still extremely complicated, since several processes occur simultaneously. For better and clearer understanding, we discuss individual processes separately here and provide a brief overview of

– wave-particle interaction within the framework of a quasilinear theory,—induced scattering of waves by particles (e.g., nonlinear Landau damping),

– aspects of wave-wave interaction within the framework of weak turbulence.

For formal reasons, we formulate the corresponding approaches only in one spatial dimension.

Quasilinear Theory of Wave-Particle Interaction

We begin with the simplest idea of how nonlinear turbulent processes can be taken into account in the development of unstable modes. If we consider, for example, the instability criterion

$$\gamma_k \sim \left.\frac{\partial f}{\partial v}\right|_{v=\frac{w_r}{k}} > 0 \tag{8.808}$$

within the framework of a linear Vlasov theory, it is noticeable that—as long as the single-particle distribution function does not change with time—the excitation rate of resonant waves ($\omega_r = kv$) is constant in time.

The unstable modes grow exponentially due to the instability. Because of the law of energy conservation, a feedback effect on the plasma particles (and their thermal energy) is inevitable. Accordingly, f, which appears in (8.808), will not remain unchanged in the long term. It is precisely this—and only this—aspect that is captured by quasilinear theory.

We illustrate quasilinear theory using the example of the Vlasov equation for electrons, where the ions appear only as a stationary background that ensures all electric fields vanish to zeroth order. For high-frequency wave phenomena, this model is justified.

In the Vlasov equation for electrons ($f = f^e$)

$$\boxed{\partial_t f + v\partial_x f - \frac{e}{m_e} E\partial_v f = 0} \tag{8.809}$$

integration has already been performed over the velocities v_y and v_z, and $v_x \equiv v$ has been set. We decompose the distribution function $f = f(x, v; t)$ as

$$f(x, v; t) = f_0(v; t) + f_1(x, v; t) \tag{8.810}$$

where f_0 is the spatially averaged (background) distribution function

$$f_0(v; t) = \langle f(x, v; t)\rangle = \lim_{L\to\infty} \frac{1}{L} \int_{-L/2}^{L/2} dx f(x, v; t) \tag{8.811}$$

We assume

$$\langle E\rangle = 0, \tag{8.812}$$

i.e., the electric field is to be generated only by f_1, and that no external field is present, so that $E = E_1$ can be determined by

$$\boxed{\partial_x E = -\frac{e}{\varepsilon_0} \int_{-\infty}^{+\infty} dv f_1(x, v; t)} \tag{8.813}$$

By averaging (8.809), we obtain

$$\partial_t\langle f\rangle + v\langle\partial_x f\rangle - \frac{e}{m_e}\langle E\partial_v f\rangle = 0. \tag{8.814}$$

The second term on the left-hand side

$$v\langle\partial_x f\rangle = v\lim_{L\to\infty}\frac{1}{L}\left[f\left(x=\frac{L}{2}\right)-f\left(x=-\frac{L}{2}\right)\right]=0 \tag{8.815}$$

can be neglected with the appropriate boundary conditions, so that from (8.814)

$$\boxed{\partial_t f_0(v;t) = \frac{e}{m_e}\partial_v\langle Ef_1(x,v;t)\rangle} \tag{8.816}$$

results. Equations (8.813) and (8.816) represent the essential nonlinear contributions; the relationship between f_1 and E is established in the same way as in the linearized Vlasov theory. This means in particular that we compute a solution of the (linear) dispersion relation [cf. (4.2.65)] for $k > 0$ as

$$\omega(k;t) = \omega_{pe}\left(1+\frac{3}{2}k^2\lambda_{De}^2\right) + i\frac{\pi}{2}\frac{\omega_{pe}^3}{k^2}\left.\frac{\partial g}{\partial v}\right|_{v=\frac{w_r}{k}} \tag{8.817}$$

A word of explanation on this.

The function $g(v;t) = n_0^{-1}f_0(v;t)$ is the zeroth-order distribution function, which at time $t=0$ is supposed to produce a positive growth rate $\gamma = \text{Im}\ \omega$ according to (8.817). The excited waves act back on the distribution function on the timescale $T\sim\gamma^{-1}$. Nevertheless, under the assumption $\gamma \ll \omega_r$ we have performed a Fourier transformation (temporarily on a scale $\tau \ll T$)). The justification for this is provided by a multiple-scale formalism. If the distribution function f_0 changes slowly, this is associated with a modification of the instability criterion, which can lead to the disappearance of the instability.

A Fourier transformation of the Vlasov equation (8.809), linearized with the ansatz (8.810) around f_0, leads, using the definition

$$\begin{aligned} f_1(x,v;t) &= \int_{-\infty}^{\infty} dk\, f_1(k,v;t)e^{ikx} \\ &= \int_{-\infty}^{\infty} dk\, \tilde{f}_1(k,v)e^{-i\omega(k;t)t+ikx} \end{aligned} \tag{8.818}$$

to

$$f_1(k,v;t) = -\frac{(e/m_e)}{i[\omega(k;t)-kv]}[\partial_v f_0(v;t)]E(k;t), \tag{8.819}$$

where $E(x;t)$ has been represented in Fourier space in analogy to (8.818).

Now we calculate

$$\begin{aligned}\langle Ef_1\rangle &= \lim_{L\to\infty}\frac{1}{L}\int_{-L/2}^{L/2} dx\, E(x;t)f_1(x,v;t)\\ &= \lim_{L\to\infty}\frac{1}{L}\int_{-L/2}^{L/2} dx\int_{-\infty}^{\infty} dk_1 f_1(k_1,v;t)\int_{-\infty}^{\infty} dk_2 E(k_2;t)e^{i(k_1+k_2)x}\\ &= 2\pi\frac{1}{L_\infty}\int_{-\infty}^{\infty} dk\, f_1(k,v;t)E(-k;t),\end{aligned} \tag{8.820}$$

where we have denoted the length of the system by $L_\infty(\to\infty)$. Taking into account (8.819), this yields (8.816) and (8.820)

$$\partial_t f_0(v;t) = -\frac{e^2}{m_e^2}\frac{2\pi}{L_\infty}\partial_v\int_{-\infty}^{+\infty} dk\, E(-k;t)E(k;t)\frac{\partial_v f_0(v;t)}{i(\omega-kv)}. \tag{8.821}$$

We write this equation in a more suggestive form as

$$\boxed{\partial_t f_0 = \partial_v D\partial_v f_0} \tag{8.822}$$

with

$$\boxed{D = \frac{2\pi e^2}{m_e^2 L_\infty}\int_{-\infty}^{+\infty} dk\frac{|E(k;t)|^2}{(-i\omega+ikv)}}\,. \tag{8.823}$$

Note that, due to the real field $E(x;t)$ and the properties of the Fourier transformation, $\omega_r(-k)=-\omega_r(k)$, $\omega_i(-k)=\omega_i(k)$ and $E(k;t)=E^*(-k;t)$ hold.

We define the spectral energy density of the field,

$$\mathcal{E}(k;t) := \frac{1}{4L_\infty}|E(k;t)|^2\,, \tag{8.824}$$

from which the averaged energy density of the field

$$\begin{aligned}\left\langle\frac{E^2}{8\pi}\right\rangle &= \frac{1}{8\pi}\lim_{L\to\infty}\frac{1}{L}\int_{-L/2}^{L/2} dx\, E^2(x;t)\\ &= \int_{-\infty}^{+\infty} dk\,\mathcal{E}(k;t)\end{aligned} \tag{8.825}$$

can be obtained by integration. Instead of (8.823) we now use

$$D = D(v;t) = -\frac{8\pi e^2}{m_e^2}\int_{-\infty}^{+\infty} dk\frac{\mathcal{E}(k;t)}{i[\omega(k;t)-kv]} \tag{8.826}$$

in (8.822). The temporal evolution of $\mathcal{E}(k;t)$ is, due to

$$E(k; t) = \tilde{E}(k) \exp[-i\omega(k; t)t], \tag{8.827}$$

$$\partial_t E(k; t) \approx -i\omega(k; t)E(k; t) \ , \tag{8.828}$$

$$\omega(k; t) = -\omega^*(-k; t) \tag{8.829}$$

given by

$$\partial_t \mathcal{E}(k; t) = 2\omega_i(k; t)\mathcal{E}(k; t) \tag{8.830}$$

($\gamma = \omega_i$) with the solution

$$\boxed{\mathcal{E}(k; t) = \mathcal{E}(k; t = 0) \exp\left\{2 \int_0^t \omega_i(k; t')dt'\right\}} \tag{8.831}$$

provided.

Equations (8.822) and (8.831) constitute the fundamental equations of quasilinear theory, where at each time D is calculated from (8.826) and $\omega_i(k; t)$ is obtained in each case from the Vlasov equation linearized with respect to f_0.

Example 8.29 (Real Diffusion Coefficient)
The diffusion coefficient (8.826) is real, since we can write

$$D = \frac{8\pi e^2}{m_e^2} \int_{-\infty}^{+\infty} \frac{i\mathcal{E}(k; t)[\omega_r - kv - i\omega_i]}{(\omega_r - kv)^2 + \omega_i^2} dk \tag{8.832}$$

and $\mathcal{E}(k; t) = \mathcal{E}(-k; t)$ holds. Together with (8.829) it then follows that

$$D = \frac{8\pi e^2}{m_e^2} \int_{-\infty}^{+\infty} dk \frac{\mathcal{E}(k; t)\omega_i(k; t)}{[\omega_r(k; t) - kv]^2 + [\omega_i(k; t)]^2} . \tag{8.833}$$

■

Example 8.30 (Contributions to the Diffusion Coefficient)
Another remark arises from the evaluation of (8.826) for $\omega_i \to 0$.. Using the Plemelj formula

$$\lim_{\varepsilon \to 0} \frac{1}{a + i\varepsilon} = \mathcal{P}\frac{1}{a} - i\pi\delta(a) \tag{8.834}$$

we can separate the contributions into resonant particles (due to the Dirac function) and non-resonant particles (contribution of the principal value $\mathcal{P}$). We find for $\omega_i \to 0$

$$D = \frac{8\pi e^2}{m_e^2} \int_{-\infty}^{+\infty} dk \left[i\mathcal{P} \left\{ \frac{\mathcal{E}(k;t)}{\omega_r - kv} \right\} + \pi\delta(\omega_r - kv)\mathcal{E}(k;t) \right]$$
$$\approx \frac{16\pi^2 e^2}{m_e^2} \frac{1}{v} \mathcal{E}(k = \frac{\omega_r}{v}; t) \quad \text{for} \quad \omega_i \to 0 \,. \tag{8.835}$$

The non-resonant part [$\sim \omega_i$, cf. (8.833)] vanishes for $\omega_i \to 0$.. However, it should be noted in balances that, in general, there are significantly more non-resonant particles than resonant ones. As a result, an effect due to the non-resonant part can occur, even though the resonant part of the diffusion coefficient dominates.

The quasilinear description accomplishes what we expect of it. We illustrate this statement with the following examples. ■

If we multiply (8.822) for an unstable situation by f_0,, we obtain for the positive functional

$$H = \frac{1}{2} \int dv (f_0)^2 \tag{8.836}$$

$$\frac{dH}{dt} = - \int dv \; D(\partial_v f_0)^2 = - \int dv \frac{8\pi e^2}{m_e^2} \int dk \frac{\mathcal{E}(k;t)\omega_i}{[\omega_r - kv]^2 + w_i^2} (\partial_v f_0)^2 . \tag{8.837}$$

Since $H > 0$ and $dH/dt < 0$ [because $\mathcal{E} > 0$ for $\omega_i > 0$ and $\mathcal{E} = 0$ otherwise], H decreases continuously as long as the system is unstable. In the time-asymptotic limit, $dH/dt = 0$ must hold, which corresponds to the marginally stable case. [For this, we must also exclude $E(k)\partial_v f_0 = 0$, but also $\omega_i \neq 0$: According to (8.819), however, this would result in $f_1(k, v; t) = 0$ and thus no excited waves in this region.]

The quasilinear equations satisfy the conservation of particles, momentum, and energy.

Example 8.31 (Conservation of Momentum)
Because $\langle f_1 \rangle = 0$ it follows from

$$\frac{d}{dt} \int dv f_0 = \int dv \partial_v D \partial_v f_0 = 0 \tag{8.838}$$

the conservation of $\langle n \rangle$. To demonstrate momentum conservation, we calculate

$$m_e \int dv v \frac{\partial f_0}{\partial t} = -m_e \int dv D \partial_v f_0 = \frac{8\pi e^2}{m_e} \int dk \; \mathcal{E} \int \frac{\partial_v f_0}{i[\omega - kv]} dv$$
$$= 2i \int_{-\infty}^{+\infty} dk \mathcal{E}(k;t) k = 0, \tag{8.839}$$

where the penultimate transformation results from the dispersion relation (4.168). Since $\mathcal{E}(k;t)$ is even in k, the expression vanishes and thus

$$\boxed{\frac{d}{dt}\langle p\rangle = 0}\,. \tag{8.840}$$

[Note: the electrostatic modes carry no momentum!] ∎

Example 8.32 (Conservation of Energy)
To demonstrate energy conservation, we proceed as follows:

$$\begin{aligned}\frac{d}{dt}\int dv\frac{1}{2}m_e v^2 f_0 &= -m_e\int dv\; vD\;\partial_v f_0\\ &= \frac{8\pi e^2}{m_e}\int dv\int_{-\infty}^{+\infty} dk\frac{1}{k}\mathcal{E}(k;t)\frac{(\partial_v f_0)[kv-\omega+\omega]}{i[\omega-kv]}\\ &= -2\omega_i\int_{-\infty}^{+\infty} dk\;\mathcal{E}(k;t),\end{aligned} \tag{8.841}$$

where again the linear dispersion relation was used. Taking into account (8.830), we can also write the result just obtained as

$$\boxed{\frac{d}{dt}\left[\int dv\frac{1}{2}m_e v^2 f_0 + \int dk\mathcal{E}(k;t)\right] = 0} \tag{8.842}$$

the sum of (kinetic) particle energy and (electrostatic) wave energy is therefore constant. ∎

Example 8.33 ("bump on tail" instability)
We now apply quasilinear theory to a situation ("weak bump on tail instability") in which a deformation of the (see Fig. 8.18) gives rise to a weak instability ($\gamma/\omega_r \ll 1$). The velocity interval $v_1 < v < v_2$ in Fig. 8.18 corresponds, due to $\omega_r \approx \omega_{pe}$ in k-space, to the intervals $k_2 < k < k_1$ and $-k_1 < k < -k_2$; with $\omega_{pe}/k_1 \approx v_1$ and $\omega_{pe}/k_2 \approx v_2$. holding. Due to the growth rate

$$\boxed{\omega_i \approx \frac{\pi}{2}\omega_k\frac{\omega_{pe}^2}{|k|^3}k\frac{1}{n_0}\left.\frac{\partial f_0}{\partial v}\right|_{v=\omega_k/k}} \tag{8.843}$$

modes grow in the aforementioned k-intervals.

The diffusion coefficient D (8.826) can, based on the Plemelj formula (8.834), be split into a resonant and a non-resonant component. The resonant component [cf. (8.835)]

$$D^r \approx \frac{16\pi^2 e^2}{m_e^2}\frac{1}{v}\mathcal{E}\left(k=\frac{\omega_r}{v};t\right) \tag{8.844}$$

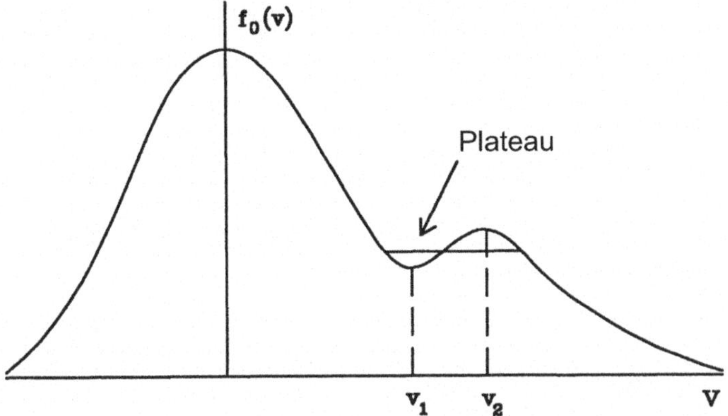

Fig. 8.18 Schematic representation of an unstable distribution function with plateau formation in the framework of quasilinear theory

leads in (8.822) to a "smearing out" of the gradients of the distribution function, i.e., one can expect $\omega_i \to 0$ for $t \to \infty$ to occur.

The non-resonant component [cf. (8.833)]

$$\begin{aligned} D^{nr} &\approx \frac{8\pi e^2}{m_e^2} \int_{-\infty}^{+\infty} dk \frac{\mathcal{E}(k;t)\omega_i}{[\omega_r - kv]^2} \\ &\approx \frac{2}{n_o m_e} \int_{-\infty}^{+\infty} dk \; \omega_i \mathcal{E}(k;t) \\ &\approx \frac{1}{n_0 m_e} \frac{\partial}{\partial t} \int_{-\infty}^{+\infty} dk \mathcal{E}(k;t) \end{aligned} \tag{8.845}$$

for $v \ll \omega_r/k$ at $k^2\lambda_{De}^2 \ll 1$ leads to

$$\partial_t f_0 = \frac{1}{n_0 m_e} \left[\frac{\partial}{\partial t} \int_{-\infty}^{+\infty} dk \; \mathcal{E}(k;t) \right] \partial_v^2 f_0. \tag{8.846}$$

With the new variable

$$\tau(t) = \frac{2}{n_0} \int_{-\infty}^{+\infty} dk \; \mathcal{E}(k;t) \tag{8.847}$$

(8.846) can be written as

$$\partial_\tau f_0(v;\tau) = \frac{1}{2m_e} \partial_v^2 f_0(v;\tau),$$

i.e., as the standard diffusion equation. With the help of the Green's function G, we write the solution to the initial value problem as

$$f_0(v; \tau(t)) = \int dv' f_0(v'; \tau(0)) G(v', v; \tau(t)) \tag{8.848}$$

with

$$G(v', v; \tau(t)) = \left\{ \frac{m_e}{2\pi[\tau(t) - \tau(0)]} \right\}^{1/2} \exp\left\{ \frac{m_e(v' - v)^2}{2[\tau(t) - \tau(0)]} \right\}. \tag{8.849}$$

For the non-resonant particles, we can assume a Maxwell distribution for f_0, so that (8.848) can be easily evaluated with the result

$$\boxed{f_0(v; t) = \left\{ \frac{m_e}{2\pi[k_B T_e + \tau(t) - \tau(0)]} \right\}^{1/2} \exp\left\{ \frac{m_e v^2}{2[k_B T_e + \tau(t) - \tau(0)]} \right\}} \tag{8.850}$$

We observe a broadening of the distribution function, as schematically shown in Fig. 8.19. The transition

$$\frac{1}{2} k_B T_e \rightarrow \frac{1}{2} k_B T_e + \Delta W \tag{8.851}$$

shows us that the non-resonant particles *gain* the energy

$$\Delta W \equiv \frac{1}{n_0} \int_{-\infty}^{+\infty} dk[\mathcal{E}(k; \infty) - \mathcal{E}(k; 0)] \tag{8.852}$$

According to the energy theorem (8.842), this means that the resonant particles *lose* the energy $-2\Delta W$. In non-resonant diffusion, we again have an example where the individual process per particle is vanishingly small, but due to the large number of particles, a summation leads to finite contributions. ■

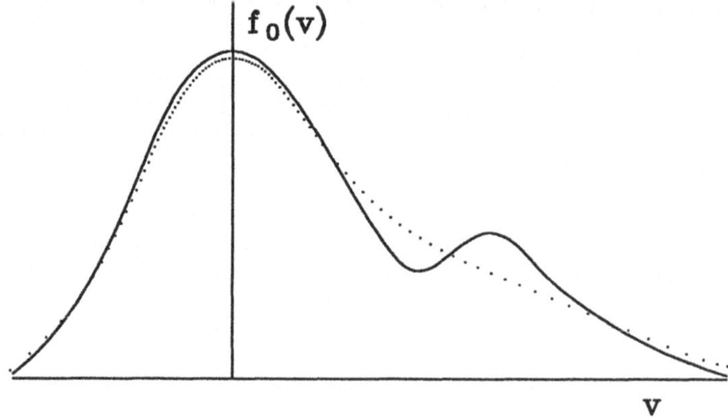

Fig. 8.19 Schematic representation of the influence of quasilinear diffusion in velocity space on the distribution function

Induced Scattering of Waves by Particles

The process of quasilinear diffusion just examined is based on the particle-wave interaction and its feedback on the averaged distribution function f_0.. However, it is easy to imagine further nonlinear processes that may play a role in a (weakly) turbulent plasma. In this section, we discuss whether, under certain circumstances, the interaction of a particle with two waves can give rise to qualitatively new effects.

For such a (turbulent) elementary process, the resonance condition

$$\boxed{\omega_1 - \omega_2 = (k_1 - k_2)v} \tag{8.853}$$

will be given, if ω_1, k_1 belongs to the first and ω_2, k_2 to a second wave. Resonances, for example with the ions, can be expected to occur to a greater extent if

$$\frac{\omega_1 - \omega_2}{k_1 - k_2} \sim \mathcal{O}(v_{ti}) \tag{8.854}$$

lies in the range of the thermal velocity of the ions. Equation (8.854) is, for example, satisfied for two (high-frequency) Langmuir oscillations. From previous chapters, we know in principle how to treat such processes. We describe the Langmuir oscillations using a hydrodynamic model, while the low-frequency perturbations (due to the particle-wave interaction) must be treated kinetically.

In detail, this means

$$\boxed{\partial_t n_e + \partial_x(n_e v_e) = 0,} \tag{8.855}$$

$$\boxed{m_e n_e[\partial_t v_e + v_e \partial_x v_e] = -\gamma_e k_B T_e \partial_x n_e - e n_e E,} \tag{8.856}$$

$$\boxed{\partial_x E = -4\pi e n_e,} \tag{8.857}$$

$$\boxed{\partial_t \delta f_i + v \partial_x \delta f_i + \frac{e}{m_i} E \partial_v f_{i0} = 0.} \tag{8.858}$$

With the following explanations.

(8.855)–(8.857) are the hydrodynamic basic equations for (high-frequency) Langmuir oscillations, in which the ions play only the role of a "smeared-out background." We also use them to describe the electron component of the low-frequency perturbations. As far as the electrons are concerned, no resonances are to be taken into account, while (8.854) requires a kinetic description of the (low-frequency) ion component. The ions are described by the Vlasov equation (8.858), where in the splitting $f_i = f_{i0}(v) + \delta f_i$ the term $\partial_v \delta f_i$ was neglected compared to $\partial_v f_{i0}$.

The (two) high-frequency fields are represented by

$$E = E_1 e^{-i\omega_1 t + ik_1 x} + E_2 e^{-i\omega_2 t + ik_2 x} + c.c. \tag{8.859}$$

Let E_1 be large compared to E_2; the frequency ω_1 is assumed to be real. For the perturbation, we make the ansatz $\sim \exp(-i\omega_3 t + ik_3 x) + c.c.$, and assume

$$\omega_1 - \omega_2 = \omega_3^*, \tag{8.860}$$

$$k_1 - k_2 = k_3 \tag{8.861}$$

with $\omega_1 \approx \omega_{2r} \gg \omega_{3r}$.

We first discuss mode 2 with frequency ω_2 and the wave vector k_2. We denote the corresponding contributions by the index 2. Equations (8.855) and (8.856) yield for the harmonic $\exp(-i\omega_2 t + ik_2 x)$

$$-i\omega_2 n_2 + ik_2(n_0 v_2 + n_1 v_3^* + n_3^* v_1) = 0, \tag{8.862}$$

$$-i\omega_2 v_2 - ik_3 v_3^* v_1 + ik_1 v_1 v_3^* = -3ik_2 \frac{k_B T_e}{m_e n_0} n_2 - \frac{e}{m_e} E_2 \,. \tag{8.863}$$

As in previous chapters, nonlinearities in the pressure term have been neglected based on similar arguments. Furthermore, only the three-mode interaction has been taken into account.

If we substitute (8.863) into (8.862), we obtain

$$-i\omega_2 n_2 = -ik_2 \left\{ v_3^* \left[n_1 - n_0 \frac{k_3}{\omega_2} v_1 + n_0 \frac{k_1}{\omega_2} v_1 \right] + n_3^* v_1 + \frac{3k_2 v_{te}^2 n_2}{\omega_2} + \frac{n_0 e}{i\omega_2 m_e} E_2 \right\}. \tag{8.864}$$

Example 8.34 (Orders of Magnitude)

On the right-hand side of this equation, for the first term $v_3^*[\ldots] \ll n_3^* v_1$ applies. The reason is

$$n_1 \approx n_0 \frac{k_1}{\omega_1} v_1 \approx n_0 \frac{k_1}{\omega_2} v_1, \tag{8.865}$$

and therefore

$$v_3^*[\ldots] \approx \frac{v_3^* v_1 n_0}{\omega_2}(k_1 - k_3 + k_1)$$
$$= \frac{v_3^* v_1 n_0}{\omega_2}(k_1 + k_2). \tag{8.866}$$

On the other hand, we have $n_3^* \approx n_0 v_3^* k_3/\omega_3^*$, from which

$$n_3^* v_1 \approx n_0 v_3^* v_1 k_3/\omega_3^* \tag{8.867}$$

follows. If we compare (8.866) and (8.867) under the assumption $\omega_3 \ll \omega_1, \omega_2$ and $k_1 \approx k_2 \approx k_3$,, the neglect stated above is suggested. ■

The high-frequency electric field E_2 is obtained from (8.857) as

$$ik_2 E_2 = -4\pi e n_2. \tag{8.868}$$

Simple rearrangements then yield from (8.864)

$$\boxed{(\omega_2^2 - 3k_2^2 v_{te}^2)\varepsilon(\omega_2, k_2) n_2 = \omega_2 k_2 n_3^* v_1}\,; \tag{8.869}$$

ε is the (high-frequency) dielectric constant

$$\varepsilon(\omega, k) \approx 1 - \frac{\omega_{pe}^2}{\omega^2 - 3k^2 v_{te}^2}. \tag{8.870}$$

For further evaluation of (8.869) we need a relation between the (low-frequency) density perturbation n_3 and n_2, v_1. To lowest order, for v_1 it follows from the momentum balance of the electrons

$$v_1 \approx \frac{eE_1}{i\omega_1 m_e}. \tag{8.871}$$

The quantity $n_3 = n_{e3} \approx n_{i3}$ is determined for low-frequency (isothermal: $\gamma_e = 1$) changes from

$$-i\omega_3 v_3 - v_1 ik_2 v_2^* + ik_1 v_1 v_2^* = -i\frac{k_B T_e}{n_0 m_e} k_3 n_3 - \frac{e}{m_e} E_3 \tag{8.872}$$

or

$$E_3 \approx -\frac{k_B T_e}{e n_0} ik_3 n_3 - \frac{im_e}{n_0 e}\frac{k_3}{k_2} v_1 \omega_2^* n_2^*. \tag{8.873}$$

Ultimately, the Vlasov equation for ions yields the low-frequency contribution (note the normalization in the one-dimensional Vlasov equation)

$$n_3 = \int_{-\infty}^{+\infty} dv\, \delta f_{i3} = E_3 \int_{-\infty}^{+\infty} dv \frac{e/m_i}{i(\omega_3 - k_3 v)} \partial_v f_{i0}$$
$$= -\frac{m_e v_1 \omega_2^* n_2^*}{k_2 k_B T_e} \frac{W}{1+W}, \tag{8.874}$$

if we express E_3 in terms of (8.873) and define

$$W = \frac{k_3 k_B T_e}{m_i n_0} \int_{-\infty}^{+\infty} dv \frac{\partial_v f_{i0}}{\omega_3 - k_3 v} \tag{8.875}$$

If we insert (8.874) into (8.869), we obtain a nonlinear dispersion relation

$$\boxed{\varepsilon_{NL} \equiv \varepsilon(\omega_2, k_2) + \frac{|v_1|^2}{v_{te}^2} \frac{\omega_2^2}{\omega_2^2 - 3k_2^2 v_{te}^2} \frac{W^*}{1+W^*} = 0}\,. \tag{8.876}$$

It is now interesting that for finite $|v_1|^2 \sim |E_1|^2$ instability can occur. To this end, we calculate the imaginary part of W (using the Plemelj formula for $\omega_{3i} \ll \omega_{3r}$)

$$\text{Im } W = W_i = -\frac{c_s^2}{n_0} \pi \partial_v f_{i0} \bigg|_{\omega_3/k_3}. \tag{8.877}$$

Because

$$\text{Im } \frac{W^*}{1+W^*} \approx -\frac{W_i}{(1+W_r)^2} \tag{8.878}$$

we find

$$\text{Im } \varepsilon_{NL} = \frac{1}{(1+W_r)^2} \frac{|v_1|^2}{v_{te}^2} \frac{\pi c_s^2}{n_0} \partial_v f_{i0} \bigg|_{\omega_3/k_3}. \tag{8.879}$$

Together with

$$\frac{\partial \varepsilon_r}{\partial \omega}\bigg|_{\omega_{2r}} \approx \frac{2}{\omega_{pe}} \tag{8.880}$$

this yields

$$\boxed{\omega_{3i} = \omega_{2i} \approx \frac{-\text{ Im } \varepsilon_{NL}}{\partial \varepsilon_r / \partial \omega|_{\omega_{2r}}} \approx -\frac{1}{(1+W_r)^2} \frac{\omega_{pe}}{2} \frac{|v_1|^2}{v_{te}^2} \frac{\pi c_s^2}{n_0} \partial_v f_{i0} \bigg|_{\omega_{3r}/k_3}}\,. \tag{8.881}$$

ω_{3r}/k_3 is the phase velocity of the difference wave and, due to (8.860) and (8.861), is equal to $(\omega_1 - \omega_{2r})/(k_1 - k_2)$.

Formula (8.881) teaches us that for two Langmuir waves 1 and 2, for which

$$\frac{\omega_1 - \omega_{2r}}{k_1 - k_2} \sim \mathcal{O}(v_{ti}) \tag{8.882}$$

applies, an instability occurs if the slope $\partial_v f_{i0}$ of the ion distribution is negative (as is the case, for example, with a Maxwellian distribution).

So far, the process of induced scattering of Langmuir waves (1) by ions, in which a Langmuir wave (2) is generated [provided (8.882) is satisfied], has been considered in a rather isolated manner. It is clear that we must include such processes in a complete (weak) turbulence theory—similar to the first section. For Langmuir turbulence, this has important consequences. With each scattering of a Langmuir wave by the ions, part of the energy of the Langmuir wave is transferred to an ion. At the same time, there is a transport of Langmuir energy into the long-wavelength range, where direct resonant absorption of the wave energy by the particles becomes fundamentally impossible. This problem of "Bose condensation" can be resolved by Langmuir collapse.

However, it would go beyond the scope of this book to discuss this in further detail. Therefore, we will now leave this aspect and turn to another important elementary process.

Wave–Wave Interaction

Since we have not yet explicitly considered the nonlinear interaction of the waves with each other, we now turn to this aspect. In the following, we assume that the basic equations contain at most quadratic terms; a typical system is given by the Zakharov equations (8.255) and (8.256). For demonstration purposes, we choose a simpler form that does not involve a system of equations.

The (only) fundamental equation has the model structure

$$\boxed{\partial_t \tilde{C}(x;t) + L\tilde{C}(x;t) = A\tilde{C}^2(x;t)}\,. \tag{8.883}$$

Here, L is a linear operator that, in the absence of nonlinearities, $(A \equiv 0)$ accounts for the linear dispersion. The restriction to the first time derivative is not significant, since after a Fourier transformation in space we will in any case make the assumption of slowly varying amplitudes ("slowly varying envelope approximation"). This is reflected in the fact that the Fourier transform (in space) of $\tilde{C}(x;t)$ is assumed as

$$\tilde{C}(k;t) = C(k;t)e^{-i\omega(k)t} \tag{8.884}$$

where the "fast" time variation is accounted for by the factor $\exp[-i\omega(k)t]$ and the amplitude $C(k;t)$ is still assumed to depend weakly on time. The nonlinear term on the right-hand side of (8.883) leads to a convolution integral in the Fourier transformation, so that (8.883) becomes

$$\boxed{\begin{aligned}\partial_t C(k;t) = \int_{-\infty}^{\infty} & dk' V(k,k',k-k')C(k';t)C(k-k';t) \\ & \times \exp\{i[\omega(k)-\omega(k')-\omega(k-k')]t\}\end{aligned}} \tag{8.885}$$

.

This equation, with a general matrix element V, is representative of many more complicated model equations that arise from specific problems. We are not seeking *a coherent* solution, but rather intend to make a statement for a (statistical) turbulent ensemble of waves.

Within the framework of weak turbulence theory, we assume the coupling via the matrix element V to be small and expand

$$\boxed{C(k;t) = C^0(k;t) + C^{(1)}(k;t) + C^{(2)}(k;t) + \dots} \tag{8.886}$$

in terms of the interaction. To zeroth order, from (8.885) follows

$$\boxed{\partial_t C^{(0)}(k;t) = 0}\,, \tag{8.887}$$

i.e., $C^{(0)}(k;t) = C^{(0)}(k)$ is constant in time. The first-order equation

$$\begin{aligned}\partial_t C^{(1)}(k;t) = \int_{-\infty}^{+\infty} & dk' V(k,k',k-k')C^{(0)}(k')C^{(0)}(k-k') \\ & \times e^{i[\omega(k)-\omega(k')-\omega(k-k')]t}\end{aligned} \tag{8.888}$$

can be integrated directly. Due to the time-independence of the $C^{(0)}$ we obtain

$$\boxed{C^{(1)}(k;t) = \int_{-\infty}^{+\infty} dk' \int_{-\infty}^{+\infty} dk'' C^{(0)}(k')C^{(0)}(k'') \int_0^t dt' F(k,k',k'',t')}\,, \tag{8.889}$$

where

$$F(k,k',k'',t) = V(k,k',k'')\delta(k-k'-k'')e^{i[\omega(k)-\omega(k')-\omega(k'')]t} \tag{8.890}$$

was introduced for reasons of symmetry. The second order in V (or F) is (in a somewhat abbreviated notation)

$$
\partial_t C^{(2)}(k;t) = \int_{-\infty}^{\infty} dk' \int_{-\infty}^{\infty} dk'' F(k,k',k'';t) \Big\{ C^{(0)}(k') \int_{-\infty}^{\infty} dk''' \int_{-\infty}^{\infty} dk'''' \\
\times C^{(0)}(k''')C^{(0)}(k'''') \int_0^t dt' F(k'',k''',k'''';t') \\
+ k' \leftrightarrow k'' \Big\} . \tag{8.891}
$$

The solution to this first-order differential equation is

$$
\boxed{\begin{aligned} C^{(2)}(k;t) = \int dk' ... dk'''' \int_0^t dt' \int_0^{t'} dt'' F(k,k',k'';t')\{F(k'',k''',k'''';t'') \\ \times C^{(0)}(k')C^{(0)}(k''')C^{(0)}(k'''') + k' \leftrightarrow k''\}. \end{aligned}} \tag{8.892}
$$

In principle, we can thus successively determine all higher orders. However, we interrupt this calculation at this point in order to introduce the additional information of statistically distributed fields. Formally, we write

$$
C^{(0)}(k) = |C^{(0)}(k)| e^{i\varphi(k)}, \tag{8.893}
$$

where $\varphi(k)$ is an arbitrary value in the interval $0 \leq \varphi(k) < 2\pi$ distributed phase. This approach means that we consider a particular configuration at time t as a realization within an ensemble of systems, where the individual ensemble members differ by the (arbitrarily chosen and statistically independent) phases. The amplitudes $|C^{(0)}|$ do not vary. The symbol $\langle \ldots \rangle = \int_0^{2\pi} d\varphi(k) \ldots /2\pi$ is intended to denote an averaging over the ensemble, where for $k \neq k'$

$$
\langle e^{i\varphi(k) \pm i\varphi(k')} \rangle = \frac{1}{4\pi^2} \int_0^{2\pi} d\varphi(k) \int_0^{2\pi} d\varphi(k') e^{i\varphi(k) \pm i\varphi(k')} = 0 \tag{8.894}
$$

applies and for $k = k'$

$$
\langle e^{i\varphi(k) - i\varphi(k')} \rangle = \frac{1}{2\pi} \int_0^{2\pi} e^{i0} d\varphi(k) = 1 \tag{8.895}
$$

can be set. This allows us to introduce an intensity $n^{(0)}$ via

$$
\langle C^{(0)}(k) C^{(0)^*}(k') \rangle = n^{(0)}(k) \delta(k - k') \tag{8.896}
$$

Analogously, we have

$$
\boxed{\langle C(k;t) C^*(k';t) \rangle = n(k;t) \delta(k - k')} \, . \tag{8.897}
$$

We now systematically calculate this last expression up to second order by substituting (8.886) and the subsequent results. A somewhat lengthy, but in principle straightforward calculation leads to

$$\langle C(k;t)C^*(k;t)\rangle = |C^{(0)}(k)|^2 + \langle C^{(1)}(k;t)C^{(1)^*}(k;t)\rangle \tag{8.898}$$

or

$$\begin{aligned}\left[n(k;t) - n^{(0)}(k)\right]\delta((k-k') = 0)&\\ = \int dk'dk''dk'''dk'''' \int_0^t dt' F(k,k',k'';t') \int_0^t dt'' F^*(k,k''',&\\ \times \langle C^{(0)}(k')C^{(0)}(k'')C^{(0)^*}(k''')C^{(0)^*}(k'''')\rangle \, .&\end{aligned} \tag{8.899}$$

In the integrand on the right-hand side, we only obtain contributions for $k' = k'''$ and $k'' = k''''$ or $k' = k''''$ and $k'' = k'''$. This means

$$\begin{aligned}&\langle C^0(k')C^{(0)}(k'')C^{(0)^*}(k''')C^{(0)^*}(k'''')\rangle\\ &= n^{(0)}(k')n^{(0)}(k'')[\delta(k'-k''')\delta(k''-k'''') + \delta(k'-k'''')\delta(k''-k''')].\end{aligned} \tag{8.900}$$

If we take this into account in (8.899), it follows that

$$\begin{aligned}[n(k;t) - n^{(0)}(k)]\delta(0) = \textstyle\int_{-\infty}^{+\infty} dk' \int_{-\infty}^{+\infty} dk'' \int_0^t dt' \int_0^t dt'' \; F(k,k',k'';t')&\\ \times n^{(0)}(k')n^{(0)}(k'') \left\{F^*(k,k',k'';t'') + F^*(k,k'',k';t'')\right\}.&\end{aligned} \tag{8.901}$$

If the matrix element V possesses the symmetry

$$V(k,k',k'') = V(k,k'',k') \, , \tag{8.902}$$

then it is also found in F, and (8.901) can be written in the form

$$\begin{aligned}[n(k;t) - n^{(0)}(k;t)]\delta(0) = 2\int dk' \int dk'' n^{(0)}(k')n^{(0)}(k'')&\\ \times \left|\int_0^t dt' F(k,k',k'';t')\right|^2&\end{aligned} \tag{8.903}$$

Similar to Fermi's golden rule for transitions in quantum electrodynamics, we note

$$\lim_{t\to\infty}\left|\int_0^t dt' e^{i\Delta\omega t'}\right|^2 = 2\pi t\delta(\Delta\omega), \tag{8.904}$$

so that for large t

$$\boxed{\left|\int_0^t dt' F\right|^2 = \left|V(k,k',k'')\right|^2 2\pi t\delta[\omega(k) - \omega(k') - \omega(k'')]\delta(k-k'-k'')\delta(0)} \tag{8.905}$$

can be set.

Equation (8.903) can therefore also be written as

$$\frac{n(k;t) - n^{(0)}(k)}{t} = 4\pi \int dk' \int dk'' |V(k,k',k'')|^2$$
$$\times n^{(0)}(k')n^{(0)}(k'')\delta[\omega(k) - \omega(k') - \omega(k'')]\delta(k - k' - k'') \qquad (8.906)$$

Although t can become large, in the limit $V \to 0 \quad |n - n^{(0)}| \ll n^{(0)}$, and we can therefore replace the left-hand sides by the derivative. Furthermore, we consider (8.906) as an iteration rule in time. We start at time $t = 0$ with $n(k; t = 0) = n^{(0)}(k)$. Then, from (8.906), we obtain $n(k;t)$, which we can in turn use as a new initial value [for $n^{(0)}(k)$ on the right-hand side of (8.906)] to calculate $n(k,T)$ for $T > t$, and so on.

This leads to the fundamental equation of weak turbulence for wave–wave interaction

$$\partial_t n(k;t) = 4\pi \int dk' \int dk'' |V(k,k',k'')|^2 n(k';t) n(k'';t)$$
$$\times \delta[\omega(k) - \omega(k') - \omega(k'')]\delta(k - k' - k''). \qquad (8.907)$$

The Dirac function with the argument $\Delta\omega$ expresses energy conservation, and the Dirac function with the argument Δk embodies momentum conservation.

In this simple formulation [with only one wave type and the frequency $\omega(k)$], Eq. (8.907), if it is not to be empty, presupposes a decay-type spectrum ("decay spectrum") $\omega(k' + k'') = \omega(k') + \omega(k'')$. This is not fulfilled for all wave types. However, we should keep in mind that we have only dealt with a simple model equation. Coupled systems (e.g., the Zakharov equations) allow for couplings between different wave types, which can lead to extremely interesting results. The "$k \to 0$ condensation" of weak Langmuir turbulence is one example of this.

Stability of MHD Systems

9

Abstract

In this chapter, we examine the stability of plasma configurations based on a magnetohydrodynamic description. This is particularly important for a wide range of configurations, since, in the pursuit of stability, the first step is to rule out rapid instability mechanisms that occur on the global magnetohydrodynamic scale. We begin by developing a general stability criterion, which takes the form of a variational principle. In the subsequent sections, we apply this principle to various configurations as illustrative examples. We start with simple pinch configurations, then, after discussing general aspects of interchange instabilities, we conclude with resistive instabilities as the final example to be addressed.ies to be discussed.

9.1 Hydrodynamic Stability Criterion

In this section, we address the stability of magnetohydrodynamic systems in the ideal approximation ($\eta = 0$). The corresponding fundamental equations have already been derived. Specifically, we now address, within the framework of ideal MHD, the problem of when an equilibrium ($\partial_t = 0$) is stable, by introducing a variational principle (energy principle).

Let the equilibrium be given by $u_0 = 0$ and

$$\boxed{\nabla p_0 = \frac{1}{\mu_0}(\nabla \times \mathbf{B}_0) \times \mathbf{B}_0} \tag{9.1}$$

K.-H. Spatschek, *Theoretical Plasma Physics*,
https://doi.org/10.1007/978-3-662-72828-4_9

Small perturbations satisfy (in linear approximation) the equation of motion

$$\rho_0 \partial_t \mathbf{u} = -\nabla p + \frac{1}{\mu_0}(\nabla \times \mathbf{B}_0) \times \mathbf{B} + \frac{1}{\mu_0}(\nabla \times \mathbf{B}) \times \mathbf{B}_0 \,. \tag{9.2}$$

All quantities without an index represent the perturbations, while the index 0 denotes the equilibrium values. By differentiating once more with respect to time, we obtain

$$\boxed{\begin{aligned} \rho_0 \partial_t^2 \mathbf{u} = &- \nabla \dot{p} + \frac{1}{\mu_0}(\nabla \times \mathbf{B}_0) \times \nabla \times (\mathbf{u} \times \mathbf{B}_0) \\ &+ \frac{1}{\mu_0}[\nabla \times \nabla \times (\mathbf{u} \times \mathbf{B}_0)] \times \mathbf{B}_0 \,. \end{aligned}} \tag{9.3}$$

Using the equation of state

$$\frac{d}{dt}\left(\frac{p}{\rho^\gamma}\right) = 0 \tag{9.4}$$

and the continuity equation for ρ, $\dot{p}$ can, to lowest order, be expressed by

$$\boxed{\dot{p} = -\mathbf{u} \cdot \nabla p_0 - \gamma p_0 \nabla \cdot \mathbf{u}} \tag{9.5}$$

so that, after substituting into (9.3), we obtain a closed equation for $\mathbf{u}$. The latter is often written in terms of the Lagrangian variable ξ instead of $\mathbf{u}$. Within the framework of linear theory, we can approximately set $\mathbf{u} = \dot{\xi}$, as the following discussion will show.

Example 9.1 (Eulerian versus Lagrangian Description)
The Eulerian description of the velocity history at a fixed location $\mathbf{r}$, i.e., $\mathbf{u}(\mathbf{r}, t)$, is, for small deviations from equilibrium, related to the Lagrangian velocity $\dot{\xi}(\mathbf{r}_0, t)$ of a particular fluid element that was at time t_0 at the position $\mathbf{r}_0$, via the Taylor expansion

$$\begin{aligned} \dot{\xi}(\mathbf{r}_0, t) = \mathbf{u}(\mathbf{r}, t) &\approx \mathbf{u}(\mathbf{r}_0, t) + \xi \cdot \nabla \mathbf{u} + \ldots \\ &\approx \mathbf{u}(\mathbf{r}_0, t) \end{aligned} \tag{9.6}$$

Thus, in (9.3) we can set $\mathbf{u} = \dot{\xi}$ and integrate once with respect to time. ■

From the equilibrium condition it follows that $\xi(\mathbf{r}_0, 0) = \dot{\xi}(\mathbf{r}_0, 0) = 0$, and the equation of motion for the deviation from equilibrium reads

$$\begin{aligned} \rho_0 \ddot{\xi} = &\nabla(\xi \cdot \nabla p_0 + \gamma p_0 \nabla \cdot \xi) + \frac{1}{\mu_0}\{\nabla \times [\nabla \times (\xi \times \mathbf{B}_0)]\} \times \mathbf{B}_0 \\ &+ \frac{1}{\mu_0}(\nabla \times \mathbf{B}_0) \times [\nabla \times (\xi \times \mathbf{B}_0)]. \end{aligned} \tag{9.7}$$

Usually, the (complicated) time-independent linear operator **F** is introduced, with which (9.7) can be written in the form

$$\boxed{\rho_0 \ddot{\xi} = \mathbf{F}(\xi)} \tag{9.8}$$

This notation should not obscure the fact that the determining equation for ξ is not easy to solve.

We can draw initial conclusions after a separation ansatz $\xi(\mathbf{r}_0, t) = \xi_k(r_0)\tau_k(t)$. Because

$$\rho_0 \xi_k \, \ddot{\tau}_k = \mathbf{F}(\xi_k)\tau_k \tag{9.9}$$

it is then possible, for example, to write

$$\tau_k \sim \exp[i(\omega_k t + \varphi_k)] \tag{9.10}$$

, where ω_k^2 is obtained from the eigenvalue equation

$$\boxed{-\rho_0 \omega_k^2 \, \xi_k = \mathbf{F}(\xi_k)} \tag{9.11}$$

Owing to the linearity of the problem, we can sum over all k to obtain

$$\xi(\mathbf{r}_0, t) = \sum_k a_k \, \xi_k(\mathbf{r}_0) \exp[i(\omega_k t + \varphi_k)] \tag{9.12}$$

This procedure yields the general solution, provided that **F** is a complete operator. Unfortunately, this requirement is not always met, and a more general approach is necessary. However, if we linger a moment longer with (9.11) and (9.12), it is easy to see that, under the already mentioned assumption of completeness, the stability criterion holds: *A hydromagnetic equilibrium is stable if and only if all eigenvalues of* ρ_0^{-1} **F** *are negative.*

However, caution is also warranted because equation $\mathbf{F}(\xi_0) = 0$ can be satisfied by arbitrarily many eigenfunctions

$$\xi_0 = \alpha(p_0)\, \mathbf{B}_0 + \beta(p_0)\, \nabla \times \mathbf{B}_0 \tag{9.13}$$

and, due to degeneracy, algebraic modes in t may occur. (Note that the above formulation of the stability criterion assumes exponential instability.)

Within the framework of ideal MHD, it is plausible to suspect that the operator **F** is Hermitian. Since we have neither dissipation nor external sources in the (conservative) ideal MHD system, ω_k^2 should be real. This general statement for the eigenvalues ω_k^2 in conservative systems excludes both purely damped and oscillatory growing modes. This is generally justified by considering a conservative system with total energy $E = T + V$. For an oscillatory motion, at the turning points $V = E$. holds. If the oscillation amplitude now increases, the turning point shifts continuously over time away from the minimum of the potential energy V, in contrast to the fixed turning points that result from $V = E$ for purely position-dependent potential energies. Hermitian operators have real eigenvalues, hence the suspicion that **F** is Hermitian.

Example 9.2 (Hermitian Operator F)

A relatively simple plausibility argument supports this suspicion. If we interpret (9.8) as a "Newtonian equation" with the "force density" **F**, we obtain, due to linearity,

$$-\int_0^{\xi} \mathbf{F}(\eta) \cdot d\eta = -\frac{1}{2}\xi \cdot \mathbf{F}(\xi) \tag{9.14}$$

as the change in the potential energy density. For the temporal change of the total energy,

$$\frac{\partial}{\partial t}\int \left[\frac{1}{2}\rho_0\, \dot{\xi}^2 - \frac{1}{2}\xi \cdot \mathbf{F}(\xi)\right] d^3r = 0\,, \tag{9.15}$$

we require it to vanish according to the law of energy conservation. Simple rearrangements of the left-hand side, together with (9.8), yield

$$\int \dot{\xi} \cdot \mathbf{F}(\xi)\, d^3r = \int \xi \cdot \mathbf{F}(\dot{\xi})\, d^3r\,. \tag{9.16}$$

The differential equation (9.8) is equivalent to a system of two first-order differential equations, so we can regard $\dot{\xi}$ and ξ as independent. With appropriate boundary conditions, (9.16) suggests Hermiticity. ■

We will provide a more rigorous proof of this mathematical property below. But first, a word about the potential energy

$$W = -\frac{1}{2}\int \xi \cdot \mathbf{F}(\xi)\, d^3r\,. \tag{9.17}$$

If W can become negative, instability is present. This is a preliminary form of the variational principle, which will be formulated more precisely later. Let us expand ξ in the form (9.12), or, using $\tau_k \sim \cos(\omega_k t + \varphi_k)$, to emphasize the real character of ξ, in terms of eigenfunctions of **F**, then, from the orthonormality of the eigenfunctions of Hermitian operators, it follows that

$$W = \sum_k a_k^2 \cos^2(\omega_k t + \varphi_k)\, \omega_k^2\,. \tag{9.18}$$

If W is negative, at least one ω_k^2 must be negative, and that means instability. The converse of the above statement (for stability) would again require the completeness of the eigenfunctions.

We now prove more generally the energy principle, according to which *a necessary and sufficient condition for stability is that the potential energy W cannot take negative values.*

The justification for positive values of W is relatively straightforward. Eq. (9.15) suggests using

$$K = \int \frac{1}{2}\rho_0\, \dot{\xi}^{\,2}\, d^3r = E - W \tag{9.19}$$

as a positive definite functional for a norm, which must be introduced within the framework of a rigorous concept of stability. For $W > 0$ it follows from

$$K < E \tag{9.20}$$

the boundedness of the norm of the perturbation. Conversely, we can show that

$$\boxed{I := \frac{1}{2}\int \rho_0 \xi^{\,2} d^3r} \tag{9.21}$$

grows exponentially with time if W can be made negative. To this end, we differentiate (9.21) twice with respect to time; a short calculation yields

$$\ddot{I} = 2K - 2W\,. \tag{9.22}$$

With this, we can write

$$\frac{d^2}{dt^2}\ln I = \frac{1}{I}\left[2K - 2W - \frac{1}{I}\left(\int \rho_0 \xi \cdot \dot{\xi}\, d^3r\right)^2\right], \tag{9.23}$$

where, based on the Schwarz inequality,

$$\left(\int \rho_0\, \xi \cdot \dot{\xi}\, d^3r\right)^2 \leq 4\, I\, K \tag{9.24}$$

can be estimated. Substituting into (9.23), it follows that

$$\boxed{\frac{d^2}{dt^2}\ln I \geq -\frac{2E}{I}}\,. \tag{9.25}$$

Now, if there exists a $\xi := \mathbf{v}_0(\mathbf{r})$, for which $W < 0$ holds, we consider the initial value problem $\xi(t = 0) = \mathbf{v}_0(\mathbf{r})$ and $\dot{\xi}(t = 0) = 0$. Obviously, for this it follows that $E < 0$; we formally write

$$E := -2\nu^2 I(0). \tag{9.26}$$

With the auxiliary quantity

$$y := \ln\left[I(t)/I(0)\right] \tag{9.27}$$

(9.25) then reads

$$\ddot{y} \geq 2\nu^2 e^{-y}. \tag{9.28}$$

The initial conditions are then given by $y(0) = \dot{y}(0) = 0$.

The function $Y(t)$ satisfies the same initial conditions and obeys the differential equation

$$\ddot{Y} = 2\nu^2 e^{-Y}. \tag{9.29}$$

Obviously, at all times, the following holds: $y \geq Y$. The differential equation (9.29) has the solution

$$Y(t) = \ln(\cosh \nu t)^2, \tag{9.30}$$

from which it now immediately follows that

$$\boxed{I(t) \geq I(0)\frac{e^{2\nu t} + 2 + e^{-2\nu t}}{4} \to \infty} \tag{9.31}$$

for $t \to \infty$. The unstable behavior (for $W < 0$) becomes evident due to the definition (9.21).

The weak point in the previous argument lies in the assumption of Hermiticity [see (9.16)] of $\mathbf{F}$. We now aim to address this with a more precise calculation.

Example 9.3 (Structure of F)

In the following, we will rewrite the term $\xi' \cdot \mathbf{F}(\xi)$ taking into account the relevant boundary conditions. This requires rather intricate vector operations, which we will nevertheless present here in detail to provide insight into this fairly typical problem.

First, we introduce the abbreviations

$$\mathbf{Q} := \nabla \times (\xi \times \mathbf{B}_0) \tag{9.32}$$

and

$$\mathbf{Q}' := \nabla \times (\xi' \times \mathbf{B}_0) \tag{9.33}$$

Furthermore, if one uses (9.1), the vector identity

$$\begin{aligned}
&\nabla(\xi \cdot \nabla p_0) + (1/\mu_0)(\nabla \times \mathbf{B}_0) \times \mathbf{Q} \\
&= (\nabla\xi) \cdot \nabla p_0 + \xi \cdot \nabla\nabla p_0 + (1/\mu_0)(\nabla \times \mathbf{B}_0) \times \mathbf{Q} \\
&= [(\nabla p_0) \times \nabla] \times \xi + (\nabla \cdot \xi)\nabla p_0 + \xi \cdot \nabla\nabla p_0 + (1/4\pi)(\nabla \times \mathbf{B}_0) \times \mathbf{Q} \\
&= (1/\mu_0)\,\{[(\nabla \times \mathbf{B}_0) \times \mathbf{B}_0] \times \nabla\} \times \xi + (\nabla \cdot \xi)\nabla p_0 \\
&\quad + \xi \cdot \nabla\nabla p_0 + (1/4\pi)(\nabla \times \mathbf{B}_0) \times \mathbf{Q} \\
&= (1/4\pi)[\mathbf{B}_0(\nabla \times \mathbf{B}_0) \cdot \nabla - (\nabla \times \mathbf{B}_0)\mathbf{B}_0 \cdot \nabla] \times \xi + \nabla \cdot \xi\nabla p_0 \\
&\quad + \xi \cdot \nabla\nabla \rho_0 + (1/\mu_0)(\nabla \times \mathbf{B}_0) \times [\mathbf{B}_0 \cdot \nabla\xi - \xi \cdot \nabla\mathbf{B}_0 - \mathbf{B}_0\nabla \cdot \xi] \\
&= (1/\mu_0)\,\{\mathbf{B}_0 \times [(\nabla \times \mathbf{B}_0) \cdot \nabla\xi] - \mathbf{B}_0 \times [\xi \cdot \nabla(\nabla \times \mathbf{B}_0)] \\
&\quad - \xi \cdot \nabla[(\nabla \times \mathbf{B}_0) \times \mathbf{B}_0] + \mathbf{B}_0 \times (\nabla \times \mathbf{B}_0)\nabla \cdot \xi\} \\
&\quad + (\nabla \cdot \xi)\nabla p_0 + \xi \cdot \nabla\nabla p_0 \\
&= (1/\mu_0)\mathbf{B}_0 \times \{\nabla \times [\xi \times (\nabla \times \mathbf{B}_0)]\} + (1/\mu_0)\mathbf{B}_0 \times (\nabla \times \mathbf{B}_0)\nabla \cdot \xi \\
&= (1/\mu_0)\mathbf{B}_0 \times \{\nabla \times [\xi \times (\nabla \times \mathbf{B}_0)]\} - (\nabla \cdot \xi)\nabla p_0
\end{aligned} \tag{9.34}$$

can be proven. This allows the following transformation to be carried out:

$$\begin{aligned}
&\xi' \cdot \mathbf{F}\{\xi\} = \xi' \cdot [\nabla(\gamma p_0\nabla \cdot \xi + \xi \cdot \nabla p_0) + (1/\mu_0)(\nabla \times \mathbf{Q}) \times \mathbf{B}_0 \\
&\quad + (1/\mu_0)(\nabla \times \mathbf{B}_0) \times \mathbf{Q}] \\
&= \xi' \cdot [\nabla(\gamma p_0\nabla \cdot \xi) + (1/\mu_0)(\nabla \times \mathbf{Q}) \times \mathbf{B}_0 \\
&\quad + (1/\mu_0)\mathbf{B}_0 \times \{\nabla \times [\xi \times (\nabla \times \mathbf{B}_0)]\} - (\nabla \cdot \xi)\nabla p_0] \\
&= \nabla \cdot \big\{\xi'\gamma p_0\nabla \cdot \xi - (1/\mu_0)\mathbf{Q} \times (\xi' \times \mathbf{B}_0) \\
&\quad + (1/\mu_0)[\xi \times (\nabla \times \mathbf{B}_0)] \times (\xi' \times \mathbf{B}_0)]\big\} \\
&\quad - \gamma p_0(\nabla \cdot \xi)(\nabla \cdot \xi') - (1/\mu_0)\mathbf{Q} \cdot \mathbf{Q}' \\
&\quad + (1/\mu_0)[\xi \times (\nabla \times \mathbf{B}_0)] \cdot \mathbf{Q}' - (\nabla \cdot \xi)\xi' \cdot \nabla p_0 \\
&= \nabla \cdot (\xi'\gamma p_0\nabla \cdot \xi - (1/\mu_0)\mathbf{Q} \times (\xi' \times \mathbf{B}_0) \\
&\quad + \xi'(1/\mu_0)\xi \cdot (\nabla \times \mathbf{B}_0) \times \mathbf{B}_0 - (1/\mu_0)\mathbf{B}_0\xi' \times \xi \cdot \nabla \times \mathbf{B}_0\big\} \\
&\quad - (1/\mu_0)\mathbf{Q} \cdot \mathbf{Q}' + \xi \cdot (1/\mu_0)(\nabla \times \mathbf{B}_0) \times \mathbf{Q}' - (\nabla \cdot \xi)(\gamma p_0\nabla \cdot \xi' + \xi' \cdot \nabla p_0) \\
&= \nabla \cdot \big\{\xi'\gamma p_0\nabla \cdot \xi - \xi\gamma p_0\nabla \cdot \xi' - (1/\mu_0)\mathbf{Q} \times (\xi' \times \mathbf{B}_0) \\
&\quad + (1/\mu_0)\mathbf{Q}' \times (\xi \times \mathbf{B}_0) + \xi'\xi \cdot \nabla p_0 - \xi\xi' \cdot \nabla p_0 - \xi' \times \xi \cdot \nabla p_0\big\} \\
&\quad + \nabla \cdot [\xi(\gamma p_0\nabla \cdot \xi' + \xi' \cdot \nabla p_0) - (1/\mu_0)\mathbf{Q}' \times (\xi \times \mathbf{B}_0)] \\
&\quad - (1/\mu_0)\mathbf{Q} \cdot \mathbf{Q}' + \xi \cdot (1/\mu_0)(\nabla \times \mathbf{B}_0) \times \mathbf{Q}' \\
&\quad - (\nabla \cdot \xi)(\gamma p_0\nabla \cdot \xi' + \xi' \cdot \nabla p_0)
\end{aligned} \tag{9.35}$$

or

$$
\begin{aligned}
\xi' \cdot \mathbf{F}\{\xi\} = \nabla \cdot \{&\xi'[\gamma p_0 \nabla \cdot \xi - (1/\mu_0)\mathbf{Q} \cdot \mathbf{B}_0 + \xi \cdot \nabla p_0] \\
&+\mathbf{B}_0((1/\mu_0)(\xi' \cdot \mathbf{Q} - \xi \cdot \mathbf{Q}') + \xi \times \xi' \cdot \nabla p_0) \\
&-\xi\,[\gamma p_0 \nabla \cdot \xi' - (1/\mu_0)\mathbf{Q}' \cdot \mathbf{B}_0 + \xi' \cdot \nabla p_0]\} \\
&+\xi \cdot [\nabla(\gamma p_0 \nabla \cdot \xi' + \xi' \cdot \nabla p_0) + (1/\mu_0)(\nabla \times \mathbf{B}_0) \times \mathbf{Q}' \\
&+(1/\mu_0)(\nabla \times \mathbf{Q}') \times \mathbf{B}_0] \\
= \nabla \cdot \{&\xi'\,[\gamma p_0 \nabla \cdot \xi - (1/\mu_0)(\mathbf{Q} \cdot \mathbf{B}_0 + \xi \cdot (\nabla \mathbf{B}_0) \cdot \mathbf{B}_0) + \xi \cdot \nabla(p_0 + B_0^2/2\mu_0)] \\
&-\xi[\gamma p_0 \nabla \cdot \xi' - (1/\mu_0)(\mathbf{Q}' \cdot \mathbf{B}_0 + \xi' \cdot (\nabla \mathbf{B}_0) \cdot \mathbf{B}_0) + \xi' \cdot \nabla(p_0 + B_0^2/2\mu_0)] \\
&+\mathbf{B}_0((1/\mu_0)(\xi' \cdot \mathbf{Q} - \xi \cdot \mathbf{Q}') + \xi \times \xi' \cdot \nabla p_0)\} + \xi \cdot \mathbf{F}\{\xi'\}\ .
\end{aligned}
\tag{9.36}
$$

This makes the structure clearer for the first time. ■

The left-hand side of (9.36), together with the last term on the right-hand side of (9.36), leads directly to Hermiticity. The remaining terms on the right-hand side, after a volume integration, result in surface contributions, whose vanishing we now need to demonstrate. The boundary conditions on the surface S, which encloses the volume V, will be important here.

Example 9.4 (Boundary Contributions)
We assume that the plasma (volume V) is confined by a vacuum magnetic field $\mathbf{B}_v$. Due to $\nabla \cdot \mathbf{B} = 0$, the continuity of the normal component of $\mathbf{B}$, follows, whose vanishing we can assume on S in the form $\mathbf{n} \cdot \mathbf{B}_0 = 0$, where $\mathbf{n}$ is the normal vector of S. (This also holds for moving surfaces; see below.) From (9.36) we can therefore derive

$$
\begin{aligned}
&\int d^3r\,[\xi' \cdot \mathbf{F}\{\xi\} - \xi \cdot \mathbf{F}\{\xi'\}] \\
&= \int d^2r\,\mathbf{n} \cdot \left\{\xi'\left[\gamma p_0 \nabla \cdot \xi - \frac{1}{\mu_0}(\mathbf{Q} \cdot \mathbf{B}_0 + \xi \cdot (\nabla \mathbf{B}_0) \cdot \mathbf{B}_0)\right.\right. \\
&\quad \left. + \xi \cdot \nabla\left(p_0 + \frac{B_0^2}{2\mu_0}\right)\right] \\
&\quad -\xi\left[\gamma p_0 \nabla \cdot \xi' - \frac{1}{\mu_0}(\mathbf{Q}' \cdot \mathbf{B}_0 + \xi' \cdot (\nabla \mathbf{B}_0) \cdot \mathbf{B}_0)\right. \\
&\quad \left.\left. +\xi' \cdot \nabla\left(p_0 + \frac{B_0^2}{2\mu_0}\right)\right]\right\}
\end{aligned}
\tag{9.37}
$$

For vanishing displacement $\mathbf{n} \cdot \xi$ at the boundary, Hermiticity has already been demonstrated.
■

Unfortunately, this boundary condition is not very meaningful for plasma applications. The plasma-vacuum boundary is generally variable; only further out is the vacuum limited by a fixed conducting surface S'. Under these circumstances, we discuss the formulas

$$\boxed{\mathbf{n} \cdot \mathbf{B}_0|_S = 0} \tag{9.38}$$

and

$$\boxed{-\gamma p_0 \nabla \cdot \xi + \frac{\mathbf{B}_0}{\mu_0} \cdot [\mathbf{Q} + (\xi \cdot \nabla)\mathbf{B}_0] = \frac{\mathbf{B}_v}{\mu_0} \cdot [\delta \mathbf{B}_v + (\xi \cdot \nabla)\mathbf{B}_v]} \tag{9.39}$$

on S in detail.

Example 9.5 (Proof of (9.38))

Eq. (9.38) was used in the transformation to (9.37). First, on the surface S we have

$$\mathbf{n} \cdot [\mathbf{B}] = 0, \tag{9.40}$$

where the square brackets [...] in the usual way denote the difference between plasma (exact: without index, linear: index o) and vacuum values (index v) at the surface. If initially $\mathbf{n} \cdot \mathbf{B}_v = 0$, then initially $\mathbf{n} \cdot \mathbf{B}_0 = 0$ also holds. Furthermore, (9.38) then follows at all times with the following arguments. We compute

$$\frac{d}{dt}(\mathbf{n} \cdot \mathbf{B}_0) = \mathbf{B}_0 \cdot \frac{d\mathbf{n}}{dt} + \mathbf{n} \cdot \frac{d\mathbf{B}_0}{dt} \tag{9.41}$$

and use

$$\begin{aligned} \frac{d\mathbf{B}_0}{dt} &= \frac{\partial \mathbf{B}_0}{\partial t} + \mathbf{u} \cdot \nabla \mathbf{B}_0 \\ &= \nabla \times \mathbf{E} + \mathbf{u} \cdot \nabla \mathbf{B}_0 \\ &= \nabla \times (\mathbf{u} \times \mathbf{B}_0) + \mathbf{u} \cdot \nabla \mathbf{B}_0 \\ &= \mathbf{B}_0 \cdot \nabla \mathbf{u} - \mathbf{B}_0 \nabla \cdot \mathbf{u}\,. \end{aligned} \tag{9.42}$$

On the other hand, we have

$$\frac{d}{dt}\mathbf{n} = -\mathbf{n}(\nabla \cdot \mathbf{u}) + \mathbf{n}\mathbf{n} \cdot (\nabla \mathbf{u}) \cdot \mathbf{n}\,. \tag{9.43}$$

To prove this relation, we start from a surface element $d^2\mathbf{r} = \mathbf{n} d^2 r = d\mathbf{r} \times d\mathbf{r}\,'$ on S. A displacement by ξ changes $d\mathbf{r}$ by $\delta d\mathbf{r} = d\mathbf{r} \cdot \nabla \xi$ (and $d\mathbf{r}\,'$ accordingly). This yields

$$\begin{aligned} \delta d^2\mathbf{r} &= (d\mathbf{r} \cdot \nabla \xi\,) \times d\mathbf{r}\,' + d\mathbf{r} \times (d\mathbf{r}\,' \cdot \nabla \xi\,) \\ &= -[(d\mathbf{r} \times d\mathbf{r}') \times \nabla] \times \xi \end{aligned} \tag{9.44}$$

to linear order in ξ. The left-hand side can be rewritten using the definition of $d^2\mathbf{r}$ and we obtain

$$d^2r\,\delta\mathbf{n} + \mathbf{n}\,\delta d^2r = -d^2r\,(\mathbf{n}\times\nabla)\times\xi\ . \tag{9.45}$$

Scalar multiplication with $\mathbf{n}$ yields, due to $\mathbf{n}\cdot\delta\mathbf{n} = 0$, the equation

$$\delta d^2r = -d^2r\,\mathbf{n}\cdot(\mathbf{n}\times\nabla)\times\xi = -d^2r\,\mathbf{n}\times(\mathbf{n}\times\nabla)\cdot\xi\ . \tag{9.46}$$

If we insert this into (9.45), we obtain

$$\begin{aligned}\delta\mathbf{n} &= -(\mathbf{n}\times\nabla)\times\xi + \mathbf{nn}\times(\mathbf{n}\times\nabla)\cdot\xi\\ &= -(\nabla\xi)\cdot\mathbf{n} + \mathbf{nn}\cdot(\nabla\xi)\cdot\mathbf{n}\ .\end{aligned} \tag{9.47}$$

With $\xi = \mathbf{u}\delta t$ and $\dot{\mathbf{n}}\delta t = \delta\mathbf{n}$, it follows from (9.41)–(9.47) the linear first-order differential equation

$$\frac{d}{dt}(\mathbf{n}\cdot\mathbf{B}_0) = \mathbf{n}\cdot\mathbf{B}_0\;\mathbf{n}\times(\mathbf{n}\times\nabla)\cdot\mathbf{u} \tag{9.48}$$

for $n\cdot\mathbf{B}_0$. The initial value $\mathbf{n}\cdot\mathbf{B}_0\,|_S = 0$ then leads to (9.38) for all times. ■

Example 9.6 (Proof of (9.39))

So much for the proof of the first auxiliary formula (9.38); we now turn to the second Eq. (9.39). This involves a jump condition as a consequence of the momentum balance. With

$$\sigma := \rho\mathbf{u} \tag{9.49}$$

and

$$\mathcal{G} := \rho\mathbf{uu} + \left(p + \frac{B^2}{2\mu_0}\right)I - \frac{\mathbf{BB}}{\mu_0} \tag{9.50}$$

we write the momentum conservation in the framework of ideal MHD as

$$\frac{\partial\sigma}{\partial t} + \nabla\cdot\mathcal{G} = 0\ . \tag{9.51}$$

The last equation can be written in a completely equivalent form as

$$\begin{aligned}\mathbf{n}\cdot\nabla[\mathbf{n}\cdot(\mathcal{G}-\mathbf{u}\sigma)] = &-\frac{d\sigma}{dt} - \sigma\mathbf{n}\cdot\nabla(\mathbf{n}\cdot\mathbf{u})\\ &+\mathbf{n}\times(\mathbf{u}\times\mathbf{n})\cdot\nabla\sigma + \mathbf{n}\cdot(\nabla\mathbf{n})\cdot\mathcal{G}\\ &+\mathbf{n}\times(\mathbf{n}\times\nabla)\cdot\mathcal{G}\end{aligned} \tag{9.52}$$

This form is suitable for calculating a jump condition, since on the right-hand side there are no gradients of the quantities $\mathbf{n}$ parallel to σ and $\mathcal{G}$ may occur. Integration along ds with $\mathbf{n} \cdot \nabla = \partial_s$ leads to

$$[\mathbf{n} \cdot (\mathcal{G} - \mathbf{u}\sigma)] = 0 \ . \tag{9.53}$$

Because of $\mathbf{n} \cdot \mathbf{B} = 0$ it follows from the explicit expressions for σ and $\mathcal{G}$

$$\left[p + \frac{B^2}{2\mu_0}\right] = 0 \ ; \tag{9.54}$$

for the tangential components we therefore require

$$\mathbf{n} \times \nabla \left(p_0 + \frac{B_0^2}{2\mu_0} - \frac{B_v^2}{2\mu_0} \right) = 0 \ . \tag{9.55}$$

The linearization of the continuity condition (9.54) for the pressure yields in the plasma, due to

$$\begin{aligned} \frac{d\mathbf{B}}{dt} &= \frac{\partial \mathbf{B}}{\partial t} + \mathbf{u} \cdot \nabla \mathbf{B} \\ &= \nabla \times (\mathbf{u} \times \mathbf{B}) + \mathbf{u} \cdot \nabla \mathbf{B}, \end{aligned} \tag{9.56}$$

$$\mathbf{B}(\mathbf{r}, t) - \mathbf{B}(\mathbf{r}_s, 0) \approx \mathbf{Q}(\mathbf{r}_s) + (\xi \cdot \nabla)\mathbf{B}(\mathbf{r}_s) \tag{9.57}$$

for displacements $\mathbf{r} = \mathbf{r}_s + \xi$ in time t. Analogously, we obtain from

$$\frac{dp}{dt} = -\gamma p \nabla \cdot \mathbf{u} \tag{9.58}$$

$$p(\mathbf{r}, t) - p(\mathbf{r}_s, 0) = -\gamma p(r_s, 0) \ \nabla \cdot \xi \ . \tag{9.59}$$

In the vacuum we set $p = 0$ and

$$\mathbf{B}_v(\mathbf{r}, t) = \mathbf{B}_v(\mathbf{r}_s, 0) + \delta\mathbf{B}_v + \xi \cdot \nabla\mathbf{B}_v(\mathbf{r}_s) \ . \tag{9.60}$$

Note that in the last formula, for a displacement by $\xi\,'$ (instead of ξ) ξ we must replace by $\xi\,'$ and $\delta\mathbf{B}_v$ by $\delta\mathbf{B}\,'_v$. Thus, to first order (9.39) it follows from (9.54) and a corresponding formula for the primed quantities. ■

We are now finally in a position to further address the surface contributions in (9.37). We substitute for

$$p_0 + \frac{B_0^2}{2\mu_0} = \left(p_0 + \frac{B_0^2}{2\mu_0} - \frac{B_v^2}{2\mu_0}\right) + \frac{B_v^2}{2\mu_0} \tag{9.61}$$

and make use of (9.55) as well as (9.39). From (9.37) we then obtain

$$\boxed{\begin{aligned}&\mu_0 \int d^3r\, [\xi' \cdot \mathbf{F}\{\xi\} - \xi \cdot \mathbf{F}\{\xi'\}] \\ &= \int_S d^2r\, \mathbf{n}\left[-\xi' \mathbf{B}_v \cdot \nabla \times \delta\mathbf{A} + \xi \mathbf{B}_v \cdot \nabla \times \delta\mathbf{A}'\right],\end{aligned}} \tag{9.62}$$

where $\delta\mathbf{B}_v = \nabla \times \delta\mathbf{A}$ or $\delta\mathbf{B}'_v = \nabla \times \delta\mathbf{A}'$ was set, respectively.

Example 9.7 (Vector Potential $\delta\mathbf{A}$)

To better understand this vector potential $\delta\mathbf{A}$, let us return to the boundary conditions of electrodynamics. The statement that the tangential component of the electric field is continuous across the boundary holds in this form only for $\mathbf{u} = 0$ at the boundary; for $\mathbf{u} \neq 0$ we have

$$\mathbf{n} \times [\mathbf{E}] = \mathbf{n} \cdot \mathbf{u}\, [\mathbf{B}] \,. \tag{9.63}$$

Rewritten for our purposes, this means

$$\mathbf{n} \times \delta\mathbf{E}_v + \mathbf{n} \times (\mathbf{u} \times \mathbf{B}) = (\mathbf{n} \cdot \mathbf{u})\, \mathbf{B}_v - (\mathbf{n} \cdot \mathbf{u})\, \mathbf{B} \,, \tag{9.64}$$

or, to first order in the perturbations, for $\mathbf{n} \cdot \mathbf{B} = 0$:

$$\mathbf{n} \times \delta\mathbf{E}_v = (\mathbf{n} \cdot \mathbf{u})\mathbf{B}_v \,. \tag{9.65}$$

Within the framework of the Coulomb gauge, we have

$$\delta\mathbf{E}_v = -\delta\dot{\mathbf{A}}, \tag{9.66}$$

so that ultimately

$$-\,\mathbf{n} \times \delta\mathbf{A} = (\mathbf{n} \cdot \xi)\mathbf{B}_v \tag{9.67}$$

follows; a corresponding (primed) equation applies for displacements ξ'. ■

If we use this in (9.62), we obtain

$$\begin{aligned}&\mu_0 \int d^3r [\xi' \cdot \mathbf{F}\{\xi\} - \xi \cdot \mathbf{F}\{\xi'\}] \\ &= \int_S d^2r \big[(\mathbf{n} \times \delta\mathbf{A}') \cdot (\nabla \times \delta\mathbf{A}) - (\mathbf{n} \times \delta\mathbf{A}) \cdot (\nabla \times \delta\mathbf{A}')\big].\end{aligned} \tag{9.68}$$

On the right-hand side, we can add a corresponding integral over the (outer, conducting) surface S' (with the same integrand) without changing the value, since on S' the tangential component of the electric field ($\sim \mathbf{n} \times \delta\mathbf{A}$) vanishes.

It thus follows from (9.68)

$$\begin{aligned}
&\mu_0 \int d^3r \left[\xi' \cdot \mathbf{F}\{\xi\} - \xi \cdot \mathbf{F}\{\xi'\}\right] \\
&= \int_{S+S'} \left[(d\mathbf{F} \times \delta\mathbf{A}') \cdot (\nabla \times \delta\mathbf{A}) - (d\mathbf{F} \times \delta\mathbf{A}) \cdot (\nabla \times \delta\mathbf{A}')\right] \\
&= \int_{S+S'} d\mathbf{F} \cdot \left[\delta\mathbf{A}' \times (\nabla \times \delta\mathbf{A}) - \delta\mathbf{A} \times (\nabla \times \delta\mathbf{A}')\right] \\
&= -\int_{V_v} d^3r \left[\delta\mathbf{A}' \cdot \nabla \times \nabla \times \delta\mathbf{A} - \delta\mathbf{A} \cdot \nabla \times \nabla \times \delta\mathbf{A}'\right].
\end{aligned} \tag{9.69}$$

Here, V_v is the volume of the vacuum located between the plasma and the conducting wall. When applying Gauss's theorem, the change of sign of the normal vector on the surface must be taken into account. The vanishing of the right-hand side of (9.69) now follows from the current-free nature of the vacuum. This completes the proof of the Hermiticity of $\mathbf{F}$.

The extensive calculations also allow us to reformulate the expression (9.17) for the potential energy accordingly. In addition to the volume contribution in the plasma, there are also surface and vacuum contributions, which we will suitably transform.

Because

$$\begin{aligned}
\nabla \cdot \left[(\xi \cdot \nabla p_0 + \gamma p_0 \nabla \cdot \xi)\xi\right] &= \xi \cdot \nabla(\xi \cdot \nabla p_0 + \gamma p_0 \nabla \cdot \xi) \\
&+ (\nabla \cdot \xi)(\xi \cdot \nabla p_0 + \gamma p_0 \nabla \cdot \xi),
\end{aligned} \tag{9.70}$$

$$\begin{aligned}
\nabla \cdot [(\xi \times \mathbf{B}_0) \times \mathbf{Q}] &= \mathbf{Q} \cdot \nabla \times (\xi \times \mathbf{B}_0) - (\xi \times \mathbf{B}_0) \cdot \nabla \times \mathbf{Q} \\
&= \mathbf{Q}^2 + \xi \cdot [(\nabla \times \mathbf{Q}) \times \mathbf{B}_0],
\end{aligned} \tag{9.71}$$

$$\xi[(\nabla \times \mathbf{B}_0) \times \mathbf{Q}] = -(\nabla \times \mathbf{B}_0) \cdot (\xi \times \mathbf{Q}) \tag{9.72}$$

the volume contribution in the plasma is

$$\boxed{\begin{aligned}
W_F = \frac{1}{2} \int_{V_p} \Bigg[&\frac{Q^2}{\mu_0} + \frac{1}{\mu_0} (\nabla \times \mathbf{B}_0) \cdot (\xi \times \mathbf{Q}) \\
&+ (\nabla \cdot \xi)\xi \cdot \nabla p_0 + \gamma p_0 (\nabla \cdot \xi)^2 \Bigg] d^3r
\end{aligned}} \tag{9.73}$$

and the difference terms are

$$W - W_F = -\frac{1}{2}\int_S \left[\frac{1}{\mu_0}(\xi \times \mathbf{B}_0) \times \mathbf{Q} + \xi(\xi \cdot \nabla p_0 + \gamma p_0 \nabla \cdot \xi)\right] \cdot d^2\mathbf{r} . \tag{9.74}$$

The latter can be transformed using similar arguments as in the proof of Hermiticity in

$$\begin{aligned} W - W_F = & -\frac{1}{2}\int_S (\xi \cdot \mathbf{n})^2 \, \mathbf{n} \cdot \nabla \left(p_0 + \frac{B_0^2}{2\mu_0} - \frac{B_v^2}{2\mu_0}\right) d^2 r \\ & + \frac{1}{2\mu_0}\int_S \mathbf{B}_v \cdot \delta\mathbf{B}_v \, \xi \cdot d^2\mathbf{r} \equiv W_S + W_v \end{aligned} \tag{9.75}$$

W_v is the contribution to the potential energy due to changes in the vacuum magnetic field. With similar steps as in (9.69), we can write

$$\boxed{W_v = \frac{1}{2\mu_0}\int_{V_v} (\delta\mathbf{B}_v)^2 d^3 r} , \tag{9.76}$$

where V_v is the vacuum volume.

9.2 Stability of a Pinch Discharge

In this section, we want to demonstrate the practicality of the stability criterion just derived using a few very simple examples. These include pinch-like configurations.

First, we choose a situation in which the plasma is in direct contact with a conducting wall. Vacuum contributions are then absent. Furthermore, we set $\gamma = 0$, which, in the case of stability, is not so restrictive: If the plasma is stable for $\gamma = 0$, then $\gamma \neq 0$ does not alter this fundamental statement. The equilibrium configuration is assumed to be force-free, i.e., we choose

$$\boxed{\nabla \times \mathbf{B}_0 = \alpha \mathbf{B}_0} \tag{9.77}$$

for $\nabla p_0 = 0$. Because of $\mathbf{n} \cdot \xi = 0$ at the fixed surface, it suffices to discuss only W_F. Our starting point in this simple example is therefore

$$\boxed{W_F = \frac{1}{2}\int \left[\frac{\mathbf{Q}^2}{\mu_0} + \frac{1}{\mu_0}(\nabla \times \mathbf{B}_0) \cdot (\xi \times \mathbf{Q})\right] d^3 r} . \tag{9.78}$$

With (9.77) and

$$\mathbf{R} := \xi \times \mathbf{B}_0 \tag{9.79}$$

one obtains

$$W_F = \frac{1}{2\mu_0} \int \left[(\nabla \times \mathbf{R})^2 - \alpha \mathbf{R} \cdot \nabla \times \mathbf{R} \right] d^3 r \,. \tag{9.80}$$

The variation with respect to **R** is quite natural. However, we must keep in mind that, due to (9.79), only the contributions perpendicular to $\mathbf{B}_0$ are included.

Another simplifying assumption can be made regarding the normalization. At first, one is inclined to choose the obvious normalization

$$\int |\xi|^2 d^3 r = 1 \tag{9.81}$$

However, it is advisable to fully exploit the available freedom. We can also use

$$\boxed{\int \mathbf{R} \cdot \nabla \times \mathbf{R} \, d^3 r = const} \tag{9.82}$$

The constant in (9.82) can be positive, negative, or zero. In the latter case, it immediately follows that $W_F > 0$, so that we can exclude this singular case without loss of generality in the following. The normalization (9.82) has the advantage of allowing the variation of (9.80) under a very suitably formulated constraint [namely (9.82)] with the Lagrange multiplier $\alpha\lambda$ to be carried out easily. We require

$$\delta \int L \, d^3 r \equiv \delta \int \left[(\nabla \times \mathbf{R})^2 - (\lambda + 1)\alpha \mathbf{R} \cdot \nabla \times \mathbf{R} \right] d^3 r = 0 \tag{9.83}$$

and determine the extremum from the Euler equations. The latter is possible because, due to the normalization (9.82), the functional L is bounded from below. For brevity, we write the Euler equations in the form

$$\frac{\partial L}{\partial R_i} = \sum_k \frac{\partial}{\partial x_k} \frac{\partial L}{\partial R_{i,k}} \,, \tag{9.84}$$

where $R_{i,k} := \partial R_i / \partial x_k$ has been set. The calculation of the individual derivatives is somewhat cumbersome, but can be carried out without difficulty. The result of the algebra is

$$\boxed{\nabla \times \nabla \times \mathbf{R}_{min} = (\lambda + 1)\alpha \nabla \times \mathbf{R}_{min}} \,. \tag{9.85}$$

In the following, we again omit the index *"min"* for the minimizing test function **R**. In (9.85), λ plays the role of an eigenvalue. We gain insight into the stability behavior by using (9.85) to determine the minimum of W_F. To do this, we multiply (9.85) scalarly by **R**, in order to obtain after a short calculation

$$\alpha \mathbf{R} \cdot \nabla \times \mathbf{R} = \frac{1}{\lambda + 1} \{ (\nabla \times \mathbf{R})^2 + \nabla \cdot [(\nabla \times \mathbf{R}) \times \mathbf{R}] \} \quad (9.86)$$

Substituting into (9.80) yields

$$\min W_F = \frac{1}{2\mu_0} \int \left[\frac{\lambda}{\lambda + 1} (\nabla \times \mathbf{R})^2 - \frac{1}{\lambda + 1} \nabla \cdot ((\nabla \times \mathbf{R}) \times \mathbf{R}) \right] d^3 r. \quad (9.87)$$

Example 9.8 (Disappearance of the Surface Term)

The surface term, which arises from the last term in (9.87), disappears for the following reason. In the corresponding surface integral, the integrand $\mathbf{n} \cdot [(\nabla \times \mathbf{R}) \times \mathbf{R}]$ appears on the surface. Recalling the definition (9.79) and the convention that ξ and $\mathbf{B}_0$ are tangential to the surface, it follows on the surface that $\mathbf{R} \parallel \mathbf{n}$ and thus the integrand vanishes. ■

We find that

$$\boxed{\min W_F = \frac{\lambda}{2\mu_0(\lambda + 1)} \int (\nabla \times R)^2 \, d^3 r} \quad (9.88)$$

is positive for $\lambda > 0$ and for $\lambda < -1$.

Now, an analysis of (9.85) is necessary in order to determine the eigenvalue spectrum, taking into account the appropriate boundary conditions.

Example 9.9 (Evaluation for Box Geometry)

In the spirit of an exercise, we choose periodic boundary conditions for a box. From (9.85), it then follows for $\mathbf{Q} = \nabla \times \mathbf{R}$

$$\frac{i}{\alpha}(k_z Q_y - k_y Q_z) = (\lambda + 1) Q_x, \quad (9.89)$$

$$\frac{i}{\alpha}(k_x Q_z - k_z Q_x) = (\lambda + 1) Q_y, \quad (9.90)$$

$$\frac{i}{\alpha}(k_y Q_x - k_x Q_y) = (\lambda + 1) Q_z. \quad (9.91)$$

Nontrivial solutions exist for

$$(\lambda + 1)^2 = \frac{k^2}{\alpha^2}. \quad (9.92)$$

From this we see that for small k, $\mathbf{k} = (k_x, k_y, k_z)$, solutions with

$$-1 < \lambda < 0 \tag{9.93}$$

exist. Thus, if the system is large enough ($L \sim 2\pi/k$), an instability can develop. ■

Unstable Pinch Discharge

As a second, more realistic example, we discuss the stability of a pinch discharge. Let us begin with the naive question of whether a pure B_θ field can confine a cylindrical plasma. Assume that no current flows in the plasma and that it is in a field-free state. The B_θ field is generated by surface currents along the cylindrical shell ($r = r_0$),

$$\boxed{B_\theta = B_0 \frac{r_0}{r}}\;. \tag{9.94}$$

Then, from the equation of motion (9.8), only

$$-\rho_0 \omega^2 \xi = \gamma p_0 \nabla \nabla \cdot \xi \tag{9.95}$$

remains. This equation is so simple that we can solve it directly without further ado. We can perform a Fourier transform in θ and z. Thus, we set

$$\xi \sim \xi(r) e^{im\theta + ikz} \tag{9.96}$$

into (9.95), and after some elementary manipulations, we obtain

$$\boxed{\frac{d^2 \xi_z}{dr^2} + \frac{1}{r}\frac{d\xi_z}{dr} - \left(\alpha^2 + \frac{m^2}{r^2}\right)\xi_z = 0} \tag{9.97}$$

with

$$\alpha^2 = \frac{-\omega^2 \rho_0}{\gamma p_0} + k^2 . \tag{9.98}$$

For the moment, let us assume $\alpha^2 > 0$ at $\omega^2 < 0$ and check whether consistent solutions exist. The differential equation (9.97) has modified Bessel functions as solutions,

$$\xi_z = A_{mk}\, I_m(\alpha r) e^{im\theta + ikz} . \tag{9.99}$$

To determine the eigenvalues (ω^2), we need the boundary conditions for the transition to the vacuum field $\mathbf{B}_\theta + \delta\mathbf{B}$. In the vacuum, no current flows, so that $\nabla \times \delta\mathbf{B} = 0$ holds. As long as we do not choose a path through the current-carrying shell, we can therefore

$$\delta\mathbf{B} = \nabla\psi \tag{9.100}$$

express this by a scalar potential ψ, which is determined by

$$\nabla^2 \psi = 0 \tag{9.101}$$

With the ansatz

$$\psi \sim \psi(r) e^{i(kz+m\theta)} \tag{9.102}$$

we find the decaying solutions

$$\psi = C_{mk}\, K_m(kr)\, e^{im\theta+ikz}. \tag{9.103}$$

From this, $\delta\mathbf{B}$ follows directly via (9.100).

Now we can explicitly formulate the boundary conditions. For this, we need the position of the boundary, which is given by ξ in the form [cf. (9.95)]

$$\begin{aligned} r =& r_0 + \xi_r = r_0 + \frac{1}{ik}\frac{d}{dr}\xi_z \\ =& r_0 + \frac{\alpha}{ik} A_{mk}\, I'_m(\alpha r_0)\, e^{ikz+im\theta} \end{aligned} \tag{9.104}$$

That is, the equation for the surface transitions from $r - r_0 = 0$ to $\varphi(\mathbf{r}) = 0$, where φ is defined by

$$\varphi := r_0 + r_1\, e^{im\theta+ikz} - r \tag{9.105}$$

with $r_1 = (\alpha/ik) A_{mk}\, I'_m(\alpha r_0)$ from (9.104). The vector (in a cylindrical coordinate system)

$$\nabla\varphi = \left(-1,\ \frac{imr_1}{r_0}\, e^{im\theta+ikz},\ ikr_1\, e^{im\theta+ikz}\right) \tag{9.106}$$

is parallel to the normal vector of the surface. We then calculate the total magnetic field on the vacuum side using the formula

$$\mathbf{B}(\mathbf{r}, t) = \mathbf{B}(r_0) + \delta\mathbf{B} + (\xi \cdot \nabla)\mathbf{B}(\mathbf{r}_0) \tag{9.107}$$

and, using (9.94), obtain

$$\mathbf{B}(\mathbf{r}, t) = B_0\hat{\theta} + \delta\mathbf{B} + \xi_r(-1)\frac{B_0}{r_0}\,\hat{\theta}. \tag{9.108}$$

The boundary condition that $\mathbf{B}(\mathbf{r}, t)$ must be perpendicular to the surface is written as

$$\nabla\varphi \cdot \mathbf{B} \;=\; 0 \;\approx\; -\delta B_r + \frac{im}{r_0} B_0 r_1\, e^{im\theta+ikz}\,. \tag{9.109}$$

With the previous solutions (9.103) and (9.104), (9.109) immediately yields

$$\frac{C_{mk}}{A_{mk}} = \frac{B_0}{r_0}\frac{\alpha m}{k^2}\frac{I'_m(\alpha r_0)}{K'_m(kr_0)} . \tag{9.110}$$

Furthermore, if we use the continuity of the total pressure, which we have already written earlier in (9.39) as

$$-\gamma p_0 \nabla \cdot \xi + \frac{\mathbf{B}}{\mu_0} \cdot [\mathbf{Q} + (\xi \cdot \nabla)\mathbf{B}] = \frac{\mathbf{B}_v}{\mu_0} \cdot [\delta \mathbf{B}_v + (\xi \cdot \nabla)\mathbf{B}_v] , \tag{9.111}$$

now for $\mathbf{B} = 0$ in the plasma, then at the surface this yields

$$\frac{\rho_0 \omega^2}{ik} I_m(\alpha r_0) = \frac{B_0}{\mu_0 r_0}\left[\frac{C_{mk}}{A_{mk}} imK_m(kr_0) + iB_0 \frac{\alpha}{k} I'_m(\alpha r_0)\right] . \tag{9.112}$$

Together with (9.110), we obtain the dispersion relation

$$\boxed{-\frac{\omega^2 r_0 \rho_0}{2\alpha p_0}\frac{I_m(\alpha r_0)}{I'_m(\alpha r_0)} = 1 + \frac{m^2}{kr_0}\frac{K_m(kr_0)}{K'_m(kr_0)}} . \tag{9.113}$$

The simplest way to see that instabilities occur is to set $m = 0$, i.e., to consider modes that are independent of θ. There then exist solutions $\omega^2 < 0$ of the dispersion relation

$$-\frac{\omega^2 r_0 \rho_0}{2\alpha p_0}\frac{I_0(\alpha r_0)}{I_1(\alpha r_0)} = 1. \tag{9.114}$$

This result is obtained through a curve analysis, assuming that $I_0 > 0$ and $I'_0 = I_1 > 0$ can be assumed for variations of $-\omega^2$ between 0 and $+\infty$. The left-hand side of (9.114) takes on values between 0 and $+\infty$ in this range of $-\omega^2$. A detailed analysis shows that solutions are possible for all k. This type of instability is known as the "sausage" instability. Furthermore, if one analyzes for $m = 1$, one finds the so-called "kink" instability.

An illustration of both instabilities can be found in Fig. 9.1

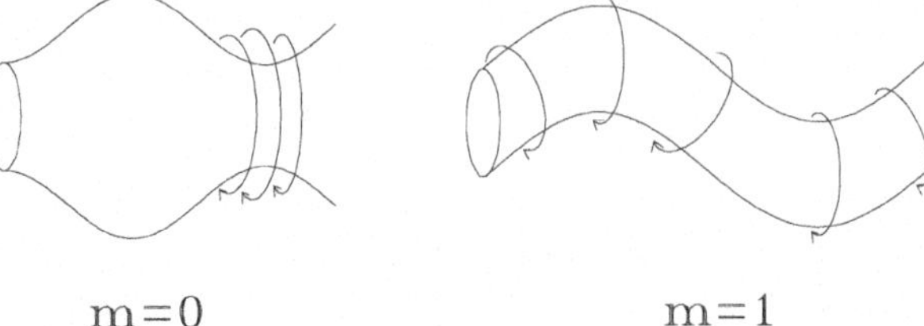

Fig. 9.1 Illustration of the $m = 0$ ("sausage") and $m = 1$ ("kink") instabilities

Stabilized Pinch Discharge

The following calculation will now show that the pinch configuration just described can be stabilized by an internal axial B_z field and an external conductor.

The internal (in the plasma) B_z field is intended to prevent the $m = 0$ instability, and the external conductor surrounding the entire arrangement is to suppress the "kink" instability. For simplicity, we introduce the following notations:

$$\mathbf{B}_{aussen} = B_\theta \hat{\theta} + B_{z,Vakuum}\, \hat{z} = B_0 \frac{r_0}{r} \hat{\theta} + b_v \hat{z}, \tag{9.115}$$

$$\mathbf{B}_{innen} = b_i \hat{z}. \tag{9.116}$$

The external conducting cylinder is located at $R_0 = \Upsilon r_0, \Upsilon > 1$. In the following, we use the energy principle rather than the equation of motion corresponding to (9.95).

Fluid component

Under the assumptions $p_0 = const$ and $\nabla \times b_i \hat{z} = 0$ in the plasma interior, the potential energy of the fluid component is given by

$$\boxed{W_F = \frac{1}{2\mu_0} \int [\mathbf{Q}^{\,2} + \mu_0 \gamma p_0 (\nabla \cdot \xi)^2] d^3 r}\,. \tag{9.117}$$

This expression cannot become negative. This statement is consistent with our earlier conclusions, according to which instability—if it occurs at all—can only arise from a negative contribution of the remaining terms. However, in that case, $\mathbf{n} \cdot \xi$ must not vanish. A look at the remaining terms convinces us that for $\mathbf{n} \cdot \xi = 0$ only positive terms are added. (Note that for the "kink" and "sausage" instabilities $\xi_r \neq 0$ holds!) We therefore allow all possible states with $\mathbf{n} \cdot \xi \neq 0$. If the total energy integral remains positive for all arbitrarily prescribed distributions of $\mathbf{n} \cdot \xi$ on the surface, we can conclude stability. Therefore, we now calculate the individual contributions W_F, W_v and W_s for a given $\mathbf{n} \cdot \xi$. The form (9.117) can be written, using the notation already employed in (9.84), as

$$\begin{aligned} W_F = \frac{1}{2\mu_0} \int \sum_{i,k} \Big\{ & b_k\, \xi_{i,k}\, Q_i - \sum_l b_l \xi_{i,k} Q_l \delta_{ik} \\ & - \xi_i\, b_{k,i} Q_k + \mu_0 \gamma p_0 \delta_{ik} \xi_{i,k} \nabla \cdot \xi \Big\} \end{aligned} \tag{9.118}$$

where, for the moment, the index i for "inside" has been omitted. To determine the minimum, the Euler equations are used. After some simple transformations, the compact form follows

$$\frac{1}{\mu_0}(\nabla \times \mathbf{Q}) \times \mathbf{B}_{innen} + \gamma p_0 \nabla \nabla \cdot \xi = 0 \,. \tag{9.119}$$

This is exactly equation $\mathbf{F}(\xi) = 0$, since we have not yet used any normalization for ξ. The latter is also not necessary in this calculation, since we keep $\mathbf{n} \cdot \xi$ fixed at the boundary and therefore no factor in ξ is still free. With the help of the Euler equations (9.119) for the minimizing ξ we can calculate the minimum of W_F. We have

$$\min W_F = \frac{1}{2} \int \left(\gamma p_0 \nabla \cdot \xi - \frac{\mathbf{B}_{innen} \cdot \mathbf{Q}}{\mu_0} \right) \xi \cdot d\mathbf{F}, \tag{9.120}$$

because $\mathbf{B}_{innen}$ is perpendicular to the normal of the plasma surface.

So far, we have only used Eq. (9.119), and not its solution, in the transformations. For the solution, we make the ansatz

$$\xi = \xi(r) e^{im\theta + ikz}. \tag{9.121}$$

Because $\mathbf{B}_{innen} \| \hat{z}$ follows for $k \neq 0$ from (9.119) directly

$$\nabla \cdot \xi = 0. \tag{9.122}$$

The case $k = 0$ must be treated separately. Due to (9.122), (9.119) becomes, after taking the divergence,

$$\nabla^2 (b_i Q_z) = 0 \tag{9.123}$$

With a constant A yet to be determined, the solution is

$$Q_z = \frac{A}{b_i} I_m(kr) e^{im\theta + ikz}. \tag{9.124}$$

Substituting into (9.120) then yields

$$\min W_F = -\frac{A}{2\mu_0} I_m(kr_0) \int \xi_r(r = r_0)\, e^{im\theta + ikz}\, dz\, r_0\, d\theta \,. \tag{9.125}$$

To determine $\xi_r(r_0)$ we proceed as follows. $\mathbf{Q}$ depends, by its definition and due to (9.116) and (9.122), on ξ in the form

$$\mathbf{Q} = ikb_i \xi \tag{9.126}$$

together. The relation (9.119) yields

$$\nabla (b_i Q_z) = ikb_i \mathbf{Q}. \tag{9.127}$$

If we combine the last two equations, it follows that

$$\begin{aligned}\xi &= -\frac{1}{k^2 b_i^2}\nabla(b_i Q_z)\\ &= -\frac{A}{k^2 b_i^2}\nabla\left[I_m(kr)e^{im\theta+ikz}\right].\end{aligned} \tag{9.128}$$

From this, $\xi_r(r = r_0) \equiv r_1 \exp[-i(m\theta + ikz)]$ can be calculated directly and inserted into the integrand of (9.125). Eq. (9.125) then yields

$$\boxed{\min W_F = \frac{\pi r_0 L k b_i^2 r_1^2}{2\mu_0}\frac{I_m(kr_0)}{I'_m(kr_0)}} \tag{9.129}$$

with $r_1 = -A\ I'_m(kr_0)/kb_i^2$. Two remarks are necessary. First, of course, only the harmonics $\xi_r(r_0) \sim \exp(-im\theta - ikz)$ are relevant in the integrand on the right-hand side of (9.125). The θ -integration yields the factor 2π; however, since we must only consider the real parts of ξ in complex notation, an additional factor $\frac{1}{2}$ appears. Second, we have replaced the integral over z by the length L.

Vacuum contribution

Next, we turn to the vacuum contribution W_v. With

$$\delta \mathbf{B}_v = \nabla\phi \tag{9.130}$$

and

$$(\nabla\phi)^2 = \nabla\cdot(\phi\nabla\phi) - \phi\nabla^2\phi \tag{9.131}$$

only the surface contribution

$$\boxed{W_v = \frac{1}{2\mu_0}\int \phi\nabla\phi\cdot d\mathbf{F}} \tag{9.132}$$

remains to be calculated. The minimizing function ϕ follows from the Euler equations associated with (9.130). The solutions of

$$\frac{\partial}{\partial x_k}\frac{\partial}{\partial\phi_{,k}}\frac{1}{2}\phi_{,k}\,\phi_{,k} = \frac{\partial}{\partial x_k}\phi_{,k} = \nabla^2\phi = 0 \tag{9.133}$$

in the vacuum region bounded on both sides (by plasma or conducting wall) $r_0 < r < R_0$ are given by

$$\phi = C I_m(kr)e^{im\theta+ikz} + D K_m(kr)e^{im\theta+ikz} \tag{9.134}$$

with still undetermined coefficients C and D. These constants must be determined by the boundary conditions.

First, at the conductor-vacuum boundary, the entire **B** field should have no normal component,

$$\left.\frac{\partial \phi}{\partial r}\right|_{r=R_0} = 0 = CkI'_m(kR_0)e^{im\theta+ikz} + DkK'_m(kR_0)e^{im\theta+ikz} . \tag{9.135}$$

Furthermore, at the plasma-vacuum boundary, the entire **B** field should be tangential. Using (9.106) and (9.115), we formulate this condition as

$$\nabla\varphi \cdot \mathbf{B} = 0 = -\left.\frac{\partial \phi}{\partial r}\right|_{r_0} + \frac{imr_1}{r_0}B_0 + ikr_1\, b_v . \tag{9.136}$$

When evaluated, this provides us with a second equation for the coefficients C and D,

$$CkI'_m(kr_0) + DkK'_m(kr_0) = ikr_1 b_v + \frac{imr_1}{r_0}B_0 . \tag{9.137}$$

Equations (9.135) and (9.137) form an inhomogeneous system of equations, whose solutions for C and D can be substituted into (9.134). The minimizing solution thus found ϕ can be used to calculate the minimum of (9.132). A short calculation yields

$$\boxed{\begin{aligned}\min W_v = {} & \frac{\pi r_0 L k}{2\mu_0}\left(b_v + \frac{mB_0}{kr_0}\right)^2 \\ & \times \frac{K'_m(kR_0)I_m(kr_0) - I'_m(kR_0)K_m(kr_0)}{K'_m(kR_0)I'_m(kr_0) - I'_m(kR_0)K'_m(kr_0)} r_1^2 .\end{aligned}} \tag{9.138}$$

Surface contribution

Finally, we still have to minimize the original surface term W_S. At the boundary between plasma and vacuum, a force equilibrium must exist, which, due to the continuity of pressure, can be written in the form

$$\nabla\left(p_0 + \frac{b_i^2}{2\mu_0}\right) = -\nabla\frac{B_0^2}{2\mu_0}\frac{r_0^2}{r^2} = \frac{B_0^2}{\mu_0 r_0}\hat{r} \tag{9.139}$$

Interestingly, this means that

$$\boxed{\min W_S = -\frac{B_0^2 L}{2\mu_0}\pi r_1^2} \tag{9.140}$$

is always negative. Only the combination of W_F, W_v and W_S can lead to positive values and thus to stability.

The results of numerical evaluations show that stabilization can be achieved. The most difficult stability criteria to satisfy are those for the $m = 0$ and $m = 1$ modes. For details, we must refer to the specialized literature. Only one remark should be made here. In the limit $R_0 \to \infty$ the stability criterion is called the Kruskal-Shafranov condition.

9.3 The Interchange Instability and Its Significance

One of the most direct applications of the energy principle is the interchange instability of magnetically confined plasmas. We will demonstrate that concavely curved plasma surfaces, in contrast to convexly curved plasma surfaces, tend to develop instabilities under magnetic confinement.

Let us consider a rather plausible change in which two neighboring flux tubes (see below) are interchanged without disturbing the rest of the system. In this simple case, one can quickly calculate whether the system can thereby transition to a state of lower energy, i.e., whether it is unstable. It is assumed that the plasma dynamics can actually proceed in such a way that the energetically more favorable state is reached. These remarks are meant to indicate that a complete analytical calculation of the minimizing perturbations is not always necessary. Often, it is appropriate to first find plausibility arguments through global considerations. For quantitative purposes (thresholds, etc.), however, one must always resort to more detailed calculations.

Within the framework of ideal magnetohydrodynamics, one finds the potential energy

$$\boxed{W = \int \left(\frac{p}{\gamma - 1} + \frac{B^2}{2\mu_0} \right) d^3 r}\,, \tag{9.141}$$

which, for an incompressible fluid ($\gamma \to \infty$), reduces to the magnetic energy

$$W_m = \int \frac{B^2}{2\mu_0}\, d^3 r \tag{9.142}$$

In the following, we will mainly discuss this contribution, especially since we already know that a compressible fluid is at least as unstable.

Two flux tubes, including the plasma contained within them, are interchanged; that is, we consider two infinitesimally neighboring states that result from the interchange of two regions.

By a flux tube, we mean a volume $V = \int F dl$, enclosed by an "end face" F and the "lateral surface," in which the magnetic field lines lie that pass through the boundary of F.

Let the "height element" above F be dl. If we denote two adjacent flux tubes by $F_1 = A_1$ and dl_1 or, respectively, $F_2 = A_2$ and dl_2,, then for incompressibility we require

$$\boxed{A_1 dl_1 = A_2 dl_2} \; . \tag{9.143}$$

The magnetic flux ϕ is defined by $\phi = BA$. It is constant within a flux tube. Due to infinite conductivity (within the framework of ideal MHD), we also have

$$\frac{d\phi}{dt} = 0 \; , \tag{9.144}$$

i.e., the flux remains constant. By interchanging two flux tubes, the magnetic energy (9.142) changes as follows:

$$\delta W_m = \delta \int \frac{B^2}{2\mu_0} A dl = \delta \left[\frac{\phi^2}{2\mu_0} \int \frac{dl}{A} \right] . \tag{9.145}$$

In calculating the last expression, we take into account that the flux ϕ_2 previously passed through A_2 and afterwards through A_1; for ϕ_1 the reverse is true. From this it follows that

$$\begin{aligned} \delta W_m =& \frac{1}{2\mu_0} \left[\phi_2^2 \left(\int \frac{dl_1}{A_1} - \int \frac{dl_2}{A_2} \right) + \phi_1^2 \left(\int \frac{dl_2}{A_2} - \int \frac{dl_1}{A_1} \right) \right] \\ =& \frac{1}{2\mu_0} (\phi_2^2 - \phi_1^2) \left(\int \frac{dl_1}{A_1} - \int \frac{dl_2}{A_2} \right) \equiv -\frac{\delta\phi^2}{2\mu_0} \delta \int \frac{dl}{A} \; . \end{aligned} \tag{9.146}$$

Because of (9.143), we can also write

$$\boxed{\delta W_m = \frac{1}{2\mu_0} (\phi_2^2 - \phi_1^2) \int \left[1 - \left(\frac{dl_2}{dl_1} \right)^2 \right] \frac{dl_1}{A_1}} \; . \tag{9.147}$$

If, for example, in a machine the magnetic field in the outer plasma edge region increases rapidly enough outward so that both $\phi_2^2 > \phi_1^2$ and $dl_2/dl_1 > 1$ are satisfied, then we can infer $\delta W_m < 0$. [In the following, index 2 always characterizes the flux tube located further outward (compared to 1).] This means instability, since there exists a neighboring state with lower potential energy.

Under this restriction, the statement is valid that concave configurations ($dl_2/dl_1 > 1$) are unstable. However, it is also evident that there can be concave configurations in which, despite $dl_2/dl_1 > 1$ for the fluxes $\phi_2 < \phi_1$ holds. For these, we therefore expect (for $\gamma \to \infty$)) stability. However, it turns out that in general, for finite values of γ, instability occurs.

Curvature Instabilities

To gain a more detailed insight into the situation with finite γ, we must (for finite γ) additionally take into account the free energy

$$\boxed{W_p = \int \frac{p}{\gamma - 1} dV} \tag{9.148}$$

For adiabatic changes $[d(pdV^\gamma)/dt = 0]$ the following holds for infinitesimal volume elements of the flux tubes

$$\begin{aligned} \delta dW_p &= \frac{1}{\gamma - 1}\delta(pdV) = \frac{1}{\gamma - 1}\delta\left(\frac{pdV^\gamma}{dV^{\gamma-1}}\right) \\ &= \frac{1}{\gamma - 1}\left\{p_2 dV_2^\gamma\left[\frac{1}{dV_1^{\gamma-1}} - \frac{1}{dV_2^{\gamma-1}}\right] + p_1 dV_1^\gamma\left[\frac{1}{dV_2^{\gamma-1}} - \frac{1}{dV_1^{\gamma-1}}\right]\right\} \\ &= \frac{1}{\gamma - 1}\left[p_2 dV_2^\gamma - p_1 dV_1^\gamma\right]\left(\frac{1}{dV_1^{\gamma-1}} - \frac{1}{dV_2^{\gamma-1}}\right) \end{aligned} \tag{9.149}$$

or

$$\delta dW_p = \frac{\delta(pdV^\gamma)}{dV^\gamma}\delta dV . \tag{9.150}$$

If we interchange two flux tubes near the plasma edge, where $p \to 0$ approaches zero, we can approximate

$$\delta(pdV^\gamma) \approx \delta p dV^\gamma \tag{9.151}$$

provided that

$$\left|\frac{\delta p}{p}\right| \gg \left|\frac{\delta dV}{dV}\right| \tag{9.152}$$

holds there. For $\delta dV > 0$ and $\delta p = p_2 - p_1 < 0$ we recognize, due to $\delta dW_p \approx \delta p \delta dV$, the destabilizing tendency.

The following statements can therefore be made in summary:

1. For $\gamma \to \infty$ (and therefore all the more so for finite values of γ)), we have already found, in the context of $dl_2/dl_1 > 1$, instability for $\phi_2 > \phi_1$ or $B_2/B_1 > A_1/A_2 = dl_2/dl_1$, i.e., for

$$\delta \int \frac{dl}{B} < 0 \tag{9.153}$$

2. If for the two flux tubes $\phi_2 = \phi_1$ (i.e., $\delta W_m = 0$) holds, the system is, under the conditions $\delta p < 0$, unstable for

$$\delta \int \frac{dl}{B} > 0 \tag{9.154}$$

The criterion (9.154) follows from (9.150) and (9.151) if we require $\delta W_p < 0$ at $\delta p < 0$; then it must be that

$$\delta dV = \delta \int A dl = \phi\, \delta \int \frac{dl}{B} > 0 \tag{9.155}$$

holds.

Thus, if (9.153) is satisfied, we have instability for $dl_2 > dl_1$, since we can interchange flux tubes with $\phi_2 > \phi_1$. On the other hand, if (9.154) holds, we can perform interchanges of flux tubes for which $\phi_2 = \phi_1$ is satisfied at $p_2 < p_1$. Incompressibility is no longer assumed here. From $p_2 < p_1$ we infer $B_2 > B_1$ (due to a pressure equilibrium to zeroth order), so that, because of $dl_2/B_2 > dl_1/B_1$, according to (9.154), $dl_2/dl_1 > 1$ must also be satisfied for these interchanges. The instability criterion

$$\boxed{dl_2 > dl_1} \tag{9.156}$$

thus holds quite generally. It only breaks down in the case of sheared magnetic fields. Furthermore, the instability criterion (9.156) requires a separate treatment of the case

$$\delta \int \frac{dl}{B} = 0\,. \tag{9.157}$$

In this case, we can write

$$\delta W = \delta W_m + \delta W_p = \frac{1}{2\mu_0}(\phi_2^2 - \phi_1^2) \int \left(\frac{dl_1}{A_1} - \frac{dl_2}{A_2}\right) + \delta p\, \delta \int A dl\,. \tag{9.158}$$

Let us now consider the interchange of flux tubes for which

$$\frac{dl_1}{A_1} = \frac{dl_2}{A_2} \tag{9.159}$$

applies. Together with (9.157), it follows that

$$\frac{A_1^2}{\phi_1} = \frac{A_2^2}{\phi_2} \tag{9.160}$$

and $\delta W_m = 0$. We can now rewrite Eq. (9.158) as

$$\boxed{\delta W = \delta p \int \left[\left(\frac{A_2}{A_1} \right)^2 - 1 \right] A_1 dl_1 = \delta p \int \left[\left(\frac{dl_2}{dl_1} \right)^2 - 1 \right] A_1 dl_1} \tag{9.161}$$

so that now, too, for $\delta p < 0$ ($B_2 > B_1$) and $dl_2 > dl_1$ instability follows due to $\delta W < 0$.

In simple cases, the results of the interchange instability just derived can also be easily justified using the hydromagnetic stability criteria derived in Sect. 9.1. Let us choose the model of a plasma confined by magnetic fields, for which (approximately) the pressure and the magnetic field are constant inside. Due to $\nabla \times \mathbf{B} = 0$ and $\nabla p = 0$ in the plasma, we can estimate

$$W \equiv W_F + W_S + W_v > W_s = -\frac{1}{2} \int dF \, (\mathbf{n} \cdot \xi)^2 \, \mathbf{n} \cdot \nabla \left(p + \frac{B^2}{2\mu_0} - \frac{B_v^2}{2\mu_0} \right) \tag{9.162}$$

Using the radius of curvature $\mathbf{R}$ for magnetic fields, it follows that

$$\mathbf{n} \cdot \nabla \left(-\frac{B_v^2}{2\mu_0} \right) = \frac{B_v^2}{\mu_0} \frac{(\mathbf{R} \cdot \mathbf{n})}{R^2} \, . \tag{9.163}$$

In the following estimates, we assume the total pressure in the plasma to be constant. For

$$\mathbf{n} \cdot \mathbf{R} < 0 \tag{9.164}$$

it immediately follows *stability*. This condition states that the normal vector $\mathbf{n}$ (pointing from the plasma into the vacuum) must have the opposite direction to the radius of curvature $\mathbf{R}$ of the magnetic field lines (pointing from the center of the curvature circle to the magnetic field lines). Such (stable) configurations are convexly curved ("curved away from the fluid"); see Fig. 9.2.

For understanding the interchange instability, the analogy to the Rayleigh-Taylor instability is very helpful. We know that a heavy fluid located above a lighter fluid in a gravitational field is unstable according to Rayleigh and Taylor. A curved magnetic field can be assigned an effective force due to the centrifugal acceleration. The force is parallel to $\mathbf{R}$. If $\mathbf{n} \cdot \mathbf{R} > 0$, then, in this picture, the heavier plasma is above the vacuum.

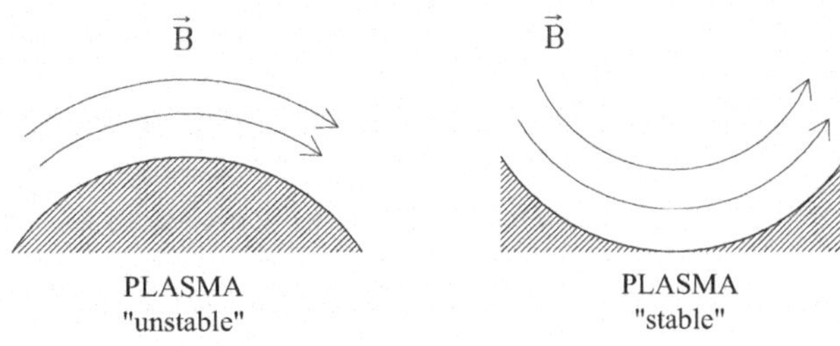

Fig. 9.2 Concave and convex arrangements, which are unstable or stable, respectively, in the context of MHD

The instability due to incorrect curvature ("bad curvature instability") thus also becomes understandable in this way.

Hydromagnetic Rayleigh-Taylor Instability

For the sake of completeness, we now calculate the growth rate for the hydromagnetic Rayleigh-Taylor instability; see Fig. 9.3

We consider the case of a plasma in the external magnetic field $\mathbf{B}_0 = \hat{z}B_0(y)$, where an (effective) gravitational field with y-direction and $\mathbf{g}_0 = -\hat{y}g_0$ is supposed to act. The equilibrium situation is then characterized by

$$\frac{\partial}{\partial y}\left(p_0 + \frac{B_0^2}{2\mu_0}\right) = -\rho_0\, g_0 \tag{9.165}$$

In the equation of motion (9.7) for the perturbations, we now also have to include the gravitational force. This can be easily accomplished if the following intermediate steps are taken into account: First, the force density appears in the equation of motion

$$\rho_0 \dot{\mathbf{u}} = \cdots + \rho \mathbf{g}_0 \tag{9.166}$$

where the contributions marked by ... on the right-hand side of (9.166) have already been treated explicitly. When differentiating with respect to time, we set for $\dot{\rho} = -\nabla(\rho_0 \mathbf{u})$, and thus it becomes clear that on the right-hand side of (9.7) the additional term

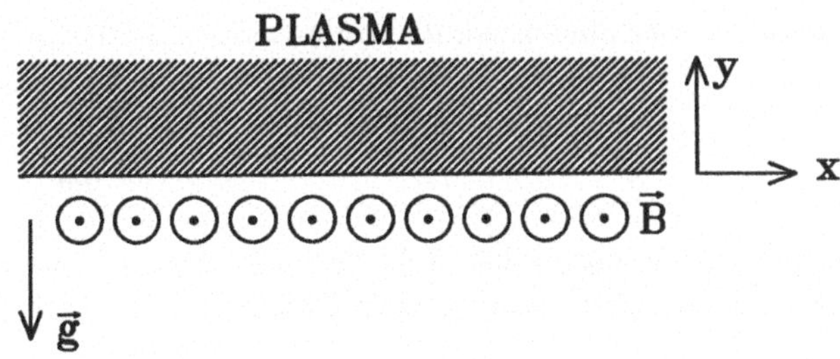

Fig. 9.3 Geometry of the hydromagnetic Rayleigh-Taylor instability

$$-\mathbf{g}_0\nabla\cdot(\rho_0\xi)=\hat{y}g_0\nabla\cdot(\rho_0\xi)=\hat{y}g_0[\xi\cdot\nabla\rho_0+\rho_0\nabla\cdot\xi] \tag{9.167}$$

must also appear. Therefore, instead of (9.7), we write

$$\begin{aligned}-\rho_0\omega^2\xi =&\nabla(\xi\cdot\nabla p_0+\gamma p_0\nabla\cdot\xi)+\frac{1}{\mu_0}[(\nabla\times\mathbf{Q})\times\mathbf{B}_0\\ &+(\nabla\times B_0)\times Q\big]+\hat{y}g_0\nabla\cdot(\rho_0\xi)\ .\end{aligned} \tag{9.168}$$

To simplify the further discussion, we make the following restrictions. First, we consider only perturbations that do not explicitly depend on the z-coordinate, and second, we select only incompressible perturbations ($\nabla\cdot\xi=0$).

These and the previously mentioned specifications lead to

$$\mathbf{Q}=-\xi_y\frac{dB_0}{dy}\hat{z}\ , \tag{9.169}$$

$$(\nabla\times\mathbf{Q})\times\mathbf{B}_0+(\nabla\times\mathbf{B}_0)\times\mathbf{Q}=-\nabla(\mathbf{B}_0\cdot\mathbf{Q})\ , \tag{9.170}$$

$$\xi:=\xi(y)e^{ikx}\ , \tag{9.171}$$

$$\rho_0\omega^2\xi_x=ik\left(\xi\cdot\nabla p_0+\frac{1}{\mu_0}\mathbf{B}_0\cdot\mathbf{Q}\right), \tag{9.172}$$

$$\rho_0\omega^2\xi_y=\frac{d}{dy}\left(\xi\cdot\nabla p_0+\frac{1}{\mu_0}\mathbf{B}_0\cdot\mathbf{Q}\right)-g_0\xi_y\frac{d\rho_0}{dy}\ . \tag{9.173}$$

First, we combine the last two equations to obtain

$$\rho_0\omega^2\xi_y=\frac{\omega^2}{ik}\frac{d}{dy}(\rho_0\xi_x)-g_0\xi_y\frac{d\rho_0}{dy}\ . \tag{9.174}$$

If both sides of this equation are differentiated with respect to x and incompressibility is used, ξ_x can be eliminated; the result is

$$\boxed{\frac{d^2\xi_y}{dy^2}+\left(\frac{1}{\rho_0}\frac{d\rho_0}{dy}\right)\frac{d\xi_y}{dy}-k^2\left(1+\frac{g_0}{\omega^2}\frac{1}{\rho_0}\frac{d\rho_0}{dy}\right)\xi_y=0}\ . \tag{9.175}$$

In general, the eigenvalue problem is not easy to solve. However, for two simple cases, the solutions can be written down directly.

Example 9.10 (Inhomogeneous Plasma)
For an exponential density variation, $d \ln \rho_0/dy = \kappa = const$, one possibility is

$$\frac{d^2\xi_y}{dy^2} = \frac{d\xi_y}{dy} = 0 \,. \tag{9.176}$$

Nontrivial solutions follow for

$$\boxed{\omega^2 = -g_0\kappa} \,. \tag{9.177}$$

For $\kappa > 0$, that is, when the density decreases in the direction of the gravitational force, we obtain instability. This is the well-known Rayleigh-Taylor instability. ■

Example 9.11 (Kruskal-Schwarzschild Instability)
A similar result is obtained for a homogeneous plasma with a sharp boundary,

$$\rho_0 = \begin{cases} \rho_{00} = const & \text{for} \quad y \geq 0, \\ 0 & \text{for} \quad y < 0. \end{cases} \tag{9.178}$$

For $y > 0$ the differential equation

$$\frac{d^2\xi_y}{dy^2} = k^2\xi_y \tag{9.179}$$

has the solution

$$\xi_y = \xi_y^{(0)} e^{-ky} \,. \tag{9.180}$$

For $y = 0$ we have a significant contribution from the derivative of ρ_0,

$$\frac{d\rho_0}{dy} = \rho_0\delta(y), \tag{9.181}$$

which is discontinuous. Therefore, if we integrate the differential equation (9.175) over y from $-\varepsilon$ to $+\varepsilon$ and then let $\varepsilon \to 0$ go to zero, we obtain from

$$\int_{-\varepsilon}^{+\varepsilon} \frac{d}{dy}\left(\rho_0 \frac{d\xi_y}{dy}\right) dy = \int_{-\varepsilon}^{+\varepsilon} k^2 \left(\rho_0 + \frac{g_0}{\omega^2}\frac{d\rho_0}{dy}\right) \xi_y \, dy \tag{9.182}$$

the relation

$$\frac{k^2 g_0}{\omega^2} \rho_0 \, \xi_y \bigg|_{y=0} = \lim_{\varepsilon \to +0} \rho_0 \frac{d\xi_y}{dy}\bigg|_{y=\varepsilon} \,. \tag{9.183}$$

If we require continuity of ξ_y at the point $y = 0$, then from (9.180) it follows

$$\boxed{\omega^2 = -kg_0}\,. \tag{9.184}$$

■

The choice (9.178) for the density distribution at $\mathbf{g}_0 = -\hat{y}g_0$ again allows us to recognize this instability as an analogue of the Rayleigh-Taylor instability. In plasma physics, it is also referred to—specifically for certain plasma configurations in a magnetic field—as the Kruskal-Schwarzschild instability.

Such instabilities play a crucial role in plasma physics. The first thing that comes to mind is magnetic confinement for achieving nuclear fusion. Many seemingly simple configurations must be ruled out directly due to the curvature argument, or rather ingenious modifications must be devised for stabilization. Further details can be found in the specialized literature on magnetically confined fusion plasmas ("magnetic confinement fusion").

Significant stability problems also arise in the inertial confinement fusion that has been pursued for several years. In the planned fusion using laser or particle beams, the goal is to compress a highly symmetric pellet to high densities so that fusion processes can occur inside. Essential plasma physical processes take place in the shell ("corona"), whose symmetry, however, can quickly be disrupted by a Rayleigh-Taylor instability. The associated changes in the absorption of the laser or particle beams pose a serious problem for the realization of controlled fusion.

In conclusion, it should perhaps be emphasized once again that magnetohydrodynamics represents only a very rough model for a plasma. This makes statements about instabilities all the more important, as they are quite robust and cannot be reversed by marginal changes.

9.4 Resistive Instabilities

We now leave ideal magnetohydrodynamics and allow for finite conductivities (non-vanishing resistances or resistivities). In this area, the theory regarding variational principles is less developed, and accordingly, in most of the following calculations, a normal mode analysis is used.

The MHD model has already been introduced. It consists of a continuity equation for ρ, Ohm's law, and the momentum balance, supplemented by Maxwell's equations for div $\mathbf{B}$, curl $\mathbf{B}$, and curl $\mathbf{E}$. For simplicity, we assume incompressibility,

$$\nabla \cdot \mathbf{u} = 0. \tag{9.185}$$

The resistivity η is assumed to depend only on the plasma variables, so that we set

$$\boxed{\frac{d}{dt}\eta = \dot{\eta} + \mathbf{u} \cdot \nabla \eta = 0} \tag{9.186}$$

For the sake of clarity, we summarize the basic equations here once again:

$$\dot{\rho} + \mathbf{u} \cdot \nabla \rho = 0, \tag{9.187}$$

$$\boxed{\eta \mathbf{j} = \mathbf{E} + \mathbf{u} \times \mathbf{B}}\,, \tag{9.188}$$

$$\rho \dot{\mathbf{u}} = \mathbf{j} \times \mathbf{B} - \nabla p + \rho \mathbf{g}, \tag{9.189}$$

$$\nabla \cdot \mathbf{B} = 0, \tag{9.190}$$

$$\nabla \times \mathbf{E} = -\dot{\mathbf{B}}, \tag{9.191}$$

$$\nabla \times \mathbf{B} = \mu_0 \mathbf{j}. \tag{9.192}$$

In the momentum balance (9.189), we have included the force density $\rho \mathbf{g}$ in an (effective) gravitational field, in order to later simulate the effects of curved magnetic field lines.

Compared to the stability analysis in ideal systems ($\eta = 0$), new effects can arise because the plasma is no longer rigidly coupled to the magnetic field lines ("frozen in magnetic field lines"). Modes that are otherwise stable can become unstable due to finite resistivity.

Let us begin with a plausibility consideration.

For $\mathbf{E} = 0$, (9.188) together with (9.189) yields the force density

$$\mathbf{F}_s = \frac{(\mathbf{u} \times \mathbf{B}) \times \mathbf{B}}{\eta} = \frac{\mathbf{B}(\mathbf{u} \cdot \mathbf{B}) - \mathbf{u}B^2}{\eta}, \tag{9.193}$$

which, for $\eta \to 0$ (bei $\mathbf{u} \cdot \mathbf{B} \neq uB$), becomes infinite and prevents the decoupling of the plasma flow from the magnetic field lines, i.e., it enforces $\mathbf{u} \cdot \mathbf{B} = uB$. However, if a small but finite resistivity $\eta \neq 0$ is present, F_s becomes small in regions where B becomes small. The following calculation will show that $\mathbf{k} \cdot \mathbf{B}$ is the decisive quantity (instead of B) for perturbations with the wave vector $\mathbf{k}$, i.e., for $\mathbf{k} \cdot \mathbf{B} \to 0$ new effects can arise.

Example 9.12 Rippling Instability

If we linearize, for example, (9.186) and (9.188) and make an exponential ansatz $\exp(st)$ for the temporal evolution, we obtain for $\mathbf{E} = 0,\ p \approx 0,\ g = 0$ *in addition* to (9.193)

$$\eta_1 \approx -\frac{1}{s}\mathbf{u}_1 \cdot \nabla \eta_0, \tag{9.194}$$

$$\mathbf{j}_1 \approx -\frac{\eta_1}{\eta_0}\mathbf{j}_0\,, \tag{9.195}$$

$$\mathbf{F}_r \approx \frac{\mathbf{u}_1 \cdot \nabla \eta_0}{s\eta_0}\,\mathbf{j}_0 \times \mathbf{B}_0\,. \tag{9.196}$$

If we illustrate the directions (see Fig. 9.4), it follows that if $x = 0$ denotes the line $B_{0y} = 0$,

$$F_{rx} \sim \begin{cases} -u_{1x} \text{ for } x > 0, \\ u_{1x} \text{ for } x < 0. \end{cases} \tag{9.197}$$

this means that perturbations for $x < 0$ can be amplified. The exact calculation leads to the "rippling instability." ■

Example 9.13 (Gravitational Instability)

A similar consideration is possible for gravitational instability if, in (9.189), the term $\rho\mathbf{g}$ on the right-hand side dominates. Analogously, one finds

$$\mathbf{F} \approx \rho_1 \mathbf{g} \approx -\frac{\mathbf{u} \cdot \nabla \rho_0}{s}\mathbf{g}, \tag{9.198}$$

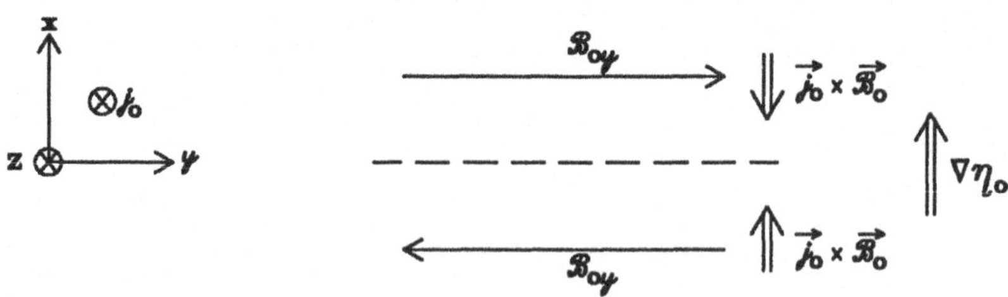

Fig. 9.4 Magnetic field topology and force directions in the "rippling instability"

which, for $\mathbf{g} \cdot \nabla \rho_0 < 0$, leads to an amplification of initial perturbations. ■

Example 9.14 (Tearing Instability)
The instability associated with the breaking of magnetic field lines ("tearing instability") is more complicated, since for long-wavelength perturbations, electric fields are induced that must be taken into account in Ohm's law (9.188). From (9.191), for Fourier modes, one obtains $\sim \exp(ik_y y)$

$$E_{1z} \approx -\frac{s B_{1x}}{ik_y}. \tag{9.199}$$

Together with $j_{1z} \approx E_{1z}/\eta_0$ from (9.188), this yields from (9.189) an additional driving force

$$F_{tx} \approx \frac{s B_{oy} B_{1x}}{ik_y \eta_0}, \tag{9.200}$$

which favors island formation in nearly symmetric configurations. This is schematically illustrated in Fig. 9.5. ■

The following stationary state ($\partial_t = 0$, (index 0) can be used as the initial solution:

$$\mathbf{u}_0 = 0, \quad \rho_0 = \rho_0(x), \quad \eta_0 = \eta_0(x), \quad \mathbf{g} = g\hat{x}, \tag{9.201}$$

where the magnetic field may have a shear of the form

$$\boxed{\mathbf{B}_0 = B_{0y}(x)\hat{y} + B_{0z}(x)\hat{z}} \tag{9.202}$$

In order to avoid a contradiction between (9.188) and (9.191), it must hold that

$$\nabla \times [\eta_0 \nabla \times \mathbf{B}_0] = 0 \tag{9.203}$$

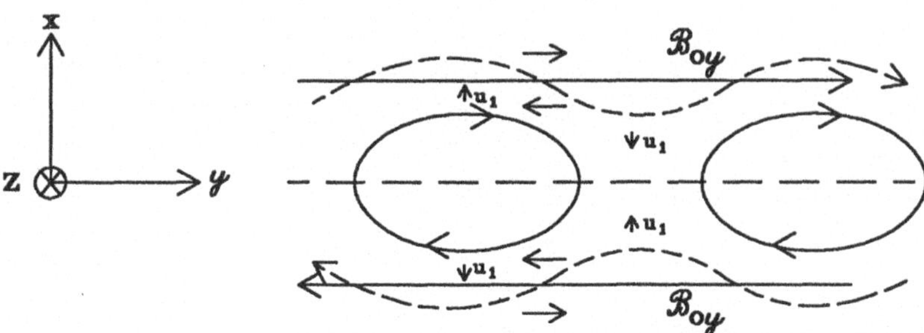

Fig. 9.5 Island formation due to the breaking of magnetic field lines in the "tearing" instability

This, together with the ansatz (9.202), has the consequence

$$\eta_0 \partial_x \mathbf{B}_0 = const = \mathbf{c}\ . \tag{9.204}$$

If we now perturb this initial state and then linearize the perturbations (index 1), we obtain for $\mathbf{B}_1$ from (9.191) and (9.188)

$$\boxed{\dot{\mathbf{B}}_1 = \nabla \times \left\{ \mathbf{u}_1 \times \mathbf{B}_0 - \frac{1}{\mu_0}[\eta_1(\nabla \times B_0) + \eta_0(\nabla \times \mathbf{B}_1)] \right\}}\ . \tag{9.205}$$

The curl of the momentum balance (9.189), on the other hand, leads to

$$\boxed{\nabla \times (\rho_0 \dot{\mathbf{u}}_1) = \nabla \times \left\{ \frac{1}{\mu_0}[(\mathbf{B}_0 \cdot \nabla)\mathbf{B}_1 + B_{1x}(\partial_x \mathbf{B}_0)] + \rho_1 g \hat{x} \right\}}\ . \tag{9.206}$$

Since the stationary state depends only on x, we can Fourier transform in y and z. Altogether, we make the normal mode ansatz for a perturbed quantity where

$$\mathbf{k} = \begin{pmatrix} 0 \\ k_y \\ k_z \end{pmatrix} \tag{9.208}$$

applies. From (9.205), after a short calculation, we obtain for the x-component of $\mathbf{B}_1$

$$sB_{1x} = i(\mathbf{k} \cdot \mathbf{B}_0)u_{1x} + \frac{\eta_0}{\mu_0}\left(\partial_x^2 B_{1x} - k^2 B_{1x}\right) - i\frac{\eta_1}{\mu_0}(\mathbf{k} \cdot \partial_x \mathbf{B}_0). \tag{9.209}$$

A second equation for B_{1x} is obtained from (9.206) by the following transformations. We first take the curl and use the vector-analytical relation $\nabla \times \nabla \times = \nabla\nabla \cdot - \nabla^2$. We then discuss the x-component of the equation. In doing so, one finds

$$\hat{x} \cdot [\nabla \times \nabla \times (\rho_0 \mathbf{u}_1)] = -\partial_x[\rho_0(\partial_x u_{1x})] + k^2 \rho_0 u_{1x} \tag{9.210}$$

as well as

$$\begin{aligned} &\hat{x} \cdot \{\nabla \times \nabla \times [(\mathbf{B}_0 \cdot \nabla)\mathbf{B}_1 + B_{1x}(\partial_x \mathbf{B}_0)]\} \\ &\quad = -i\varphi\left(\partial_x^2 B_{1x}\right) + i\left(\partial_x^2 \varphi\right)B_{1x} + ik^2\varphi\, B_{1x}, \end{aligned} \tag{9.211}$$

where we have introduced the auxiliary quantity

$$\boxed{\varphi := \mathbf{k} \cdot \mathbf{B}_0} \tag{9.212}$$

If we also take into account the linearized form of (9.186), i.e.,

$$s\eta_1 + u_{1x}(\partial_x \eta_0) = 0 \tag{9.213}$$

and

$$\hat{x} \cdot \left[\nabla \times \nabla \times \left(\rho_1 \hat{x}\right)\right] = k^2 \rho_1, \tag{9.214}$$

then from (9.206) and (9.209)–(9.214) we can obtain the following system:

$$\boxed{\eta_0 \partial_x^2 B_{1x} = \left(\mu_0 s + k^2 \eta_0\right) B_{1x} - i\left[\mu_0 \varphi + \frac{1}{s}(\partial_x \eta_0)(\partial_x \varphi)\right] u_{1x},} \tag{9.215}$$

$$\boxed{\begin{aligned}\varphi\, \partial_x^2 B_{1x} = &\left[(\partial_x^2 \varphi) + k^2 \varphi\right] B_{1x} + \mu_0 i k^2 \left[s \rho_0 + \frac{g}{s}\,(\partial_x \rho_0)\right] u_{1x} \\ &- \mu_0 i s \partial_x \left[\rho_0 \partial_x (u_{1x})\right] .\end{aligned}} \tag{9.216}$$

From these two equations, we eliminate $\partial_x^2 B_{1x}$, by multiplying the first equation by φ and the second by η_0 and then subtracting. The result is

$$\begin{aligned} B_{1x}\left(\mu_0 s \varphi - \eta_0 \varphi''\right) =& i \mu_0 s \left[u_{1x} \varphi^2 s^{-1} \left(1 + \frac{\eta_0' \varphi'}{\mu_0 s \varphi}\right)\right. \\ &\left. + u_{1x} k^2 \eta_0 \left(\rho_0 + g \rho_0' s^{-2}\right) - \eta_0 \left(\rho_0 u_{1x}'\right)' \right]. \end{aligned} \tag{9.217}$$

The prime denotes differentiation with respect to x.

We discuss the eigenvalue problem using this equation, together with (9.215).

First, we note that the property (9.204) of the stationary solution can be written in the form

$$\eta_0 \varphi' = c_0 \tag{9.218}$$

In the following, we consider perturbations of the stationary solutions that satisfy

$$\varphi(x = 0) = 0 \quad , \quad \varphi' > 0 \tag{9.219}$$

The characteristic length L_s ("shear length") of $\varphi = \mathbf{k} \cdot \mathbf{B}_0$ is used in the following step of nondimensionalization. Let

$$\begin{aligned} &\psi := \frac{B_{1x}}{B_0} \quad , \quad W := -i u_{1x}\, k \tau_R \quad , \quad \boxed{F := \frac{\varphi}{k B_0}} \quad , \\ &\alpha := k L_s \quad , \quad \xi := \frac{x}{L_s} \quad , \quad L_s \approx \left(\frac{\partial F}{\partial x}\right)^{-1} , \end{aligned} \tag{9.220}$$

where a resistive time τ_R ("diffusion time") is defined by

$$\tau_R := \mu_0 \frac{L_s^2}{\langle \eta_0 \rangle} \tag{9.221}$$

The analogous hydrodynamic time τ_H ("Alfvén-wave transit time") is given by

$$\tau_H = \frac{L_s}{B_0}\sqrt{\mu_0 \langle \rho_0 \rangle} \tag{9.222}$$

The ratio

$$\boxed{S = \frac{\tau_R}{\tau_H}} \tag{9.223}$$

can be interpreted as the magnetic Reynolds number. The quantities $\langle \eta_0 \rangle$ and $\langle \rho_0 \rangle$ represent mean values of η_0 and ρ_0, respectively, which can usefully be employed for normalization. It is natural to nondimensionalize the growth rate s with τ_R and to introduce for this purpose

$$p = s\tau_R \tag{9.224}$$

Furthermore, we abbreviate

$$\tilde{\rho} = \rho_0 / \langle \rho_0 \rangle \quad , \quad \tilde{\eta}_0 = \eta_0 / \langle \eta_0 \rangle \tag{9.225}$$

and choose $\langle \eta_0 \rangle$ such that in (9.218) $c_0 = 1$ can be set. This is possible because F' is of order 1; it results in

$$\tilde{\eta}_0 = F'^{-1} , \tag{9.226}$$

where we assume $B_0 = |\mathbf{B}_0| \approx const.$

The two equations (9.215) and (9.217) then yield, after some rearrangement,

$$\boxed{p\psi'' - p\psi\left(\alpha^2 + \frac{F''}{F}\right) = (p\psi + WF)\left(pF' - \frac{F''}{F}\right),} \tag{9.227}$$

$$\boxed{\frac{p^2}{\alpha^2 S^2 F}\left[(\tilde{\rho}_0 W')' + \alpha^2 W\left(\frac{S^2 G}{p^2} - \tilde{\rho}_0\right)\right] = (p\psi + WF)\left(pF' - \frac{F''}{F}\right),} \tag{9.228}$$

where differentiation with respect to ξ is now denoted by a prime, and

$$\boxed{G := -\tau_H^2\, g \frac{d\rho_0}{dx} \frac{1}{\langle \rho_0 \rangle}} \tag{9.229}$$

has been abbreviated.

The two eigenvalue equations (9.227) and (9.228) are still quite complicated. In further analysis, one takes advantage of the fact that S is generally a very large number.

Let us now take a closer look at the limiting case $S \to \infty$, $p \to \infty$, $p/S \to 0$. Outside a small neighborhood around $\xi = 0$ (only in a small neighborhood around $\xi = 0$ does F become very small), it follows that

$$\psi'' - \psi\left(\alpha^2 + \frac{F''}{F}\right) \approx (p\psi + WF)F', \tag{9.230}$$

$$W\frac{G}{F} \approx (p\psi + WF)F'p\,. \tag{9.231}$$

To leading order in p^0 this means

$$WF \approx -p\psi. \tag{9.232}$$

From (9.231) it then follows directly

$$(p\psi + WF)F' \approx -\psi\frac{G}{F^2}. \tag{9.233}$$

If we substitute this into the right-hand side of (9.230), we obtain (outside the small neighborhood around $\xi = 0$, where $F \approx 0$ can be used)

$$\psi'' - \left(\alpha^2 + \frac{F''}{F} - \frac{G}{F^2}\right)\psi = 0. \tag{9.234}$$

In the vicinity of $F = 0$ we approximate $F \approx F'\xi$ and solve the resulting differential equations from (9.227) and (9.228). If we disregard the technical details for the moment, we may assume the existence of solutions in complementary spatial regions.

In the outer region ($F \neq 0$) we find solutions ψ_1 with $\psi_1 \to 0$ for $\xi \to -\infty$ and ψ_2 with $\psi_2 \to 0$ for $\xi \to +\infty$. In order to meaningfully connect these solutions in the region $\xi \approx 0$, the slope of ψ must vary very sharply there. More precisely, if we were to extend the outer solutions up to $\xi = 0$, we would find a discontinuous change of ψ' at $\xi = 0$. Only continuity $\psi_1 = \psi_2$ at $\xi = 0$ could be required without loss of generality. A formal solution of (9.234) is helpful in this context. With the boundary conditions just discussed, we have

$$\psi_1 = e^{\alpha\xi}\left[\int_{-\infty}^{\xi} dx\, e^{-2\alpha x}\int_{-\infty}^{x} dy\left(\frac{F''}{F} - \frac{G}{F^2}\right)\psi_1\, e^{\alpha y} + A\right], \tag{9.235}$$

$$\psi_2 = e^{-\alpha\xi}\left[\int_{+\infty}^{\xi} dx\, e^{2\alpha x}\int_{\infty}^{x} dy\left(\frac{F''}{F} - \frac{G}{F^2}\right)\psi_2\, e^{-\alpha y} + B\right]. \tag{9.236}$$

Example 9.15 (Solution for $G = 0$ and $F = \tanh\xi$)
To gain a somewhat more concrete idea, we choose as a demonstration example $G = 0$ (for $g = 0$) and

$$\boxed{F = \tanh\xi}\,. \tag{9.237}$$

Then the solutions ψ_1 and ψ_2 are explicitly

$$\psi_1 = (\tanh\xi - \alpha)e^{\alpha\xi}, \tag{9.238}$$

$$\psi_2 = -(\tanh\xi + \alpha)e^{-\alpha\xi}. \tag{9.239}$$

At the point $\xi = 0$ (i.e., already outside the actual range of validity) we have $\psi_1 = \psi_2 = -\alpha$ and

$$\psi_1' = 1 - \alpha^2\,,\ \psi_2' = -1 + \alpha^2. \tag{9.240}$$

The quantity

$$\Delta' := \left.\frac{\psi_2' - \psi_1'}{\psi_1}\right|_{\xi=0} = 2\left(\frac{1}{\alpha} - \alpha\right) \tag{9.241}$$

takes on all values for $0 < \alpha < \infty$; the variation $\infty > \Delta' > -\infty$ is in this case monotonic with α. This statement holds for practically all relevant examples. ■

In the general case (9.235) and (9.236) we have

$$\Delta' \approx -2\alpha - \frac{1}{\psi_1}\int_{-\infty}^{+\infty} dx\, e^{-\alpha|x|}\,\psi\frac{F''}{F} + \mathcal{O}\left[\frac{G}{(F')^2\xi_0}\right], \tag{9.242}$$

where ξ_0 is the approximate width of the region in which ψ' (the exact solution) changes rapidly.

At this point, it is appropriate to say something about the general strategy for the further procedure. As already indicated, two length scales are relevant throughout the entire problem. A large length scale, which determines the variation in the outer region, and a small scale, which is characteristic for the variation around $\xi = 0$.

The inner and outer solutions are "matched" at an intermediate point in order to obtain an approximate solution to the entire problem.

With regard to the outer solution, we do not make a significant error if we choose the "matching point" at $\xi = 0$. (In fact, it should be at $\xi_0 \neq 0$, but between ξ_0 and 0, ψ_1 and ψ_2 do not vary significantly.) For the inner solution, we must introduce a new scale $\xi_1 = \varepsilon^{-1}\xi$. For large values of ξ_1, the inner solution must smoothly connect to the outer solution. Therefore, for the inner solution, we will also calculate Δ' and, without significant error, determine the eigenvalue p from the condition

$$\Delta'_{\text{äußere Lösung}}(\xi = 0) \stackrel{!}{=} \Delta'_{\text{innere Lösung}}(|\xi_1| \to \infty) \tag{9.243}$$

Thus, to carry out this program, it remains to calculate the inner solution and the corresponding Δ'.

In the following, we outline the solution in the inner region, under the assumption $F = F'\xi$, F' is constant.

In the vicinity of $\xi = 0$, let F'', $\tilde{\eta}$, $\tilde{\eta}'$, G and $\tilde{\rho}_0$ also be (nearly) constant. Then $\tilde{\rho}_0' W' \ll \tilde{\rho}_0 W''$ holds, and from (9.228), it follows with $\xi = \varepsilon\xi_1$

$$\begin{aligned} W'' + \varepsilon^2\alpha^2\left(\frac{S^2 G}{p^2\tilde{\rho}_0} - 1\right)W - \varepsilon^2\,\frac{\alpha^2 S^2}{p^2\tilde{\rho}_0}F'\varepsilon\xi_1\left[p(F')^2\varepsilon\xi_1 - F''\right]W \\ = \frac{\alpha^2 S^2}{p^2\tilde{\rho}_0^2}\,\varepsilon^2 p\psi\left[p(F')^2\varepsilon\xi_1 - F''\right]. \end{aligned} \tag{9.244}$$

If one chooses

$$\varepsilon = \left[\frac{p\tilde{\rho}_0\tilde{\eta}_0}{4\alpha^2 S^2 (F')^2}\right]^{1/4} \tag{9.245}$$

and

$$\xi_1 = \theta + \frac{1}{2}\,\frac{F''}{p\varepsilon(F')^2} = \theta - \frac{1}{2}\,\frac{\tilde{\eta}_0'}{p\varepsilon}\,, \tag{9.246}$$

then (9.244) simplifies to

$$W'' + \left(\Upsilon - \frac{1}{4}\theta^2\right)W = \frac{p}{4\varepsilon F'}\left(\theta + \frac{\eta_0'}{2p\varepsilon}\right)\psi \tag{9.247}$$

with

$$\Upsilon = \frac{S^2\alpha^2\varepsilon^2 G}{p^2\tilde{\rho}_0} - \varepsilon^2\alpha^2 + \frac{(\tilde{\eta}_0')^2}{16\varepsilon^2 p^2} ; \tag{9.248}$$

the prime now of course denotes differentiation with respect to θ.

A corresponding variable substitution must also be performed in (9.227). If we simultaneously define the new quantities

$$U := \frac{4\varepsilon F'}{p} W \quad , \quad \Omega := \frac{p\varepsilon}{4\tilde{\eta}_0} \quad , \quad \delta := -\frac{\tilde{\eta}_0'}{2p\varepsilon} , \tag{9.249}$$

then, after a short calculation, we obtain

$$\boxed{\psi'' - \varepsilon^2\alpha^2\psi = \varepsilon\Omega[4\psi + U(\theta - \delta)],} \tag{9.250}$$

$$\boxed{U'' + \left(\Upsilon - \frac{1}{4}\theta^2\right)U = (\theta - \delta)\psi.} \tag{9.251}$$

In the first of these two equations, the small quantity ε still appears; in comparison to (9.251), we therefore infer from (9.250) that U in θ varies significantly faster than is the case for ψ. Furthermore, (9.250) offers a further simplification due to $\varepsilon^2\alpha^2 \ll 4\varepsilon\Omega$. The difference of the gradients then follows from

$$\begin{aligned} \Delta' &= \lim_{x\to\infty}\left\{\frac{1}{\varepsilon\psi(x)}\int_{-x}^{x} d\theta\ \psi''\right\} \\ &\approx \frac{\Omega}{\psi_1}\int_{-\infty}^{+\infty}[4\psi + U(\theta - \delta)]d\theta. \end{aligned} \tag{9.252}$$

Eq. (9.251) now suggests expanding the functions ψ and U in terms of Hermite polynomials u_n. The differential equation for u_n is namely

$$u_n'' + \left(n + \frac{1}{2} - \frac{1}{4}\theta^2\right)u_n = 0. \tag{9.253}$$

If we therefore set

$$U = \sum_n a_n u_n , \tag{9.254}$$

$$\psi = \sum_n b_n u_n \tag{9.255}$$

then, due to the weak variation of ψ $[\varepsilon|\psi'/\psi| \sim \varepsilon|\Delta'| \ll 1]$

$$\frac{b_n}{\psi_1} = \frac{1}{\psi_1}\int_{-\infty}^{+\infty} d\theta\ \psi u_n \approx \int_{-\infty}^{+\infty} d\theta u_n$$

$$= \begin{cases} 2^{3/4}\left[\Gamma\left(\frac{1}{2}n+\frac{1}{2}\right)/\Gamma\left(\frac{1}{2}n+1\right)\right]^{1/2} & \text{for } n \text{ even,} \\ 0 & \text{for } n \text{ odd.} \end{cases} \tag{9.256}$$

On the other hand, (9.251) yields

$$\begin{aligned} \frac{a_n}{\psi_1} &\approx \frac{1}{\Upsilon-\left(n+\frac{1}{2}\right)}\int_{-\infty}^{+\infty} d\theta u_n(\theta-\delta) \\ &= \frac{1}{\Upsilon-\left(n+\frac{1}{2}\right)}\begin{cases} -\delta\, 2^{3/4}\left[\Gamma\left(\frac{1}{2}n+\frac{1}{2}\right)/\Gamma\left(\frac{1}{2}n+1\right)\right]^{1/2}, & n \text{ even,} \\ 2^{9/4}\left[\Gamma\left(\frac{1}{2}n+1\right)/\Gamma\left(\frac{1}{2}n+\frac{1}{2}\right)\right]^{1/2}, & n \text{ odd.} \end{cases} \end{aligned} \tag{9.257}$$

Taking all this into account, we obtain from (9.252)

$$\Delta' = 2^{7/2}\pi\ \Omega\left[\frac{\Gamma\left(\frac{3}{4}-\frac{1}{2}\Upsilon\right)}{\Gamma\left(\frac{1}{4}-\frac{1}{2}\Upsilon\right)} - \frac{\delta^2}{8}\frac{\Gamma\left(\frac{1}{4}-\frac{1}{2}\Upsilon\right)}{\left(\frac{3}{4}-\frac{1}{2}\Upsilon\right)}\right]. \tag{9.258}$$

The right-hand side of this equation must coincide with the right-hand side of (9.242), for example, in the special case (9.237) with the right-hand side of (9.241). This requirement results in a characteristic equation for the eigenvalue p.

Since Δ' according to expression (9.250) always varies between $-\infty$ and $+\infty$ when Υ remains finite and traverses the range

$$n+\frac{1}{2} < \Upsilon < n+\frac{3}{2}, \qquad n = 0, 1, 2, \ldots \tag{9.259}$$

, for a given value of Δ' one can find infinitely many eigenvalues from

$$\Upsilon \approx 1, 2, 3, \ldots \tag{9.260}$$

An illustrative sketch of this situation can be found in Fig. 9.6.

With the definition (9.248), the determination of p becomes relatively simple.

Let us now take a closer look at some special cases.

Example 9.16 (Rippling Instability)

If in (9.248) the gradient in resistivity plays the dominant role, i.e.,

$$\boxed{\Upsilon \approx \frac{(\tilde{\eta}_0')^2}{16\varepsilon^2 p^2}} \tag{9.261}$$

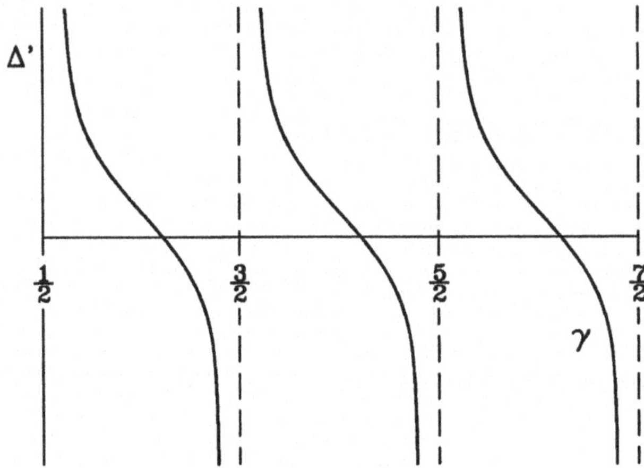

Fig. 9.6 Representation of the relationship (9.258)

and Υ remains finite, then together with (9.245) it follows

$$\boxed{p = \left(\frac{(\tilde{\eta}_0')^4\alpha^2 S^2 (F')^2}{64\Upsilon^2\tilde{\rho}_0\tilde{\eta}_0}\right)^{1/5}} . \tag{9.262}$$

The maximum growth rate p is achieved for minimal $\Upsilon\left(\frac{1}{2} \le \Upsilon \le \frac{3}{2}\right)$ and small wavelengths ($\alpha \gg 1$). This instability is called the rippling instability because of the small wavelengths. However, it should be noted that due to the various approximations α must not become arbitrarily large. For example, from (9.260), if we compare with (9.248), it follows

$$\alpha^2\varepsilon^2 \ll \Upsilon \approx 1 \tag{9.263}$$

or $\alpha \ll 1/\varepsilon$, so that the growth rate (9.261) cannot become infinitely large. The wavelength of the most unstable mode also remains finite. ■

Example 9.17 (Gravitational Interchange Instability)
A similar case arises when gravity (or the curvature of the magnetic field) in (9.248) plays the dominant role, i.e.,

$$\boxed{\Upsilon \approx S^2\alpha^2\varepsilon^2 G/p^2\tilde{\rho}_0} \tag{9.264}$$

applies. Note that for instability, $G > 0$ must hold. The growth rate for this gravitational interchange instability can be determined similarly to the "rippling mode." ■

Finally, let us consider the case of nearly symmetric ground states (around $\xi = 0$)). Then let $\tilde{\eta}_0' \approx 0,\ G \approx 0$ and $|\Upsilon| \ll \frac{1}{2}$. In this case, (9.258) simplifies to

$$\Delta' = \Omega(12 - 13\delta^2) \approx 12\Omega. \tag{9.265}$$

The instability condition is then obviously

$$\Delta' > 0\,, \tag{9.266}$$

and Ω follows from the Δ' prescribed by (9.241) or (9.242). If we now use (9.226), (9.245), and (9.249), it follows that

$$\boxed{p \approx 4\left(\Omega^2 \alpha S\, \tilde{\rho}_0^{-\frac{1}{2}}\, \tilde{\eta}_0^{\frac{1}{2}}\right)^{2/5}}\,. \tag{9.267}$$

The growth rate p is therefore proportional to $(\Delta')^{4/5}$ and $\alpha^{2/5}\left[= (kL_s)^{2/5}\right]$. From the specific example (9.241) it follows that $(\Delta')^{4/5} \sim \alpha^{-4/5}$ for $\Delta' > 0$, so that the expression (9.267) is maximal for $\alpha \ll 1$. However, there is a lower limit for α due to the approximation $\varepsilon|\Delta'| \ll 1$, which led to (9.256). With $\Delta' \approx 2/\alpha$ it follows that

$$\alpha^4 \gg \rho_0^{1/2} \tilde{\eta}_0^2 / S. \tag{9.268}$$

Of course, this does not yet cover all instabilities in which finite resistivity plays a role. In particular, those with finite $k_\parallel$ ("ballooning modes") have recently gained considerable importance. However, due to space constraints, their study must be referred to the specialized literature.

Part II
Current Research Areas

Plasma Confinement and Anomalous Transport 10

Abstract

In this chapter, we supplement the general fundamentals with some (essential) aspects that arise in magnetic confinement aimed at energy-producing nuclear fusion. Fusion in toroidal configurations such as the tokamak, stellarator, etc., is an important long-term physical and technical project that touches on many highly interesting areas of theoretical physics. Here, we address only a few of these. The selection criterion is the direct connection to the general fundamentals discussed in previous chapters. Therefore, some other important topics are left out. In this regard, we must refer to the specialized literature.

10.1 Particle Motion in Toroidal Geometry

We begin by examining how single-particle motion is influenced by toroidal geometry. This is by no means merely an exercise, as the characteristic motion allows us—as already discussed in the introductory chapter—to draw important conclusions about the fundamental confinement of particles.

A toroidal configuration is called axisymmetric if, as in a tokamak, it exhibits a certain symmetry about its axis (the z-axis). This means that physical properties such as the magnetic field, the density, and the temperature of the plasma are arranged with radial symmetry around the central axis of the tokamak. The plasma in a tokamak possesses rotational symmetry with respect to its main axis. This symmetry is important for the

K.-H. Spatschek, *Theoretical Plasma Physics*,
https://doi.org/10.1007/978-3-662-72828-4_10

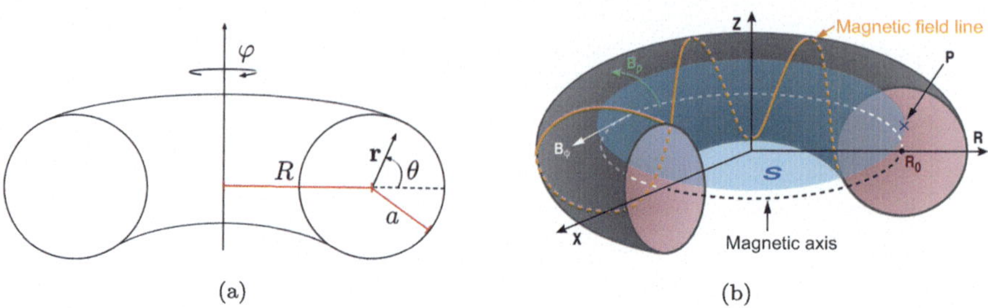

Fig. 10.1 (**a**) Toroidal coordinates, (**b**) toroidal magnetic field configuration

stability and control of the plasma. For a relatively clear description, toroidal (R, φ) and poloidal (r, θ) coordinates are introduced, as sketched in Fig. 10.1(a).

The main axis (major semi-axis) of a tokamak is referred to as the torus axis or magnetic axis. The axis runs through the center of the ring-shaped plasma vessel and forms the midpoint of the rotationally symmetric structure of the plasma.

Magnetic Surfaces

The main coils of a tokamak generate a strong magnetic field that runs around the main axis (symmetry axis). In addition, a current is induced in the plasma itself by transformer action, which also generates a magnetic field that contributes to the stabilization and confinement of the plasma. The magnetic field that runs around the main axis of the tokamak and keeps the plasma on its ring-shaped path is called the toroidal magnetic field. This field is necessary to keep the plasma away from the walls of the tokamak. The additional poloidal magnetic field runs in smaller circles within the plasma, i.e., perpendicular to the magnetic axis. Fig. 10.1(b) shows a sketch.

The magnetic field lines satisfy the equation

$$\frac{dx}{B_x} = \frac{dy}{B_y} = \frac{dz}{B_z} = \frac{dl}{B}, \quad dl = \sqrt{dx^2 + dy^2 + dz^2}\,. \tag{10.1}$$

Magnetic surfaces can be characterized in the form

$$\boxed{\psi(\mathbf{r}) = \text{ const}\,, \quad [\nabla\psi(\mathbf{r})]\cdot\mathbf{B} = 0} \tag{10.2}$$

Example 10.1 (Magnetic Field Configurations)
In the case of a straight cylinder with cylindrical coordinates r, θ, z, the magnetic field $\mathbf{B} = \nabla \times \mathbf{A}$ is obtained from the components of the vector potential $\mathbf{A}$ via

$$B_r = \frac{1}{r}\frac{\partial A_z}{\partial \theta} - \frac{\partial A_\theta}{\partial z}, \quad B_\theta = \frac{\partial A_r}{\partial z} - \frac{\partial A_z}{\partial r}, \quad B_z = \frac{1}{r}\frac{\partial (rA_\theta)}{\partial r} - \frac{1}{r}\frac{\partial A_r}{\partial \theta}, \tag{10.3}$$

and for an *axisymmetric* configuration with $\partial/\partial\theta \equiv 0$, the magnetic surfaces are determined by

$$\boxed{\psi(r,z) = rA_\theta(r,z) = \text{ const}} \tag{10.4}$$

Specifically,

$$B_r \frac{\partial(rA_\theta)}{\partial r} + 0 + B_z \frac{\partial(rA_\theta)}{\partial z} = 0 \,. \tag{10.5}$$

For a *translational* configuration with $\partial/\partial z \equiv 0$, we have

$$\boxed{\psi(r,\theta) = A_z(r,\theta)} \,. \tag{10.6}$$

For *helical* symmetry,

$$\boxed{\psi(r,\theta - \alpha z) = A_z(r,\theta - \alpha z) + \alpha r A_\theta(r,\theta - \alpha z)} \,, \tag{10.7}$$

where α is the pitch angle. ■

If we want to describe a torus as in Fig. 10.1 (i.e., with a circular cross-section) analytically, we can use various coordinate systems:

$$R\,,\ z\,,\ \varphi \qquad \text{cylindrical CS,} \tag{10.8}$$

$$r\,,\ \theta\,,\ \varphi \qquad \text{toroidal CS.} \tag{10.9}$$

We introduce $h(r,\theta)$ via

$$R = h(r,\theta)R_0\,, \quad h(r,\theta) = 1 + \frac{r}{R_0}\cos\theta \equiv 1 + \varepsilon\cos\theta \tag{10.10}$$

The toroidal plasma current generates a toroidal vector potential and a poloidal magnetic field:

$$\mathbf{A} = A_\varphi \hat{\varphi} \quad \rightsquigarrow \quad B_\theta = (\nabla \times \mathbf{A}) \cdot \hat{\theta} = \frac{1}{h}\frac{\partial}{\partial r}(hA_\varphi) \,. \tag{10.11}$$

In the appendix, we present vector analysis relations and provide examples for both cylindrical and toroidal coordinates.

For example, in toroidal coordinates we have

$$\nabla \cdot \mathbf{B}_\theta = \frac{R_0}{r}\frac{\partial}{\partial \theta}(hB_\theta) = 0 \quad \rightsquigarrow \quad B_\theta(r,\theta) = \frac{B_\theta^0(r)}{h(r,\theta)} \quad \rightsquigarrow A_\varphi(r,\theta) = \frac{A_\varphi^0(r)}{h(r,\theta)} \,. \tag{10.12}$$

The previously anticipated vanishing of the radial component follows from

$$B_r = (\nabla \times \mathbf{A}) \cdot \hat{r} = -\frac{1}{hr}\frac{\partial}{\partial \theta}(hA_\varphi) = 0 \,. \tag{10.13}$$

The magnetic flux surfaces are simply given by

$$\boxed{\psi(r,\theta) = h(r,\theta)R_0 A_\varphi(r,\theta) = R_0 A_\varphi^0(r) = \text{const}} \,. \tag{10.14}$$

As also expected, these are concentric circles. To prove this, one calculates

$$\mathbf{B} \cdot \nabla\psi = (B_\varphi \hat{\varphi} + B_\theta \hat{\theta}) \cdot \left(R_0 \frac{\partial A_\varphi^0}{\partial r}\hat{r} + \frac{R_0}{r}\frac{\partial A_\varphi^0}{\partial \theta}\hat{\theta} \right) = 0 \,. \tag{10.15}$$

Example 10.2 (Explicit Form of Magnetic Flux Surfaces)
We can obtain further insights into the structure of the magnetic flux surfaces approximately as follows. We have

$$\mu_0 j_\varphi(r,\theta) = (\nabla \times \mathbf{B}) \cdot \hat{\varphi} = \frac{1}{r}\frac{\partial (rB_\theta)}{\partial r} = \frac{1}{r}\frac{\partial}{\partial r}\left[\frac{r}{h}\frac{\partial}{\partial r}(hA_\varphi)\right] . \tag{10.16}$$

If we insert B_θ, we find

$$\mu_0 j_\varphi(r,\theta) = \frac{1}{h(r,\theta)}\left[\frac{B_\theta^0(r)}{r} + \frac{\partial B_\theta^0(r)}{\partial r} - \frac{B_\theta \cos\theta}{R_0}\right] . \tag{10.17}$$

If we neglect, for small aspect ratios a/R_0, where a is the minor radius, i.e., for

$$r \le a \ll R_0 , \tag{10.18}$$

the last term in the square brackets, we can formally indicate the dependence

$$j_\varphi(r,\theta) \approx \frac{j_\varphi^0(r)}{h(r,\theta)} \tag{10.19}$$

Now, if we determine A_φ from (10.16), we assume for this calculation that the toroidal current over the cross-sectional area with minor radius $r \le a$ is practically constant. Then the integration yields

$$A_\varphi(r,\theta) \approx \frac{\mu_0}{4}\frac{j_\varphi^0(r)r^2}{h(r,\theta)} \quad \rightsquigarrow \quad B_\theta(r,\theta) \approx \frac{\mu_0}{2}\frac{rj_\varphi^0(r)}{h(r,\theta)} \,. \tag{10.20}$$

The magnetic flux surfaces are determined by

$$\boxed{\psi(r,\theta) = RA_\varphi = R_0 h A_\varphi \approx \frac{\mu_0}{4} j^0_\varphi(r) r^2 R_0 \sim R_0 r^2 = \text{const}} \tag{10.21}$$

■

Drift in the Torus

We now turn to the (approximate) particle motion. We continue to refer to Fig. 10.1, i.e., a toroidal magnetic field configuration

$$B_r = 0\,, \quad B_\varphi = \frac{B_0 R_0}{R}\,, \tag{10.22}$$

which is extended by the poloidal field

$$\mathbf{B}_{pol} \equiv B_\theta \hat{\theta} \tag{10.23}$$

So much for the representation in toroidal coordinates, which we have also used for the magnetic surfaces.

Now we return to cylindrical coordinates R, φ, z and use the fact that for axisymmetric situations the φ coordinate is cyclic. Then, in Hamiltonian theory, the associated canonical particle momentum is

$$P_\varphi = mR^2\dot{\varphi} + qRA_\varphi = \text{const}\,. \tag{10.24}$$

An important result of the following consideration will be that the particles move "essentially" on the magnetic surfaces. The deviations are of the order of the gyroradius, formed with the poloidal magnetic field.

A magnetic surface is represented in cylindrical coordinates R, φ, z by the coordinate relation

$$R^* A_\varphi(R^*, z^*) = C_m \equiv \text{const} \tag{10.25}$$

for the coordinates $R = R^*$ and $z = z^*$. Now, if we choose $C_m = \frac{P_\varphi}{q}$, it follows that

$$RA_\varphi(R, z) - R^* A_\varphi(R^*, z^*) = -\frac{m}{q} R^2 \dot{\varphi}\,. \tag{10.26}$$

With the definition

$$\boxed{\delta = (R - R^*)\hat{R} + (z - z^*)\hat{z}} \tag{10.27}$$

we write (10.26) as the first-order Taylor approximation of

$$\delta \cdot \nabla(RA_\varphi) \approx -\frac{m}{q}R^2\dot{\varphi} \,. \tag{10.28}$$

Since

$$RB_R \approx \frac{\partial(RA_\varphi)}{\partial z} \,, \quad RB_z \approx \frac{\partial(RA_\varphi)}{\partial R} \tag{10.29}$$

(10.28) becomes

$$\left[\mathbf{B}_{pol} \times \delta\right]_\varphi \approx -\frac{m}{q}R\dot{\varphi} \,. \tag{10.30}$$

In terms of magnitude, we can write

$$\boxed{B_{pol}\delta \approx \frac{m}{q}v_\varphi \quad \rightsquigarrow \quad \delta = \mathcal{O}\left(\rho_{pol}\right)} \,, \tag{10.31}$$

with the poloidal gyroradius ρ_{pol} for the deviation. The sign of the deviation from the flux surface depends—as at least suggested—on the sign of v_φ.

The particles thus move "almost" on flux surfaces, with deviations on the order of the poloidal gyroradius. The flux surfaces are "approximately" determined by $r = const$. The $Grad - B$ drift and the curvature drifts do not cause any net radial deviations.

More or less, the particles follow the helical magnetic field lines (for deviations, see the following sections) and thus move through regions of different magnetic field strength (inner vs. outer side). The magnetic field strength is greater on the inner side ($\theta = \pi$) than on the outer side ($\theta = 0$). This leads to magnetic trapping for a portion of the particles.

Which particles are trapped is determined by the conservation of energy, involving the kinetic energy W and the electrostatic potential ϕ. For the sake of clarity, we will no longer explicitly write out the latter, as we intend to neglect its poloidal variation. In the case of a static field, the total energy E is constant, so that we can obtain the parallel velocity component $v_\parallel$ from

$$\boxed{v_{\parallel}(\theta) = \sqrt{\frac{2}{m}\left(E - \frac{\bar{\mu}B^0}{h(\theta)} - q\phi\right)} \quad \text{with } r \approx const.} \tag{10.32}$$

The average magnetic moment is denoted by $\bar{\mu}$. For the magnetic field strength, we approximate

$$B^0 = \sqrt{(B_\theta^0)^2 + (B_\varphi^0)^2} \approx B_\varphi^0 \,. \tag{10.33}$$

Thus, if

$$\frac{\bar{\mu}B^0}{E - q\phi} \geq h(\theta = \pi) = 1 - \varepsilon \,, \quad \varepsilon = \frac{r}{R_0} \,, \tag{10.34}$$

holds, the particle is reflected at the position $v_{\parallel} = 0$. On the outside, at $\theta = \pi$, the magnetic field strength is smaller, namely $B \approx B^0(1 - \varepsilon)$, while on the inside, at $\theta = 0$, it is larger, namely $B \approx B^0(1 + \varepsilon)$.

We now discuss in more detail (for $\phi = 0$) and use $v_\perp^2 \sim \bar{\mu}B$ as well as $\bar{\mu} \approx const$ for the magnetic moment. This directly yields

$$v_\perp(\theta) = v_\perp(\theta = 0)\sqrt{\frac{B(\theta)}{B(\theta = 0)}} \,. \tag{10.35}$$

The angle in velocity space (pitch angle) α_θ is defined at the outer edge by

$$\boxed{\tan\alpha_0 = \left.\frac{v_\perp}{v_\parallel}\right|_{\theta=0}} \,. \tag{10.36}$$

We abbreviate the contributions to the kinetic energy with $W = W_\perp + W_\parallel = const$. Then $\sin\alpha = \frac{\tan\alpha}{\sqrt{1+\tan^2\alpha}}$ leads to

$$\boxed{\sin\alpha_0 = \sqrt{\frac{W_\perp(\theta = 0)}{W}} = \sqrt{\frac{B(\theta = 0)}{B(\theta)}}\sin\alpha_\theta} \,. \tag{10.37}$$

With increasing θ from the outside inwards, $\sin\alpha_\theta$ thus becomes larger, but must not exceed one. If it does, the particle reverses direction beforehand. The latter applies to particles with

$$\sin\alpha_0 > \sqrt{\frac{B_{min}}{B_{max}}} = \sqrt{\frac{1 - \varepsilon}{1 + \varepsilon}} \tag{10.38}$$

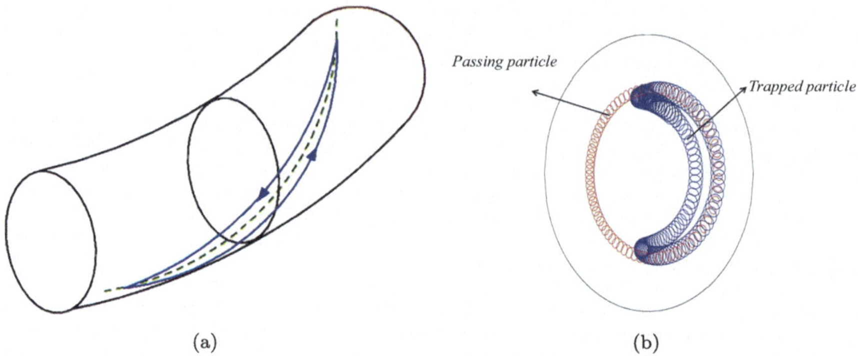

Fig. 10.2 (**a**) Complete orbit of trapped particles in the torus (**b**) Banana orbit in projection

Together with the already discussed deviation of the motion from the exact flux surfaces, the trajectory of such a trapped particle is as sketched in Fig. 10.2. Because of its shape, especially in projection (b), it is called a banana orbit.

10.2 Nuclear Fusion

Thermonuclear fusion is the essential energy source of stars. Nuclear fusion is a process in which light atomic nuclei merge to form heavier nuclei as the end product. In general, energy is released in the form of kinetic energy of the newly formed nucleus and other reaction products (neutrons, γ radiation).

Due to the reduction in energy, the reaction partners transition from a less stable to a more stable state. The mass of the new atomic nucleus is less than the sum of the masses of its components. The mass difference Δm (mass defect) corresponds to the amount of energy $\Delta E \equiv B = \Delta m c^2$. The energy released during the formation of atomic nuclei from protons and neutrons is the energy that must be supplied to separate the nucleons of a nucleus from each other. The binding energy is a function of the number (mass number) A of the nucleons. If one plots the binding energy per nucleon B/A against the nucleon number, the result is the diagram shown in Fig. 10.3.

The binding energy per nucleon shows a maximum. This means that, on the one hand, energy can be gained by fusion of light elements and, on the other hand, by fission of heavy elements.

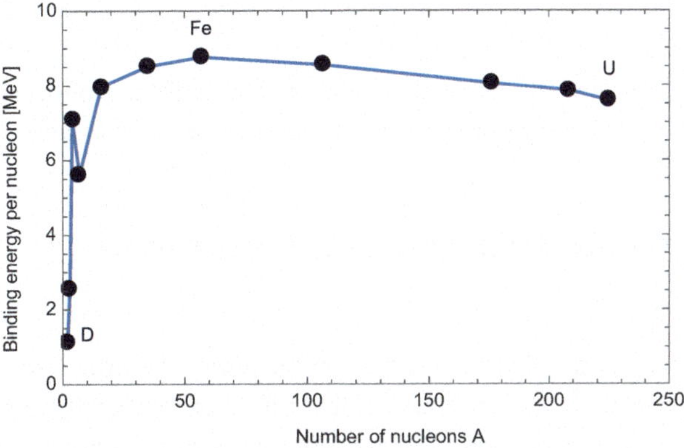

Fig. 10.3 Binding energy per nucleon *B*/*A* as a function of nucleon number A

Since atomic nuclei can only be disassembled again by supplying their binding energy, the nuclei in the vicinity of $A \approx 56$ are particularly stable. These are chromium, manganese, iron, nickel, cobalt, and copper.

Positively charged nuclei repel each other. To approach to the distance r_c we require the energy

$$\boxed{E = \frac{Z_A Z_B e^2}{4\pi\varepsilon_0 r_c} \approx \frac{1.4\ Z_A Z_B}{(r_c \text{ in Fermi})}\ \text{MeV}}\ , \tag{10.39}$$

where 1 fermi $= 10^{-15}$ m. Near the atomic nuclei, repulsion turns into attraction. Figuratively speaking: the atoms fall into a "potential well" if the velocity is right.

We have already estimated the central temperature of the Sun to be 1.7×10^7 K. At a temperature of 10^7 K ($\hat{=}1$ keV) it is trivial that only extremely few MeV particles are available. The few fast particles in the tail of the Maxwell distribution represent only a tiny fraction of the total number of particles. We can estimate this fraction using exp(-1000).

Nevertheless, fusion in the keV range is possible because quantum mechanically, tunneling through the Coulomb barrier is possible.

We will discuss this effect, which is due to the Gamow factor

$$e^{-\sqrt{E_G/E}} \approx e^{-22} \quad \text{at} \quad \text{kT} \approx 1 \text{ eV} \tag{10.40}$$

with

$$E_G = (\pi\alpha Z_A Z_B)^2 2\,mc^2 \tag{10.41}$$

in detail later. Overall, the probability of fusion is proportional to

$$e^{-E/kT} \times e^{-(E_G/E)^{1/2}} ; \tag{10.42}$$

it has a maximum at 2×10^7 K for hydrogen burning. We will discuss more details—especially regarding fusion cross sections—a bit later in this section.

"Terrestrial" Deuterium-Tritium Fusion

In controlled nuclear fusion on Earth, the aim is to utilize deuterium-tritium fusion to form helium (plus a neutron). The reason for choosing this reaction, for example as opposed to deuterium-deuterium fusion, is mainly the lower reaction cross section. Deuterium can be obtained from seawater. The tritium must be bred from ^6Li. The positive prospects of controlled nuclear fusion on Earth can be illustrated by the fact that only a few liters of seawater and a few rocks (for the deuterium and lithium, respectively) would be sufficient to meet the energy needs of a family of four for one year.

The relatively large binding energy of about 28 MeV makes ^4He a fusion product that is particularly attractive for energy generation. In practice, however, it is unlikely that two neutrons and two protons will fuse directly with each other, as the simultaneous collision is too improbable. However, there are nuclear reactions that produce ^4He successively.

In "terrestrial" nuclear fusion, deuterium and tritium are used as starting materials, in order to then generate fusion energy via the $D - T$ reaction

$$\boxed{D + T \rightarrow {}^4He\ (3.517\ \text{MeV}) + n\ (14.069\ \text{MeV})} \tag{10.43}$$

to obtain fusion energy.

The numbers in parentheses indicate the share of the released energy. In total, we obtain approximately 17.6 MeV. Compared to the 28 MeV binding energy in ^4He, we already have about 10 MeV of binding energy in D and T. The advantage is that we are now only dealing with a (probable) two-body collision of D and T nuclei. The disadvantage is that we first have to breed tritium. This is done via

$$\boxed{{}^6Li + n \rightarrow {}^4He + T} . \tag{10.44}$$

Ernest Rutherford had already discussed the first fusion processes in 1919, much earlier than Otto Hahn and Fritz Strassmann discovered nuclear fission in 1938. The prospects for controlled nuclear fusion on Earth have continuously improved over the past decades. Nevertheless, we are still far from its large-scale use. In contrast, Enrico Fermi realized a nuclear chain reaction as early as 1942, and fission reactors have been operating for decades. This shows the major challenges that must be overcome for controlled nuclear fusion by magnetic or inertial confinement on Earth. Nature has solved nuclear fusion by gravitational confinement for billions of years.

Fusion temperature

Fusion processes occur at lower temperatures than a naive estimate (incorrectly) suggests. The naive estimate calculates the energy required to overcome the Coulomb repulsion up to a distance of 1 fermi $= 10^{-15}$ m. The strong interaction, which is attractive and leads to fusion, becomes effective at distances ≤ 1 fermi. Naively, to overcome the Coulomb barrier, one needs the value already given

$$E = \frac{1.4\ Z_A Z_B}{(r_N \text{ in fermi})} \text{ MeV} . \tag{10.45}$$

Example 10.3 (Approach of Charged Particles)

The idea behind a better estimation can be illustrated using the interaction energy U of two atomic nuclei as a function of their distance r; see Fig. 10.4. For very large distances $r \to \infty$ approaches $U \to 0$, since no interaction takes place. At smaller distances, the Coulomb interaction initially dominates

$$U \sim \frac{q_1 q_2}{r} . \tag{10.46}$$

Because $q_1, q_2 > 0$ holds $dU/dr < 0$, and the nuclei repel each other. The Coulomb interaction dominates up to a minimum distance r_K,

$$r_K \approx r_0\, A^{1/3} \quad , \quad r_0 \approx 1.3 \times 10^{-13} \text{ cm} . \tag{10.47}$$

For radii smaller than the nuclear radius r_K the attraction due to nuclear forces prevails.

A nucleus approaching another thus faces an "energy barrier," whose peak for $D - T$ interaction lies at approximately

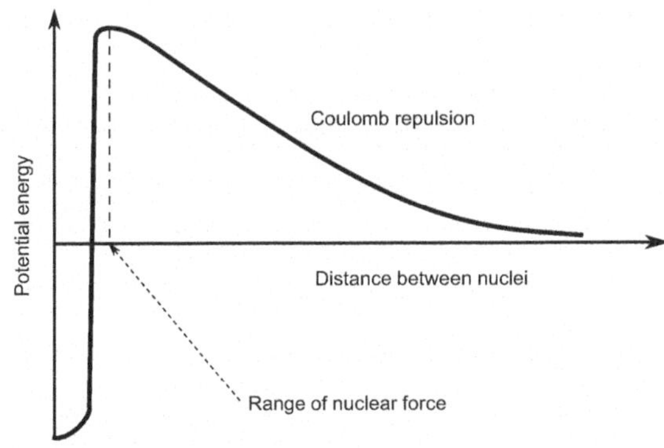

Fig. 10.4 Schematic representation of the interaction energy U of two atomic nuclei as a function of their distance r

$$r_m \approx 3.7 \text{ Fermi}$$

The peak is about $U_m \approx 0.4$ MeV high. Most nuclei are repelled from each other by their Coulomb barriers. Thus, the nuclei are usually scattered elastically. An estimate shows that at $T \approx 10$ keV, on average, there are 8000 elastic scatterings for every fusion reaction. ■

Fusion already occurs at 10 million K $= 10^7$ K $\hat{=} 1$ keV , and not only at MeV. The main reason for this is the tunneling effect.

To estimate the order of magnitude, we assume that atomic nuclei fuse as soon as their wave functions overlap. This occurs when their separation is on the order of their de Broglie wavelength λ_{dB}. From equating

$$\frac{1}{4\pi\varepsilon_0}\frac{Z_A Z_B e^2}{\lambda_{dB}} \approx \frac{(h/\lambda_{dB})^2}{2\mu_m}, \quad \lambda_{dB} = \frac{h}{\sqrt{mkT}}, \tag{10.48}$$

and solving for $T = T_{\text{Tunnel}}$

$$\boxed{T_{\text{Tunnel}} = \frac{Z_A^2 Z_B^2 e^4 \mu_m}{12\pi^2\varepsilon_0^2 h^2 k} \approx 10^7 \text{ K} \quad \text{for } Z_A = Z_B = 1,\ \mu_m = \frac{m_p}{2}}. \tag{10.49}$$

The more precise calculation for the probability of penetrating the Coulomb barrier yields

$$P \sim e^{-\sqrt{E_G/E}} \tag{10.50}$$

with the Gamow energy

$$\begin{aligned} E_G &= \frac{m_r Z_A^2 Z_B^2 \alpha^2 c^2}{8\varepsilon_0^2} \hat{=} (\pi\alpha Z_A Z_B)^2\, 2m_r c^2 \\ &\approx 493 \text{ keV} \end{aligned} \tag{10.51}$$

for protons. A small fraction T_{12} tunnels through the Coulomb wall,

$$\boxed{T_{12} \sim \exp\left[\frac{-4\pi^2 Z_1 Z_2 e^2}{2h}\sqrt{\frac{2m_r}{E}}\right]}. \tag{10.52}$$

The fine-structure constant is

$$\alpha = \frac{e^2}{\hbar c}; \tag{10.53}$$

m_r is the reduced mass,

$$\mu_m \equiv m_r = \frac{m_1 m_2}{m_1 + m_2} , \tag{10.54}$$

where m_1 and m_2 are the masses of the nuclei, respectively.

The tunneling probability T_{12} decreases sharply with the atomic numbers Z_1 and Z_2, respectively. If we set

$$E = \frac{1}{2} m_r v^2 , \tag{10.55}$$

we see that T_{12} increases with increasing relative velocity v. Fusion reactions therefore occur practically only with light nuclei (small Z) and at high relative velocities.

For two protons at 1 keV, the penetration probability is approximately e^{-22}. Using this probability and a Maxwell distribution for the energy (= kinetic energy) of the particles, fusion cross sections can be calculated. After tunneling, fusion of the nuclei does not always occur, but only with an (additional) probability p_{12}, which depends on the details of the nuclear structure.

Reaction parameters and fusion power density

We now calculate the number R_{12} of fusion reactions per unit volume and time. As a model, we represent a nucleus as a sphere, which is "seen" with the cross-sectional area Q_{12}. In one unit of time, nucleus 1 sweeps out with this area the volume $Q_{12}v$. Nuclei 2 within this volume interact with nucleus 1. With the spatial density n_2 we find per unit time $n_2 Q_{12} v$ interactions. Only the fraction $T_{12} p_{12}$ of these lead to fusion. If we consider not just nucleus 1, but all n_1 nuclei per unit volume, the reaction rate is given by

$$\boxed{R_{12} = n_1\, n_2\, Q_{12}\, T_{12}\, p_{12}\, v\ ,} \tag{10.56}$$

i.e., the number of fusions per unit time and volume. The quantity

$$\boxed{\sigma = Q_{12}\, T_{12}\, p_{12}} \tag{10.57}$$

is called the cross section for fusion.

Example 10.4 (Cross Section and Mean Free Path)
The relationship between cross section and mean free path can be found as follows: The probability of a reaction after a distance Δx is $\sigma n \Delta x$, where n is the density of the target particles. No reaction occurs with probability $1 - \sigma n \Delta x$. This holds for infinitesimal segments. For finite distances $x = N \Delta x$ the probabilities are multiplied, i.e.,

$$\lim_{N\to\infty} \left[1 - \sigma n \frac{x}{N}\right]^N = e^{-\sigma n x} . \tag{10.58}$$

From this, a mean free path can be defined

$$\boxed{l = \int_0^\infty x\, e^{-\sigma n x}\, dx = \frac{1}{n\sigma}} \tag{10.59}$$

■

To determine the cross section σ at a relative energy E of the interacting partners, we make the ansatz

$$\boxed{\sigma(E) = \frac{S(E)}{E}\, e^{-\sqrt{E_G/E}} ,} \tag{10.60}$$

which agrees well with measured values. The factor $S(E)$ accounts for the nuclear physics with the corresponding interactions. In many cases, however, $S(E)$ can be assumed constant. The $1/E$ dependence reflects the experimental finding that the cross section is proportional to the square of the de Broglie wavelength ($\lambda^2 = h^2/2m_r E$).

Example 10.5 (Mean Reaction Rate)
From the reaction rate R_{12} we arrive at the mean reaction rate. For $\langle R_{12}\rangle$ one must average over all relative velocities v. We first obtain the contributions

$$dR_{12} = dn_1\, dn_2\, \sigma(v) v \tag{10.61}$$

with $v = |\mathbf{v}_1 - \mathbf{v}_2| = \sqrt{2E/m}$; dn_1 is the density of particles in the velocity interval $\left[\mathbf{v}_1, \mathbf{v}_1 + d^3 v_1\right]$. The same applies for dn_2.

The two quantities dn_1 and dn_2 can be easily calculated when Maxwell distributions are valid. $\langle R_{12}\rangle$ then follows from integration over all velocities,

$$\boxed{\langle R_{12}\rangle = n_1 n_2 \langle \sigma v\rangle} . \tag{10.62}$$

This relation will be further evaluated below. ■

The mean time that elapses between two fusion processes is proportional to $l/v \sim 1/n\sigma v$, where we average over the relative velocity distribution.

We define the reaction parameter

$$\boxed{\langle \sigma v_r \rangle = \int_0^\infty \sigma\, v_r f(v_r)\, dv_r} \tag{10.63}$$

and then find

$$\tau_A = \frac{1}{n_B \langle \rho v_r \rangle} \,. \tag{10.64}$$

This is the mean time for a selected nucleus 1. The Maxwell distribution f yields a factor

$$f(v)d^3v = \left[\frac{m_r}{2\pi kT}\right]^{3/2} \exp\left[-\frac{m_r v^2}{2kT}\right] 4\pi\, v^2\, dv \,, \tag{10.65}$$

where we still set $E = \frac{1}{2}\, mv^2$.

The conversion to $E = p^2/2m$ is performed according to

$$f(E)dE \sim f(p)p^2 dp \sim f(p)E\frac{dp}{dE}dE \sim \sqrt{E}\exp(-E/kT)dE \,. \tag{10.66}$$

Since $v_r \sim \sqrt{E}$, we obtain in total the integral

$$\boxed{\langle R_{12} \rangle = \left(\frac{2}{kT}\right)^{3/2} \frac{n_1 n_2}{\pi \mu_m} \int_0^\infty S(E) \exp\left[-\sqrt{\frac{E_G}{E}}\right] \exp\left[-\frac{E}{kT}\right] dE} \,. \tag{10.67}$$

The two factors

$$e^{-E/kT} \times e^{-\sqrt{E_G/E}} \tag{10.68}$$

yield, around the maximum value $E_{max} = \left(\frac{\sqrt{E_G}kT}{2}\right)^{2/3}$, a fusion window of the size

$$\boxed{\Delta = \frac{4}{3^{1/2}2^{1/3}}\, E_G^{1/6}\, (kT)^{5/6}} \,. \tag{10.69}$$

This "window" is explained in Fig. 10.5

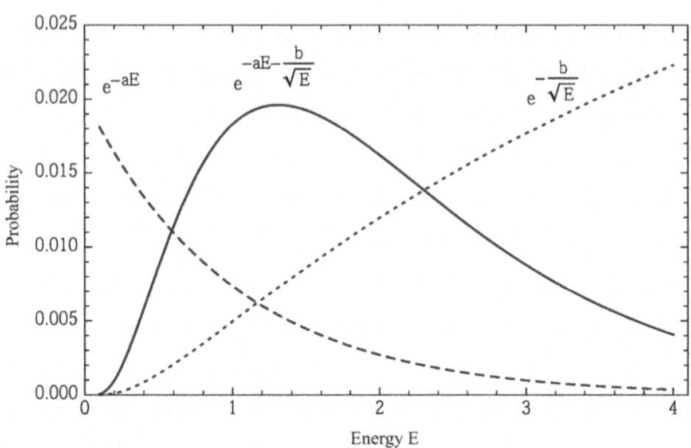

Fig. 10.5 Determination of a fusion window from tunneling probability and Maxwell distribution

The fusion rate (reactions per particle and per unit time and volume)

$$\langle R_{12}\rangle = n_1\, n_2\, \langle \sigma v_r\rangle \tag{10.70}$$

then essentially depends on three parameters: $S(E)$, kT and E_G. For constant $S(E)$, the essential dependence is

$$R_{AB} \sim n_A\, n_B\, S(E_0)\, e^{-3(E_G/4kT)^{1/3}}\ . \tag{10.71}$$

Often, a Taylor expansion is also performed in the integrand, yielding

$$R_{AB} \approx r_0 X_A X_B \rho^2 T^{\beta}\ ; \tag{10.72}$$

r_0 is a constant,

$$X_i = n_i m_i / \rho \tag{10.73}$$

captures the relative fraction of the particle species in the total mass, and the coefficient $1 \le \beta \le 40$ approximates the actual distribution.

With the help of this formula (or formulas), it is, for example, very easy to estimate that the reaction

$$p + d \rightarrow {}^3He + \gamma \tag{10.74}$$

is clearly favored over the reaction

$$p + {}^{12}C \rightarrow {}^{13}N + \gamma \tag{10.75}$$

at low temperatures (2×10^7 K).

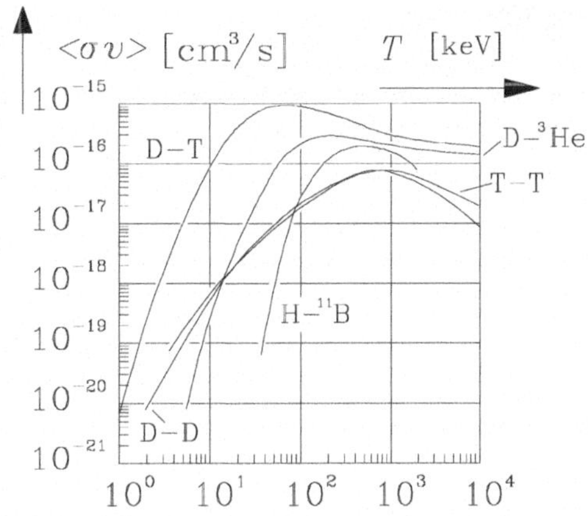

Fig. 10.6 Reaction parameter $\langle \sigma v \rangle$ as a function of temperature T for various fusion reactions

If in each reaction the energy Q_{ij} is released, then the energy generation rate per unit mass follows as

$$\boxed{\varepsilon \equiv \varepsilon_{nuc}^{ij} \approx Q_{ij} r_0 X_i X_j \rho T^{\beta}} \tag{10.76}$$

in erg g^{-1} s^{-1}.

Fig. 10.6 shows the reaction parameter $\langle \sigma v \rangle$ as a function of temperature T for various fusion reactions. A comparison of the different reactions shows that the D-T reaction has by far the largest reaction parameter at comparatively low temperature. This is the reason why, on Earth, the $D - T$ reaction is preferred for controlled nuclear fusion. Since fusion reactions are enabled by the thermal motion of ions in a hot plasma, this is also referred to as thermonuclear fusion.

In magnetically confined plasmas, the pressure

$$P = P_1 + P_2 \approx n_1 k T_1 + n_2 k T_2 \approx (n_1 + n_2) k T \tag{10.77}$$

is balanced by magnetic forces. In such systems,

$$\boxed{n_i \sim \frac{1}{T}} \tag{10.78}$$

applies, and thus

$$\langle R_{12} \rangle \sim \frac{\langle \sigma v \rangle}{T^2} \,. \tag{10.79}$$

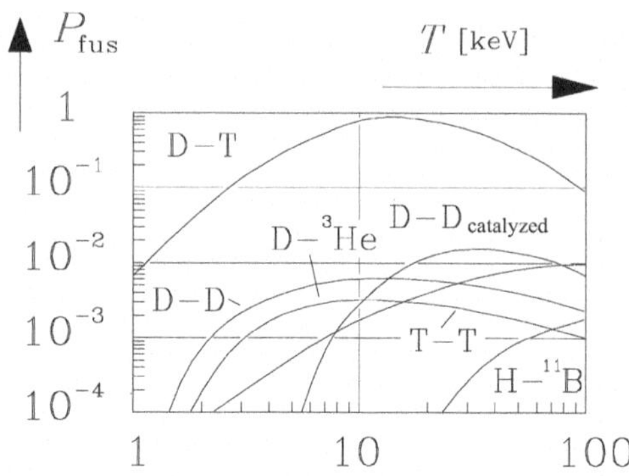

Fig. 10.7 Power densities for various fusion reactions as a function of temperature T. The curves are shown relative to the maximum for $D-T$ reactions

The fusion power density P_{fus} is obtained by multiplying with the reaction energy E_{fus} ($\approx$ 16.6 MeV bei $D-T$-Reaktionen),

$$\boxed{P_{fus} = n_1 n_2 \langle \sigma v \rangle \, E_{fus} \sim \frac{\langle \sigma v \rangle}{T^2} \, E_{fus}} \, . \tag{10.80}$$

Fig. 10.7 shows the power densities for various fusion reactions as a function of temperature T. The values are normalized to the maximum value, which occurs for the $D-T$ reaction at $T = 15$ keV. This reference value is $P_{fus} \approx 1.85$ W cm^{-3} for the typical particle densities $n_D = n_T = 5 \times 10^{13}$ cm^{-3}. Expressed in terms of the total ion density

$$n = n_0 + n_T = 2n_D \tag{10.81}$$

we obtain

$$P_{fus} = \frac{n^2}{4} \langle \sigma v \rangle E_{fus} \, . \tag{10.82}$$

Example 10.6 (Power Densities)
Let us compare $P_{fus} \approx 1.85$ W cm^{-3} with power densities in other systems. In the combustion chamber of a coal-fired power plant boiler, the value is 0.2 W cm^{-3}, for a light water reactor the value is 90 W cm^{-3}. Current estimates for the center of the sun yield fractions of mW cm^{-3} (averaged over the entire solar volume, further reduced by a factor of 10^{-1}). ■

Ignition condition
In a plasma with equal numbers of electrons and ions, the total thermal energy is calculated as

$$W = \int dV \; 3nT \equiv 3 \, \overline{nT} \; V \, . \tag{10.83}$$

The energy loss rate P_L defines the energy confinement time τ_E by

$$\boxed{P_L = \frac{W}{\tau_E}} . \tag{10.84}$$

If we balance the energy loss by additional heating, it follows that

$$\tau_E = \frac{W}{P_H} . \tag{10.85}$$

Only a fraction of the released energy is transferred to the ^4He nuclei (α particles). The α particles are charged (in contrast to the neutrons, which leave the plasma unhindered) and interact with the other plasma particles. Through collisions, they can transfer (parts of) the 3.5 MeV ($= E_\alpha$) to the other constituents. For this reason, the so-called α particle heating is defined as

$$\boxed{P_\alpha = \frac{1}{4} \overline{n^2 \langle \sigma v \rangle} \, V E_\alpha} . \tag{10.86}$$

Within the framework of the fusion program, the aim is to use additional heating P_H to bring a plasma to fusion conditions, so that ultimately the α particle heating alone compensates for the (unavoidable) losses. With additional heating, the burn condition is found to be

$$P_H + P_\alpha = P_L \tag{10.87}$$

or, for the additional heating power,

$$P_H = \left(\frac{3nT}{\tau_E} - \frac{1}{4} n^2 \langle \sigma v \rangle E_\alpha \right) V , \tag{10.88}$$

where, for simplicity, we have omitted the bars denoting volume averaging.

Self-sustained burning (ignition at $P_H = 0$) can occur for

$$n\tau_E > \frac{12}{\langle \sigma v \rangle} \frac{T}{E_\alpha} \tag{10.89}$$

The right-hand side is a function of temperature, which is shown in Fig. 10.8. From the minimum at $T \approx 30$ keV, one finds

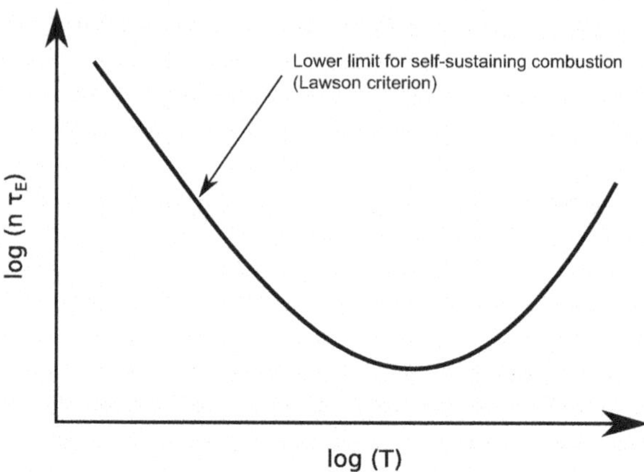

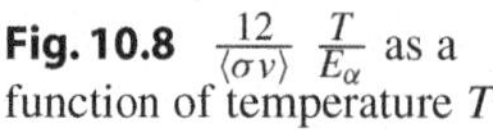
Fig. 10.8 $\frac{12}{\langle\sigma v\rangle}\frac{T}{E_\alpha}$ as a function of temperature T

$$\boxed{n\tau_E \geq 1.5 \times 10^{20}\ m^{-3}\ \ \mathrm{s}\ .} \tag{10.90}$$

Such an estimate is also known as the Lawson criterion.

Since τ_E itself depends on the plasma parameters (thus also on the temperature, or more generally on the transport coefficients), its usefulness is limited. A reformulation takes advantage of the fact that in the temperature range

$$10\ \mathrm{keV} \leq T \leq 20\ \ \mathrm{keV} \tag{10.91}$$

$$\langle\sigma v\rangle \approx 1.1 \times 10^{-24}\ \ \mathrm{T}^2\ \ \mathrm{m}^3\ \ \mathrm{s}^{-1} \tag{10.92}$$

applies, when T is also given in keV. Then, the ignition criterion can be written as

$$\boxed{n\ T\ \tau_E \geq 3 \times 10^{21}\ \ \mathrm{m}^{-3}\ \ \mathrm{keV}\ \ \mathrm{s}} \tag{10.93}$$

Example 10.7

For $n = 10^{20}\ \ \mathrm{m}^{-3}$ and $T = 10\ \ \mathrm{keV}$ a required energy confinement time of approximately 3 s results. ■

As a measure of progress in controlled nuclear fusion on Earth, the following has been introduced

$$Q = \frac{5P_\alpha}{P_H} \tag{10.94}$$

The factor 5(17.5/3.5 = 5) arises from the fact that the total energy gain, and not just the α component, is included in the calculation of Q. $Q \to \infty$ means ignition.

Fusion Processes in Stars

Stellar fusion was already discussed in detail by Atkinson and Hontermans in 1929. After the discovery of the first fusion reactions in the laboratory (1932–1934), the specific fusion reactions responsible for stellar burning, in addition to hydrogen fusion, were elucidated by von Weizsäcker and Bethe towards the end of the 1930s.

We distinguish between low-mass and high-mass stars, with the boundary lying approximately at the solar mass.

pp Reactions
In low-mass stars, fusion begins with the proton-proton reaction

$$\boxed{p + p \to D + e^+ + \nu\,,} \tag{10.95}$$

and then proceeds most frequently via

$$\boxed{D + p \to {}^3He + \gamma} \tag{10.96}$$

into the reaction

$$\boxed{{}^3He + {}^3He \to {}^4He + 2p} \tag{10.97}$$

see Fig. 10.9.

Carbon Cycle
In more massive stars, the energy released by the $p - p$ fusion process is not sufficient to maintain the necessary internal pressure. The stars continue to contract, and their temperature rises further due to the virial theorem according to

$$E_{kin} \approx -\frac{1}{2}\,E_{gr} > 0 \tag{10.98}$$

until finally the so-called carbon cycle provides a significantly higher fusion output. The carbon cycle is schematically illustrated in Fig. 10.10.

In the carbon cycle, a carbon nucleus serves as a catalyst for the successive capture of four protons. The overall result is again the fusion of four protons into ${}^4He$, albeit in the

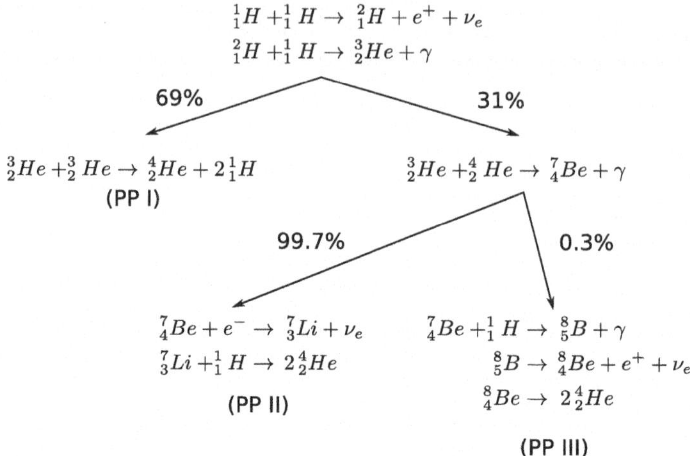

Fig. 10.9 Three branches of the pp reactions with probabilities for solar parameters

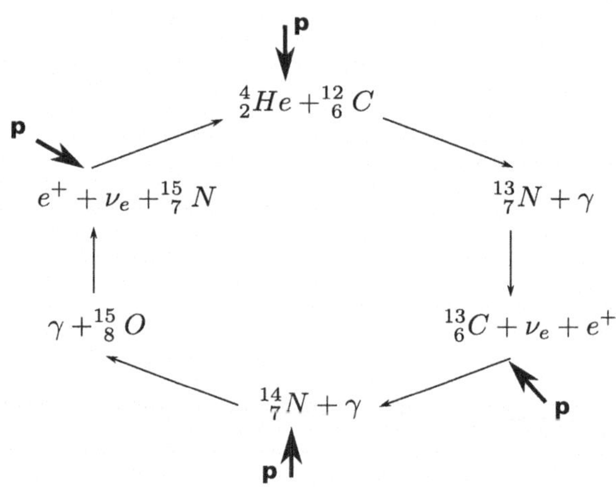

Fig. 10.10 Schematic representation of the carbon cycle. The reaction chain is to be read clockwise starting from the top

form shown in Fig. 10.10. The cycle begins at the top with ^{12}C, proceeds clockwise, and ends again at ^{12}C. The application of the carbon cycle to more massive stars goes back to Bethe (in 1938). Its use at higher temperatures, in contrast to the pp-chain, is further illustrated in Fig. 10.11.

Example 10.8 (Production rates ε)
Approximately 90% of all stars (main sequence stars) burn hydrogen into helium. In this case, the production rate can be approximated by

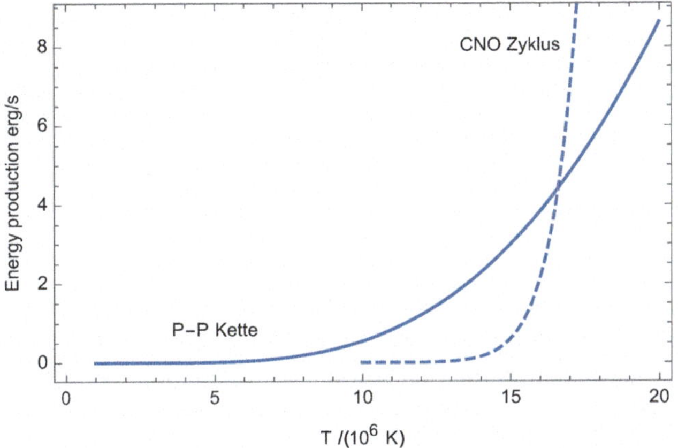

Fig. 10.11 Temperature dependencies of the fusion power of various processes

$$\varepsilon \approx 1.07 \times 10^{-5} X^2 \rho \left[\frac{\mathrm{g}}{\mathrm{cm}^3}\right] \left(\frac{T}{10^6\ \mathrm{K}}\right)^4 \tag{10.99}$$

Here, X denotes the relative hydrogen fraction. The formula has been adjusted for solar parameters.

For the CNO cycle, one approximates

$$\varepsilon \approx 8.24 \times 10^{-24} X X_{CNO} \rho \left[\frac{\mathrm{g}}{\mathrm{cm}^3}\right] \left(\frac{T}{10^6\ \mathrm{K}}\right)^{19.9} . \tag{10.100}$$

■

The various fusion cycles proceed at different rates. For example, the D-T reaction is fast compared to the carbon cycle. This, together with the different temperatures required to achieve the "right" pressures in stars, is crucial for the proper selection of the relevant fusion cycles. For further details on the relevant fusion cycles in stars, the specialized literature is recommended.

In summary, it can be stated that a star passes through various stages of burning, as shown in the following overview.

$T = 10^7$ K	Hydrogen	→	Helium
$M > 0.08\,M_\odot$			
$T = 10^8$ K	Helium	→	Carbon, oxygen
$M > \frac{1}{2} M_\odot$			
$T = 5 \times 10^8$ K	Carbon	→	Oxygen, neon, etc.

(continued)

(continued)

$M > 8\,M_\odot$			
$T = 10^9$ K	Neon	→	Oxygen, magnesium
$T = 2 \times 10^9$ K	Oxygen	→	Magnesium to sulfur
$T = 3 \times 10^9$ K	Silicon	→	Iron, etc.
$M > 11\,M_\odot$			

Nuclear burning phases can be separated in time and space. The spatial separation is evident in the so-called "onion structure" of the star in later stages of evolution. The ash of one burning phase becomes the fuel for the subsequent one. Whether another burning phase occurs depends on the maximum temperature that can be reached and thus on the stellar mass. For $M \geq 8\,M_\odot$ practically all possible thermonuclear burning phases are passed through.

It is interesting that in cosmological evolution, although the temperatures were high enough, the densities were too low to allow fusion. Only the formation of massive stars initiated fusion. The burning cycles stop at ^{56}Fe, since the binding energy per nucleon reaches its maximum there.

Heavy elements are formed by neutron capture and subsequent β decay (of the neutron into a proton and an electron) or α reactions.

Neutrinos are produced in fusion reactions. Neutrino losses play a major role in the evolution of massive stars, but can also be detected in the cooling of white dwarfs.

10.3 Equilibria in Toroidal Geometry

In this section, we discuss fundamental aspects of possible confinement configurations. The basis is magnetohydrodynamics (MHD). However, this theory is only applicable on certain length and time scales, so some phenomena are excluded from the outset.

Magnetohydrodynamics (MHD) is the fundamental theory for fusion plasmas. Therefore, fusion physicists refer to more specialized books for specific aspects, such as Hazeltine and Meiss (1992), Wesson (2004), or Stacey (2005) [134–136], to name just a few. Fundamental aspects of MHD are comprehensively covered, for example, in the works of Biskamp (1997) and Goedbloed (2010) [66, 137]. In the following, we will illustrate the capabilities of MHD using several examples.

The evolution of a magnetic field in a plasma with constant electrical conductivity is described by the following equation:

$$\boxed{\frac{\partial \mathbf{B}}{\partial t} - \nabla \times (\mathbf{u} \times \mathbf{B}) \approx \frac{1}{\mu_0 \sigma} \nabla^2 \mathbf{B}} \,. \tag{10.101}$$

The right-hand side represents a diffusion term. For $\mathbf{u} \approx 0$ we can estimate the characteristic time for diffusion as

$$\tau \sim \mu_0 \sigma L^2 \,, \tag{10.102}$$

where L is the characteristic length for variations of the magnetic field $\mathbf{B}$. If τ is large (e.g., in the limiting case $\sigma \to \infty$), we can neglect the diffusion term. A large τ means that it is large compared to the characteristic time of the phenomenon under consideration.

Within ideal MHD ($\sigma \to \infty$) we have

$$\frac{\partial \mathbf{B}}{\partial t} \approx \nabla \times (\mathbf{u} \times \mathbf{B}) \,, \tag{10.103}$$

The latter case describes frozen-in magnetic field lines.

To see this, we integrate over a surface that moves with the plasma (magnetofluid) to obtain

$$\int \frac{\partial \mathbf{B}}{\partial t} \cdot d\mathbf{F} - \int (\mathbf{u} \times \mathbf{B}) \cdot d\mathbf{l} = \int \frac{\partial \mathbf{B}}{\partial t} \cdot d\mathbf{F} + \int \mathbf{B} \cdot (\mathbf{u} \times d\mathbf{l}) = \frac{d}{dt} \int \mathbf{B} \cdot d\mathbf{F} = 0. \tag{10.104}$$

The magnetic flux is therefore constant.

Let us consider ideal MHD equilibria in the static case ($\partial_t = 0,\ \mathbf{u} = 0$):

$$\boxed{\nabla p = (\mathbf{j} \times \mathbf{B}) \,, \quad \nabla \times \mathbf{B} = \mu_0 \mathbf{j} \,, \quad \nabla \cdot \mathbf{B} = 0} \,. \tag{10.105}$$

For obvious reasons, this case is referred to as ideal magnetohydrostatics. Eliminating $\mathbf{j}$ leads to

$$(\nabla \times \mathbf{B}) \times \mathbf{B} = \mu_0 \nabla p \,. \tag{10.106}$$

If we write the last equation in the form

$$\nabla \left(p + \frac{B^2}{2\mu_0} \right) = \mathbf{B} \cdot \nabla \mathbf{B} \,, \tag{10.107}$$

we can easily interpret the relation as a pressure balance. In addition to the kinetic pressure p, a magnetic pressure also appears. The right-hand side shows that a magnetic tension also arises.

The case $\nabla p = 0$ is referred to as a force-free equilibrium, for which

$$(\nabla \times \mathbf{B}) \times \mathbf{B} = 0 . \tag{10.108}$$

In a force-free equilibrium, $\nabla \times \mathbf{B} \sim \mathbf{B}$ should hold. We express this by

$$\nabla \times \mathbf{B} = \alpha(\mathbf{r})\,\mathbf{B} . \tag{10.109}$$

Due to the solenoidal nature of a magnetic field, it follows that

$$\mathbf{B} \cdot \nabla \alpha = 0 . \tag{10.110}$$

This means that the magnetic field lines should lie in the surfaces α = const, which are referred to as magnetic surfaces.
If $\nabla p \neq 0$, similar considerations lead to

$$\mathbf{j} \cdot \nabla p = 0 , \quad \mathbf{B} \cdot \nabla p = 0 . \tag{10.111}$$

This means that the surfaces $p(\mathbf{r})$ = const are both magnetic and current surfaces.

Due to $\nabla \cdot \mathbf{B} = 0$ and $\nabla \cdot \mathbf{j} = 0$, both fields, $\mathbf{B}(\mathbf{r})$ and $\mathbf{j}(\mathbf{r})$, are either closed or extend to infinity. Thus, the magnetic surfaces $p(\mathbf{r})$ = const are either tubes or tori.

As an example, we now examine the possible magnetic field configuration under the assumption of axial symmetry and in the approximation of a scalar pressure. For this purpose, we introduce a cylindrical coordinate system r, ϕ, z and assume azimuthal symmetry about the z-axis; see Fig. 10.12. As we will discuss, due to $\nabla \cdot \mathbf{B} = 0$ the magnetic field $\mathbf{B}$ can be described by two Euler potentials. The most general form of an axially symmetric magnetic field is then

$$\boxed{\mathbf{B} = \frac{1}{2\pi}(\nabla \psi \times \nabla \phi + \mu_0 I \nabla \phi) \equiv \mathbf{B}_{pol} + \mathbf{B}_{tor} , \quad \mathbf{B}_{tor} = \frac{1}{2\pi}\mu_0 I \nabla \phi} . \tag{10.112}$$

Here, $I = I(r, z)$ is a current flowing in the z-direction and associated with a circle of radius r whose center lies on the axis at the axial position z. By integrating over the area of the circle, we obtain

$$\int \nabla \times \mathbf{B}_{tor} \cdot d\mathbf{F} = \oint_C \mathbf{B}_{tor} \cdot d\mathbf{l} = \mu_0 I . \tag{10.113}$$

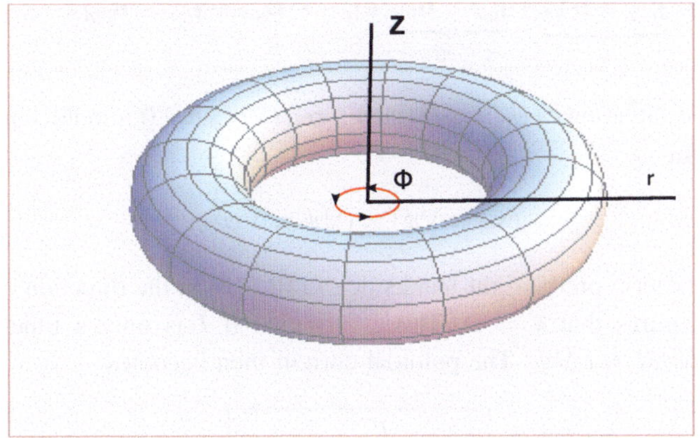

Fig. 10.12 Cylindrical coordinate system for an axially symmetric configuration

Next, we determine the poloidal flux, which we identify with ψ.

To do this, we again integrate over a surface perpendicular to the symmetry axis. In cylindrical coordinates, the azimuthal component of the poloidal magnetic field ($\mathbf{B}_{pol} \cdot \hat{\phi} = 0$) vanishes. The integration yields:

$$\int \mathbf{B}_{pol} \cdot d\mathbf{F} = \frac{1}{2\pi} \int_0^r (\nabla\psi \times \nabla\phi) \cdot \hat{\mathbf{z}} 2\pi r' dr' = \int_0^r \frac{d\psi}{dr'} dr' = \psi \,. \tag{10.114}$$

The Euler potentials allow for the straightforward identification of the poloidal flux with ψ. In cylindrical coordinates, $\nabla\phi = r^{-1}\hat{\phi}$ holds (whereas in spherical coordinates, as we will see later, $\nabla\phi = (r \sin\theta)^{-1}\hat{\phi}$ applies).

Example 10.9 (Current Densities)

After the poloidal and toroidal magnetic fields have been defined in an axisymmetric situation, we can calculate the corresponding current densities using $\mu_0 \mathbf{j} = \nabla \times \mathbf{B}$, that is,

$$\mathbf{j}_{pol} \equiv \frac{1}{\mu_0} \nabla \times \mathbf{B}_{pol} = \frac{1}{2\pi} \nabla I \times \nabla\phi \,, \tag{10.115}$$

$$\mathbf{j}_{tor} \equiv \frac{1}{\mu_0} \nabla \times \mathbf{B}_{tor} = -\frac{r^2}{2\pi\mu_0} \nabla \cdot \left(\frac{1}{r^2} \nabla\psi \right) \nabla\phi \,. \tag{10.116}$$

As expected, the toroidal current density lies in the azimuthal direction, while the poloidal current density lies in the r, z plane. ■

Next, we substitute the results into $\nabla p = \mathbf{j} \times \mathbf{B}$, which leads to the following:

$$\nabla p = \underbrace{\mathbf{j}_{pol} \times \mathbf{B}_{pol}}_{*} + \underbrace{\mathbf{j}_{tor} \times \mathbf{B}_{tor}}_{=0} + \mathbf{j}_{pol} \times \mathbf{B}_{tor} + \mathbf{j}_{tor} \times \mathbf{B}_{pol} \ . \tag{10.117}$$

In axisymmetric situations ($\frac{\partial}{\partial\phi} = 0$), therefore, $(\nabla p)_\phi = 0$ should hold. Let us now consider the term

$$* \sim (\nabla I \times \nabla\phi) \times (\nabla\psi \times \nabla\phi) \sim \nabla\phi \tag{10.118}$$

which is the only term on the right-hand side of (10.117) in the direction of $\hat{\phi}$. Therefore, axisymmetry requires that $* = 0$. This is satisfied if I is only a function of ψ, i.e., $I = I(\psi)$, where $\nabla I = I'\nabla\psi$. The poloidal current then becomes

$$\boxed{\mathbf{j}_{pol} = \frac{I'}{2\pi}\nabla\psi \times \nabla\phi} \ . \tag{10.119}$$

The evaluation of (10.117) is now straightforward,

$$\nabla p = -\frac{1}{(2\pi)^2}\left[\frac{\mu_0 I I'}{r^2} + \frac{1}{\mu_0}\nabla\cdot\left(\frac{1}{r^2}\nabla\psi\right)\right]\nabla\psi \ . \tag{10.120}$$

Thus, $\nabla p \| \nabla\psi$ holds, and for an isotropic pressure p, this is only a function of ψ, i.e., $p = p(\psi)$.

By introducing $\nabla p = p', \nabla\psi$, we obtain from (10.120)

$$\nabla\cdot\left(\frac{1}{r^2}\nabla\psi\right) + \frac{\mu_0^2 I I'}{r^2} + 4\pi^2\mu_0 p' = 0 \ , \tag{10.121}$$

what is called the Grad-Shafranov equation [138, 139] (in cylindrical coordinates). The Grad-Shafranov equation contains the unknown functions $I = I(\psi)$ and $p = p(\psi)$, which must be specified in order to solve it.

In general, the Grad-Shafranov equation is a nonlinear partial differential equation. Solovev [140] showed that even the linear Grad-Shafranov equation provides useful solutions for tokamak physics.

Example 10.10 (Linear Grad-Shafranov Equation)
Suppose the plasma pressure is a linear function of ψ, e.g., $p = p_0 + \lambda\psi$. In addition, the current I is modeled as a conduction current along the z-axis, so that $I' = 0$ within the plasma. Then, for $r > 0$, the (linear) Grad-Shafranov equation is

$$\boxed{\frac{\partial^2\psi}{\partial r^2} - \frac{1}{r}\frac{\partial\psi}{\partial r} + \frac{\partial^2\psi}{\partial z^2} + 4\pi^2 r^2 \mu_0 \lambda = 0} \ . \tag{10.122}$$

It has the exact solution

$$\psi(r,z) = \psi_0 \frac{r^2}{r_0^4}(2r_0^2 - r^2 - 4\alpha^2 z^2) \ , \tag{10.123}$$

with constants ψ_0, r_0 and α, which yield

$$\lambda = \frac{2\psi_0}{\pi^2 r_0^4 \mu_0}(1+\alpha^2) \tag{10.124}$$

A typical contour plot in the r, z diagram is shown in Fig. 10.13. Closed contours, i.e., closed magnetic surfaces ψ = const, exist. The closed contours are separated from the contours that go to infinity by a separatrix $r^2 + 4\alpha^2 z^2 = 2r_0^2$. The common center of all closed surfaces is called the magnetic axis.

For $I = 0$ there exists an analogue to (10.122) in spherical coordinates:

$$\frac{\partial^2\psi}{\partial r^2} + \frac{1}{r^2}\frac{\partial^2\psi}{\partial\theta^2} - \frac{\text{cotg }\theta}{r^2}\frac{\partial\psi}{\partial\theta} = -4\pi^2\mu_0 r^2 \sin^2\theta \frac{dp}{d\psi} \ . \tag{10.125}$$

The homogeneous equation ($p = 0$) has a solution with a dipole structure $\psi \sim -r^{-1}\sin^2\theta$. ■

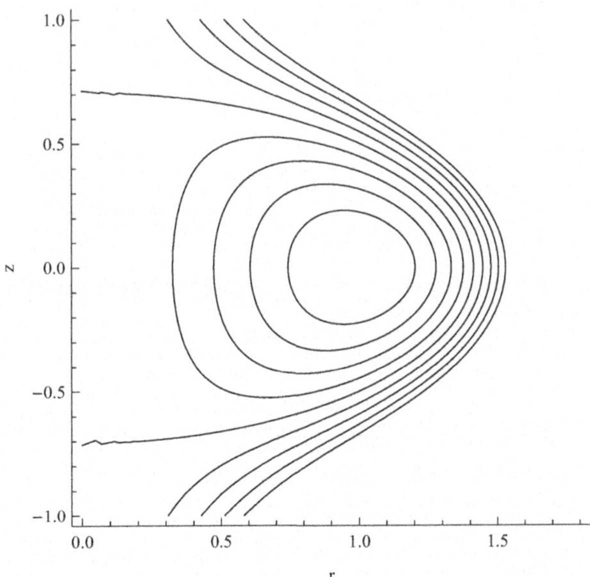

Fig. 10.13 Contour plot of the Solovev solution [140] of the Grad-Shafranov Eq. (10.121)

Example 10.11 (Pinch Discharge)

Now let us consider a straight (linear) pinch discharge configuration with infinite extension in the z-direction. We assume cylindrical symmetry, with an electric current in the z-direction, i.e., $\mathbf{j} = j(r)\hat{z}$. The current generates an azimuthal magnetic field $\mathbf{B} = B(r)\hat{\phi}$. Again, we consider the stationary case ($\partial_t = 0$) without mean flow ($\mathbf{u} = 0$). The pressure equilibrium (10.107) simplifies to (note that $[(\mathbf{B} \cdot \nabla)\mathbf{B}]_r = -B^2(r)/r$)

$$\frac{dp}{dr} = -\frac{1}{2\mu_0 r^2}\frac{d}{dr}(r^2 B^2) \, . \tag{10.126}$$

The azimuthal magnetic field component is then

$$B(r) = \frac{\mu_0}{r}\int_0^r rj(r)dr \, . \tag{10.127}$$

For a simple application, we specify $j(r)$ as constant for $r < R$ and zero for $r > R$. Then

$$B(r) = \frac{\mu_0}{2\pi}\frac{Ir}{R^2} \quad \text{for } r < R \, , \qquad B(r) = \frac{\mu_0}{2\pi}\frac{I}{r} \qquad \text{for } r > R \, , \tag{10.128}$$

where I is the total pinch current. By integration, the pressure $p(r)$ follows,

$$p(r) = \frac{\mu_0 I^2}{4\pi^2 R^2}\left(1 - \frac{r^2}{R^2}\right) \quad \text{for } r < R \, , \tag{10.129}$$

if we apply the obvious boundary condition $p(r) = 0$ for $r > R$. Now, the number N of enclosed particles is related to the pinch current I. At constant temperature and with $p = nk_B T$ we have

$$N = \int_0^R n(r) 2\pi r dr \, , \tag{10.130}$$

so that

$$N k_B T = \frac{\mu_0 I^2}{8\pi} \, . \tag{10.131}$$

The magnetic surfaces $\psi = \text{const}$ are cylinders around the z-axis. ■

> In summary, we have found exact equilibrium solutions, although the terminology is not entirely correct from a thermodynamic perspective. The term stationary solutions is more appropriate. The reason is that the solutions do not necessarily have to be stable.

10.4 Nonlinear Transport

Nonlinear signatures of particle and heat transport are typical, for example, in fluid dynamics, plasma physics, and astrophysics. Transport anomalies range from Bohm-like diffusion in gas discharges to the penetration of low-energy cosmic rays into the heliosphere. In the classical picture, strong magnetic guide fields significantly reduce perpendicular diffusion. On the one hand, collisions appear as an obstacle to free motion along the field lines, and on the other hand, they increase transport in the perpendicular direction.
The specific reason for the extraordinary interest in the mechanisms of diffusion in stochastic fields lies in the unexpectedly large losses caused by anomalous transport [12, 141]. Anomalous particle and heat losses have enormous consequences for practical applications. For example, in magnetic fusion devices, heat fluxes must be relatively low and evenly distributed over the wall in order to achieve tolerable local power densities. Therefore, it is desirable to identify the fundamental processes that determine anomalous losses in open nonlinear (chaotic) systems. determine.

The analysis of anomalous transport is one of the central problems in plasma physics with applications in nuclear fusion [29]. In recent decades, significant progress has been made in understanding the fundamental properties [142].

Initially, linear transport theory was modified to account for geometric effects and large mean free paths. The neoclassical theory (see [57] and references therein) is a major intellectual and practical achievement. Nevertheless, it is not able to solve all problems.

It is known [143] that in many cases, strong deviations of the diffusion rate from classical or neoclassical predictions are due to nonlinear effects caused by (electrostatic and electromagnetic) fluctuations. In the past, several approaches have been made toward a self-consistent theory of nonlinear transport; see [29, 144] and references therein.

Quasilinear theory and transport estimates based on weak turbulence theory are by far the most successful analytical approaches. However, it is known [143] that their fields of application are limited. Strong plasma turbulence represents an extremely complex and complicated problem. The physical kinetics of plasma turbulence, with a focus on quasi-particle models, was summarized some time ago [145]. Analytical evaluations are generally difficult, which is why numerical simulations are becoming increasingly important. These lead to an extensive database with numerous indications of fundamental transport scalings.

Anomalous transport of charged particles is also a longstanding problem in astrophysical contexts [146–159]. A variety of issues, such as the penetration of low-energy cosmic rays into the heliosphere, the transport of galactic cosmic rays into and out of the interstellar magnetic field, the Fermi acceleration mechanism, and others, are at the top of the astrophysical agenda [160]. In most cases, astrophysical transport is collisionless. Normally, there are no dominant guiding fields. Nonlinear effects are responsible for the differences between linear theoretical predictions and experimental facts.

A very simple argument illustrates the occurrence of nonlinearities in transport theory. The continuity equation for the particle current density $\mathbf{\Gamma}$ can formally be written as

$$\partial_t \mathbf{\Gamma} = -\nabla \cdot (\mathbf{u}\mathbf{\Gamma}) - \nabla P - \nabla \cdot \Pi + \frac{e}{m}\mathbf{\Gamma} \times \mathbf{B} + \frac{e}{m}\mathbf{E} + \frac{1}{m}\mathbf{R} . \tag{10.132}$$

On the right-hand side appear the convective term, contributions from pressure and fields, as well as the resistive term. Each variable is decomposed into an averaged and a fluctuating part, for example

$$X := \langle X \rangle + \delta X . \tag{10.133}$$

After a short calculation, the averaged x-component of the current density follows

$$\boxed{\langle \Gamma_x \rangle = \underbrace{\frac{1}{m\Omega}\langle R_y \rangle}_{classical} + \underbrace{\frac{1}{\Omega}\big\langle (\hat{b} \times \nabla \cdot \Pi) \cdot \hat{x} \big\rangle}_{neo-classical} + \Gamma_x^{anomalous} + \Gamma_x^{time-dependent}} . \tag{10.134}$$

The first part, caused by the friction term, is the classical one. The second part, originating from the pressure tensor, is mainly discussed in neoclassical theory. The third part on the right-hand side is the one that will be the focus of the next section, namely

$$\Gamma_x^{anomalous} = \frac{1}{B}\langle \delta n \, \delta E_y \rangle + \frac{1}{B}\big\{ \langle \delta\Gamma_\parallel \delta B_x \rangle - \langle \delta\Gamma_x \delta B_\parallel \rangle \big\} . \tag{10.135}$$

The fourth, time-dependent term is usually neglected.

It should be emphasized that this simple argument does not cover the important aspect of gyro-averaging. The latter forms the central component of gyrokinetic models.

Fluctuation Spectra and Transport

Weak turbulence theory has proven to be a very successful approach, developed specifically for plasma fluctuations with a characteristic frequency. Essentially, in its simplest form, it results from perturbation theory. It is therefore limited to weak fluctuation levels.

For the second moment of fluctuating electrostatic potentials, we can introduce the spectral density $n(\mathbf{k}, t)$,

$$\langle \delta\varphi_{\mathbf{k}} \delta\varphi^*_{\mathbf{k}'} \rangle = n(\mathbf{k}, t)\delta(\mathbf{k} - \mathbf{k}') \, , \tag{10.136}$$

for which a wave-kinetic equation is obtained [145]

$$\boxed{\frac{\partial n(\mathbf{k}, t)}{\partial t} = \mathcal{I}(\mathbf{k}, t) + \Gamma(\mathbf{k})n(\mathbf{k}, t)} \, . \tag{10.137}$$

On the right-hand side appear a nonlinear term ($\mathcal{I}$) and a damping term (Γ). If the damping can be neglected in a certain (inertial) region of $\mathbf{k}$-space, then for the spectral energy density $E(k)$ (see below) in isotropic turbulence, one can write

$$\boxed{\frac{\partial E(k)}{\partial t} + \frac{\partial P(k)}{\partial k} = 0} \, . \tag{10.138}$$

Such an equation can be used to analyze wavenumber spectra.

Dimensional arguments often prove useful. Let us repeat the simple dimensional analysis known from fluid theory. In three-dimensional space, we can write the dimensions of energy E and the spectral energy density $E(k)$ as follows:

$$[E] = \left[\frac{gL^2}{T^2}\right] = \left[\frac{\rho L^5}{T^2}\right] \quad \Rightarrow \quad [E(k)] = \left[\frac{\rho L^3}{T^2}\right] \text{ and } [P(k)] = \left[\frac{\rho L^2}{T^3}\right] , \tag{10.139}$$

where we use

$$\frac{E}{V} = \int E(k)dk \tag{10.140}$$

independent of the dimension of the system. The famous argument by Kolmogorov [161] (applied to fluid turbulence) states that the time T can be eliminated by the constant transfer function P, leading to the following expression:

$$\boxed{[E(k)] = \left[\frac{\rho L^3}{T^2}\right] \sim \rho^{1/3} \, P^{2/3} \, k^{-3+\frac{4}{3}} \sim k^{-5/3}} \, , \tag{10.141}$$

i.e., to the well-known Kolmogorov spectrum.

When a characteristic frequency appears in plasmas, the situation is different. The characteristic time results from the finite frequency $\omega = \omega(k)$, and we can formulate a dimensionless variable ξ in the form:

$$\xi = \frac{Pk^2}{\rho\omega^3} \, . \tag{10.142}$$

Now, there is more freedom, since we can introduce an arbitrary function f of the dimensionless quantity, to obtain for the spectral density

$$E(k) \sim \rho\omega^2(k)k^{-3}f(\xi) \tag{10.143}$$

If $f(\xi) \sim \xi^{1/3}$ holds, some fundamental aspects of drift wave turbulence become apparent. However, in a strict sense, this functional form is not uniquely prescribed.

There have been many attempts to generalize weak turbulence theory to strong plasma turbulence using renormalization methods [162–165]. Although these are intellectually appealing, in most cases the evaluation for practical applications is technically very demanding (and perhaps more time-consuming than solving the original gyrokinetic problem).

Especially for drift wave turbulence, there exist semi-empirical models that have been published under various names such as mixing length theory, marginal stabilization scenarios, etc. [13].

For electrostatic turbulence, the nonlinear current density in the radial direction r is given by

$$\Gamma_r \approx \langle \delta n\, \delta u_r \rangle \mathrel{\hat{=}} -D\frac{\langle n \rangle}{L_n}\,, \tag{10.144}$$

where on the right-hand side the current density is expressed as an effective diffusive flux with a diffusion coefficient D and an inhomogeneity length L_n. For drift waves, the linear growth rate, frequency, density fluctuation, and $E \times B$-velocity are

$$\frac{\gamma_k}{\omega_k} \sim \delta_k,\ \omega_k \approx \frac{kT}{eBL_n},\ \delta n_k \approx \langle n \rangle \frac{e\phi_k(1 - i\delta_k)}{T},\ \delta u_r \approx ik\frac{\phi_k}{B}\,. \tag{10.145}$$

Taking all this into account, it follows that

$$\boxed{\Gamma_r \approx \langle n \rangle \frac{T}{eB} \sum_k k\delta_k \left| \frac{\delta n_k}{\langle n \rangle} \right|^2}\,. \tag{10.146}$$

The last expression contains the as yet unknown fluctuation spectrum. For the level of this spectrum, we assume that it increases until the corresponding local density change eliminates the density gradient driving the instability,

$$\nabla \delta n \mathrel{\hat{=}} k\delta n_k \sim \frac{\langle n \rangle}{L_n}\,. \tag{10.147}$$

Thus, we ultimately obtain

$$\boxed{D \approx \frac{\gamma_{\max}}{k_{\max}^2}}, \tag{10.148}$$

if we approximate the sum by its maximum term. This expression is frequently used for estimates of diffusion processes. It is related to the mixing length. According to Prandtl, the turbulent diffusion coefficient can be written as

$$D \approx \frac{L_r^2}{\tau_c} \tag{10.149}$$

where L_r is the characteristic length (the radial correlation length of the modes) and τ_c is the eddy turnover time (autocorrelation time for the turbulent field).

Now we can have two extremely different scenarios. First, small-scale turbulence with $L_r \approx \rho_s \sim \rho_i$ and the diamagnetic frequency

$$\boxed{\tau_c^{-1} \approx \omega_T^* \equiv \frac{k_\theta T}{BL_T} \sim \frac{v_{thi}}{a}, \quad k_\theta \approx \rho_i^{-1}}. \tag{10.150}$$

In this case, we have

$$D \sim \rho_i^2\, \omega_T^* \sim D_{Bohm} \frac{\rho^*}{L_T^*} \sim T^{3/2} B^{-2}, \tag{10.151}$$

which is known as gyro-Bohm scaling.

On the other hand, for large-scale turbulence, e.g., with $L_r \approx \sqrt{a\rho_s}$, where a is the system size (the minimum radius of the tokamak), we estimate

$$\boxed{D \sim \rho_i v_{ti} \sim D_{Bohm}\left[\frac{m^2}{s}\right] = \frac{T[eV]}{16B[T]} \sim TB^{-1}}. \tag{10.152}$$

This is the famous Bohm scaling.

Other scalings have also been proposed. For example, the Goldstone regime $D \sim \sqrt{\rho_i}$ in the L-regime of a tokamak.

The main question remains: How can we predict the characteristic scales? Over the past decades, several approaches have been proposed. A rather convincing one is the following self-organization hypothesis:

Let a system be modeled by a nonlinear partial differential equation with dissipation for the field u. The system contains (at least) two quadratic or higher-order conserved quantities in the absence of dissipation. One of the conserved quantities, say $A(u)$, decays faster than the other(s), e.g., $B(u)$. It is expected that the modal cascade reaches a quasi-stationary state that minimizes A at constant B, i.e., $\delta A - \lambda\,\delta B = 0$.

This formulation, as well as arguments supporting it, can be found in the review article by Hasegawa [166].

In addition to the obvious advances in analytical theory, it should be kept in mind that (truly) self-consistent models are generally very difficult to solve and often demonstrate their strength only qualitatively, i.e., in scaling predictions. Heuristic, quasi-self-consistent models are often ad hoc, well suited for rough interpretations, but strictly speaking are not predictive. Nonlocality, intermittency, interactions with coherent structures, etc., are open problems that have not yet been solved in full generality. Many new insights are based on numerical simulations. The progress in numerical codes is enormous; unfortunately, analytical theory lags behind, but remains indispensable.

10.5 Stochastic Magnetic Fields

One possible approach to solving the problem of anomalous plasma transport is to consider the (passive) motion of (test) particles under the influence of given perturbations; see [167] and the references therein. Such an approach is widespread in fluid turbulence [168], where the passive motion of scalars, vectors, particles, etc., has been extensively studied.

In magnetic fusion research, there is an additional, qualitatively important reason to analyze particle motion in given stochastic fields. Disturbances in the magnetic field structure are more or less unavoidable due to errors in the arrangement of the device coils. Moreover, and this aspect has recently gained significant importance, additional coils are being installed in tokamaks to control the particle and heat loads on the walls by magnetically stochastizing the plasma edge

[169–173]. By "stochastic transport" we mean transport in stochastic fields generated by external means. In this section, we discuss models for magnetic field fluctuations.

The strength of magnetic field fluctuations can be very small, e.g., less than a tenth of a percent of the zeroth-order confining field (which defines the *parallel* direction). Nevertheless, magnetic field fluctuations have a strong impact on *perpendicular* transport. Particles move in the perpendicular direction significantly faster than predicted by classical theory, as has been shown in what are now classic works [146, 148, 149, 174, 175]. According to these findings, the perpendicular diffusion coefficient exhibits a crucial dependence on the parallel diffusion coefficient. This is because perpendicular fluctuations in the magnetic field create small channels for parallel diffusivity in the perpendicular direction.

The kinetics of test particle diffusion, with particular attention to the collision-dominated limit, has been studied using a modified perturbation theory [176]. Myra et al. [177] showed both analytically and by Monte Carlo simulation that the diffusion coefficient can be sensitive to the spectral width of the magnetic turbulence. A general variational principle for particle diffusion across braided magnetic surfaces was developed by Laval [178]. Models for magnetic field line configurations have been discussed by various authors, e.g., in [64, 179–181]. The statistical theory of plasma transport [182–184] has attracted considerable interest [185, 186].

The Langevin equation model [183] has been used to analyze the subdiffusive behavior of charged particles in stochastic magnetic fields [187, 188]. The connection with continuous random walks has been elucidated [189]. The transition from Euler to Liouville correlation functions proved to be a very important step [188, 190–194]. Effects of time-dependent magnetic field fluctuations [195] and of plasma motion [196] on particle diffusion have been investigated. Trajectory structures and transport have become a major topic [197–209]. Larmor radius effects have proven important in the percolative regime [210–212]. New processes such as a ratchet current in plasma devices have been discussed, see, e.g., [213–215] and the references cited therein.

To calculate the effective perpendicular transport, one needs the parallel diffusion coefficient. In the collisionless case, classical theory predicts superdiffusive transport in the parallel direction. In fact, this transport does not occur as long as magnetic perturbation fields are present as an additional source of collision-like interactions. Even in the absence of binary collisions, the parallel motion should exhibit diffusive behavior. This behavior has recently been discussed for plasmas in the context of magnetic fluctuations.

The entire field of "nonlinear transport" is vast. A global overview would go beyond the scope of a single chapter. Fortunately, several specialized books already exist, e.g., [13, 28, 144, 145], which should be consulted if all aspects of nonlinear transport are of interest.

In this chapter, we focus on a single subtopic, namely "stochastic transport theory" in fluctuating magnetic fields. This subject has several applications in nuclear fusion and astrophysics. It also offers the advantage that "simple" analytical models can be developed, which provide deeper insights into nonlinear transport processes. The following can therefore be regarded as a tutorial on fundamental aspects of nonlinear plasma transport. For advanced aspects of nonlinear transport, we refer to the continually growing literature, including the increasingly important numerical simulations.

A first approach to describing anomalous transport in stochastic magnetic fields can be based on the following fluid-dynamical picture. Suppose that magnetic field fluctuations are present in the plasma, so that the total field consists of a (strong) zeroth-order contribution $\mathbf{B}_0$ and fluctuations $\delta\mathbf{B}$,

$$\boxed{\mathbf{B} = \mathbf{B}_0 + \delta\mathbf{B}, \quad \text{with } \mathbf{b} = \frac{\delta\mathbf{B}}{B}} \; . \tag{10.153}$$

We assume that locally there exist components of the fluctuating magnetic field $\delta\mathbf{B}$ that are perpendicular to the zeroth (main) magnetic field $\mathbf{B}_0$. To avoid problems of ambipolarity, we consider heat transport. To zeroth order, heat transport occurs mainly along the zeroth-order magnetic field; $\kappa_\perp \ll \kappa_\parallel$ applies for the linear thermal conductivities. Taking into account the magnetic perturbations, we can formulate the dominant terms in the heat conduction equation as follows:

$$\frac{\partial T}{\partial t} \approx \kappa_\parallel (\hat{n} \cdot \nabla)^2 T \; . \tag{10.154}$$

The direction vector $\hat{n}$ points in the direction of the total magnetic field. We have

$$\hat{n} \equiv \frac{\mathbf{B}}{B} \, , \quad \hat{n}_0 \equiv \frac{\mathbf{B}_0}{B_0} \, , \quad \kappa_\parallel \approx \frac{v_{th}^2}{\nu_{coll}} \gg \kappa_\perp \, , \tag{10.155}$$

where ν_{coll} is the binary collision frequency. By introducing fluctuations, we obtain an additional heat transport in the direction perpendicular to the zeroth-order magnetic field. Assuming that the temperature is uniform on the zeroth-order magnetic surfaces, we can write for the change in the (over the zeroth-order magnetic surfaces) averaged temperature $\langle T \rangle$ up to second order:

$$\boxed{\frac{\partial \langle T \rangle}{\partial t} = \kappa_\parallel \langle b^2 \rangle \frac{\partial^2 \langle T \rangle}{\partial x_\perp^2} + 2\kappa_\parallel \left\langle b \frac{\partial}{\partial x_\perp} (\hat{n}_0 \cdot \nabla) \delta T \right\rangle} \; . \tag{10.156}$$

The first term on the right-hand side already shows that (perpendicular) fluctuations can lead to quite effective perpendicular transport, which is caused by an effective parallel heat conduction.

The phenomena summarized so far have led to the often-used paradigm: plasma stochasticity, which arises predominantly from plasma instabilities in the high-dimensional phase space, acts counterproductively for good confinement, i.e., increased losses occur (due to microinstabilities, MHD instabilities, ripple losses, etc.).

Hamiltonian Formalism for Stochastic Magnetic Fields

Nonlinear dynamics teaches us that even systems with few degrees of freedom can become stochastic. In plasmas, sources of magnetic stochasticity include, for example, imperfect coils for confinement, heating, shaping as well as correction coils, MHD control coils and edge region coils, ergodic divertor coils, and others.

In a "new" perspective, stochasticity can have positive aspects, since it is to some extent controllable and can contribute to the mitigation of edge localized modes (ELMs) [170, 216, 217]. Moreover, it can be useful for particle exhaust, zonal flows [218], radial electric fields and transport barriers [219–222], understanding escape velocities of fast particles [223–225], heat flow patterns [226, 227], and other phenomena. Special coil configurations have already been tested in Tore Supra, W7-AS (island divertor), DIII-D, Textor, JET, and ASDEX-U, and are planned for ITER operation.

In the following, we present the fundamental techniques for describing and investigating stochastic magnetic fields.

For the magnetic field, we use the Clebsch representation [228].

$$\boxed{\mathbf{B} = \nabla\psi \times \nabla\theta - \nabla H \times \nabla\varphi}\,. \tag{10.157}$$

This representation arises, for example, from the vector potential in cylindrical coordinates, where $\varphi \hat{=} z$ (toroidal "height"), θ (poloidal coordinate), and ρ (radial coordinate in the poloidal cross-section) are,

$$\boxed{\mathbf{A} = A_\rho \nabla\rho + A_\theta \nabla\theta + A_\varphi \nabla\varphi}\,. \tag{10.158}$$

With

$$A_\rho = \frac{\partial G(\rho, \theta, \varphi)}{\partial \rho} \tag{10.159}$$

and the definitions

$$\psi = A_\theta - \frac{\partial G(\rho, \theta, \varphi)}{\partial \theta}, \quad -H = A_\varphi - \frac{\partial G(\rho, \theta, \varphi)}{\partial \varphi}, \tag{10.160}$$

we arrive at the Clebsch form.

Example 10.12 (Clebsch Representation)
To demonstrate this, we substitute the components of the vector potential, as defined in (10.159) and (10.160), into $\mathbf{B} = \nabla \times \mathbf{A}$ and use for cylindrical coordinates

$$(\nabla \times \mathbf{A})_\rho = \frac{1}{\rho}\frac{\partial A_z}{\partial \theta} - \frac{\partial A_\theta}{\partial z}, \tag{10.161}$$

$$(\nabla \times \mathbf{A})_\theta = \frac{\partial A_\rho}{\partial z} - \frac{\partial A_z}{\partial \rho}, \tag{10.162}$$

$$(\nabla \times \mathbf{A})_z = \frac{1}{\rho}\frac{\partial (\rho A_\theta)}{\partial \rho} - \frac{1}{\rho}\frac{\partial A_\rho}{\partial \theta}. \tag{10.163}$$

■

Using the Clebsch representation, it becomes clear that the system of magnetic field lines exhibits a Hamiltonian form. Starting from the differentials

$$d\psi = \mathbf{B} \cdot \nabla\psi = -(\nabla H \times \nabla\varphi) \cdot \nabla\psi \tag{10.164}$$

$$= -\left(\frac{\partial H}{\partial \theta}\nabla\theta \times \nabla\varphi\right) \cdot \nabla\psi = -\frac{\partial H}{\partial \theta}(\nabla\psi \times \nabla\theta) \cdot \nabla\varphi, \tag{10.165}$$

$$d\theta = \mathbf{B} \cdot \nabla\theta = -(\nabla H \times \nabla\varphi) \cdot \nabla\theta \tag{10.166}$$

$$= -\left(\frac{\partial H}{\partial \psi}\nabla\psi \times \nabla\varphi\right) \cdot \nabla\theta = \frac{\partial H}{\partial \psi}(\nabla\psi \times \nabla\theta) \cdot \nabla\varphi, \tag{10.167}$$

$$d\varphi = \mathbf{B} \cdot \nabla\varphi = (\nabla\psi \times \nabla\theta) \cdot \nabla\varphi, \tag{10.168}$$

we immediately obtain the symplectic structure

$$\boxed{\frac{d\psi}{d\varphi} = -\frac{\partial H}{\partial \theta}, \quad \frac{d\theta}{d\varphi} = \frac{\partial H}{\partial \psi}}. \tag{10.169}$$

Example 10.13 (Breaking of Flux Surfaces)
In comparison to classical mechanics, φ corresponds to "time," and ψ is the associated "conjugate momentum" to the "coordinate" θ. Within the Hamiltonian formalism, action (hereafter denoted by J) and angle coordinates (hereafter denoted by θ) can be introduced. In general, the Hamiltonian consists of an integrable part H_0 (which leads to closed magnetic surfaces) and a perturbation

$$H(J,\theta) = H_0(J) + \delta H(J.\theta) \ . \tag{10.170}$$

The corresponding canonical equations are

$$\frac{\partial H}{\partial J} = \dot{`} = \mathbf{\Omega}(J) + \frac{\partial \delta H}{\partial J} \ , \qquad \frac{\partial H}{\partial `} = -\dot{J} = \frac{\partial \delta H}{\partial \theta} \ . \tag{10.171}$$

After Fourier expansion

$$\delta H(J,\theta) = \sum_{\mathbf{m}} e^{i\mathbf{m}\cdot `} H_{\mathbf{m}} \tag{10.172}$$

the following approximate solutions are obtained

$$\theta \approx \theta_0 + \mathbf{\Omega} t \ , \qquad \Delta J = -\int dt \frac{\partial \delta H}{\partial `} \approx -i \sum_{\mathbf{m}} \frac{\mathbf{m} H_{\mathbf{m}} e^{i\mathbf{m}\cdot `_0}\left(e^{i\mathbf{m}\cdot\mathbf{\Omega} t} - 1\right)}{\mathbf{m}\cdot\mathbf{\Omega}} \ , \tag{10.173}$$

with resonances at $\mathbf{m} \cdot \mathbf{\Omega}(J) = 0$. As is well known, resonances lead to the breaking of magnetic surfaces. This can result in stochastic motion of the field lines. ■

For applications in tokamaks, the perturbation terms with toroidal and poloidal mode numbers can be written in the form $\delta H \sim H_{mn} \cos(m\theta + n\varphi + \chi_{m0})$. Resonances occur at the rational magnetic surface ψ_{mn} with $q(\psi_{mn}) = m/n$, resulting in chains of islands. The width of the islands is [30]

$$W_{mn} = 4 \left| \frac{\varepsilon H_{mn}(\psi_{mn})}{dq^{-1}/d\psi} \right|^{1/2} \ , \tag{10.174}$$

and chaos sets in at

$$\sigma_{Chir} = \frac{W_{mn} + W_{m+1n}}{|\psi_{m+1n} - \psi_{mn}|} \geq 1 \ . \tag{10.175}$$

This (qualitative) theoretical prediction has been successfully tested in many numerical applications and is also very helpful for the interpretation of experimental results.

Let us now consider the cylindrical coordinate system (R, φ, Z), where R is the major radius, φ is the toroidal angle, and Z is the vertical coordinate. The field line equations in this coordinate system are

$$\boxed{\frac{1}{R}\frac{dZ}{d\varphi} = \frac{B_Z}{B_\varphi}, \qquad \frac{1}{R}\frac{dR}{d\varphi} = \frac{B_R}{B_\varphi}}. \tag{10.176}$$

The magnetic field components B_R, B_φ, B_Z can be determined by the vector potential $\mathbf{A}$. The coordinates (R, φ, Z) are related to each other via the previously defined coordinates (r, θ, φ) as follows:

$$R = R_0 + r\cos\theta, \qquad Z = r\sin\theta. \tag{10.177}$$

We choose the radial component of the vector potential A_R to be zero, i.e., $A_R = 0$. The reason for this choice is the gauge invariance of the vector potential. We assume that the Z component of the vector potential (A_Z), which determines the toroidal field $B_\varphi = -\partial A_Z/\partial R$, is given in the form $A_Z = -B_0 R_0/R$, i.e., $B_\varphi = B_0 R_0/R$, where B_0 is the strength of the magnetic field at the major radius of the torus $R = R_0$. We further assume that the equilibrium poloidal field of the plasma (B_R, B_Z) and the perturbed magnetic field are generated by external coils, so that they can be fully described by the toroidal component of the vector potential $A_\varphi(R, \varphi, Z)$:

$$B_Z = \frac{1}{R}\frac{\partial(RA_\varphi)}{\partial R}, \qquad B_R = -\frac{\partial A_\varphi}{\partial Z}. \tag{10.178}$$

We now introduce the normalized coordinate z and the canonical momentum p_z:

$$z = \frac{Z}{R_0}, \qquad p_z = \ln\frac{R}{R_0},$$

With this, Eqs. (10.176) appear in Hamiltonian form

$$\boxed{\frac{dz}{d\varphi} = \frac{\partial H}{\partial p_z}, \qquad \frac{dp_z}{d\varphi} = -\frac{\partial H}{\partial z}}, \tag{10.179}$$

where the Hamiltonian function $H = H(z, p_z, \varphi)$ is the normalized φ component of the vector potential, i.e.,

$$H \equiv H(z, p_z, \varphi) = \frac{R(p_z)A_\varphi(R(p_z), \varphi, zR_0)}{B_0 R_0^2}, \tag{10.180}$$

with $R = R_0\exp(p_z)$.

In the axisymmetric case, the magnetic field does not depend on the toroidal angle φ, $A_\varphi = A_\varphi(R, Z)$, and thus $H = H(z, p_z)$ holds. In this case, the Hamiltonian system (10.179) is completely integrable. The field lines lie on nested toroidal surfaces, which are defined by the surface function $H(z, p_z) = f(Z, R) = \text{const}$.

The intersection of a toroidal surface with the plane $\varphi = \text{const}$ is shown in Fig. 10.14. One can introduce action and angle variables (I, ϑ) via

$$I = \frac{1}{2\pi} \oint_C p_z dz, \qquad \vartheta = \frac{\partial}{\partial I} \int^z p_z(z', I) dz', \tag{10.181}$$

where the integration is performed along the closed contour C, which results from the intersection of the surface function $f(R, Z) = \text{const}$ with the poloidal plane $\varphi = \text{const}$ (see Fig. 10.14). The action variable I coincides with the normalized toroidal flux ψ, since

$$I = \frac{1}{2\pi} \oint\limits_C p_z dz = \frac{1}{2\pi} \int_S dp_z dz = \frac{1}{2\pi R_0^2 B_0} \int_S B_\varphi(R, Z) dR dZ = \psi, \tag{10.182}$$

which gives the meaning of the normalized flux of the toroidal field B_φ through the surface S enclosed by the closed contour C on the poloidal plane $\varphi = \text{const}$. The angle variable ϑ is precisely the *intrinsic* poloidal angle.

In action-angle variables (ψ, ϑ), the Hamiltonian function is $H = H(\psi)$, and the field lines obey Eq. (10.169).

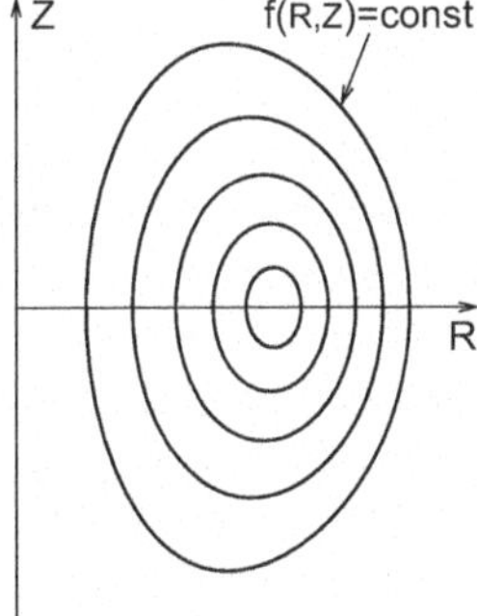

Fig. 10.14 Magnetic flux surfaces $H(z, p_z) = f(Z, R) = \text{const}$

The inverse safety factor $q(\psi)$ is defined by $dH(\psi)/d\psi$. It can also be derived from the field line Eq. (10.179). By definition, q is equal to the number of toroidal turns per poloidal turn, i.e., $q = \Delta\varphi/2\pi$, where $\Delta\varphi$ is the change in the toroidal angle φ when a field line completes one full poloidal turn.

From the first Eq. (10.179), it then follows that

$$q(\psi) = \frac{\Delta\varphi}{2\pi} = \int_C \frac{dz}{\partial H/\partial p_z}, \tag{10.183}$$

where the integration is performed along the closed contour C of $H = H_0(z, p_z) = \text{const}$.

The geometric coordinates R, Z (or r, θ) of the field lines are periodic functions of the angular variable ϑ, where the following holds: $R(\vartheta, \psi) = R(\vartheta + 2\pi, \psi)$ and $Z(\vartheta, \psi) = Z(\vartheta + 2\pi, \psi)$. Therefore, they can be represented in a Fourier series,

$$Z(\vartheta, \psi) = \sum_m Z_m(\psi)e^{im\vartheta}, \qquad R(\vartheta, \psi) = \sum_m R_m(\psi)e^{im\vartheta}, \tag{10.184}$$

or

$$r(\vartheta, \psi) = \sum_{m=0}^{\infty}\left(r_m^{(c)}(\psi)\cos m\vartheta + r_m^{(s)}(\psi)\sin m\vartheta\right), \tag{10.185}$$

$$\theta(\vartheta, \psi) = \vartheta + \sum_{m=0}^{\infty} \alpha_m(\psi)\sin m\vartheta. \tag{10.186}$$

The coefficients $R_m(\psi), Z_m(\psi)$ (or $r_m^{(c,s)}(\psi)$, $\alpha_m(\psi)$), which depend on the toroidal flux ψ, can be determined by integrating the Hamiltonian field Eq. (10.179). The field lines $R(\varphi), Z(\varphi)$ [or $r(\varphi), \theta(\varphi)$] in real geometric space (R, Z) or (r, θ) are determined by Eqs. (10.184) and (10.185), if the intrinsic poloidal angle ϑ is taken as $\vartheta = \varphi/q(\psi) + \vartheta_0$.

In the case of non-axisymmetric magnetic perturbations, the toroidal component of the vector potential $A_\varphi(R, Z, \varphi)$ can be represented as a sum,

$$A_\varphi = A_\varphi^{(0)}(R, Z) + A_\varphi^{(per)}(R, Z, \varphi). \tag{10.187}$$

In typical situations, the perturbed part of the component A_Z is small and can be neglected compared to the unperturbed part $A_Z^{(0)}(R)$, which determines the toroidal magnetic field B_φ. Then the magnetic perturbation $\delta\mathbf{B}(R, Z, \varphi)$ has only two nonzero components, namely δB_R and δB_Z. The latter, according to (10.178), are expressed by the perturbed part of the vector potential $A_\varphi^{(per)}(R, Z, \varphi)$ as

$$\delta\,\mathbf{B}(R,Z,\varphi) = \left(-\hat{e}_R\frac{1}{R}\frac{\partial}{\partial Z} + \hat{e}_Z\frac{1}{R}\frac{\partial}{\partial R}\right)RA_\varphi^{(per)}(R,Z,\varphi). \tag{10.188}$$

The introduction of the toroidal flux ψ and the intrinsic poloidal angle ϑ according to (10.181) for the equilibrium magnetic field leads to the Hamiltonian equations for the perturbed field lines, which are given as follows:

$$\boxed{\frac{d\vartheta}{d\varphi} = \frac{1}{q(\psi)} + \varepsilon\frac{\partial H_1}{\partial \psi}, \qquad \frac{d\psi}{d\varphi} = -\varepsilon\frac{\partial H_1}{\partial \vartheta}}\,. \tag{10.189}$$

The perturbed Hamiltonian function $\varepsilon H_1 \equiv H_1(\psi,\vartheta,\varphi)$ is

$$H_1(\psi,\vartheta,\varphi) = \frac{R(\psi,\vartheta)}{R_0^2 B_0}A_\varphi^{(per)}(R(\psi,\vartheta), Z(\psi,\vartheta),\varphi), \tag{10.190}$$

with $R(\psi,\vartheta) = R_0 + r(\psi,\vartheta)\cos\theta(\psi,\vartheta)$. For

$$A_\varphi^{(per)}(r,\theta,\varphi) = \varepsilon B_0 R_0 a(r,\theta,\varphi), \tag{10.191}$$

$$a(r,\theta,\varphi) = \sum_m a_{mn}(r,\theta)\cos(m\theta + n\varphi + \chi_{m0}) \tag{10.192}$$

one obtains the Fourier expansion

$$H_1(\psi,\vartheta,\varphi) = \varepsilon\sum_m H_{mn}(\psi)\cos(m\vartheta + n\varphi + \chi_{m0})\,. \tag{10.193}$$

The dimensionless perturbation parameter ε characterizes the strength of the perturbations. According to (10.191), the Fourier components $H_{m,n}(\psi)$ are given by integrals,

$$\begin{aligned} H_{mn}(\psi) &= \mathrm{Re}\iint_0^{2\pi}\frac{R(\psi,\vartheta)}{(2\pi)^2R_0^2B_0\varepsilon}A_\varphi^{(per)}(R(\psi,\vartheta), Z(\psi,\vartheta),\varphi)e^{-im\vartheta-in\varphi}d\vartheta d\varphi \\ &= \mathrm{Re}\iint_0^{2\pi}\frac{R(\psi,\vartheta)}{(2\pi)^2R_0}a(r(\psi,\vartheta),\theta(\psi,\vartheta),\varphi)e^{-im\vartheta-in\varphi}d\vartheta d\varphi\,. \end{aligned} \tag{10.194}$$

One can also expand the perturbed magnetic field $\delta\mathbf{B}$ in a Fourier series with respect to the intrinsic poloidal angle ϑ and the toroidal angle φ. This is done similarly to the expansion of the perturbation Hamiltonian H_1, as given by Eq. (10.193). We obtain

$$\delta\,\mathbf{B}(\psi,\vartheta,\varphi) = \sum_m \mathbf{B}_{mn}(\psi)\sin(m\vartheta + n\varphi + \chi_{m0}). \tag{10.195}$$

Frequently, the Fourier components $\mathbf{B}_{mn}(\psi)$ are used to estimate the width of magnetic islands that arise in the presence of magnetic perturbations.

The relationship between the Fourier components of the perturbation Hamiltonian $H_{mn}(\psi)$ and the perturbation field $\mathbf{B}_{mn}(\psi)$ can be derived using Eq. (10.188) and (10.190). Assuming $|dH_{mn}/d\psi| \ll m|H_{mn}|$, we obtain

$$| \mathbf{B}_{mn}(\psi)| \approx \varepsilon m H_{mn}(\psi) B_0 \equiv m H_{mn}(\psi) B_c \,, \quad \text{with } \varepsilon = B_c/B_0 \,. \tag{10.196}$$

Diffusion processes

We now summarize the fundamental statistical assumptions for fluctuating fields, which are frequently used in theory for both qualitative predictions and quantitative estimates. The models appear simple, and it should be clear to everyone that they cannot capture all the details that occur in complex plasma systems. The situation is ambivalent: models should not deviate too much from realistic conditions, but on the other hand, they must be simple enough to allow for analytical calculations as far as possible. In this section, the focus is on analytical tractability.

Let us begin with the current density for a three-dimensional diffusion process

$$\boxed{\mathbf{\Gamma} = -\overleftrightarrow{D}\,\nabla n}\,, \tag{10.197}$$

with the diffusion tensor $\overleftrightarrow{D}$, which in the gyrotropic case can be written as follows:

$$\overleftrightarrow{D} = \chi_\perp(\hat{e}_x\hat{e}_x + \hat{e}_x\hat{e}_x) + \chi_{zz}\hat{e}_z\hat{e}_z \,. \tag{10.198}$$

The continuity equation for the particle density leads to

$$\frac{\partial n}{\partial t} = \left[\chi_\perp(\nabla_x^2 + \nabla_y^2) + \chi_{zz}\nabla_z^2\right] n \,. \tag{10.199}$$

With the particle density n, which is normalized to the total particle number density N, we connect the probability of finding a particle at a given time t at a specific position $\mathbf{r}$. We then determine averages over

$$\langle \cdots \rangle = \frac{1}{N} \int d^3r \ \cdots \ n \,. \tag{10.200}$$

For purely diffusive processes, we have

$$\frac{d}{dt}\langle \mathbf{r} \rangle = 0 \,. \tag{10.201}$$

With the difference

$$\delta\mathbf{r} = \mathbf{r}(t) - \langle \mathbf{r}(t) \rangle \tag{10.202}$$

we introduce the mean square deviation:

$$\langle (\delta\mathbf{r})^2 \rangle = \langle (\mathbf{r})^2 \rangle - \langle \mathbf{r} \rangle^2 \,. \tag{10.203}$$

In the case of (pure) diffusion, we then have

$$\frac{d}{dt}\langle(\delta\mathbf{r})^2\rangle = \frac{d}{dt}\big[\langle(\mathbf{r})^2\rangle - \langle\mathbf{r}\rangle^2\big] = \frac{d}{dt}\langle(\mathbf{r})^2\rangle\,. \tag{10.204}$$

The right-hand side is calculated as follows:

$$\frac{d}{dt}\langle(\mathbf{r})^2\rangle = \frac{1}{N}\int d^3r\,\mathbf{r}^2\frac{\partial n}{\partial t} = \frac{1}{N}\int d^3r\,\mathbf{r}^2\Big[\chi_\perp(\nabla_x^2+\nabla_y^2)+\chi_{zz}\nabla_z^2\Big]n\,. \tag{10.205}$$

In the case of position-independent diffusion coefficients, we integrate by parts and, after a short calculation, obtain for the displacements in the various directions:

$$\chi_\perp(t) = \frac{1}{2}\frac{d}{dt}\langle x^2\rangle = \frac{1}{2}\frac{d}{dt}\langle y^2\rangle\,,\quad \chi_{zz}(t) = \frac{1}{2}\frac{d}{dt}\langle z^2\rangle\,. \tag{10.206}$$

These are called time-dependent, running diffusion coefficients

$$\boxed{\chi_\perp(t) = \frac{1}{2}\frac{d}{dt}\langle\delta x^2(t)\rangle = \frac{1}{2}\frac{d}{dt}\langle\delta y^2(t)\rangle\,,\quad \chi_{zz}(t) = \frac{1}{2}\frac{d}{dt}\langle\delta z^2(t)\rangle}\,. \tag{10.207}$$

The relation (10.207) can also be used for advection-diffusion processes in the form

$$\frac{\partial n}{\partial t} = -\mathbf{V}(t)\cdot\nabla n(\mathbf{r},t) + \Big[\chi_\perp(\nabla_x^2+\nabla_y^2)+\chi_{zz}\nabla_z^2\Big]n(\mathbf{r},t) \tag{10.208}$$

if

$$\frac{d}{dt}\langle\mathbf{r}\rangle = \mathbf{V}(t) \neq 0\,. \tag{10.209}$$

We have found the famous Einstein relations (10.207) between the mean square displacements and the running diffusion coefficients. From this, we can directly derive the Green-Kubo formula.

Let us take a closer look at the mean square displacement (10.203). We can evaluate this by integration. Let us demonstrate this using the z component of a purely diffusive process with

$$\delta z(t) = z(t) - \langle z(t)\rangle\,,\quad \langle z(t)\rangle = \text{const}\,,\quad \frac{dz(t)}{dt} = v_z(t)\,, \tag{10.210}$$

in the form

$$z(t) = \underbrace{z(t_0 = 0)}_{z_0} + \int_0^t dt' \; v_z(t') = z_0 + \delta z(t) \; . \tag{10.211}$$

Obviously, $\langle \delta z(t) \rangle = 0$ holds and

$$\langle [\delta z(t)]^2 \rangle = \int_0^t dt_1 \int_0^t dt_2 \; \underbrace{\langle v_z(t_1) v_z(t_2) \rangle}_{R_{zz}(t_1, t_2)} \; . \tag{10.212}$$

For temporally homogeneous situations, the velocity correlation function R_{zz} depends only on the time difference, i.e.:

$$\langle v_z(t_1) v_z(t_2) \rangle \equiv R_{zz}(t_1, t_2) = R_{zz}(t_2 - t_1) \; . \tag{10.213}$$

In this case, the evaluation yields

$$\begin{aligned} &\int_0^t dt_1 \int_0^t dt_2 \; R_{zz}(t_1, t_2) = 2 \int_0^t ds \int_0^s dt_2 \; R_{zz}(t_2 - s) = 2 \int_0^t ds \int_{-s}^0 d\tau \; R_{zz}(\tau) \\ &= 2 \int_0^t ds \int_0^s d\tau \; R_{zz}(\tau) = 2 \Big[t \int_0^t d\tau \; R_{zz}(\tau) - \int_0^t ds \; s \; R_{zz}(s) \Big] \\ &= 2 \int_0^t ds \; (t - s) \; R_{zz}(s) \; , \quad t > 0 \; . \end{aligned} \tag{10.214}$$

In combination with (10.207) and by differentiating with respect to t we obtain

$$\chi_{zz}(t) = \int_0^t ds \; R_{zz}(s) \; , \quad \chi_{zz} \equiv \chi_{zz}(t \to \infty) = \int_0^\infty ds \; R_{zz}(s) \; , \tag{10.215}$$

that is, the famous Green-Kubo formula, which expresses the diffusion coefficient as a time integral of the velocity autocorrelation function.

Of course, the present result can be generalized to other diffusion coefficients, e.g.

$$\chi_\perp = \int_0^\infty ds \; R_{xx}(s) \; , \quad R_{xx}(t_1, t_2) \equiv R_{xx}(t_2 - t_1) = \langle v_x(t_1) v_x(t_2) \rangle \; . \tag{10.216}$$

Obviously, the Green-Kubo formulas provide a direct approach to the diffusion coefficient χ_{zz}. We will use them to determine the parallel diffusion coefficient using perturbation theory.

A very simple form of the velocity autocorrelation function can be derived from the Langevin equations for the motion of particles in an external magnetic field $B_0\hat{z}$ in the case of collisions with a collision frequency ν and without magnetic fluctuations.

The corresponding velocity to zeroth order is denoted by $\mathbf{v} \equiv \eta$. The Langevin equations are

$$\boxed{\frac{d\mathbf{r}}{dt} = \eta \,,} \tag{10.217}$$

$$\boxed{\frac{d\eta_\perp}{dt} = \Omega\eta_\perp \times \hat{z} - \nu\eta_\perp + \mathbf{a}_\perp(t) \,,} \tag{10.218}$$

$$\boxed{\frac{d\eta_\parallel}{dt} = -\nu\eta_\parallel + a_z(t) \,.} \tag{10.219}$$

Example 10.14 (Solution for the velocity autocorrelation function)
We solve for the velocity as a function of the initial value η_0,

$$\eta(t) = G(t) \cdot \eta_0 + \int_0^t d\theta G(\theta) \cdot \mathbf{a}(t - \theta) \,, \tag{10.220}$$

where

$$G(t) \equiv \begin{pmatrix} \exp(-\nu t)\cos(\Omega t) & \exp(-\nu t)\sin(\Omega t) & 0 \\ -\exp(-\nu t)\sin(\Omega t) & \exp(-\nu t)\cos(\Omega t) & 0 \\ 0 & 0 & \exp(-\nu t) \end{pmatrix} . \tag{10.221}$$

As already introduced, $\Omega = \frac{qB_0}{m}$ the gyrofrequency. The stochastic force $\mathbf{a}(t)$ is assumed to be a stationary, delta-correlated Gaussian process (white noise), so that

$$\boxed{\langle a_i(t)\rangle = 0 \,, \quad \langle a_i(t_0)a_j(t_o + t)\rangle = A\delta_{ij}\delta(t) \,, \quad i,j = x,y,z} \,. \tag{10.222}$$

Due to the random, collision-induced force, the velocity η also becomes a random variable. We will decompose it into its perpendicular component $\eta_\perp$ and its parallel component $\eta_\parallel$. We define

$$\boxed{R_{ij}^{(0)} \equiv \langle \eta_i(t_1)\eta_j(t_2)\rangle} \tag{10.223}$$

as the zeroth-order velocity autocorrelation function between the components i and j. For a fixed initial velocity η_0 we average over the stochastic force to obtain

$$\langle \eta(t) \rangle = G(t) \cdot \eta_0 \,, \tag{10.224}$$

and thus for $t_1 > t_2$

$$\begin{aligned} R_{xx}^{(0)} &= \exp\left(-\nu(t_1+t_2)\right)\left[\cos(\Omega t_1)\cos(\Omega t_2)v_{0x}^2 + \sin(\Omega t_1)\sin(\Omega t_2)v_{0y}^2\right] \\ &\quad + \tfrac{A}{2\nu}\exp\left(-\nu(t_1-t_2)\right)\left[1-\exp\left(-2\nu t_2\right)\right]\cos(\Omega(t_1-t_2)) \,. \end{aligned} \tag{10.225}$$

Next, we average over the initial velocity η_0 with the probability distribution

$$\varphi(\eta_0) = \frac{1}{\pi^{\frac{3}{2}} v_{th}^3} \exp\left(-\frac{\eta_0^2}{v_{th}^2}\right) , \tag{10.226}$$

with the result

$$\begin{aligned} \overline{R_{xx}^{(0)}} =& \frac{1}{2}\left[\frac{A}{\nu}\exp\left(-\nu|t_1-t_2|\right) + \left(v_{th}^2 - \frac{A}{\nu}\right)\exp\left(-\nu(t_1+t_2)\right)\right] \\ & \times \cos(\Omega(t_1-t_2)) \,. \end{aligned} \tag{10.227}$$

Since for stationary fluctuations the dependence on $t_1 + t_2$ is unphysical, we set $A \equiv \nu v_{th}^2$ and obtain

$$\boxed{\overline{R_{xx}^{(0)}} = \frac{1}{2}v_{th}^2 \exp\left(-\nu|\tau|\right)\cos(\Omega\tau) \,, \quad \overline{R_{yy}^{(0)}} = \frac{1}{2}v_{th}^2 \exp\left(-\nu|\tau|\right)\cos(\Omega\tau) \,,} \tag{10.228}$$

$$\boxed{\overline{R_{xy}^{(0)}} = \frac{1}{2}v_{th}^2 \exp\left(-\nu|\tau|\right)\sin(\Omega\tau) \,, \quad \overline{R_{yx}^{(0)}} = -\frac{1}{2}v_{th}^2 \exp\left(-\nu|\tau|\right)\sin(\Omega\tau) \,,} \tag{10.229}$$

$$\boxed{\overline{R_{zz}^{(0)}} = \frac{1}{2}v_{th}^2 \exp\left(-\nu|\tau|\right) \,,} \tag{10.230}$$

with $\tau = t_1 - t_2$. ■

If we use $R_{xx} = \overline{R_{xx}^{(0)}}$ in the Green-Kubo formula (10.216), we obtain

$$\chi_\perp(t) = \frac{v_{th}^2}{2}\frac{\nu + [\Omega\sin(\Omega t) - \nu\cos(\Omega t)]\exp(-\nu t)}{\nu^2+\Omega^2} \,, \tag{10.231}$$

with the asymptotic value

$$\chi_\perp = \frac{v_{th}^2}{2}\frac{\nu}{\nu^2+\Omega^2} \,. \tag{10.232}$$

A similar calculation, using $R_{zz} = \overline{R_{zz}^{(0)}}$ in the Green-Kubo formula (10.215), leads to

$$\chi_\parallel(t) \equiv \chi_{zz}(t) = \frac{v_{th}^2}{2} \frac{1 - \exp(-\nu t)}{\nu} , \tag{10.233}$$

with the asymptotic value

$$\chi_\parallel \equiv \chi_{zz} = \frac{v_{th}^2}{2\nu} . \tag{10.234}$$

Correlation Functions for Magnetic Field Fluctuations

In this subsection, we examine possible forms of the magnetic fluctuation autocorrelation functions. In contrast to the previously considered constant magnetic field, we now assume that the magnetic field has a fluctuating component **b**.

We introduce

$$\mathbf{B} = B_0(b_0\hat{e}_z + \mathbf{b}) , \tag{10.235}$$

where $B_0 = const$ is factored out for dimensional reasons and $b_0 \equiv 1$ represents the constant external magnetic field. The challenge is to formulate magnetic autocorrelation functions that do not contradict the constraint

$$\nabla \cdot \mathbf{B} = B_0 \nabla \cdot \mathbf{b} = 0 \tag{10.236}$$

imposed by Maxwell's equations. We introduce the transverse and longitudinal correlation lengths $\lambda_\perp$ and $\lambda_\parallel$. We normalize the spatial coordinates with these lengths,

$$\frac{x}{\lambda_\perp} \to x , \quad \frac{y}{\lambda_\perp} \to y , \quad \frac{z}{\lambda_\parallel} \to z . \tag{10.237}$$

In the following, x, y, z are the dimensionless values, where x, y are combined into $\mathbf{x}_\perp$. Furthermore, we introduce the vector potential accordingly,

$$\mathbf{A} = \mathbf{A}_0 + \delta\mathbf{A} , \tag{10.238}$$

with

$$\boxed{\delta\mathbf{A} = \lambda_\perp B_0 \psi(x, y, z)\, \hat{e}_z , \quad \langle\psi\rangle = 0} . \tag{10.239}$$

Because of $\mathbf{B} = \nabla \times \mathbf{A}$ one immediately finds

$$b_x = \frac{\partial\psi}{\partial y} , \quad b_y = -\frac{\partial\psi}{\partial x} . \tag{10.240}$$

Note that in the present formulation we use $b_z = 0$ because $b_z \ll b_0$ holds. By specifying the statistics of the vector potential, we avoid problems with the solenoidal nature of the magnetic field.

The following notation is used for the correlation function in real space,

$$\boxed{\langle \psi(\mathbf{x}_{\perp 0}, z_0)\ \psi(\mathbf{x}_{\perp 0} + \mathbf{x}_\perp, z_0 + z)\rangle = \beta^2 C_\perp(\mathbf{x}_\perp)\ C_\parallel(z)}\ . \tag{10.241}$$

After Fourier transformation

$$\psi(\mathbf{x}_\perp, z) = \int d^2k_\perp \int dk_\parallel e^{i\mathbf{k}_\perp\cdot\mathbf{x}_\perp + ik_\parallel z}\ \hat{\psi}(\mathbf{k}_\perp, k_\parallel) \tag{10.242}$$

the correlation function in Fourier space is obtained as

$$\langle \hat{\psi}(\mathbf{k}_\perp, k_\parallel)\ \hat{\psi}(\mathbf{k}'_\perp, k'_\parallel)\rangle = \beta^2 \hat{C}_\perp(\mathbf{k}_\perp)\ \hat{C}_\parallel(k_\parallel)\ \delta(\mathbf{k}_\perp + \mathbf{k}'_\perp)\delta(k_\parallel + k'_\parallel)\ . \tag{10.243}$$

Without further information about the fluctuation field, one usually makes the ansatz

$$C_\perp(\mathbf{x}_\perp) = e^{-\frac{1}{2}(x^2+y^2)}\ , \quad C_\parallel(z) = e^{-\frac{1}{2}z^2}\ , \tag{10.244}$$

which leads to

$$\hat{C}_\perp(\mathbf{k}_\perp) = \frac{1}{2\pi}e^{-\frac{1}{2}k_\perp^2}\ , \quad \hat{C}_\parallel(k_\parallel) = \frac{1}{\sqrt{2\pi}}e^{-\frac{1}{2}k_\parallel^2} \tag{10.245}$$

In the next step, we calculate

$$\boxed{\langle b_i(\mathbf{0}, 0)\ b_j(\mathbf{x}_\perp, z) \equiv \beta^2 E_{\perp ij}(\mathbf{x}_\perp)\ C_\parallel(z) \equiv E_{ij}(\mathbf{r})\ , \quad i, j = x, y}\ . \tag{10.246}$$

Obviously, under the above assumptions, the following holds

$$\begin{aligned} E_{\perp xx}(\mathbf{x}_\perp) &= -\frac{\partial^2}{\partial y^2}C_\perp(\mathbf{x}_\perp)\ , \quad E_{\perp yy}(\mathbf{x}_\perp) = -\frac{\partial^2}{\partial x^2}C_\perp(\mathbf{x}_\perp)\ , \\ E_{\perp xy}(\mathbf{x}_\perp) &= E_{\perp yx}(\mathbf{x}_\perp) = \frac{\partial^2}{\partial x \partial y}C_\perp(\mathbf{x}_\perp)\ . \end{aligned} \tag{10.247}$$

The ansatz (10.244) then immediately leads to

$$\begin{aligned} E_{\perp xx}(\mathbf{x}_\perp) &= (1 - y^2)e^{-\frac{1}{2}(x^2+y^2)}\ , \quad E_{\perp yy}(\mathbf{x}_\perp) = (1 - x^2)e^{-\frac{1}{2}(x^2+y^2)}\ , \\ E_{\perp xy}(\mathbf{x}_\perp) &= E_{\perp yx}(\mathbf{x}_\perp) = xye^{-\frac{1}{2}(x^2+y^2)}\ . \end{aligned} \tag{10.248}$$

Returning to quantities with dimensions, we write

$$\boxed{\begin{aligned} E_{ij}(\mathbf{r}) &\equiv \langle b_i(\mathbf{r}) b_j(0)\rangle \\ &= \beta^2 \begin{pmatrix} 1-\frac{y^2}{\lambda_\perp^2} & \frac{xy}{\lambda_\perp^2} \\ \frac{xy}{\lambda_\perp^2} & 1-\frac{x^2}{\lambda_\perp^2} \end{pmatrix} \exp\left(-\frac{x^2+y^2}{2\lambda_\perp^2} - \frac{z^2}{2\lambda_\parallel^2}\right), \quad i,j=1,2. \end{aligned}} \tag{10.249}$$

where β denotes the strength, $\lambda_\perp$ the transverse, and $\lambda_\parallel$ the longitudinal correlation length.

This form of the autocorrelation matrix is compatible with the Maxwell equation $\nabla \cdot \mathbf{B} = 0$.

It should be borne in mind that the correlation tensor represents an ensemble average. For homogeneous statistical systems, we can, for example, perform the averaging as a spatial average,

$$\langle \ldots \rangle = \frac{1}{V} \int d^3x_0 \, \ldots \, . \tag{10.250}$$

After Fourier transformation of (10.249),

$$E_{ij}(\mathbf{k}) = \frac{1}{(2\pi)^3} \int E_{ij}(\mathbf{r}) \exp(-i\mathbf{k}\cdot\mathbf{r}) d^3r \, , \tag{10.251}$$

with the corresponding inverse (in asymmetric form)

$$E_{ij}(\mathbf{r}) = \int d^3k E_{ij}(\mathbf{k}) \exp(i\mathbf{k}\cdot\mathbf{r}) \, , \tag{10.252}$$

one directly obtains the result

$$E_{ij}(\mathbf{k}) = (k_\perp^2 \delta_{ij} - k_i k_j) A(\mathbf{k}) \, , \tag{10.253}$$

$$A(\mathbf{k}) = \frac{\beta^2}{(2\pi)^{3/2}} \lambda_\parallel \lambda_\perp^4 \exp\left(-\frac{1}{2}\lambda_\parallel^2 k_\parallel^2 - \frac{1}{2}\lambda_\perp^2 k_\perp^2\right) . \tag{10.254}$$

The Lagrangian form of the correlation function for magnetic field lines is

$$\boxed{L_{mn}(\zeta) = \langle b_m(\mathbf{x}_\perp(\zeta), \zeta) b_n(\mathbf{x}_\perp(0), 0)\rangle \, , \quad m,n = x,y} \, , \tag{10.255}$$

where $\mathbf{x}_\perp(\zeta)$ denotes the transverse position of the magnetic field line, which starts at $\mathbf{x}_\perp(0)$ for $\zeta = 0$.

Balescu and collaborators [192] introduced the Corrsin approximation by rewriting (10.255) as follows:

$$L_{mn}(\zeta) = \int d\rho_x d\rho_y \langle b_m(\rho_\perp, \zeta) b_n(\mathbf{x}_\perp(0), 0)\delta(\rho_\perp - \mathbf{x}_\perp(\zeta))\rangle \ , \quad m, n = x, y \ . \quad (10.256)$$

Subsequently, the exact propagator $\delta(\rho_\perp - \mathbf{x}_\perp(\zeta))$ is replaced by its ensemble average,

$$L_{mn}(\zeta) \approx \int dr_x dr_y \langle b_m(\mathbf{r}_\perp, \zeta) b_n(\mathbf{0}, 0)\rangle \langle \delta(\mathbf{r}_\perp - \delta\mathbf{x}_\perp(\zeta))\rangle \ , \quad m, n = x, y \ , \quad (10.257)$$

where $\mathbf{r}_\perp = \rho_\perp - \mathbf{x}_\perp(0)$ and $\delta\mathbf{x}_\perp(\zeta) = \mathbf{x}_\perp(\zeta) - \mathbf{x}_\perp(0)$ is. Thus, the Eulerian correlator of the magnetic field fluctuations appears on the right-hand side. Together with the averaged propagator

$$\gamma(\mathbf{r}_\perp, \zeta) \equiv \langle \delta(\mathbf{r}_\perp - \delta\mathbf{x}_\perp(\zeta))\rangle \ , \quad m, n = x, y \ , \quad (10.258)$$

the Lagrangian correlation appears in the form

$$L_{mn}(\zeta) \approx \int dr_x dr_y E_{mn}(\mathbf{r}_\perp, \zeta)\gamma(\mathbf{r}_\perp, \zeta) \ , \quad m, n = x, y \ . \quad (10.259)$$

Homogeneity and gyrotropy (i.e., isotropy with respect to the $\mathbf{B}_0$-axis) imply the properties [192]

$$\gamma(\mathbf{r}_\perp, \zeta) = \gamma(r_\perp, \zeta) \ , \quad \boxed{L_{mn} = L\delta_{mn}} \ , \quad m, n = x, y \ . \quad (10.260)$$

To calculate the averaged propagator γ we switch to the Fourier representation,

$$\gamma(\mathbf{r}_\perp, \zeta) = \int dk_x dk_y e^{i\mathbf{k}_\perp \cdot \mathbf{r}_\perp} \left\langle e^{-i\mathbf{k}_\perp \cdot \delta\mathbf{x}_\perp(\zeta)} \right\rangle \ , \quad (10.261)$$

and perform a cumulant expansion

$$\boxed{\left\langle e^{-i\mathbf{k}_\perp \cdot \delta\mathbf{x}_\perp(\zeta)} \right\rangle \approx e^{-\frac{1}{2}\sum_{r,s} k_r k_s \Gamma_{rs}(\zeta)}} \quad (10.262)$$

with

$$\Gamma_{rs}(\zeta) = \langle \delta x_r(\zeta) \delta x_s(\zeta)\rangle \ , \quad r, s = x, y \ . \quad (10.263)$$

Example 10.15 (Form of the correlation function L_{mn})
To prove $L_{mn} = L\delta_{mn}$ we recall the Green-Kubo formula

$$\Gamma_{mn} = \int_0^\zeta d\zeta_1 \int_0^\zeta d\zeta_2 \underbrace{L_{mn}(\zeta_1, \zeta_2)}_{L_{mn}(\zeta_2 - \zeta_1)} = 2\int_0^\zeta d\zeta'(\zeta - \zeta')L_{mn}(\zeta') \ . \quad (10.264)$$

Obviously, Γ_{mn} satisfies the differential equation

$$\frac{d^2\Gamma_{mn}}{d\zeta^2} = 2L_{mn} . \tag{10.265}$$

By substituting on the right-hand side of (10.259), we obtain

$$\frac{d^2\Gamma_{mn}}{d\zeta^2} = 2\int d^2k_\perp \int dk_\| e^{ik_\|\zeta} E_{mn}(\mathbf{k}_\perp, k_\|) e^{-\frac{1}{2}\sum_{r,s} k_r k_s \Gamma_{rs}(\zeta)} . \tag{10.266}$$

We assume a Gaussian form for the Euler correlation and introduce abbreviations:

$$\zeta = \lambda_\| \tau , \quad g_{mn} = \frac{\Gamma_{mn}(\lambda_\| \tau)}{\lambda_\perp^2} , \quad \alpha = \beta \frac{\lambda_\|}{\lambda_\perp} . \tag{10.267}$$

Then we obtain

$$\frac{d^2 g_{mn}}{d\tau^2} = \frac{2\alpha^2 e^{-\tau^2/2}}{\{[1+g_{xx}(\tau)][1+g_{yy}(\tau)] - g_{xy}^2(\tau)\}^{3/2}} [\delta_{mn} + g_{mn}(\tau)] . \tag{10.268}$$

The following initial values result from (10.263),

$$g_{mn}(0) = 0 , \quad \left.\frac{dg_{mn}}{d\tau}\right|_{\tau=0} = 0 . \tag{10.269}$$

For $m \neq n$ the homogeneous differential equation takes the form

$$y'' = f(\tau) y , \tag{10.270}$$

so that it can be integrated directly, with

$$y(0) = 0 , \quad y'(0) = 0 . \tag{10.271}$$

If initially $y = 0$, the variable vanishes at all times. Therefore, we conclude

$$g_{mn}(\tau) = 0 \quad \text{für } m \neq n . \tag{10.272}$$

On the other hand, for the diagonal elements we introduce $g(\tau) \equiv g_{xx} - g_{yy}$ in order to obtain for g again a differential equation of the form (10.270). The same arguments as above lead to $g_{xx} = g_{yy}$ and thus finally to (10.260). ■

Using these properties, we can simplify (10.261), with the result

$$\gamma(r_\perp, \zeta) = \frac{1}{4\pi \int_0^\zeta d\zeta'(\zeta - \zeta')L(\zeta')} \exp\left\{ - \frac{r_\perp^2}{4\int_0^\zeta d\zeta'(\zeta - \zeta')L(\zeta')} \right\} . \tag{10.273}$$

This can be substituted into (10.259) together with a Gaussian-shaped Euler correlator. Ultimately, for L one obtains the integral equation

$$L(\zeta) = \beta^2 \frac{\lambda_\perp^4 e^{-\zeta^2/2\lambda_\parallel^2}}{\left[\lambda_\perp^2 + 2\int_0^\zeta d\zeta'(\zeta - \zeta')L(\zeta')\right]^2} . \tag{10.274}$$

For $\lambda_\perp \to \infty$ the approximate Lagrangian correlation

$$L(\zeta) = \beta^2 e^{-\zeta^2/2\lambda_\parallel^2} \tag{10.275}$$

agrees with the Eulerian form.

Example 10.16 (Correlation of the Derivatives)
In the context of the divergence of magnetic field lines, the derivatives of the magnetic fluctuations become of interest. For the derivatives of the magnetic field fluctuations, we introduce a

$$\boxed{b_{m,\alpha} \equiv \frac{d}{dx_\alpha} b_m} , \quad \alpha = 1, 2 \hat{=} x, y , \quad \mathbf{r} = \begin{pmatrix} x_1 \\ x_2 \\ \zeta \end{pmatrix} \equiv \begin{pmatrix} x \\ y \\ \zeta \end{pmatrix} \equiv \begin{pmatrix} \mathbf{x}_\perp \\ \zeta \end{pmatrix} . \tag{10.276}$$

The Euler correlator for the derivatives is obtained from that for the fields

$$E_{mn} = \langle b_m(\mathbf{r}) b_n(\mathbf{0}) \rangle \tag{10.277}$$

by switching to the Fourier transform:

$$E_{mn}^{\alpha\beta} = \int d^3k e^{i\mathbf{k}\cdot\mathbf{r}} k_\alpha k_\beta E_{mn}(\mathbf{k}) . \tag{10.278}$$

In the integrand, we use (10.254) and (10.253). We immediately recognize the symmetries

$$E_{mn}^{\alpha\beta} = E_{mn}^{\beta\alpha} = E_{nm}^{\alpha\beta} = E_{nm}^{\beta\alpha} . \tag{10.279}$$

Five independent components remain, namely

$$E_{xx}^{xx} = E_{yy}^{yy} = -E_{xy}^{xy} = \frac{\beta^2}{\lambda_\perp^2}\left(1 - \frac{x^2}{\lambda_\perp^2}\right)\left(1 - \frac{y^2}{\lambda_\perp^2}\right)\mathcal{E} , \tag{10.280}$$

$$E_{xx}^{yy} = \frac{\beta^2}{\lambda_\perp^2}\left(3 - 6\frac{y^2}{\lambda_\perp^2} + \frac{y^4}{\lambda_\perp^4}\right)\mathcal{E} , \quad E_{yy}^{xx} = \frac{\beta^2}{\lambda_\perp^2}\left(3 - 6\frac{x^2}{\lambda_\perp^2} + \frac{x^4}{\lambda_\perp^4}\right)\mathcal{E} , \tag{10.281}$$

$$E_{yy}^{xy} = -E_{yx}^{xx} = -\frac{\beta^2 xy}{\lambda_\perp^4}\left(3 - \frac{x^2}{\lambda_\perp^2}\right)\mathcal{E} , \quad E_{xx}^{xy} = -E_{yx}^{yy} = -\frac{\beta^2 xy}{\lambda_\perp^4}\left(3 - \frac{y^2}{\lambda_\perp^2}\right)\mathcal{E} , \tag{10.282}$$

where

$$\mathcal{E} \equiv \exp\left(-\frac{x^2+y^2}{2\lambda_\perp^2} - \frac{\zeta^2}{2\lambda_\parallel^2}\right) . \tag{10.283}$$

The Lagrange correlator

$$\boxed{E_{mn}^{\alpha\beta} = \langle b_{m,\alpha}(\mathbf{x}_\perp(\zeta),\zeta) b_{n,\beta}(\mathbf{0},0)\rangle} \tag{10.284}$$

can be calculated in a similar way as we have already determined the Lagrange correlator for the magnetic field fluctuations. With the averaged propagator, it follows that

$$L_{mn}^{\alpha\beta}(\zeta) \approx \int dr_x dr_y E_{mn}^{\alpha\beta}(\mathbf{r}_\perp,\zeta)\gamma(\mathbf{r}_\perp,\zeta) , \quad m,n = x,y . \tag{10.285}$$

All non-vanishing components can be formed from a function $K(\zeta)$:

$$L_{xx}^{xx}(\zeta) = L_{yy}^{yy}(\zeta) = -L_{xy}^{xy}(\zeta) = -L_{xy}^{yx}(\zeta) \equiv K(\zeta) , \tag{10.286}$$

$$L_{xx}^{yy}(\zeta) = L_{yy}^{xx}(\zeta) = 3K(\zeta) , \quad L_{xx}^{xy}(\zeta) = L_{xy}^{yy}(\zeta) = L_{yy}^{xy}(\zeta) = L_{yx}^{xx}(\zeta) = 0 . \tag{10.287}$$

Therefore, we only need to explicitly calculate one element. Let us choose $L_{xx}^{xx}(\zeta)$, which after integration over the angles reads

$$K(\zeta) = \beta^2 \frac{2\pi}{\lambda_\perp^2} \frac{e^{-\zeta^2/2\lambda_\parallel^2}}{4\int_0^\zeta d\zeta'(\zeta-\zeta')L(\zeta')} \int dr_\perp r_\perp \left[1 - \frac{r_\perp^2}{\lambda_\perp^2} + \frac{r_\perp^4}{8\lambda_\perp^4}\right] e^{-\frac{r_\perp^2}{4\int_0^\zeta d\zeta'(\zeta-\zeta')L(\zeta')}} e^{-\frac{r_\perp^2}{2\lambda_\perp^2}} . \tag{10.288}$$

Here, L is the Lagrange correlator for magnetic field lines. We perform the $r_\perp$ integration, with the result

$$\boxed{K(\zeta) = \beta^2 e^{-\zeta^2/2\lambda_\parallel^2} \frac{\lambda_\perp^4}{\left[\lambda_\perp^2 + 2\int_0^\zeta d\zeta'(\zeta-\zeta')L(\zeta')\right]^3}} . \tag{10.289}$$

■

As an application, we calculate the diffusion associated with the spatial variation of magnetic field lines. The equation for a magnetic field line (10.299) can be written as

$$\frac{dx}{dz} = b_x . \tag{10.290}$$

This equation must be supplemented by the statistical properties of the magnetic fluctuations. It is an equation of the Langevin type. The solution for the mean square displacement follows the considerations of the Green-Kubo formula and yields

$$\langle \delta x^2(z)\rangle = \int_0^z dz' \int_0^z dz'' \langle b_x(z')b_x(z'')\rangle = 2\int_0^z dz'(z-z')E_{xx}(z') \, . \tag{10.291}$$

In the quasilinear limit $\lambda_\perp \to \infty$ we use

$$E_{xx} = E_{xx}(z) = \beta^2 \exp\left(-\frac{z^2}{2\lambda_\parallel^2}\right) , \tag{10.292}$$

to ultimately obtain

$$\langle \delta x^2(z)\rangle = 2\lambda_\parallel^2\beta^2 \left\{\sqrt{\frac{\pi}{2}}\frac{z}{\lambda_\parallel}\text{erf}\left(\frac{z}{\sqrt{2}\lambda_\parallel}\right) + \left[\exp\left(-\frac{z^2}{2\lambda_\parallel^2}\right) - 1\right]\right\} . \tag{10.293}$$

For large z we have

$$\langle \delta x^2(z)\rangle \sim 2\sqrt{\frac{\pi}{2}}\lambda_\parallel\beta^2 z \, , \tag{10.294}$$

which leads to the following magnetic diffusion coefficient:

$$\boxed{D = \sqrt{\frac{\pi}{2}}\beta^2\lambda_\parallel} \, . \tag{10.295}$$

The factor $\sqrt{\frac{\pi}{2}}$ arises from the assumed exponential form of the Eulerian correlation function. Therefore, for a less specific form of the correlator, the form of the magnetic diffusion coefficient is assumed as follows:

$$D_{m(agnetic)} = \beta^2\lambda_\parallel \, . \tag{10.296}$$

Divergence of Magnetic Field Lines

The Kolmogorov length characterizes the exponential divergence of neighboring magnetic field lines. Standard works on nonlinear dynamical systems, see e.g. [32], provide an overview of expressions for L_K.

Example 10.17 (The Standard Map for Stochastic Magnetic Fields)
We use a simple approach for a mapping that allows some analytical estimates to be carried out. We will perform evaluations in the fully stochastic regime. The magnetic field configuration has small deviations in the transverse direction compared to a leading-order

field,

$$\mathbf{B} = \mathbf{B}_0 + \delta\mathbf{B}_\perp \,, \quad B_0 \gg |\delta\mathbf{B}_\perp| \,. \tag{10.297}$$

As before, $\mathbf{B}_0$ is a constant magnetic field in the z-direction, and $\delta\mathbf{B}_\perp$ is the transverse perturbation field with components in the x and y directions. The relative components are defined via $\mathbf{b}_\perp = \delta\mathbf{B}_\perp / B_0$ as b_x and b_y.

When we proceed one step along a magnetic field line, the components dx, dy, and dz of the infinitesimal path are approximately related by the following equation:

$$\boxed{\frac{dy}{\delta B_{\perp y}} \approx \frac{dz}{B_0} \approx \frac{dx}{\delta B_{\perp x}}} \,. \tag{10.298}$$

We can introduce a parameter t (which will later be identified with the time for particle motion), for example in the simple form

$$\frac{dx}{dt} = b_x \frac{dz}{dt} \,, \tag{10.299}$$

$$\frac{dy}{dt} = b_y \frac{dz}{dt} \,, \tag{10.300}$$

$$\frac{dz}{dt} = \eta_\parallel \equiv const \,. \tag{10.301}$$

These equations already show some formal similarities with the Langevin equations. However, at this point, no stochastic source has yet been introduced. For tokamak-like configurations, several simple forms of the perturbation field have been discussed [229–232]. The starting point is the Fourier decomposition of the x-component,

$$b_x = \sum_{m,n} b_{mn}(x) e^{i(2\pi my - 2\pi nz)} + c.c., \tag{10.302}$$

We can identify x with a normalized radial coordinate, y with the normalized poloidal, and z with the normalized toroidal (angular) coordinate. The radial coordinate is measured in the shear length L_s, while y and z each change by 1 after a poloidal or toroidal revolution, respectively. It is assumed that the y-component is responsible for the shear,

$$\delta B_{\perp y} = B_0 \frac{x}{L_s} \quad \rightarrow \quad b_y = x \,. \tag{10.303}$$

In the next step, we simplify to

$$m = 1, \qquad b_{1n} = \varepsilon \frac{e^{i2\pi\varphi_{10}}}{2i} \equiv \frac{\varepsilon}{2i} \,, \quad b_{mn} = 0 \;\text{ für } m \neq 1 \,, \tag{10.304}$$

with constant phase $\varphi_{10} \equiv 0$ (see below). ε determines the strength of the perturbation. With

$$\sum_n b_{1n} e^{i(2\pi y - 2\pi nz)} + c.c. = \varepsilon \sum_{n \in N} \sin(2\pi y - 2\pi nz) \tag{10.305}$$

and $\sin(2\pi y - 2\pi nz) = \sin(2\pi y)\cos(2\pi nz) - \cos(2\pi y)\sin(2\pi nz)$ we obtain

$$b_x = \varepsilon \sin(2\pi y) \left[1 + 2 \sum_{n \in N^+} \cos(2\pi nz) \right] . \tag{10.306}$$

If we take into account the identity (Poisson summation formula)

$$\sum_{n \in Z} \cos(2\pi nz) = \sum_{k \in Z} \delta(z - k) , \tag{10.307}$$

we obtain

$$b_x = \varepsilon \sin(2\pi y) \sum_{k \in Z} \delta(z - k) . \tag{10.308}$$

Physically, the Dirac delta function δ produces "kicks" each time z changes by $\Delta z \equiv 1$. Due to (10.299) and (10.300), a "kick" leads to changes

$$\Delta x = \varepsilon \sin(2\pi y) , \quad \Delta y = x . \tag{10.309}$$

In other words, we can interpret the result in a poloidal cross-section (Poincaré section) as the changes in the penetration points of the magnetic field lines that occur with each toroidal revolution. Eq. (10.309) is equivalent to the standard map, written for the $(t+1)$-th iteration y_{t+1} and x_{t+1} in terms of the t-th iteration. For simplicity, we can set $2\pi x \to x$, $2\pi y \to y$ and $2\pi\varepsilon \to \varepsilon$ to obtain the form from [32],

$$\boxed{x_{t+1} = x_t + \varepsilon \sin(y_t),} \tag{10.310}$$

$$\boxed{y_{t+1} = y_t + x_{t+1}.} \tag{10.311}$$

Fig. 10.15 shows a Poincaré plot of the standard map for $\varepsilon = 2.4$ in Figs. (10.310) and (10.311). We have chosen 10 initial points; the number of iterations was 1000. It can be seen that for this relatively large control parameter ε a rather large stochastic region already exists. Embedded in the stochastic sea are islands, including a central one at $(0, \pi)$ and four significant ones surrounding the central island. Some regions with irrational winding numbers are indicated by solid lines.

Several interesting features of nonlinear dynamics, which are hidden in the standard map, are discussed in the literature; see, e.g., [32, 233]. ■

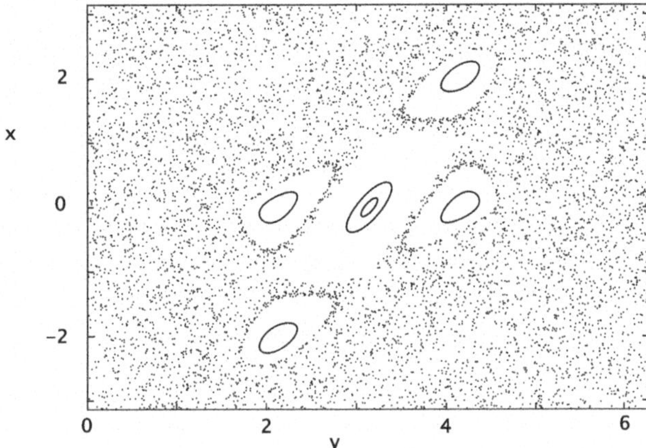

Fig. 10.15 Poincaré plot of the standard map (10.310) and (10.311) for $\varepsilon = 2.4$

In the present context, the divergence of field lines, the Lyapunov exponent, and the Kolmogorov length are of interest. The general theory of these tools can be found in standard works on nonlinear dynamics [234].

Example 10.18 (Kolmogorov Length for the Standard Map)
If, within the framework of the standard map model, we consider the deviations of two initially neighboring field lines in the stochastic region, a deviation formula is obtained

$$d\mathbf{I}_k = \begin{pmatrix} dx_k \\ dy_k \end{pmatrix}, \qquad d\mathbf{I}_{k+1} = J_k d\mathbf{I}_k, \qquad J_k = \begin{pmatrix} \frac{\partial x_{k+1}}{\partial x_k} & \frac{\partial x_{k+1}}{\partial y_k} \\ \\ \frac{\partial y_{k+1}}{\partial x_k} & \frac{\partial y_{k+1}}{\partial y_k} \end{pmatrix}, \tag{10.312}$$

which is determined by the Jacobian matrix J_k. For the determinant, we have $\det J_k = 1$. The eigenvalues $\lambda_1^{(k)}$ and $\lambda_2^{(k)}$ of the Jacobian matrix

$$\begin{vmatrix} \frac{\partial x_{k+1}}{\partial x_k} - \lambda_{1,2}^{(k)} & \frac{\partial x_{k+1}}{\partial y_k} \\ \\ \frac{\partial y_{k+1}}{\partial x_k} & \frac{\partial y_{k+1}}{\partial y_k} - \lambda_{1,2}^{(k)} \end{vmatrix} = 0\,, \qquad \lambda^{(k)} = \max\left(\lambda_1^{(k)}, \lambda_2^{(k)}\right), \tag{10.313}$$

determine the Lyapunov exponent. We can define a global Lyapunov exponent for unstable trajectories,

$$\boxed{\lambda = \lim_{N \to \infty} \frac{1}{N} \ln \prod_{k=1}^{N} \lambda^{(k)} > 0}\,. \tag{10.314}$$

From the latter, we can define a local e-folding length (Kolmogorov length):

$$L_K = \frac{1}{\lambda} \,. \tag{10.315}$$

In general, the Kolmogorov length is determined numerically, starting from the Lyapunov exponent λ. Various numerical methods have been proposed [235–237]. For the standard map, the Chirikov method [235] is very effective. Here, the differences between the field lines (x_t, y_t) and (x_t', y_t') are considered, with

$$\eta_t = x_t' - x_t \,, \quad \xi_t = y_t' - y_t \,. \tag{10.316}$$

We determine the differences at each step by iteration (see (10.310) and (10.311)). For neighboring field lines we use

$$\lim_{y_t' \to y_t} \left(\frac{\sin(y_t') - \sin(y_t)}{y_t' - y_t} \right) = \cos(y_t) \tag{10.317}$$

and obtain

$$\eta_{t+1} \approx \eta_t + \varepsilon \xi_t \cos(y_t) \,, \tag{10.318}$$

$$\xi_{t+1} \approx \xi_t + \eta_{t+1} \,. \tag{10.319}$$

The distance

$$\Delta r(t) = \sqrt{\eta_t^2 + \xi_t^2} \quad \text{with initial value } \Delta r(0) = \Delta r_0 \tag{10.320}$$

leads to the Lyapunov exponent via

$$\lambda = \lim_{t \to \infty} \frac{1}{t} \ln \left(\frac{|\Delta r(t)|}{|\Delta r_0|} \right) . \tag{10.321}$$

After a sufficient number of iterations of (10.318) and (10.319), the exponential divergence can be written as

$$\langle [\Delta r(t)]^2 \rangle \sim [\Delta r(0)]^2 \exp\left\{ \frac{2t}{L_K} \right\} . \tag{10.322}$$

However, Eqs. (10.318) and (10.319) are only valid for small values of η and ξ. Therefore, rescaling is required from time to time during the iteration. As a typical value, a total of 10^5 iterations is sufficient in most cases for determining the Lyapunov exponent. For relatively large control parameters ε, Chirikov [235] has calculated the analytical limit:

$$\boxed{L_K = \frac{1}{\ln\left(\frac{\varepsilon}{2}\right)}} \,. \tag{10.323}$$

■

We have just determined the Kolmogorov length (10.323) for the standard map in the fully stochastic parameter regime. This is a highly simplified model. In the following, we do not simplify to a discrete map, but instead choose a continuum description for the magnetic field trajectories.

We determine the separation of magnetic field lines in the quasilinear regime. Two different field lines start at different positions. The (longitudinal) coordinate ζ is used as a parameter to determine the (lateral) positions $\mathbf{x}_{\perp 1}$ and $\mathbf{x}_{\perp 2}$. The components Δx and Δy of the distance $\Delta \mathbf{x}_\perp = \mathbf{x}_{\perp 2} - \mathbf{x}_{\perp 1}$ obey the equations:

$$\frac{d\Delta x}{d\zeta} = b_x(\mathbf{x}_{\perp 2}(\zeta), \zeta) - b_x(\mathbf{x}_{\perp 1}(\zeta), \zeta) \,, \tag{10.324}$$

$$\frac{d\Delta y}{d\zeta} = b_y(\mathbf{x}_{\perp 2}(\zeta), \zeta) - b_y(\mathbf{x}_{\perp 1}(\zeta), \zeta) \,. \tag{10.325}$$

After linearization, it follows that

$$\boxed{\frac{d\Delta x}{d\zeta} \approx b_{x,x}(\mathbf{x}_\perp(\zeta), \zeta)\Delta x + b_{x,y}(\mathbf{x}_\perp(\zeta), \zeta)\Delta y \,,} \tag{10.326}$$

$$\boxed{\frac{d\Delta y}{d\zeta} \approx b_{y,x}y(\mathbf{x}_\perp(\zeta), \zeta)\Delta x + b_{y,y}(\mathbf{x}_\perp(\zeta), \zeta)\Delta y \,.} \tag{10.327}$$

The derivatives $b_{m,n}$ are functions of the coordinates. Due to the ζ-dependence of the coefficients, even the linearized equations for the separation do not possess *simple* exponential solutions. As already discussed, the derivatives $b_{m,n}$ are also random fields, and a statistical description is appropriate. From now on, we use (10.326) and (10.327) as the starting equations (where $\approx$ is replaced by $=$). Simple algebraic manipulations (using the notation $(\Delta x)^2 \equiv \Delta x^2$ and so on) lead to

$$\frac{d\Delta x^2}{d\zeta} = 2b_{x,x}(\mathbf{x}_\perp(\zeta), \zeta)\Delta x^2 + 2b_{x,y}(\mathbf{x}_\perp(\zeta), \zeta)\Delta x\Delta y \,, \tag{10.328}$$

$$\frac{d\Delta y^2}{d\zeta} = 2b_{y,x}(\mathbf{x}_\perp(\zeta), \zeta)\Delta x\Delta y + 2b_{y,y}(\mathbf{x}_\perp(\zeta), \zeta)\Delta y^2 \,, \tag{10.329}$$

$$\frac{d\Delta x\Delta y}{d\zeta} = b_{x,x}(\mathbf{x}_\perp(\zeta), \zeta)\Delta x\Delta y + b_{x,y}(\mathbf{x}_\perp(\zeta), \zeta)\Delta y^2 + b_{y,x}(\mathbf{x}_\perp(\zeta), \zeta)\Delta x\Delta y + b_{y,y}(\mathbf{x}_\perp(\zeta), \zeta)\Delta y^2 . \tag{10.330}$$

For an iterative, quasilinear solution, we formally integrate these equations over ζ, in order to replace the squares on the right-hand side by the formal solutions. After averaging, we obtain

$$\frac{d\langle\Delta x^2\rangle}{d\zeta} \approx 4\bar{L}^{xx}_{xx}\langle\Delta x^2\rangle + 4\bar{L}^{xy}_{xx}\langle\Delta x\Delta y\rangle + 2\bar{L}^{yx}_{xx}\langle\Delta x\Delta y\rangle + 2\bar{L}^{yy}_{xx}\langle\Delta y^2\rangle + 2\bar{L}^{yx}_{xy}\langle\Delta x^2\rangle + 2\bar{L}^{yy}_{xy}\langle\Delta x\Delta y\rangle , \tag{10.331}$$

and similar equations for $\frac{d\langle\Delta y^2\rangle}{d\zeta}$ and $\frac{d\langle\Delta x\Delta y\rangle}{d\zeta}$. The coefficients

$$\bar{L}^{\alpha\beta}_{mn} = \int_0^\infty d\zeta\, L^{\alpha\beta}_{mn}(\zeta) \tag{10.332}$$

follow by integrating the correlation function for the derivatives of the magnetic field fluctuations. Implicitly, we have assumed that

$$\lambda_\parallel \ll L_K , \tag{10.333}$$

where L_K is the characteristic (exponentiation) length for the mean squares. Making use of the properties of the correlation functions$L^{\alpha\beta}_{mn}(\zeta)$ and by introducing

$$\mathcal{K} = \int_0^\infty K(\zeta)d\zeta \equiv \frac{1}{4L_K} \tag{10.334}$$

we obtain

$$\boxed{\frac{d\langle\Delta x^2\rangle}{d\zeta} = 2\mathcal{K}\langle\Delta x^2\rangle + 6\mathcal{K}\langle\Delta y^2\rangle ,} \tag{10.335}$$

$$\boxed{\frac{d\langle\Delta y^2\rangle}{d\zeta} = 6\mathcal{K}\langle\Delta x^2\rangle + 2\mathcal{K}\langle\Delta y^2\rangle ,} \tag{10.336}$$

$$\boxed{\frac{d\langle\Delta x\Delta y\rangle}{d\zeta} = -4\mathcal{K}\langle\Delta x\Delta y\rangle .} \tag{10.337}$$

The solution of this linear system of first-order ordinary differential equations with constant coefficients is straightforward. The analysis shows that two eigenvalues $-L_K^{-1}$ are negative (and degenerate), while one is positive, leading to exponential growth,

$$\langle \Delta x^2 \rangle \sim \langle \Delta y^2 \rangle \sim \exp\left[2\frac{\zeta}{L_K}\right] ; \tag{10.338}$$

L_K is the exponentiation length (Kolmogorov length). In general, it is not easy to determine. In the limiting case of very large perpendicular correlation lengths $\lambda_\perp$, with

$$K(\zeta) = \beta^2 e^{-\zeta^2/2\lambda_\parallel^2} \frac{\lambda_\perp^4}{\left[\lambda_\perp^2 + 2\int_0^\zeta d\zeta'(\zeta - \zeta')L(\zeta')\right]^3} \approx \frac{\beta^2}{\lambda_\perp^2} e^{-\zeta^2/2\lambda_\parallel^2} , \tag{10.339}$$

integration over ζ leads to

$$\boxed{L_K \approx \sqrt{\frac{2}{\pi} \frac{\lambda_\perp^2}{4\beta^2 \lambda_\parallel}}} . \tag{10.340}$$

The condition (10.333) determines the range of applicability

$$4\sqrt{\frac{\pi}{2}} \beta^2 \frac{\lambda_\parallel^2}{\lambda_\perp^2} \ll 1 , \tag{10.341}$$

which is satisfied for small Kubo numbers.

10.6 Stochastic Particle Transport

The following subsections present heuristic theories of test particle diffusion in defined stochastic magnetic fields. We distinguish between the spatial diffusion of magnetic field lines, characterized by a magnetic diffusion coefficient, and the motion of particles along as well as the decorrelation of particles from a given line, for example due to collisions.
We begin with negligible binary collisions and then discuss models in partial regimes with increasing collisional effects. The results can be used to estimate the thermal electron conductivity in defined, static, chaotic tokamak edge regions. Ion transport and ambipolarity are not addressed. This is the reason why the models are only applicable to thermal electron conductivity, even though we refer to the expressions as particle diffusion coefficients.

In a brief historical overview, we summarize some pioneering works to which we partly refer. Jokipii [146] and Rosenbluth et al. [238] discussed the diffusive motion of magnetic field lines in perturbed systems. Stix [239] was probably the first to apply the ideas of magnetic turbulence to tokamaks [240]. Jokipii and Parker [149] emphasized the importance of magnetic stochasticity in astrophysical transport problems. Rechester and Rosenbluth [174] calculated electron heat transport in a tokamak with destroyed magnetic surfaces. At the same time, Kadomtsev and Pogutse [175] derived a diffusion coefficient for stochastic plasmas with strong collisions. Following these pioneering works, many authors, e.g., [176, 178, 191, 197–199, 241, 242] and many others, further developed the theory, although to this day no complete, consistent theory of transport in stochastic plasmas is available.

The starting point for a simple but systematic stochastic theory of the perpendicular transport of electrons and ions is stochastic differential equations. The details of this method have been extensively presented in the excellent monograph [144] by Balescu, and for this reason we do not repeat these calculations here. The following remarks are merely intended to give an impression of how this type of transport theory works.

Balescu and collaborators [144] made significant contributions to the progress in the field of stochastic transport by using the so-called V-Langevin equations:

$$\boxed{\frac{dx_p(t)}{dt} = b_x[x_p(t), y_p(t), z_p(t)]\frac{dz_p(t)}{dt} + \eta_{\perp x}(t) ,} \tag{10.342}$$

$$\boxed{\frac{dy_p(t)}{dt} = b_y[x_p(t), y_p(t), z_p(t)]\frac{dz_p(t)}{dt} + \eta_{\perp y}(t) ,} \tag{10.343}$$

$$\boxed{\frac{dz_p(t)}{dt} = \eta_{\parallel}(t) .} \tag{10.344}$$

The V-Langevin equations use the guiding center approach for small Larmor radii of the particles. In the absence of collisions ($\eta_\perp = \eta_\parallel = 0$), the particle positions $\mathbf{r}_p(t)$ follow the magnetic field lines. Collisions lead to a diffusive motion along the zeroth-order magnetic field (parallel direction) as well as to deviations from the perturbed magnetic field lines in the perpendicular direction.
The V-Langevin equations are simplifications of the A-Langevin equation (Acceleration Langevin Equation)

$$\frac{d\mathbf{v}}{dt} = \frac{Ze}{m}\mathbf{v} \times \mathbf{B} - \nu\mathbf{v} + \mathbf{a} , \tag{10.345}$$

which must be used when finite Larmor radii become important.

The V-Langevin equations allow for simple estimates of particle diffusion. Using the Taylor-Green-Kubo formula [243–245] in 3D

$$D = \frac{1}{3} \int_0^\infty d\tau \langle \mathbf{u}[\mathbf{x}(\tau), \tau] \cdot \mathbf{u}[0, 0] \rangle \ , \tag{10.346}$$

we obtain for the magnetic diffusion coefficient in the x-direction the estimate already used in the evaluation of the quasilinear expression, namely

$$\boxed{D_{m(agnetic)} \sim \int_0^\infty d\zeta \langle b_x[\mathbf{x}_\perp(\zeta), \zeta] b_x[\mathbf{x}_\perp(0), 0] \rangle \sim b^2 L_{corr}} \ . \tag{10.347}$$

It should be noted that the Taylor-Green-Kubo formula contains the Lagrangian correlation function,

$$L_{rs}[\zeta] = \langle b_r[\mathbf{x}_\perp(\zeta), \zeta] b_s[\mathbf{x}_\perp(0), 0] \rangle \ , \tag{10.348}$$

and not the (simpler) Eulerian correlation function. Reasonable assumptions for the latter were discussed in the previous section.

The relationship between the Lagrangian correlation function and the Eulerian correlation function is a difficult problem that in principle requires knowledge of the exact dynamics. All discussions on the following pages focus on this problem. Drawing on arguments from fluid dynamics, the Corrsin approximation [190] is often applied, in which the exact propagator is replaced by the averaged propagator. The averaged propagator can be further evaluated by using a cumulant expansion [168].

Within the Corrsin approximation, Balescu and collaborators were able to derive an integral equation for the Lagrangian form of the correlation function. The solution of the integral equation leads, for example, to

$$D_m \approx \begin{cases} b^2 \lambda_\parallel & \text{for } \lambda_\perp \to \infty \\ b \lambda_\perp & \text{for } \lambda_\parallel \to \infty \end{cases} \ . \tag{10.349}$$

for the magnetic diffusion process. These expressions, as well as more general results on particle diffusion perpendicular to a strong external magnetic field, will be discussed below on the basis of heuristic arguments.

Perpendicular Particle Diffusion

Using the quasilinear magnetic diffusion coefficient, we can apply random walk estimates for the particle diffusion coefficient. In the collisionless case, we obtain the particle diffusion coefficient as a function of the magnetic (field line) diffusion coefficient $D_m \equiv D_{m(agnetic)}$ in the form

$$\boxed{D_{\perp particle} \sim \frac{\langle(\Delta x)^2\rangle}{\Delta\tau} \approx \frac{D_{m(agnetic)}l}{l/v_{th}} = D_{m(agnetic)}\, v_{th} \approx b^2\lambda_{\parallel} v_{th}} \ . \tag{10.350}$$

Example 10.19 (Particle diffusion coefficient for $\lambda_\perp \to \infty$)
This collisionless quasilinear result, which occurs for large (infinite) transverse correlation lengths ($\lambda_\perp \to \infty$), is based on the following picture. For a particle that travels a distance $l \approx \lambda_\parallel$ along a magnetic field line with a small lateral perturbation δB compared to the zeroth-order field B_0, we obtain from the equations of motion of the field lines for the transverse displacement Δr

$$\frac{\Delta r}{\delta B} \approx \frac{\lambda_\parallel}{B_0} \ . \tag{10.351}$$

If the particle has an approximately constant, typical velocity $v \approx v_{th}$, the time step is estimated by the following equation:

$$\Delta\tau \approx \frac{\lambda_\parallel}{v_{th}} \ . \tag{10.352}$$

In the random walk approximation, this leads to

$$D_{\perp particle} \sim \frac{(\Delta r)^2}{\Delta\tau} \approx v_{th}\left(\frac{\delta B}{B_0}\right)^2 \lambda_\parallel \sim v_{th} D_m, \tag{10.353}$$

as already summarized in (10.350). ■

Collisions affect the quasilinear result. For weak collisions ($\lambda_{coll} \gg \lambda_\parallel$) and large transverse correlation lengths $\lambda_\perp \to \infty$, the result will approximately correspond to the collisionless case, i.e., (10.350).

In the next step, we explicitly take into account finite transverse correlation lengths and low collision rates. We begin with diffusion caused by the stochasticity of the magnetic field lines, as described by the formula for magnetic cross-field diffusion after a distance $L_\parallel$:

$$(\Delta r)^2 \approx 2 D_m L_\parallel \ . \tag{10.354}$$

Parallel to the strong magnetic field line with $L_\parallel$, decorrelation is achieved. The decorrelation length $L_\parallel$ should be greater than the Kolmogorov length L_K. Then, a small initial cross-field displacement is significantly amplified. The collision-induced motion along the magnetic field lines links the characteristic time $\Delta\tau$ with the (parallel) length $L_\parallel$ in the form

$$L_\parallel^2 \approx 2\chi_\parallel \Delta\tau \ . \tag{10.355}$$

The time step $\Delta\tau$ is the decorrelation time during which the perpendicular deviation grows from 0 to $\lambda_\perp$. By combining (10.354) with (10.355), we obtain the perpendicular particle diffusion coefficient

$$\boxed{D_\perp = \frac{1}{2}\frac{(\Delta r)^2}{\Delta\tau} \approx \frac{D_m\sqrt{2\chi_\parallel}}{\sqrt{\Delta\tau}}}\,, \tag{10.356}$$

which will serve as the starting point for the following estimates.

When we apply the magnetic diffusion coefficient D_m, we are in the limiting case

$$\frac{\lambda_\parallel}{\lambda_\perp} \ll 1\,. \tag{10.357}$$

Furthermore, we assume that the (classical) parallel diffusion coefficient is much greater than the perpendicular one,

$$\frac{\chi_\perp}{\chi_\parallel} \ll 1\,, \quad \leftrightarrow \quad \frac{\nu_{coll}}{\Omega} \ll 1\,. \tag{10.358}$$

For the characteristic lengths, we require [144]

$$\lambda_{coll} \ll L_K \lesssim L_\parallel\,. \tag{10.359}$$

Detailed theories [192] further show that

$$\lambda_\parallel < L_K \tag{10.360}$$

is required for the Markovian approximation when evaluating integrals over correlation functions.

Within the Kadomtsev-Pogutse model [175], the time step $\Delta\tau \equiv \Delta\tau_{KP}$ is determined by

$$\lambda_\perp^2 \approx 2\chi_\perp\Delta\tau_{KP}\,. \tag{10.361}$$

In Fig. 10.16, this situation is illustrated schematically.

Collision broadening occurs around a magnetic field line. When the lateral displacement reaches the order of the perpendicular correlation length, the particle becomes decorrelated from the magnetic field line. This defines $L_\parallel$. Due to strong field line wandering, the mean displacement Δr can be larger than the perpendicular correlation length. In this regime, one obtains the perpendicular particle diffusion coefficient

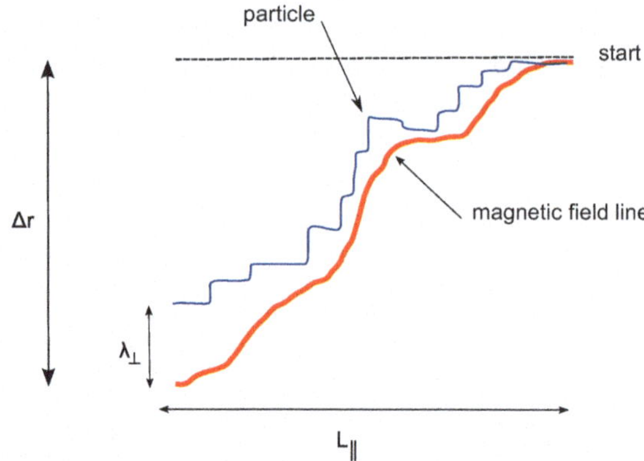

Fig. 10.16 Sketch of the physical conditions for the validity of the Kadomtsev-Pogutse diffusion coefficient

$$D_{KP} \approx \frac{2D_m\sqrt{\chi_\parallel \chi_\perp}}{\lambda_\perp} \approx \frac{D_m v_{th}\rho}{\lambda_\perp}\ , \tag{10.362}$$

i.e., the so-called Kadomtsev-Pogutse diffusion coefficient, which in this form is independent of the collision frequency ν_{coll}.

As we will see, it can be regarded as the limiting case of the Rechester-Rosenbluth diffusion coefficient D_{RR} (see below) for large collision frequencies.

The present form of the Kadomtsev-Pogutse diffusion coefficient should not be confused with $D_{KP}^{(II)} \approx b\lambda_\perp v_{th}$, which arises from the second form of (10.362) and is sometimes also referred to as the Kadomtsev-Pogutse diffusion coefficient. The form $D_{KP}^{(II)}$ was derived in the percolation limit without taking particle trapping into account. Therefore, it is not correct.

In the present case, the decorrelation length has the form

$$L_\parallel \approx \lambda_\perp \sqrt{\frac{\chi_\parallel}{\chi_\perp}} \equiv L_{KP}\ ; \tag{10.363}$$

L_{KP} is called the characteristic length according to Kadomtsev-Pogutse. We note the order of magnitude

$$L_{KP} \approx \frac{\Omega}{\nu_{coll}} \lambda_\perp \gg \lambda_\perp\ . \tag{10.364}$$

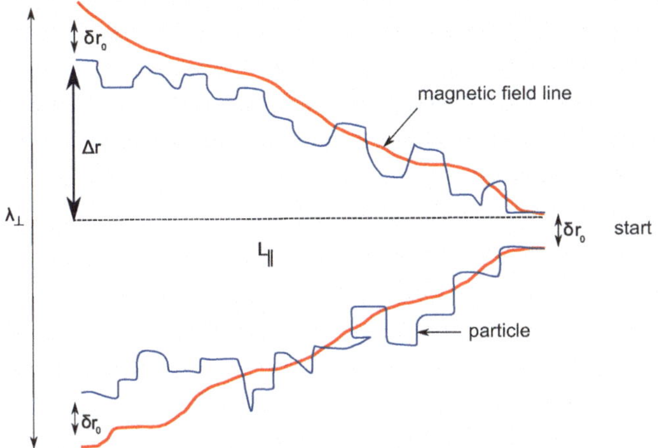

Fig. 10.17 Sketch of the scenario for the Rechester-Rosenbluth diffusion coefficient

The determination of the time step $\Delta\tau \equiv \Delta\tau_{RR}$ within the Rechester-Rosenbluth model [174] is somewhat more complex, since it is now assumed that the main decorrelation occurs due to the exponential divergence of neighboring field lines, as sketched in Fig. 10.17.

The condition that the decorrelation mechanism due to the chaotic field dominates over cross-field displacement by collisions is

$$L_{\|} < L_{KP} \ . \tag{10.365}$$

The decorrelation due to exponential field line separation (with the Kolmogorov length L_K) ends when the perpendicular distance reaches the perpendicular correlation length,

$$\lambda_{\perp} \approx \delta r_0 \exp\left\{\frac{L_{\|}}{L_K}\right\} . \tag{10.366}$$

At that point, the particle has traversed a longitudinal distance $L_{\|}$ during the time $\Delta\tau_{RR}$; both quantities are approximately related by (10.355). The initial width δr_0 is determined self-consistently by taking into account that the repeating process (see Fig. 10.17) requires

$$\delta r_0^2 \approx 2\chi_{\perp}\Delta\tau_{RR} \ . \tag{10.367}$$

Here it is assumed that the collisional broadening distributes different particles onto different field lines. By combining the last two equations together with (10.355), an equation for $\Delta\tau_{RR}$ is obtained, namely

$$\lambda_\perp \approx \sqrt{2\chi_\perp}\sqrt{\Delta\tau_{RR}}\exp\left\{\frac{\sqrt{2\chi_\parallel}\sqrt{\Delta\tau_{RR}}}{L_K}\right\}. \quad (10.368)$$

This guarantees self-consistency on the one hand, but on the other hand cannot be solved explicitly for $\Delta\tau_{RR}$. We assume that $\lambda_\perp$, $\chi_\perp$, $\chi_\parallel$ and L_K are either known or can be easily determined for a specific situation.

After introducing the new variable μ via [232]

$$\sqrt{\frac{\chi_\parallel}{\Delta\tau_{RR}}} = \frac{1}{\sqrt{2}}\frac{v_{th}}{\sqrt{\mu}}, \quad (10.369)$$

we can rewrite (10.368) as:

$$L_K \ln\left[\frac{\lambda_\perp}{\rho\sqrt{\mu}}\right] \approx \lambda_{coll}\sqrt{\mu}. \quad (10.370)$$

This is an equation for the variable μ.

Its solution determines the Rechester-Rosenbluth diffusion coefficient in the form

$$D_{RR} \approx \frac{D_m v_{th}}{\sqrt{\mu}}. \quad (10.371)$$

Example 10.20 (Rechester-Rosenbluth Diffusion Coefficient)
Other reformulations of the Rechester-Rosenbluth formula may be useful. In its original form [174], it was written as

$$D_{RR} \approx \frac{2D_m\chi_\parallel}{L_K \ln\left(\frac{\lambda_\perp}{L_K}\sqrt{\frac{\chi_\parallel}{\chi_\perp}}\right)} \sim \frac{D_m\chi_\parallel}{L_K}, \quad (10.372)$$

which follows from Eq. (10.356) and the iterative solution

$$\sqrt{\Delta\tau_{RR}} \approx \frac{L_K}{\sqrt{2\chi_\parallel}} \ln\sqrt{\frac{\lambda_\perp}{\sqrt{2\chi_\perp\Delta\tau_{RR}}}} \approx \frac{L_K}{\sqrt{2\chi_\parallel}} \ln\sqrt{\frac{\lambda_\perp\sqrt{2\chi_\parallel}}{\sqrt{2\chi_\perp}L_K}} \quad (10.373)$$

With (10.355) we obtain

$$L_\parallel \gg L_K, \quad (10.374)$$

however, for estimates, the somewhat questionable approximation $L_\parallel \sim \mathcal{O}(L_K)$ is often used. ■

Now we introduce the Kubo number. In general, the Kubo number is the ratio between the distance traveled ℓ during an autocorrelation time and the correlation length ℓ_{corr},

$$K \approx \frac{\ell}{\ell_{corr}} \,. \tag{10.375}$$

Let us apply this definition to the deviations in the perpendicular direction. Propagating in the parallel direction over the distance $\lambda_{\|}$, for $b \ll 1$ the perpendicular displacement is $b\lambda_{\|}$, which we should compare with $\lambda_{\perp}$. Therefore, we set $\ell \approx b\lambda_{\|}$ and $\ell_{corr} \approx \lambda_{\perp}$, and obtain

$$\boxed{K \approx \frac{b\lambda_{\|}}{\lambda_{\perp}}} \,. \tag{10.376}$$

This leads to the scaling

$$D_{RR} \sim \chi_{\|}\, b^2\, K^2 \,, \tag{10.377}$$

i.e., the diffusion coefficient increases with the square of the Kubo number.

If a field line (or a particle) is trapped, it remains stuck to an island in the Poincaré section. With island-like structures in phase space, the system is not fully stochastic. Nevertheless, for motion along a trapped field line, the distance traveled along the field line [unless it propagates into the (perpendicular) uncorrelated region at a perpendicular distance $\lambda_{\perp}$] is very large, so that $K \gg 1$. On the other hand, if a field line does not explore an uncorrelated perpendicular region as it advances a parallel distance $\lambda_{\|}$, then $K \approx \frac{\ell}{\lambda_{\perp}} \approx \frac{b\lambda_{\|}}{\lambda_{\perp}} \ll 1$ holds.

For the Kadomtsev-Pogutse formula, we obtain the scaling

$$D_{KP} \approx \frac{b^2\lambda_{\|}}{\lambda_{\perp}} \sqrt{\chi_{\|}\chi_{\perp}} \sim b\, K\, D_{Bohm} \,, \tag{10.378}$$

where the Bohm diffusion coefficient

$$D_{Bohm} \approx \frac{1}{16} \sqrt{\chi_{\|}\chi_{\perp}} \sim \frac{T}{eB} \tag{10.379}$$

has been introduced. We observe the linear dependence on the Kubo number.

We will return to the role of the Kubo number after a brief comparison of the two diffusion coefficients.

By introducing the time step in the Kadomtsev-Pogutse model from (10.361),

$$\Delta\tau_{KP} \approx \frac{\lambda_\perp^2}{2\chi_\perp}\,, \tag{10.380}$$

we find from (10.368) for the decorrelation time $\Delta\tau_{RR}$ the implicit equation

$$\sqrt{\Delta\tau_{KP}} \approx \sqrt{\Delta\tau_{RR}}\exp\left[\frac{\sqrt{\Delta\tau_{RR}}\sqrt{2\chi_\parallel}}{L_K}\right] \approx \sqrt{\Delta\tau_{RR}}\exp\left[\frac{\sqrt{\Delta\tau_{RR}}v_{th}}{\sqrt{v_{coll}}L_K}\right] \tag{10.381}$$

From this, two conclusions follow. First, it always holds that

$$\Delta\tau_{RR} \leq \Delta\tau_{KP} \tag{10.382}$$

and therefore

$$\boxed{D_{RR} \geq D_{KP}}\,, \tag{10.383}$$

i.e., the Rechester-Rosenbluth coefficient is always (as long as the model assumptions hold) the dominant transport process. Second, we have the limit

$$\boxed{D_{RR} \to D_{KP} \quad \text{for} \quad v_{coll} \to \infty}\,; \tag{10.384}$$

i.e., D_{KP} is the lower bound for D_{RR} at high collision frequencies.

In Fig. 10.18 we compare the Kadomtsev-Pogutse and Rechester-Rosenbluth diffusion formulas. Several parameters of the stochastic system are required for the evaluations.

Since we have not explicitly introduced any other system so far, we take them from the standard mapping. This means that we identify D_m with the diffusion coefficient

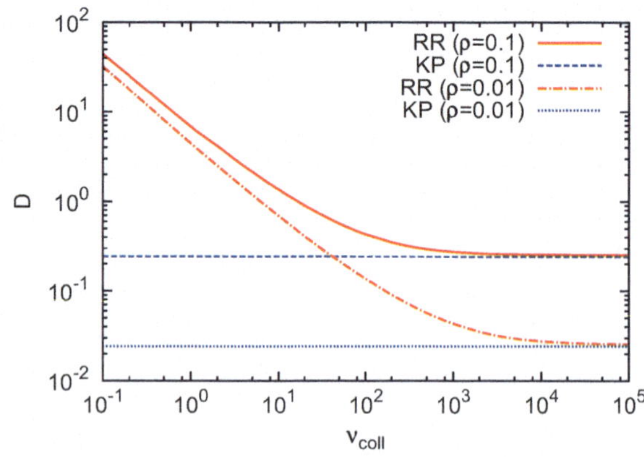

Fig. 10.18 Diffusion coefficients as a function of collision frequency for the stochastic parameter $\varepsilon = 10$. For illustration, two different Larmor radii, $\rho = 0.1$ and $\rho = 0.01$, were chosen. The limiting process $D_{RR} \to D_{KP}$ is shown for $v_{coll} \to \infty$

of the standard mapping (10.310) and (10.311) for $\varepsilon = 10$. The Kolmogorov length is also numerically evaluated from the standard mapping for $\varepsilon = 10$. The other parameters are $\lambda_\perp = 2\pi$ and $v_{th} = 1$. For illustration, two different Larmor radii, $\rho = 0.1$ and $\rho = 0.01$, were chosen. For smaller Larmor radii, higher collision frequencies are required so that $D_{KP} \sim \mathcal{O}(D_{RR})$. Furthermore, the following holds: The smaller ρ, the smaller the (perpendicular) diffusion coefficients. The numerical results confirm the predictions (10.383) and (10.384).

Example 10.21 (Kadomtsev-Pogutse Regime)
The parameter values at which the Kadomtsev-Pogutse regime begins can be estimated as follows. For D_{KP} we need for the decorrelation time

$$\Delta\tau_{KP} \approx \Delta\tau_{RR} \quad \leftrightarrow \quad \frac{\lambda_\perp^2}{2\chi_\perp} \gtrsim \frac{L_K^2}{2\chi_\parallel} \,. \tag{10.385}$$

By using [192]

$$L_K \approx \sqrt{\frac{2}{\pi}} \frac{\lambda_\perp^2}{4\beta^2 \lambda_\parallel} \tag{10.386}$$

we can rewrite the condition as follows:

$$\boxed{\frac{\nu_{coll}}{\Omega} \gtrsim \beta K} \,, \tag{10.387}$$

where K is still the Kubo number. ■

Example 10.22 (Collision-Dominated Fluid Regime)
Another collision-dominated case has been discussed in the literature [176, 239–242]. In the so-called collision-dominated fluid limit for strong collisions, it is assumed that a fluid element is transported along the actual magnetic field line, which, based purely on geometric considerations, leads to the following result:

$$\Delta x \sim b \Delta z. \tag{10.388}$$

Therefore,

$$D_{\perp particle} \sim \frac{\langle (\Delta x)^2 \rangle}{\Delta\tau} \sim b^2 \frac{\langle (\Delta z)^2 \rangle}{\Delta\tau} \sim b^2 \frac{D^{classical}_{\parallel particle} \Delta\tau}{\Delta\tau} \sim b^2 \chi_\parallel \,. \tag{10.389}$$

the classical (collision-dominated) diffusion coefficients

$$D^{classical}_{\perp particle} \equiv \chi_\perp = \frac{1}{2} \rho^2 \nu_{coll} \,, \quad D^{classical}_{\parallel particle} \equiv \chi_\parallel = \frac{1}{2} \lambda^2_{coll} \nu_{coll} \tag{10.390}$$

are used, with the Larmor radius $\rho = \frac{v_{th}}{\Omega}$ and the mean free path $\lambda_{coll} \approx \frac{v_{th}}{\nu_{coll}}$. The result (10.389) also follows from the perspective that, in the case of strong collisions, particles moving along stochastic field lines are effectively decorrelated from the field lines after a distance of $\lambda_{coll} \approx v_{th}/\nu_{coll}$:

$$\boxed{D_{\perp particle} \sim \frac{\langle(\Delta x)^2\rangle}{\Delta\tau} \sim \frac{b^2\lambda_{coll}\lambda_{coll}}{\tau_{coll}} \sim b^2\lambda_{coll}^2\nu_{coll} \sim b^2\, D_{\| particle}^{classical}} \,. \tag{10.391}$$

■

For strong collisions $\lambda_{coll} \ll \lambda_\|$ and negligible Larmor radii (i.e., when no significant perpendicular displacements occur apart from the divergence of the magnetic field lines), the (collision-dominated) diffusive motion along the main magnetic field (z-direction) leads to a broadening

$$z^2 \approx 2\chi_\| t \tag{10.392}$$

during the time $t \hat{=} \Delta\tau$. If the collisional broadening in the perpendicular direction is neglected, we obtain the perpendicular motion solely due to the progression of the field lines, i.e., for a (parallel) length z, the (quadratic) perpendicular deviation is

$$(\Delta r)^2 \approx 2D_m z \approx 2D_m\sqrt{2\chi_\|}\sqrt{t}\,. \tag{10.393}$$

An estimate using the random walk approximation yields

$$\frac{(\Delta r)^2}{t} \sim D_{RR}^{sub}(t) \sim \frac{D_m\sqrt{\chi_\|}}{\sqrt{t}} \to 0 \quad \text{for} \quad t \to \infty\,. \tag{10.394}$$

This is the famous subdiffusive behavior in the strongly (infinitely) magnetized case ($\rho = 0$), which was first formulated by Rechester and Rosenbluth [174].

In general, however, the deviations of the particles from the magnetic field lines are important. The prediction of the subdiffusive nature is well founded. However, the value 1/2 for the exponent ν in $(\Delta r)^2 \sim t^\nu$ is not without reservations, because it is based on rather rough arguments that we have used for demonstration purposes.

Parallel Test Particle Diffusion Coefficient

The motion of particles along a zeroth-order magnetic field is also influenced by fluctuations of the magnetic field in the perpendicular direction. The estimation of the longitudinal diffusion coefficient (and the corresponding parallel mean free path) is the subject of the following discussion.

We calculate the particle diffusion in the direction of a strong external magnetic field. This physical process is abbreviated as "parallel diffusion." In principle, the same methods as in the previous section can be applied to this problem. However, it has been shown that more elegant methods are available for parallel diffusion. These will be used in the following.

Relation to Pitch Angle Diffusion

We begin with the single-particle distribution function $f = f(z, \mu, t)$, where μ is the cosine of the pitch angle and z is the spatial coordinate along the external magnetic field. Isotropy is assumed in the plane perpendicular to the external magnetic field. Due to the symmetry of the problem, the distribution function f depends only on the coordinate z, the time t, the velocity v, and the pitch angle μ. The variable v can be hidden, as it does not change during interaction with magnetic fluctuations. Therefore, we formally write $f = f(z, \mu, t)$. The kinetic equation for f is a two-dimensional Fokker-Planck equation (see below), which can be derived from the relativistic Vlasov equation [160] by neglecting momentum diffusion due to the assumption of purely magnetic fluctuations. From f we obtain the pitch-angle averaged particle density

$$\boxed{M(z, t) = \frac{1}{2} \int_{-1}^{+1} d\mu f} \, . \tag{10.395}$$

The normalization is given by

$$\int_{-\infty}^{+\infty} dz \, M(z, t) = 1 \, . \tag{10.396}$$

Then the particle current density

$$j_{\parallel} = \frac{1}{2} \int_{-1}^{+1} d\mu v \mu f \tag{10.397}$$

can be calculated, where $\parallel$ denotes the direction of the external magnetic field $\mathbf{B} \approx B_0 \hat{z}$. Here, $\mathbf{v}$ stands for the velocity of the particle, the parallel velocity component is $v_{\parallel}$. Obviously, for the parallel velocity component we have

$$v_{\parallel} = v\mu \ , \quad \boxed{\mu = \cos(\vartheta)} \ , \tag{10.398}$$

where $\vartheta = \sphericalangle(\mathbf{v}, \mathbf{B})$ is the angle between $\mathbf{v}$ and $\mathbf{B}$. If only magnetic fields act on the particle, the velocity v remains unchanged, although the direction of $\mathbf{v}$ changes.

The total magnetic field consists of the large background field $\mathbf{B}_0$ and the fluctuations,

$$\mathbf{B} = B_0\hat{z} + \delta\mathbf{B} \widehat{=} B_0(b_0\hat{z} + \mathbf{b}) \ , \quad b_0 \equiv 1 \ . \tag{10.399}$$

A simple indexing in the form $b_\nu, \ \nu = 0, 1, 2, 3$ can be used, where $b_x = b_1, b_y = b_2, b_z = b_3$ applies. For strong background fields, we can neglect $b_3 \ll b_0$. In principle, we can set $b_0 = 1$; however, we sometimes retain b_0 to make some formulas clearer. The dimension of the magnetic field is always contained in B_0, and we should keep in mind that $\delta\mathbf{B} = B_0\mathbf{b}$ holds. The non-relativistic equations of motion are

$$\boxed{\dot{v}_x = \Omega v_y + \Omega\left(v_y\frac{\delta B_z}{B_0} - v_z\frac{\delta B_y}{B_0}\right) ,} \tag{10.400}$$

$$\boxed{\dot{v}_y = -\Omega v_x + \Omega\left(v_z\frac{\delta B_x}{B_0} - v_x\frac{\delta B_z}{B_0}\right) ,} \tag{10.401}$$

$$\boxed{\dot{v}_z = \Omega\left(v_x\frac{\delta B_y}{B_0} - v_y\frac{\delta B_x}{B_0}\right) ,} \tag{10.402}$$

with the gyrofrequency $\Omega = \frac{qB_0}{m}$. q is the charge and m the (rest) mass of the particle. Eq. (10.400) can be written using the Larmor radius $\rho_L = \frac{v}{\Omega}$ as

$$\dot{\mu} = \frac{1}{\rho_L}\left(v_x\frac{\delta B_y}{B_0} - v_y\frac{\delta B_x}{B_0}\right) . \tag{10.403}$$

Now we come to the calculation of the pitch-angle diffusion coefficient

$$\boxed{D_{\mu\mu} = \lim_{t\to\infty}\int_0^t dt' \langle\dot{\mu}(t')\dot{\mu}(0)\rangle} \ . \tag{10.404}$$

With a Fokker-Planck description, the change in the distribution function $f(z, \mu, t)$ results from

$$\boxed{\frac{\partial f}{\partial t} + v\mu\frac{\partial f}{\partial z} = \frac{\partial}{\partial \mu}\left(D_{\mu\mu}\frac{\partial f}{\partial \mu}\right)} \ . \tag{10.405}$$

The following derivation will show that (quasilinear) Fokker-Planck scattering in velocity space leads to the diffusion approximation in configuration space.

We have already mentioned that the magnetic fluctuations do not change the magnitude of the particle velocity. However, they can lead to complete isotropization over long timescales:

$$t \to \infty : \quad f \to M(z,t) = \frac{1}{2}\int_{-1}^{+1} d\mu f \ . \tag{10.406}$$

From the Fokker-Planck Eq. (10.405) we obtain

$$\boxed{\frac{\partial M}{\partial t} + \frac{\partial j_{\|}}{\partial z} = 0} \ . \tag{10.407}$$

With (10.405) we find

$$\frac{\partial f}{\partial \mu} = \frac{1}{D_{\mu\mu}}\int_{-1}^{\mu} d\nu \left[\frac{\partial f}{\partial t} + v\nu\frac{\partial f}{\partial z}\right] , \tag{10.408}$$

which can be inserted into the equation for the current density,

$$j_{\|} \equiv -\frac{v}{4}\int_{-1}^{+1} d\mu \frac{\partial(1-\mu^2)}{\partial \mu} f = \frac{v}{4}\int_{-1}^{+1} d\mu (1-\mu^2)\frac{\partial f}{\partial \mu} \ . \tag{10.409}$$

Note that $D_{\mu\mu}(\mu = \pm 1) = 0$ due to (10.404) and $\dot{\mu} \sim v_x$ (or v_y) $\sim v\sqrt{1-\mu^2}$. By substituting (10.408) into (10.409) we obtain

$$j_{\|} = \frac{v}{4}\int_{-1}^{+1} d\mu \frac{1-\mu^2}{D_{\mu\mu}}\int_{-1}^{\mu} d\nu \frac{\partial f}{\partial t} + \frac{v^2}{4}\int_{-1}^{+1} d\mu \frac{1-\mu^2}{D_{\mu\mu}}\int_{-1}^{\mu} d\nu\nu \frac{\partial f}{\partial z} \ . \tag{10.410}$$

For large times t we assume that $f \to M$, and therefore, asymptotically, by replacing f with M, we obtain

$$\begin{aligned} j_{\|} &\approx \frac{v}{4}\frac{\partial M}{\partial t}\int_{-1}^{+1} d\mu \frac{(1-\mu^2)(1+\mu)}{D_{\mu\mu}} - \frac{v^2}{8}\frac{\partial M}{\partial z}\int_{-1}^{+1} d\mu \frac{(1-\mu^2)^2}{D_{\mu\mu}} \\ &\equiv \chi_{zt}\frac{\partial M}{\partial t} - \chi_{zz}\frac{\partial M}{\partial z} \ . \end{aligned} \tag{10.411}$$

We have defined two coefficients χ_{zt} and χ_{zz}. If we compare the two terms on the right-hand side of (10.411), we can argue that the second term dominates. To justify this, we rewrite (10.411) as

$$j_\parallel \approx \chi_{zt}\frac{\partial M}{\partial t} - \chi_{zz}\frac{\partial M}{\partial z} = -\frac{\partial}{\partial z}\left[\chi_{zt} j_\parallel + \chi_{zz} M\right]. \tag{10.412}$$

Because of the tendency toward isotropization ($f \to M$), for large times the following should hold

$$t \to \infty: \quad j_\parallel = \frac{v}{2}\int_{-1}^{+1} d\mu \mu f \to \frac{Mv}{2}\int_{-1}^{+1} \mu d\mu = 0. \tag{10.413}$$

The smaller $j_\parallel$ is, the less important the first term in (10.411) becomes. Therefore, (after a sufficiently long waiting time) we can approximate

$$\boxed{j_\parallel(z,t) \approx -\chi_{zz}\frac{\partial M(z,t)}{\partial z}}. \tag{10.414}$$

We have arrived at the famous diffusion equation. The diffusion coefficient is

$$\chi_{zz} = \frac{v^2}{8}\int_{-1}^{+1} \frac{\left(1-\mu^2\right)^2}{D_{\mu\mu}} d\mu. \tag{10.415}$$

This is an important relationship between the diffusion coefficient in the parallel direction and the pitch-angle diffusion coefficient $D_{\mu\mu}$.

By substituting (10.403) into the definition of $D_{\mu\mu}$, we can decompose the overall result into the following contributions:

$$\boxed{\begin{aligned} D_{\mu\mu} &= \int_0^\infty dt \langle \dot\mu(t)\dot\mu(0)\rangle \\ &= \frac{1}{\rho_L^2}\int_0^\infty dt \left\{\langle v_x(t)v_x(0)b_y(t)b_y(0)\rangle + \langle v_y(t)v_y(0)b_x(t)b_x(0)\rangle \right. \\ &\quad \left. -\langle v_x(t)v_y(0)b_y(t)b_x(0)\rangle - \langle v_y(t)v_x(0)b_x(t)b_y(0)\rangle\right\} \\ &\equiv D_{\mu\mu}^{(I)} + D_{\mu\mu}^{(II)} + D_{\mu\mu}^{(III)} + D_{\mu\mu}^{(IV)}. \end{aligned}} \tag{10.416}$$

The four terms on the right-hand side will be calculated separately below.

Properties of χ_{zz}

By evaluating the parallel diffusion coefficient χ_{zz}, we obtain direct access to the (parallel) mean free path $\lambda_\parallel^{mfp}$. The relationship

$$\boxed{\chi_{zz} = \frac{1}{3}\langle v\rangle \lambda_\parallel^{mfp}} \tag{10.417}$$

holds and will be justified below.

Diffusion is the physical process by which particles spread from a region of higher concentration to a region of lower concentration. The average distance a particle travels between collisions is called the mean free path. The relationship

$$\frac{\partial M}{\partial t} = \chi_{zz}\frac{\partial^2 M}{\partial z^2} \tag{10.418}$$

is known as the second Fick's law of diffusion. The diffusion coefficient χ_{zz} has the unit $m^2\ s^{-1}$ and provides a measure of the number of particles that move through a given cross-sectional area per unit time.

The diffusion coefficient is related to the mean free path via the Einstein-Smoluchowski equation:

$$\chi_{zz} \sim \frac{\lambda_{\parallel}^{mfp\ 2}}{\tau} , \tag{10.419}$$

where τ is the average time between collisions. Assuming that this can be calculated from the average velocity $\langle v \rangle$ and the mean free path $\lambda_{\parallel}^{mfp}$, we have:

$$\tau \sim \frac{\lambda_{\parallel}^{mfp}}{\langle v \rangle} . \tag{10.420}$$

We obtain

$$\chi_{zz} \sim \lambda_{\parallel}^{mfp} \langle v \rangle . \tag{10.421}$$

To determine the proportionality constant in (10.420) precisely, we will discuss in more detail the physical statistics underlying the relationships just mentioned. We consider self-diffusion in a system of identical particles (gas) under the assumption that collisions in the gas occur randomly. The probability that a particle can travel a distance z without a collision can be assumed to have the form

$$\boxed{P_0(z) \approx \frac{1}{\lambda} e^{-z/\lambda}} , \tag{10.422}$$

with the mean free path $\lambda \equiv \lambda_{\parallel}^{mfp}$.

Example 10.23 (Collision Probability)
If we calculate the average distance a particle travels on average without a collision,

$$\langle z \rangle = \int_0^{\infty} z P_0(z) dz = \lambda , \tag{10.423}$$

this is referred to as the mean free path. The distribution (10.422) results from the independence of collisions according to the following argument. The average number of collisions per unit length is $1/\lambda$, and the probability that a collision occurs in an interval dz is dz/λ. The probability that no collision occurs in the interval $z + dz$ is

$$P_0(z+dz) \approx P_0(z) + dz\frac{dP_0}{dz} \approx \underbrace{P_0(z)}_{\text{no collision in z}} \underbrace{\left(1-\frac{dz}{\lambda}\right)}_{\text{no collision in dz}} , \tag{10.424}$$

which leads to the following differential equation:

$$\frac{dP_0}{dz} \approx -\frac{1}{\lambda}P_0 \, . \tag{10.425}$$

It has the normalized solution (10.422). ■

Next, we introduce the average speed $\langle v \rangle$ and the average relative speed $\langle v_r \rangle$ for a gas of identical particles. The two velocities are defined as

$$\langle v \rangle = \int d^3v f_M(\mathbf{v}) \, |\mathbf{v} - \mathbf{V}| \, , \tag{10.426}$$

$$\langle v_r \rangle = \int d^3v_A d^3v_B f_M(\mathbf{v}_A) f_M(\mathbf{v}_B) \, |\mathbf{v}_A - \mathbf{v}_B| \, , \tag{10.427}$$

where $\mathbf{V}$ is the mean directed velocity and f_M is assumed to follow a Maxwell distribution,

$$f_M(\mathbf{v}) = \left(\frac{m}{2\pi k_B T}\right)^{3/2} \exp\left(-\frac{m(\mathbf{v}-\mathbf{V})^2}{2k_B T}\right) . \tag{10.428}$$

T is the temperature. Integration yields

$$\langle v_r \rangle = \sqrt{\frac{16 k_B T}{\pi m}} = \sqrt{2}\langle v \rangle \, . \tag{10.429}$$

Example 10.24 (Mean Free Path)
In the center-of-mass system, the reduced mass $m_r = m/2$ appears for identical particles. Therefore, when calculating $\langle v_r \rangle$ we can use the formula for $\langle v \rangle$ by replacing $m \to m_r$, which in the above formula leads to the factor $\sqrt{2}$. Assuming a collision cross section σ for binary collisions, the collision frequency is

$$\nu_{\text{coll}} \equiv \tau_{\text{coll}}^{-1} \equiv n\sigma \langle v_r \rangle, \tag{10.430}$$

which yields for the mean free path

$$\lambda = \langle v \rangle \tau_{\text{coll}} = \frac{1}{\sqrt{2} n\sigma} \, . \tag{10.431}$$

■

Example 10.25 (Self-Diffusion)
For the relation to the self-diffusion coefficient χ_{zz} we identify some of the (otherwise identical) particles as "tracer particles." Self-diffusion describes the transport of "tracer particles" through a gas of otherwise identical particles. The rate at which spatial inhomogeneities of the "tracer particles" are evened out is determined by the self-diffusion coefficient χ. We assume that the density of the "tracer particles" $n^T(z)$ varies in the z-direction, while the total particle density n is kept constant. We determine the net flux through an imaginary wall at $z = 0$. Let dS be a segment of the wall. The origin of the spatial coordinates should be at the center of dS. We consider a volume element dV containing "tracer particles" that strike the wall from above. It is to be centered in polar coordinates at r, ϑ, φ. The average number of "tracer particles" per unit time in the volume element dV that undergo collisions is

$$\frac{n^T(z)\,dV}{\tau_{\mathrm{coll}}} = \frac{\langle v\rangle n^T}{\lambda}dV\;. \tag{10.432}$$

After the collisions, the "tracer particles" leave the volume element in random directions. The fraction moving toward dS is determined by the solid angle under which dS is seen from dV, i.e.,

$$\frac{d\Omega}{4\pi} = \frac{dS|\cos\vartheta|}{4\pi r^2}\;. \tag{10.433}$$

The probability of reaching dS before a new collision is $P_0(r)$. Therefore, the number of "tracer particles" that collide in dV and reach dS per unit time is

$$dn^T = \frac{\langle v\rangle\, n^T\, dV}{\lambda}\frac{dS|\cos\vartheta|}{4\pi r^2}e^{-r/\lambda}\;. \tag{10.434}$$

The total number of "tracer particles" striking a unit area of the wall from above per unit time is

$$\dot{N}_+ = \frac{\langle v\rangle}{4\pi\lambda}\int_0^\infty dr r^2 \int_0^{\pi/2} d\vartheta\,\sin\vartheta \int_0^{2\pi} d\varphi\, n^T(z)\,\cos\vartheta\frac{e^{-r/\lambda}}{r^2}\;. \tag{10.435}$$

Note that for constant n^T we would obtain the well-known result, namely

$$\dot{N}_+ = \frac{n\langle v\rangle}{4}\quad \text{for } n^T = n = \text{const}\;. \tag{10.436}$$

If we consider the particles moving from below towards dS, we obtain a similar expression, except that we must integrate over ϑ from $\pi/2$ to π, and $|\cos\vartheta| = -\cos\vartheta$ is, which leads to the following result:

$$\dot{N}_+ = -\frac{\langle v\rangle}{4\pi\lambda}\int_0^\infty dr r^2 \int_{\pi/2}^{\pi} d\vartheta\,\sin\vartheta \int_0^{2\pi} d\varphi\, n^T(z)\,\cos\vartheta\frac{e^{-r/\lambda}}{r^2}\;. \tag{10.437}$$

The net rate is therefore

$$\dot{N}_+ - \dot{N}_- = \frac{\langle v \rangle}{4\pi\lambda} \int_0^\infty dr r^2 \int_0^\pi d\vartheta \sin\vartheta \int_0^{2\pi} d\varphi \, n^T(z) \cos\vartheta \frac{e^{-r/\lambda}}{r^2} . \tag{10.438}$$

In evaluating the right-hand side, we expand to second order using the Taylor series

$$n^T(z) \approx n^T(0) + z \frac{\partial n^T}{\partial z} + \frac{z^2}{2} \frac{\partial^2 n^T}{\partial z^2} . \tag{10.439}$$

We obtain only a contribution from the second term of the Taylor expansion (10.439), with the result

$$\dot{N}_+ - \dot{N}_- \approx \frac{\langle v \rangle \lambda}{3} \frac{\partial n^T}{\partial z} . \tag{10.440}$$

If the density increases in the z-direction, i.e., $\dot{N}_+ - \dot{N}_- > 0$, more particles will propagate in the negative z-direction (from above). Therefore, we can use as the particle flux density

$$\Gamma(z) \approx -\left(\dot{N}_+ - \dot{N}_-\right) \equiv -\chi_{zz} \frac{\partial n^T(z)}{\partial z} \tag{10.441}$$

and (10.417) follows.

The continuity equation

$$\frac{\partial n^T}{\partial t} + \frac{\partial \Gamma}{\partial z} = 0 \tag{10.442}$$

finally leads to

$$\boxed{\frac{\partial n^T}{\partial t} - \chi_{zz} \frac{\partial^2 n^T}{\partial z^2} = 0} , \tag{10.443}$$

i.e., (10.418). ■

Evaluation of (10.416)

The "pitch angle" is the angle between the velocity of a particle and the direction of a specified reference system, typically the direction of a magnetic field. In plasma physics and astrophysics, the pitch angle describes how strongly the motion of a charged particle is aligned along or across a magnetic field. A pitch angle of 0 means that the particle moves parallel to the magnetic field line, while a pitch angle of 90 degrees means that the particle moves perpendicular to the magnetic field line. we are now ready to calculate the pitch angle diffusion coefficient.

According to (10.416), we have four contributions; we begin with the first. We evaluate it under the following assumption: only first-order corrections due to magnetic field fluctuations are considered. This means we approximate

$$D_{\mu\mu}^{(I)} \equiv \frac{1}{\rho_L^2}\int_0^\infty dt\langle v_x(t)v_x(0)b_y(t)b_y(0)\rangle \approx \frac{1}{\rho_L^2}\int_0^\infty dt\langle \eta_x(t)\eta_x(0)b_y(t)b_y(0)\rangle \,. \quad (10.444)$$

Since the integrand already explicitly contains the magnetic field fluctuations up to second order, only collisions are considered for the velocity correlations. The velocity component η_x must be calculated at a time t, where the path of the particle is traced under the influence of the random collision force **a**. As already mentioned, the force has a perpendicular ($\perp$) and a parallel ($\|$) component. In addition, the total correlation expresses an averaging over the magnetic field fluctuations (b). Therefore, we can speak of a triple stochastic process, which we denote by the notation:

$$\langle \eta_x(t)\eta_x(0)b_y(t)b_y(0)\rangle \equiv \langle\langle\langle \eta_x(t)\eta_x(0)b_y(t)b_y(0)\rangle_\perp\rangle_\|\rangle_b \,. \quad (10.445)$$

A part of the total averaging can be performed separately. For example, the averaging over the magnetic field fluctuations does not affect the lowest-order velocity component η_x. If we denote the path of the particle by $\mathbf{R}(t)$, which in the present approximation is influenced only by collisions, we first consider

$$\langle b_i(t)b_j(0)\rangle_b = \left\langle \int_\infty^\infty E_{ij}(\mathbf{r})\delta[\mathbf{r}-\mathbf{R}(t)]d^3r \right\rangle_b \,. \quad (10.446)$$

We evaluate in the Corrsin approximation,

$$\left\langle \int_\infty^\infty E_{ij}(\mathbf{r})\delta[\mathbf{r}-\mathbf{R}(t)]d^3r \right\rangle_b \approx \int_\infty^\infty E_{ij}(\mathbf{r})\langle\delta[\mathbf{r}-\mathbf{R}(t)]\rangle_b d^3r \,. \quad (10.447)$$

For the averaged propagator $\langle\delta[\mathbf{r}-\mathbf{R}(t)]\rangle_b$ we use the Fourier representation of the delta function

$$\delta[\mathbf{r}-\mathbf{R}(t)] = \frac{1}{(2\pi)^3}\int e^{-ik\cdot[\mathbf{r}-\mathbf{R}(t)]}d^3k \,. \quad (10.448)$$

Furthermore, we also use the Fourier transform of $E_{ij}(r)$, to obtain

$$\langle b_i(t)b_j(0)\rangle_b = \int E_{ij}(\mathbf{k})\langle\exp[i\mathbf{k}\cdot\mathbf{R}(t)]\rangle_b \, d^3k \,. \quad (10.449)$$

To lowest order, the trajectory is

$$\mathbf{R}(t) \approx \int_0^t \eta(t')dt' \, . \tag{10.450}$$

Thus,

$$\boxed{D_{\mu\mu}^{(I)} \approx \frac{1}{\rho_L^2} \int E_{yy}(\mathbf{k}) \langle\langle \eta_y(t)\eta_y(0) \ \mathcal{E}\rangle_\perp\rangle_\| d^3k dt} \, . \tag{10.451}$$

In a similar way, it follows that

$$\boxed{D_{\mu\mu}^{(II)} \approx \frac{1}{\rho_L^2} \int E_{xx}(\mathbf{k}) \langle\langle \eta_y(t)\eta_y(0) \ \mathcal{E}\rangle_\perp\rangle_\| d^3k dt \, ,} \tag{10.452}$$

$$\boxed{D_{\mu\mu}^{(III)} \approx -\frac{1}{\rho_L^2} \int E_{yx}(\mathbf{k}) \langle\langle \eta_x(t)\eta_y(0) \ \mathcal{E}\rangle_\perp\rangle_\| d^3k dt \, ,} \tag{10.453}$$

$$\boxed{D_{\mu\mu}^{(IV)} \approx -\frac{1}{\rho_L^2} \int E_{xy}(\mathbf{k}) \langle\langle \eta_y(t)\eta_x(0) \ \mathcal{E}\rangle_\perp\rangle_\| d^3k dt \, ,} \tag{10.454}$$

with the definition

$$\mathcal{E} \equiv \exp\left[ik_x \int_0^t \eta_x(t')dt' + ik_y \int_0^t \eta_y(t')dt' + ik_z \int_0^t \eta_z(t')dt' \right] . \tag{10.455}$$

We will first focus on the (simpler) collisionless case for $\nu = 0$. In this case, the velocities in the external magnetic field can easily be determined in the following form:

$$\eta_x = v_\perp \cos(\phi_0 - \Omega t) = \frac{v_\perp}{2}\left[e^{i(\phi_0 - \Omega t)} + e^{-i(\phi_0 - \Omega t)} \right] , \tag{10.456}$$

$$\eta_y = v_\perp \sin(\phi_0 - \Omega t) = \frac{v_\perp}{2i}\left[e^{i(\phi_0 - \Omega t)} - e^{-i(\phi_0 - \Omega t)} \right] , \tag{10.457}$$

$$\eta_z = v_\| \, . \tag{10.458}$$

Here, ϕ_0 is an arbitrary phase that can be used to perform an ensemble average. It is easy to see that upon averaging

$$\langle \cdots \rangle = \frac{1}{2\pi} \int_0^{2\pi} d\phi_0 \cdots \tag{10.459}$$

the correlation functions (10.228)–(10.230) for $\nu = 0$ recovered. In the following, we use $v_\perp = v\sqrt{1-\mu^2}$ and $v_\| = v\mu$, since in the absence of collisions the velocity v does not change. The factor $\mathcal{E}$, as defined in (10.455), requires the calculations

$$\int_0^t \eta_x(t')dt' = x_0 - \frac{v_\perp}{\Omega}\{\sin(\phi_0 - \Omega t) - \sin\phi_0\}, \tag{10.460}$$

$$\int_0^t \eta_y(t')dt' = y_0 + \frac{v_\perp}{\Omega}\{\cos(\phi_0 - \Omega t) - \cos\phi_0\}, \tag{10.461}$$

$$\int_0^t \eta_z(t')dt' = z_0 + v_\parallel t\ . \tag{10.462}$$

Thus,

$$\boxed{\mathcal{E} = \exp\{i\frac{k_\perp v_\perp}{\Omega}[\sin(\psi + \omega t - \phi_0) - \sin(\psi - \phi_0)] + ik_\parallel v_\parallel t\}}\ . \tag{10.463}$$

Here we have introduced the azimuthal angle ψ in k-space according to $k_x = k_\perp \cos\psi$, $k_y = k_\perp \sin\psi$, $k_z \equiv k_\parallel$. The initial position is assumed to be $\mathbf{r}_0 = (x_0, y_0, z_0)$. Next, we combine the trigonometric functions. To perform the averaging, we use the representation

$$\boxed{e^{iU\sin\alpha} = \sum_{n=-\infty}^{\infty} J_n(U)e^{in\alpha}} \tag{10.464}$$

with the Bessel functions J_n of order n. After a short calculation, we obtain

$$\mathcal{E} = \sum_{n=-\infty}^{\infty}\sum_{m=-\infty}^{\infty} J_n(W)J_m(W)e^{i(n-m)(\psi-\phi_0)+in\Omega t+ik_\parallel v_\parallel t} \tag{10.465}$$

with $W := \frac{k_\perp v_\perp}{\Omega}$. Now we can further simplify,

$$D_{\mu\mu}^{(I)} = \frac{1}{\rho_L^2}\int_0^\infty dt \int_{-\infty}^{\infty} d^3k E_{yy}(\mathbf{k})\langle\langle \eta_x(t)\eta_x(0)\mathcal{E}\rangle_\perp\rangle_\parallel\ , \tag{10.466}$$

where we substitute

$$\eta_x(t)\eta_x(0) = \frac{v_\perp^2}{4}\left[e^{i(2\phi_0-\Omega t)} + e^{-i\Omega t} + e^{i\Omega t} + e^{-i(2\phi_0-\Omega t)}\right]. \tag{10.467}$$

The averaging over the phase is performed after multiplication with $\frac{1}{2\pi}\int_0^{2\pi} d\phi_0$. This yields

$$\begin{aligned} D_{\mu\mu}^{(I)} =& \frac{v_\perp^2}{4\rho_L^2}\int_0^\infty dt \int_{-\infty}^{\infty} d^3k\ E_{yy}(\mathbf{k}) \sum_{n=-\infty}^{\infty} e^{-in\Omega t+ik_\parallel v_\parallel t} \\ &\times \left[J_{n+1}^2(W) + J_{n-1}^2(W) + J_{n+1}(W)J_{n-1}(W)(e^{i2\psi} + e^{-i2\psi})\right]. \end{aligned} \tag{10.468}$$

Note that the integration over t can be easily performed, provided the magnetic field fluctuation spectrum does not depend on time. We use

$$\int_0^\infty dte^{i(k_\| v_\| - n\Omega)t} = \pi\delta(ik_\| v_\| - n\Omega) \equiv R_n(\mathbf{k}) = R_n(k_\|) \; . \tag{10.469}$$

For later generalizations, we have the $k_\|$ resonance function abbreviated with R_n. With this definition, we obtain

$$D_{\mu\mu}^{(I)} = \frac{v_\perp^2}{4\rho_L^2} \int_{-\infty}^{\infty} d^3kE_{yy}(\mathbf{k}) \sum_{n=-\infty}^{\infty} R_n(k_\|)\Big[J_{n+1}^2 + J_{n-1}^2 + J_{n+1}J_{n-1}(e^{i2\psi} + e^{-i2\psi})\Big] \; . \tag{10.470}$$

Obviously, the other contributions $D_{\mu\mu}^{(II)}, D_{\mu\mu}^{(III)}$ and $D_{\mu\mu}^{(VI)}$ can be calculated in a similar way, yielding

$$\eta_y(t)\eta_y(0) = -\frac{v_\perp^2}{4}\Big[e^{i(2\phi_0-\Omega t)} - e^{-i\Omega t} - e^{i\Omega t} + e^{-i(2\phi_0-\Omega t)}\Big], \tag{10.471}$$

$$\eta_x(t)\eta_y(0) = \frac{v_\perp^2}{4i}\Big[e^{i(2\phi_0-\Omega t)} - e^{-i\Omega t} + e^{i\Omega t} - e^{-i(2\phi_0-\Omega t)}\Big], \tag{10.472}$$

$$\eta_y(t)\eta_x(0) = \frac{v_\perp^2}{4i}\Big[e^{i(2\phi_0-\Omega t)} + e^{-i\Omega t} - e^{i\Omega t} - e^{-i(2\phi_0-\Omega t)}\Big] \; . \tag{10.473}$$

The results are

$$D_{\mu\mu}^{(I)} = \frac{v_\perp^2}{4\rho_L^2} \int_{-\infty}^{\infty} d^3kE_{yy}(\mathbf{k}) \sum_{n=-\infty}^{\infty} R_n(k_\|) \times \Big[J_{n+1}^2(W) + J_{n-1}^2(W) + J_{n+1}(W)J_{n-1}(W)(e^{2i\psi} + e^{-2i\psi})\Big] \, , \tag{10.474}$$

$$D_{\mu\mu}^{(II)} = \frac{v_\perp^2}{4\rho_L^2} \int_{-\infty}^{\infty} d^3kE_{xx}(\mathbf{k}) \sum_{n=-\infty}^{\infty} R_n(k_\|) \times \Big[J_{n+1}^2(W) + J_{n-1}^2(W) - J_{n+1}(W)J_{n-1}(W)(e^{2i\psi} + e^{-2i\psi})\Big] \, , \tag{10.475}$$

$$D_{\mu\mu}^{(III)} = \frac{v_\perp^2}{i4\rho_L^2} \int_{-\infty}^{\infty} d^3kE_{yx}(\mathbf{k}) \sum_{n=-\infty}^{\infty} R_n(k_\|) \times \Big[-J_{n+1}^2(W) + J_{n-1}^2(W) + J_{n+1}(W)J_{n-1}(W)(e^{2i\psi} - e^{-2i\psi})\Big] \, , \tag{10.476}$$

$$D_{\mu\mu}^{(IV)} = \frac{v_\perp^2}{i4\rho_L^2} \int_{-\infty}^{\infty} d^3kE_{xy}(\mathbf{k}) \sum_{n=-\infty}^{\infty} R_n(k_\|) \times \Big[J_{n+1}^2(W) - J_{n-1}^2(W) + J_{n+1}(W)J_{n-1}(W)(e^{2i\psi} - e^{-2i\psi})\Big] \, . \tag{10.477}$$

Example 10.26 (Calculation without Corrsin Approximation)
Let us again consider the collisionless case $\nu = 0$, without explicitly using the Corrsin approximation. For the velocities, we use (10.456) and (10.457). Then (10.403) can be written as follows:

$$\dot{\mu} = \frac{\Omega\sqrt{1-\mu^2}}{B_0}\left[\delta B_y \cos\phi - \delta B_x \sin\phi\right], \tag{10.478}$$

with $\phi = \phi_0 - \Omega t$. For the magnetic field fluctuations, we use the representation

$$\delta B_L(\mathbf{r}, t) := \frac{1}{\sqrt{2}}\big(\delta B_x(\mathbf{r}, t) + i\delta B_y(\mathbf{r}, t)\big) \tag{10.479}$$

$$\delta B_R(\mathbf{r}, t) := \frac{1}{\sqrt{2}}\big(\delta B_x(\mathbf{r}, t) - i\delta B_y(\mathbf{r}, t)\big). \tag{10.480}$$

Then (10.403) can be written as

$$\dot{\mu} = \frac{i\Omega}{\sqrt{2}B_0}\sqrt{1-\mu^2}\left[\delta B_R e^{i\phi} - \delta B_L e^{-i\phi}\right]. \tag{10.481}$$

From this follows the pitch-angle diffusion coefficient (10.404) in the form

$$D_{\mu\mu} = \frac{\Omega^2(1-\mu^2)}{2B_0^2}\Re\int_0^\infty dt\left[\mathcal{E}_{RR}e^{-i\Omega t} - \mathcal{E}_{RL}e^{2i\phi_0 - i\Omega t} - \mathcal{E}_{LR}e^{-2i\phi_0 + i\Omega t} + \mathcal{E}_{LL}e^{i\Omega t}\right], \tag{10.482}$$

with

$$\mathcal{E}_{XY}(\mathbf{r}, t) = \left\langle \delta B_X(\mathbf{r}, t)\delta B_Y^*(\mathbf{r_0}, 0)\right\rangle, \quad \text{with } X, Y = R, L. \tag{10.483}$$

Next, we apply a Fourier decomposition of the magnetic field fluctuations,

$$\delta B_{L,R}(\mathbf{r}, t) = \int d^3k \delta B_{L,R}(\mathbf{k}, t) e^{i\mathbf{k}\cdot\mathbf{r}(t)}, \tag{10.484}$$

which leads to

$$\mathcal{E}_{XY} = \mathcal{E}_{XY}(\mathbf{r}, t, \mathbf{r}_0, 0) = \int d^3k \int d^3k' \left\langle \delta B_X(\mathbf{k}, t)\delta B_Y(\mathbf{k}', 0)\right\rangle e^{i[\mathbf{k}\cdot\mathbf{r}(t) - \mathbf{k}'\cdot\mathbf{r}_0]} \tag{10.485}$$

Now, using cylindrical coordinates in k-space,

$$k_x = k_\perp \cos\psi, \; k_y = k_\perp \sin\psi, \; k_z = k_\parallel, \tag{10.486}$$

we quickly obtain

$$\mathbf{k}\cdot\mathbf{r}(t)=\mathbf{k}\cdot\mathbf{r}_0+\frac{k_\perp v_\perp}{\Omega}[\sin(\psi-\phi_0+\Omega t)-\sin(\psi-\phi_0)]+k_\parallel v_\parallel t. \tag{10.487}$$

All together, in real space this yields

$$\mathcal{E}_{XY}(\mathbf{r},t,\mathbf{r}_0,0)=\int d^3k\mathcal{E}_{XY}(\mathbf{k},t)e^{iW[\sin(\psi-\phi_0+\Omega t)-\sin(\psi-\phi_0)]+iv_\parallel k_\parallel t}\,, \tag{10.488}$$

with

$$W=\frac{k_\perp v_\perp}{\Omega}=k_\perp\rho_L\sqrt{1-\mu^2} \tag{10.489}$$

and

$$\mathcal{E}_{XY}(\mathbf{k},t)=\frac{1}{(2\pi)^3}\int d^3r\mathcal{E}_{XY}(\mathbf{r},t)e^{-i\mathbf{k}\cdot\mathbf{r}}\,. \tag{10.490}$$

With (10.464) we can further write

$$\begin{aligned}D_{\mu\mu}=&\frac{\Omega^2(1-\mu^2)}{2B_0^2}\Re\int_0^\infty dt\int d^3k\sum_{n,m=-\infty}^{\infty}J_n(W)J_m(W)e^{in(\psi-\phi_0+\Omega t)-im(\psi-\phi_0)+iv_\parallel k_\parallel t}\\&\times\Big[\mathcal{E}_{RR}(\mathbf{k})e^{-i\Omega t}-\mathcal{E}_{RL}(\mathbf{k})e^{2i\phi_0-i\Omega t}-\mathcal{E}_{LR}(\mathbf{k})e^{-2i\phi_0+i\Omega t}+\mathcal{E}_{LL}(\mathbf{k})e^{i\Omega t}\Big]\,.\end{aligned} \tag{10.491}$$

Then we average over all initial angles by multiplying with $\frac{1}{2\pi}\int_0^{2\pi}d\phi_0$. The averaging yields

$$\begin{aligned}D_{\mu\mu}=&\frac{\Omega^2(1-\mu^2)}{2B_0^2}\Re\sum_{n=-\infty}^{\infty}\int d^3kR_n(k_\parallel)[J_{n+1}^2(W)\mathcal{E}_{RR}(\mathbf{k})+J_{n-1}^2(W)\mathcal{E}_{LL}(\mathbf{k})\\&-J_{n+1}(W)J_{n-1}(W)\Big(\mathcal{E}_{RL}(\mathbf{k})e^{2i\psi}+\mathcal{E}_{LR}(\mathbf{k})e^{-2i\psi}\Big)]\,,\end{aligned} \tag{10.492}$$

where, similar to (10.469),

$$R_n(k_\parallel)=\int_0^\infty dte^{i(k_\parallel v_\parallel+n\Omega)t}=\pi\delta(k_\parallel v_\parallel+n\Omega)\,. \tag{10.493}$$

With the help of (10.490), a short calculation yields

$$\mathcal{E}_{RR}(\mathbf{k},t)=\frac{B_0^2}{2}\big[E_{xx}(\mathbf{k})-iE_{yx}(\mathbf{k})+iE_{xy}(\mathbf{k})+E_{yy}(\mathbf{k})\big]\,, \tag{10.494}$$

where $E_{ij}(\mathbf{k})$ from (10.249) is used. Furthermore, $b_i=\frac{\delta B_i}{B_0}$ also holds. As before, we obtain

$$\mathcal{E}_{LL}(\mathbf{k},t)=\frac{B_0^2}{2}\big[E_{xx}(\mathbf{k})+iE_{yx}(\mathbf{k})-iE_{xy}(\mathbf{k})+E_{yy}(\mathbf{k})\big]\,, \tag{10.495}$$

$$\mathcal{E}_{RL}(\mathbf{k},t)=\frac{B_0^2}{2}\big[E_{xx}(\mathbf{k})-iE_{yx}(\mathbf{k})-iE_{xy}(\mathbf{k})-E_{yy}(\mathbf{k})\big]\,, \tag{10.496}$$

$$\mathcal{E}_{LR}(\mathbf{k},t)=\frac{B_0^2}{2}\big[E_{xx}(\mathbf{k})+iE_{yx}(\mathbf{k})+iE_{xy}(\mathbf{k})-E_{yy}(\mathbf{k})\big]\,. \tag{10.497}$$

By inserting the results into $D_{\mu\mu}$, we can, with respect to the contributions from E_{ij}, perform a separation to obtain the following expressions:

$$\begin{aligned} D_{\mu\mu}^{(I)} =&\frac{v_\perp^2}{4\rho_L^2}\int d^3kE_{yy}(\mathbf{k})\sum_{n=-\infty}^{\infty}R_n(k_\parallel)\\ &\times[J_{n+1}^2(W)+J_{n-1}^2(W)+J_{n+1}(W)J_{n-1}(W)\Big(e^{2i\psi}+e^{-2i\psi}\Big)\,, \end{aligned} \tag{10.498}$$

$$\begin{aligned} D_{\mu\mu}^{(II)} =&\frac{v_\perp^2}{4\rho_L^2}\int d^3kE_{xx}(\mathbf{k})\sum_{n=-\infty}^{\infty}R_n(k_\parallel)\\ &\times[J_{n+1}^2(W)+J_{n-1}^2(W)-J_{n+1}(W)J_{n-1}(W)\Big(e^{2i\psi}+e^{-2i\psi}\Big)\,, \end{aligned} \tag{10.499}$$

$$\begin{aligned} -D_{\mu\mu}^{(III)} =&\frac{iv_\perp^2}{4\rho_L^2}\int d^3kE_{yx}(\mathbf{k})\sum_{n=-\infty}^{\infty}R_n(k_\parallel)\\ &\times[-J_{n+1}^2(W)+J_{n-1}^2(W)+J_{n+1}(W)J_{n-1}(W)\Big(e^{2i\psi}-e^{-2i\psi}\Big)\,, \end{aligned} \tag{10.500}$$

$$\begin{aligned} -D_{\mu\mu}^{(IV)} =&\frac{iv_\perp^2}{4\rho_L^2}\int d^3kE_{xy}(\mathbf{k})\sum_{n=-\infty}^{\infty}R_n(k_\parallel)\\ &\times[J_{n+1}^2(W)-J_{n-1}^2(W)+J_{n+1}(W)J_{n-1}(W)\Big(e^{2i\psi}-e^{-2i\psi}\Big)\,, \end{aligned} \tag{10.501}$$

with $\left(1-\mu^2\right)=\frac{v_\perp}{v}$ and $\rho_L=\frac{v}{\Omega}$. These formulas agree with the previously derived expressions. ■

Parallel diffusion in the quasilinear limit $\lambda_\perp\to\infty$

The so-called quasilinear case with a slab model for the magnetic field fluctuation spectrum is frequently considered in the theory of anomalous transport. It applies in the limiting case $\lambda_\perp\to\infty$ and for $\nu=0$. In this case, the magnetic field fluctuations are greatly simplified.

For an exponential dependence in the parallel direction (note that especially in astrophysics, different forms for the parallel spectrum are used), we make the ansatz

$$\boxed{E_{ij}(\mathbf{k}) \equiv E_{ij}^{slab}(\mathbf{k}) = \frac{1}{\sqrt{2\pi}}\beta^2\lambda_\parallel e^{-\frac{1}{2}\lambda_\parallel^2 k_\parallel^2}\delta(k_x)\delta(k_y)\,\delta_{ij}} \,. \tag{10.502}$$

In the literature, a somewhat different ansatz can also be found,

$$E_{ij}(\mathbf{k}) = g^{slab}(k_\parallel)\frac{1}{k_\perp}\delta(k_\perp)\,\delta_{ij}, \tag{10.503}$$

where different functions $g^{slab}(k_\parallel)$ are commonly used [151, 156, 246]. We should note that in this case, cylindrical coordinates are used in k-space and that $dk_x\,dk_y \to k_\perp\,dk_\perp\,d\varphi$ holds, as well as

$$\int_{-\infty}^{\infty} dk_x \int_{-\infty}^{\infty} dk_y \delta(k_x)\delta(k_y) = 1 \to \int_0^{2\pi} d\varphi \int_0^{\infty} dk_\perp \frac{1}{k_\perp \pi}\delta(k_\perp)k_\perp d\varphi = 1 \,. \tag{10.504}$$

The spectrum should be normalized,

$$E_{ij}(\mathbf{r} = \mathbf{0}) = \beta^2 = \int d^3k\; E_{ij}(\mathbf{k}) \,. \tag{10.505}$$

First, we use (10.498)–(10.501), although in the present special case $\nu = 0$ and $\lambda_\perp \to \infty$ a more direct approach is available (see below). By inserting (10.502) into (10.498), we obtain

$$\begin{aligned} D_{\mu\mu}^{(I)} =& \frac{\sqrt{\pi}\lambda_\parallel\beta^2 v_\perp^2}{4\sqrt{2}\rho_L^2} \sum_{n=-\infty}^{\infty} \int_{-\infty}^{\infty} d^3k e^{-\frac{1}{2}\lambda_\parallel^2 k_z^2}\delta(k_x)\delta(k_y)\delta(k_z v_\parallel - n\Omega) \\ & \times \left[J_{n-1}^2(W) + 2\cos(2\psi)J_{n-1}(W)J_{n+1}(W) + J_{n+1}^2(W)\right] , \end{aligned} \tag{10.506}$$

where $k_\perp = \sqrt{k_x^2 + k_y^2}$ and $k_\parallel = k_z$ are. The integration over k_z leads to

$$\begin{aligned} D_{\mu\mu}^{(I)} =& \frac{\sqrt{\pi}\lambda_\parallel\beta^2 v_\perp^2}{4\sqrt{2}\rho_L^2|v_\parallel|} \sum_{n=-\infty}^{\infty} \int_{-\infty}^{\infty} dk_x \int_{-\infty}^{\infty} dk_y e^{-\frac{1}{2}\lambda_\parallel^2\left(\frac{n\Omega}{v_\parallel}\right)^2}\delta(k_x)\delta(k_y) \\ & \times \left[J_{n-1}^2(W) + 2\cos(2\psi)J_{n-1}(W)J_{n+1}(W) + J_{n+1}^2(W)\right] . \end{aligned} \tag{10.507}$$

Due to the properties of Bessel functions, e.g.,

$$J_n(0) = 0 \quad \text{for } n \neq 0\,, \quad J_0(0) = 1\,, \tag{10.508}$$

we obtain

$$D_{\mu\mu}^{(I)} = \frac{\sqrt{\pi}}{2\sqrt{2}}\frac{\beta^2\lambda_\parallel v_\perp^2}{\rho_L^2|v_\parallel|}e^{-\frac{1}{2}\lambda_\parallel^2\frac{\Omega^2}{v_\parallel^2}} \,. \tag{10.509}$$

In a similar way, we obtain

$$\boxed{D_{\mu\mu}^{(II)} = D_{\mu\mu}^{(I)} \equiv D_{\mu\mu}\,, \quad D_{\mu\mu}^{(III)} = D_{\mu\mu}^{(IV)} = 0}\,, \tag{10.510}$$

with

$$\boxed{D_{\mu\mu} = \frac{\sqrt{\pi}\beta^2\lambda_\parallel v_\perp^2}{\sqrt{2}\rho_L^2|v_\parallel|} e^{-\frac{1}{2}\lambda_\parallel^2\frac{\Omega^2}{v_\parallel^2}}}\,. \tag{10.511}$$

Example 10.27 (Direct Integration)
Before we discuss the result, let us return to the point that, under the present assumptions, the diffusion coefficient can be calculated directly from (10.451)–(10.454). For example, by substituting (10.502) into (10.451) we obtain

$$D_{\mu\mu}^{(I)} \approx \frac{1}{\rho_L^2}\beta^2\lambda_\parallel \frac{v_\perp^2}{2\sqrt{2}\sqrt{\pi}} \int dt \int dk_\parallel \tfrac{1}{2}\left[e^{-i\Omega t} + e^{i\Omega t}\right] e^{ik_\parallel v_\parallel t} e^{-i\frac{1}{2}\lambda_\parallel^2 k_\parallel^2}\,. \tag{10.512}$$

Because

$$\int_0^\infty dt e^{i(k_\parallel v_\parallel \pm n\Omega)t)} = \pi\delta(k_\parallel v_\parallel \pm n\Omega) \tag{10.513}$$

and

$$\int_{-\infty}^\infty dk_\parallel \delta(k_\parallel v_\parallel \pm n\Omega)\cdots = \frac{1}{|v_\parallel|}\int_{-\infty}^\infty dk_\parallel \delta\left(k_\parallel \pm \frac{n\Omega}{v_\parallel}\right)\cdots \tag{10.514}$$

the integration can be performed directly, yielding the result

$$D_{\mu\mu}^{(I)} = \frac{\sqrt{\pi}}{2\sqrt{2}}\frac{\beta^2\lambda_\parallel v_\perp^2}{\rho_L^2|v_\parallel|} e^{-\frac{1}{2}\lambda_\parallel^2\frac{\Omega^2}{v_\parallel^2}} \tag{10.515}$$

In the assumed case, this agrees with the previous result. The other expressions (10.452)–(10.454) are obtained in a similar manner. ■

According to a notation that is especially common in plasma astrophysics, we next introduce the rigidity.

$$\boxed{R = \frac{\rho_L}{l_{slab}} \sim \frac{\rho_L}{\lambda_\parallel}}\,, \tag{10.516}$$

where the so-called "slab bendover length" l_{slab} is defined such that the area $l_{slab}\beta^2$ corresponds to the area under the correlation function, i.e.,

$$l_{slab} = \int_0^\infty dz e^{-\frac{1}{2}\frac{z^2}{\lambda_\parallel^2}} = \sqrt{\frac{\pi}{2}}\lambda_\parallel\,. \tag{10.517}$$

The rigidity provides a measure of a particle's momentum. For large values of rigidity,

$$D_{\mu\mu} \approx \frac{\sqrt{\pi}\beta^2\lambda_\| v_\perp^2}{\sqrt{2}\rho_L^2 |v_\||} = \frac{\sqrt{\pi}\beta^2\lambda_\| v}{\sqrt{2}\rho_L^2} \frac{1-\mu^2}{|\mu|} . \tag{10.518}$$

By substituting into (10.415), we can integrate analytically and obtain

$$\chi_{zz} = \frac{v^2}{8}\int_{-1}^{+1} \frac{\left(1-\mu^2\right)^2}{D_{\mu\mu}} d\mu \approx \frac{1}{8\sqrt{2}\sqrt{\pi}} \frac{v^3}{\Omega^2\beta^2\lambda_\|} . \tag{10.519}$$

Apart from a numerical factor, this agrees with the result derived on the basis of a small gyro expansion in [247].

In the cited work, the "magnetic collision frequency" is defined as

$$\nu_{\text{mag}} = \sqrt{2}\Omega^2\beta^2 \frac{\lambda_\|}{v_{th}} , \tag{10.520}$$

so that the quasilinear parallel and perpendicular diffusion coefficients can each be written in forms corresponding to the classical forms, namely,

$$\boxed{D_\| = \frac{v_{th}^2}{2\nu_{\text{mag}}} , \quad D_\perp = \frac{v_{th}^2 \nu_{\text{mag}}}{2\Omega^2}} . \tag{10.521}$$

Eq. (10.417) allows the calculation of the parallel mean free path as

$$\lambda_\|^{mfp} = 3\frac{\chi_{zz}}{v} \approx \frac{3}{8\sqrt{2}\sqrt{\pi}} \frac{v^2}{\Omega^2\beta^2\lambda_\|} \sim R^2 . \tag{10.522}$$

The mean free path in the quasilinear approximation is proportional to the square of the rigidity.

Starting from the relation (10.415) between the parallel diffusion coefficient and the pitch-angle diffusion coefficient, as well as the expressions (10.451)–(10.454) for the latter, more general cases can be considered, e.g., $\nu \neq 0$ and finite large $\lambda_\perp$. Further details would go beyond the scope of this book.

Laser Plasma Physics at High Intensities 11

Abstract

This chapter builds upon the previously discussed fundamentals of laser-plasma interaction. However, it now outlines modern developments that have emerged following the enormous increase in short-pulse laser intensities. We begin with a brief account of the history and perspectives of laser development. Subsequently, we focus primarily on two current branches of research and development: plasma accelerators and plasma optics. We could add further topics, such as plasma mirrors, but must omit them for reasons of space.

11.1 Laser Development and Applications

In this section, we outline how the idea and realization of CPA ("chirped-pulse amplification") revolutionized laser development.

The interaction of high-intensity lasers with matter is a broad field of research that has experienced explosive scientific development over the past three decades. Wavelength, energy, pulse duration, and focus size are the four "magic numbers" [248] that characterize laser development with respect to nonlinear laser plasma physics. The interaction of high-intensity lasers with long pulses and plasma is primarily relevant for inertial fusion. However, since the invention of the chirped-pulse amplification (CPA) technique [249], the interaction of lasers with short pulses has opened up an entirely new field of research, now known as relativistic optics [250]. Femtosecond and even attosecond timescales (1 fs

K.-H. Spatschek, *Theoretical Plasma Physics*,
https://doi.org/10.1007/978-3-662-72828-4_11

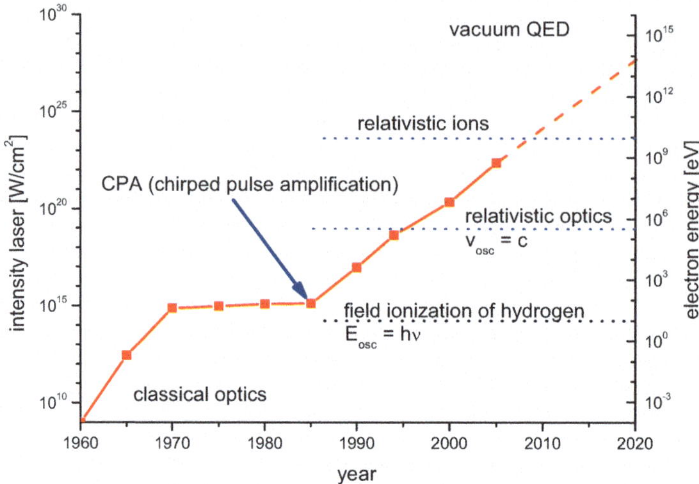

Fig. 11.1 Schematic representation of the historical development of the most intense laser pulses

$= 10^{-15}$ s, 1 as $= 10^{-18}$ s) are becoming increasingly accessible at relatively high powers. Figure 11.1 provides an overview of the currently available high-intensity laser pulses and the associated new areas of physical research that can be explored.

A new field of research in laser-plasma interaction began about 40 years ago, when laser fields became strong enough to directly ionize matter. With lasers operating at wavelengths from 0.25 μm to 13.4 μm, the photoelectric effect in ordinary materials cannot be effective, since the photon energy $\hbar\omega$ is much smaller than the atomic potential barrier experienced by the electron near the nucleus (e.g., 13.6 eV for the ground state of hydrogen). With more powerful laser systems in the 1960s and 1970s, multiphoton ionization became possible. The rich physics that already occurs in the weakly nonlinear low-intensity regime is summarized, for example, in [24].

Before the CPA era, direct amplification of pulses was limited, since intensities of several GW/cm^2 would damage the gain medium. Stretching the pulse in a controlled (reversible) manner by adding a chirp (temporal variation of the frequency) is the first stage in the CPA technique. The chirped pulse is then amplified. For this, one or more conventional laser amplifier stages can generally be used to increase the pulse energy by a factor of 10^7 to 10^9. Finally, a compressor performs the exact optical inversion of the stretcher to deliver an amplified pulse with the same duration as at the beginning. Optical parametric amplification (OPCPA) in a nonlinear crystal even promises higher intensities than conventional CPA.

Example 11.1 (Data from two existing laser systems)
An example: The "magic numbers" of the Jena *TW* laser are [248]: wavelength 800 nm, energy 0.8 J, pulse duration 80 fs and focus size 5 μm², which leads to an approximate intensity of 10^{20} W/cm². In this regime, the generated laser plasma becomes relativistic. It is convenient to characterize the laser field by the dimensionless parameter $a_0 = eA_0/m_e c$, where A_0 is the peak vector potential of the laser pulse. Since a_0 corresponds to the normalized (by $m_e c$) maximum transverse momentum of an electron in the laser field, $a_0 \geq 1$ is often referred to as the threshold for entering relativistic optics (see below). Facilities such as the ARCTURUS laser at the University of Düsseldorf can generate $a_0 \sim 10$. ■

A next major step will be reached at $a_0 \sim 1000$, when ions are also accelerated to relativistic velocities. At even higher intensities, it will become possible to investigate the generation of electron-positron pairs and other quantum electrodynamics (QED) effects.

Intensities of 10^{19} W/cm² are now available in tabletop terawatt laser systems (Tera $\hat{=}\, 10^{12}$), which are based on chirped-pulse amplification. Let us take as a typical example one joule of light compressed into a picosecond (pico $\hat{=}\, 10^{-12}$) to produce a peak power of one terawatt. The laser can be focused to a few squares of the wavelength, e.g., 10 μm². This results in an intensity of

$$I_0 \sim \frac{1\ \text{Joule}}{1\ \text{ps}\ 10\,\mu\text{m}^2} \sim \frac{1}{10^{-12}\ 10\ 10^{-8}} \sim 10^{19} \left[\frac{\text{W}}{\text{cm}^2}\right] . \tag{11.1}$$

The photon density $n_{\text{photon}} = N/V$ follows, with the photon energy

$$\hbar\omega \equiv \frac{hc}{\lambda} \approx \frac{6.6 \times 10^{-34}\,\text{Js}\ 3 \times 10^{8}\,\text{ms}^{-1}}{10^{-6}\,\text{m}} \approx 2 \times 10^{-19}\,\text{J} \quad \text{at}\ \lambda = 1\ \mu\text{m} \tag{11.2}$$

to the number $N = 5 \times 10^{18}$ and thus

$$n_{\text{photon}} = \frac{5 \times 10^{18}}{10 \times 10^{-8}\,\text{cm}^2\ 3 \times 10^{10}\,\text{cm}\,\text{s}^{-1} 10^{-12}\,\text{s}} \approx 1.6 \times 10^{27}\ \frac{\text{photons}}{\text{cm}^3} . \tag{11.3}$$

Regimes of Relativistic Optics

In the study of the motion of a single particle (electron) in a plane electromagnetic wave, we have already begun to discuss the significance of relativistic effects during laser–particle interaction.

Equation (2.286) is central in this context. For clarity, we repeat it here:

$$I_0\lambda^2 = \zeta\left[1.37 \times 10^{18}\,\frac{\text{W}}{\text{cm}^2}\,\mu\text{m}^2\right]a_0^2\,. \tag{11.4}$$

If we specify the circular polarization $\zeta = 2$ and the wavelength $\lambda = 1\ \mu m$, we obtain

$$I_0 \approx 2.7 \times 10^{18}\,a_0^2\left[\frac{\text{W}}{\text{cm}^2}\right] \quad \text{for } \lambda = 1\ \mu\text{m}\,,\ \zeta = 2\,. \tag{11.5}$$

For $\lambda = 1\ \mu$m this leads to the electric field strength E_{field} expressed in a_0,

$$E_{\text{field}} \approx 3.14 \times 10^{12}\ a_0\left[\frac{\text{V}}{\text{m}}\right] \quad \text{for } \lambda = 1\ \mu\text{m}\,. \tag{11.6}$$

This allows the intensity to also be written as

$$I_0 \approx \frac{(E_{\text{field}}\,[\text{V/cm}])^2}{377\,[\Omega]}\left[\frac{\text{W}}{\text{cm}^2}\right] \quad \text{for } \lambda = 1\ \mu\text{m}\,,\ \zeta = 2\,. \tag{11.7}$$

Since a_0^2 is proportional to $I_0\lambda^2$, relativistic effects are favored at large wavelengths λ (or low frequencies ω). Higher frequencies, which can be focused onto smaller spots, are therefore not as well suited for observing relativistic effects in ions, but could be well utilized to reach the QED regime. For the latter, I_0 itself is important. Higher harmonics at solid surfaces are advantageous for reaching the QED regime much more easily.

The weakly relativistic regime

From the approximate solution of the motion of a single electron in a laser field, for $|\mathbf{v}| \ll c$ the condition is obtained

$$a_0 = \frac{eA_0}{m_e c} \ll 1\,. \tag{11.8}$$

In this limit, the γ factor can be expanded,

$$\gamma = \sqrt{1 + \frac{|\mathbf{p}_e|^2}{m_e^2 c^2}} \approx \sqrt{1 + a_0^2} \approx 1 + \frac{a_0^2}{2}\,. \tag{11.9}$$

In the non-relativistic case, $\gamma = 1$.

The high-intensity regime

The following terminology was introduced by Mourou et al. [250]. The high-intensity regime (for electrons) is defined as the range in which

$$\hbar\omega \ll m_e c^2 \left[\sqrt{1+a_0^2}-1\right] < m_e c^2 \,, \tag{11.10}$$

when we use $\gamma \approx \sqrt{1+a_0^2}$. Relativistic effects are very important here, but in this range the physics is not yet ultrarelativistic. Obviously, the upper limit is at $a_0 \approx 1$. For $\lambda = 1\ \mu\text{m}$, the estimate follows from the condition $a_0 = 1$ as

$$I_{0e} \approx 2.7 \times 10^{18}\,\frac{\text{W}}{\text{cm}^2}\,. \tag{11.11}$$

Higher intensities on the order of

$$I_{0i} \approx 10^{25}\,\frac{\text{W}}{\text{cm}^2} \tag{11.12}$$

are required for the ion motion to become relativistic. A first estimate leads to the range

$$5 \times 10^{-6} \ll a_0^2 \le 1\,. \tag{11.13}$$

If we introduce a safety factor of 100 for the lower limit, we obtain

$$2.2 \times 10^{-2} \le a_0 \le 1\,. \tag{11.14}$$

Expressed in terms of intensities, we have

$$5 \times 10^{14}\left[\frac{\text{W}}{\text{cm}^2}\right] \le I_0 \le 2.7 \times 10^{18}\left[\frac{\text{W}}{\text{cm}^2}\right] \quad \text{for } \lambda = 1\ \mu\text{m}\,. \tag{11.15}$$

Note that the numerical values depend on the laser wavelength. We have used $\lambda = 1\ \mu\text{m}$, keeping in mind that for an excimer laser $\lambda \approx 248$ nm an upper limit on the order of 10^{19} W/cm^2 results, while for a CO_2 laser $\lambda \approx 10.6\ \mu\text{m}$ an upper limit on the order of 10^{16} W/cm^2 applies for entering the so-called ultra-high intensity regime, which will be discussed below.

Ultra-high intensities and the ultrarelativistic regime

The ultra-high intensity regime—above the high-intensity regime (11.13)—covers the range $a_0 > 1$ or is also referred to as the *ultrarelativistic regime.* There,

$$eE_{\text{field}}\lambda > 2\pi m_e c^2\,, \tag{11.16}$$

applies for the electric field E_{field} of the laser. Depending on the dominant physical processes, we distinguish between different regions:

The relatively low-amplitude ultrarelativistic regime

An accelerated relativistic electron moving in a laser field loses energy due to synchrotron radiation. Using the standard relativistic formula [251, 252] for the radiation intensity,

$$P = \frac{2}{3}\frac{e^2}{4\pi\varepsilon_0 m^2 c^3}\gamma^2\omega^2|\mathbf{p}|^2 \;, \tag{11.17}$$

one can, for relatively low laser amplitudes $|\mathbf{p}| \approx mca_0$, make an approximate assumption (see below). The radius of the electron trajectory is approximately $R = c/\omega_0 = \lambda/2\pi$, where a_0 is the normalized amplitude of the vector potential, $m \equiv m_e$ is the rest mass of the electron, and ω_0 is the laser frequency. In addition, for $a_0 \gg 1$ and as long as $|\mathbf{p}| \approx m_e c a_0$, it holds that $\gamma \approx a_0$. Using this, we obtain

$$P \approx \underbrace{\frac{4\pi r_e}{3\lambda}}_{\varepsilon_{\text{rad}}} \omega_0 mc^2 a_0^4 \;, \tag{11.18}$$

with

$$\varepsilon_{\text{rad}} = \frac{4\pi r_e}{3\lambda} \approx 1.2 \times 10^{-8} \quad \text{at } \lambda = 1\,\mu\text{m} \tag{11.19}$$

and the classical electron radius

$$r_e = \frac{1}{4\pi\varepsilon_0}\frac{e^2}{mc^2} \approx 2.82 \times 10^{-13}\,\text{cm} \;. \tag{11.20}$$

The electron energy

$$E = \gamma mc^2 \approx mc^2 a_0 \tag{11.21}$$

depends on the laser amplitude. Based on the energy formula, we approximate the temporal change by

$$\frac{\partial E}{\partial t} \sim \omega_0 mc^2 a_0 \;. \tag{11.22}$$

Comparing (11.18) with (11.22), we find that radiation becomes dominant for

$$\varepsilon_{\text{rad}} a_0^3 \gg 1 \;. \tag{11.23}$$

In the region below the upper limit, it then holds that

$$1 \le a_0 \le \frac{1}{\varepsilon_{\text{rad}}^{1/3}} \equiv a_{\text{radiation}} \approx 440 \quad \text{for} \quad \lambda = 1\,\mu\text{m} \;. \tag{11.24}$$

This is referred to as the relatively low-amplitude regime. Above this regime, radiation effects become important. At the upper edge, the electric field is on the order of

$$E_{\text{field}}^{\text{radiation}} \approx 1.4 \times 10^{15} \left[\frac{\text{V}}{\text{m}}\right] ; \tag{11.25}$$

the corresponding intensity is

$$I_0^{\text{radiation}} \approx 5.2 \times 10^{23} \left[\frac{\text{W}}{\text{cm}^2}\right] \quad \text{at } \lambda = 1\,\mu\text{m}\,. \tag{11.26}$$

To summarize:

$$2.7 \times 10^{18} \left[\frac{\text{W}}{\text{cm}^2}\right] \leq I_0 \leq 5.2 \times 10^{23} \left[\frac{\text{W}}{\text{cm}^2}\right] \quad \text{at } \lambda = 1\,\mu\text{m} \tag{11.27}$$

characterizes the relatively low-intensity region of the ultrarelativistic regime.

The intermediate ultrarelativistic regime

Before we address the new physical effects that occur at even larger amplitudes, let us return to the relationship between $|\mathbf{p}|$ and a_0. Without proof, we state (see [250])

$$|\mathbf{p}| \approx \begin{cases} m_e c a_0 & \text{for} \quad 1 < a_0 < a_{\text{radiation}}\,, \\ m_e c (a_0/\varepsilon_{\text{rad}})^{1/4} & \text{for } a_{\text{radiation}} < a_0 < a_{\text{quantum}}\,. \end{cases} \tag{11.28}$$

Quantum effects become important when the photons produced as a result of Compton scattering have energies on the order of (or greater than) the electron energy $E_e = \gamma m_e c^2$. The energy of a photon is

$$E_{\text{photon}} = \hbar\omega_m\,. \tag{11.29}$$

The frequency of the photon produced by an electron rotating with frequency ω is [253]

$$\omega_m = \gamma^3 \omega\,. \tag{11.30}$$

Equating the electron energy with the photon energy $E = E_{\text{photon}}$ leads to the critical γ value

$$\gamma_{\text{quantum}} \equiv \sqrt{\frac{m_e c^2}{\hbar\omega}}\,, \tag{11.31}$$

so that for

$$\gamma > \gamma_{\text{quantum}} \approx 600 \quad \text{for} \quad \lambda = 1\,\mu\text{m} \tag{11.32}$$

quantum effects come into play. Using (11.28) and

$$\gamma \approx (a_0/\varepsilon_{\text{rad}})^{1/4} \approx \left(\frac{3a_0\lambda}{4\pi r_e}\right)^{1/4} \tag{11.33}$$

we can also reformulate the condition,

$$a_0 > a_{\text{quantum}} = \frac{1}{4\pi\varepsilon_0}\frac{2e^2 m_e c}{3\hbar^2\omega} = \frac{1}{3\pi}\frac{r_e\lambda}{\bar{\lambda}_c^2} \approx 2000 \quad \text{for} \quad \lambda = 1\ \mu\text{m}\,, \tag{11.34}$$

where

$$\bar{\lambda}_C = \frac{\hbar}{m_e c} \approx 3.86 \times 10^{-11}\ \text{cm} \tag{11.35}$$

is the Compton wavelength.

The range

$$a_{\text{radiation}} \leq a_0 \leq a_{\text{quantum}} \tag{11.36}$$

is referred to as the intermediate (intermediary) ultrarelativistic region. At the upper end, the electric field is on the order of

$$E_{\text{field}}^{\text{quantum}} \approx \frac{1}{4\pi\varepsilon_0}\frac{2em_e^2c^2}{3\hbar^2} \approx 6.5 \times 10^{15}\left[\frac{\text{V}}{\text{m}}\right]; \tag{11.37}$$

the corresponding intensity is

$$I_0^{\text{quantum}} \approx 10^{25}\left[\frac{\text{W}}{\text{cm}^2}\right] \quad \text{at } \lambda = 1\ \mu\text{m}\,. \tag{11.38}$$

Let us summarize:

$$5.2 \times 10^{23}\left[\frac{\text{W}}{\text{cm}^2}\right] \leq I_0 \leq 10^{25}\left[\frac{\text{W}}{\text{cm}^2}\right] \quad \text{at } \lambda = 1\ \mu\text{m} \tag{11.39}$$

characterizes the intermediate region (transition region) of the ultrarelativistic regime.

The ultrarelativistic QED regime

For even greater intensities, we reach the regime in which vacuum breakdown with spontaneous pair production becomes possible. It should be noted that the field can perform enough work on a virtual electron-positron pair to induce vacuum breakdown, i.e.,

$$eE_{\text{field}}\bar{\lambda}_C > 2m_e c^2 \tag{11.40}$$

should hold, and we find the Schwinger field

$$E_{\text{field}}^{\text{Schwinger}} = \frac{m_e^2 c^3}{e\hbar} \approx 1.3 \times 10^{16} \left[\frac{\text{V}}{\text{cm}}\right] . \tag{11.41}$$

The corresponding normalized vector potential is

$$a_{\text{Schwinger}} = \frac{m_e c^2}{\hbar\omega} \approx 4 \times 10^5 . \tag{11.42}$$

Such fields are reached at laser intensities for which

$$I_0^{\text{Schwinger}} \approx 4.5 \times 10^{29} \left[\frac{\text{W}}{\text{cm}^2}\right] \quad \text{for } \lambda = 1\ \mu\text{m} . \tag{11.43}$$

applies. At this intensity, we enter the nonlinear QED regime.

Hawking-Unruh Radiation

The original works that initiated the discussion on Unruh radiation are the following references: [254–257]. Subsequently, many papers on this interesting topic appeared, e.g., [258].

The existence of Unruh radiation is not universally accepted. Some claim that it has already been observed, while others assert that it is not emitted at all. While the skeptics accept that an accelerated object thermalizes at the Unruh temperature, they do not believe this leads to the emission of photons, since the emission and absorption rates of the accelerated particle are in equilibrium.

Here we follow the interpretation of K.T. McDonald [259]. According to Hawking, an observer outside a strong gravitational field, such as that found in a black hole, experiences a bath of thermal radiation with a temperature T, given by

$$k_B T_{\text{Hawking}} = \frac{\hbar g}{2\pi c} , \tag{11.44}$$

where g is the local acceleration due to gravity. The gravitational field interacts with the quantum fluctuations of the electromagnetic field, allowing energy to be transferred to an "observer." If the temperature is equivalent to 1 MeV or more, virtual electron-positron pairs emerge from the vacuum as real particles.

The idea of W.G. Unruh was that, similar to Hawking radiation, a uniformly accelerated detector also experiences a bath of thermal radiation. An accelerated observer in a gravity-free environment is (locally) subject to the same physics as a stationary observer in a gravitational field. The temperature should be given by

$$k_B T_{\text{Unruh}} = \frac{\hbar a^*}{2\pi c} , \tag{11.45}$$

where a^* is the acceleration measured in the observer's instantaneous rest frame.

The "observer" could be an electron that is accelerated by an electromagnetic field E_{field}. Suppose the characteristic energy $k_B T_{\text{Unruh}}$ is much smaller than 1 MeV. Then, the Thomson scattering of the electron on photons in the apparent thermal bath would be interpreted by a laboratory observer as an additional contribution to the radiation rate of the accelerated charge. The power of the additional (Unruh) radiation is then calculated from

$$\frac{dU_{\text{Unruh}}}{dt} = F_{\text{energy flux}} \times \sigma_{\text{Thomson}} , \tag{11.46}$$

where

$$F_{\text{Energiefluss}} = U_{\text{Unruh}}\, c , \quad \sigma_{\text{Thomson}} = \frac{8\pi r_e^2}{3} . \tag{11.47}$$

Assume the radiation is in thermal equilibrium, so that the energy density of the thermal radiation is given by the Planck expression,

$$\frac{dU_{\text{Unruh}}}{d\nu} = \frac{8\pi}{c^3} \frac{h\nu^3}{e^{h\nu/k_B T_{\text{Unruh}}} - 1} , \tag{11.48}$$

where ν is the frequency. Note that these relations apply in the instantaneous rest frame of the electron. Combining (11.46) with (11.48) yields

$$\frac{dU_{\text{Unruh}}}{dt d\nu} = \frac{8\pi}{c^2} \frac{h\nu^3}{e^{h\nu/k_B T_{\text{Unruh}}} - 1} \frac{8\pi r_e^2}{3} . \tag{11.49}$$

After integration over ν we find a Stefan–Boltzmann-like relation

$$\frac{dU_{\text{Unruh}}}{dt} = \frac{8\pi^3 \hbar r_e^2}{45 c^2} \left(\frac{k_B T_{\text{Unruh}}}{\hbar} \right)^4 . \tag{11.50}$$

With (11.45) we ultimately obtain

$$\boxed{\frac{dU_{\text{Unruh}}}{dt} = \frac{\hbar r_e^2 a^{*4}}{90\pi c^6} \approx \frac{e^4 E_{\text{field}}^4 \hbar r_e^2}{90\pi m^4 c^6}} . \tag{11.51}$$

What is important is the proportionality to the fourth power of $a^* \approx eE_{\text{field}}/m$. If we compare this with other radiation, e.g., Larmor radiation

$$\frac{dU_{\text{Larmor}}}{dt} = \frac{1}{4\pi\varepsilon_0} \frac{2e^2 a^{*2}}{3c^3} , \tag{11.52}$$

we find

$$\frac{dU_{\text{Unruh}}}{dt} \geq \frac{dU_{\text{Larmor}}}{dt} \tag{11.53}$$

for

$$E_{\text{field}} \geq E_{\text{field}}^{\text{Unruh}} \equiv \sqrt{\frac{60\pi}{\alpha} \frac{m^2c^3}{e\hbar}} \approx 3 \times 10^{18} \left[\frac{\text{V}}{\text{cm}}\right]. \tag{11.54}$$

Here, $\alpha = e^2/4\pi\varepsilon_0\hbar c$ is the fine-structure constant. The field required for significant Unruh radiation is much greater than the Schwinger field. The required acceleration is also extremely large, $a^{*\text{Unruh}} \approx \mathcal{O}(10^{31})g$.

Such fields are rare in nature, but are believed to occur at the surface of neutron stars and could play a role in pulsar physics. Critical fields can be generated temporarily in the laboratory by the superposition of Coulomb fields during the collision of two heavy atomic nuclei.

It is known that plasma wakefields, which are excited either by a laser pulse or an intense electron beam, can theoretically enable an acceleration rate of up to 100 GeV/cm or 10^{23} g. This acceleration is based on the collective perturbations of the plasma density, which are excited by the driving pulse and restored by the immobile ions, and is thus an effect over a plasma period.

However, another aspect of laser-driven electron acceleration is of significance: When a laser is ultrarelativistic, an electron can be instantaneously accelerated (and decelerated) under the direct influence of the laser in each laser cycle. This leads to intermittent acceleration that is much more violent than that provided by plasma wakefields.

For petawatt-class lasers (peta $\hat{=}10^{15}$), which, for example, are focused to $10\,\mu\text{m}^2$, resulting in $I_0 \approx 10^{22} - 10^{23}$ W/cm^2 for $\lambda \sim \mathcal{O}(1\ \mu\text{m})$, fields on the order of magnitude of $E_{\text{field}} \approx 6 \times 10^{12}$ V/cm are obtained. The corresponding acceleration is on the order of 2×10^{25} g.

Successful Applications [260]

A very nice summary of the developments over the past two decades was recently written by Riconda and Weber [260], which we will now partially reproduce verbatim. References to some older literature can be found in Ref. [91]. Here first is the general appreciation of the significance:

"In the past two decades, the importance of fully ionized plasmas for the controlled manipulation of high-power coherent light has increased significantly. Numerous ideas have been developed on how to control or modify the properties of laser pulses—such as their frequency, spectrum, intensity, and polarization. The corresponding interaction with a plasma can be either self-organized or achieved through prior adjustment. While theoretical studies and simulations have made extensive progress, there is currently a need for experimental verification and the associated detailed characterization of plasma-optical elements. Existing feasibility studies need to be extended to higher power levels. There is little doubt that plasmas offer enormous potential for future use in high-power optics."

The following is a chronological overview, in bullet points, of the development, taken from the aforementioned article by Riconda and Weber [260]:

- **1999:** A theoretical analysis of plasma amplification based on stimulated Raman scattering (SRS) is presented [87]. This identifies a self-similar solution (or a nonlinear regime) that leads to ultra-intense and ultra-short pulses based on electronic time scales. This is an example of how parametric instabilities can be advantageously utilized [in contrast to inertial confinement fusion (ICF), where they are considered detrimental to the implosion process].
- **2005:** A remarkable experiment is conducted demonstrating Raman amplification in the nonlinear regime [261].
- **2006:** Plasma amplification is analyzed using ion-based gratings in the so-called strong coupling regime of stimulated Brillouin scattering (sc-SBS). A self-similar solution is also identified in this regime [91].
- **2009:** A multicolor scheme for controlled energy transfer between beams to fine-tune implosion symmetry in ICF experiments is proposed [262].
- **2010:** It is shown that ellipsoidal plasma mirrors can be used to refocus high-power laser beams for intensity enhancement [263, 264].
- **2014:** The generation of spatial structures in overdense plasmas at the surface of originally smooth solids by optical lasers is experimentally demonstrated. The interaction of these transient structures with an ultra-intense laser pulse is also reported [265]. Plasma mirrors and their optical properties, including harmonic generation, are presented [266].
- **2016:** A scheme for generating transient photonic crystals for high-power lasers by tuning counter-propagating laser beams in a plasma in the sc-SBS regime is proposed [267]. An experimental demonstration of a plasma wave plate, based on laser-induced birefringence, is reported, in which a large-area, low-intensity ($I = 1013$ W/cm^2) laser beam interacts with a second beam in a gas-jet plasma [268].
- **2017:** Extremely high gain and significant energy transfer via Raman amplification are reported, with 170 mJ extracted from a 70-J pump laser [269]. It is shown that a

holographic prepulse beam focused onto a flat solid target generates modulations that persist for picoseconds and can be used as plasma holograms [270].

- **2018:** It is proposed that energy redistribution between multiple energy beams interacting in a plasma can produce a well-collimated beam. Such a plasma-based beam combiner is experimentally demonstrated at the National Ignition Facility [271]. Frequency conversion in ionized media (plasmas) is proposed. The evolution of wave frequency, amplitude, and energy density in a plasma with a temporally decreasing refractive index is investigated [272].
- **2019:** Record energy transfer beyond the joule level and very high efficiency of up to 20% in laser-plasma-based amplification in the sc-SBS regime are experimentally demonstrated [273]. A fluid model is proposed to predict the nonlinear dynamics and properties of quasi-neutral gratings in a spatially periodic ponderomotive potential. Such gratings can be generated by two intense lasers of the same frequency and exhibit characteristic growth times that depend on the laser amplitude [274].
- **2020:** An overview of experimental results on plasma-based amplification via SRS and sc-SBS is presented. By analyzing the nonlinear (or self-similar) regime, criteria are proposed to improve the efficiency of the scheme [275].
- **2021:** A new scheme for plasma-based frequency conversion and broadening via dynamic plasma gratings is proposed [276].
- **2022:** A theoretical scheme for generating a holographic plasma lens that can focus or collimate a probe pulse in an underdense plasma is proposed and experimentally demonstrated [277]. Plasma-based CPA is proposed, based on a compact high-power laser system utilizing plasma transmission gratings with currently achievable parameters [278].

11.2 Plasma Accelerators

Since the innovative ideas of Dawson and Tajima [279] and colleagues, confidence has grown that plasma accelerators will soon enable the construction of much more compact machines that—compared to the current state—can withstand electric accelerating fields up to a factor of 1000 higher. Two different approaches to electron acceleration with plasma accelerators are presented, and the requirements for the plasma and laser regime are briefly discussed.

In addition to the "plasma beat wave accelerator", the "laser wakefield accelerator" in the "bubble regime" is currently of particular interest. We begin with the older concept.

Plasma Beat Wave Accelerator (PBWA)

Before the CPA technique ("chirped-pulse amplification technique") was invented, it was not possible to generate short laser pulses of very high intensity. Therefore, in the 1970s, Dawson and Tajima proposed allowing two long laser pulses with frequencies ω_1 and ω_2 and wavenumbers k_1 and k_2 to interact and, through resonance, to excite a plasma oscillation that propagates with a high phase velocity.

If we describe the normalized vector potential of the lasers by

$$a = a_1 \cos(k_1 z - \omega_1 t) + a_2 \cos(k_2 z - \omega_2 t) , \tag{11.55}$$

then the square a^2 in the ponderomotive force $\nabla a^2/2$ yields contributions

$$(a^2)_{res} = a_1 a_2 \cos(\Delta k\, z - \Delta\omega\, t) \tag{11.56}$$

with

$$\Delta k = k_1 - k_2 , \quad \Delta\omega = \omega_1 - \omega_2 . \tag{11.57}$$

The goal is to achieve

$$\Delta\omega \approx \omega_{pe} \quad \text{and} \quad \Delta k \approx k_p \equiv \frac{\omega_{pe}}{c} \tag{11.58}$$

Then, the phase velocity of the beat approaches the order of magnitude of the group velocity of the lasers, i.e.,

$$\boxed{\frac{\Delta\omega}{\Delta k} \sim \mathcal{O}\left(v_g = c\left[1 - \frac{\omega_{pe}^2}{\omega_{1,2}^2}\right]\right) \sim \mathcal{O}(c)} . \tag{11.59}$$

Essentially, the theoretical description that follows is similar to that for stimulated Raman scattering. We can use a fluid-Maxwell system,

$$\frac{\partial n_e}{\partial t} + \nabla \cdot (n_e \mathbf{v}_e) = 0 , \tag{11.60}$$

$$\left(\frac{\partial}{\partial t} + \mathbf{v}_e \cdot \nabla\right)(\gamma_e \mathbf{v}_e) = -\frac{e}{m_e}(\mathbf{E} + \mathbf{v}_e \times \mathbf{B}) - \frac{3k_B T_e}{n_0 m_e} \nabla n_e , \tag{11.61}$$

$$\nabla \times \mathbf{E} = -\frac{\partial \mathbf{B}}{\partial t} , \tag{11.62}$$

$$\nabla \times \mathbf{B} = \mu_0 \mathbf{j} + \mu_0 \varepsilon_0 \frac{\partial \mathbf{E}}{\partial t} , \tag{11.63}$$

$$\nabla \cdot \mathbf{E} = \frac{e}{\varepsilon_0}(n_e - n_0) \, . \tag{11.64}$$

We discuss the response only on the fast electron timescale (e) and, for larger amplitudes (and the resulting nonlinear effects), take into account the relativistic factor

$$\gamma_e = \frac{1}{\sqrt{1 - \frac{v_e^2}{c^2}}} \, . \tag{11.65}$$

Since we have already performed similar calculations several times, we will abbreviate the derivation of a reduced model under the assumption of slowly varying amplitudes and refer directly to Esarey, Schroeder, and Leemans [280].

For the normalized electrostatic potential $\phi \sim \delta n_e$, neglecting electron nonlinearity, one obtains

$$\left(\frac{\partial^2}{\partial t^2} + \omega_{pe}^2\right)\phi = \frac{\omega_{pe}^2}{2} a_1 a_2 \cos(\Delta k\, z - \Delta\omega\, t) \, . \tag{11.66}$$

If the resonance condition $\Delta\omega = \omega_{pe}$ is exactly satisfied, the solution of the second-order differential equation exhibits secular growth $\sim t$.

With Rosenbluth and Liu [281], we write the solution as

$$\phi = -\frac{1}{4} a_1 a_2 k_p |\underbrace{z - ct}_{\zeta}| \sin(\Delta k\, z - \Delta\omega\, t) \, . \tag{11.67}$$

Here, ζ can be interpreted as the distance behind the laser front. The phase velocity of the "electric wave" is

$$v_{ph} = c\left(1 - \frac{\omega_{pe}^2}{\omega_1 \omega_2}\right) . \tag{11.68}$$

With a view to the LWFA in the following section, we can venture the following interpretation: The laser beat wave essentially acts like a series of laser pulses, each of which has an amplitude $a_1 a_2$ and a duration $\Delta\tau = 2\pi/\Delta\omega$. Each of these pulses generates a wakefield with amplitude $\sim \frac{\pi}{2} a_1 a_2$. Over a length $L = N\lambda_p$ the field sums to an amplitude $\sim N\frac{\pi}{2} a_1 a_2$, where N is the number of periods of the laser beat wave within the pulse.

If nonlinear effects due to relativistic mass change are taken into account, (11.66) extends to

$$\left(\frac{\partial^2}{\partial t^2} + \omega_{pe}^2\right)\phi = \frac{3}{8}\omega_{pe}^2 \phi^3 + \frac{\omega_{pe}^2}{2} a_1 a_2 \cos(\Delta k\, z - \Delta\omega\, t) \, . \tag{11.69}$$

Even before wave breaking, for the density perturbation normalized with n_0 at $\delta n \sim \mathcal{O}(1)$ saturation occurs at

$$\boxed{\phi_{sat} = (2\pi a_1 a_2)^{1/3}} \tag{11.70}$$

This saturation is explained by the nonlinear increase of the wave period over larger distances.

Example 11.2 (Field Strength Values)
If one considers theoretically possible values, for $n \approx 10^{19}$ cm^{-3} a maximum field strength of

$$E \sim \mathcal{O}\big(10^9\big)\frac{\mathrm{V}}{\mathrm{cm}}\ , \tag{11.71}$$

is obtained, which would accelerate electrons to GeV energies over a distance of 1 cm. The Coulomb field strength of a proton at the Bohr radius is (however, only) about five times as large. ■

Two limitations must be mentioned. First, the above calculations assume that, despite the relative length, everything happens quickly, i.e., before the ions come into play. And second, even more importantly, that effective injection of particles into the wave can be achieved. The latter is a major experimental challenge.

Laser Wakefield Accelerator (LWFA)

The LWFA is particularly relevant today, since, with the help of CPA technology, short, high-intensity laser pulses ($\geq 10^{19}$ W cm^{-2} at focus) can be generated. The wakefield behind a short laser pulse has already been discussed. Now, the basic equations for simulation practice [282].

We start with the well-known momentum balance for relativistic electrons, which we now write without the pressure term for a spatially one-dimensional electron motion:

$$\frac{\partial \mathbf{p}}{\partial t} + v_z \frac{\partial \mathbf{p}}{\partial z} = -e(\mathbf{E} + \mathbf{v} \times \mathbf{B})\ , \tag{11.72}$$

where

$$\mathbf{p} = m_e \gamma_e \mathbf{v}\,, \quad \gamma_e = \sqrt{1 + \frac{p^2}{m_e^2 c^2}}\,. \tag{11.73}$$

The direction of propagation is z. Since the laser pulse with vector potential $\mathbf{A}_\perp$ generates an ambipolar field ϕ in the plasma, we write

$$\mathbf{E} = -\frac{\partial \mathbf{A}_\perp}{\partial z} - \hat{z}\frac{\partial \phi}{\partial z}\,, \quad \mathbf{A}_\perp = \hat{x}A_x + \hat{y}A_y\,, \quad \mathbf{B} = \nabla \times \mathbf{A}_\perp\,. \tag{11.74}$$

As discussed earlier, we can deduce

$$\mathbf{p}_\perp = e\mathbf{A}_\perp \equiv m_e c\mathbf{a}(z,t) \tag{11.75}$$

. This also allows us to transform γ_e. A straightforward calculation yields

$$\gamma_e = \gamma_a \gamma_\parallel\,, \quad \gamma_a \equiv \sqrt{1 + \mathbf{a}^2}\,, \quad \gamma_\parallel \equiv \sqrt{1 - \frac{v_z^2}{c^2}}\,. \tag{11.76}$$

The longitudinal component of the momentum balance, the electron continuity equation, the Poisson equation for $\varphi = \frac{e\phi}{m_e c^2}$, and the wave equation for $\mathbf{a}$ yield, for $\varepsilon_0 \mu_0 = c^{-2}$,

$$\boxed{\frac{1}{c}\frac{\partial}{\partial t}\left(\gamma_a\sqrt{\gamma_\parallel^2 - 1}\right) + \frac{\partial}{\partial z}(\gamma_a \gamma_\parallel) = \frac{\partial \varphi}{\partial z}}\,, \tag{11.77}$$

$$\boxed{\frac{1}{c}\frac{\partial n}{\partial t} + \frac{\partial}{\partial z}\left(n\frac{\sqrt{\gamma_\parallel^2 - 1}}{\gamma_\parallel}\right) = 0}\,, \tag{11.78}$$

$$\boxed{\frac{\partial^2 \varphi}{\partial z^2} = \frac{\omega_{p0}^2}{c^2}\left(\frac{n}{n_0} - 1\right)}\,, \tag{11.79}$$

$$\boxed{c^2\frac{\partial^2 \mathbf{a}}{\partial z^2} - \frac{\partial^2 \mathbf{a}}{\partial t^2} = \omega_{p0}^2 \frac{n}{n_0}\frac{\mathbf{a}}{\gamma_a \gamma_\parallel}}\,. \tag{11.80}$$

With an ansatz for slowly varying amplitudes

$$\mathbf{a}(z,t) = \frac{1}{2}\mathbf{a}_0(\xi,\tau)e^{-i\theta} + c.c.\,, \quad \xi = z - v_g t\,, \; v_g = \frac{\partial \omega_0}{\partial k_0}\,, \; \theta = \omega_0 t - k_0 z\,, \tag{11.81}$$

we obtain, after some calculation, for $\frac{\partial^2}{\partial \tau^2} \ll \omega_0^2$ from (11.80)

$$\left[2\frac{\partial}{\partial \tau}\left(i\omega_0 a_0 + v_g\frac{\partial a_0}{\partial \xi}\right) + c^2\left(1 - \frac{v_g^2}{c^2}\right)\frac{\partial^2 a_0}{\partial \xi^2} + 2i\omega_0\left(\frac{c^2 k_0}{\omega_0} - v_g\right)\frac{\partial a_0}{\partial \xi}\right]e^{-i\theta} + c.c.$$

$$= \left[c^2 k_0^2 - \omega_0^2 + \frac{n}{n_0} \omega_{p0}^2 \gamma_a \gamma_\| \right] a_o e^{-i\theta} + c.c. \tag{11.82}$$

This equation shows the change of the pump pulse due to the excitation of a wakefield. Of course, it is coupled to the rest of the system of equations, which can be seen, for example, from the fact that n and $\gamma_\|$ appear in it. Only the higher time derivatives of a_0 have been neglected.

The quasistationary approximation (in the reference frame of the group velocity v_g) refers to the assumption that the dependence on z and t in Eqs. (11.77) and (11.78) occurs only through the combination $z - v_g t$. In this case, each equation can be integrated once. Using the boundary condition

$$\text{at } |a_0|^2 = 1 \text{ and } \gamma_a = 1: \quad n = n_0\,, \; \gamma_\| = 1\,, \; \varphi = 0\,, \tag{11.83}$$

one obtains, together with $\beta_0 = \frac{v_g}{c}$, the "conservation laws"

$$\gamma_a \left(\gamma_\| - \beta_0 \sqrt{\gamma_\|^2 - 1} \right) = \varphi + 1\,, \tag{11.84}$$

$$n \left(\beta_0 \gamma_\| - \sqrt{\gamma_\|^2 - 1} \right) = n_0 \beta_0 \gamma_\|\,. \tag{11.85}$$

In the Poisson equation (11.79), n appears, which we can relate to $\gamma_\|$ using (11.85). Equation (11.84) links $\gamma_\|$ with φ (where we always retain a_0 as a variable). Against this background, in the literature the coupled system of equations is often given in the form of the pulse equation (11.80) in the "slowly varying envelope approximation" together with the Poisson equation, with (11.84) being implicitly assumed:

$$\boxed{2i\omega_0\frac{\partial a_0}{\partial \tau}+2c\beta_0\frac{\partial^2 a_0}{\partial \tau \partial \xi}+\frac{c^2\omega_{p0}^2}{\omega_0^2}\frac{\partial^2 a_0}{\partial \xi^2}=-\omega_{p0}^2\mathcal{H}a_0}\,, \tag{11.86}$$

$$\boxed{\frac{\partial^2 \varphi}{\partial \xi^2}=\frac{\omega_{p0}^2}{c^2}\mathcal{G}} \tag{11.87}$$

with (11.84) and

$$\mathcal{H}=1-\frac{\beta_0}{\gamma_a\left(\beta_0\gamma_\parallel-\sqrt{\gamma_\parallel^2-1}\right)}\,,\quad \mathcal{G}=\frac{\sqrt{\gamma_\parallel^2-1}}{\beta_0\gamma_\parallel-\sqrt{\gamma_\parallel^2-1}}\,. \tag{11.88}$$

Unfortunately, the two equations for the laser pulse a_0 and the wakefield φ cannot be solved analytically. However, numerical methods are part of the standard repertoire in the accelerator community.

Significant distortion of the rear end of the laser pulse occurs. The distortions arise where the wakefield potential reaches a minimum and the density a maximum. A spike forms because the photons interact with the inhomogeneity of the plasma density, with some photons being accelerated (decelerated) as they move down (up) the density gradient; this effect is known as photon acceleration. The distortion of the rear end increases with increasing values of ω_{p0}/ω_0.

The longitudinally varying potential is significantly greater than the fields achieved in the plasma beat wave accelerator (PBWA), where they are limited by relativistic detuning. Such saturation does not occur in the laser wakefield accelerator (LWFA).

A special regime emerges for high laser pulse intensities. It has been shown—initially in numerical simulations—that all electrons can be completely expelled from the vicinity of the axis. This regime is referred to as the "blow-out," "bubble," or "cavitation regime." Some of the plasma electrons can be recaptured and accelerated to high energies; see Fig. 11.2. A regime in which the plasma electrons are completely expelled from a region around the axis has been observed for both laser and electron beam drivers and is referred to as a *nonlinear* plasma wakefield accelerator.

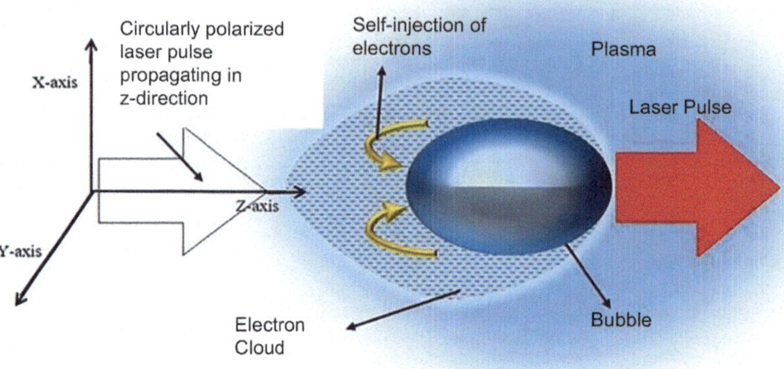

Fig. 11.2 Schematic representation of an LWFA in the (nonlinear) bubble regime

In the blow-out regime, all plasma electrons can be expelled from the vicinity of the driver and immediately behind it. The blow-out region of the wake is characterized by an accelerating field that is constant as a function of radius and changes linearly with the distance behind the driver. In addition, there is a focusing field that varies linearly with the radius. This regime can exhibit advantageous acceleration properties, as the focusing forces are linear; as a result, the normalized emittance of an accelerated electron bunch is preserved.

In the experiments at SLAC, the blow-out wakefield led to an energy gain of more than 40 GeV for a portion of the electrons in the rear part of the bubble. However, the majority of the electrons in the main part of the bubble lost energy.

11.3 Plasma Gratings

In this section, we revisit ideas that were already outlined in the context of parametric instabilities and also in LBWA. The plasma is influenced by the laser via the ponderomotive force. When laser pulses interact, waves (oscillations) can arise in the plasma. If these structures move little (if at all), and ions come into play on a longer timescale, we are led to new plasma-optical components, such as plasma gratings.

Particle Motion with Ponderomotive Force

We start with the equation of motion for an electron in electromagnetic fields,

$$\frac{d\mathbf{p}}{dt} = -e(\mathbf{E} + \mathbf{v} \times \mathbf{B}) \,. \tag{11.89}$$

In an electromagnetic wave, the electron motion is dominated by the oscillations in the transverse electric field. Within a nonrelativistic model, we immediately obtain

$$\mathbf{p}_\perp \approx e\mathbf{A}_\perp \,; \tag{11.90}$$

[see, e.g., Sect. 2.3 of Ref. [48]]. The $\mathbf{v} \times \mathbf{B}$ term produces, on average, a ponderomotive force (radiation pressure). Introducing the dimensionless form

$$\mathbf{a} = \frac{e}{m_e c}\mathbf{A} \tag{11.91}$$

we find

$$-e\mathbf{v} \times \mathbf{B} = -m_e c^2 \mathbf{a} \times \nabla \times \mathbf{a} = -m_e c^2 \left[\frac{1}{2}\nabla(\mathbf{a} \cdot \mathbf{a}) - (\mathbf{a} \cdot \nabla)\mathbf{a}\right] \approx -\frac{m_e c^2}{2}\nabla \mathbf{a}^2 \tag{11.92}$$

and thus the average force

$$\left\langle \frac{d\mathbf{p}}{dt} \right\rangle \approx -\nabla \phi_p \,, \tag{11.93}$$

with the ponderomotive potential

$$\boxed{\phi_p = \frac{m_e c^2}{2}\langle \mathbf{a}^2 \rangle} \,. \tag{11.94}$$

We now decompose the total vector potential into contributions from two pulses: (0) and (1). We then obtain

$$\mathbf{a} = \mathbf{a}_0 + \mathbf{a}_1 \,, \quad \langle (\mathbf{a}_0 + \mathbf{a}_1)^2 \rangle = \langle \mathbf{a}_0^2 \rangle + \langle \mathbf{a}_1^2 \rangle + 2\langle \mathbf{a}_0 \cdot \mathbf{a}_1 \rangle \,. \tag{11.95}$$

Circular Polarization
In the case of circular polarization, we start with the ansatz

$$\mathbf{a}_1 = \frac{1}{2}(\mathbf{e}_y + \mathrm{i}\mathbf{e}_z)a_1 e^{\mathrm{i}\theta_1} + c.c., \quad \mathbf{a}_0 = -\frac{1}{2}(\mathbf{e}_y + \mathrm{i}\mathbf{e}_z)a_0 e^{\mathrm{i}\theta_0} + c.c., \tag{11.96}$$

where $\theta_1 = k_1 x - \omega_1 t$ and $\theta_0 = -k_0 x - \omega_0 t$. In this notation, one (pump) wave propagates from right to left, while the other moves in the opposite direction. We can assume the amplitude of the laser pulse $a_0 = |a_0|$ to be real, while pulse (1) may have an additional slowly varying phase φ, i.e.,

$$a_1 = |a_1| e^{i\varphi} \,. \tag{11.97}$$

Then a short calculation yields [283]

$$\boxed{\frac{\phi_p}{m_e c^2} = \frac{|a_1|^2}{2} + \frac{a_0^2}{2} - |a_1| a_0 \cos(\psi + \varphi)} \,, \tag{11.98}$$

where

$$\psi \equiv \theta_1 - \theta_0 \approx 2k_1 x + \Delta\omega t \quad \text{for} \quad k_1 \approx k_0 \,, \ \Delta\omega = \omega_0 - \omega_1 \ll \omega_1 \,. \tag{11.99}$$

From momentum conservation, we obtain $k = k_0 + k_1 \approx 2k_0$. Obviously, we average over the rapid motion with frequency $\omega_0 \sim \omega_1$, but not over slow variations characterized by $\Delta\omega \equiv \omega_0 - \omega_1$.

Note that the ponderomotive force on ions is smaller by a factor of m_e/m_i.

Linear Polarization

For linear polarization, we start with

$$\mathbf{a}_1 = \mathbf{e}\left(\frac{a_1^*}{2} e^{i\theta_1} + c.c.\right), \quad \mathbf{a}_0 = -\mathbf{e}\left(\frac{a_0^*}{2} e^{i\theta_0} + c.c.\right), \tag{11.100}$$

which leads to the following ponderomotive potential:

$$\boxed{\frac{\phi_p}{m_e c^2} = \frac{|a_1|^2}{4} + \frac{a_0^2}{4} - \frac{1}{2}|a_1| a_0 \cos(\psi - \varphi)} \,. \tag{11.101}$$

The equation of motion for a single electron is then

$$\ddot{\psi}_j + \omega_b^2 \sin(\psi_j - \varphi) = 0 \,, \tag{11.102}$$

with the oscillation frequency (hereafter called the bounce frequency)

$$\omega_b = \sqrt{2}\sqrt{|a_1| a_0}\, \omega_1 \,. \tag{11.103}$$

We now consider counterpropagating, linearly polarized pulses in plasma at wavelengths around 800 nm. The identical frequencies are $\omega_0 = 2.35 \times 10^{15}\ \text{s}^{-1}$. The normalized vector potentials $\mathbf{a} = \frac{e}{m_e c}\mathbf{A}$ are written using (11.100) with $\theta_1 = k_1 x - \omega_1 t$ and $\theta_0 = -k_0 x - \omega_0 t$. Again: We are dealing with two laser beams propagating in opposite directions, and to illustrate the basic procedure, we use linear polarization of both laser fields with the polarization vector **e**.

A ponderomotive potential arises from the two oppositely propagating laser pulses; its (spatially relatively rapidly) varying component is

$$\boxed{\frac{\phi_p}{m_e c^2} \approx -\frac{1}{2}|a_1|a_0 \cos(\psi - \varphi)}\,, \tag{11.104}$$

with $\psi = 2k_1 x$; φ is a phase difference. We assume $\omega_0 = \omega_1$, i.e., $\Delta\omega \equiv 0$.

As we will see, the plasma's response depends on the laser intensity, plasma density, as well as the electron and ion temperature. However, the plasma always responds on two timescales. Initially, for short times, we observe a response exclusively from the electrons. This response can occur either collectively or individually.

Collective electron response

When electrons respond collectively, they can be modeled in macroscopic terms. For times on the order of the inverse electron plasma frequency ω_{pe}^{-1} we keep the ions fixed.

The simplest *linear* model (1D, vanishing temperature) for the electrostatic electron density fluctuations n_e and the velocity fluctuations v_e is given by

$$\frac{\partial n_e}{\partial t} + n_0 \frac{\partial v_e}{\partial x} = 0\,, \tag{11.105}$$

$$\frac{\partial v_e}{\partial t} = \frac{e}{m_e}\frac{\partial \phi}{\partial x} - \frac{1}{m_e}\frac{\partial \phi_p}{\partial x}\,, \tag{11.106}$$

$$\frac{\partial^2 \phi}{\partial x^2} = 4\pi e n_e\,, \tag{11.107}$$

$$\phi_p = -\frac{1}{2}a_0|a_1|\cos(\psi - \varphi) m_e c^2\,. \tag{11.108}$$

Here, ϕ is the electrostatic and ϕ_p the ponderomotive potential. If we introduce

$$\delta n_e \equiv \frac{n_e}{n_0} \tag{11.109}$$

we obtain the following equation for the normalized density variations:

$$\boxed{\frac{\partial^2 \delta n_e}{\partial t^2} + \omega_{pe}^2 \delta n_e = \omega_b^2 \cos(\psi - \varphi)}\,. \tag{11.110}$$

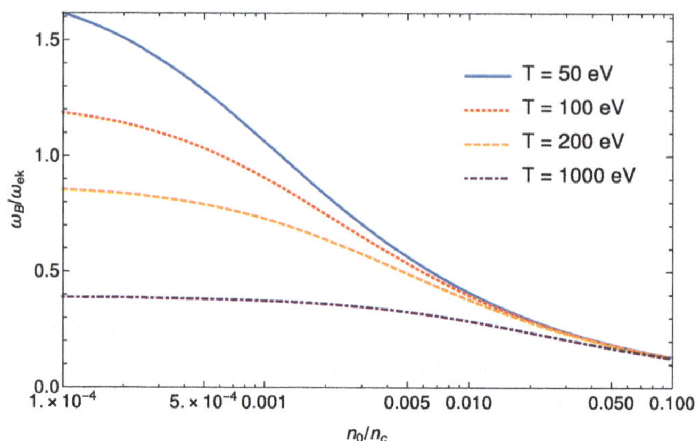

Fig. 11.3 The ratio of the bounce frequency $\omega_b = \sqrt{2a_0a_1}\omega_0$ to the Bohm-Gross frequency $\omega_{ek} = \sqrt{\omega_{pe}^2 + \frac{3}{2}v_{the}^2 k^2}$ with $k = 2k_0\sqrt{1 - n_0/n_c}$ as a function of plasma density n_0 (in units of the critical density n_c) for electron temperatures $T_e = 50$ eV, 100 eV, 200 eV, and 1000 eV. The amplitudes of the linearly polarized (pump) lasers are $a_0 = |a_1| = 0.03$

In this equation, the plasma frequency $\omega_{pe} = \sqrt{\frac{4\pi n_e e^2}{m_e}}$ competes with the bounce frequency $\omega_b = \sqrt{2|a_1|a_0}\,\omega_1$. It should be noted that the model can easily be generalized to finite electron temperatures T_e. In this case, the electron plasma frequency must be replaced by the Bohm-Gross frequency:

$$\boxed{\omega_{pe} \quad \rightarrow \quad \omega_{ek} = \sqrt{\omega_{pe}^2 + \frac{3}{2}v_{the}^2 k^2}}\,, \tag{11.111}$$

where v_{the} is the electron thermal velocity $\sqrt{\frac{T_e}{m_e}}$ is. A comparison of ω_b with ω_{ek} is shown in Fig. 11.3.

Several parameters determine the ratio between the two frequencies, namely the background density n_0, the electron temperature T_e, and the pump amplitudes a_0, $|a_1|$. For example: the lower the density and the stronger the (pump) lasers, the earlier we enter the trapping regime. For $\omega_b/\omega_{ek} < 1$ we expect a collective electron response, whereas for $\omega_b/\omega_{ek} > 1$ an individual electron response is to be expected.

Individual electron response

When ponderomotive effects dominate, i.e., when the bounce frequency becomes greater than the plasma frequency, a trapping regime for the electrons should set in immediately. This estimate results from a comparison of the (collective) oscillations with the motion induced by the ponderomotive force.

Oscillating particles occupy different regions at different times. At and near the turning points, the velocity is negligible, and therefore the probability density reaches its maximum at these points. Perfect bunching means that the density distribution $n(x, t)$ should be sharply peaked with a period of $\Delta x = \frac{\lambda_1}{2}$.
To calculate the density distribution, we start with the position x_j of the jth particle. By summing over all particles, we obtain the discrete particle density:

$$n_e(x, t) = \frac{1}{F} \sum_j \delta(x - x_j) \; , \tag{11.112}$$

where F is the unit area perpendicular to the x-axis.

For later applications (transparent matching condition within a three-wave model), the density is expanded in a Fourier series with respect to the phase difference $\psi = \theta_1 - \theta_0 \approx 2k_1 x$,

$$\boxed{n_e(\psi, t) = n_0 + \sum_{l=-\infty}^{\infty} \hat{n}_l e^{il\psi} \; , \; l \neq 0} \; . \tag{11.113}$$

In fact, this means that in the original density expression we replace $x \to \frac{\psi}{2k_1}$, with the result

$$n_e(\psi, t) = \frac{1}{F} \sum_j \delta\left(\frac{\psi}{2k_1} - x_j\right) . \tag{11.114}$$

The coefficients are calculated as

$$\hat{n}_l(\psi_0, t) = \frac{1}{2\pi} \int_{\psi_0-\pi}^{\psi_0+\pi} n_e(\psi, t) e^{-il\psi} d\psi \; . \tag{11.115}$$

Note that the period length in x results from the periodicity of ψ, i.e.

$$\Delta\psi \equiv 2k_1 \Delta x \stackrel{!}{=} 2\pi \quad \to \quad \Delta x = \frac{\lambda_1}{2} \; . \tag{11.116}$$

Here, $\frac{\lambda_1}{2}$ is the period of the ponderomotive potential used for averaging; ψ_0 is a position in the potential well. For an infinite and strictly periodic beat pattern, $\hat{n}_l$ would not depend on ψ_0. However, for pulses of finite duration, a dependence on ψ_0 arises.

By substituting the expression for the density, we obtain

$$\hat{n}_l(\psi_0, t) = \frac{2k_1}{2\pi F} \sum_j \int_{\psi_0-\pi}^{\psi_0+\pi} \delta(\psi - 2k_1 x_j) e^{-il\psi} d\psi$$
$$= \frac{1}{\frac{\lambda_1}{2} F n_0} n_0 \sum_{j, |\psi_j - \psi_0| < \pi} e^{-il\psi_j} \equiv n_0 \langle e^{-il\psi_j} \rangle_{\frac{\lambda_1}{2}} , \qquad (11.117)$$

where $\psi_j = 2k_s x_j(t)$ and $\frac{\lambda_1}{2} F n_0$ are the number of particles with index j in the potential well.

In summary:

$$\boxed{\hat{n}_l(\psi_0, t) = n_0 \langle e^{-il\psi_j} \rangle_{\frac{\lambda_1}{2}}} . \qquad (11.118)$$

In the case of infinite extent and ideal periodicity, the coefficients will be constant.

Example 11.3 (Field Expansion)
Analogous to the density, we expand the longitudinal electric field in a Fourier series:

$$E_x(\psi, t) = E_{x0} + \sum_{l=-\infty}^{\infty} \hat{E}_{xl} e^{il\psi} , \quad \text{for } l \neq 0 . \qquad (11.119)$$

Via the Poisson equation, we find the relation for the Fourier coefficients,

$$2ilk_1 \hat{E}_{xl} = -4\pi e \hat{n}_l = -\frac{m_e \omega_{pe}^2}{e} \langle e^{-il\psi_j} \rangle_{\frac{\lambda_1}{2}} . \qquad (11.120)$$

■

The force exerted by these fields must be added to the ponderomotive force. The resulting equation of motion for an electron is

$$\boxed{\underbrace{\ddot{\psi}_j + \omega_b^2 \sin(\psi_j + \varphi)}_{\text{bounce}} = \underbrace{-i\omega_{pe}^2 \sum_{l=-\infty}^{\infty} \frac{\hat{n}_l}{l n_0} e^{il\psi_j}}_{\text{collective}}} . \qquad (11.121)$$

In the simplest version, the equation of motion could be written as follows:

$$\ddot{\psi}_j + \omega_b^2 \sin \psi_j = \omega_{pe}^2 \frac{\hat{n}_1}{n_0} \sin \psi_j . \qquad (11.122)$$

By comparing the prefactors, the criterion for the dominant bounce motion follows.

The collective oscillations are limited by wave breaking. In its simplest form, the criterion for wave breaking leads to a maximum density fluctuation [284]:

$$\left.\frac{\delta n_e}{n_0}\right|_{max} \approx \frac{1}{2} . \tag{11.123}$$

This should lead to the more precise criterion for superradiance:

$$\omega_b^2 \geq \frac{1}{2}\omega_{pe}^2 . \tag{11.124}$$

Many effects, such as temperatures, velocities, variations of the envelope, etc. [133] influence wave breaking. We should expect a transition only at approximately $\omega_b \approx \omega_{pe}/\sqrt{2}$ from the collective motion of the plasma to the motion of individual, independently trapped electrons in the ponderomotive potential.

Ion motion at a later time

On the second timescale, which is appropriate for later times in both models, we consider the ion motion. Bunched electrons generate an electrostatic field E_x, which follows from the Poisson equation:

$$\frac{\partial}{\partial x}E_x = -4\pi e n_0 \delta n_e . \tag{11.125}$$

Rapid variations on the timescale of the electrons (inverse electron plasma frequency) do not affect the ions. The latter respond only to the average field $\langle E_x \rangle$.

Neglecting the ponderomotive force on the ions, we have a force in the x-direction

$$\frac{dp_i}{dt} = Ze\langle E_x \rangle . \tag{11.126}$$

After a typical time t_g (see below), the ions will approximately compensate the electron charge distribution.

Once the adjustment to the electrons in the density maxima is nearly complete, the electrons can, however, be further compressed. In the optimal phase, we have bunched electrons together with ions. The maximum density reached in the density maxima during this phase is determined by the ion temperature. The higher the temperature, the lower the maximum density.

Particle distribution in the collective electron regime
We consider the case $\omega_{pe} > \omega_b \gg \omega_{pi}$ (where we always assume $\Delta\omega = 0$) again in detail, using typical parameters.

In the present case, we distinguish between fast electron oscillations with the electron plasma frequency ω_{pe} and slower bounce behavior in the ponderomotive potential wells.

Starting from $\psi = 2k_s x$, a particular solution of the inhomogeneous Eq. (11.110) is given by

$$\delta n_e(x) = \frac{\omega_b^2}{\omega_{pe}^2}\cos(2k_s x - \varphi) , \tag{11.127}$$

while the general solution of the homogeneous equation represents a plasma oscillation:

$$\delta n_e(x,t) = A(x)\cos(\omega_{pe} t + \varphi_1(x)) . \tag{11.128}$$

For $t = 0$ the density variation should vanish, which points to the following general solution:

$$\delta n_e(x) = \frac{\omega_b^2}{\omega_{pe}^2}\cos(2k_s x - \varphi)[1 - \cos(\omega_{pe} t)] , \tag{11.129}$$

which satisfies (11.110).

If we (e.g., for $\varphi \equiv 0$) analyze the variation of $\delta n_e(x)$ by tracking $\cos(2k_s x)[1 - \cos(\omega_{pe} t)]$ at different times t, we clearly find a resulting density distribution whose minimum lies at the maximum of the ponderomotive potential distribution ϕ_p; see Fig. 11.4(a).

When averaged over the fast oscillations, a variation remains that is proportional to $\cos(2k_s x)$, i.e., an average bunching of the electrons is present. The bunching as well as the oscillation with the Bohm-Gross frequency are clearly visible in numerical Vlasov simulations, which are shown in Fig. 11.5. The driven electron plasma oscillations become apparent when the ions also begin to move.

Using the Poisson equation (11.125), we obtain the longitudinal electric field.

$$E_x = -4\pi e n_0 \frac{\omega_b^2}{2k_s\omega_{pe}^2}\sin(2k_s x + \varphi)[1 - \cos(\omega_{pe} t)] . \tag{11.130}$$

Its mean value

$$\langle E_x\rangle = -4\pi e n_0 \frac{\omega_b^2}{2k_s\omega_{pe}^2}\sin(2k_s x + \varphi) \tag{11.131}$$

drives the ions. In Fig. 11.4(a), the forces would point from the center toward the edges; see also Fig. 11.4(b). For simplicity, in the following discussion we assume that $\varphi \equiv 0$. The electric field is exactly zero at the center of Fig. 11.4(a) at $x = \frac{\lambda}{4}$, as it should be due

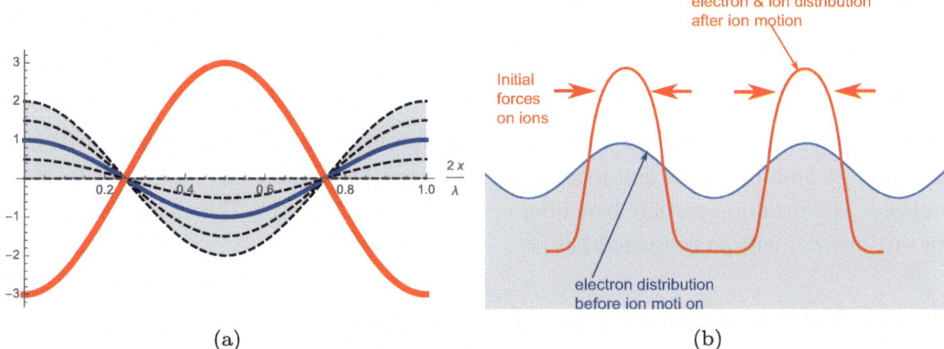

Fig. 11.4 (**a**) Sketch of the electron density oscillation in the early phase of the temporal evolution. The thick blue curve represents the averaged electron density distribution (averaged over fast electron plasma oscillations). The corresponding distribution of the ponderomotive potential ϕ_p is shown in the figure by the (thick) red curve (not to scale). (**b**) The situation after the initial electron bunching is shown. The averaged electron distribution from (a) is represented by the blue line. The electrostatic forces (arrows) accelerate ions, which in turn attract and drag along electrons. This ultimately leads to enhanced bunching of both electrons and ions, sketched by the red line

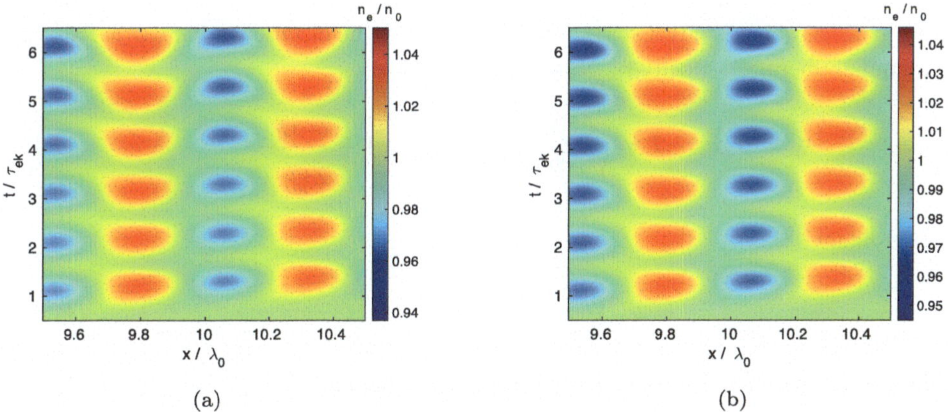

Fig. 11.5 Vlasov simulations show the temporal evolution of the electron density n_e/n_0 over two spatial periods of the ponderomotive lattice, which is generated by two linearly polarized lasers with amplitudes $a_0 = |a_1| = 0.03$. The initial homogeneous plasma density is $n_0 = 0.1n_c$, the electron temperature is (**a**) $T_e = 50$ eV and (**b**) $T_e = 200$ eV. Density oscillations with the Bohm-Gross period $t_{ek} = 2\pi/\omega_{ek}$ are clearly observable

to symmetry. Individually, we have the equation of motion

$$\boxed{\frac{dp_i}{dt} \approx Ze\langle E_x \rangle} \tag{11.132}$$

for an ion when the ponderomotive force on the ion is neglected. This leads to enhanced bunching of both electrons and ions at the locations where electrons have already accumulated. The resulting particle bunching is shown in Fig. 11.4(b). The characteristic time for this process will be estimated below.

Example 11.4 (Estimation of the Ion Response Time)
The ions move from both sides toward the maxima. The plasma becomes nearly quasineutral (since the electrons compensate the ions), except at the maxima, where complicated dynamics occur. There, a narrow and sharp electrostatic field can extend over an electron Debye length.

By integrating the equation of motion once (after multiplying by $\frac{dx_i}{dt}$), we obtain

$$\left(\frac{dx_i}{dt}\right)^2 = 2Z\frac{\omega_{pi}^2}{\omega_{pe}^2}\frac{\omega_b^2}{4k_s^2}[\cos(2k_s x_i) + 1] + v_0^2 \,, \tag{11.133}$$

when considering an ion that starts at the maximum of the electric field at $x_i(t=0) = \frac{3}{8}\lambda_s$ with initial (thermal) velocity v_0. Now, let us introduce:

$$X_i = 2k_s x_i \,, \quad T = \sqrt{2Z\frac{m_e}{m_i}}\omega_b t \,, \quad V_0^2 = 1 + \frac{v_0^2}{Z\frac{m_e}{m_i}\frac{\omega_b^2}{2k_s^2}} \,, \tag{11.134}$$

so we obtain

$$\left(\frac{dX_i}{dT}\right)^2 = \cos X_i + V_0^2 \,. \tag{11.135}$$

Obviously, separation of variables is possible. The equation

$$T = \int_{\frac{3\pi}{2}}^{X_i} \frac{dX'}{\sqrt{V_0^2 + \cos X'}} \equiv E(X_i) \tag{11.136}$$

implicitly determines the position X_i of the ions as a function of time T.

The integral on the right-hand side is directly related to the elliptic integral of the first kind

$$F(\phi, k) \equiv \int_0^{\phi} \frac{d\theta}{\sqrt{1 - k^2 \sin^2\theta}} \,, \tag{11.137}$$

Since in (11.136) the identity $\cos X' = 1 - 2\sin^2\left(\frac{1}{2}X'\right)$ can be used, the right part of (11.136) varies between 0 and 1.2.

Then, using Eq. (11.134), we can estimate the (dimensional) time t_g for the formation of the ion lattice,

$$t_g \sim \mathcal{O}\left(\sqrt{\frac{m_i}{Zm_e}}\omega_b^{-1}\right). \tag{11.138}$$

■

Fluid model for the temporal evolution of the lattice
We now discuss a simple analytical model that explains the main findings of the previous subsection. Very detailed studies of the nonlinear dynamics of *homogeneous* laser-generated ion plasma gratings already exist [274]. For the discussion of *inhomogeneous* plasma gratings, we start from a simplified homogeneous model [285]. It has turned out that this latter model describes the initial phase of the homogeneous situation quite well.

We will normalize the frequency ω with the pump frequency ω_0, the time t with $2\pi/\omega_0$ (approx. 2.67 fs), distances with the laser wavelength λ_0 in vacuum (800 nm), and wavenumbers k with $k_0 \equiv 2\pi/\lambda_0$. In the plasma, the pump wavenumber is $k_1 = k_0 N_0$. The mean density n_0 is used for density normalization, while the speed of light c serves as the velocity unit. Then, $2k_1 x \rightarrow 4\pi N_0 x$ holds in dimensionless form.

A laser pulse propagates in the plasma with the group velocity $v_{g0} = cN_0$. Normalized vector potentials $\mathbf{a} = \frac{e}{m_e c}\mathbf{A}$ are used.

The individual pulse envelopes are assumed as

$$\text{pump envelope} \sim \exp\left[-\frac{(x \mp x_0 \pm v_{g0}t)^2}{2\langle x^2\rangle}\right], \tag{11.139}$$

where $\langle x^2\rangle$ denotes the root mean square width of the pump pulses. Initially, the two counterpropagating pump pulses are well separated when they are centered at $\pm x_0$. The overlap at later times leads to the factor

$$\mathcal{E}_{a_0a_0} = e^{-\frac{-x^2}{\langle x^2\rangle}}\, e^{-\frac{-v_{g0}^2 t^2}{\langle x^2\rangle}} \tag{11.140}$$

for the combined effect, when we reset the zero point of the time axis accordingly. Then the factor $\mathcal{E}_{a_0a_0}$ appears in the ponderomotive potential during the head-on collision of two oppositely propagating single pump pulses.

In the following, it is assumed that the variation of the envelopes on the λ_0 scale is slow. Generated by two pump pulses with amplitudes a_0, the rapidly varying part of the ponderomotive potential is ϕ_p (here still in dimensional form) [286]

$$\frac{\phi_p}{m_e c^2} \approx -\frac{1}{2} a_0^2 \mathcal{E}_{a_0 a_0} \cos(2k_1 x - \varphi) , \tag{11.141}$$

where φ represents the phase difference. Space charges gènerate an averaged electric field

$$\langle E_x \rangle \approx -4\pi e n_0 \frac{\omega_{b0}^2}{2k_1 \omega_{pe}^2} \mathcal{E}_{a_0 a_0} \sin(2k_1 x + \varphi) , \tag{11.142}$$

which we will describe with the bounce frequency of the electrons $\omega_{b0} = \sqrt{2} a_0 \omega_0$ have written. According to Ma et al. [285], we can determine a fluid velocity u_i of the ions from [285, 286],

$$m_i \frac{\partial u_i}{\partial t} \approx Ze \langle E_x \rangle . \tag{11.143}$$

We set $Z = 1$. The dimensionless formulation using the units mentioned above and introducing

$$b = 2\pi a_0^2 N_0^{-1} \frac{m_e}{m_i} , \quad h = 4\pi N_0 , \tag{11.144}$$

follows as

$$\boxed{\frac{\partial u_i}{\partial t} = -b\, e^{-N_0^2 t^2/\langle x^2 \rangle} \sin(hx)\, e^{-x^2/\langle x^2 \rangle}} . \tag{11.145}$$

In fact, the pump lasers are effective as long as $t \lesssim \sqrt{\langle x^2 \rangle}/N_0$. However, since we are interested in the ion response on longer time scales, on the order of $t \sim \mathcal{O}\left(\sqrt{\frac{m_i}{m_e}\frac{n_c}{n_0}}\right)$, we can make the simplified assumption that the ion velocity remains constant at later times, at least in the simplest approximation. Using $\int_{-\infty}^{\infty} e^{-N_0^2 t^2/\langle x^2 \rangle} dt = \frac{\sqrt{\pi \langle x^2 \rangle}}{N_0}$, the zeroth-order result is

$$u_i^{(0)} \approx -b_0 \sin(hx)\, e^{-x^2/\langle x^2 \rangle} , \quad b_0 = 2\pi \sqrt{\pi} a_0^2 \sqrt{\langle x^2 \rangle} N_0^{-2} \frac{m_e}{m_i} . \tag{11.146}$$

A more precise calculation is, however, possible, starting from

$$u_i = -b \sin(hx)\, e^{-x^2/\langle x^2 \rangle} \int_{-\infty}^{t} e^{-N_0^2 t'^2/\langle x^2 \rangle} dt' . \tag{11.147}$$

We are guided by the formulation of Ma et al. [285] and use the density continuity equation

$$\boxed{\frac{\partial n}{\partial t} + u_i \frac{\partial n}{\partial x} = -n \frac{\partial u_i}{\partial t}} . \tag{11.148}$$

Note that the exponential function $e^{-x^2/\langle x^2\rangle}$ varies slowly in space compared to $\sin(hx)$ and $\cos(hx)$. We will use this fact to determine an approximate solution to the initial value problem. In addition, we introduce a new time τ defined by

$$d\tau = dt \int_{-\infty}^{t} e^{-N_0^2 t'^2/\langle x^2\rangle} dt' , \tag{11.149}$$

so that

$$\tau = \int_0^t dt'' \int_{-\infty}^{t''} e^{-N_0^2 t'^2/\langle x^2\rangle} dt' \tag{11.150}$$

applies for $t \geq 0$, which leads to the following relation:

$$\tau(t) = \frac{\sqrt{\pi\,\langle x^2\rangle}}{2N_0}\, t \left[1 + \operatorname{erf}\left(\frac{N_0 t}{\sqrt{\langle x^2\rangle}}\right)\right] + \frac{\langle x^2\rangle}{2N_0^2}\left[e^{-N_0^2 t^2/\langle x^2\rangle} - 1\right] . \tag{11.151}$$

Written out explicitly with respect to the time τ, the approximate equation for ion continuity becomes

$$\boxed{\frac{\partial n}{\partial \tau} - b\,\sin(hx)\,e^{-x^2/\langle x^2\rangle}\,\frac{\partial n}{\partial x} \approx b\,h\,\cos(hx)\,e^{-x^2/\langle x^2\rangle}\,n} . \tag{11.152}$$

To eliminate the coefficients h and b, in this section we modify the already dimensionless variables t and x to

$$\tilde{t} = hb\,\tau \;, \quad \tilde{x} = hx \; . \tag{11.153}$$

Then follows

$$\frac{\partial n}{\partial \tilde{t}} - \tilde{\mathcal{E}}\,\sin(\tilde{x})\,\frac{\partial n}{\partial \tilde{x}} \approx \tilde{\mathcal{E}}\,\cos(\tilde{x})\,n \;, \tag{11.154}$$

with

$$\tilde{\mathcal{E}} = e^{-\frac{\tilde{x}^2}{\langle \tilde{x}^2\rangle}} \;, \quad \langle \tilde{x}^2\rangle = h^2\langle x^2\rangle \; . \tag{11.155}$$

Next, we should familiarize ourselves with the following intuitive picture. We have already assumed that the envelope changes only slowly in space. This means we can introduce a local (position-dependent) time T (and, for aesthetic reasons, also $X \equiv \tilde{x}$), i.e.,

$$\boxed{T = \tilde{t}e^{-\frac{\tilde{x}^2}{\langle \tilde{x}^2\rangle}} \;, \quad X \equiv \tilde{x}} \; . \tag{11.156}$$

This *local* time reflects the fact that pulses at different locations effectively interact for different durations. The fundamental equation now becomes

$$\boxed{\frac{\partial n}{\partial T} - \sin(X)\,\frac{\partial n}{\partial X} = \cos(X)\,n}\ . \tag{11.157}$$

We have replaced the approximation symbol with an equals sign, fully aware that we are only seeking an approximate solution. As long as we do not concern ourselves with the initial condition, the solution to this quasilinear differential equation can be written with an arbitrary function F as follows:

$$n(X,T) = \frac{F(-\ln[\csc(X) + \cot(X)] + T)}{\sin(X)}\ . \tag{11.158}$$

At the time $T = 0$ we assume the condition $n(X,0) = 1$. With $z \equiv \sin(X)$ we obtain the relation

$$z = F\left(-\ln\left[\frac{1+\sqrt{1-z^2}}{z}\right]\right) \equiv F(f(z))\ . \tag{11.159}$$

Therefore, $f(z)$ is the inverse of $F(z)$. From

$$f(z) = F^{-1}(z) \quad \text{with} \quad f^{-1}(z) = \frac{2e^z}{(e^z)^2+1} \tag{11.160}$$

we obtain

$$F(Z) = \frac{2e^Z}{e^{2Z}+1}\ , \tag{11.161}$$

and thus

$$n(X,T) = \frac{1}{\sin(X)}\,\frac{2e^Z}{e^{2Z}+1}\bigg|_{Z=T-\ln[\csc(X)+\cot(X)]}\ . \tag{11.162}$$

From a direct evaluation of the right-hand side, we obtain the final expression

$$n(X,T) = 2e^T\,\frac{1+\cos(X)}{[1+\cos(X)]^2 + e^{2T}\sin^2(X)}\ . \tag{11.163}$$

Recall (11.153) and (11.156), which define the inhomogeneous envelope at time t. This leads to

$$\boxed{n(x,t) = 2e^T\,\frac{1+\cos(X)}{[1+\cos(X)]^2 + e^{2T}\sin^2(X)}\bigg|_{X=hx,\ T=hb\,\tau(t)\,e^{-x^2/\langle x^2\rangle}}}\ . \tag{11.164}$$

Table 11.1 Fourier coefficients for the lowest-order harmonics of the analytical solution for the plasma density (11.164) at $t = 560$. Note that all β_m vanish due to the even symmetry of the plasma density $n(x, t)$

m	0	1	2	3	4	5
$\alpha_m(560)$	2.000	0.89	0.40	0.18	0.08	0.03

Before we compare the analytical solution with a numerical one, we want to discuss an approximation that is used to evaluate the simple models presented in the following sections. In these models, we neglect deviations from purely harmonic behavior, i.e., we ignore higher harmonics.

Obviously, the result (11.164) consists of several harmonic contributions $\sim \cos(mhx)$, $m = 1, 2 \ldots$ [and possibly additional terms of odd parity $\sim \sin(mhx)$]. To estimate the significance of each Fourier component, we define time-dependent coefficients, namely

$$\alpha_m(t) = \frac{2}{L^*} \int_0^{L^*} \cos(mx)\, n(x, t)\, dx \;, \tag{11.165}$$

$$\beta_m(t) = \frac{2}{L^*} \int_0^{L^*} \sin(mx)\, n(x, t)\, dx \;, \tag{11.166}$$

where $L^* = 1/2N_0$ is the periodicity of $n(x, t)$ when x is measured in λ_0. These coefficients are taken from the homogeneous case, i.e., from $n(x, t)$ for $\langle x^2 \rangle \to \infty$. We assess their significance, for example, at the time $t = 560$ (corresponding to 1.5 ps) using Table 11.1.

Using the Fourier coefficients, we construct the following simple model for the inhomogeneous lattice density:

$$n^M_{approx}(x, t) = \frac{1}{2}\alpha_0(t) + \sum_{m=1}^{M} [\alpha_m(t) \cos(mhx) + \beta_m(t) \sin(mhx)]\, e^{-x^2/\langle x^2 \rangle} \;. \tag{11.167}$$

Even for small values of M, Eq. (11.167) provides an excellent approximation for $n(x, t)$. Equation (11.167) is used to evaluate the coupled mode equations and to apply the effective medium approach.

For a fixed value of n_0/n_c as well as given pump amplitudes a_0 and widths $\langle x^2 \rangle$, there remains one free parameter, which is the time t. Figure 11.6 shows the density distributions for $n_0/n_c = 0.05$, $a_0 = 0.0216$ and $\langle x^2 \rangle = 2300$. The left subfigure shows the analytical solution (11.164) for $t = 560$, while the right subfigure shows the approximate solution (11.167) for $M = 3$ at the same time.

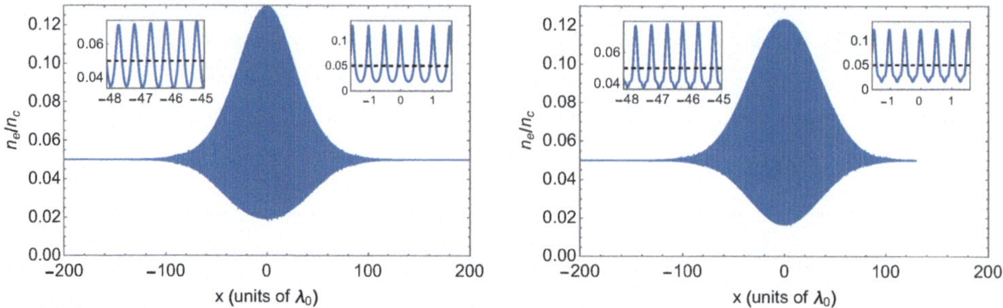

Fig. 11.6 Plasma lattice structures for $n_0 = 0.05\,n_c$ and $\langle x^2 \rangle = 2300$. (**a**) The analytical result (11.164) at $t = 560$. (**b**) For comparison, the simpler model (11.167) for $M = 3$. The dashed line in the insets represents the initial density of $0.05\,n_c$

Lattice Models and Possible Applications

Photonic crystals are systems in which the dielectric function is periodically modulated with a period on the order of an optical wavelength. The modulation leads to optical band gaps. The latter are similar to the electronic band gaps in solids with discrete atomic lattices [287].

Originally, photonic crystals in plasmas were introduced for microwave radiation. In this case, thin layers of plasma and dielectric material are arranged periodically to form a frequency filter [288]. Implementation was achieved in the form of an array of microplasmas [289]. A combination of metamaterial and plasma enables cloaking and nonlinear effects in the microwave range [290]. The transient plasma-based photonic crystal (TPPC) considered here self-organizes in the presence of two counter-propagating (and interacting) laser beams with frequency ω_0. The resulting lattice is transient on a timescale of a few picoseconds. Nevertheless, it can be used to manipulate short laser pulses (of a few femtoseconds duration) with wavelengths in the μm range. The intensity of the probe pulse can be large (e.g., 10^{17} W/cm^2).

The fundamental idea for a TPPC was published in references [267, 291]. Starting with two counter-propagating, linearly polarized laser beams, a standing wave can be formed. The ponderomotive force initially displaces the electrons and leads to the formation of an electrostatic field. The ions respond with a time delay. Eventually, the ions drive a ballistic evolution of the plasma density. As soon as the laser interference disappears, the electric field also vanishes almost completely, and the

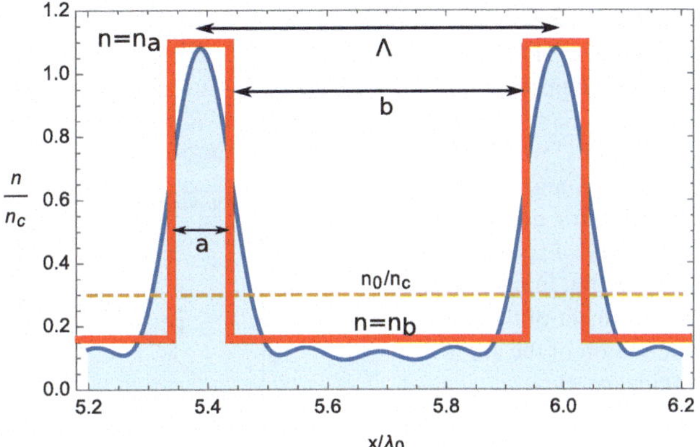

Fig. 11.7 Sketch of a simple model for the density distribution in a crystal, approximated by a stepwise density variation along the x-axis. The dashed line shows the original density

electron and ion densities become nearly equal. The electrons remain trapped in the nodes of the ponderomotive potential, and the fastest counter-propagating ions begin to cross their trajectories.

Figure 11.7 shows, in the shaded areas, the density pattern of a typical one-dimensional (1D) crystal [see, e.g., Fig. 3 of Ref. [291]], when only the first four Fourier coefficients are taken into account. The figure shows the density variation in the x-direction, with neighboring peaks separated by distances of about $\lambda_0/2$, where λ_0 is the laser wavelength. The original density distribution n_0 was constant and significantly below the critical density n_c; in the case considered, $n_0 = 0.3\, n_c$ applies. The critical density n_c is defined by the condition $\omega_0 = \omega_{p0}$, where $\omega_{p0} = \sqrt{n_0 e^2/\varepsilon_0 m_e}$ is the electron plasma frequency and ω_0 is the laser frequency. The system is homogeneous in the y- and z-directions.

Next, we construct a simple model for the TPPC. Figure 11.7 serves as a guide to reduce the density to an ideal lattice, as depicted in Figs. 11.7 and 11.8(a). Based on such a simplification, we can easily estimate the potential of the TPPC for various applications. For example, we will investigate the behavior of the TPPC for different angles of incidence θ of the laser light, as sketched in the schematic in Fig. 11.8(c).

However, before we present new predictions of the model, we should first substantiate the general approach. To validate the method, we adopt parameter values from numerical simulations and compare the band structure predictions of the simple model [see Fig. 11.8] with the numerical simulations presented in references [267, 291]. Considering a lattice period $\Lambda = a + b$, where a and b represent the regions of high (n_a) and low density (n_b), respectively, Fig. 11.8 suggests that

$$n_a = 1.1\, n_c\ , \quad n_b = 0.1736\, n_c\ , \quad a = 0.13644\, \Lambda\ , \quad b = 0.86356\, \Lambda\ . \tag{11.168}$$

These values satisfy the normalization condition

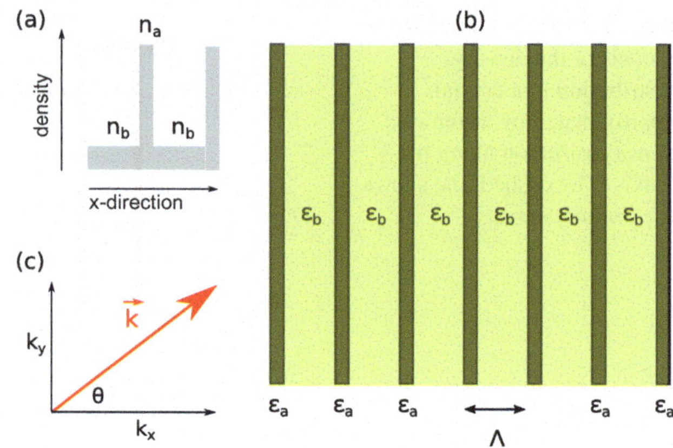

Fig. 11.8 Sketch of a simple model for a TPPC. (**a**) Stepwise density variation along the x-axis; (**b**) Top view of the TPPC with lattice constant Λ and variations of the dielectric constant ε; (**c**) Geometry of laser light propagation in relation to the arrangement of the crystal (definition of the angle of incidence θ)

$$\frac{an_a + bn_b}{\Lambda} = n_0 \, . \tag{11.169}$$

As discussed in references [267, 291], the base length Λ is given by $\Lambda = \pi/k_1$, where k_1 is the wavenumber in the plasma for a wave entering with frequency ω_0 and wavenumber k_0 from vacuum. Since $k_1 = k_0\sqrt{1 - \frac{n_0}{n_c}}$, we have

$$\Lambda = \frac{\lambda_0}{2} \frac{1}{\sqrt{1 - \frac{n_0}{n_c}}} \, . \tag{11.170}$$

For $\frac{n_0}{n_c} = 0.3$ we obtain $\Lambda = 0.6\,\lambda_0$. We give frequencies in $2\pi c/\Lambda$ on. Then the average electron plasma frequency is given by

$$\Omega_{p0} = \frac{\omega_{p0}\Lambda}{2\pi c} = 0.6\frac{\omega_{p0}}{\omega_0} = 0.6\sqrt{\frac{n_0}{n_c}} \approx 0.3286 \, , \tag{11.171}$$

while the local plasma frequencies are

$$\Omega_{pb} = \frac{\omega_{pb}\Lambda}{2\pi c} \approx 0.25 \, , \quad \Omega_{pa} = \frac{\omega_{pa}\Lambda}{2\pi c} \approx 0.63 \tag{11.172}$$

The latter are each calculated using the densities n_b and n_a. The wave numbers k are normalized either by $2\pi/\Lambda$ or by k_1; both normalizations are simply related by

$$K \equiv \frac{k\Lambda}{2\pi} = \frac{1}{2}\frac{k}{k_1} \, . \tag{11.173}$$

Similarly, we obtain

$$\Omega \equiv \frac{\omega\Lambda}{2\pi c} = \frac{\Lambda}{\lambda_0}\frac{\omega}{\omega_0} \, . \tag{11.174}$$

Within the individual layers, the dielectric constants are

$$\varepsilon_a = 1 - \frac{\omega_{pa}^2}{\omega^2} = 1 - \frac{\Omega_{pa}^2}{\Omega^2}\,, \quad \varepsilon_b = 1 - \frac{\omega_{pb}^2}{\omega^2} = 1 - \frac{\Omega_{pb}^2}{\Omega^2} \tag{11.175}$$

for a wave with frequency ω approaching the crystal.

Transmission and Reflection

We distinguish between two polarization directions of the incident wave. The s-wave is also known as the TE wave, where the electric field vector $\mathbf{E} = E\hat{z}$ is perpendicular to the plane of incidence. Maxwell's equations yield

$$\left(\frac{\partial^2}{\partial x^2} + \frac{\partial^2}{\partial y^2}\right)E + \varepsilon\frac{\omega^2}{c^2}E = 0\,. \tag{11.176}$$

For the solutions, we require the continuity of E_y and H_z at the interfaces.

The p-wave is also known as the TM wave, where the magnetic field vector $\mathbf{H} = H\hat{z}$ is perpendicular to the plane of incidence. Maxwell's equations yield

$$\frac{\partial}{\partial x}\left(\frac{1}{\varepsilon}\frac{\partial H}{\partial x}\right) + \frac{\partial}{\partial y}\left(\frac{1}{\varepsilon}\frac{\partial H}{\partial y}\right) + \frac{\omega^2}{c^2}H = 0\,. \tag{11.177}$$

For the solutions, we require the continuity of E_z and H_y at the interfaces.

The problem of reflection and transmission of electromagnetic radiation through a multilayer system can be analyzed using the matrix method. We follow the procedure described in Ref. [292] and apply the geometry shown in Fig. 11.8.

For TE modes (s-polarization), the band structure of the periodic layered medium is given by the dispersion relation

$$\boxed{\begin{aligned}\cos(2\pi K_x) = {} & \cos\left(2\pi\frac{a}{\Lambda}K_{xa}\right)\cos\left(2\pi\frac{b}{\Lambda}K_{xb}\right) \\ & -\frac{1}{2}\left(\frac{K_{xa}}{K_{xb}} + \frac{K_{xb}}{K_{xa}}\right)\sin\left(2\pi\frac{a}{\Lambda}K_{xa}\right)\sin\left(2\pi\frac{b}{\Lambda}K_{xb}\right),\end{aligned}} \tag{11.178}$$

where

$$K_{xa} = \sqrt{\Omega^2 - \Omega_{pa}^2 - K_y^2}\,, \quad K_{xb} = \sqrt{\Omega^2 - \Omega_{pb}^2 - K_y^2}\,. \tag{11.179}$$

Equation (11.178) establishes a relationship between the normalized frequency $\Omega = \frac{\omega\Lambda}{2\pi c}$ and the wave vector components $K_x = \frac{k_x\Lambda}{2\pi}$ and $K_y = \frac{k_y\Lambda}{2\pi}$.

For TM modes (p-polarization), we obtain the slightly different dispersion relation

$$\boxed{\begin{aligned}\cos(2\pi K_x) = &\cos\left(2\pi\frac{a}{\Lambda}K_{xa}\right)\cos\left(2\pi\frac{b}{\Lambda}K_{xb}\right)\\ &-\frac{1}{2}\left(\frac{K_{xa}}{K_{xb}}\frac{\Omega^2-\Omega_{pb}^2}{\Omega^2-\Omega_{pa}^2}+\frac{K_{xb}}{K_{xa}}\frac{\Omega^2-\Omega_{pa}^2}{\Omega^2-\Omega_{pb}^2}\right)\\ &\times\sin\left(2\pi\frac{a}{\Lambda}K_{xa}\right)\sin\left(2\pi\frac{b}{\Lambda}K_{xb}\right)\,.\end{aligned}} \tag{11.180}$$

It is immediately apparent that for $K_y = 0$, i.e., for normal incidence on the array, the two dispersion relations (11.178) and (11.180) are identical. These can be easily solved, for example, with MATHEMATICA. The result is shown in Fig. 11.9 for $K_y = 0$. We have plotted $\frac{\omega}{\omega_0}$ versus K_x to allow a direct comparison with Fig. 6(a) from [291]. The agreement between the model presented here (stepwise layered material) and the numerical results is excellent. Only modes with $\omega/\omega_0 > \sqrt{n_0/n_c}$ can propagate.

In examining oblique propagation, we begin with the TE modes. Numerical simulations are also available for this case, for example those published in Fig. 8 of Ref. [291]. Evaluating the dispersion relation (11.178) for the parameters (11.168) yields the band structure shown in Fig. 11.10. The shaded areas are zones of allowed bands, in which $|\cos(2\pi K_x)| < 1$ holds. These match exactly with the band structure observed in the numerical simulations [267, 291]. The analytical model (11.178) now effortlessly explains the shape of the bands.

Since in (11.178) the frequency and the transverse wave number appear only in the combination $\Omega^2 - K_y^2$, the analytical model for TE modes clearly predicts the dependence

$$\Omega^2 = \Omega_0^2 + K_y^2 \tag{11.181}$$

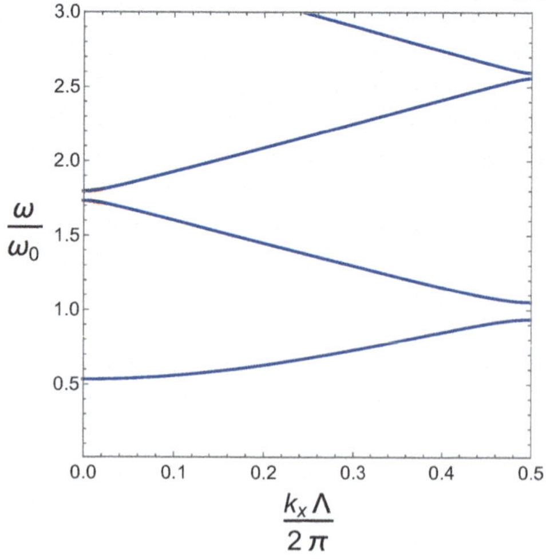

Fig. 11.9 Model calculation based on the dispersion relations (11.178) and (11.180) for propagation with $k_y = 0$. TE and TM modes yield the same result. This matches exactly with Fig. 6(a) from Ref. [291]

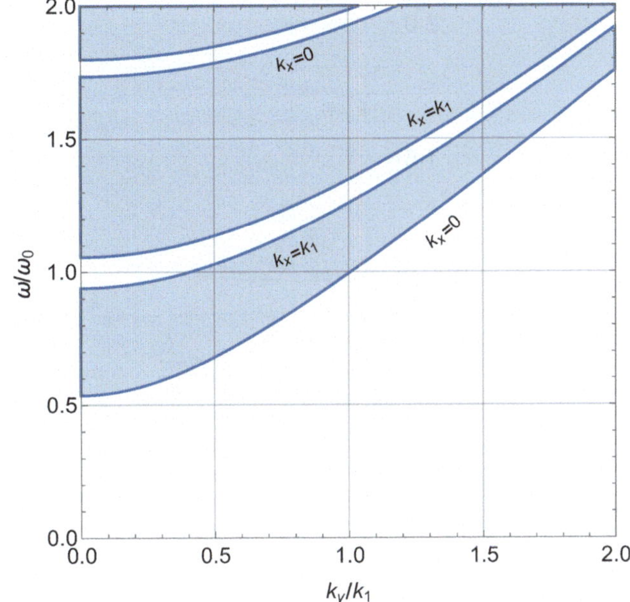

Fig. 11.10 Band structure for oblique incidence ($k_y \neq 0$) of TE modes, calculated using the dispersion relation (11.178). For TE modes, we find complete agreement with the results previously published in Fig. 8 of Ref. [291]

that appears in Fig. 11.10. The constants Ω_0 for the band edges are obtained from Fig. 11.9 (which applies for $K_y = 0$), when we set $|\cos(2\pi K_x)| = 1$.

In conclusion, the discovery of the dependence (11.181) in the simulations for oblique propagation of TE modes provides another successful test of our analytical model.

The different forms of the dispersion relations (11.178) and (11.180) for $K_y \neq 0$ indicate that oblique propagation becomes polarization-dependent (birefringent). Indeed, analysis of the TM waves with (11.180) for $K_y \neq 0$ shows a different behavior. Figure 11.11 shows the band structure of the TM modes.

Several points are noteworthy. First, we find significant differences in the phase velocity between TE and TM modes.

A band gap then arises around $\omega = \omega_0$. In this frequency range, TM-polarized waves cannot propagate in the crystal, regardless of the angle of incidence. This feature can be used to employ the plasma crystal as a polarizer.

Finally, the dispersion relation predicts the propagation of TM-polarized light for frequencies $\omega/\omega_0 < \sqrt{n_0/n_c}$. The lower frequency limit for TM modes is now determined by the density n_b ($< n_0$), i.e., ω/ω_0 merely needs to be greater than $\sqrt{n_b/n_c}$. For such modes, the initially homogeneous plasma appears overdense. However, the lower density regions in the lattice allow propagation as long as they are underdense.

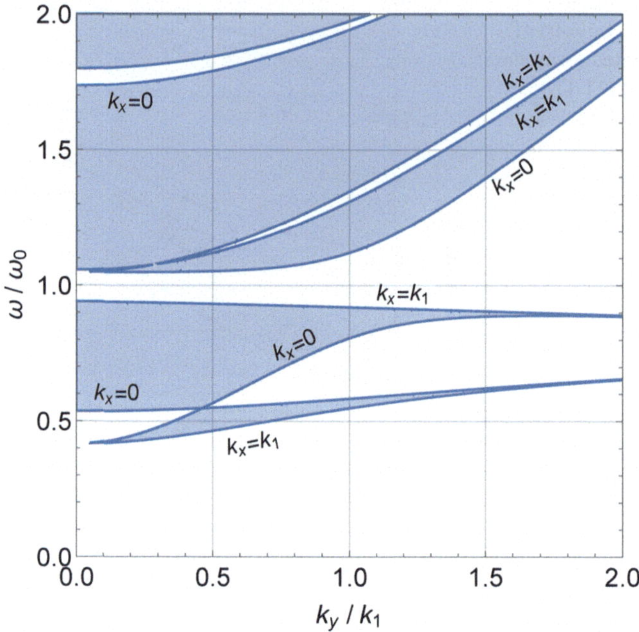

Fig. 11.11 Band structure for oblique propagation ($k_y \neq 0$) of TM modes, as obtained from the dispersion relation (11.180)

11.4 Plasma Compressor

A particularly important application of a plasma lattice could be the plasma compressor, which should demonstrate its advantages at very high intensities. Essentially, this plasma compressor is a chirped plasma lattice. In this section, we also introduce the coupled mode equations as a suitable tool for easily calculating the effectiveness of plasma optical components.

Laser pulse compression is a technology used in laser physics and ultrashort pulse laser research to shorten the duration of laser pulses and increase their energy. This process plays an important role in various scientific, industrial, and medical applications. The compression of laser pulses is crucial for generating intense, ultrashort laser pulses that can last only femtoseconds (1 fs $= 10^{-15}$ s) or picoseconds (1 ps $= 10^{-12}$ s). Such ultrashort pulses make it possible to achieve extremely high power densities, making them ideal for many applications, including

- Material processing: Ultrashort pulse lasers can cut, drill, and mark materials with precision, without transferring heat to the surrounding area;
- medical applications: Ultrashort pulse lasers are used in eye surgery and other medical procedures;
- research: In basic research, ultrashort laser pulses enable the study of ultrafast processes in atoms and molecules.

The plasma-based laser pulse compressor is intended to utilize the properties of a plasma to obtain even more intense short pulses. Successful development of plasma-based laser pulse compressors will lead to enormous advances in laser technology and make it possible to carry out a wide range of fascinating applications in various fields even more intensively.

To avoid damage to optical components in high-power laser systems, the CPA (chirped-pulse amplification) technique was developed, which was without doubt highly deserving of awards [249, 293].

In the CPA method, a laser pulse is temporally stretched before amplification, typically by a chirped grating. The stretched pulse is then amplified by an amplifier (e.g., a solid-state amplifier) without damaging the material of the amplifier. After amplification, the pulse is recompressed to its original short duration by a second chirped element, resulting in a dramatic increase in intensity. This is the fundamental principle behind almost all high-power laser sources.

To this day, pulse stretching, amplification, and compression are carried out exclusively by pairs of conventional solid-state components [294]. The highest peak power is limited by the optical threshold of the compressor components. In comparison to solid-state materials, plasma, which is already ionized, does not suffer from breakdown at extreme light intensities. Therefore, the high-intensity laser community is increasingly focusing on plasma-based components to overcome the limitations imposed by intensity-limited solid-state components [295–297]. Particular interest is directed at the final compressor stage in a CPA chain [298].

Obviously, an increase in intensity in basic research, material processing [299], laser medicine [300, 301], diagnostics of ultrafast processes in atoms and molecules [302], laser fusion [303], and so on will open up new perspectives. In this context, a plasma-based laser pulse compressor should utilize the properties of a plasma lattice. As already discussed, a plasma lattice is a purely optically generated plasma structure that only changes on the ion time scale. Therefore, it can be used to manipulate short, high-intensity laser pulses. The formation of electron and ion density lattices by the interaction of two counter-propagating laser pulses has been known for at least 20 years [304–307]. Since then, many fundamental properties have been worked out [265, 267, 274, 285, 286, 291, 308–315].

The plasma lattice is tunable, as its period can be varied by changing the angle between the pump pulses. On the other hand, a plasma does not allow the creation of a lattice with sharp boundaries and a fairly homogeneous amplitude. Typically, a lattice is generated by the superposition of pump laser pulses. Their profiles determine the spatial envelope of the generated lattice. This results in lattice structures with constant lattice periods but spatially dependent envelopes [316, 317]. A strictly homogeneous plasma lattice with constant amplitude is an idealization, the limitations of which have been discussed [318]. Understanding the differences makes it possible to extend many model statements obtained with constant lattice amplitudes to inhomogeneous lattices.

For high-power laser physics, the creation of chirped plasma lattices is a major challenge for the future. Much can be learned from the development of fiber optics [294, 319–325].

In the field of fiber optics, refractive index modifications for volume-chirped Bragg gratings have been proposed [294] by considering the interaction of a focused and a defocused writing laser. It is still an open question whether this technique can be transferred to the production of chirped plasma-based gratings. Currently, another idea is preferred. A chirped plasma-based grating can occur when two counter-propagating chirped pump lasers interact within a plasma layer. There is also a third idea [326], which is based on the fact that light reflection from an inhomogeneous plasma occurs at different positions for different frequencies. We will briefly comment on this when we analyze the coupled mode equations for inhomogeneous plasma. The coupled mode equations [327] show a direct analogy between a quadratically chirped and a linearly inhomogeneous grating. However, this correspondence is based on some simplifying assumptions. Therefore, it remains to be seen whether the predictions actually apply to plasma situations. If so, this could potentially provide a simple way to realize a plasma laser pulse compressor.

The present discussion was strongly inspired by an article by Edwards and Michel [277]. They proposed a realistic scenario for a compact CPA system in which the final stage consists of a homogeneous plasma lattice. This is intended to compensate for a small angular dispersion. Here, we pose the question of whether and how a chirped plasma lattice can be produced and subsequently used in a CPA system.

Chirped plasma gratings could offer a way to compress chirped high-power laser pulses in reflection, similar to conventional solid-state gratings in CPA schemes. Solid-state gratings typically have to be operated at energy fluences below 0.1 J/cm^2 for 30-fs pulses, which corresponds to peak intensities of about 10^{12} W/cm^2. Such intensities are close to the ionization threshold of solid-state materials, which is why traditional compressor gratings must be correspondingly large (several hundred cm^2 for petawatt systems). At the same time, the repetition rates of CPA-based laser systems are currently limited to about 1 Hz (for PW lasers) to 10 Hz (for 0.1-PW lasers). Higher repetition rates are desirable to increase the average power, which in turn would raise the average power of, for example, laser-driven radiation or particle sources. The current limitations on repetition rate are mainly due to the laser pumping processes, amplifier cooling, but also to thermally induced deformations of the compressor gratings [328–330].

Plasma compression gratings could provide a solution to two problems. Their damage threshold is generally determined by the requirement that the density modulation should not be altered by the ponderomotive potential of the probe pulse. For typical underdense plasmas, this corresponds to intensities on the order of 10^{17} W/cm^2, that is, five orders of magnitude higher than for solid-state gratings. At the same time, a fresh plasma grating could be used for each laser shot, which would allow repetition rates far beyond a few hertz.

Shape of a Chirped Plasma Grating

A chirped plasma grating is generated by chirped laser pulses (pump pulses). We describe the electric field of a chirped laser pulse as

$$E(\tau) = Ae^{-B\tau^2} e^{i\omega_0(\tilde{\tau}+b\tilde{\tau}^2)} \tag{11.182}$$

with

$$A = \frac{\sigma}{\sqrt{4iD_2 + \sigma^2}}\,, \quad B = \frac{\sigma^2}{16D_2^2 + \sigma^4}\,, \quad b = \frac{1}{\omega_0}\frac{4D_2}{16D_2^2 + \sigma^4}\,. \tag{11.183}$$

Here, D_2 is the coefficient of a quadratic phase, which corresponds to $\exp\left[-iD_2(\omega - \omega_0)^2\right]$, and σ characterizes a Gaussian envelope, which is proportional to $\frac{\sigma}{\sqrt{2}} \exp\left[-\frac{\sigma^2(\omega-\omega_0)^2}{4}\right]$. Depending on the sign of D_2, a positive ($b > 0$) or negative ($b < 0$) chirp results.

The minimum pulse duration $\tau_{\min} = 2\sigma\sqrt{\ln(2)}$ is achieved for $D_2 = 0$. For finite D_2, the FWHM duration of the chirped pulse is

$$\tau_{\mathrm{ch}} = \tau_{\min}\sqrt{1 + \left(\frac{16D_2\ln(2)}{\tau_{\min}^2}\right)^2}\,. \tag{11.184}$$

Pulse 1 is to propagate from left to right; its retarded time is

$$\tau \to \tau_1 = \bar{t} - \frac{1}{v_g}(x + x_0)\,, \quad \tilde{\tau} \to \tilde{\tau}_1 = \bar{t} - \frac{k_0}{\omega_0}(x + x_0)\,, \tag{11.185}$$

when the starting time $\bar{t} = 0$ is at $x = -x_0 < 0$. Here, v_g is the group velocity and $\frac{\omega_0}{k_0}$ is the phase velocity. On the other hand, pulse 2 should propagate from right to left; its retarded time is

$$\tau \to \tau_2 = \bar{t} + \frac{1}{v_g}(x - x_0) \ , \quad \tilde{\tau} \to \tilde{\tau}_2 = \bar{t} + \frac{k_0}{\omega_0}(x - x_0) \ , \tag{11.186}$$

when the starting time $\bar{t} = 0$ is at $x = x_0$. We could simplify the representation (which we do not wish to do) by assuming that $n_0 \ll n_c$ holds, so that $\omega_0 \approx ck_0$ applies. In this case, a distinction between τ and $\tilde{\tau}$ is no longer necessary, and the calculation would become more straightforward.

The ponderomotive potential is proportional to the product $E_1 E_2^* +$ c.c. We first calculate the contribution of the envelope, starting with the cases $B_1 = B_2 \equiv B$:

$$S_1 := e^{-B\tau_1^2} e^{-B\tau_2^2} = \exp\left[-\frac{\sigma^2}{16D_2^2 + \sigma^4}\left(\frac{2}{v_g^2}x^2 + 2t^2\right)\right] \tag{11.187}$$

for $t = \bar{t} - \frac{1}{v_g}x_0$. On the other hand, for $b_2 = 0$ and $b_1 \equiv b$

$$S_1 := e^{-B\tau_1^2} e^{-\sigma^{-2}\tau_2^2} = \exp\left[-\frac{16D_2^2 + 2\sigma^4}{(16D_2^2 + \sigma^4)\sigma^2}\left(\frac{1}{v_g^2}x^2 + t^2\right)\right] \times \exp\left[\frac{-16D_2^2}{(16D_2^2 + \sigma^4)\sigma^2}\frac{2}{v_g}x\,t\right] . \tag{11.188}$$

For $t = 0$ the pulses overlap optimally.

Next, we calculate the product of the phase factors

$$S_2 := e^{i\omega_0(\tilde{\tau}_1 + b_1\tilde{\tau}_1^2)} e^{-i\omega_0(\tilde{\tau}_2 + b_2\tilde{\tau}_s^2)} . \tag{11.189}$$

A short calculation yields

$$S_2 = \begin{cases} \exp\{-2ik_0x[1 + 2b(t + t_\varepsilon)]\} & \text{for } b = b_1 = +b_2 \ , \\ \exp\left\{-2ik_0x\left(1 - \frac{k_0}{\omega_0}bx\right)\right\}e^{2ib\omega_0(t+t_\varepsilon)^2} & \text{for } b = b_1 = -b_2 \\ \exp\left\{-2ik_0x\left[1 - \frac{k_0}{\omega_0}\frac{b}{2}x + b(t + t_\varepsilon)\right]\right\}e^{ib\omega_0(t+t_\varepsilon)^2} & \text{for } b = b_1,\ b_2 = 0 \ , \end{cases} \tag{11.190}$$

where

$$t_\varepsilon = \left(\frac{1}{v_g} - \frac{k_0}{\omega_0}\right)x_0 \approx 0 \ . \tag{11.191}$$

In most cases, the limit $16D_2^2 \gg \sigma^4$ will be of particular interest, as this limit covers the situation with significant chirp.

Example 11.5 (The Case $b = b_1 = -b_2 \neq 0$)
The prediction for this case of interacting pulsed pump lasers with opposite chirp is clear. A spatially chirped grating with linear chirp should form. The grating should exhibit a spatial density variation:

$$\frac{\delta n}{n_0} \sim a(x) \cos\left[2k_0 x\left(1 - \frac{k_0}{\omega_0} bx\right)\right]. \tag{11.192}$$

For a Gaussian envelope

$$a(x) \sim \exp\left(-\frac{2}{v_g^2}\frac{\sigma^2}{16D_2^2 + \sigma^4}x^2\right) \approx \exp\left(-\frac{1}{v_g^2}\frac{\sigma^2}{8D_2^2}x^2\right) \tag{11.193}$$

the width increases with the chirp parameter.

In summary, a chirped grating with variable wave vector $K = 2k_0 - \frac{2k_0^2 b}{\omega_0}x$ appears. The grating has a Gaussian shape with a width (FWHM), e.g., for $16D_2^2 \gg \sigma^4$

$$W = \sqrt{2\ln 2}\, v_g \frac{\sqrt{16D_2^2 + \sigma^4}}{\sigma} \approx 4\sqrt{2\ln 2}\, v_g \frac{|D_2|}{\sigma}. \tag{11.194}$$

The width increases with the chirp of the pump lasers. ■

Example 11.6 (The Case $b = b_1 = +b_2 \neq 0$)
In this case, a grating with a fixed grating constant is formed,

$$\frac{\delta n}{n_0} \sim a(x) \cos[2k_0 x]. \tag{11.195}$$

For the shape of the envelope, two factors from (11.187) and (11.190) are relevant, namely

$$\exp\left[-\frac{2\sigma^2 t^2}{16D_2^2 + \sigma^4}\right] \quad \text{and} \quad \exp[-4ik_0 bx(t + t_\varepsilon)]. \tag{11.196}$$

In the following, we use the integral

$$\begin{aligned} &\int e^{-2Bt^2} \exp[-4ik_0 b(t + t_\varepsilon)x]\, dt \\ &= \frac{1}{2\sqrt{2}}\sqrt{\frac{\pi}{B}} \exp\left\{-\frac{2k_0^2 b^2 x^2}{B}\right\} \operatorname{erf}\left(\sqrt{2B}t + i\frac{2k_0 bx}{\sqrt{2B}}\right) \exp(-4ik_0 bt_\varepsilon x). \end{aligned} \tag{11.197}$$

The superposition of the pulses occurs over a short time compared to the total existence time of the grating. Therefore, we average over time, so that for $t_\varepsilon \approx 0$ the x-dependence $\exp\left\{-\frac{2k_0^2 b^2 x^2}{B}\right\}$ is preserved together with a constant factor. We combine this with $\exp\left(-\frac{\sigma^2}{16D_2^2+\sigma^4}\frac{2}{v_g^2}x^2\right)$ from (11.187) to obtain the following for the product:

$$\exp\left(-\frac{2k_0^2b^2x^2}{B}\right)\exp\left(-\frac{\sigma^2}{16D_2^2+\sigma^4}\frac{2}{v_g^2}x^2\right)=\exp\left(-\frac{2x^2}{\sigma^2}\frac{1}{v_g^2}\right)e^{-\varepsilon x^2} \tag{11.198}$$

with

$$\varepsilon=\frac{2}{\sigma^2}\frac{16D_2^2}{16D_2^2+\sigma^4}\left(\frac{k_0^2}{\omega_0^2}-\frac{1}{v_g^2}\right)\approx 0 \text{ for } n_0 \ll n_c\,. \tag{11.199}$$

In summary, no chirp appears in the grating. The grating has a Gaussian shape with a width (FWHM)

$$W=\sqrt{2\ln 2}\,v_g\sigma \tag{11.200}$$

for $t_\varepsilon \sim \varepsilon \approx 0$. Its width does not depend on the chirp of the pump waves. ■

Example 11.7 The Case $b_1 = b,\ b_2 = 0$
This case encompasses both effects discussed in the previous subsections, namely chirp and length change. Initially, there is a spatial density variation with chirp,

$$\frac{\delta n}{n_0}\sim a(x)\,\cos\left[2k_0x\left(1-\frac{k_0}{\omega_0}\frac{b}{2}x\right)\right]. \tag{11.201}$$

For the calculation of the width in x we need the integral

$$\int e^{-(A_1-iA_4)t^2-(A_2+iA_3)tx}\,dt$$
$$=\frac{1}{2}\sqrt{\frac{\pi}{A_1-iA_4}}\exp\left\{\frac{(A_2+iA_3)^2x^2}{4(A_1-iA_4)}\right\}\operatorname{erf}\left(\sqrt{A_1-4A_4t}-\frac{A_2+iA_3}{2\sqrt{A_1-iA_4}}x\right), \tag{11.202}$$

with appropriate definitions of A_1, A_2, A_3 and A_4. These can be done using (11.188) and (11.190).

Without repeating similar calculations as before, we summarize that the envelope of the grating has a Gaussian shape. For $16D_2^2 \gg \sigma^4$ its spatial width (FWHM) is

$$W=4\sqrt{\ln 2}\,v_g\frac{|D_2|}{\sigma}\,. \tag{11.203}$$

The situation in which only one pulse is chirped and the other is not corresponds to the case of two oppositely chirped pulses with half the value of b. ■

Example from a numerical simulation

As an example for a chirped grating, we show a numerical simulation in which only the laser field E_1 is chirped, i.e., stretched to a longer duration. The second laser pulse E_2 is assumed to be bandwidth-limited, i.e., with an FWHM duration of 30 fs ($b_2 = 0$).

The pulse E_1 is chirped such that it is 100 times longer than the short pulse ($b_1 = 1.31 \times 10^{10}\,\mathrm{s}^{-1}$, i.e., $D_2 = 8.1 \times 10^{-27}\,\mathrm{s}^2$). Both pulses have the same peak intensity of $2.5 \times 10^{14}\,\mathrm{W/cm}^2$, which means that the long pulse carries 100 times more energy than the short pulse. The length of the entire interaction region between the two pulses is determined mainly by the length of the long pulse and is approximately $600\,\lambda_0$.

The ponderomotive field of the overlapping laser pulses initiates the formation of an electron density grating. Due to the chirp of the long pulse, the wavelength and frequency of the grating vary spatially.

Figure 11.12 (top) shows the electron density 12.6 ps after the interaction. In the interaction region, the plasma density is modulated by rapid oscillations with wavenumbers $k = 2k_1$, where $k_1 = k_0\sqrt{1 - n_0/n_c}$ applies. Due to the temporal Gaussian envelopes of the two laser pulses, the envelope of the density modulation is also Gaussian [318]. Ion and electron densities are nearly identical.

Figure 11.12 (bottom) shows a spectrogram of the density modulation, which makes the changes in the local wavenumber visible. The interaction between the two pulses E_1 and E_2 begins at $x = 400\,\lambda_0$, where the low-frequency components of E_1 interfere with the unchirped pulse E_2. By the end of the interaction, E_2 has propagated up to $x = -400\,\lambda_0$ and there interferes with the high-frequency components of E_1. The result is a linearly chirped plasma density grating.

Mode Coupling Equations for a Chirped Grating

A coupled mode system can be used to interpret the propagation of test pulses in chirped plasma gratings. The results obtained in this way can be compared with numerical PIC simulations. First, we establish the fundamental equations. We then discuss suitable boundary conditions to determine the transmission and reflection coefficients. As a byproduct of the general formulation, we can derive the influence

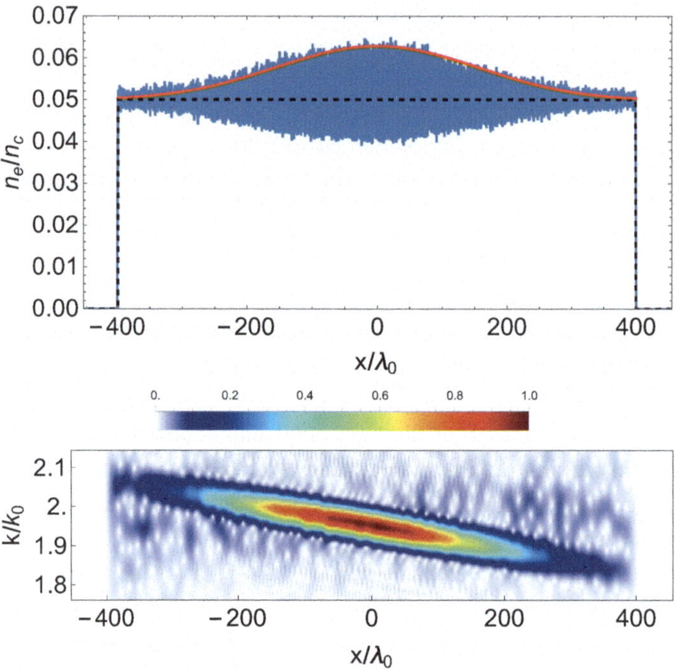

Fig. 11.12 Top: Electron density (blue) n_e at $t = 12.6\,\text{ps}$. Rapid oscillations on the scale of $\lambda_0/2$ are not visible due to the scale. The envelope of the density fluctuation is shown as a solid red line, the initial density profile as a black dashed line. bottom: Normalized spectrogram of the variation of the electron density $\delta n_e = n_e - n_0$ for n_e, as shown above. It shows which wavenumber component contributes to the density oscillation at which position in the plasma grating. The corresponding Fourier amplitudes are shown on a linear scale

of inhomogeneous density variations in comparison to the chirp effect. The investigation benefits from earlier work [305, 306, 321, 331, 332] on group velocity in Bragg gratings with linear chirp.

Within a coupled mode analysis, we start with incident plane waves. The individual waves can be interpreted as spectral contributions to pulses. Therefore, the reflections shown here are not directly applicable to the reflected pulses in a PIC simulation. However, they will be extremely helpful for interpreting PIC results.

We consider a plane test wave $E \sim e^{-i\omega t}$ with frequency ω, so that (prior to normalization) the stationary amplitudes are obtained from the following equation:

$$\boxed{\frac{d^2E}{dx^2} + k^2\frac{N^2}{N_0^2}E = 0} \tag{11.204}$$

with

$$k \equiv \frac{\omega N_0}{c}\,, \quad N = \sqrt{1 - \frac{\omega_{pe}^2}{\omega^2}}\,. \tag{11.205}$$

N is the refractive index, and N_0 is a reference index. Obviously, as the test wave frequency approaches the pump frequency, i.e., for $\omega \to \omega_0$

$$k \to k_1 \equiv \frac{\omega_0 N_0}{c}\,, \quad N \to N_0 = \sqrt{1 - \frac{\omega_{pe}^2}{\omega_0^2}}\,. \tag{11.206}$$

From the (dimensioned) wave vector k_1 and the (dimensional) spatial coordinate x, we can construct a dimensionless variable:

$$\xi = k_1 x = \frac{k_1}{k_0} \underbrace{k_0 \lambda_0}_{2\pi} \frac{x}{\lambda_0} \hat{=} 2\pi N_0 x\,. \tag{11.207}$$

Here, in the last term on the right-hand side, x is dimensionless, i.e., normalized with λ_0. The normalization is performed using the following units.

We normalize the frequency ω by the pump frequency ω_0, the time t by $2\pi/\omega_0$, distances by the laser wavelength λ_0 in vacuum, and wavenumbers k by $k_0 \equiv \frac{2\pi}{\lambda_0}$. In the plasma, the pump wavenumber is $k_1 = k_0 N_0$ with $N_0 = \sqrt{1 - n_0/n_c}$. The (constant) mean density n_0 is used for density normalization, while the speed of light c serves as the unit of velocity. Then, $2k_1 x \to 4\pi N_0 x$ holds in dimensionless form. For the (normalized) frequency deviation Δ we obtain

$$\Delta = \omega - 1 \quad \rightsquigarrow \quad k^2 \approx k_1^2(1 + 2\Delta) \tag{11.208}$$

for $|\Delta| \ll 1$. Likewise,

$$N^2 \approx N_0^2\left(1 + \left[1 - \frac{1}{N_0^2}\right]\delta n_e - \left[1 - \frac{1}{N_0^2}\right]2\Delta\right), \tag{11.209}$$

and

$$k^2 \frac{N^2}{N_0^2} \approx k_1^2\left(1 + \left[1 - \frac{1}{N_0^2}\right]\delta n_e + \frac{2}{N_0^2}\Delta\right). \tag{11.210}$$

The variation of the electron density δn_e is driven by the ponderomotive force, and (at least in the first part of the present discussion) there can be an inhomogeneous contribution $\delta n_e^{inh} \sim D$. We write the ansatz in generalized form as

$$\delta n_e = \frac{1}{2} C(x)\left(e^{i\psi} + e^{-i\psi}\right) + D(x)\,, \tag{11.211}$$

where the coefficients C and D may still depend on position. Furthermore, let

$$\psi = 2\xi + \varphi(\xi) \,. \tag{11.212}$$

We allow for a (nonlinear) phase φ, which becomes essential for chirped gratings. For example, a linear chirp is equivalent to a quadratic phase

$$\varphi(\xi) = \beta(\xi - \xi_0)^2 \,. \tag{11.213}$$

The normalized wave equation (11.204) reads

$$\begin{aligned} \frac{d^2E}{d\xi^2} &+ \{1 + \frac{2}{N_0^2}\Delta + \left[1 - \frac{1}{N_0^2}\right] D(\xi) \\ &+ \frac{1}{2}\left[1 - \frac{1}{N_0^2}\right] C(\xi)\left(e^{2i\xi + i\varphi} + e^{-2i\xi - i\varphi}\right)\}E = 0 \,. \end{aligned} \tag{11.214}$$

For the electric field E we make the ansatz

$$\boxed{E(\xi) = a_+(\xi)e^{i\xi} + a_-(\xi)e^{-i\xi}} \,, \tag{11.215}$$

with slowly varying envelopes $a_\pm$. Later, we will generalize to carrier wave numbers $k \neq k_1$.

In the following, we assume that the envelopes vary only weakly, so that

$$\frac{da_+}{dx} = i\pi N_0\left[\frac{2}{N_0^2}\Delta + \left(1 - \frac{1}{N_0^2}\right)D\right]a_+ + i\frac{\pi N_0}{2}\left(1 - \frac{1}{N_0^2}\right)Ce^{i\varphi}a_- \,, \tag{11.216}$$

$$\frac{da_-}{dx} = -i\pi N_0\left[\frac{2}{N_0^2}\Delta + \left(1 - \frac{1}{N_0^2}\right)D\right]a_- - i\frac{\pi N_0}{2}\left(1 - \frac{1}{N_0^2}\right)Ce^{-i\varphi}a_+ \,. \tag{11.217}$$

As already mentioned in several places, x is dimensionless (normalized with λ_0). For the amplitudes u and v, which are defined via

$$\boxed{a_+(\xi) = u(\xi)e^{i\varphi/2} \,, \quad a_-(\xi) = v(\xi)e^{-i\varphi/2}} \tag{11.218}$$

the mode coupling equations are obtained as

$$\boxed{\frac{du(\xi)}{d\xi} = i[\sigma(\xi)\,u(\xi) + \kappa(\xi)\,v(\xi)]} \,, \tag{11.219}$$

$$\boxed{\frac{dv(\xi)}{d\xi} = -i[\sigma(\xi)\,v(\xi) + \kappa(\xi)\,u(\xi)]} \,. \tag{11.220}$$

Here,

$$\sigma(\xi) = \frac{1}{N_0^2}\Delta + \frac{1}{2}\left(1 - \frac{1}{N_0^2}\right)D(\xi) - \frac{1}{2}\frac{d\varphi}{d\xi}\,, \tag{11.221}$$

$$\kappa(\xi) = \frac{1}{4}\left(1 - \frac{1}{N_0^2}\right)C(\xi)\,. \tag{11.222}$$

Example 11.8 (Density Inhomogeneity)
In this example, we would like to point out an interesting general aspect. Equation (11.221) shows that two terms appear side by side, namely a possible inhomogeneity in the density and the derivative of the phase. If we establish the correspondence

$$\frac{1}{2}\left(1 - \frac{1}{N_0^2}\right)D(\xi) \mathrel{\hat{=}} -\frac{1}{2}\frac{d\varphi}{d\xi} \tag{11.223}$$

we can observe that a linear chirp is equivalent to a linear density variation.

Since plasma gratings are generated by pump pulses with non-constant envelopes, the generation of a purely linear density grating could be difficult (or even only theoretically conceivable). In experiments, exponential variations usually (additionally) occur, i.e.

$$D(\xi) \sim \exp\left(-\frac{\xi^2}{\xi_0^2}\right)\,. \tag{11.224}$$

These then correspond to a phase change

$$\varphi(\xi) \sim -\int^{\xi} D(\xi')d\xi' \sim \frac{\sqrt{\pi}}{2}\,\mathrm{erf}\left(\frac{\xi}{\xi_0}\right)\,. \tag{11.225}$$

For the Taylor expansion of the error function, the following holds

$$\mathrm{erf}\,(x) \approx \frac{2}{\sqrt{\pi}}x - \frac{2}{3\sqrt{\pi}}x^3 + \mathcal{O}(x^5)\,. \tag{11.226}$$

Therefore, a Gaussian inhomogeneity is expected to contribute a third-order phase term, which should be minimized. In other words, it will be difficult to generate a grating with a purely linear chirp. Corrections due to a quadratic chirp are always to be expected because of the inhomogeneous envelopes. ■

Boundary Conditions
Now, a few remarks on the boundary conditions and the definition of reflection and transmission coefficients. Since in dimensionless form

$$2\pi N_0(1+\Delta)x = 2\pi kx \,, \tag{11.227}$$

we can introduce

$$\sigma = \Delta - \frac{1}{2}\frac{d\varphi}{d\xi} + \bar{\sigma} \,, \quad \tilde{u} = u\, e^{-i\Delta\xi + i\varphi/2} \,, \quad \tilde{v} = v\, e^{i\Delta\xi - i\varphi/2} \,, \tag{11.228}$$

to obtain the electric field of a test wave (in dimensionless form) with the appropriate carrier wavenumber k as

$$E = \tilde{u}(\xi)\, e^{2\pi ikx} + \tilde{v}(\xi)\, e^{-2\pi ikx} \tag{11.229}$$

The modified form of the mode coupling equations is then

$$\frac{d\tilde{u}(\xi)}{d\xi} = i\Big[\bar{\sigma}(\xi)\,\tilde{u}(\xi) + \kappa(\xi)\, e^{-2i\Delta\xi + i\varphi}\,\tilde{v}(\xi)\Big] \,, \tag{11.230}$$

$$\frac{d\tilde{v}(\xi)}{d\xi} = -i\Big[\bar{\sigma}(\xi)\,\tilde{v}(\xi) + \kappa(\xi)\, e^{2i\Delta\xi - i\varphi}\,\tilde{u}(\xi)\Big] \tag{11.231}$$

with

$$\bar{\sigma}(\xi) = 2\pi N_0\left(1 - \frac{1}{N_0^2}\right)\left[\frac{1}{2}D(\xi) - \Delta\right] . \tag{11.232}$$

When solving the coupled mode equations for a finite grating, we recall that ω_0 is the fixed frequency of the pump lasers that generate the grating. The variable probe frequency is ω. Therefore, the transformation $\xi = 2\pi N_0 x$ is useful with respect to the boundary conditions and the spatial variable; for fixed n_0/n_c the factor N_0 remains constant.

For the region $-L \le x \le L$ we use the boundary conditions $\tilde{u}(x=-L) = 1$ and $\tilde{v}(x=L) = 0$. Then, the reflection coefficient R and the transmission coefficient T are given by

$$\boxed{r = v(x=-L)e^{-i2\pi N_0\Delta L + i\varphi(x=-L)/2} \rightarrow R \equiv |r|^2 = |v(x=-L)|^2 \,,} \tag{11.233}$$

$$\boxed{t = u(L)e^{-i2\pi N_0\Delta L + i\varphi(x=L)/2} \rightarrow T \equiv |t|^2 = |u(L)|^2 \,.} \tag{11.234}$$

Spectral Properties of a Homogeneous Grating with Linear Chirp

To analyze the spectral properties of a chirped grating, we start with the standard equations for coupled modes (11.219) and (11.220) under the assumptions

$$D \equiv 0 \,, \quad C = \text{const} \,, \quad \varphi(\xi) = \beta(\xi - \xi_0)^2 \,. \tag{11.235}$$

We assume propagating plane waves (Fourier modes) within the slowly varying envelope approximation. For these, the following holds

$$\boxed{\frac{du(\xi)}{d\xi} = i\left[\frac{1}{N_0^2}\Delta - \beta(\xi - \xi_0)\right] u(\xi) + iC_0\, v(\xi) \,,} \tag{11.236}$$

$$\boxed{\frac{dv(\xi)}{d\xi} = -i\left[\frac{1}{N_0^2}\Delta - \beta(\xi - \xi_0)\right] v(\xi) - iC_0\, u(\xi) \,,} \tag{11.237}$$

where β and $C_0 = \frac{1}{4}\left(1 - \frac{1}{N_0^2}\right)C < 0$ are constants. The variable u corresponds to the envelope of the incident (and transmitted) wave, while v describes the reflected wave.

As before, the frequency difference Δ is a fixed parameter. In this case, analytical solutions are possible.

We rewrite (11.236) and (11.237), with

$$Z(\xi) = \frac{1}{N_0^2}\Delta\xi - \frac{1}{2}\varphi(\xi) \,, \quad Z'(\xi) = \frac{1}{N_0^2}\Delta - \beta(\xi - \xi_0) \,, \quad Z''(\xi) = -\beta \,, \tag{11.238}$$

and introduce

$$\bar{u} = ue^{-iZ} \,, \quad \bar{v} = ve^{iZ} \,. \tag{11.239}$$

Then it follows that

$$\frac{d^2\bar{u}(\xi)}{d\xi^2} + 2iZ'\frac{d\bar{u}(\xi)}{d\xi} - C_0^2\bar{u} = 0 \,, \tag{11.240}$$

$$\frac{d^2\bar{v}(\xi)}{d\xi^2} - 2iZ'\frac{d\bar{v}(\xi)}{d\xi} - C_0^2\bar{v} = 0 \,. \tag{11.241}$$

We can transform this system of equations into the standard forms for known polynomials by defining the dimensionless spatial variable

$$\zeta = C_0(\xi - \xi_0) - \frac{C_0}{\beta N_0^2}\Delta \rightsquigarrow \xi = \frac{\zeta}{C_0} + \frac{\Delta}{N_0^2\beta} + \xi_0 \,, \tag{11.242}$$

and rewriting the dependent variables:

$$U(\zeta) \equiv \bar{u}(\xi) \,, \quad V(\zeta) \equiv \bar{v}(\xi) \,. \tag{11.243}$$

Note that

$$V(\zeta) = -i\frac{dU(\zeta)}{d\zeta}\, e^{2iZ} , \tag{11.244}$$

$$U(\zeta) = -i\frac{dV(\zeta)}{d\zeta}\, e^{-2iZ} . \tag{11.245}$$

We obtain the set of equations

$$\boxed{\frac{d^2U}{d\zeta^2} = i\chi_1\zeta\frac{dU}{d\zeta} + U ,} \tag{11.246}$$

$$\boxed{\frac{d^2V}{d\zeta^2} = -i\chi_1\zeta\frac{dV}{d\zeta} + V ,} \tag{11.247}$$

with

$$\chi_1 = \frac{2\beta}{C_0^2} . \tag{11.248}$$

The system of differential equations can be solved in various ways. Here we present a solution using Kummer functions. Alternatively, Hermite polynomials can also be used.

Starting from Eqs. (11.246) and (11.247), we introduce a new coordinate,

$$z = \frac{i}{2}\chi_1\zeta^2 . \tag{11.249}$$

Then we obtain

$$z\frac{d^2U}{dz^2} + \left(\frac{1}{2} - z\right)\frac{dU}{dz} + i\frac{1}{2\chi_1}U = 0 , \tag{11.250}$$

$$z\frac{d^2V}{dz^2} + \left(\frac{1}{2} + z\right)\frac{dV}{dz} + i\frac{1}{2\chi_1}V = 0 . \tag{11.251}$$

For the first equation, two independent solutions can be written in terms of confluent hypergeometric functions of the first kind, also known as Kummer functions $M(a, b, z)$ [321]:

$$U_1(\zeta) = M\left(-i\frac{1}{2\chi_1}, \frac{1}{2}, \frac{i}{2}\chi_1\zeta^2\right) , \tag{11.252}$$

$$U_2(\zeta) = \zeta\, M\left(\frac{1}{2} - i\frac{1}{2\chi_1}, \frac{3}{2}, \frac{i}{2}\chi_1\zeta^2\right). \tag{11.253}$$

Similarly, the second equation can be treated as follows:

$$V_1(\zeta) = M\left(i\frac{1}{2\chi_1}, \frac{1}{2}, -\frac{i}{2}\chi_1\zeta^2\right), \tag{11.254}$$

$$V_2(\zeta) = \zeta\, M\left(\frac{1}{2} + i\frac{1}{2\chi_1}, \frac{3}{2}, -\frac{i}{2}\chi_1\zeta^2\right). \tag{11.255}$$

From this, we construct the general solutions with initially arbitrary coefficients A_1, A_2, B_1 and B_2, i.e.

$$U(\zeta) = A_1\, U_1(\zeta) + A_2\, U_2(\zeta)\,, \tag{11.256}$$

$$V(\zeta) = B_1\, V_1(\zeta) + B_2\, V_2(\zeta)\,. \tag{11.257}$$

Additionally, we use (11.244) and (11.245) with

$$Z = \frac{\Delta}{N_0^2}\left[\frac{1}{2}\frac{\Delta}{N_0^2\beta} + \xi_0\right] - \frac{\beta}{2C_0^2}\zeta^2\,. \tag{11.258}$$

The boundary condition $V(\zeta_+) = 0$ leads to

$$\rho \equiv \frac{B_2}{B_1} = -\frac{M\left(i\frac{1}{2\chi_1}, \frac{1}{2}, -\frac{i}{2}\chi_1\zeta_+^2\right)}{M\left(\frac{1}{2} + i\frac{1}{2\chi_1}, \frac{3}{2}, -\frac{i}{2}\chi_1\zeta_+^2\right)}, \tag{11.259}$$

for $\xi_0 = 2\pi N_0 x_0$. For further relations, we do not use normalized input for $\zeta = \zeta_-$. Instead, we compare the series expansion of $U(\zeta)$ with that of $V(\zeta)$ using (11.244) and (11.245). A short calculation yields

$$\frac{A_1}{B_2} = ie^{-i\theta}\,, \qquad \frac{B_1}{A_2} = -ie^{i\theta}\,, \qquad \theta = 2\frac{\Delta}{N_0^2}\left[\frac{1}{2}\frac{\Delta}{N_0^2\beta} + \xi_0\right], \qquad \rho = \frac{B_2}{B_1} = \frac{A_1}{A_2}\,. \tag{11.260}$$

To calculate reflection and transmission, we use the two quantities

$$r = \frac{\tilde{v}(\xi_-)}{\tilde{u}(\xi_-)} = \exp\left[-4\pi i N_0\Delta + i\frac{\Delta^2}{N_0^4\beta} - i\frac{\pi}{2}\right]\frac{V_1(\zeta_-) + \rho V_2(\zeta_-)}{\rho U_1(\zeta_-) + U_2(\zeta_-)}, \tag{11.261}$$

$$t = \frac{\tilde{u}(\xi_+)}{\tilde{u}(\xi_-)} = \exp\left[-4\pi i N_0\Delta\left(1 - \frac{1}{N_0^2}\right)L\right]\frac{\rho U_1(\zeta_+) + U_2(\zeta_+)}{\rho U_1(\zeta_-) + U_2(\zeta_-)}, \tag{11.262}$$

which are obtained using (11.260). Then, the reflection coefficient R and the transmission coefficient T are given by

$$\boxed{R = |r|^2 = \left|\frac{V_1(\zeta_-) + \rho\, V_2(\zeta_-)}{\rho\, U_1(\zeta_-) + U_2(\zeta_-)}\right|^2} \tag{11.263}$$

and

$$\boxed{T = |t|^2 = \left|\frac{\rho\, U_1(\zeta_+) + U_2(\zeta_+)}{\rho\, U_1(\zeta_-) + U_2(\zeta_-)}\right|^2} \,. \tag{11.264}$$

These formulas yield results that are identical to (11.275) and (11.276), respectively.

Example 11.9 (Solution using Hermite polynomials)
Here we present an alternative method for solving the fundamental Eqs. (11.246) and (11.247). Hermite polynomials of imaginary order are used:

$$U(\zeta) = H_{i/\chi_1}\left(\pm\sqrt{i\frac{\chi_1}{2}}\zeta\right), \quad V(\zeta) = H_{-i/\chi_1}\left(\pm\sqrt{i\frac{\chi_1}{2}}\zeta\right). \tag{11.265}$$

They are applied in the region $-L \le x \le L$, which for ζ lies between the following limits:

$$\zeta_\pm = C_0 2\pi N_0(\pm L - x_0) - \frac{C_0}{\beta N_0^2}\Delta\,. \tag{11.266}$$

As already discussed, the corresponding boundary conditions result from $\tilde{u}(x = -L) = 1$ and $\tilde{v}(x = L) = 0$. These must be evaluated for $U(\zeta)$ and $V(\zeta)$ taking into account the definition (11.243). Based on

$$\tilde{u}(\xi) = U(\zeta)e^{-i\left(1-\frac{1}{N_0^2}\right)\Delta\xi} \tag{11.267}$$

$$\tilde{v}(\xi) = V(\zeta)e^{i\left(1-\frac{1}{N_0^2}\right)\Delta\xi} \tag{11.268}$$

we obtain

$$U(\zeta_-) = e^{-i\left(1-\frac{1}{N_0^2}\right)\Delta 2\pi N_0 L}, \quad V(\zeta_+) = 0\,. \tag{11.269}$$

Now we write the general solution as

$$U(\zeta) = c_1\, H_{i/\chi_1}\left(\sqrt{i\frac{\chi_1}{2}}\zeta\right) + c_2\, H_{i/\chi_1}\left(-\sqrt{i\frac{\chi_1}{2}}\zeta\right), \tag{11.270}$$

and using (11.244) and (11.245) it follows that

$$V(\zeta) = \sqrt{\frac{2i}{\chi_1}} \left\{ c_1 H_{i/\chi_1 - 1}\left(\sqrt{i\frac{\chi_1}{2}}\zeta\right) - c_2 H_{i/\chi_1 - 1}\left(-\sqrt{i\frac{\chi_1}{2}}\zeta\right) \right\} e^{2iZ} . \qquad (11.271)$$

The boundary conditions (11.269) lead to an inhomogeneous linear system of equations for the coefficients c_1 and c_2, which must be solved in order to find the correct analytical solution. We obtain

$$c_2 = \underbrace{\frac{H_{i/\chi_1 - 1}\left(\sqrt{i\frac{\chi_1}{2}}\zeta_+\right)}{H_{i/\chi_1 - 1}\left(-\sqrt{i\frac{\chi_1}{2}}\zeta_+\right)}}_{c_{211}} c_1 \equiv c_{211} c_1 , \qquad (11.272)$$

$$c_1 = \frac{e^{-i\left(1 - \frac{1}{N_0^2}\right)\Delta 2\pi N_0 L}}{H_{i/\chi_1}\left(\sqrt{i\frac{\chi_1}{2}}\zeta_-\right)} - c_{211}\, c_1 \underbrace{\frac{H_{i/\chi_1}\left(-\sqrt{i\frac{\chi_1}{2}}\zeta_-\right)}{H_{i/\chi_1}\left(\sqrt{i\frac{\chi_1}{2}}\zeta_-\right)}}_{c_{212}} , \qquad (11.273)$$

or, respectively,

$$c_1 = \frac{1}{1 + c_{211}\, c_{212}} \frac{e^{-i\left(1 - \frac{1}{N_0^2}\right)\Delta 2\pi N_0 L}}{H_{i/\chi_1}\left(\sqrt{i\frac{\chi_1}{2}}\zeta_-\right)} . \qquad (11.274)$$

Once the coefficients c_1 and c_2 have been determined, we can calculate the transmission coefficient T as well as the reflection coefficient R from (11.270). The result is

$$\boxed{T = |\tilde{u}(x = +L)|^2 = |U(\zeta_+)|^2 ,} \qquad (11.275)$$

$$\boxed{R = |\tilde{v}(x = -L)|^2 = |V(\zeta_-)|^2 .} \qquad (11.276)$$

The formulas suggest two remarks. First, we can estimate to what extent the chirp changes the effectiveness of the grating in reflection compared to an unchirped grating.

Furthermore, an examination of the argument in the Hermite polynomials is insightful. It generally reveals the dependence of the spatial variation on the chirp parameter of the grating. Consequently, an incident plane wave is reflected with a slight change in the effective wavelength. ■

Example 11.10 (Resonance Frequency in the Chirped Grating)
Note that in (11.208) the normalized frequency difference Δ is defined with respect to ω_0. However, if $x_0 = -L$ is chosen, the resonance frequency ω_0 occurs at the entrance of the grating, i.e., at $x = -L$.

Very often, the frequency deviation is defined with respect to the resonance frequency at the center of the grating, i.e., at $x = 0$. In this case, we shift Δ by $2\pi N_0|\beta|L$.

This shift becomes evident from (11.212) and (11.213). We can write in dimensionless form

$$\psi = 4\pi N_0 x + 4\pi^2 N_0^2 \beta (x - x_0)^2 , \qquad (11.277)$$

which, via $k_{\text{eff}} = \partial\Psi/\partial x$, leads to the effective wavenumber

$$k_{\text{eff}} = 4\pi N_0[1 + 2\pi N_0 \beta (x - x_0)] \qquad (11.278)$$

of the grating. Therefore, we introduce

$$\boxed{\tilde{\Delta} = \Delta + 2\pi N_0 |\beta| L} \qquad (11.279)$$

for the following plots, in which the resonance frequency occurs at the center of the grating. ■

Parameter dependencies
We now discuss the specific properties of a more realistic, chirped, plasma-based grating. We address the question of effectiveness by varying different parameters.

The grating (for $b_1 = b$ and $b_2 = 0$) can be written in the form:

$$\delta n_e = C \cos\left[4\pi N_0 x + 4\pi^2 N_0^2 \beta (x - x_0)^2\right] , \qquad (11.280)$$

where the density perturbation δn_e is normalized with n_0 and x is measured in λ_0. For the theoretical prediction, it is assumed that the grating is spatially homogeneous and exists in the range $-L \leq x \leq L$.

We now begin with the parameter set

$$L \equiv L_0 = 270 , \quad \beta \equiv \beta_0 = -2.35 \times 10^{-5} , \quad C \equiv C_0 = 0.2 , \qquad (11.281)$$

and $n_0 = 0.05 n_c$. The initial position is $x_0 = -L$, where the grating starts with the wavenumber $2k_0$. The resonance is shifted to the center of the grating by changing Δ to $\tilde{\Delta}$. The sign of the grating chirp is chosen such that the probe pulse arrives at the side of the grating where the large wavenumbers are located. We use (11.263) and (11.264) to investigate how parameter variations affect reflection and transmission. For clarity, we restrict the discussion to reflection, since the transmission follows from $T = 1 - R$.

Although in a realistic situation grating amplitude, length, and chirp are interdependent, in this section we treat them as independent in order to analyze the influence of each individual parameter separately.

Let us begin by varying the chirp parameter β; the results are shown in Fig. 11.13(a). We observe a clear dependence on the strength of the chirp parameter for both the bandwidth and the strength of the reflection behavior. Within the reflection windows, different frequencies are reflected with nearly identical amplitudes. The oscillations observed in the reflection coefficient resemble previous predictions for Bragg gratings with linear chirp [321, 325]. Clearly, the chirp increases the window width (bandwidth), while the reflectivity decreases.

Next, we vary the strength of the grating, i.e., the parameter C. The results are shown in Fig. 11.13(b). Increasing the grating amplitude obviously does not change the bandwidth in Δ for the reflection. However, it leads to the expected increase in the reflection rate.

Finally, we vary the total length $2L$ of the grating. The results are shown in Fig. 11.13(c). Varying the grating length leads to a change in the reflection window.

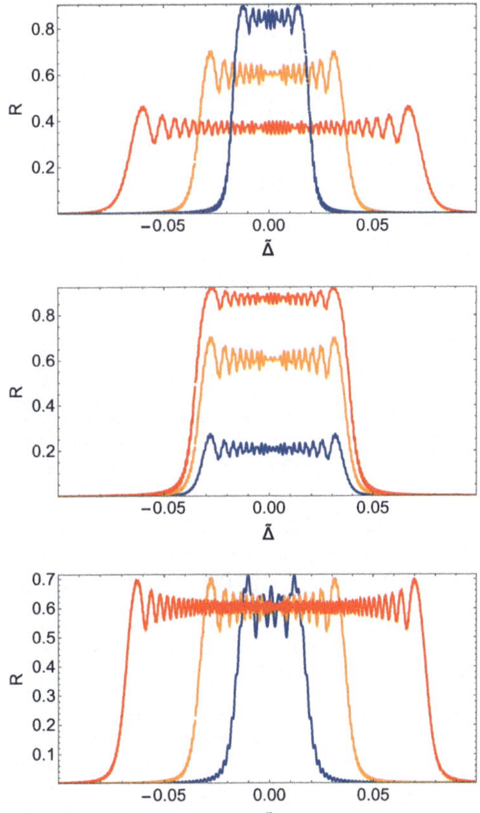

Fig. 11.13 (**a**) Solutions, using Kummer functions, of (11.246) and (11.247) for chirp rates $\beta_0/2$ (blue line), β_0 (orange line), and $2\beta_0$ (red line). The grating amplitude $C = 0.2$ and the length $L = 270$ are fixed. (**b**) Results for a fixed chirp rate β_0 and a fixed grating length L_0, but with amplitudes $C = 0.1$ (blue line), $C = 0.2$ (orange line), and $C = 0.3$ (red line). (**c**) Results for variation of the grating length $L = L_0/2$ (blue line), $L = L_0$ (orange line), and $L = 2L_0$ (red line). The grating amplitude $C = 0.2$ and the chirp rate β_0 are fixed. In all diagrams, the reflection coefficient R is plotted as a function of the frequency detuning $\Delta = \tilde{\Delta}$

Longer gratings allow for a greater bandwidth in Δ. However, the length does not significantly affect the strength of the reflection.

Simulation results

To demonstrate that pulse compression is possible, we investigate the reflection properties of the chirped grating in a numerical simulation. Once the grating is fully formed, we send a chirped laser pulse onto the grating. The chirp rate of this probe pulse matches that of the chirped driver pulse. If the compression were perfect, the reflected pulse would have a duration 100 times shorter and, at the same time, an intensity 100 times higher. For demonstration, we use a probe intensity of 10^{14} W/cm^2.

Figure 11.14 shows the electric field of the incident probe pulse (propagating from left to right) as well as that of the reflected portion of the pulse. The fields in Fig. 11.14 are normalized to the maximum of the incident pulse. The reflected pulse is clearly shorter than the incident pulse; however, the maximum electric field is only about five times greater than that of the incident pulse. The intensity of the reflected pulse is about 27 times higher. The main reason that an intensity increase by a factor of 100 is not achieved is that about 50% of the incident laser energy is transmitted through the grating. In particular, the energy of the frequencies in the wings of the probe spectrum is not sufficiently reflected. This reduces the effective bandwidth of the laser pulse. Accordingly, we measure a FWHM duration of the reflected pulse of about 50 fs.

The spectrum of the reflected pulse exhibits an almost flat phase, i.e., only a small residual chirp, which is mainly quadratic and is due to the spatial inhomogeneity of the

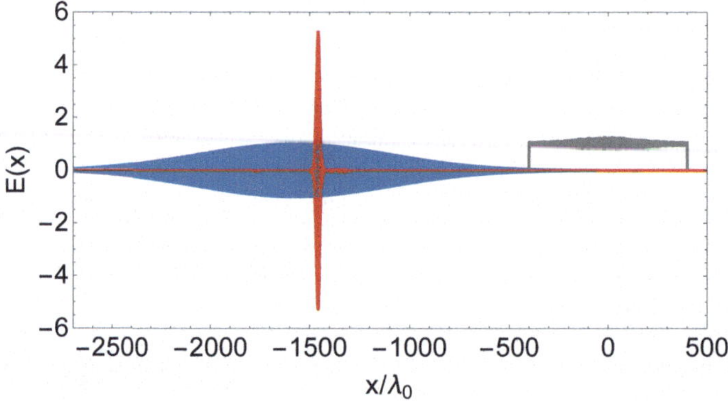

Fig. 11.14 Reflection of an incident chirped pulse (blue) by a chirped plasma grating (position marked by gray line), resulting in a strongly compressed reflected pulse (red line). Shown are the electric fields of the laser pulses, normalized to the maximum of the incident laser pulse. The plasma density n_e/n_c was scaled by a factor of 20 for better visibility. The incident laser pulse has a maximum intensity of 10^{14} W/cm^2

grating amplitude. Overall, the grating reflects 50% of the pulse energy and leads to a spectral narrowing, which in turn results in a lengthening of the pulse duration.

The coupled mode equations show that both the reflection and the effective bandwidth of chirped plasma gratings are optimizable parameters. The spectral bandwidth of the grating can be increased at the expense of reflection efficiency.

Complex Plasmas 12

Abstract

This chapter aims to fill a gap concerning a significant pillar of plasma physics. This can only be achieved very incompletely, as complex plasmas (also referred to as dusty, non-ideal, or colloidal) would deserve a much broader scope if one were to do justice to their rapid development, both scientifically and technologically. But as with magnetic confinement (Chap. 10) and laser plasmas (Chap. 11), this is a vast field that can only be fully covered in specialized books. (The fact that the chapter on "Complex Plasmas" is shorter than the two preceding ones is also—unintentionally—due to the fact that complex plasmas are not the author's main research focus. However, the brevity is by no means intended to signal lesser importance.) The idea behind including it here is to show how such a research field builds upon the fundamentals presented in Part I. We choose three examples for demonstration: charging of heavy dust particles, a new oscillation mode that appears due to the presence of a third component in the plasma, and crystallization, when we can no longer speak of ideal plasma situations.

12.1 Overview

We begin with a brief overview, referring to authors such as Bonitz [333], Melzer [334], and Ishihara [335], who have shaped the field over the past decades.

A dusty plasma consists of at least three components: electrons, ions, and heavy dust particles. This can be seen as an extension of the two-component electron-ion plasmas we have mainly considered so far. Here, the third component is not formed by additional

K.-H. Spatschek, *Theoretical Plasma Physics*,
https://doi.org/10.1007/978-3-662-72828-4_12

ions, but by (usually negatively) charged heavy dust particles, which have not arisen through direct elementary processes such as ionization in a gas mixture.

The charges of the dust particles can be approximately ten thousand times that of an elementary charge. This brings us into a regime that we described at the very beginning of this book as non-ideal. The charging process is now largely understood. We will return to a variant of it in the next section. However, the absolute charge of a dust particle is not necessarily constant over time, which can certainly add an extra layer of complexity.

The mass of the dust particles, measured for example in proton masses, is orders of magnitude greater than one. This introduces a completely new timescale in the dusty plasma. The question of screening must be reconsidered. It is also relevant which new phenomena gain additional significance due to the qualitatively different third component in a multi-component, non-ideal plasma.

The interest in dusty plasmas is driven by quite different reasons. Originally motivated by astrophysical considerations (e.g., by L. Spitzer [336] or H. Alfvèn [337]), technological interest soon developed, for example from the semiconductor industry. By now, dusty plasmas, not least accompanied by experiments, have established their own place in basic research.

A special role in all considerations is played by the coupling parameter

$$\Gamma = \frac{\text{mean potential (Coulomb) energy}}{\text{mean kinetic energy}} . \tag{12.1}$$

For an electron-ion plasma, we set

$$\Gamma_{e-i} \sim \frac{e^2}{4\pi \varepsilon_0 d k_B T_e} = 3^{-1/3}(4\pi n \lambda_D^3)^{-2/3} , \tag{12.2}$$

where d is the mean interparticle distance and λ_D is the electron Debye length:

$$d \sim n^{-1/3} , \quad \lambda_D = \sqrt{\frac{\varepsilon_0 k_B T_e}{n e^2}} . \tag{12.3}$$

If the number of particles in a Debye volume is large, then the plasma coupling parameter is small and we speak of an ideal plasma.

Three parameters enter into the coupling parameter: temperature, density, and charge. Large coupling parameters signal the transition to a crystalline state.

In a two-component plasma, we can "tune" the density and temperature to increase the coupling parameter by increasing the density or lowering the temperature. However, when lowering the temperature, recombination "throws a wrench in the works" when aiming to achieve a crystalline plasma state. Non-neutral plasmas may allow for exceptions here.

The alternative for Coulomb crystallization of two-component plasmas is to increase the particle density. However, the required orders of magnitude

$$\boxed{\Gamma_{e-i} \geq \mathcal{O}(135)} \tag{12.4}$$

require very high densities, which *classically* can hardly be achieved even in astrophysical environments.

We have already discussed in Part I that *quantum mechanically* the situation is different and the Brueckner parameter plays a decisive role. Roughly, we distinguish classical and quantum mechanical regimes using the degeneracy parameter χ_a for the particle species a, where χ_a is defined with the thermal de Broglie wavelength Λ_a:

$$\chi_a \equiv n_a \Lambda_a^3 , \quad \Lambda_a = \sqrt{\frac{h^2}{2\pi m_a k_B T_a}} . \tag{12.5}$$

If the thermal de Broglie wavelength is greater than the mean interparticle distance, quantum mechanics must be used. When estimating the potential energy relative to the kinetic energy, we must use the quantum mechanical expression for the kinetic energy. This leads to the already discussed Brueckner parameter

$$\boxed{r_S \equiv \lambda_B \sim \frac{\text{mean particle distance}}{\text{Bohr radius}}} . \tag{12.6}$$

Example 12.1

We obtain the quantum mechanical Brueckner parameter of order 1 for electron densities of the order of $n \approx 1.6 \times 10^{30}\ \text{m}^{-3}$. These are solid-state densities. ■

Back to the classical case. We do not need to enter the quantum mechanical regime to observe strong non-ideality effects in dusty plasmas. The reason is the third selectable parameter, in addition to density and temperature, namely the charge.

The coupling parameter for dusty plasmas is

$$\Gamma_{dust} \sim \frac{Z_d^2 e^2}{4\pi \varepsilon_0 d k_B T_d} \tag{12.7}$$

or

$$\Gamma_{dust} \quad \longrightarrow \quad \gamma_{dust}^{eff} \equiv \Gamma_{dust}\, e^{-\kappa} \ , \quad \kappa \approx \frac{d}{\lambda_D}, \tag{12.8}$$

if we also take screening into account. $d = \left(\frac{3}{4\pi n}\right)^{1/3}$ is called the Wigner-Seitz radius.

Example 12.2
For a typical dusty laboratory plasma, we assume $Z_d \approx 1000$, $d \approx 10^{-4}$ m and $k_B T_d \approx 0.03$ eV. This yields $\Gamma_{dust} \approx 500 \gg 1$. ■

The resulting and relatively easily observable **Coulomb crystallization** allows significant phenomena to be investigated and understood in a particularly vivid way. This includes not only the formation itself, but also the geometric structure, the dynamics, possible phase transitions, and much more.

In addition to crystallization, the **effective interaction** of dust particles is an extremely interesting statistical problem. Naturally, one expects screening of charged dust particles at greater distances, similar to what we have already found for ions in two-component plasmas. In addition, the influence of the surrounding plasma can, under certain circumstances, lead to a change in the **repulsive or attractive behavior**.

As already mentioned, the **charging processes** of dust particles are of particular interest. Here, we will present only one essential process.

Another major research focus is on the **eigenmodes** of complex plasmas. Due to the enormous differences in charge and mass of the dust particles, not only do generalizations occur that are already known from multicomponent plasmas. If crystallization is present, the collective behavior is further extended by **lattice modes**.

All this demonstrates the **broad spectrum** by which theoretical plasma physics is expanded through dusty plasmas. In addition, there is enormous **technological relevance**. For reasons of space, we can only highlight three aspects below—similar to what was done for magnetic confinement and laser plasmas—and must then refer to the extensive further literature.

The specialized literature on complex plasmas has now become very extensive. Here is a brief selection:

The article by Ikezi [338] is one of the earliest in the field. It investigates the formation of Coulomb solid structures from small particles in plasmas. Ikezi describes theoretical models that explain the interactions between charged particles in a plasma and the conditions for the formation of a solid Coulomb system. The results provide important insights into the physics of complex plasmas, particularly with regard to self-organization and the emergence of solid-state structures under certain conditions. Later articles [339–341] take up this topic.

There is now a wide range of books that present the research field of "complex plasmas" excellently for students and researchers. From the abundance, the following books are particularly noteworthy:

The book by Shukla and Mamun [342] offers a comprehensive introduction to the physics of dusty and self-gravitating plasmas in space. It covers theoretical concepts and mathematical models for describing such systems, particularly with regard to collective effects, wave propagation, and instabilities. In addition, astrophysical applications such as plasma processes in planetary rings, interstellar clouds, and accretion disks are discussed.

The book by Huber [343] covers the physical fundamentals and experimental methods for investigating complex plasmas and colloidal dispersions. It combines concepts from plasma physics and statistical physics to analyze collective phenomena, phase transitions, and self-organization at the microscopic level. A particular focus is placed on particle-resolved experiments, which make it possible to model liquids and solids at the atomic level.

The book by Thomas [344] deals with the experimental investigation of the dynamics of Yukawa systems in complex plasmas. It analyzes the fundamental physical mechanisms that determine the interactions and collective phenomena in these systems. Special emphasis is placed on the experimental methodology for capturing particle dynamics as well as on the analysis of phase transitions and structure formation.

The book "Complex and Dusty Plasmas: From Laboratory to Space" by Tsytovich et al. [345] provides an in-depth analysis of complex and dusty plasmas, a special class of plasmas containing charged dust particles. It covers both theoretical foundations and experimental and astrophysical applications. The book combines a systematic presentation of the theory with an overview of experimental findings and real-world applications. The mathematical treatment is demanding but well structured, and numerous application examples from laboratory and space plasmas provide practical relevance.

The book by Franz [346] offers a comprehensive introduction to low-pressure plasmas and their application in microstructure technology. It covers the physical fundamentals of plasmas, their interactions with solid surfaces, as well as various technological applications such as etching and coating processes. The presentation is both theoretically sound and practice-oriented, making the book a valuable resource for scientists and engineers in plasma technology and semiconductor manufacturing.

This book by Fortov and Morfill [347] provides a comprehensive account of complex and dusty plasmas, which occur both in laboratory environments and in space. It covers the theoretical foundations, experimental methods, and astrophysical applications of such plasmas. Special emphasis is placed on collective effects, nonlinear phenomena, and technological applications.

The book by Ivlev et al. [348] offers a detailed study of complex plasmas and colloidal dispersions with a focus on particle-resolved methods. It combines concepts from plasma physics with the statistical physics of classical liquids and solids. Experimental techniques as well as theoretical models for describing collective phenomena, phase transitions, and self-organization are discussed.

The book by Bonitz et al. [349] provides a comprehensive introduction to the physics of complex plasmas and covers both theoretical and experimental aspects. It addresses fundamental physical concepts, interactions between charged particles, as well as collective phenomena and self-organization. In addition to classical plasma models, modern developments and applications in research and technology are discussed.

The book by Melzer [350] offers a detailed analysis of dynamic processes in complex plasmas, with particular focus on the interactions and behavior of charged particles in non-ideal systems. Both theoretical models and experimental techniques are discussed, enabling an understanding of particle dynamics in complex plasmas.

The book by Thoma [351] provides a comprehensive overview of complex plasmas and their scientific as well as technological significance. It covers fundamental physical concepts, experimental methods, and modern applications. Special emphasis is placed on collective phenomena, non-equilibrium effects, and innovative technologies based on complex plasmas.

This book by Ivlev [352] offers a comprehensive analysis of complex plasmas, highlighting both fundamental scientific questions and technological applications. It covers theoretical concepts, experimental methods, and modern developments in plasma physics. Special emphasis is placed on collective phenomena, non-equilibrium effects, and innovative applications in research and industry.

12.2 Charging

In this section, we discuss a simple model for the charging of a dust particle in a plasma composed (otherwise) of electrons and ions. This is the so-called OML (orbital motion limit) model.

The model now presented represents the simplest version of currents that can impinge on the surface of a dust particle. For this, little more than knowledge of the classical version of the two-body problem is required. We neglect many additional effects, such as collisions along the trajectories of incoming electrons or ions, absorption of radiation, secondary electron emission, etc.

The idea is that large particles in an electron-ion system are struck by incoming charge carriers and thus become charged. Let us consider a state with the charge Q_d of a dust particle. Then, in the stationary case, the following holds

$$\boxed{\frac{dQ_d}{dt} = \sum_l I_l = 0}\ , \tag{12.9}$$

where I_l denotes the various currents of particle species l. Let us start with a (positive) ion current onto an already negatively charged dust particle. The dust particle then has a (floating) potential ϕ_p. The concept of the floating potential originates from probe theory. The "floating potential" describes the electric potential that an electrically isolated surface (such as a probe) assumes when it is placed in a plasma and no external currents are allowed. The floating potential is the potential of an isolated, non-conductively connected surface in the plasma. This surface can freely adjust to a potential at which the sum of the currents of the various plasma components (electrons and ions) is zero.

If a (singly charged) ion comes from infinity with initial velocity v_{i0}, then for the respective velocity v_i the following holds according to the law of energy conservation

$$\frac{1}{2}m_i v_{i0}^2 = \frac{1}{2}m_i v_i^2 + e\phi_p\ . \tag{12.10}$$

The (as usual defined) impact parameter (measured from the center of a dust particle, here assumed to be spherical) is b. There is a critical value b_c, such that for $b < b_c$ the ion strikes the dust particle. The angular momentum at the critical impact parameter is

$$L_c = |\mathbf{r} \times \mathbf{p}| = m_i v_{i0} b_c\ . \tag{12.11}$$

At the moment of contact with the dust particle of radius a, for the critical value b_c, that is, when the incoming ion just grazes the dust particle, we have on the surface of the sphere

$$L_c = m_i v_i a\ , \tag{12.12}$$

since we assume conservation of angular momentum. A short calculation, making use of the law of energy conservation, yields

$$\frac{1}{2}m_i v_{i0}^2 = \frac{1}{2}m_i v_{i0}^2\left(\frac{v_i^2}{v_{i0}^2} + \frac{e\phi_p}{\frac{1}{2}m_i v_{i0}^2}\right) = \frac{1}{2}m_i v_{i0}^2\left(\frac{b_c^2}{a^2} + \frac{e\phi_p}{\frac{1}{2}m_i v_{i0}^2}\right) \tag{12.13}$$

as well as

$$b_c^2 = a^2\left(1 - \frac{2e\phi_p}{m_i v_{i0}^2}\right)\ . \tag{12.14}$$

With this, we can define a cross section,

$$\boxed{\sigma_c = \pi b_c^2 = \pi a^2 \left(1 - \frac{2e\phi_p}{m_i v_{i0}^2}\right)} . \tag{12.15}$$

Since $\phi_p < 0$, the cross section is larger than for collisions with uncharged particles. This is, of course, understandable due to the attraction.

Next, we calculate the charge current densities (I_i for ions) by averaging over all initial configurations. If we interpret $v_i \hat{=} v_{i0}$, we write

$$dI_i = \sigma_c(v_i) n_i e v_i f(v_i) dv_i \tag{12.16}$$

with an isotropic Maxwell distribution

$$f(v_i) = 4\pi v_i^2 \left(\frac{m_i}{2\pi k_B T_i}\right)^{3/2} e^{-\frac{m_i v_i^2}{2k_B T_i}} . \tag{12.17}$$

We obtain the integral expression

$$I_i = \int dI_i = 4\pi^2 a^2 n_i e \left(\frac{m_i}{2\pi k_B T_i}\right)^{3/2} \int_0^\infty \left(1 - \frac{2e\phi_p}{m_i v_i^2}\right) e^{-\frac{m_i v_i^2}{2k_B T_i}} dv_i . \tag{12.18}$$

The integral can be evaluated analytically with the result

$$\boxed{I_i = \pi a^2 n_i e \sqrt{\frac{8 k_B T_i}{\pi m_i}} \left(1 - \frac{e\phi_p}{k_B T_i}\right)} . \tag{12.19}$$

Example 12.3 (Evaluation with a shifted Maxwell distribution)
The evaluation with an isotropic Maxwell distribution is only accurate if the flow velocity v_f of the ions in normalized form

$$u = \frac{v_f}{v_{Ti}} \quad \text{with } v_{Ti} = \sqrt{\frac{8 k_B T_i}{\pi m_i}} \tag{12.20}$$

is negligible. However, we can also carry out the calculation analytically for a shifted Maxwell distribution. The result is then

$$I_i = \pi a^2 n_i e \sqrt{\frac{8 k_B T_i}{\pi m_i}} e^{-u^2} \left\{ \frac{1}{2} + \frac{\sqrt{\pi}}{2} \left[u + \frac{1}{2u}\left(1 - \frac{e\phi_p}{k_B T_i}\right)\right] e^{u^2} \operatorname{erf}(u) \right\} , \tag{12.21}$$

where for the error function the following holds

$$\operatorname{erf}(u) \underset{u \to 0}{\longrightarrow} \frac{2}{\sqrt{\pi}} \left(u - \frac{u^3}{3}\right) , \tag{12.22}$$

$$\operatorname{erf}(u) \xrightarrow[u\to\infty]{} 1 \,. \tag{12.23}$$

It is easy to see that for $u \to 0$ the expression (12.21) reduces to (12.19). ■

Without recalculating, however, we cannot simply use the substitutions $e \to -e$, $n_i \to n_e$, $T_i \to T_e$, $m_i \to m_e$ to state the result for the corresponding electron current. The reason is that electrons with low velocity ($v_e \to 0$) do not reach the dust particle at all. The electrons require a minimum velocity

$$v_{min} = \sqrt{\frac{-2e\phi_p}{m_e}} \,, \tag{12.24}$$

and the integral is

$$I_e = -4\pi^2 a^2 n_e e \left(\frac{m_e}{2\pi k_B T_e}\right)^{3/2} \int_{v_{min}}^{\infty} \left(1 + \frac{2e\phi_p}{m_e v_e^2}\right) e^{-\frac{m_e v_e^2}{2k_B T_e}}\, dv_e \,. \tag{12.25}$$

The evaluation yields

$$\boxed{I_e = -\pi a^2 n_e e \sqrt{\frac{8k_B T_e}{\pi m_e}}\, e^{\frac{e\phi_p}{k_B T_e}}} \,. \tag{12.26}$$

If we were to calculate the currents for a positively charged dust particle, the functional forms for electrons and ions would be exactly reversed.

Now we can turn to the calculation (estimation) of the charge of a dust particle in the stationary case. From the condition $I_e + I_i = 0$ we find an equation for ϕ_p as a function of the plasma parameters:

$$1 - \frac{e\phi_p}{k_B T_i} = \sqrt{\frac{m_i}{m_e}\frac{T_e}{T_i}\frac{n_e}{n_i}}\, e^{\frac{e\phi_p}{k_B T_e}} \,. \tag{12.27}$$

The equation is not analytical, but can be easily solved numerically.

Example 12.4

A relevant estimate for an electron-proton plasma in astrophysical situations goes back to Spitzer:

$$T_e \approx T_i \,, \quad n_e \approx n_i \,, \quad \frac{m_i}{m_e} \approx 1836 \,, \quad z \equiv -\frac{e\phi_p}{k_B T_e} \,. \tag{12.28}$$

The equation for z

$$(1+z)\frac{1}{\sqrt{1836}} \approx e^{-z} \tag{12.29}$$

has the quite well-fitting solution $z \approx 2.5$ as a result. ■

From ϕ_p we calculate $Q_d \equiv Z_d e$. For this, we treat the dust particle as a spherical capacitor with capacitance C. The following holds

$$\boxed{C = \frac{Q_d}{\phi_p}} . \tag{12.30}$$

From electrodynamics, we know the capacitor formula in vacuum, which can usually also be used with shielding in plasma because of

$$C \approx 4\pi\varepsilon_0 a\left(1+\frac{a}{\lambda_D}\right) \quad \rightarrow \quad C \approx 4\pi\varepsilon_0 a \tag{12.31}$$

for $a \ll \lambda_D$. Here, a is the inner radius; the outer radius has been taken to infinity.

For a typical value $\phi_p \sim \mathcal{O}(-2k_B T_e/e)$ we thus obtain

$$Q_d \approx -8\pi\frac{\varepsilon_0}{e} a k_B Te . \tag{12.32}$$

Using

$$\frac{\varepsilon_0}{e} \approx 8.854 \times 10^{-12}\frac{\mathrm{As}}{\mathrm{eV\ m}} , \quad 1\mathrm{As} \approx \frac{10^{19}}{1.6}e, \tag{12.33}$$

we find after a short calculation

$$Q_d \approx -1400\, a[\mu\mathrm{m}]\, T_e[\mathrm{eV}]\, e . \tag{12.34}$$

Finally, a word on the temporal evolution of charging. Due to the higher mobility of electrons, the characteristic charging time by electrons is significantly shorter than that for ions. The overall duration of the charging process is therefore determined by the ion currents.

If we replace in the ion current ϕ_p by $\frac{Q_d}{4\pi\varepsilon_0 a}$, we obtain for $Q \equiv Q_d$ the differential equation

$$\frac{dQ}{dt} = \pi a^2 n_i e \sqrt{\frac{8k_B T_i}{\pi m_i}} \left(1 - \frac{eQ}{4\pi\varepsilon_0 a k_B T_i}\right) . \quad (12.35)$$

Without explicitly solving the differential equation, we can find a characteristic charging time from

$$\boxed{\tau_i = \frac{4\pi\varepsilon_0 a k_B T_i}{e\pi a^2 e n_i v_{thi}} = \sqrt{2\pi}\frac{\lambda_{Di}}{a}\omega_{pi}^{-1}} \quad (12.36)$$

Example 12.5 (Solution of the Differential Equation)
A differential equation of the form

$$\dot{y} = C - \frac{y}{\tau} \quad (12.37)$$

is solved by separation of variables. Integration then leads to the solution

$$y = (y_0 - \tau C)e^{-t/\tau} + \tau C \quad \text{with } y(t=0) = y_0 . \quad (12.38)$$

■

A calculation for charging by electrons yields the characteristic time τ_e on the order of

$$\boxed{\tau_e = \mathcal{O}\left(\frac{\lambda_{De}}{\lambda_{Di}}\frac{\omega_{pi}}{\omega_{pe}}\right)\tau_i} . \quad (12.39)$$

It is therefore significantly shorter, which explains the predominantly negative charge of the dust particles. However, the charging process is only completed on the timescale τ_i.

12.3 Forces between Dust Particles

In this section, we first compare the electromagnetic interaction with the gravitational interaction for dust particles. In contrast to "normal" electron-ion plasmas, the gravitational attraction can become significant due to the size and mass of the additional dust particles. We estimate the relevant regimes and briefly mention the astrophysical structure formation from dust clouds. Finally, we briefly address the non-gravitational attraction between dust particles caused by plasma effects.

The electric potential of strongly negatively charged dust particles will—as we can justifiably surmise based on our knowledge of the general principles—be shielded outwardly.

We have already discussed this effect, with reference to ions, in detail in the general section. A Debye shielding, primarily by positive ions, is to be expected for dust particles. The characteristic shielding length in a Debye-Hückel potential will again be the Debye length λ_D. Without an in-depth discussion, we set

$$\frac{1}{\lambda_D^2} = \frac{1}{\lambda_{De}^2} + \frac{1}{\lambda_{Di}^2}, \tag{12.40}$$

where, for a quick estimate, we again provide the calculation rules for Debye lengths

$$\lambda_{De} \approx 74.3\left(\frac{10^{16}\,\mathrm{m}^{-3}}{n_e}\right)^{1/2}\left(\frac{k_B T_e}{1\,\mathrm{eV}}\right)^{1/2}\,\mu\mathrm{m}, \tag{12.41}$$

$$\lambda_{Di} \approx 23.5\frac{1}{z_i}\left(\frac{10^{16}\,\mathrm{m}^{-3}}{n_i}\right)^{1/2}\left(\frac{k_B T_i}{0.1\,\mathrm{eV}}\right)^{1/2}\,\mu\mathrm{m} \tag{12.42}$$

The potential of a dust particle shielded according to Debye is written in the form

$$\phi(r) = \frac{Q}{4\pi\varepsilon_0 r}\exp\left(-\frac{r}{\lambda_D}\right) \tag{12.43}$$

It should be mentioned that the Debye-Hückel potential in dusty plasmas is also commonly referred to as the Yukawa potential.

Assuming radial symmetry, we calculate the electric field

$$E_r = -\frac{\partial\phi}{\partial r} = \left(\frac{1}{r} + \frac{1}{\lambda_D}\right)\phi\,. \tag{12.44}$$

We now easily find the relation between charge and floating potential already discussed in the previous section. With the relationship between surface charge ($r = a$) and the normal component of the electric field

$$Q = 4\pi\varepsilon_0 E_r \tag{12.45}$$

we find

$$\boxed{Q_d = 4\pi\varepsilon_0 a\left(1 + \frac{a}{\lambda_D}\right)\phi_p(a)}\,. \tag{12.46}$$

This is familiar from (12.30) and (12.31).

As a reminder: The floating potential $\phi_p(a)$ is to be understood as the difference between the potential of the surface charge and the bulk plasma potential (bulk = main volume).

Coulomb repulsion and gravitational attraction
Now let us compare, for (isolated) dust particles, the electromagnetic and gravitational interactions. Let us take two dust particles (indices 1 and 2) at a distance d; the forces are equal under the following condition:

$$\frac{|Q_1 Q_2|}{4\pi\varepsilon_0 d^2} e^{-d/\lambda_D} = G\frac{m_1 m_2}{d^2} , \tag{12.47}$$

which we can rewrite using the dust mass densities $\rho_{1,2} = 3m_{1,2}/(4\pi a_{1,2}^3)$ as

$$\boxed{a_1 a_2 = \frac{3}{4\pi}\sqrt{\frac{4\pi\varepsilon_0}{G}\frac{\phi_1\phi_2}{\rho_1\rho_2}e^{-d/\lambda_D}}} \tag{12.48}$$

Now we can distinguish two cases: Let us begin with $\boldsymbol{d \gg \lambda_D}$. Gravitational attraction is always present. However, the dust particles do not fall directly toward each other, but rather orbit one another and occasionally merge when the angular momentum slowly decreases (due to whatever influences). A dust cloud forms.

In the opposite case $\boldsymbol{d \ll \lambda_D}$ we can neglect the shielding factor and attraction occurs provided that

$$a \geq \left(\frac{9\varepsilon_0}{4\pi G}\right)^{1/4}\sqrt{\frac{\phi_p}{\rho}} \tag{12.49}$$

is satisfied. For simplicity, we have assumed identical dust particles.

Example 12.6 (Size of Dust Particles)
For $\phi_p \approx 1$ V and $\rho \approx 10^4$ kg m^{-3} a lower value of $a \approx 6$ mm is obtained. This is relatively large, so in this second case gravity can usually be neglected. ■

Example 12.7 (Jeans Criterion from a Fluid Model)
We now turn to the astrophysically interesting question of whether a dust cloud caused by self-gravity is stable or will continue to collapse. This question is known as the Jeans instability. An exact instability analysis is generally complicated. First, one must know the initial state precisely, which is to be examined for stability. Then, one linearizes around the state to be investigated, which leads to an eigenvalue problem. The latter is generally not easy to solve. However, if one finds an unstable mode, instability can be inferred. Proving stability is even more difficult, since in that case one must exclude *all* unstable modes.

We now attempt to derive an instability criterion for a dust cloud using a fluid model.

To do this, we start from the fluid equations

$$\frac{\partial \rho}{\partial t} + \nabla \cdot (\rho \mathrm{v}) = 0 \,, \tag{12.50}$$

$$\frac{\partial \mathrm{v}}{\partial t} + \mathrm{v} \cdot \nabla \mathrm{v} = -\frac{1}{\rho} \nabla P - \nabla \phi \,, \tag{12.51}$$

$$\nabla^2 \phi = 4\pi \, G \rho \tag{12.52}$$

for the mass density ρ, the fluid velocity $\mathbf{v}$, and the gravitational potential ϕ of the dust component. Charges are (initially) neglected. We will discuss the pressure P shortly.

It is now natural to linearize the seemingly simple equations around a stationary solution. One problem is determining a stationary solution. For example, if we use

$$\rho_0 = \text{ const} \,, \quad \mathrm{v}_0 = 0 \,, \quad P_0 = \text{ const} \,, \quad \phi_0 = \text{ const} \tag{12.53}$$

the field equation for the gravitational potential is not satisfied, except if one sets $\rho_0 = 0$, which, however, would completely miss our original physical question. If one nevertheless uses a solution ($\rho_0 \neq 0$) for linearization, this is referred to as the so-called Jeans swindle *("Jeans swindle")*. This only shows how difficult stability problems are. For more detailed investigations, better methods (e.g., variational principles) are used.

The "Jeans swindle" can be reasonably accepted in two cases. First, when the length scales of the perturbations are much shorter than the extent of the stationary solution. And second, when equilibrium is maintained not by pressure gradients but by rotation. In that case, a homogeneous state can also be considered as the initial value.

A note on the first case: One starts from $\rho_0 = \rho_0(\varepsilon \mathbf{r})$ and, to lowest order, the spatial variation can be neglected in each case (e.g., $\mathrm{v}_1 \cdot \nabla \rho_0 \ll \rho_0 \nabla \cdot \mathrm{v}_1$). The equilibrium is then determined by

$$\mathrm{v}_0 = 0 \tag{12.54}$$

together with

$$0 = -\frac{1}{\rho_0} \nabla P_0 - \nabla \phi_0 \,, \tag{12.55}$$

$$\nabla^2 \phi_0 = 4\pi \, G \rho_0 \tag{12.56}$$

Linearization about the state (12.54)–(12.56) leads to

$$\frac{\partial \rho_1}{\partial t} + \rho_0 \nabla \cdot \mathrm{v}_1 = 0 \,, \tag{12.57}$$

$$\frac{\partial \mathrm{v}_1}{\partial t} = -\nabla h_1 - \nabla \phi_1 \,, \tag{12.58}$$

$$\nabla^2 \phi_1 = 4\pi G \rho_1 \,. \tag{12.59}$$

Here, we have expressed the terms

$$\frac{\rho_1}{\rho_0^2} \nabla P_0 - \frac{1}{\rho_0} \nabla P_1 \equiv -\nabla h_1 \tag{12.60}$$

in terms of the enthalpy perturbation h_1. We assume that the pressure depends only on ρ, and write

$$\begin{aligned} \frac{1}{\rho} \nabla P(\rho) &= \frac{1}{\rho} \frac{dP}{d\rho} \nabla \rho = \nabla \int\limits_0^{\rho(\mathbf{r})} \frac{1}{\rho} \frac{dP}{d\rho} d\rho \\ &= \nabla \int_0^{\rho} \frac{dP}{\rho} \equiv \nabla h \,. \end{aligned} \tag{12.61}$$

The adiabatic speed of sound is denoted by $\mathrm{v}_s = \sqrt{\gamma P/\rho}$.

Thus, for h_1

$$h_1 = \int\limits_{\rho_0}^{\rho_0+\rho_1} \frac{1}{\rho} \frac{dP}{d\rho} d\rho \approx \left.\frac{dP}{d\rho}\right|_0 \frac{\rho_1}{\rho_0} \equiv \mathrm{v}_s^2 \frac{\rho_1}{\rho_0} \,. \tag{12.62}$$

the equations for the perturbations can then be combined into

$$\frac{\partial^2 \rho_1}{\partial t^2} - \mathrm{v}_s^2 \nabla^2 \rho_1 - 4\pi G \rho_0 \rho_1 = 0 \tag{12.63}$$

After Fourier transformation, we obtain the dispersion relation

$$\omega^2 = \mathrm{v}_s^2 k^2 - 4\pi G \rho_0 \,, \tag{12.64}$$

where, due to the ansatz

$$\rho \sim e^{i\mathbf{k}\cdot\mathbf{r} - i\omega t} \tag{12.65}$$

a positive imaginary part of ω signifies exponential growth (instability). This case occurs for

$$\boxed{k^2 < k_J^2 \equiv \frac{4\pi G \rho_0}{v_s^2}} . \tag{12.66}$$

The new frequency that arises is often called the Jeans frequency ω_J, for which the following holds:

$$\omega_J^2 = 4\pi G \rho_0 . \tag{12.67}$$

In light of the previous results, we can rewrite the instability criterion as

$$\boxed{\lambda^2 \equiv \left(\frac{2\pi}{k}\right)^2 > \lambda_J^2 \equiv \frac{\pi v_s^2}{G \rho_0}} \tag{12.68}$$

A Jeans mass can be defined as

$$\begin{aligned} M_J =& \frac{4\pi}{3}\rho_0\left(\frac{1}{2}\lambda_J\right)^3 = \frac{1}{6}\pi\rho_0\left(\frac{\pi v_s^2}{G\rho_0}\right)^{3/2} \\ \sim& 1.37 \times 10^5 M_\odot \left(\frac{T}{10^2\,\mathrm{K}}\right)^{3/2}\left(\frac{\rho}{10^{-24}\,\mathrm{g\,cm^{-3}}}\right)^{-1/2} \mu^{-3/2} . \end{aligned} \tag{12.69}$$

Here, μ is the mean molecular weight, with which the particle density $n = \rho/\mu m_p$ is determined. If we raise both sides of (12.68) to the third power and multiply both sides by ρ_0^2, we immediately obtain the statement: If the mass M of the cloud,

$$M = \frac{4\pi}{3}\rho_0 R^3 , \tag{12.70}$$

is greater than the Jeans mass, an instability can develop. The timescale is

$$\tau_d \approx \frac{1}{\sqrt{4\pi G \rho_0}} , \tag{12.71}$$

thus practically identical to the collapse time (free-fall time). ■

The Jeans instability provides us with a way to understand the formation of structures in astrophysical systems. The *stellar* condensations that develop as a result of the Jeans instability are called protostars.

Back to our current topic. In the case $d \ll \lambda_D$, (12.47) leads to the equality

$$\omega_J^2 = \omega_d^2 , \tag{12.72}$$

where ω_d is the dust plasma frequency,

$$\omega_d^2 = \frac{n_d Z_d^2 e^2}{\varepsilon_0 m_d} . \tag{12.73}$$

If (for $d < \lambda_D$ and without considering collective effects due to the influence of the plasma) the electrostatic attraction dominates, then $\omega_J < \omega_d$.

The dust particles will then tend to disperse until the shielding becomes more effective and gravitational effects increase again. The dust plasma frequency can be written using the Havnes parameter

$$\boxed{P \equiv \frac{|Z_d| n_d}{n_e} \geq 1 \text{ for collective effects}} \tag{12.74}$$

and the quasineutrality condition

$$n_d Z_d = z_i n_i - n_e \tag{12.75}$$

in the form

$$\omega_d^2 = \frac{m_i}{m_e} \frac{Z_d}{z_i} \frac{P}{1+P} \omega_{pi}^2 \tag{12.76}$$

Collective attraction of negatively charged dust particles
Collective effects play a role when the Havnes order parameter P becomes greater than 1.

Electrically charged bodies with the same sign of charge repel each other. This is, of course, universally valid. However, in a streaming (complex) plasma, it can be observed that, for example, two negatively charged dust particles can also attract each other.

One of the best-known examples results from the flow of ions around dust particles in the plasma. When ions flow around a dust particle with sufficiently high flow velocity (Mach numbers), they generate a strongly oscillating electrostatic potential behind an oppositely charged dust particle. A similar phenomenon, namely the wakefield, has already been observed with short laser pulses. Behind the dust particles, the sign of the space charge potential alternates. Therefore, not only oppositely charged particles, but—at suitable locations—also particles with the same sign of charge can be trapped in it. If we move to the center-of-mass frame of the ions, the dust particles (we are considering streaming ions in the laboratory frame) move with the velocity v_d, and we define the

Mach number as

$$M = \frac{|v_d|}{C_S} \tag{12.77}$$

with the ion sound speed $C_s = \sqrt{k_B T_e / m_i}$. The attractive phenomena occur for $M > 1$.

The theory was developed by Ishihara, Tsytovich, and many others. Due to the already advanced length of this section, we will forgo a detailed presentation here. More information can be found in original papers or review articles, e.g., [335, 345, 353].

12.4 Plasma Crystal

In this section, we discuss the formation of a plasma crystal in dusty plasmas. Theoretical predictions—based on Wigner models in solid-state physics—predicted regions for Coulomb crystallization in complex plasmas. Experiments confirmed the idea. In contrast to the presentation in other sections, here we rely heavily, in large part even verbatim, on (two) works or authors who have significantly shaped the field: H. Ikezi [338] and H. Thomas, G.E. Morfill, V. Demmel, J. Goree, B. Feuerbacher, and D. Möhlmann [354]. The original quotations nicely illustrate the interplay between theory and experiment. The following overview is also strongly influenced by the summary script by A. Melzer [334].

The search for model systems for crystalline structures to study phase transitions was initiated in the 1930s by Wigner [355] with the theory of the Wigner crystal. Since then, experimental confirmations have been achieved for several specific systems. On the atomic level, these are ion and electron crystals; on the macroscopic level, colloidal crystals in aqueous solutions. Each of these systems has advantages and disadvantages for the detailed study of the phase transition of interest, such as the formation, growth, and melting of crystalline structures [354].

Theoretical prediction by Ikezi [338]

As early as 1986, Ikezi correctly predicted Coulomb crystallization in dusty plasmas. In essence, he writes: "When the ratio between the Coulomb energy and the kinetic energy of a system of charged particles

$$\boxed{\Gamma = \frac{q^2}{4\pi\varepsilon_0 b k_B T}} \tag{12.78}$$

exceeds a critical value, $\Gamma_c \simeq 170$, a Coulomb lattice is formed. Here, q is the (negative) electric charge of a (dust) particle, b is the Wigner-Seitz distance between the particles, and T is the temperature. The Wigner-Seitz distance b is defined as:

$$b = \left(\frac{3}{4\pi N}\right)^{1/3}, \tag{12.79}$$

where $N \equiv n_d$ is the particle density (of the charged dust particles). Compared to a gaseous plasma, when q corresponds to the charge *of a single* electron e, in the latter the crystallization condition

$$\boxed{\Gamma > \Gamma_c} \tag{12.80}$$

is only fulfilled in systems with extremely high density and very low temperature, which is difficult to achieve. In dusty plasmas, however, relatively small particles (with a radius of, for example, $a \approx 0.1\,\mu\text{m}$) can, for example, acquire 10^3 electron charges, and Γ can become greater than Γ_c."

We have already estimated the typical charge of a dust particle. A small, negatively charged dust particle with a radius a smaller than the Debye length

$$\lambda_D = \left(\frac{1}{\lambda_{D,e}^2} + \frac{1}{\lambda_{D,i}^2}\right)^{-1/2}, \quad \lambda_{De,i} = \sqrt{\frac{\varepsilon_0 k_B T_{e,i}}{n_{e,i} e^2}}, \tag{12.81}$$

has the approximate potential $q/4\pi\varepsilon_0 a$. This potential should be equal to the floating potential $\phi_p \sim -2(k_b T_e/e)$. This yields an approximation for the charge:

$$\boxed{q = q_1 \approx -8\pi\varepsilon_0 \frac{a k_B T_e}{e}}. \tag{12.82}$$

Here, $n = n_i$ is the plasma density (for singly charged ions) and T_e and T_i are the electron and ion temperatures, respectively. Ikezi continues: "If $a > \lambda_D$, then the potential of the particle is not $q/4\pi\varepsilon_0 a$, so that (12.82) is not a suitable approximation. However, this case will not be considered further here. Furthermore, it must be taken into account that the system should be quasi-neutral. A comparison of the ion density $n = n_i$ (equal to the original plasma density) with the dust density $N = n_d$ of the charged dust particles shows that $n \leq |q/e|N$ must hold. For

$$q = q_2 = \frac{e n_i}{N} \tag{12.83}$$

the system consists only of negatively charged dust particles and positive ions with density n_i (and no free electrons at all). Larger charges than

$$Z_{c,\text{lim}} = e\frac{n_i}{n_d} \tag{12.84}$$

are therefore not possible on average.

The shielding effect must still be taken into account as an extension of (12.78).

Since the plasma screens the field of the particles, the crystallization condition is only valid when $b \ll \lambda_D$. To take the Debye screening effect into account, we replace q^2/b with the screened Coulomb potential $q^2 \exp(-b/\lambda_D)/b$ and introduce the quantity

$$\Gamma_{\text{eff}} = \frac{Z_d^2 e^2 \exp(-b/\lambda_D)}{4\pi\varepsilon_0 b k_B T_d} \equiv \frac{q^2 \exp(-b/\lambda_D)}{4\pi\varepsilon_0 b k_B T} \equiv \Gamma \exp(-\kappa) \tag{12.85}$$

The critical value $\Gamma_{\text{c,eff}}$ is still defined by

$$\Gamma_{\text{c,eff}} \approx 170 \qquad \text{(melting line)} \tag{12.86}$$

For $\Gamma_{\text{eff}} \geq \Gamma_{\text{c,eff}}$ we expect Coulomb crystallization. The screening is essentially provided by the ions, i.e.,

$$\lambda_D \approx \lambda_{D,i} \; . \tag{12.87}$$

The critical value $\Gamma_{\text{c,eff}}$ may not be exactly 170, but is used here for the following discussion as long as no better values are available for the case of screened interactions."

By combining

$$\Gamma_{\text{eff}} > \Gamma_{\text{c.eff}} \tag{12.88}$$

with the expression for q [which is given either by (12.82) or (12.83) (depending on the relationship between n and N)] and the functional forms $\lambda_D(n)$ as well as $b(N)$ for parameters a and T, a range of n and N can be found in which crystallization occurs.

An example of such a diagram is shown in Fig. 1 in Ikezi [338]. The parameters used there are $T_e = 3\,\text{eV}$, the particle temperature $T = T_i = 0.03\,\text{eV}$ and $m_i/m_e = 40 \times 1800$.

Now, at what values of ion and dust density can Wigner crystallization occur in dusty plasmas? Melzer [334] has described the situation in detail: "The ion density influences the screening length λ_D and the charge limit of the dust particles. A high ion density means a high maximum dust charge, but also strong screening. The dust density also influences the maximum dust charge and the distance between the particles b. High dust densities lead to a low maximum dust charge, but also to small distances between the particles, resulting in strong coupling. Both parameters therefore have opposing effects.

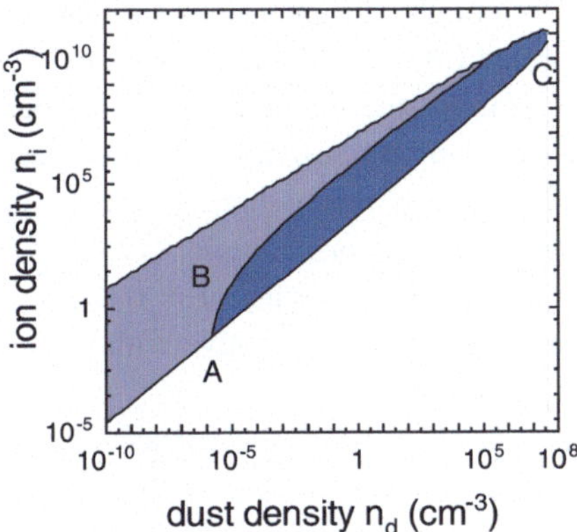

Fig. 12.1 Existence diagram of Wigner crystals in dusty plasmas. In the dark region, Coulomb crystallization should be possible for particles with a radius of 10 μm. The parameters used are $T_e = 3\,\text{eV}$, the particle temperature $T = T_i = 0.03\,\text{eV}$ and $m_i/m_e = 40 \times 1800$. In the fully shaded region, crystallization can occur for particles of any size. Note the wide logarithmic scale on both axes. The diagram is based on the work of Ikezi and was published by Melzer for demonstration purposes

From these considerations, it follows that Wigner crystallization should be possible in the dark region ABC of Fig. 12.1. The boundary of this region is dominated by different mechanisms. From A to B, the dust density remains nearly constant. The charge Z_d also remains constant; there are no depletion effects due to the low dust density. Along the entire upper boundary from A via B to C, the particle is charged to its single-particle charge (it is assumed here that $a = 10\,\mu\text{m}$ applies). From A to B, the Debye length is much greater than the distance between the particles, so screening effects do not play a role.

Near point B, the situation changes. The ion density becomes so high that the Debye length is now on the order of the distance between the particles, and shielding effects become dominant. Therefore, the boundary bends toward much higher dust densities and thus smaller particle separations, until point C is reached. On the boundary from C to A, there are relatively high dust densities and low ion densities. Here, depletion effects become dominant. The charge of the dust particles is determined by the available free electrons, which limits the coupling parameter."

Ikezi's considerations show that Coulomb crystallization in dusty plasmas is possible over a range of ion and dust densities spanning several orders of magnitude. For typical plasma discharges with $n_i = 10^9$ to $10^{10}\,\text{cm}^{-3}$, plasma crystals should exist for dust densities in the range from $n_d = 10^3$ to $10^5\,\text{cm}^{-3}$.

Ikezi also discussed concrete possibilities in laboratory plasmas, well before the subsequent realization described below: "A partially ionized plasma with the parameters $T_e = 3\,\text{eV}$ and $n = 10^7 - 10^{10}\,\text{cm}^{-3}$ can easily be produced in a neutral gas with a density of $n_n = 10^{12} - 10^{13}\,\text{cm}^{-3}$. The plasma is formed by collisional ionization of neutral gas by electrons injected from hot filaments or by RF fields. When small particles are introduced into such a plasma, the temperature of the particles corresponds to the temperature of the neutral gas, which is usually room temperature. This is because the particle-neutral gas relaxation time is much shorter than the particle-electron relaxation time. In the case of an argon plasma, for example, (12.82) $q = 1.6 \times 10^{-6}\,\text{esu}$, which corresponds to 3.6×10^3 electron charges when $a = 0.3\,\mu\text{m}$ is given. Since the plasma vessel, which is at the floating potential, is negative with respect to the plasma potential, the particles are electrostatically confined.

The plasma that fills the space between the particles is generally not in a state of equilibrium. The electron temperature corresponds to the kinetic energy that electrons acquire during collisional ionization. In the device described here, it is in the range of a few electronvolts. The kinetic energy of the ions at their formation corresponds to the neutral gas temperature. The charged particle system generates potential waves with an amplitude of about $k_B T_e/e$. The ions are accelerated and decelerated by the waves. Since the ion birth rate is uniform in space, the average kinetic energy of the ions is about $k_B T_e$. Some of the ions are trapped by the particle potential, while others are not. The trapped ions have a mean lifetime of about

$$\tau_T = \frac{1}{3}\left(\frac{b}{a}\right)^2 \frac{1}{v_i}, \tag{12.89}$$

where τ_T is determined by

$$\tau_T = \frac{\text{number of ions in a unit cell}}{\text{ion influx onto the surface of a particle}} \tag{12.90}$$

Here, v_i is the mean velocity. On the other hand, the ions are cooled by charge exchange. The ratio between τ_T and the charge exchange time τ_{ex} is approximately

$$\frac{\tau_T}{\tau_{\text{ex}}} = \left(\frac{n_n}{4\pi N}\right)\left(\frac{\sigma_{\text{ex}}}{a^2}\right). \tag{12.91}$$

Here, σ_{ex} is the charge exchange cross section, which is about $4 \times 10^{-15}\,\text{cm}^2$ for argon. If $\tau_T/\tau_{\text{ex}} \gg 1$, the ion temperature corresponds to room temperature. In the opposite case, $T_i \simeq T_e$ applies. If the neutral particle density $n_n = 10^{13}\,\text{cm}^{-3}$, $N \equiv n_d == 4 \times 10^6\,\text{cm}^{-3}$ and $a = 3000\,\text{Å}$ are used, then $\tau_T/\tau_{\text{ex}} = 1$ results. The temperature of the untrapped ions is determined by the ratio between the size of the plasma vessel and the mean free path for charge exchange. Therefore, the temperature range of trapped and untrapped ions can be regulated by the neutral gas density between $T < T_i < T_e$.

The plasma crystal is subject to the influence of gravity. The original crystal structure falls to the bottom of the vessel if the gravitational energy $\sim a^3 \rho g l$ is greater than the Coulomb energy $\sim q^2 \exp(-b/\lambda_D)/b$. Here, ρ is the mass density of the particle material and l is the characteristic length scale of the plasma. These conditions set an upper limit for the particle size and a lower limit for N. If the experiment is conducted in microgravity, there is no restriction on the particle size. A crystal with a large lattice constant can be produced in plasmas with low density, such as interplanetary plasma, since according to (12.82) q is proportional to $a\sqrt{\rho}$ and $\Gamma \propto q^2 N^{1/3}$, so that the lower limit of N, which satisfies the crystallization condition (segment A-B in Fig. 12.1), is proportional to a^{-6}. Like a colloidal lattice, a plasma crystal in laboratory plasmas should be observed both by optical microscopy and by light scattering."

Pioneering experiment by Thomas et al. [354]
In 1994, Thomas et al. [354] published the first experimental realization of a plasma crystal. They wrote (Fig. 12.2): "A macroscopic Coulomb crystal of solid particles in a plasma has been observed. Images of a cloud of 7-μm dust particles, which are charged and levitated in a weakly ionized argon plasma, show a hexagonal crystal structure. The crystal is visible to the naked eye. The particles are cooled to 310 K by the neutral gas and carry a charge of $q \geq 9800\ e$, corresponding to a Coulomb coupling parameter of $\Gamma \geq 20.700$. At such a high Γ value, the theory of strongly coupled plasmas predicts that the particles should organize into a Coulomb solid, which is consistent with our observations.

In the present experiment, the structure of a cloud of charged particles levitating in a weakly ionized plasma is investigated. A low-energy argon plasma at 2.05 ± 0.05 mbar was generated by applying a 13.56 MHz signal to the lower electrode of a parallel plate reactor. The lower electrode is a disk with a diameter of 8 cm, while the upper electrode is a ring electrode with inner and outer diameters of 3 and 10 cm, respectively. The electrode

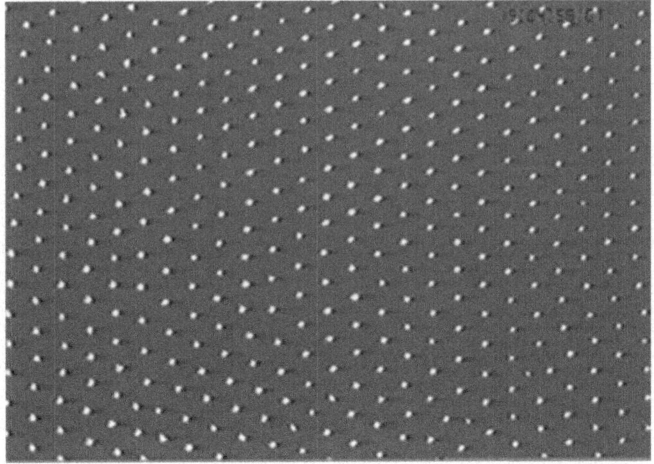

Fig. 12.2 View of a two-dimensional plasma crystal in the MPE laboratories. Due to electrical interactions, the charged microparticles in the plasma form regular structures, such as the lattice shown here. Courtesy of MPE

gap is 2 cm. The DC voltage at the lower electrode was -14.5 ± 0.5 V, measured at the electrical feedthrough. The RF power was 4.5 ± 0.2 W (forward minus reflected), measured between the RF generator and the matching network. This measurement does not account for losses in the matching network, connecting cables, or reactor hardware. These losses could amount to 90% or more; therefore, we roughly estimate the power coupled into the plasma to be 0.4 ± 0.3 W.

12.5 Acoustic Dust Mode

In this section, we derive the dispersion relation for a new mode [356] in a plasma with negatively charged dust particles. The appearance of an additional mode with an increasing number of components is, in principle, not surprising. In parts of the plasma community, the discovery of a "dust acoustic mode" was still regarded as a breakthrough decades later [357], because on the one hand, simple observations were possible, and on the other hand, astrophysical applications are under consideration.

The theoretical model is simple. With the particle densities n_e for electrons, n_i for ions, and n_d for dust particles, we determine the spatially one-dimensional dynamics of the dust particles from

$$\boxed{\frac{\partial n_d}{\partial t} + \frac{\partial}{\partial x}(n_d\, v_d) = 0}\,, \tag{12.92}$$

$$\boxed{\frac{\partial v_d}{\partial t} + v_d \frac{\partial v_d}{\partial x} = \frac{Z_d}{m_d}\frac{\partial \phi}{\partial x}}\,, \tag{12.93}$$

$$\boxed{\frac{\partial^2 \phi}{\partial x^2} = -\frac{e}{\varepsilon_0}(n_i - n_e - Z_d n_d)}\,. \tag{12.94}$$

Due to the slow dynamics of the dust particles, we can assume equilibrium distributions (Boltzmann distributions) for electrons and ions:

$$n_i = n_{i0} \exp\left(-\frac{e\phi}{k_B T_i}\right)\,, \tag{12.95}$$

$$n_e = n_{e0} \exp\left(\frac{e\phi}{k_B T_e}\right)\,. \tag{12.96}$$

Quasineutrality for negatively charged dust particles is expressed by the relation

$$n_{i0} = n_{e0} + Z_d n_{d0} \tag{12.97}$$

For the dispersion equation, we may linearize. Strictly speaking, we introduce

$$n_d = n_{d0} + \tilde{n}_d \,, \quad v_d = 0 + \tilde{v}_d \,, \quad \phi = 0 + \tilde{\phi} \tag{12.98}$$

for this purpose, but for clarity, we omit the tilde for the perturbations in the following. Thus, we start with

$$\frac{\partial n_d}{\partial t} = n_{d0} \frac{\partial v_d}{\partial x} \,, \tag{12.99}$$

$$\frac{\partial v_d}{\partial t} = \frac{Z_d}{m_d} \frac{\partial \phi}{\partial x} \,, \tag{12.100}$$

which immediately leads to

$$\frac{\partial^2 n_d}{\partial t^2} = -\frac{e Z_d n_{d0}}{m_d} \frac{\partial^2 \phi}{\partial x^2} \tag{12.101}$$

If we include the change in density of the dust particles from the Poisson equation,

$$n_d = \frac{1}{4\pi e Z_d} \frac{\partial^2 \phi}{\partial x^2} - \frac{e}{Z_d} \left(\frac{n_{i0}}{T_i} + \frac{n_{e0}}{T_e} \right) \phi \,, \tag{12.102}$$

we find the dispersion relation

$$\omega^2 \left[\frac{e}{Z_d} \left(\frac{n_{i0}}{T_i} + \frac{n_{e0}}{T_e} \right) + \frac{1}{4\pi e Z_d} k^2 \right] = \frac{e Z_d n_{d0}}{m_d} k^2 \,. \tag{12.103}$$

With the parameters

$$\delta \equiv \frac{n_{i0}}{n_{e0}} \,, \quad \eta \equiv \frac{T_e}{T_i} \,, \quad \beta^2 \equiv \frac{Z_d(\delta - 1)}{1 + \delta\eta} \quad C_s^2 \equiv \frac{T_e}{m_d} \tag{12.104}$$

we can rewrite the dispersion relation as:

$$\boxed{\omega^2 = \beta^2 C_s^2 k^2 \frac{1}{1 + \frac{k^2 \lambda_{de}^2}{1 + \eta\delta}}} \,. \tag{12.105}$$

The dust acoustic modes do not propagate with C_s, but instead have the phase velocity

$$\boxed{C_{DA} \equiv \beta C_s} \,. \tag{12.106}$$

For $\eta \gg 1$ we find

$$C_{DA} = P\sqrt{\frac{n_{i0}}{n_{d0}}\frac{T_i}{m_d}} \tag{12.107}$$

with the Havnes parameter

$$P \equiv Z_d \frac{n_{d0}}{n_{i0}} \ . \tag{12.108}$$

Example 12.8 (Orders of Magnitude)
If we choose

$$T_e = 2.5 \text{ eV} = 100\ T_i\,, \quad T_d = 0.025 \text{ eV}\,, \quad n_{i0} = 10^{14} \text{ m}^{-3}\,, \quad n_{d0} = 10^{10} \text{ m}^{-3} \tag{12.109}$$

as well as

$$a = 0.5\,\mu\text{m}\,, \quad m_d = 10^{-15}\text{ kg}\,, \quad Z_d = 2000\,, \tag{12.110}$$

then it follows

$$C_{DA} \approx 4\,\frac{\text{cm}}{\text{s}} \text{ ateat } \lambda = 0{,}5\,\text{cm}\,. \tag{12.111}$$

The frequency is then

$$\omega \approx 8 \text{ Hz}\ . \tag{12.112}$$

■

The low frequency range led to the name "dust *acoustic* mode." Although first derived by Rao et al. [356], the mode is primarily associated with Padma Shukla († 2013).

Part III
Appendix

13 Useful Basic Equations

Abstract

This appendix provides a few basic equations that should be particularly helpful for physical formulation and mathematical analysis. In general, it is highly recommended to consult a more comprehensive collection of formulas. A very concise formula collection tailored to plasma physics applications is published by the Naval Research Laboratory [33].

13.1 Unit Systems

Since a plasma consists of electrically charged particles, in this section we summarize the electromagnetic field equations (Maxwell's equations). For the unit systems, we choose the SI unit system in the main text. In the field of electrodynamics, it coincides with the (earlier) MKSA system. The SI system is not always the preferred system in plasma physics; often the Gaussian unit system is favored. This section deals with the differences and conversions between these systems.

The most commonly used unit systems in electrodynamics are the **SI system (or MKSA system)** and the **Gaussian unit system.**

The abbreviation MKSA stands for meter, kilogram, second, and ampere. The SI unit system is essentially a further development of the MKSA system, was introduced in 1960, and also includes additional base units for, e.g., temperature (kelvin), amount of substance (mole), and luminous intensity (candela). In classical electrodynamics (Maxwell's

K.-H. Spatschek, *Theoretical Plasma Physics*,
https://doi.org/10.1007/978-3-662-72828-4_13

equations), there are no substantive differences between the MKSA system and the SI system.

In the **Gaussian system**, Maxwell's equations (in vacuum) take the form

$$\nabla \cdot \mathbf{E} = 4\pi\rho \,, \tag{13.1}$$

$$\nabla \times \mathbf{B} - \frac{1}{c}\frac{\partial \mathbf{E}}{\partial t} = \frac{4\pi}{c}\mathbf{j} \,, \tag{13.2}$$

$$\nabla \times \mathbf{E} + \frac{1}{c}\frac{\partial \mathbf{B}}{\partial t} = 0 \,, \tag{13.3}$$

$$\nabla \cdot \mathbf{B} = 0 \,. \tag{13.4}$$

In the **SI system** we have correspondingly

$$\nabla \cdot \mathbf{E} = \frac{1}{\varepsilon_0}\rho \,, \tag{13.5}$$

$$\nabla \times \mathbf{B} = \mu_0\varepsilon_0\frac{\partial \mathbf{E}}{\partial t} + \mu_0\mathbf{j} \,, \tag{13.6}$$

$$\nabla \times \mathbf{E} = -\frac{\partial \mathbf{B}}{\partial t} \,, \tag{13.7}$$

$$\nabla \cdot \mathbf{B} = 0 \,. \tag{13.8}$$

The electric field constant or permittivity of vacuum ("electric permittivity of free space") ε_0 is

$$\varepsilon_0 \approx 8.854 \times 10^{-12}\,\frac{A\,s}{V\,m} \quad \text{(MKSA)} \,. \tag{13.9}$$

The magnetic field constant or permeability of vacuum ("permeability of free space") μ_0 is

$$\mu_0 = 4\pi \times 10^{-7}\,\frac{kg\,m}{A^2\,s^2} \approx 1.257 \times 10^{-6}\,\frac{H}{m} \quad \text{(MKSA)} \,. \tag{13.10}$$

In general, the following holds

$$\frac{1}{\varepsilon_0\mu_0} = c^2 \,, \tag{13.11}$$

where c is the speed of light in vacuum (approx. 3×10^8 meters per second in the MKSA system).

For the sake of completeness, the macroscopic forms (in matter) are also given. In the **Gaussian system** the following applies

$$\nabla \cdot \mathbf{D} = 4\pi\rho \; , \tag{13.12}$$

$$\nabla \times \mathbf{H} = \frac{4\pi}{c}\mathbf{j} + \frac{1}{c}\frac{\partial \mathbf{D}}{\partial t} \; , \tag{13.13}$$

$$\nabla \times \mathbf{E} = -\frac{1}{c}\frac{\partial \mathbf{B}}{\partial t} \; , \tag{13.14}$$

$$\nabla \cdot \mathbf{B} = 0 \; . \tag{13.15}$$

The dielectric displacement $\mathbf{D}$ is related to the electric field strength $\mathbf{E}$ and the polarization $\mathbf{P}$ by

$$\mathbf{D} = \mathbf{E} + 4\pi\mathbf{P} \tag{13.16}$$

is linked. The magnetic field strength $\mathbf{H}$ is related to the magnetic induction $\mathbf{B}$ via the magnetization $\mathbf{M}$ in the form

$$\mathbf{H} = \mathbf{B} - 4\pi\mathbf{M} \tag{13.17}$$

Note that in the sets of equations in vacuum and in matter, ρ and $\mathbf{j}$ do not have the same meaning. The free charge density ρ in matter is obtained from the total charge density after subtracting the polarization charge density $-\nabla \cdot \mathbf{P}$; the free electric current density $\mathbf{j}$ in matter differs from the total current density in vacuum by the magnetization current density $c\nabla \times \mathbf{M}$ and the polarization current density $\partial\mathbf{P}/\partial t$.

In the **SI system** the macroscopic Maxwell equations are

$$\nabla \cdot \mathbf{D} = \rho \; , \tag{13.18}$$

$$\nabla \times \mathbf{H} = \mathbf{j} + \frac{\partial \mathbf{D}}{\partial t} \; , \tag{13.19}$$

$$\nabla \times \mathbf{E} = -\frac{\partial \mathbf{B}}{\partial t} \; , \tag{13.20}$$

$$\nabla \cdot \mathbf{B} = 0 \; , \tag{13.21}$$

where

$$\mathbf{D} = \varepsilon_0 \mathbf{E} + \mathbf{P} , \quad \mathbf{H} = \frac{1}{\mu_0}\mathbf{B} - \mathbf{M} \tag{13.22}$$

applies. The Lorentz force is written in the SI system (MKSA system) as

$$\mathbf{F}_L = q(\mathbf{E} + \mathbf{v} \times \mathbf{B}) , \tag{13.23}$$

while in the Gaussian system we have

$$\mathbf{F}_L = q\left(\mathbf{E} + \frac{\mathbf{v}}{c} \times \mathbf{B}\right) \tag{13.24}$$

The relativistically correct equation of motion for a particle of rest mass m_0 is thus

$$\frac{d}{dt}\left(\frac{m_0 \mathbf{v}}{\sqrt{1 - \frac{v^2}{c^2}}}\right) = \mathbf{F}_L . \tag{13.25}$$

One can introduce a scalar potential ϕ and a vector potential $\mathbf{A}$. In the MKSA system, this is done via

$$\mathbf{B} = \nabla \times \mathbf{A} , \quad \mathbf{E} = -\nabla\phi - \frac{\partial \mathbf{A}}{\partial t} . \tag{13.26}$$

The potentials are not unique. Through

$$\mathbf{A}' = \mathbf{A} - \nabla\psi , \quad \phi' = \phi + \frac{\partial \psi}{\partial t} \tag{13.27}$$

we can perform a gauge transformation with a gauge function ψ without changing the physically relevant statements.

Wave equations in vacuum initially take the form

$$\nabla \times \nabla\mathbf{A} + \frac{1}{c^2}\nabla\frac{\partial \phi}{\partial t} + \frac{1}{c^2}\frac{\partial^2 \mathbf{A}}{\partial t^2} = \mu_0 \mathbf{j} , \tag{13.28}$$

$$\nabla^2\phi + \nabla\frac{\partial \mathbf{A}}{\partial t} = -\frac{1}{\varepsilon_0}\rho\,. \tag{13.29}$$

If we choose the Lorentz gauge $\frac{1}{c^2}\frac{\partial\phi}{\partial t} + \nabla\cdot\mathbf{A} = 0$, the wave equations simplify to the symmetric form

$$\nabla^2\mathbf{A} - \frac{1}{c^2}\frac{\partial^2\mathbf{A}}{\partial t^2} = -\mu_0\mathbf{j}\,, \tag{13.30}$$

$$\nabla^2\phi - \frac{1}{c^2}\frac{\partial^2\phi}{\partial t^2} = -\frac{1}{\varepsilon_0}\rho\,. \tag{13.31}$$

Another common gauge is the Coulomb gauge $\nabla\cdot\mathbf{A} = 0$.

An advantage of the SI (MKSA) system is the *easy* convertibility, for example, to the Gaussian system. For the transition from Gaussian to SI, the following relations can be used, where the quantities marked with a prime are measured in the SI (MKSA) system and the quantities without a prime apply in the Gaussian system:

$$\mathbf{E} \mathrel{\widehat{=}} \sqrt{4\pi\varepsilon_0}\,\mathbf{E}'\,, \tag{13.32}$$

$$\mathbf{D} \mathrel{\widehat{=}} \sqrt{\frac{4\pi}{\varepsilon_0}}\,\mathbf{D}'\,, \tag{13.33}$$

$$\rho \mathrel{\widehat{=}} \frac{1}{\sqrt{4\pi\varepsilon_0}}\,\rho'\,, \tag{13.34}$$

$$\mathbf{j} \mathrel{\widehat{=}} \frac{1}{\sqrt{4\pi\varepsilon_0}}\,\mathbf{j}'\,, \tag{13.35}$$

$$\mathbf{P} \mathrel{\widehat{=}} \frac{1}{\sqrt{4\pi\varepsilon_0}}\,\mathbf{P}'\,, \tag{13.36}$$

$$\mathbf{B} \mathrel{\widehat{=}} \sqrt{\frac{4\pi}{\mu_0}}\,\mathbf{B}'\,, \tag{13.37}$$

$$\mathbf{H} \mathrel{\widehat{=}} \sqrt{4\pi\mu_0}\,\mathbf{H}'\,, \tag{13.38}$$

$$\mathbf{M} \mathrel{\widehat{=}} \sqrt{\frac{\mu_0}{4\pi}}\,\mathbf{M}'\,. \tag{13.39}$$

In this book, we use SI units, although Gaussian units are especially widespread in plasma physics. The use of SI units is based on international recommendations. Any formula in the SI system can be *easily* converted into Gaussian units by using the following conversion table (quantities on the left are in SI, quantities on the right are in the Gaussian system):

$$\text{SI} \;\Rightarrow\; \text{Gauss} \tag{13.40}$$

$$\varepsilon_0 \Rightarrow \frac{1}{4\pi}\,, \qquad \mu_0 \Rightarrow \frac{4\pi}{c^2}\,, \tag{13.41}$$

$$\mathbf{B} \Rightarrow \frac{1}{c}\mathbf{B}\,, \qquad \mathbf{H} \Rightarrow \frac{c}{4\pi}\mathbf{H}\,, \tag{13.42}$$

$$\mathbf{D} \Rightarrow \frac{1}{4\pi}\mathbf{D}\,, \qquad \mathbf{M} \Rightarrow c\mathbf{M}\,, \tag{13.43}$$

$$\Phi \Rightarrow \Phi\,, \qquad \mathbf{A} \Rightarrow \frac{1}{c}\mathbf{A}\,. \tag{13.44}$$

All other symbols such as ρ, $\mathbf{j}$, $\mathbf{E}$ and $\mathbf{P}$ remain unchanged in the conversion.

13.2 Fourier and Laplace Transformations

In this section, we briefly summarize the essential definitions of the Fourier and Laplace transformations, as they are used repeatedly in the text for solving linear problems. It is always necessary to check whether a symmetric or asymmetric form is chosen for the forward and inverse transformation. For further questions, please refer to the literature.

A non-periodic function $f(x)$ cannot be expanded into a Fourier series, but under certain conditions it can be represented by a Fourier integral. The function $f(x)$ must be absolutely integrable, i.e., it must satisfy

$$\int_{-\infty}^{+\infty} |f(x)|dx < \infty \tag{13.45}$$

and the Dirichlet conditions must be fulfilled. The latter state that every finite interval can be divided into a finite number of subintervals in which $f(x)$ is continuous and monotonic; if x_0 is a discontinuity of $f(x)$, then $f(x_0+0)$ and $f(x_0-0)$ must exist. Under these conditions, for all x the following holds

$$f(x) = \int_0^{\infty} [a(k) \cos kx + b(k) \sin kx] \, dk \tag{13.46}$$

with

$$a(k) = \frac{1}{\pi} \int_{-\infty}^{\infty} f(x') \cos kx' \, dx', \tag{13.47}$$

$$b(k) = \frac{1}{\pi} \int_{-\infty}^{+\infty} f(x') \sin kx' \, dx'. \tag{13.48}$$

The right-hand side of (10.2.2) is called the Fourier integral of $f(x)$. When calculating the coefficients a and b at the discontinuities of f

$$f(x) = \frac{1}{2}[f(x+0) + f(x-0)] \tag{13.49}$$

is set. The Fourier integral can also be written as

$$f(x) = \frac{1}{\pi} \int_0^{\infty} dk \int_{-\infty}^{+\infty} f(x') \cos[k(x'-x)] \, dx' \tag{13.50}$$

which suggests the complex form

$$f(x) = \frac{1}{2\pi} \int_{-\infty}^{+\infty} dk \int_{-\infty}^{+\infty} f(x') e^{ik(x'-x)} dx' \tag{13.51}$$

This formula can be interpreted as a superposition of

$$F(k) = \frac{1}{\sqrt{2\pi}} \int_{-\infty}^{+\infty} f(x) e^{ikx} dx, \tag{13.52}$$

$$f(x) = \frac{1}{\sqrt{2\pi}} \int_{-\infty}^{+\infty} F(k) e^{-ikx} dk \tag{13.53}$$

$F(k)$ is called the Fourier transform of $f(x)$, and the operation $f \to F$ is referred to as the Fourier transformation. The inverse Fourier transformation is described by (13.53). In a Fourier transformation in real space (x-coordinate), k is called the wave number, since the Fourier integral represents the function $f(x)$ as a sum of infinitely many oscillations with continuously varying wavelength $\lambda = 2\pi/k$. If a Fourier transformation is performed in time, the following representation

$$F(\omega) = \frac{1}{\sqrt{2\pi}} \int_{-\infty}^{+\infty} f(t) e^{-i\omega t} dt \tag{13.54}$$

has become standard, since—motivated by physics—we write the phases as

$$\theta = kx - \omega t \tag{13.55}$$

Several remarks are in order:

1. The decomposition of (13.51) into (13.52) and (13.53) is not unique. The Fourier transformation can also be defined asymmetrically by distributing the factor $1/2\pi$ differently. We make use of this in the text at several points.
2. The generalization to R^3 is obvious; for example, we write a Fourier transformation in space (position vector $\mathbf{r}$) and in time t as

$$F(\mathbf{k}, \omega) = \frac{1}{\sqrt{2\pi}^3} \int_{R^3} d^3r \frac{1}{\sqrt{2\pi}} \int_{-\infty}^{+\infty} dt\, f(\mathbf{r}, t) e^{i(\mathbf{k}\cdot\mathbf{r} - \omega t)} . \tag{13.56}$$

3. In general, F is complex even for real f. The original function f can also be a complex-valued function of the real argument x. In this case, the integral in (13.52) is to be understood in the sense of the Cauchy principal value.
4. One can also define Fourier cosine and Fourier sine transforms, for example,

$$F_c(k) = \sqrt{\frac{2}{\pi}} \int_0^{\infty} f(x) \cos kx\, dx, \tag{13.57}$$

$$f(x) = \sqrt{\frac{2}{\pi}} \int_0^{\infty} F_c(k) \cos kx\, dk \; ; \tag{13.58}$$

   for an even function, $F(k) = F_c(k)$ holds.
5. Derivatives are essentially reflected in the Fourier transform domain as multiplications by the independent variable. For example, the Fourier transform of df/dx is $-ikF(k)$. We therefore often write symbolically

$$\nabla \to -i\mathbf{k}, \tag{13.59}$$

$$\partial_t \to +i\omega. \tag{13.60}$$

The Laplace transformation is closely related to the Fourier transformation. A complex function $f(t)$ of the real variable t is called Laplace transformable if it is defined for $t \geq 0$ and integrable over $(0, \infty)$, and is subject to an exponential growth restriction

$$| f(t) | \leq Ke^{ct} \tag{13.61}$$

If p is a complex variable, the function

$$F(p) = \int_0^\infty e^{-pt} f(t)\, dt \tag{13.62}$$

is called the Laplace transform of $f(t)$. The integral converges absolutely for $\mathrm{Re}\, p > c$, so that $F(p)$ remains bounded in the half-plane $\mathrm{Re}\, p \geq c_0$ with $c_0 > c$. The inverse transformation is defined by

$$f(t) = \frac{1}{2\pi i} \int_{c_0 - i\infty}^{c_0 + i\infty} e^{pt}\, F(p)\, dp \tag{13.63}$$

where $c_0 > c$ holds. The path of integration is thus a straight line parallel to the imaginary axis in the complex p-plane, running to the right of the imaginary axis shifted by c.

The representation (13.63) is unique, since $F(p)$ is analytic in the half-plane $\mathrm{Re}\, p > c$. The inverse transformation (13.63) can be understood via the relation

$$\lim_{\varepsilon \to \infty} \frac{1}{\pi} \frac{\sin[(t' - t)/\varepsilon]}{t' - t} = \delta(t' - t) \tag{13.64}$$

A major advantage of the Laplace transformation lies in solving initial value problems for ordinary linear differential equations. The desired particular solution is obtained directly, without first having to adapt the general solution to the given initial values.

Example 13.1

This important fact is best illustrated by an example. The solution of the differential equation

$$f'(t) + 2f(t) = te^{-2t} \tag{13.65}$$

is sought for the initial condition $f(0) = 1$. After applying the Laplace transformation, we obtain

$$pF(p) - 1 + 2F(p) = \frac{1}{(2 + p)^2} \tag{13.66}$$

or

$$F(p) = \frac{1}{2 + p} + \frac{1}{(2 + p)^3}\,. \tag{13.67}$$

The subsequent inverse transformation yields the solution

$$f(t) = e^{-2t} + \frac{t^2}{2} e^{-2t} . \tag{13.68}$$

■

Example 13.2
In the example just discussed, we can also immediately find the solution using the method of variation of parameters. The homogeneous solution

$$f = Ae^{-2t} \tag{13.69}$$

yields, with $A = A(t)$ after substituting into (13.65)

$$A(t) = A_0 + \frac{t^2}{2} . \tag{13.70}$$

With this, we have the general solution

$$f(t) = A_0 e^{-2t} + \frac{t^2}{2} e^{-2t} , \tag{13.71}$$

where the constant A_0 is only determined at the end by the initial condition $f(0) = 1$ to $A_0 = 1$. ■

13.3 Vector Analytic Relations

In this section, we present some important formulas for vectors and vector operators. However, the explicit result of a calculation depends on the chosen coordinate system. In this regard, consulting relevant formula collections is recommended. Only for the coordinate systems describing a torus do we provide two simple examples.

For vectors **A**, **B** and **C** the following holds

$$\mathbf{A} \cdot \mathbf{B} \times \mathbf{C} = \mathbf{A} \times \mathbf{B} \cdot \mathbf{C} = \mathbf{B} \cdot \mathbf{C} \times \mathbf{A} = \mathbf{B} \times \mathbf{C} \cdot \mathbf{A} = \mathbf{C} \cdot \mathbf{A} \times \mathbf{B} = \mathbf{C} \times \mathbf{A} \cdot \mathbf{B} , \tag{13.72}$$

$$\mathbf{A} \times (\mathbf{B} \times \mathbf{C}) = (\mathbf{C} \times \mathbf{B}) \times \mathbf{A} = (\mathbf{A} \cdot \mathbf{C})\mathbf{B} - (\mathbf{A} \cdot \mathbf{B})\mathbf{C} , \tag{13.73}$$

$$\mathbf{A} \times (\mathbf{B} \times \mathbf{C}) + \mathbf{B} \times (\mathbf{C} \times \mathbf{A}) + \mathbf{C} \times (\mathbf{A} \times \mathbf{B}) = 0 , \tag{13.74}$$

$$(\mathbf{A} \times \mathbf{B}) \cdot (\mathbf{C} \times \mathbf{D}) = (\mathbf{A} \cdot \mathbf{C})(\mathbf{B} \cdot \mathbf{D}) - (\mathbf{A} \cdot \mathbf{D})(\mathbf{B} \cdot \mathbf{C}) , \tag{13.75}$$

$$(\mathbf{A} \times \mathbf{B}) \times (\mathbf{C} \times \mathbf{D}) = (\mathbf{A} \times \mathbf{B} \cdot \mathbf{D})\mathbf{C} - (\mathbf{A} \times \mathbf{B} \cdot \mathbf{C})\mathbf{D} . \tag{13.76}$$

If f and g are scalars and $\mathbf{A}$ as well as $\mathbf{B}$ are vector-valued functions, then the following relations hold:

$$\nabla(fg) = \nabla(gf) = f\nabla g + g\nabla f , \tag{13.77}$$

$$\nabla \cdot (f\mathbf{A}) = f\nabla \cdot \mathbf{A} + \mathbf{A} \cdot \nabla f , \tag{13.78}$$

$$\nabla \times (f\mathbf{A}) = f\nabla \times \mathbf{A} + \nabla f \times \mathbf{A} , \tag{13.79}$$

$$\nabla \cdot (\mathbf{A} \times \mathbf{B}) = \mathbf{B} \cdot \nabla \times \mathbf{A} - \mathbf{A} \cdot \nabla \times \mathbf{B} , \tag{13.80}$$

$$\nabla \times (\mathbf{A} \times \mathbf{B}) = \mathbf{A}(\nabla \cdot \mathbf{B}) - \mathbf{B}(\nabla \cdot \mathbf{A}) + (\mathbf{B} \cdot \nabla)\mathbf{A} - (\mathbf{A} \cdot \nabla)\mathbf{B} , \tag{13.81}$$

$$\mathbf{A} \times (\nabla \times \mathbf{B}) = (\nabla \mathbf{B}) \cdot \mathbf{A} - (\mathbf{A} \cdot \nabla)\mathbf{B} , \tag{13.82}$$

$$\nabla(\mathbf{A} \cdot \mathbf{B}) = \mathbf{A} \times (\nabla \times \mathbf{B}) + \mathbf{B} \times (\nabla \times \mathbf{A}) + (\mathbf{A} \cdot \nabla)\mathbf{B} + (\mathbf{B} \cdot \nabla)\mathbf{A} , \tag{13.83}$$

$$\nabla^2 f = \nabla \cdot \nabla f , \tag{13.84}$$

$$\nabla^2 \mathbf{A} = \nabla(\nabla \cdot \mathbf{A}) - \nabla \times \nabla \times \mathbf{A} , \tag{13.85}$$

$$\nabla \times \nabla f = 0 , \tag{13.86}$$

$$\nabla \cdot \nabla \times \mathbf{A} = 0 . \tag{13.87}$$

If a volume V is bounded by S, with $d\mathbf{S} = \mathbf{n}dS$, where $\mathbf{n}$ is the outward-pointing normal vector, then the following holds

$$\int_V dV \nabla f = \int_S d\mathbf{S} f , \tag{13.88}$$

$$\int_V dV \nabla \cdot \mathbf{A} = \int_S d\mathbf{S} \cdot \mathbf{A} , \tag{13.89}$$

$$\int_V dV \nabla \cdot \underline{\underline{T}} = \int_S d\mathbf{S} \cdot \underline{\underline{T}} , \tag{13.90}$$

$$\int_V dV \nabla \times \mathbf{A} = \int_S d\mathbf{S} \times \mathbf{A} , \tag{13.91}$$

$$\int_V dV (f \nabla^2 g - g \nabla^2 f) = \int_S d\mathbf{S} \cdot (f \nabla g - g \nabla f) , \tag{13.92}$$

$$\int_V dV (\mathbf{A} \cdot \nabla \times \nabla \times \mathbf{B} - \mathbf{B} \cdot \nabla \times \nabla \times \mathbf{A})$$

$$= \int_S d\mathbf{S} \cdot (\mathbf{B} \times \nabla \times \mathbf{A} - \mathbf{A} \times \nabla \times \mathbf{B}) . \tag{13.93}$$

Here, $\underline{\underline{T}}$ is a second-rank tensor.

For open surfaces S, which are bounded by C (with the line element $d\mathbf{l}$), the following integral theorems exist

$$\int_S d\mathbf{S} \times \nabla f = \oint_C d\mathbf{l} f , \tag{13.94}$$

$$\int_S d\mathbf{S} \cdot \nabla \times \mathbf{A} = \oint_C d\mathbf{l} \cdot \mathbf{A} , \tag{13.95}$$

$$\int_S (d\mathbf{S} \times \nabla) \times \mathbf{A} = \oint_C d\mathbf{l} \times \mathbf{A} , \tag{13.96}$$

$$\int_S d\mathbf{S} \cdot (\nabla f \times \nabla g) = \oint_C f dg = - \oint_C g df . \tag{13.97}$$

Example 13.3 (Torus in Cylindrical Coordinates)
In Fig. 10.1 we introduced cylindrical coordinates R, z and φ for a torus. As an example, we provide the components of $\nabla \times \mathbf{A}$ in these coordinates:

$$(\nabla \times \mathbf{A})_R = \frac{1}{R} \frac{\partial A_z}{\partial \varphi} - \frac{\partial A_\varphi}{\partial z} , \tag{13.98}$$

$$(\nabla \times \mathbf{A})_\varphi = \frac{\partial A_R}{\partial z} - \frac{\partial A_z}{\partial R} , \tag{13.99}$$

$$(\nabla \times \mathbf{A})_z = \frac{1}{R} \frac{\partial (R A_\varphi)}{\partial R} - \frac{1}{R} \frac{\partial A_R}{\partial \varphi} . \tag{13.100}$$

■

Example 13.4 (Torus in Toroidal Coordinates)
In Fig. 10.1 we introduced toroidal coordinates r, θ and φ for a torus. R is now the constant radius to the center of the torus. Again, as an example, we provide the components of $\nabla \times \mathbf{A}$ in these coordinates:

$$(\nabla \times \mathbf{A})_r = \frac{1}{R + r\cos\theta}\left\{\frac{\partial A_\theta}{\partial \varphi} - \frac{1}{r}\frac{\partial[(R + r\cos\theta)A_\varphi]}{\partial \theta}\right\}, \tag{13.101}$$

$$(\nabla \times \mathbf{A})_\varphi = \frac{1}{r}\left\{\frac{\partial A_r}{\partial \theta} - \frac{\partial (rA_\theta)}{\partial r}\right\}, \tag{13.102}$$

$$(\nabla \times \mathbf{A})_\theta = \frac{1}{R + r\cos\theta}\left\{\frac{\partial\left[(R + r\cos\theta)A_\varphi\right]}{\partial r} - \frac{\partial A_r}{\partial \varphi}\right\}. \tag{13.103}$$

■

References

1. F. Chen. *Introduction to Plasma Physics*. Plenum, New York, 1984.
2. L. Spitzer, Jr. *The Physics of Fully Ionized Gases*. Interscience, 1956.
3. T.H. Stix. *The Theory of Plasma Waves*. McGraw-Hill, 1962.
4. R. Balescu. *Statistical Mechanics of Charged Particles*. Interscience, New York, 1963.
5. Yu.L. Klimontovich. *Kinetic Theory of Nonideal Gases and Nonideal Plasmas*. Pergamon Press, Oxford, 1982.
6. D.R. Nicholson. *Introduction to Plasma Physics*. Wiley, New York, 1983.
7. R.A. Cairns. *Plasma Physics*. Blackie, Glasgow, Scotland, 1985.
8. N.A. Krall and A.W. Trivelpiece. *Principles of Plasma Physics*. San Francisco Press, 1986.
9. K. Nishikawa and M. Wakatani. *Plasma Physics: Basic Theory with Fusion Applications*. Springer, Berlin, 1990.
10. K.H. Spatschek. *Theoretische Plasmaphysik*. Teubner, Stuttgart, 1990.
11. S. Ichimaru. *Statistical Plasma Physics*. Addison-Wesley, New York, 1992.
12. R. J. Goldston and P. H. Rutherford. *Introduction to Plasma Physics*. Institute of Physics, Philadelphia, 1995.
13. K. Itoh, S.-I. Itoh, and A. Fukuyama. *Transport and structural formation in plasmas*. Institute of Physics Publishing, Bristol, 1999.
14. L. C. Woods. *Physics of Plasmas*. Wiley-VCH, 2004.
15. W. M. Stacey. *Fusion Plasma Physics*. Wiley-VCH, Weinheim, 2005.
16. P.M. Bellan. *Fundamentals of plasma physics*. Cambridge UP, 2006.
17. Karl-Heinz Spatschek. *High Temperature Plasmas*. WILEY-VCH, 2012.
18. Karl-Heinz Spatschek. *Astrophysik (3rd edition)*. Springer, 2021.
19. U. Stroth. *Plasmaphysik: Phänomene, Modelle, Werkzeuge*. Springer, Berlin, 2018.
20. Karl-Heinz Spatschek. *Astrophysics*. Springer, 2024.
21. A. Hasegawa. *Plasma Instabilities and Nonlinear Effects*. Springer, Berlin, 1975.
22. B.W. Carroll and D.A. Ostlie. *Modern Astrophysics*. Addison-Wesley, Reading, 1996.
23. K.H. Spatschek. *Astrophysik*. Teubner, Stuttgart, 2003.
24. W. Kruer. *The Physics of Laser-Plasma Interaction*. Addison-Wesley, 1988.
25. S. Atzeni and J. Meyer ter Vehn. *The Physics of Inertial Fusion. Beam Plasma Interaction, Hydrodynamics, Hot Dense Matter*. Oxford Univ. Press, Oxford, 2004.
26. Ralph d'Agostino, Riccardo d' Agostino, Pietro Favia, Yoshinobu Kawai, Hideo Ikegami, Noriyoshi Sato, and Farzaneh Arefi-Khonsari. *Advanced Plasma Technolog*. Wiley, Weinheim, 2007.
27. B.B. Kadomtsev. *Plasma Turbulence*. Academic Press, New york, 1965.
28. B.B. Kadomtsev. *Tokamak Plasma: a Complex Physical System*. IoP, 1993.

K.-H. Spatschek, *Theoretical Plasma Physics*,
https://doi.org/10.1007/978-3-662-72828-4

29. R. D. Hazeltine and J. D. Meiss. *Plasma confinement*. Addison-Wesley, Redwood City, Calif., 1992.
30. R.B. White. *The Theory of Toroidally Confined Plasmas*. Imperial College Press, London, 2001.
31. Meghnad Saha. On a physical theory of stellar spectra. *Proc. Roy.Soc. London, Series A*, 99:135–153, 1921.
32. L.E. Reichl. *A modern course in statistical physics*. Edward Arnold, 1980.
33. D. L. Book. *NRL Plasma Formulary*. Naval Research Lab, Washington, 1990.
34. R.W.P. McWhirter. *Plasma Diagnostic Techniques, R.H. Huddlestone and S.L. Leonard, eds.*, chapter Spectral Intensities. Academic Press, New York, 1965.
35. H.R. Griem. *Plasma Spectroscopy*. Academic Press, Newe York, 1966.
36. J.D. Lawson. Some criteria for a power producing thermonuclear reactor. *Proc. Phys. Soc.*, 70:6–10, 1957.
37. Herbert Goldstein. *Klassische Mechanik*. Akademische Verlagsgesellschaft, Frankfurt, 1963.
38. W. Nolting. *Grundkurs: Theoretische Physik – 3 Elektrodynamik*. Zimmermann-Neufang, Ulmen, 1993.
39. T. G. Northrop. *The adiabatic motion of charged particles*. Wiley, New York, 1963.
40. R. G. Littlejohn. *J. Math. Phys.*, 20:2445, 1979.
41. R. G. Littlejohn. *Phys. Fluids*, 24:1730, 1981.
42. R. G. Littlejohn. *J. Plasma Phys.*, 29:111, 1983.
43. R. G. Littlejohn. Differential forms and canonical variables for drift motion in toroidal geometry. *Phys. Fluids*, 28:2015, 1985.
44. R. Balescu. *Transport Processes in Plasmas Vol. 1: Classical Transport*. North Holland, Amsterdam, 1988.
45. R. Balescu. *Transport processes in plasmas: 2. Neoclassical transport theory*. North-Holland, Amsterdam, 1988.
46. R. Balescu. *Aspects of anomalous transport in plasmas*. Institute of Physics Publishing, Bristol, 2005.
47. E. Fermi. On the origin of the cosmic radiation. *Phys. Rev.*, 75:1169–1174, 1949.
48. K. H. Spatschek. *High Temperature Plasmas*. Wiley-VCH, 2012.
49. H. L. Friedman. *Ionic Solution Theory*. Wiley, 1962.
50. G. Ecker. *Theory of fully ionized plasmas*. Academic Press, New York, 1972.
51. Joseph Edward Mayer and Maria Goeppert-Mayer. *Statistical Mechanics*. Wiley, 2nd edition, 1977.
52. M. Baranger and B. Mozer. *Phys. Rev.*, 115:521, 1959.
53. J. Holtsmark. *Ann. Phys. (Leipzig)*, 58:577, 1919.
54. G. Ecker and K. G. Müller. *Z. Phys.*, 153:317, 1958.
55. B. Scott and J. Smirnov. *Phys. Plasmas*, 17:112302, 2010.
56. B. Scott. *Phys. Plasmas*, 17:102306, 2010.
57. R. Balescu. *Transport processes in plasmas: 2. Neoclassical transport theory*. North-Holland, Amsterdam, 1988.
58. R. D. Hazeltine and J. D. Meiss. *Plasma confinement*. Addison-Wesley, Redwood City, 1992.
59. A. Hasegawa and K. Mima. Pseudo-three-dimensional turbulence in magnetized nonuniform plasma. *Phys. Fluids*, 21:87, 1978.
60. V. Naulin, K. H. Spatschek, and A. Hasegawa. Selective decay within a one-field model of dissipative drift-wave turbulence. *Phys. Fluids B*, 4:2672–2674, 1992.
61. K. H. Spatschek and J. Uhlenbusch, editors. *Contributions to high-temperature plasma physics*. Akademie Verlag, Berlin, 1994.
62. V. Naulin, K. H. Spatschek, S. Musher, and L.I. Piterbarg. Properties of a two-nonlinearity model of driftwave turbulence. *Phys. Plasmas*, 2:2640–2652, 1995.

63. S. I. Braginskii. In M.A. Leontovich, editor, *Reviews of Plasma Physics*, volume 1, page 205. Consultants Bureau, 1965.
64. K. H. Spatschek, M. Eberhard, and H. Friedel. On models for magnetic field line diffusion. *Physicalia Mag.*, 20:85–93, 1998.
65. T.G. Cowling. *Magnetohydrodynamics*. Wiley, New York, 1957.
66. D. Biskamp. *Nonlinear Magnetohydrodynamics*. Cambridge UP, Cambridge, 1997.
67. R.C. Davidson. *Handbook of Plasma Physics*. North-Holland, Amsterdam, 1984.
68. A. B. Mikhailovskii. *Handbook of Plasma Physics*, page 587. North-Holland, Amsterdam, 1984.
69. K. Mima, M. S. Jovanović, Y. Sentoku, Z.-M. Sheng, M. M. Škorić, and T. Sato. Stimulated photon cascade and condensate in a relativistc laser-plasma interaction. *Phys. Plasmas*, 8(5):2349–2356, 2001.
70. L. M. Gorbunov and V. I. Kirsanov. Excitation of plasma waves by an electromagnetic wave packet. *Zh. Eksp. Teor. Fiz.*, 93:509, 1987.
71. V. E. Zakharov. *Sov. Phys. JETP*, 35:908–914, 1972.
72. V. E. Zakharov and E. A. Kuznetsov. *Sov. Phys. JETP*, 39:285, 1974.
73. V. E. Zakharov and A. M. Rubenchik. *Sov. Phys. JETP*, 38:494, 1975.
74. V. E. Zakharov and V. S. Synakh. *Sov. Phys. JETP*, 41:465, 1975.
75. V. E. Zakharov, E. A. Kuznetsov, and S. L. Musher. *JETP Lett.*, 41:154, 1985.
76. K. H. Spatschek. Parametrische Instabilitäten in Plasmen. *Fortschritte der Physik*, 24:687–729, 1976.
77. C. S. Liu and V. K. Tripathi. *Interaction of electromagnetic waves with electron beams and plasmas*. World Scientific, Singapore, 1994.
78. J. F. Drake, P. K. Kaw, Y. C. Lee, G. Schmidt, C. S. Liu, and M. N. Rosenbluth. Parametric instabilities of electromagnetic waves in plasmas. *Phys. Fluids*, 17:778, 1974.
79. D. W. Forslund, J. M. Kindel, and E. L. Lindman. Theory of stimulated scattering processes in laser-irradiated plasmas. *Phys. Fluids*, 18:1002, 1975.
80. C. S. McKinstrie and R. Bingham. Stimulated raman scattering and the relativistic modulation instability of light waves in rarefied plasma. *Phys. Fluids B*, 4:2626, 1992.
81. A. S. Sakharov and V. I. Kirsanov. Theory of Raman scattering for a short ultrastrong laser pulse in a rarefied plasma. *Phys. Rev. E*, 49:3274–3282, 1994.
82. S. Guerin, G. Laval, P. Mora, J. C. Adam, A. Heron, and A. Bendip. Modulational and Raman instabilities in the relativistic regime. *Phys. Plasmas*, 2:2807, 1995.
83. Z. Toroker, V. M. Malkin, and N. J. Fisch. Seed laser chirping for enhanced backward Raman amplification in plasmas. *Phys. Rev. Lett.*, 109:085003, 2012.
84. Phillippe Mounaix, Denis Pesme, Wojcech Rozmus, and Michel Casanova. Space and time behavior of parametric instabilities for a finite pump duration in a bounded plasma. *Phys. Fluids B*, 5:3304, 1993.
85. G. Shvets, J. S. Wurtele, and B. A. Shadwick. Analysis and simulation of Raman backscatter in underdense plasmas. *Phys. Plasmas*, 4:1872, 1997.
86. N. A. Yampolsky, V. M. Malkin, and N. J. Fisch. Finite-duration seeding effects in powerful backward Raman amplifiers. *Phys. Rev. E*, 69:036401, 2004.
87. V. Malkin, G. Shvets, and N. J. Fisch. Fast compression of laser beams to highly overcritical powers. *Phys. Rev. Lett.*, 82:4448–4451, 1999.
88. V. M. Malkin, G. Shvets, and N. J. Fisch. Ultra-powerful compact amplifiers for short laser pulses. *Phys. Plasmas*, 7:2232, 2000.
89. R. M. G. M. Trines, F. Fiuza, R. Bingham, R. A. Fonseca, L. O. Silva, R. A. Cairns, and P. A. Norreys. Production of picosecond, kilojoule, and petawatt laser pulses via Raman amplification of nanosecond pulses. *Phys. Rev. Lett.*, 107:105002, 2011.

90. G. L. Lamb. Pi-pulse propagation in lossless amplifier. *Phys. Lett. A*, 29:507, 1969.
91. A. A. Andreev, C. Riconda, V. T. Tikhonchuk, and S. Weber. Short light pulse amplification and compression by stimulated brillouin scattering in plasmas in the strong coupling regime. *Phys. Plasmas*, 13:053110, 2006.
92. P. N. Guzdar, C. S. Liu, and R. H. Lehmberg. Stimulated Brillouin scattering in the strong couplin regime. *Phys. Plasmas*, 3:3414–3419, 1996.
93. B. L. Bobroff and H. A. Haus. Impulse response of active coupled wave systems. *J. Appl. Phys.*, 38:390, 1967.
94. A. Mančić, Lj. Hadžievski, and M. Škorić. Dynamics of electromagnetic solitons in a relativistic plasma. *Phys. Plasmas*, 13:052305, 2006.
95. Lj. Hadžievski, M. S. Jovanović, M. M. Škorić, and K. Mima. Stability of one-dimensional electromagnetic solitons in relativistic laser plasmas. *Phys. Plasmas*, 9:2569, 2002.
96. R. Bingham, U. de Angelis, M. R. Amin, R. A. Cairns, and B. McNamara. Relativistic langmuir waves generated by ultra-short pulse lasers. *Plasma Phys. Control. Fusion*, 34:557–567, 1992.
97. J. Faure, Y. Glinec, A. Pukhov, S. Kiselev, S. Gordienko, E. Lefebvre, J.-P. Rousseau, F. Burgy, and V. Malka. A laser-plasma accelerator producing monoenergetic electron beams. *Nature*, 431:541, 2004.
98. J. Faure, C. Rechatin, A. Norlin, A. Lifschitz, Y. Glinec, and V. Malka. Controlled injection and acceleration of electrons in plasma wakefields by colliding laser pulses. *Nature*, 444:737, 2006.
99. V. Malka, J. Faure, Y. Glinec, and A. F. Lifschitz. Laser-plasma accelerators: a new tool for science and for society. *Plasma Phys, Control. Fusion*, 47:481, 2005.
100. C. G. R. Geddes, Cs. Toth, J. van Tillborg, E. Esarey, C. B. Schroeder, D. Bruhwller, C. Nieter, J. Cary, and W. P. Leemans. High-quality electron beams from a laser wakefield accelerator using plasma-channel guiding. *Nature*, 431:538, 2004.
101. F. Amiranoff, A. Antonietti, P. Audebert, D. Bernard, B. Cros, F. Dorchies, J. C. Gauthier, J. P. Geindre, G. Grillon, F. Jacquet, G. Matthieussent, J. R. Marquès, P. Mine, P. Mora, A. Modena, J. Morillo, F. Moulin, Z. Najmudin, A. E. Specka, and C. Stenz. Laser particle acceleration: beat-wave and wakefield experiments. *Plasma Phys. Control. Fusion*, 38:295, 1996.
102. R. Kodama, Y. Sentoko, Z. L. Chen, G. R. Kumar, S. P. Hatchett, Y. Toyama, T. E. Cowan, R. R. Freeman, J. Fuchs, Y. Izawa, M. H. Key, Y. Kitagawa, K. Kondo, T. Matsuoka, H. Nakamura, M. Nakatsutsumi, P. A. Norreys, T. Norimatsu, R. A. Snavely, R. B. Stephens, M. Tampo, K. A. Tanaka, and T. Yabuuchi. Plasma devices to guide and collimate a high density of mev electrons. *Nature*, 432:1005, 2004.
103. S. V. Bulanov. New epoch in the charged particle acceleration by relativistically intense laser radiation. *Plasma Phys. Control. Fusion*, 48:29, 2006.
104. V. Malka, S. Fritzler, E. Lefebvre, M.-M. Aleonard, F. Burgy, J.-P. Chambaret, J.-F. Chemin, K. Krushelnick, G. Malka, S. P. D. Mangles, Z. Najmudin, M. Pittman, J.-P. Rousseau, J.-N. Sheurer, B. Walton, and A. E. Dangor. Electron acceleration by a wakefield forced by ab intense ultrashort laser pulse. *Science*, 298:1596, 2002.
105. D. Kaganovich, A. Ting, D. F. Gordon, R. F. Hubbard, T. G. Jones, A. Zigler, and P. Sprangle. First demonstration of a staged all-optical laser wakefield acceleration. *Phys. Plasmas*, 12:100702, 2005.
106. K. Krushelnik, E. L. Clark, F. N. Beg, A. E. Dangor, Z. Najmudin, P. A. Norreys, M. Wei, and M. Zepf. High intensity laser-plasa sources of ions-physics and future applications. *Plasma Phys. Control. Fusion*, 47:451–463, 2005.

107. A. Modena, Z. Najmudin, A. E. Dangor, C. E. Clayton, K. A. Marsh, C. Joshi, V. Malka, C. B. Darrow, C. Danson, D. Neely, and F. N. Walsh. Electron acceleration from the breaking of relativistic plasma waves. *Nature*, 377:606–608, 1995.
108. A. Pukhov. Strong field interaction of laser radiation. *Rep. Prog. Phys.*, 66:47–101, 2003.
109. I. Kostyukov, A. Pukhov, and S. Kiselev. Phenomenological theory of laser-plasma interaction in "bubble" regime. *Phys. Plasmas*, 11(11):5256, 2004.
110. T. Katsouleas and W. B. Mori. Wave-breaking amplitude of relativistic oscillations in a thermal plasma. *Phys. Rev. Lett.*, 61(1):90–93, July 1988.
111. D. Teychenné, G. Bonnaud, and J.-L. Bobin. Wave-breaking limit to the wake-field effect in an underdense plasma. *Phys. Rev. E*, 48(5):3248–3251, 1993.
112. Eric Esarey and Mark Pilloff. Trapping and acceleration in nonlinear plasma waves. *Phys. Plasmas*, 2(5):1432–1436, 1995.
113. S. V. Bulanov, F. Pegoraro, A. M. Pukhov, and A. S. Sakharov. Transverse-wake wave breaking. *Phys. Rev. Lett.*, 78(22):4205–4208, 1997.
114. E. Esarey, R. F. Hubbard, W. P. Leemans, A. Ting, and P. Sprangle. Electron injection into plasma wake fields by colliding laser pulses. *Phys. Rev. Lett.*, 79:2682, 1997.
115. S. Bulanov, N. Naumova, F. Pegoraro, and J. Sakai. Particle injection into the wave acceleration phase due to nonlinear wake wave breaking. *Phys. Rev. E.*, 58:5257, 1998.
116. N. E. Andreev, B. Cros, L. M. Gorbunov, G. Matthieussent, P. Mora, and R. R. Ramazashvili. Laser wakefield structure in a plasma column created in capillary tubes. *Phys. Plasmas*, 9(9):3999–4009, 2002.
117. L. M. Gorbunov, P. Mora, and R. R. Ramazashvili. Laser surface wakefield in a plasma column. *Phys. Plasmas*, 10(11):4563–4566, 2003.
118. L. M. Gorbunov, P. Mora, and A. A. Solodov. Dynamics of a plasma channel created by the wakefield of a short laser pulse. *Phys. Plasmas*, 10(4):1124–1134, 2003.
119. P. Tomassini, M. Galimberti, A. Giulietti, L. A. Gizzi, and L. Labate. Production of high-quality electron beams in numerical experiments of laser wakefield acceleration with longitudinal wave breaking. *PRST-AB*, 6:121301, 2003.
120. G. Fubiani, E. Esarey, C. B. Schroeder, and W. P. Leemans. Beat wave injection of electrons into plasma using two interfering laser pulses. *Phys. Rev. E*, 70:016402, 2004.
121. N. J. Sircombe, T. D. Arber, and R. O. Dendy. Accelerated electron populations formed by langmuir wave-caviton interactions. *Phys. Plasmas*, 12:012303, 2005.
122. C. S. Liu and V. K. Tripathi. Ponderomotive effect on electron acceleration by plasma wave and betatron resonance in short pulse laser. *Phys. Plasmas*, 12:043103, 2005.
123. T. Ohkubo, S. V. Bulanov, A. G. Zhidkov, T. Esirkepov, J. Koga, M. Uesaka, and T. Tajima. Wave-breaking injection of electrons to a laser wake field in plasma channels at the strong focusing regime. *Phys. Plasmas*, 13:103101, 2006.
124. S. Yu. Kalmykov, L. M. Gorbunov, P. Mora, and G. Shvets. Injection, trapping, and acceleration of electrons in a three-dimensional nonlinear laser wakefield. *Phys. Plasmas*, 13:113102, 2006.
125. C. Du and Z. Xu. Positron acceleration by a laser pulse in a plasma. *Phys. Plasmas*, 7:1582, 2000.
126. M. Borghesi, S. V. Bulanov, T. Zh. Esirkepov, S. Fritzler, S. Kar, T. V. Liseikina, V. Malka, F. Pegoraro, L. Romagnani, J. P. Rousseau, A. Schiavi, O. Willi, and A. V. Zayats. Plasma ion evolution in the wake of a high-intensity ultrashort laser pulse. *Phys. Rev. Lett.*, 94:195003, 2005.
127. T. Esirkepov, S. V. Bulanov, and M. Yamagiwa nd T. Tajima. Electron, positron, and photon wakefield acceleration: trapping, wake overtaking, and ponderomotove acceleration. *Phys. Rev. Lett.*, 96:014803, 2006.

128. J. M. Dawson. Nonlinear electron oscillations in a cold plasma. *Phys. Rev.*, 113:383, 1959.
129. R. C. Davidson and P P. Schram. *Nucl. Fusion*, 8:183, 1968.
130. R. J. England, J. B. Rosenzweig, and N. Barov. Plasma electron fluid motion and wave breaking near a density transition. *Phys. Rev. E*, 66:016501, 2002.
131. R. M. G. M. Trines and P. A. Norreys. Wave-breaking limits for relativistic electrostatic waves in a one-dimensional warm plasma. *Phys. Plasmas*, 13:123102, 2006.
132. A. I. Akhiezer and R. V. Polovin. Theory of wave motion of an electron plasma. *Sov. Phys. JETP [Zh. Eksp. Teor. Fiz. 30, 915 (1956)]*, 3:696, 1956.
133. G. Lehmann, E. W. Laedke, and K. H. Spatschek. Localized wake-field excitation and relativistic wave-breaking. *Phys. Plasmas*, 14:103109, 2007.
134. R. D. Hazeltine and J. D. Meiss. *Plasma confinement*. Addison-Wesley, Redwood City, Calif., 1992.
135. J. Wesson. *Tokamaks*. Clarendon Press, Oxford, 2004.
136. W. M. Stacey. *Fusion Plasma Physics*. Wiley-VCH, Weinheim, 2005.
137. J. P. Goedbloed, R. Keppens, and S. Poedts. *Advanced Magnetohydrodynamics*. Cambridge U.P., 2010.
138. H. Grad and H. Rubin. Mhd equilibrium in an axisymmetric toroid. In *Proceedings of the 2nd UN Conf. on the Peaceful Uses of Atomic Energy, Vol. 31, Vienna*, 1958.
139. V. D. Shafranov. On magnetohydrodynamical equilibrium configurations. *Sov. Phys. JETP*, 6:545, 1958.
140. L. S. Solovev. Hydromagnetic stability of claosed magnetic configurations. *Rev. Plasma Phys.*, 6:239, 1976.
141. J. Wesson. *Tokamaks*. Clarendon Press, Oxford, 2004.
142. J. W. Connor. Transport in tokamaks: Theoretical models and comparsion with experimental results. *Plasma Phys. Control. Fusion*, 37:119–133, 1995.
143. A. A. Galeev and R. Z. Sagdeev. *in: Handbook of Plasma Physics, Basic Plasma Physics I, A.A. Galeev and R.N. Sudan, eds., p. 679*. North-Holland, Amsterdam, 1984.
144. R. Balescu. *Aspects of anomalous transport in plasmas*. Institute of Physics Publishing, Bristol, 2005.
145. P. H. Diamond, S.-I. Itoh, and K. Itoh. *Modern Plasma Physics, Vol. 1: Physical Kinetics of Turbulent Plasmas*. Cambridge U.P., 2010.
146. J.R. Jokipii. Cosmic ray propagation: I. charged particles in a random magnetic field. *Astrophys. J.*, 146:480, 1966.
147. J. R. Jokipii. Addendum and erratum to cosmic-ray propagation. i. *Astrophys. Journal*, 152:671, 1968.
148. J. R. Jokipii and E. N. Parker. Cosmic-ray life and the stochastic nature of the galactic magnetic field. *Astrophys.l Journal*, 155:799, 1969.
149. J. R. Jokipii. The rate of separation of magnetic lines of force in a random magnetic field. *Astrophys. J.*, 183:1029, 1973.
150. F. Casse, M. Lemoine, and G. Pelletier. Transport of cosmic rays in chaotic magnetic fields. *Phys. Rev. D*, 65:023002, 2001.
151. G. Qin, W. H. Matthaeus, and J. W. Bieber. Perpendicular transport of charged particles in composite model turbulence: recovery of diffusion. *Astrophys. J*, 578:L117, 2002.
152. G. Qin, W. H. Matthaeus, and J. W. Bieber. Subdiffusive transport of charged particles perpendicular to the large scale magnetic field. *Geophys. Res. Lett.*, 29:7, 2002.
153. W. H. Matthaeus, G. Qin, J. W. Bieber, and G. P. Zank. Nonlinear collisionless perpendicular diffusion of charged particles. *Astrophys. J.*, 590:L000, 2003.
154. D. Ruffolo, W. H. Matthaeus, and P. Chuychai. Separation of magnetic field lines in two-component turbulence. *Astrophys. J.*, 614:420–434, 2004.

155. S. A. Khrapak and G. E. Morfill. Dust diffusion across a magnetic field due to random charge fluctuations. *Phys. Plasmas*, 9:619, 2002.
156. A. Shalchi, J. W. Bieber, W. H. Matthaeus, and G. Qin. Nonlinear parallel and perpendicular diffusion of charged cosmic rays in weak turbulence. *Astrophys. J.*, 616:617, 2004.
157. A. Shalchi, J. W. Bieber, W. H. Matthaeus, and R. Schlickeiser. Parallel and perpendicular transport of heliospheric cosmic rays in an improved dynamical turbulence model. *Astrophys. J.*, 642:230, 2006.
158. J. W. Bieber, W. H. Matthaeus, A. Shalchi, and G. Qin. Nonlinear guiding center theory of perpendicular diffusion: General properties and comparion with observation. *Geophys. Res. Lett.*, 31:101029, 2004.
159. P. Chuychai, D. Ruffolo, W. H. Matthaeus, and G. Rowlands. Suppressed diffusive escape of topologically trapped magnetic field lines. *Astrophys. J.*, 633:L49–L52, 2005.
160. Reinhard Schlickeiser. *Cosmic Ray Astrophysics*. Springer, Berlin, 2002.
161. A. N. Kolmogorov. The local structure of turbulence in incompressible viscous fluid for very large reynolds number. *Doklady Akademii Nauk SSSR*, 30:301, 1941.
162. T. H. Dupree. Perturbation theory of strong plasma turbulence. *Phys. Fluids*, 9:1773, 1966.
163. T. H. Dupree. Nonlinear theory of drift-wave turbulence and enhanced diffusion. *Phys. Fluids*, 10:1049, 1967.
164. J. Weinstock. Formulation of a statistical theory of strong plasma turbulence. *Physics of Fluids*, 12:1045, 1969.
165. J. Weinstock. Turbulent plasmas in a magnetic field—a statistical theory. *Physics of Fluids*, 13:2308, 1970.
166. A. Hasegawa. Self-organization in continuous media. *Advances in Physics*, 34:1–42, 1985.
167. R. Balescu. *Statistical Dynamics: Matter out of Equilibrium*. Imperial College Press, London, 1977.
168. W. D. McComb. *The Physics of Fluid Turbulence*. Clarendon Press, Oxford, 1990.
169. T.E. Evans, R.A. Moyer, P.R. Thomas, J.G. Watkins, T.H. Osborne, J.A. Boedo, E.J. Doyle, M.E. Fenstermacher, K.H. Finken, R.J. Groebner, M. Groth, J.H. Harris, R.J. LaHaye, C.J. Lasnier, S. Masuzaki, N. Ohyabu, D. G. Pretty, T.L. Rhodes, H. Reimerdes, D.L. Rudakov, M.J. Schaffer, G. Wang, and L. Zeng. Suppression of large edge-localized modes in high-confinement DIII-D plasmas with a stochastic magnetic boundary. *Phys. Rev. Lett.*, 92:235003, 2004.
170. T. E. Evans, R. A. Moyer, K. H. Burrell, M. E. Fenstermacher, I. Joseph, A. W. Leonard, T. H. Osborne, G. D. Porter, M. J. Schaffer, P.B. Snyder, P.R. Thomas, J.G. Watkins, and W.P. West. *Nature Physics*, 2:419, 2006.
171. M.W. Jakubowski, O. Schmitz, S.S. Abdullaev, S. Brezinsek, K.H. Finken, A. Kraemer-Flecken, M. Lehnen, U. Samm, K.H. Spatschek, B. Unterberg, R. C. Wolf, and the TEXTOR team. Effect of the change in magnetic-field topology due to an ergodic divertor on the plasma structure and transport. *Phys. Rev. Lett.*, 96:035004, 2006.
172. K. H. Finken, S. S. Abdullaev, M. F. M. de Bock, M. von Hellermann, M. Jakubowski, R. Jaspers, H. R. Koslowski, A. Kraemer-Flecken, M. Lehnen, Y. Liang, A. Nicolai, R. C. Wolf, O. Zimmermann, M. de Baar, G. Bertschinger, W. Biel, S. Brezinsek, C. Busch, A. J. H. Donnacutee, H. G. Esser, E. Farshi, H. Gerhauser, B. Giesen, D. Harting, J. A. Hoekzema, G. M. D. Hogeweij, P. W. Huettemann, S. Jachmich, K. Jakubowska, D. Kalupin, F. Kelly, Y. Kikuchi, A. Kirschner, R. Koch, M. Korten, A. Kreter, J. Krom, U. Kruezi, A. Lazaros, A. Litnovsky, X. Loozen, N. J. Lopes Cardozo, A. Lyssoivan, O. Marchuk, G. Matsunaga, Ph. Mertens, A. Messiaen, O. Neubauer, N. Noda, V. Philipps, A. Pospieszczyk, D. Reiser, D. Reiter, A. L. Rogister, M. Sakamoto, A. Savtchkov, U. Samm, O. Schmitz, R. P. Schorn,

B. Schweer, F. C. Schueller, G. Sergienko, K. H. Spatschek, G. Telesca, M. Tokar, R. Uhlemann, B. Unterberg, G. Van Oost, T. Van Rompuy, G. Van Wassenhove, E. Westerhof, R. Weynants, S. Wiesen, and Y. H. Xu. Toroidal plasma rotation induced by the dynamic ergodic divertor in the textor tokamak. *Phys. Rev. Lett*, 94:015003, 2005.

173. K. H. Finken, S. S. Abdullaev, M. W. Jakubowski, M. F. M. de Bock, S. Bozhenkov, C. Busch, M. von Hellermann, R. Jaspers, Y. Kikuchi, A. Kraemer-Flecken, M. Lehnen, D. Schege, O. Schmitz, K. H. Spatschek, B. Unterberg, A. Wingen, R. C. Wolf, O. Zimmermann, and the TEXTOR Team. Improved confinement due to open ergodic field lines imposed by the dymanic ergodic divertor in textor. *Phy. Rev. Lett.*, 98:065001, 2007.
174. A. B. Rechester and M. N. Rosenbluth. Electron heat transport in a tokamak with destroyed magnetic surfaces. *Phys. Rev. Lett.*, 40:38–41, 1978.
175. B. B. Kadomtsev and O. P. Pogutse. Electron heat conductivity of the plasma across a braided magnetic field. *in: Plasma Physics and Controlled Nuclear Fusion Research. Proc. 7th. Int. Conf. (Innsbruck, Austria, August 23–30, 1978)*, 1:649–662, 1979.
176. J. A. Krommes, C. Oberman, and R. G. Kleva. Plasma transport in stochastic magnetic fields. part 3. kinetics of test particle diffusion. *J. Plasma Phys.*, 30:11, 1983.
177. J. R. Myra, P. J. Catto, H. E. Mynick, and R. E. Duvall. Quasilinear diffusion in stochastic magnetic fields: Reconciliation of drift-orbit modification calculations. *Phys. Fluids B*, 5(5):1160–1163, 1993.
178. G. Laval. Particle diffusion in stochastic magnetic fields. *Phys. Fluids B*, 5:711, 1993.
179. M. De Rover, N. J. Lopes Cardozo, and A. Montvai. Hamiltonian description of the topology of drift orbits of relativistic particles in a tokamak. *Phys. Plasmas*, 3:4468–4477, 1996.
180. M. DeRover, N. J. LopesCardozo, and A. Montvai. Motion of relativistic particles in axially symmetric and perturbed magnetic fields in a tokamak. *Phys. Plasmas*, 3:4478, 1996.
181. M. DeRover, A. M. Schilham, A. Montvai, and N. J. LopesCardozo. *Phys. Plasmas*, 6:2443–2451, 1999.
182. R. Kubo. *J. Math. Phys.*, 4:174, 1963.
183. N.G. VanKampen. Stochastic differential equations. *Phys. Reports*, 24:171–228, 1976.
184. N. VanKampen. Stochastic processes in physics and chemistry. *Phys. Fluids*, 19:11, 1996.
185. R. Balescu. *Statistical Dynamics, Matter out of Equilibrium.* Imperial College Press, London, 2000.
186. Y. Elskens and D. Escande. *Microscopic dynamics of plasmas and chaos.* Institute of Physics Publishing, Bristol, 2003.
187. R. Balescu, H. D. Wang, and J. H. Misguich. Langevin equation versus kinetic equation: Subdiffusive behavior of charged particles in a stochastic magnetic field. *Phys. Plasmas*, 1:3826–3842, 1994.
188. E. Vanden-Eijnden and R. Balescu. Statistical description and transport in stochastic magnetic fields. *Phys. Plasmas*, 3:874, 1996.
189. R. Balescu. Anomalous transport in turbulent plasmas and continuous time random walk. *Phys. Rev. E*, 51:4807–4822, 1995.
190. S. Corrsin. Progress report on some turbulent diffusion research. *Atmospheric Diffusion and Air Pollution, Advances in Geophysics*, 6, 1959.
191. E. VandenEijnden and R. Balescu. Liouvillian theory of magnetic fluctuations. *J. Plasma Phys.*, 54:185–199, 1995.
192. H. D. Wang, E. Vanden-Eijnden, F. Spineanu, J. H. Misguich, and R. Balescu. Diffusive processes in a stochastic magnetic field. *Phys. Rev. E*, 51:4844, 1995.
193. E. Vanden-Eijnden and R. Balescu. Strongly anomalous diffusion in sheared magnetic configurations. *Phys. Plasmas*, 3:815–823, 1996.
194. M. Vlad and F. Spineanu. Trajectory structures and transport. *Phys. Rev. E*, 70:056304–1, 2004.

195. M. Vlad, F. Spineanu, and J. H. Misguich. Effects of stochastic drifts and time variation on particle diffusion in magnetic turbulence. *Phys. Rev. E*, 53:5302–5314, 1996.
196. M. Vlad, F. Spineanu, J. H. Misguich, and R. Balescu. Effects of plasma flow on particle diffusion in stochastic magnetic fields. *Phys. Rev. E*, 54:791, 1996.
197. M. B. Isichenko. Effective plasma heat conductivity in braided magnetic field-i: Quasilinear limit. *Plasma Phys. Controlled Fusion*, 33:795, 1991.
198. M. B. Isichenko. Effective plasma heat conductivity in braided magnetic field – II: Percolation limit. *Plasma Phys. Control. Fusion*, 33:809, 1991.
199. M. B. Isichenko. Percolation, statistical topography, and transport in random media. *Rev. Mod. Phys.*, 64:961, 1992.
200. J.-D. Reuss and J. H. Misguich. Low-frequency percolation scaling for particle diffusion in electrostatic turbulence. *Phys. Rev. E*, 54:1857–1869, 1996.
201. M. Vlad, F. Spineanu, J. H. Misguich, and R. Balescu. Diffusion with intrinsic trapping in two-dimensional incompressible stochastic velocity fields. *Phys. Rev. E*, 58:7359–7368, 1998.
202. M. Vlad, F. Spineanu, J. H. Misguich, and R. Balescu. Collisional effects on diffusion scaling laws in electrostatic turbulence. *Phys. Rev. E*, 61:3023–3032, 2000.
203. M. Vlad, F. Spineanu, J. H. Misguich, and R. Balescu. Diffusion in biased turbulence. *Phys. Rev. E*, 63:066304–1, 2001.
204. M. Vlad, F. Spineanu, J. H. Misguich, and R. Balescu. Electrostatic turbulence with finite parallel correlation length and radial diffusion. *Nucl. Fusion*, 42:157–164, 2002.
205. M. Vlad, F. Spineanu, J. H. Misguich, and R. Balescu. Magnetic line trapping and effective transport in stochastic magnetic fields. *Phys. Rev. E*, 67:026406, 2003.
206. R. Balescu, M. Vlad, F. Spineanu, and J. Misguich. Anomalous transport in plasmas. *Int. J. Quantum Chem.*, 98:125–130, 2004.
207. M. Vlad, F. Spineanu, J. H. Misguich, J.-D. Reuss, R. Balescu, K. Itoh, and S.-I. Itoh. Lagrangian versus eulerian correlations and transport scaling. *Plasma Phys. Control. Fusion*, 46:1051–1063, 2004.
208. O. G. Bakunin. Correlation effects and turbulent diffusion scalings. *Rep. Prog. Phys.*, 67:1, 2004.
209. O. G. Bakunin. Percolation models of turbulent transport and scaling estimates. *Chaos, Solitons and Fractals*, 23:1703, 2005.
210. M. Vlad and F. Spineanu. Larmor radius effects on impurity transport in turbulent plasmas. *Plasma Phys. Control. Fusion*, 47:281–294, 2005.
211. M. Neuer and K. H. Spatschek. Diffusion of test particles in stochastic magnetic fields for small Kubo numbers. *Phys. Rev. E*, 73:026404, 2006.
212. M. Neuer and K. H. Spatschek. Diffusion of test particles in stochastic magnetic fields in the percolative regime. *Phys. Rev. E*, 74:036401, 2006.
213. M. Vlad, F. Spineanu, and S. Benkadda. Impurity pinch from the ratchet effect. *Phys. Rev. Lett.*, 96:085001, 2006.
214. M. Vlad, F. Spineanu, and S. Benkadda. Collision and average velocity effects on the ratchet pinch. *Phys. Plasmas*, 15:032306, 2008.
215. A. B. Schelin and K. H. Spatschek. Directed chaotic transport in the tokamap with mixed phase space. *Phys. Rev. E*, 81:016205, 2010.
216. T. E. Evans, A. Wingen, J Watkins, and K.H. Spatschek. *Nonlinear Dynamics*, chapter A conceptual model for the nonlinear dynamics of edge-localized modes in tokamak plasmas, pages 59 –78. INTECH, Vukovar, Croatia, ISBN: 978-953-7619-61-9, 2010.
217. A. Wingen, T. E. Evans, C.J. Lasnier, and K.H. Spatschek. Numerical modeling of edge-localized-mode filaments on divertor plates based on thermoelectric currents. *Phys. Rev. Lett.*, 104:175001, 2010.

218. Z. Lin, T.S. Hahm, W.W. Lee, W.M. Tang, and R.B. White. Turbulent transport reduction by zonal flows: Massively parallel simulations. *Science*, 281:1835–1837, 1998.
219. K.H. Spatschek. Basic principles of stochastic transport. In S. Benkadda, editor, *Turbulent transport in fusion plasmas: First ITER International Summer School, AIP Conf. Proc. 1013, 250*, volume 1013 of *AIP Conf. Proc.*, page 250, Aix. France, 16-20 July 2007, 2008.
220. A. Wingen and K.H. Spatschek. Ambipolar stochastic particle diffusion and plasma rotation. *Phys. Plasmas*, 15:052305, 2008.
221. A. Wingen and K.H. Spatschek. Sheared plasma rotation in stochastic magnetic fields. *Phys. Rev. Lett.*, 102:185002, 2009.
222. A. Wingen and K. K. H. Spatschek. Influence of different ded base mode configurations on the radial electric field at the plasma edge of textor. *Nucl. Fusion*, 50:034009, 2010.
223. A. Wingen, S.S. Abdullaev, K.H. Finken, and K.H. Spatschek. Influence of stochastic magnetic fields on relativistic electrons. *Nucl. Fusion*, 46:941–952, 2006.
224. A. Wingen, S. Abdullaev, K. H. Finken, M. Jakubowski, and K. H. Spatschek. Influence of stochastic magnetic fields on relativistic electrons. *Nucl. Fusion*, 46:941–952, 2006.
225. A. Wingen, K.H. Spatschek, S.S. Abdullaev, and M. Jakubowski. Interpretation of heat losses from open chaotic systems. *Physics AUC*, 17:44–58, 2007.
226. A. Wingen, M. Jakubowski, S.S. Abdullaev, K.H. Spatschek, and K.H. Finken et al. Analysis of wall patterns and transport mechanisms in open chaotic systems by stable and unstable manifolds with applications to the textor-ded. 2006.
227. A. Wingen, M.A. Jakubowski, K.H. Spatschek, S.S. Abdullaev, K.H. Finken, M. Lehnen, and the TEXTOR team. Traces of stable and unstable manifolds in heat flux patterns. *Phys. Plasmas, submitted*, 2007.
228. A. H. Boozer. Physics of magnetically confined plasmas. *Rev. Mod. Phys.*, 76:1071–1141, 2004.
229. A. B. Rechester, M. N. Rosenbluth, and R. B. White. Calculation of the klomogorov entropy for motion along a stochastic magnetic field. *Phys. Rev. Lett.*, 42:1247, 1979.
230. A. B. Rechester and R. B. White. Calculation of turbulent diffusion for the chirikov-taylor model. *Phys. Rev. Lett.*, 44:1586, 1980.
231. A. B. Rechester, M. N. Rosenbluth, and R. B. White. Fourier-space paths applied to the calculation of diffusion for the Chirikov-Taylor model. *Phys. Rev. A*, 23:2664–2672, 1981.
232. J. M. Rax and R. B. White. Effective diffusion and nonlocal heat transport in a stochastic magnetic field. *Phys. Rev. Lett.*, 68:1523–1526, 1992.
233. E. Ott. *Chaos in Dynamical Systems.* Cambridge UP, Cambridge, 1994.
234. Heinz Georg Schuster. *Deterministisches Chaos.* VCH, 1994.
235. B.V. Chirikov. A universal instability of many-dimensional oscillator systems. *Phys. Reports*, 52:263–379, 1979.
236. A. Wolf, J. B. Swift, H. L. Swinney, and J. A. Vastano. Determining lyapunov exponents from a time series. *Physica D: Nonlinear Phenomena*, 16(3):285–317, 1985.
237. S.-C. Zhang and J. Elgin. Application of kolmogorov entropy to the self-amplified spontaneous emission free-electron lasers. *Phys. Plasmas*, 11(4):1663–1668, 2004.
238. M.N. Rosenbluth, R.Z. Sagdeev, G.B. Taylor, and G.M Zaslavsky. Destruction of magnetic surfaces by magnetic irregularities. *Nucl. Fusion*, 6:297, 1966.
239. T. H. Stix. *Phys. Rev. Lett.*, 30:833, 1973.
240. R.J. Bickerton. Magnetic turbulence and the transport of energy and part in tokamaks. *Plasma Phys. Control. Fusion*, 39:339–365, 1997.
241. J. F. Drake, N. T. Gladd, C. S. Liu, and C. L. Chang. *Phys. Rev. Lett.*, 44:994, 1980.
242. H. A. Rose. *Phys. Rev. Lett.*, 48:260, 1982.
243. G. I. Taylor. Diffusion by continuous movement. *Proc. London Math. Soc.*, 20:196, 1922.

244. M. S. Green. Brownian motion in a gas of noninteracting molecules. *J. Chem. Phys.*, 19:1036, 1951.
245. R. Kubo. Statistical-mechanical theory of irreversible processes. I. General theory and simple applications to magnetic and conduction problems. *J. Phys. Soc. Jpn.*, 12:570, 1957.
246. G. Qin, W.H. Matthaeus, and J.W. Bieber. Parallel diffusion of charged particles in strong two-dimensional turbulence. *Astrophys. Journal Lett.*, 640(1):L103–L106, 2006.
247. M. Neuer and K.H. Spatschek. Pitch angle scattering and effective collision frequency caused by stochastic magnetic fields. *Phys. Plasmas*, 15:022304, 2008.
248. P. Gibbon. *Short pulse laser interactions with matter*. Imperial College Press, 2005.
249. D. Strickland and G. Mourou. Compression of amplified chirped optical pulses. *Optics Communications*, 55:447–449, 1985.
250. Gerard A. Mourou, Toshiki Tajima, and Sergei V. Bulanov. Optics in the relativistic regime. *Reviews of Modern Physics*, 78:309, 2006.
251. W. K. H. Panofsky and M. Phillips. *Classical Electricity and Magnetism*. Addison-Wesley, Reading, 2nd edition edition, 1977.
252. J. D. Jackson. *Classical Electrodynamics*. Wiley, New York, 1999.
253. L. D. Landau and E. M. Lifshitz. *Lehrbuch der Theoretischen Physik II: Klassische Feldtheorie*. Akademie Verlag, Berlin, 1977.
254. S. W. Hawking. Black hole explosions? *Nature*, 248:30, 1974.
255. S. W. Hawking. Particle creation by black hole evaporation. *Comm. Math. Phys.*, 43:199, 1975.
256. W. G. Unruh. Notes on black hole evaporation. *Phys. Rev. D*, 14:870, 1976.
257. W. G. Unruh. Particle detectors and black hole evaporation. *Ann. N.Y. Acad. Sci.*, 302:186, 1977.
258. Pisin Chen and Toshi Tajima. Testing Unruh radiation with ultraintense lasers. *Phys. Rev. Lett.*, 83:256, 1999.
259. K. T. McDonalds. Positron production by laser light.Joseph Henry Laboratories, Princeton University, Seminar presented September 30, 1998.
260. C. Riconda and S. Weber. Plasma optics: A perspective for high-power coherent light generation and manipulation. *Matter and Radiation at Extremes*, 8:02300110.1063/1.4943200, 2023.
261. W. Cheng, Y. Avitzour, Y. Ping, S. Suckewer, N. J. Fisch, M. S. Hur, and J. S. Wurtele. Reaching the nonlinear regime of Raman amplification of ultrashort laser pulses. *Phys. Rev. Lett.*, 94:045003, 2005.
262. P. Michel, L. Divol, E. A. Williams, S. Weber, C. A. Thomas, D. A. Callahan, S. W. Haan, J. D. Salmonson, S. Dixit, D. E. Hinkel, M. J. Edwards, B. J. MacGowan, J. D. Lindl, S. H. Glenzer, and L. J. Suter. Tuning the implosion symmetry of iCF targets via controlled crossed-beam energy transfer. *Phys. Rev. Lett.*, 102:025004, Jan 2009.
263. M. Nakatsutsumi, A. Kon, S. Buffechoux, P. Audebert, J. Fuchs, and R. Kodama. Fast focusing of short-pulse lasers by innovative plasma optics toward extreme intensity. *Opt. Lett.*, 35:2314, 2010.
264. R. Wilson, M. King, R. J. Gray, D. C. Carroll, R. J. Dance, C. Armstrong, S. J. Hawkes, R. J. Clarke, D. J. Robertson, D. Neely, and P. McKenna. Ellipsoidal plasma mirror focusing of high power laser pulses to ultra-high intensities. *Physics of Plasmas*, 23(3), March 2016.
265. S. Monchoce, S. Kahaly, A. Leblanc, L. Videau, P. Combis, F. Reau, D. Garzella, P. D'Oliveira, Ph. Martin, and F. Quere. Optically controlled solid-density transient plasma gratings. *Phys. Rev. Lett.*, 112:145008, 2014.
266. H. Vincenti, S. Monchoce, S. Kahaly, G. Bonnaud, Ph. Martin, and Quere F. Optical properties of relativistic plasma mirrors. *Nature Communications*, 5(1), March 2014.

267. G. Lehmann and K. H. Spatschek. Transient plasma photonic crystal for high-power lasers. *Phys. Rev. Lett.*, 116:225002, 2016.
268. D. Turnbull, P. Michel, T. Chapman, E. Tubman, B. B. Pollock, C. Y. Chen, C. Goyon, J. S. Ross, L. Divol, N. Woolsey, and J. D. Moody. High power dynamic polarization control using plasma photonics. *Phys. Rev. Lett.*, 116:205001, May 2016.
269. G. Vieux, S. Cipiccia, D. W. Grant, N. Lemos, P. Grant, C. Ciocarlan, B. Ersfeld, M. S. Hur, P. Lepipas, G. G. Manahan, G. Raj, D. Reboredo Gil, A. Subiel, G. H. Welsh, S. M. Wiggins, S. R. Yoffe, J. P. Farmer, C. Aniculaesei, E. Brunetti, X. Yang, R. Heathcote, G. Nersisyan, C. L. S. Lewis, A. Pukhov, J. M. Dias, and D. A. Jaroszynski. An ultra-high gain and efficient amplifier based on raman amplification in plasma. *Scientific Reports*, 7(1), May 2017.
270. A. Leblanc, A. Denoeud, L. Chopineau, G. Mennerat, Ph. Martin, and F. Quéré. Plasma holograms for ultrahigh-intensity optics. *Nature Phys.*, 13:440, 2017.
271. R. K. Kirkwood, D. P. Turnbull, T. Chapman, S. C. Wilks, M. D. Rosen, R. A. London, L. A. Pickworth, W. H. Dunlop, J. D. Moody, D. J. Strozzi, P. A. Michel, L. Divol, O. L. Landen, B. J. MacGowan, B. M. Van Wonterghem, K. B. Fournier, and B. E. Blue. Plasma-based beam combiner for very high fluence and energy. *Nature Phys.*, 14:80, 10 2017.
272. K. Qu, Q. Jia, M. R. Edwards, and N. J. Fisch. Theory of electromagnetic wave frequency upconversion in dynamic media. *Phys. Rev. E*, 98:023202, 2018.
273. J.-R. Marquès, L. Lancia, T. Gangolf, M. Blecher, S. Bolanos, J. Fuchs, O. Willi, F. Amiranoff, R. L. Berger, M. Chiaramello, S. Weber, and C. Riconda. Joule-level high effiency energy transfer to sub-picosecond laser pulses by a plasma-based amplifier. *Phys. Rev. X*, 9:021008, 2019.
274. H. Peng, C. Riconda, M. Grech, J.-Q. Su, and S. Weber. Nonlinear dynamics of laser-generated ion-plasma gratings: A unified description. *Phys. Rev. E*, 100:061201, 2019.
275. R. M. G. M. Trines, E. P. Alves, E. Webb, J. Vieira, F. Fiuza, R. A. Fonseca, L. O. Silva, R. A. Cairns, and R. Bingham. New criteria for efficient raman and brillouin amplification of laser beams in plasma. *Scientific Reports*, 10(1):19875, 2020.
276. H. Peng, C. Riconda, S. Weber, C.T. Zhou, and S.C. Ruan. Frequency conversion of lasers in a dynamic plasma grating. *Phys. Rev. Applied*, 15:054053, May 2021.
277. Matthew R. Edwards and Pierre Michel. Plasma transmission gratings for compression of high-intensity laser pulses. *Phys. Rev. Appl.*, 18:024026, Aug 2022.
278. M. R. Edwards, V. R. Munirov, A. Singh, N. M. Fasano, E. Kur, N. Lemos, J. M. Mikhailova, J. S. Wurtele, and P. Michel. Holographic plasma lenses. *Phys. Rev. Lett.*, 128:065003, Feb 2022.
279. T. Tajima and J. M. Dawson. Laser electron accelerator. *Phys. Rev. Lett.*, 43:267, 1979.
280. E. Esarey, C. B. Schroeder, and W. P. Leemans. Physics of laser-driven plasma-based electron accelerators. *Reviews of Modern Physics*, 81:1229, 2009.
281. M. N. Rosenbluth and C. S. Liu. *Phys. Rev. Lett.*, 29:701, 1972.
282. R. Bingham and R. Trines. Introduction to plasma accelerators: The basics. In *Proc. of the CAS-CERN accelerator school : plasma wake acceleration*, page 67, 2016.
283. M. Dreher. Superradiante Verstärkung ultrakurzer Laserpulse in Plasmen. Technical Report MPQ 250, Max-Planck, 2000.
284. R. C. Davidson. *Methods in Nonlinear Plasma Theory*. Academic, New York, 1972.
285. H. Ma, Su-Ming Weng, P. Li, X. Li, Y. Wang, S. Yew, Min Chen, Paul McKenna, and Z. Sheng. Growth, saturation and breaking down of laser-driven plasma density gratings. *Phys. Plasmas*, 27:073105, 2020.
286. G. Lehmann and K. H. Spatschek. Plasma photonic crystal growth in the trapping regime. *Phys. Plasmas*, 26:013106, 2019.

287. K. Ohtaka. Energy band of photons and low-energy photon diffraction. *Phys. Rev. B*, 19:5057, 1979.
288. H. Hojo and A. Mase. Dispersion relation of electromagnetic waves in one-dimensional plasma photonic crystals. *J. Plasma Fusion Res.*, 80:89, 2004.
289. Osamu Sakai, Takui Sakaguchi, and Kunihide Tachibana. Verification of a plasma photonic crystal for microwaves of millimeter wavelength range using two-dimensional array of columnar microplasmas. *Appl. Phys. Lett.*, 87:241505, 2005.
290. O Sakai, S Yamaguchi, A Bambina, A Iwai, Y Nakamura, Y Tamayama, and S Miyagi. Plasma metamaterials as cloaking and nonlinear media. *Plasma Phys. Control. Fusion*, 59:014042, 2017.
291. G. Lehmann and K. H. Spatschek. Laser-driven plasma photonic crystals for high-power lasers. *Phys. Plasmas*, 24:056701, 2017.
292. Pochi Yeh. *Optical waves in layered media.* Wiley Intersience, New York, 2005.
293. Gerard A. Mourou, Christopher P. J. Barry, and Michael D. Perry. Ultrahigh-intensity lasers: Physics on the extreme on a tabletop. *Phys. Today*, 51:22–28, 1998.
294. Leonid Glebov, Vadim Smirnov, Eugeniu Rotari, Ion Cohanoschi, Larissa Glebova, Oleg Smolski, Julien Lumeau, Christopher Lantigua, and Alexei Glebov. Volume-chirped Bragg gratings: Monolithic components for stretching and compression of ultrashort laser pulses. *Optical Engineering*, 53(5):051514, feb 2014.
295. H. Peng, J.-R. Marquès, L. Lancia, F. Amiranoff, R. L. Berger, S. Weber, and C. Riconda. Plasma optics in the context of high intensity lasers. *Matter Radiat. Extremes*, 4:065401, 2019.
296. Y. Michine and H. Yoneda. Ultra high damage threshold optics for high power lasers. *Communications Physics*, 3:24, 2020.
297. P. Michel. Plasma photonics: Manipulating light using plasmas. Technical report, Lawrence Livermore Nat. Lab., 2021.
298. Matthew R. Edwards and Pierre Michel. Plasma transmission gratings for compression of high-intensity laser pulses. *Phys. Rev. Appl.*, 18:024026, Aug 2022.
299. J. Smith and A. Johnson. High-intensity lasers in advanced material processing. *J. Laser Applications*, 10:123–135, 2022.
300. Victor Malka, Jerome Faure, Yann A. Gauduel, Erik Lefebvre, Antoine Rousse, and Kim Ta Phuoc. Principles and applications of compact laser-plasma accelerators. *Nature Physics*, 4(6):447–453, June 2008.
301. Qisheng Peng, Asta Juzeniene, Jiyao Chen, Lars Svaasand, Trond Warloe, Karl-Erik Giercksky, and Johan Moan. Lasers in medicine. *Reports on Progress in Physics*, 71:056701, 04 2008.
302. Ference Krausz and Misha Ivanov. Attosecond physics. *Rev. Mod. Phys.*, 81:162, 2009.
303. H. Abu-Shawareb et al. (Indirect Drive ICF Collaboration). Lawson criterion for ignition exceeded in an inertial fusion experiment. *Physical Review Letters*, 129(7):075001, aug 2022.
304. Z.-M. Sheng, J. Zhang, and D. Umstadter. Plasma density gratings induced by intersecting laser pulses in underdense plasmas. *Appl. Phys. B*, 77:673, 2003.
305. H.-C. Wu, Z.-M. Sheng, Q.-J. Zhang, Y. Cang, and J. Zhang. Controlling ultrashort intense laser pulses by plasma Bragg gratings with ultrahigh damage threshold. *Laser and Particle Beams*, 23:417–421, 2005.
306. Hui-Chun Wu, Zheng-Ming Sheng, Qiu-Ju Zhang, Yu Cang, and Jie Zhang. Manipulating ultrashort intense laser pulses by plasma Bragg gratings. *Phys. Plasmas*, 12(11):113103, nov 2005.
307. Hui-Chu Wu, Zheng-Ming Sheng, and Jie Zhang. Chirped pulse compression in nonuniform plasma Bragg gratings. *Appl. Phys. Lett.*, 87:201502, 2005.

308. S. Suntsov, D. Abdollahpour, D. G. Papazoglou, and S. Tzortzakis. Femtosecond laser induced plasma diffraction gratings in air as photonic devices for high intensity laser applications. *Appl. Phys. Lett.*, 94:251104, 2009.
309. Magali Durand, Yi Liu, Benjamin Forestier, Aurélien Houard, and André Mysyrowicz. Experimental observation of a traveling plasma grating formed by two crossing filaments in gases. *Applied Physics Letters*, 98(12):121110, mar 2011.
310. L.-L. Yu, Y. Zhao, L.-J. Qian, M. Chen, S.-M. Weng, Z.-M. Sheng, D.A. Jaroszynski, W.B. Mori, and J. Zhang. Plasma optical modulators for intense lasers. *Nat. Commun.*, 7:11839, 2016.
311. H. Peng, C. Riconda, M. Grech, C.-T. Zhou, and S. Weber. Dynamical aspects of plasma gratings driven by a static ponderomotive potential. *Plasma Phys. Contr. Fusion*, 62(11):115015, November 2020.
312. Chaojie Zhang, Zan Nie, Yipeng Wu, Mitchell Sinclair, Chen-Kang Huang, Ken A Marsh, and Chan Joshi. Ionization induced plasma grating and its applications in strong-field ionization measurements. *Plasma Phys. Control. Fusion*, 63:095011, 2021.
313. G. Lehmann and K.H. Spatschek. Reflection and transmission properties of a finite-length electron plasma grating. *Matter Radiat. Extremes*, 7:054402, 2022.
314. Gregory Vieux, Silvia Cipiccia, Gregor H. Welsh, Samuel R. Yoffe, Felix Gartner, Matthew P. Tooley, Bernhard Ersfeld, Enrico Brunetti, Bengt Eliasson, Craig Picken, Graeme McKendrick, MinSup Hur, Joao M. Dias, Thomas Kuehl, Goetz Lehmann, and Dino A. Jaroszynski. The role of transient plasma photonic structures in plasma-based amplifiers. *Communications Physics*, 6:9, 2023.
315. Y. X. Wang, X. L. Zhu, S. M. Weng, P. Li, X. F. Li, H. Ai, H. R. Pan, and Z. M. Sheng. Fast efficient photon deceleration in plasmas by using two laser pulses at different frequencies. *Matter and Radiation at Extremes*, 9:037201, 2024.
316. J.E. Sipe, L. Poladian, and M. deSterke. Propagation through nonuniform grating structures. *J. Opt. Soc. Am. A*, 11:1307, 1994.
317. L. Poladian. Graphical and WKB analysis of nonuniform Bragg gratings. *Phys. Rev. E*, 48:4758, 1993.
318. Goetz Lehmann and Karl H. Spatschek. Formation and properties of spatially inhomogeneous plasma density gratings. *Phys. Rev. E*, 108:055204, 2023.
319. R. Szipocs, K. Ferencz, Chr. Spielmannn, and F. Krausz. Chirped multilayer coatings for broadband dispersion control in femtosecond lasers. *Optics Lett.*, 19:201, 1994.
320. M. Sumetsky, B.J. Eggleton, and C.M. deSterke. Theory of group delay ripple generated by chirped fiber gratings. *Optics Express*, 10:332, 2002.
321. O.V. Belai, E.V. Podivilov, and D.A. Shapiro. Group delay in Bragg grating with linear chirp. *Optics Communications*, 266(2):512–520, oct 2006.
322. S. Kaim, S. Mokhov, B.Y. Zeldovich, and L.B. Glebov. Stretching and compressing of short laser pulses by chirped Bragg gratings: analytic and numerical modeling. *Optical Engineering*, 53:05150, 2014.
323. Zhuang Rongrong and Cai Ping. Analysis on the reflection characteristic and the dispersion compensation performance of linear chirped fiber grating. *Information Technology Journal*, 13(11):1868–1872, may 2014.
324. Zhenhua Tian and Lingyu Yu. Rainbow trapping of ultrasonic guided waves in chirped phononic crystal plates. *Scientific Reports*, 7(1):1, jan 2017.
325. Ivan Ulyanov. Theoretical analysis of the stretched optical pulse ripple and novel chirped pulse retrieving algorithm. *Oprics Express*, 27:28166, 2019.
326. Min Sup Hur, Bernhard Ersfeld, Hyojeong Lee, Hyunsuk Kim, Kyungmin Roh, Yunkyu Lee, Hyung Seon Song, Manoj Kumar, Samuel Yoffe, Dino A. Jaroszynski, and Hyyong Suk.

Laser pulse compression by a density gradient plasma for exawatt to zettawatt lasers. *Nature Photonics*, 17:1074, 2023.

327. Herwig Kogelnik. Coupled wave theory for thick hologram gratings. *The Bell System Technical Journal*, 48:2909, 1969.
328. Luis Roso. High repetition rate petawatt lasers. *EPJ Web of Conferences*, 167:01001, 2018.
329. David A. Alessi, Paul A. Rosso, Hoang T. Nguyen, Michael D. Aasen, Jerald A. Britten, and Constantin Haefner. Active cooling of pulse compression diffraction gratings for high energy, high average power ultrafast lasers. *Optics Express*, 24(26):30015, 2016.
330. Vincent Leroux, Timo Eichner, and Andreas R. Maier. Description of spatio-temporal couplings from heat-induced compressor grating deformation. *Optics Express*, 28(6):8257, March 2020.
331. Z. Wu, Q. Chen, A. Morozov, and S. Suckewer. Compression of laser pulses by near-forward Raman amplification in plasma. *Physics of Plasmas*, 27(1):013104, 2020.
332. Zhaohui Wu, Yanlei Zuo, Xiaoming Zeng, Zhaoli Li, Zhimeng Zhang, Xiaodong Wang, Bilong Hu, Xiao Wang, Jie Mu, Jingqin Su, Qihua Zhu, and Yaping Dai. Laser compression via fast-extending plasma gratings. *Matter and Radiation at Extremes*, 7(6):064402, nov 2022.
333. M. Bonitz, C. Henning, and D. Block. Complex plasmas: a laboratory for strong correlations. *Rep. Prog. Phys.*, 73:066501, 2010.
334. A. Melzer. Introduction to colloidal (dusty) plasmas. Technical report, EMA Universität Greifswald, 2016.
335. Osamu Ishihara and Noriyoshi Sato. Attractive force on like carges in a complex plasma. *Phys. Plasmas*, 12:070705, 2005.
336. L. Spitzer. *Physical Processes in the Interstellar Medium.* Wiley, New York, 1st edition, 1978.
337. Hannes Alfvén. *Cosmic Plasma.* D. Reidel Publishing Company, Dordrecht, 1981.
338. H. Ikezi. Coulomb solid of small particles in plasmas. *Phys. Fluids*, 29:1764, 1986.
339. Frank Verheest. Linear and nonlinear electrostatic waves in dusty plasmas: A review. *Physics Reports*, 361(3-4):157–259, 2000.
340. Osamu Ishihara. Complex plasma, dusty plasma: An overview and future perspective. *Journal of Physics D: Applied Physics*, 40(8):R121–R147, 2007.
341. Michael Bonitz, C. Henning, and D. Block. Dusty plasmas: The state of understanding from an experimental, theoretical, and computational perspective. *Plasma Physics and Controlled Fusion*, 54(12):124001, 2012.
342. P. K. Shukla and A. A. Mamun. *Dusty and Self-Gravitational Plasmas in Space.* Springer, Berlin, Heidelberg, 2002.
343. Peter Huber, Alexei Ivlev, and Gregor Morfill. *Complex Plasmas and Colloidal Dispersions: Particle-Resolved Studies of Classical Liquids and Solids.* Springer, Berlin, Heidelberg, 2007.
344. M. K. Thomas. *Complex Plasmas: Experimental Studies of Dynamics of Yukawa Systems.* VDM Verlag Dr. Müller, Saarbrücken, 2007.
345. V. N. Tsytovich, G. E. Morfill, H. Thomas, and S. V. Vladimirov. *Complex and Dusty Plasmas: From Laboratory to Space.* CRC Press, Boca Raton, 2008.
346. Gerhard Franz. *Niederdruckplasmen und Mikrostrukturtechnik.* Springer, Berlin, Heidelberg, 2009.
347. Vladimir E. Fortov and Gregor E. Morfill. *Complex and Dusty Plasmas: From Laboratory to Space.* CRC Press, Boca Raton, 2010.
348. Alexei V. Ivlev, Hartmut Löwen, Gregor E. Morfill, and Christoph P. Royall. *Complex Plasmas and Colloidal Dispersions: Particle-Resolved Studies of Classical Liquids and Solids.* World Scientific, Singapore, 2012.
349. Michael Bonitz, Jose Lopez, Kurt Becker, and Hauke Thomsen. *Physics of Complex Plasmas.* Springer, Cham, 2014.

350. Andre Melzer. *Dynamical Processes in Complex Plasmas*. Springer, Cham, 2019.
351. Marco Thoma. *Complex Plasmas: Scientific Challenges and Technological Opportunities*. Springer, Cham, 2021.
352. Andrey V. Ivlev. *Complex Plasmas: Scientific Challenges and Technological Opportunities*. Springer, Cham, 2022.
353. Osamu Ishihara. Complex plasma: dusts in plasma. *J. Phys. D: Appl. Phys.*, 40:R121, 2007.
354. H. Thomas, G. E. Morfill, V. Demmel, J. Goree, B. Feuerbacher, and D. Möhlmann. Plasma crystal: Coulomb crystallization in a dusty plasma. *Phys. Rev. Lett.*, 73:652, 1994.
355. E. Wigner. On the interaction of electrons in metals. *Physical Review*, 46(11):1002–1011, 1934.
356. N. N. Rao, P. K. Shukla, and M. Y. Yu. Dust-acoustic waves in dusty plasmas. *Planet. Space Sci.*, 38:543, 1990.
357. R. Merlino. 25 years of dust acoustic waves. *J. Plasma Phys.*, 80:773, 2014.

The manufacturer's authorised representative in the EU is Springer Nature Customer Service Centre GmbH, Europaplatz 3, 69115 Heidelberg, Germany. If you have any concerns regarding our products, please contact ProductSafety@springernature.com

Printed and bound by CPI Group (UK) Ltd, Croydon, CR0 4YY
07/07/2026
02160920-0008